T5-BCI-293

Basin Compartments and Seals

Edited by

Peter J. Ortoleva

Department of Chemistry

Indiana University

Bloomington, Indiana

AAPG Memoir 61

Published by

The American Association of Petroleum Geologists

Tulsa, Oklahoma, U.S.A.

Printed in the U.S.A.

ISBN: 0-89181-340-3

Association Editor: Kevin T. Biddle
Science Director: Richard Steinmetz
Publications Manager: Ken Wolgemuth
Special Projects Editor: Anne H. Thomas
Production: Custom Editorial Productions, Inc., Cincinnati, Ohio

This and other AAPG publications are available from:
The AAPG Bookstore
P.O. Box 979
Tulsa, OK 74101-0979
Telephone: (918) 584-2555; (800) 364-AAPG (USA—book orders only)
FAX: (918) 584-0469; (800) 898-2274 (USA—book orders only)

AAPG
Wishes to thank the following
for their generous contribution
to

Basin Compartments and Seals

a publication prepared as an account of work sponsored on behalf of
Gas Research Institute

Gas Research Institute

Contributions are applied against the production costs of the publication, thus directly reducing the book's purchase price and making the volume available to a greater audience.

About the Editor

Peter J. Ortoleva was born in Brooklyn, New York, in 1942. He received his B.S. in physics from Rensselaer Polytechnic Institute; he earned his Ph.D. in applied physics at Cornell University, under Professor Mark Nelkin, on the statistical mechanics of simple fluids. Dr. Ortoleva carried out postdoctoral research in theoretical physical chemistry with Professor John Ross at Massachusetts Institute of Technology. In 1975 he joined the physical chemistry faculty at Indiana University, where he is presently Professor of Chemistry and of Geological Sciences. He has received Sloan and Guggenheim awards, is the author of more than 100 scientific publications, and is the author of two monographs and the editor of two conference proceedings.

His main research interests are in the theory of self-organization and other nonlinear phenomena, and in far-from-equilibrium reaction-transport systems. Over the past ten years he has focused on problems in geochemistry, including the physico-chemical dynamics of a sedimentary basin, metamorphic differentiation, zoned crystal growth from magmas and aqueous solutions and reaction fronts. He has developed a number of general reaction-transport codes to simulate these phenomena for academic and industrial research.

Table of Contents

Preface

In the early 1970s, Powley and Bradley recognized that basins worldwide exhibited an unexpected degree of hydrologic segregation. They observed regions of a sedimentary basin that were isolated from their surroundings by a relatively thin envelope of low-permeability rock and had an interior of sufficiently high permeability to maintain a consistent internal hydrostatic fluid pressure gradient. They named these regions "fluid (or pressure) compartments."

Pressure compartments have several remarkable features. First, their internal fluid pressures may greatly exceed or be significantly less than any regional topographically controlled hydrologic head or drain. Thus, their internal pressure cannot be explained by a connection to another similarly pressured source or sink region.

Next, their bounding rock often does not follow a single stratigraphic unit but transverses several units. This contradicts conventional thinking, wherein low permeability beds such as shale, salt, or bedded anhydrite serve to separate geopressured regions above and below but there are no recognized lateral boundaries. Furthermore, this surrounding shell was such an efficient barrier to fluid flow that it could retain very large potentiometric pressure heads over geologic time—i.e., on the hundred-million-year time frame. They thereby termed the surrounding rock a "seal"—i.e., a relatively thin region of rock at the top, bottom, and all sides that can retain pressure gradients on time scales very much longer than that required for dissipating nonhydrostatic gradients within the interior.

Powley and Bradley also conjectured that the formation of the seals was, at least in part, of diagenetic character. They noted the common occurrence of alteration within the horizontal seal zones, common association of the side boundaries of compartments with faults, and that the interiors of highly pressured compartments were often naturally fractured. The fact that these compartments often contained large reserves of oil and gas made them very interesting to Powley and Bradley because they recognized that their rather curious properties could have wide-ranging implications for petroleum exploration, reserve assessment, field development, and production.

For almost two decades, Powley and Bradley used the compartment concept in their exploration and production activities at Amoco. They gathered a great wealth of data on compartments worldwide and were quite successful with this notion within Amoco.

In the late 1980s they decided that they had reached a stage in the use of the compartment concept wherein the advantages of observing external research insights into compartment properties, particularly origins, might counterbalance the loss of proprietary advantage. Furthermore, Powley, Bradley, and the Amoco administration believed that making their findings more public would help to stimulate academic research on petroleum geoscience when there was a climate of cutting back in university efforts in this area. They believed that such cut-backs could have a negative impact on the industry in the future due to loss of a well-trained technical base.

The Gas Research Institute (GRI) recognized the potential exploration and production importance of the compartment concept and undertook a two-stage program of research in the late 1980s. Phase I was designed to generate an independent test of the compartment concept. Amoco assisted greatly during the early GRI-sponsored studies by making its extensive well data files available to the researchers. In Phase II (presently ongoing), the focus is on the development of specific concepts and tools that could more effectively take advantage of the existence of compartments.

The present volume summarizes much of the findings of Phase I of the GRI compartment project. As a result of this work, the Powley and Bradley concepts were not only verified but were found to be just a part of a very fascinating new area of geoscience that bridges hydrology, sedimentology, geophysics, and geochemistry.

A variety of compartment and seal types were discovered and their origins are beginning to be understood. The development of compartment and seal genesis models not only is of interest from the viewpoint of fundamental geoscience, but also could give a degree of predictability to compartment-sequestered petroleum that would be of great practical value.

The chapters in this memoir summarize the findings of the GRI project as of the spring of 1992. The volume commences with a contribution of Powley and Bradley and then presents the findings of the GRI-supported research groups (Part I).

Bradley and Powley show that abnormal pressures in sedimentary basins are a worldwide phenomenon. They define compartment and seal nomenclature and effectiveness. Seal origin possibilities are discussed. A thermoporoelastic formulation is given to determine the relative importance of various origins of abnormal pressure for given conditions and locality. The understanding of basin fluid flow regimes, which control diagenesis and hydrocarbon migration and trapping, is based upon the knowledge of pressure compartments and seals within the basin.

A comprehensive study of reservoir fluid pressures was conducted at Oklahoma State University that included data from all producing horizons in the Anadarko basin (Part II). The geometries and pressure regimes of compartments were established, and the petrography and diagenesis of pressure seals and their importance to the Powley and Bradley compartment concept was evaluated.

The megacompartment complex (MCC) is a basin-wide overpressured domain in the Anadarko basin discovered by the Oklahoma group. Its geometric configuration was revealed through the examination of more than 4000 pressure data points. The Anadarko basin MCC is 150 miles long, 70 miles wide, and has a maximum thickness of approximately 16,000 ft. The top of the MCC is relatively flat and lies approximately 7500 to 10,000 ft below the surface. The base of the MCC appears to coincide with the Woodford Shale. The southern lateral boundary of the MCC is associated with the frontal fault zone of the Wichita Mountain uplift, while the remaining lateral boundaries are formed by the convergence of the Woodford Shale and the top seal. Its interior is composed of an array of overpressured compartments. They exhibit a wide range of pressure gradients and are isolated by seals.

In the Anadarko basin, as in many other sedimentary basins, rock/pore fluid reactions induce significant diagenetic modifications in shale and sandstone intervals with depth. One important modification is the change in shale and sandstone lithologies resulting from the formation of diagenetic banding structures between 7000 and 10,000 ft. It is important to note that these structures are absent in the same lithologies above the seal interval. This diagenetic banding coincides with pressure seal intervals that are verified by pressure depth profiles and burial history curves. Petrographic analyses of cored pressure-seal intervals suggest that such diagenetic cementation bands may be critical to seal competence.

The genesis of banded seals was studied using integrated geologic, petrographic, and burial history data to relate seal generation and compartmentation to basin evolution. The seal zones apparently developed as the result of several diagenetic episodes that are induced during basin subsidence. A quartz "protoseal" formed in sandstones as a result of compaction, dissolution, and precipitation. This seal was augmented by later phases of carbonate and silica cements.

Detailed pressure analyses within stratigraphic intervals fostered the development of a three-level compartment classification scheme. Level 1 compartments (such as the megacompartment) are the large overpressured domains that transect stratigraphic boundaries. Level 2 compartments are district- or field-sized configurations within a particular stratigraphic interval that exhibit similar pressure regimes. A Level 3 compartment is a single, small field or reservoir nested within Level 1 or 2.

In the Anadarko basin and most deep sedimentary basins, pressure compartments are generally encountered in rocks that were buried to depths greater than 10,000 ft. Predicting the boundaries of the Level 1 compartment facilitates the execution of safer drilling programs that include casing strings and adjusted drilling mud weights needed to compensate for changes in pressure encountered at the top and base of the megacompartment. Comparable pressures within Level 2 compartments allow gradient prediction across entire trends or fields. This facilitates recognition of depletion caused by production or leaking. Recognizing Level 3 compartment boundaries is essential to the development of drilling strategies. Nonproducing compartments may exist in close proximity to producing ones. Diagenetic seals may isolate reservoirs within the same interval without a significant facies change. Petroleum accumulations can be found in compartments downdip of water-bearing compartments. The compartment concept helps improve the accuracy of estimating reserves and thereby significantly enhances exploration and production strategies.

The University of Wisconsin tested the Powley and Bradley compartment concept in the Michigan basin (Part IV). This interior cratonic basin has an active gas play in the Middle Ordovician St. Peter Sandstone and Glenwood formations. A large portion of these formations is overpressured. The overpressured zone is centered just northwest of the terminus of Saginaw Bay. The pressure anomaly is too high to be caused by the modern hydrologic regime. In fact, the overpressured compartment resides within the hydrologic systems discharge area. Vertical gradients between formations are reversed compared to those expected for the present-day flow system. Seals within the St. Peter and Glenwood maintain the nonequilibrium pressure distribution. The pressure anomaly itself may have been generated by glacial loading in the Pleistocene.

The seal zones are believed to result from diagenetic banding. Banding is not controlled by lithofacies or sedimentary structure in these homogeneous quartz sandstones. The origin and timing of the bands reflect the rock's burial and fluid flow history, rather than its depositional environments. Bands of quartz cement formed early in the burial history, as a result of compaction and silica dissolution/precipitation. Petrographic evidence shows that pressure solution in the St. Peter is sufficient to supply all the silica cement in the banded intervals in the formation. Stable isotope data suggest that the quartz cements were precipitated from warm water of low salinity, probably meteoric in origin. The later-stage dolomitic cement showed isotopic values consistent with origin from a hot brine.

Diagenetic seals can be traced laterally in the Michigan basin, even in the absence of pressure record or core data, by means of wireline log signature. A new multivariate technique of log analysis that classifies log sediments into one of several "electrofacies" was developed. From comparison with core studies, the electrofacies are shown to correspond to hydraulic and diagenetic characteristics of the St. Peter. Thus, seal horizons within the sandstone can be followed in the subsurface solely on the basis of well-bore geophysical logs.

The diagenetic sequence of the St. Peter/Glenwood interval can be understood in terms of the overall basin subsidence history, as deduced from fission track and vitrinite reflectance data. Thermal history data showed higher temperatures than would be predicted from the present basin depth and geothermal gradient. These data were interpreted to result from previously greater burial depths in the basin; the additional burial was by late Paleozoic strata, which were then eroded in the early Mesozoic.

Quantitative models and simulator codes for the analysis of basin diagenesis were developed at the Laboratory for Computational Geochemistry at Indiana University to test prevailing concepts of compartment and seal genesis and dynamics and establish new ones (Part VI). The ultimate goal of these studies is to develop tools to predict the distribution and characteristics of compartments in a basin and to delineate the tectonic, sedimentary, and thermal factors that control them.

The models demonstrated the importance of diagenetic banding in establishing high-quality seals through the precipitation of locally derived solutes so that long-range silica or other import was not required. The diagenetic reaction, transport, and mechanical (RTM) processes leading to banding were identified and quantitative models and simulation codes were developed (Part VIII). Factors favoring banded seal genesis were shown to be small grain size, rapid burial, relatively low fluid pressure, and high geothermal gradient. The bands were found to form via several different processes, all involving a feedback resulting from the coupling of stress, reaction, and (sub-meter-scale) transport. It was shown that such banded seals could form spontaneously (i.e., "self-organize") in relatively uniform sediment, or could occur as the amplification of sedimentary features through diagenesis. The unified model can explain the banded compaction/cementation alternations, carbonate bands, stylolites, clay seams, and comminution bands observed in the sandstone seals of the Anadarko and Michigan basins.

A quantitative basin diagenesis simulator was developed that illustrated the RTM processes underlying the following phenomena:

- Hydrofracturing of an overpressured compartment interior and an oscillatory release of fluids results from a repetitive (temporally oscillatory) seal fracture/healing cycle.
- The top of overpressure migrates upward in a basin experiencing rapid burial.
- Top seals of compartments cut across lithologies.
- In several mechanisms "smart seals" actually form and improve quality when pressure heads are applied to them as a result of diagenetic feedback (most notably through the pressure dependence of equilibria).
- Vertical seals adjacent to (and not within) faults were shown to form via a pressure-solution mediated mechanism of "autoisolation" of a large overpressuring compartment from a fault acting as a conduit.
- Finer-grained lithologies were shown to autoisolate from coarser-grained lithologies via a pressure solution-mediated mechanism.
- Three-dimensional closure of a seal array to form a compartment can typically only develop with a strong component of diagenesis—i.e., cannot be explained simply in terms of the geometry of the lithologies and faults; inorganic RTM processes required to do this were identified.
- Phenomena on four scales (micro-, meso-, macro-, and megascopic) underlie three distinct scales of compartmentation; the genesis and distribution of compartmentation depend strongly on the interaction of the phenomena on these four scales.

Novel methods were developed for these very formidable RTM simulations that allow for the modeling of two- and three-spatial-dimensional systems and account for a wide range of physical and chemical rock and fluid interaction processes. A method was also developed to capture the meter- and sub-meter scale aspects of banded seals in basin-scale systems; this "homogenization approach" allows for the modeling of the multiple-scale dynamics of compartment genesis and dynamics. These techniques open the way for the predictive modeling of compartments and their applications to exploration, reservoir characterization, and reserve prediction. The models will be most powerful when used to aid in the interpretation of remote and down-hole observations and for extrapolating data from a well-characterized part of the basin to a less well known prospect.

Two major computer programs were developed for the practical application of the compartment concept. One focused on the role of infiltrating reactive fluids in diagenesis, petroleum engineering procedures (e.g., near borehole damage remediation and steam flooding), and the injection of toxic waste into underpressured compartment repositories. Also, a basin-scale simulator was developed for the prediction of the location and characteristics of seals and compartments in a basin of given thermal, sedimentary, and tectonic history. Both codes have a common user-friendly modeling interface; built-in thermal, kinetic, and transport databases; and graphical viewing and animation module.

The Wyoming group investigated compartments in the Powder River basin (PRB), Wyoming (Part V). Overpressured compartments were distinguished on two scales in the Cretaceous section: a basinwide pressure compartment in the shales and small, isolated pressure compartments in the sandstones.

The overpressured Cretaceous shale section in the PRB is a basinwide dynamic pressure compartment. The driving mechanism is the generation of liquid hydrocarbons that subsequently partially react to gas, converting the fluid-flow system to a multiphase regime where capillarity dominates the permeability, resulting in elevated displacement pressures within the shales below a depth of approximately 8000 ft. The top of the overpressured zone is a transitional zone 500–1000 ft thick, and typically occurs within the Steele Formation. The zone of maximum overpressure, ~2000 ft thick, begins in the Frontier Formation, persists down to the Fuson Shale (lowermost organic-rich shale in the Cretaceous), and includes the Mowry Shale. Below the Fuson Shale, the pressure regime is typically normal. The top of the anomalously pressured zone is identified by marked increases in sonic transit time, production index, clay diagenesis (smectite → illite), vitrinite reflectance, and a significant increase in bitumen-filled microfractures.

In contrast, the Cretaceous sandstones are subdivided stratigraphically and diagenetically into relatively small, isolated pressure compartments whose largest dimension is 1 to 10 miles. Compartments in sandstones are generally bounded by diagenetically modified internal and external stratigraphic elements, particularly paleosols along unconformities and transgressive shales. The stratigraphic elements are low-permeability rocks that have finite leak rates in a single-phase fluid-flow system but that evolve diagenetically (smectite→illite; kaolinite→chlorite) into capillary seals with discrete displacement pressures (and no leakage below those pressures) as the flow regime evolves into a multiphase fluid-flow system by the addition of hydrocarbons during progressive burial. These capillary seals can support overpressure indefinitely, barring diastrophic fractures or failure when their displacement pressure is exceeded; and such fractures are evidently self-sealing in some cases by precipitation of carbonate cements. Geochemical modeling suggests that rupture of boundary seals, accompanied by migration (formation water mixing and temperature and pressure drop) or degassing can cause calcite precipitation and, consequently, seal restoration. The three-dimensional closure of the capillary seals above, below, and within a sandstone results in compartments *within* the sandstone.

The Wyoming group also investigated the detection and delineation of pressure compartments and the gas accumulations within them by surface seismic methods, the organic aspects of diagenesis, particularly the contribution of carboxylic acid anions, and the modeling of the processes of overpressure generation so as to both elucidate the processes and develop useful predictive techniques.

The Whelan, Eglinton, Cathles research group postulated that precipitated bitumen plus the associated gas act together to enhance sealing such that the seals may survive large segments of geologic time (Part III). This hypothesis was shown to be consistent with the organic geochemical signatures of samples across well-defined pressure seal zones, particularly those from the deep Tuscaloosa trend in Louisiana provided by the Penn State group. The characteristic organic geochemical signatures of pressure seals include:

- a zone of increased organic maturation at the top of the pressure seal zone;
- distinct gas chromatographic signatures for C7-C8 hydrocarbons that depend on position above, within, or below the seal zone; and
- a predominance of asphaltene and pyrobitumen over less polar aliphatic and aromatic hydrocarbons in the vicinity of the seal zone.

Dissolution of oil in gas would also provide a very efficient oil and gas migration mechanism at pressures greater than 200 bars and depths greater than 10,000 ft. Calculations suggest that even relatively low TOC sediments (0.4%) may be able to provide the required methane during the maturation process.

The Penn State group documented a pressure compartment from the deep Tuscaloosa trend in Louisiana. RFT data were used to identify the pressure seal. The Tuscaloosa trend has hydrostatic fluid columns that greatly exceed the thickness of individual sandstone layers. Compartments in the Tuscaloosa trend are surrounded by seals that crosscut stratigraphy, suggesting that the upper and side seals are diagenetic and fault-bounded, respectively. The seal zone cuts through shale beds, so it is apparent that shale was not responsible for sealing the rock.

Petrography indicates that the pressure seal in the Tuscaloosa trend forms by secondary compaction. The chemical composition of water and gas produced there was investigated to determine the origin and history of brines above and below the pressure seal. Similarities in brine chemistry and observed diagenetic sequences above and below the pressure seal supports seal formation by secondary compaction. Precipitation of quartz from supersaturated brine may provide a mechanism for maintaining the seal by infilling or healing of fractures in the seal. The behavior of Ca, Mg, HCO_3, pH, and p_{CO_2} suggests that any overpressured water leaking through the seal might dissolve carbonate minerals; this is consistent with the loss of carbonate cement and secondary compaction.

The Penn State Group also investigated compartmentation in the Texas Gulf Coast (Part III). The depth to the top of the undercompacted shale was located using shale density (cuttings samples), conductivity, sonic, plus bulk density logs, and the depth to the top of abnormal pore pressure was determined using bottom-hole pressure (BHP) and RFT data. Two types of field areas in the Tertiary portion of the Gulf of Mexico are distinguished based on whether or not the top of undercompaction and the top of abnormal pressure correspond. Tertiary fields have pressure-depth profiles characterized by pressure gradients showing either three (the Alazan-type field) or four (the Ann Mag-type field) linear segments.

An attempt was made to interpret deep seismic lines that cut through overpressured sections using data from the Port Lavaca area of the Texas Gulf Coast. The presence of a high-density, high-velocity cap at the pressure transition zone was demonstrated and shown to have unusual associated AVO variations. Deeper reflections, within the overpressured zone, do not exhibit the degree of Poisson's ratio variation or the velocity gradients associated with the almost complete cementation of the pressure transition zone above.

Hydrostatic compression tests were carried out to measure the variation in mechanical properties of Devonian core cut through an ancient pressure transition zone in the Appalachian Plateau of Western New York (Part VIII). Rocks within the transition zone were found to have lower intrinsic compressibilities than those of overlying rocks. The marked change in elastic properties near the transition zone may serve as a seismic reflector, an important property in the exploration for pressure compartments.

Geophysical, hydraulic, and geochemical data were compiled from a literature search for liquid-dominated geothermal fields of sediment-filled rift valleys and several sedimentary basins. The effects of hot brines on the reservoir properties of clastic sediments in geothermal systems and in sedimentary basins were evaluated. Analogous precipitation (overgrowths), dissolution (secondary porosity), and alteration ("sealing") mechanisms were shown to take place in the seals around some pressure compartments and geothermal systems.

In summary, the main findings of Phase I of the GRI compartment project were that:

- pressure compartments exist;
- compartments appear to form with or without the presence of petroleum; and
- diagenetic alteration of seal zones has been observed in all compartments reported in this volume.

In addition, Phase I has shown a richness of new phenomena. We believe it opens up a new chapter in our understanding of the workings of the deeper parts of sedimentary basins and possibly into deeper basement rocks. The potential economic benefits of compartment research make further investigation warranted. The combination of exciting new geoscience and economic potential has fostered a new atmosphere of industrial-university cooperation. This collaboration is important for the mission of the GRI to insure a steady, long-term supply of natural gas and for the wider goal of economic security of the petroleum industry and its role in maintaining our economic viability.

Peter J. Ortoleva and Paul Westcott

I. General Considerations

Chapter 1

Pressure Compartments in Sedimentary Basins: A Review

John S. Bradley
Consultant
Tulsa, Oklahoma, U.S.A.

David E. Powley
Consultant
Tulsa, Oklahoma, U.S.A.

ABSTRACT

Pressure compartments are found in sedimentary basins throughout the world. They are defined primarily by hydraulic potentials calculated from pressure measurements but may be indicated by differing brine and hydrocarbon chemistries; by mineralogic differences; by electrical resistivity, sonic velocity, and density of the shales; and by mud weight requirements and drilling rate changes.

Pressure compartments are characterized by an effective seal, in three dimensions, that prevents pressure equilibration to normal hydrostatic pressure. A pressure seal, as opposed to a capillary seal, restricts flow of both hydrocarbon and brine and is formed where the pore throats become effectively closed, i.e., the permeability approaches zero. A leaking pressure seal, called a "rate seal," occurs when the pressure difference caused by subsidence-sedimentation or uplift-erosion or other pressure source is greater than the seal pressure leakage. When the internal fluid pressure in the compartment exceeds the fracture pressure of the seal, the seal will fracture and fluids will escape from the compartment. The fracture and resealing may occur repeatedly.

Multiple pressure seal origins must be invoked to explain their geometric and stratigraphic occurrence. Certainly some pressure seals appear to be stratigraphically controlled with perhaps more or less diagenetic enhancement. Some seals, particularly those that cross stratigraphy, appear to be entirely diagenetic. The lateral seals, which appear to be subvertical to vertical, are possibly due to faulting and fracturing or to lateral facies changes. An extensive investigation of seal origin, recognition, and duration is being undertaken by a consortium of universities under the sponsorship of the Gas Research Institute.

The cause of the abnormal fluid pressure, or flow if there is no seal, is epeirogeny (uplift or subsidence) with accompanying erosion or sedimentation. This changes the temperatures, pressure, and stress in the system leading to either changes in pressure, if sealed, or flow, if not sealed.

The knowledge and understanding of pressure compartments and compartment seals are vital to understanding the basin fluid flow regimes that control diagenesis and hydrocarbon migration and trapping.

INTRODUCTION

Many deep sedimentary basins contain a shallow hydraulic system at normal hydrostatic pressures overlying one or more hydraulic compartments that may have pressures above or below the normal hydrostatic pressures (Powley, 1990). As the hydraulic compartments are usually defined only by hydraulic potential differences calculated from pressure measurement (Chapman, 1981, pp. 62–63), it would be possible to have undetectable normally pressured compartments also. Multiple compartments may be found both vertically and laterally.

The knowledge and understanding of pressure compartments is of great importance to the drilling engineer because of the possibilities of blowouts and/or lost circulation. But the recognition of compartments is also vital to understanding the fluid flow regimes in the basin that control diagenesis and hydrocarbon migration and trapping. The trapping possibilities of compartmentation (Figure 1) indicate the exploration significance of the phenomenon.

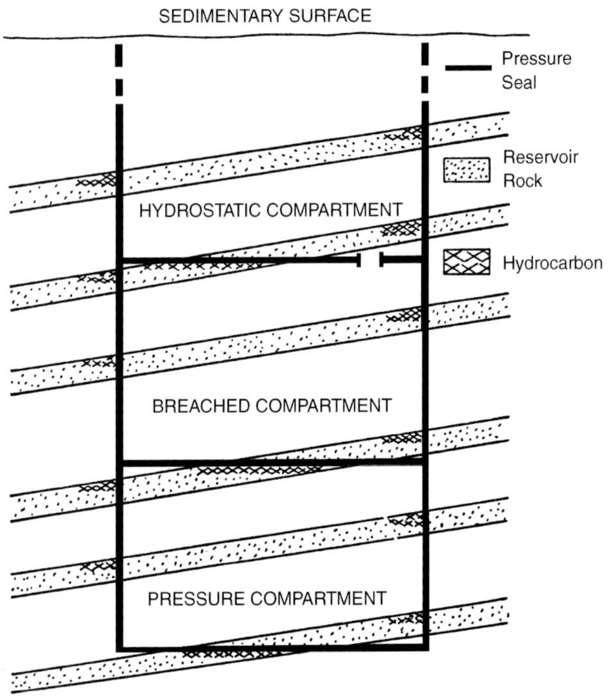

Figure 1. Compartmental seal traps.

The measurements that indicate compartmentation vary greatly in number and quality. It is expected that the compartment designation will stand through time even though additional measurements may give somewhat different pressures and areal extent. It is also important to understand that the "normal" pressure gradients are often reasonable assumptions but may not have been measured. By varying the assumed salinity of the brines, one can vary gradients (Table 1) and thus pressures by up to a maximum of approximately 1000 psi at depth (Figure 2).

The relationship between interpretations of conventional hydrodynamic profiles and sealing concept profiles is also indicated in Figure 2. The width (steepness) of the seal or transition ramp is usually determined by the spacing of the pressure measurements. Thus, it is seen that the compartmentation (effective sealing) is merely at the low-permeability limit of the spectrum of permeabilities of normal hydrodynamic models. Toward this low-permeability limit of the spectrum, the dynamic flow model tends to become a static no-flow model.

Flow is commonly inferred indirectly from differences in brine chemistry or from formation-fluid pressure measurements. Measured pressures, corrected to hydraulic potentials, infer flow from a higher to lower potential. But as Dickey (1972) pointed out, the inference leads to a gross ambiguity. "It is a mistake to assume, as has often been done, that differences in fluid potential between separate aquifers or reservoirs indicate flow between them. On the contrary, differences in potential between aquifers can be more logically interpreted as indicating lack of hydraulic connection."

This "lack of hydraulic connection" or "sealing" leads to the concept of compartmentation of a basin (Bradley, 1975).

An understanding of fluid movement in sedimentary basins is essential to understanding the diagenetic processes that change porosity and permeability. The loss of primary porosity to lithification and the creation of secondary porosity by solution affect the permeability that controls the profitability of petroleum reservoirs. By establishing physical limits on the

Table 1. Densities.

Density g/cm³	Gradient psi/ft	Salinity TDS ‰	Mud Weight lbs/gal	Oil Gravity API°
.6852	.297			70
.7201	.312			65
.7389	.320			60
.7587	.329			55
.7796	.337			50
.7883	.341			48
.8063	.349			44
.8251	.357			40
.8448	.366			36
.8654	.375			32
.8762	.389			30
.8871	.384			28
.9100	.394			24
.9340	.404			20
.9659	.418		8	15
1.00	.433	0*	8.34	10
1.01	.437	13.5		
1.02	.441	27.5		
1.029	.444	37.0**		
1.03	.445	41.4		
1.04	.459	55.4		
1.05	.454	69.4		
1.06	.459	83.7		
1.07	.463	98.4		
1.075	.465	100.0		
1.08	.467	113.2		
1.09	.471	128.3		
1.10	.476	143.5		
1.11	.480	159.5		
1.12	.485	175.8		
1.13	.489	192.4		
1.135	.491	200.0		
1.137	.493	210.0		
1.153	.500	230.0		
1.176	.510	260.0		
1.20	.520		10	
1.205	.524	300.0		
1.255	.544	350.0		
1.32	.570		11	
1.44	.625		12	
1.56	.675		13	
1.68	.725	* Fresh Water	14	
1.80	.783	** Sea Water	15	
1.92	.830		16	
2.0	.866			
2.04	.885		17	
2.1	.909			
2.16	.935		18	
2.2	.953			
2.28	.980		19	
2.3	.996			
2.4	1.070		20	

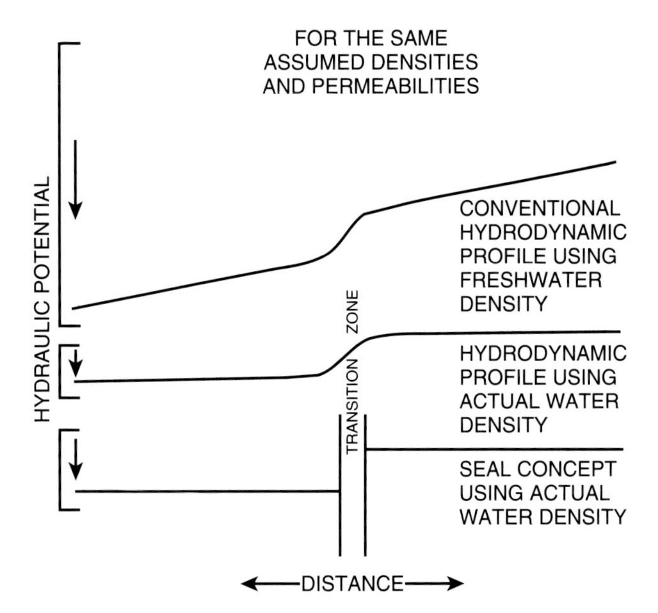

Figure 2. Hydraulic potential cross sections. After Powley (1983).

amount, duration, distance, and direction of fluid movement, some of the constraints for the physical-chemical environment in which diagenetic reactions take place can be established. Compartmentation can profoundly affect the fluid movements within a sedimentary basin.

DISCUSSION

Definitions

The hydraulic and chemical systems defined herein must be carefully delimited with regard to scale, to time, and to boundary conditions. The definitions consider a water-wet sedimentary basin. The systems of scale for cores, for reservoirs, and for basins vary widely. The permeabilities can be orders-of-magnitude different at reservoir and basin scale. Boundary condition definition can reverse a system definition—the volume *within* a pressure compartment is an open hydraulic system, but when the seals are considered the boundaries, the compartment itself is a closed hydraulic system.

"EFFECTIVE"—While the definitions are in absolute terms (no flow, no continuity, no ion exchange), naturally occurring systems are seldom absolute (minor leakage, minimal exchange, etc.). The modifier "effective" is helpful in defining natural systems as in "an effectively closed hydraulic system." This would mean that the hydraulic leakage, if any, is less than the fluid volume changes caused by orogeny, sedimentation, and erosion. Permeability or leakage is not usually measured, so pressure is the sole hydraulic parameter known.

SEAL—A *capillary seal,* which restricts the flow of hydrocarbons, is formed when the capillary pressure across the pore throats (a function of the surface tension between the wetting fluid and the migrating hydrocarbon) is greater than or equal to the buoyancy

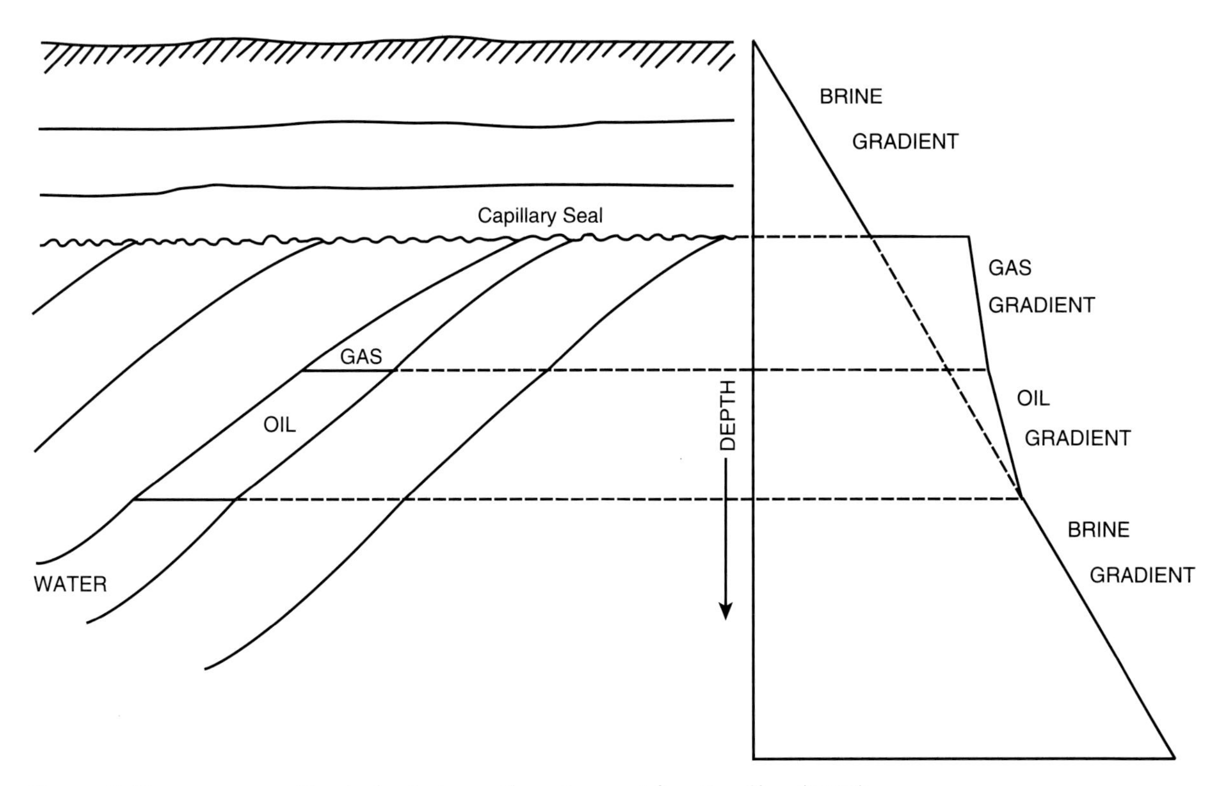

Figure 3. Buoyancy gradients for brine, oil, and gas. After Bradley (1975).

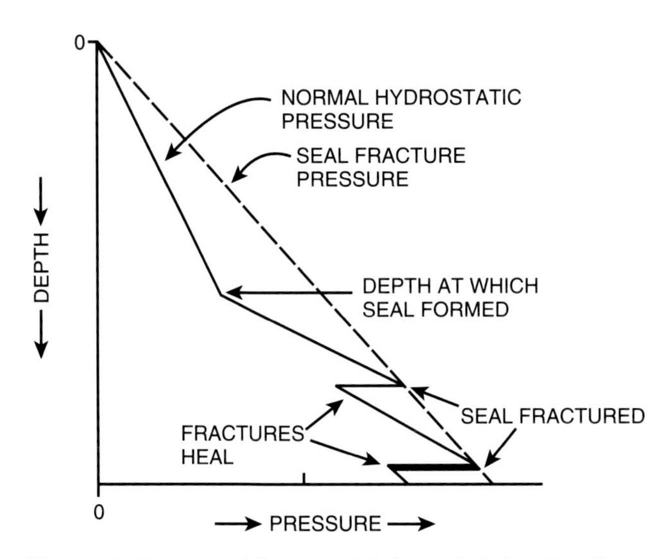

Figure 4. Pressure history with burial. After Powley (1983).

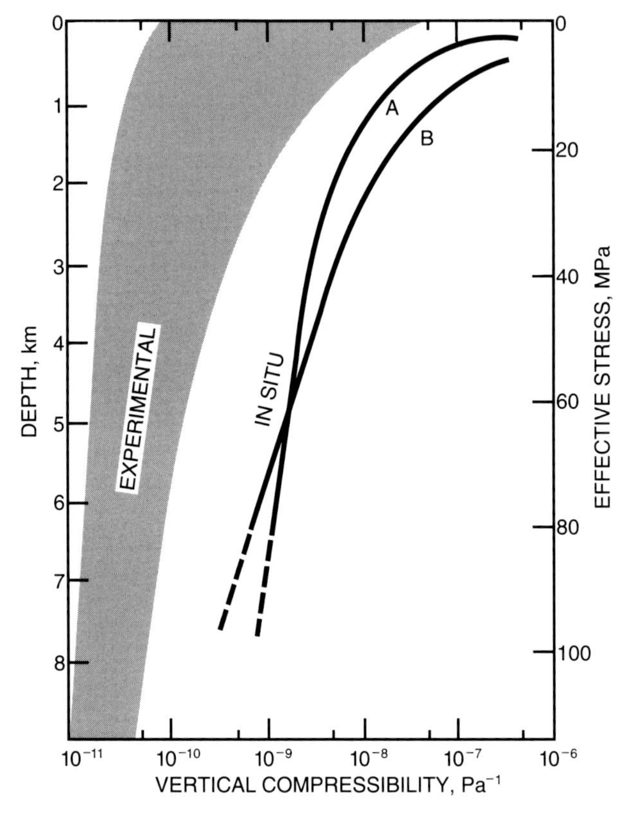

Figure 5. Rock compressibility—experimental vs. calculated. The shaded region encloses all experimental data from sources cited by Neuzil (1985). Curve A was computed from porosity data for the Gulf Coast presented by Dickinson (1953); curve B is from a representative porosity-depth relation suggested by Cathles and Smith (1983). The convention that effective stress increases at approximately 1.3×10^4 Pa/m was used to compute curves A and B and to plot the experimental data on an equivalent depth scale.

Table 2. Notation.

ε_T	total pore strain	$\frac{cm^3}{cm^3}$	
ε_B	bulk strain	$\frac{cm^3}{cm^3}$	
ε_G	grain strain	$\frac{cm^3}{cm^3}$	
ε_D	diagenetic strain	$\frac{cm^3}{cm^3}$	
ε_F	fluid strain	$\frac{cm^3}{cm^3}$	
ϕ	porosity		
β_B	bulk compressibility	$\frac{1}{psi}$	
β_G	grain compressibility	$\frac{1}{psi}$	
β_F	fluid compressibility	$\frac{1}{psi}$	
σ_H	horizontal effective stress	psi	
σ_m	mean effective stress	psi	
σ_v	vertical effective stress	psi	
P	pore pressure	psi	
T	temperature	°F	
α_G	grain cubic thermal expansivity	$\frac{1}{°F}$	
α_F	fluid cubic thermal expansivity	$\frac{1}{°F}$	
V_p	pore volume	cm^3	
Q	flow volume	cm^3	
f	empirical factor		
D	depth	ft	
M	axial modulus	psi	$M = K + \frac{4G}{3}$
S	total vertical stress	psi	overburden
K	bulk modulus	psi	
G	shear modulus	psi	

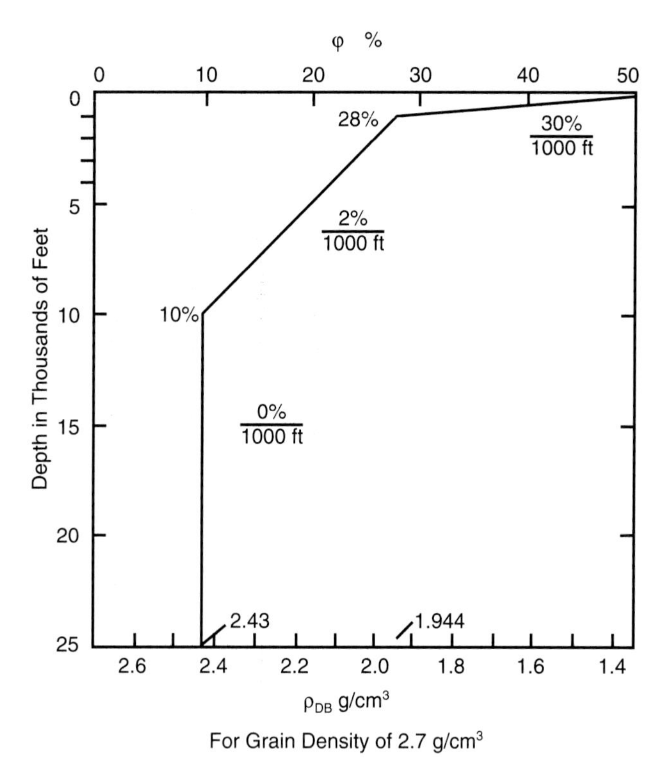

Figure 6. Shale porosity and dry bulk density. After Bradley (1986).

pressure of the hydrocarbons (a function of the height of the hydrocarbons and the density difference of the fluids). Pore throat size reduction, which increases the capillary pressure, may be due to deposition, diagenesis, deformation (including faulting), or some combination. Brine can move through the interconnected pore system of a capillary seal.

A *pressure seal,* which restricts both hydrocarbon and brine, is formed where the pore throats become effectively closed, i.e., the permeability approaches zero. A leaking pressure seal, called a "rate seal," occurs when the pressure caused by subsidence-sedimentation or uplift-erosion is greater than the seal pressure leakage. When the internal fluid pressure exceeds the fracture pressure, the seal will fracture and fluids will escape from the compartment.

TRAP—A trap is a geometrical configuration of a volume of reservoir rocks and seal in which hydrocarbons can accumulate. The trap capacity is defined by its geometry and/or the properties of the seals. A trap may or may not be filled to capacity.

PRESSURE COMPARTMENT—A pressure compartment, which is a closed hydraulic and closed chemical system, is defined *only* by deviation of pressure, either high or low, as compared with normal hydrostatic pressure. No timing or duration for the compartment is implied. An open hydraulic and chemical system exists *within* the compartment. The seals of a pressure compartment may act as a petrolum-trapping mechanism.

OPEN HYDRAULIC SYSTEM—An open hydraulic system has fluid (brine, oil, or gas) continuity throughout, resulting in normal hydrostatic pressures in the water system. There may be brine flow across a capillary seal within an open hydraulic system.

CLOSED HYDRAULIC SYSTEM—A closed hydraulic system has no fluid continuity across the

Table 3. Abnormal pressure and fluid flow calculation, Wasatch Shale, Eocene, Duchesne County, Utah.

Rock parameters:	
Bulk modulus, K	= 2.190E + 06 psi
Shear modulus, G	= 1.780E + 06 psi
Young's Modulus	= 4.202E + 06 psi
Poisson's ratio	= 0.180
Porosity	= 6.70%
Grain parameters:	
Linear expansivity	= 5.000E – 06 /°F
Linear compressibility	= 7.100E – 08 /psi
Fluid parameters:	
Cubic expansivity	= 4.000E – 04 PF
Vol. compressibility	= 3.000E – 06 /psi
Burial gradients and depth increment:	
dS/dZ	= 1.000 psi/ft
dT/dZ	= 0.012 °F/ft
dZ/dD	= 100.000 ft
Pressure compartment (pressure rise):	
ΔP:	125.32 psi
Stress:	28.70 psi
Thermal:	96.61 psi
Abnormal pressure increment: 75.32 psi	
Open system (fluid expelled, in porosity units):	
$\Delta\phi$:	0.00314 %
Rock:	0.00093 %
Fluid:	0.00221 %

Table 4. Porosity response for additional 100 ft burial at two depths.

	Shale at 7500 Ft	
	$\Delta\phi$%	% of Total Change
Rock	.00093	.465
Fluid	.00221	1.105
Diagenesis	.197	98.5
	Total	.20
	Shale at 15,000 Ft*	
	$\Delta\phi$%	% of Total Change
Rock	.00093	29.6
Fluid	.00221	70.4
Diagenesis	0	0
Total	.00314	

*Assumes the same shale moduli at the two depths.

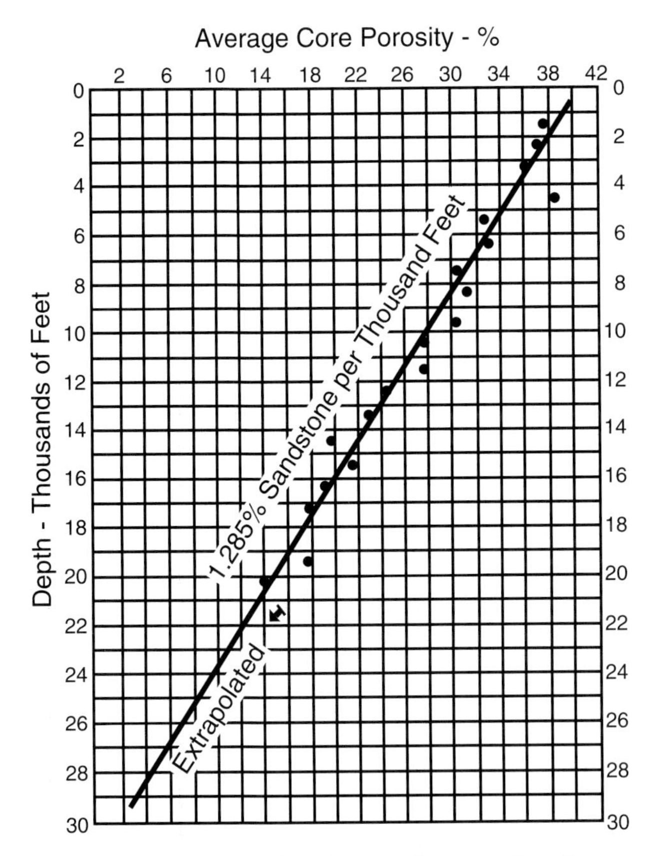

Figure 7. Porosity vs. depth—sandstones. South Louisiana sands, 17,367 samples. After Atwater and Miller (1965) and Atwater et al. (1986).

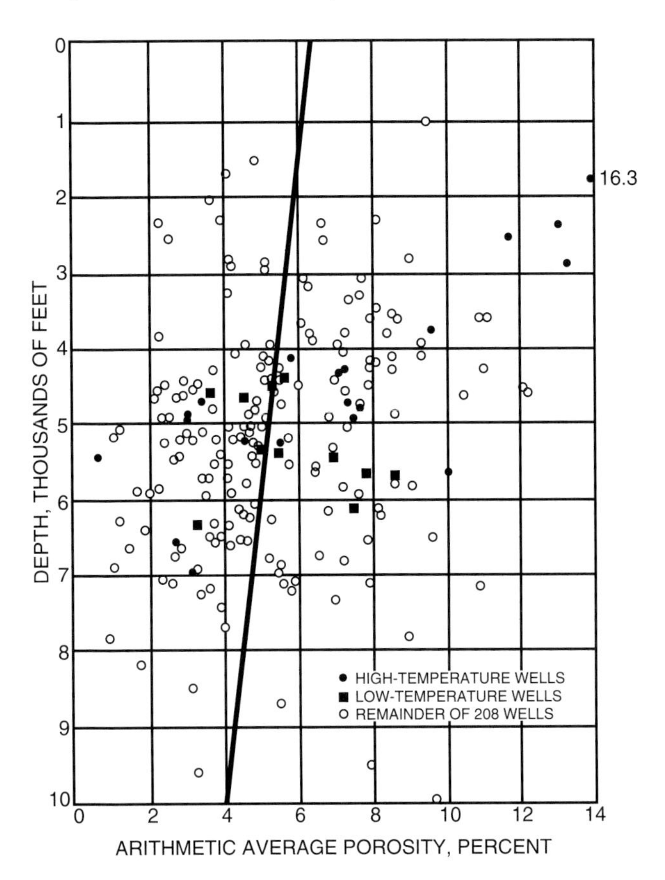

Figure 8. Porosity vs. depth—carbonates in the Alberta basin.

bounding pressure seals, so that the fluids within the system may be underpressured, normally pressured, or overpressured. There is no fluid flow (brine or hydrocarbon) across the seal. Because closed hydraulic systems are not necessarily abnormally pressured, they may not be recognized solely by pressure measurements.

CHEMICAL SYSTEMS—The chemistry within a rock-water system may change with the addition and subtraction of matter by diffusion or fluid flow. System chemistry may also change without fluid flow, due to changes of temperature, pressure, and stress with uplift or burial.

CLOSED CHEMICAL SYSTEM—A closed chemical system has no transfer of matter into or out of the system. Such a system may be hydraulically open or closed. Solution and precipitation of solids occur due to changes in temperature, pressure, and stress.

OPEN CHEMICAL SYSTEM—An open chemical system has matter transfer which, in changing the chemistry of the pore water in conjunction with changing temperature, pressure, and stress, allows the solution or precipitation of solids. An open chemical system cannot be a closed hydraulic system.

Origins of Pressure/Flow

The primary cause of either pressure change or flow is epeirogenic movement, with concomitant erosion or deposition, which changes the temperature, stress, and pore pressure in the system, which, in turn, affect the diagenesis in the system. These changes result in a change in pore volume and fluid volume, the algebraic sum of which determines the pressure/flow. In a closed hydraulic system (sealed), the pressure will change, but in an open hydraulic system the fluid will flow in or out.

The factors causing the volume changes occur throughout the basin and throughout the time of epeirogenic movement. The location and time of formation of seals determine the pressure and/or flow regimes.

Overpressures may be caused by the buoyancy effect of hydrocarbons over brine (Bradley, 1975). Note that only a capillary seal, not a pressure seal, is involved in this phenomenon (Figure 3). The buoyancy pressure is a function of the difference in density gradient of the fluids (Table 1) and the height of the hydrocarbon column.

Within an open hydraulic system, several types of flow may occur. "Compaction" flow is induced by the loss of porosity with depth (Bradley, 1986) while "gravity" flow is caused by differential topographic elevations. "Density" flow (convection) can be induced by thermal and salinity gradients (Bradley, 1986). "Potential equilibration" flow may occur episodically into or out of pressure compartments

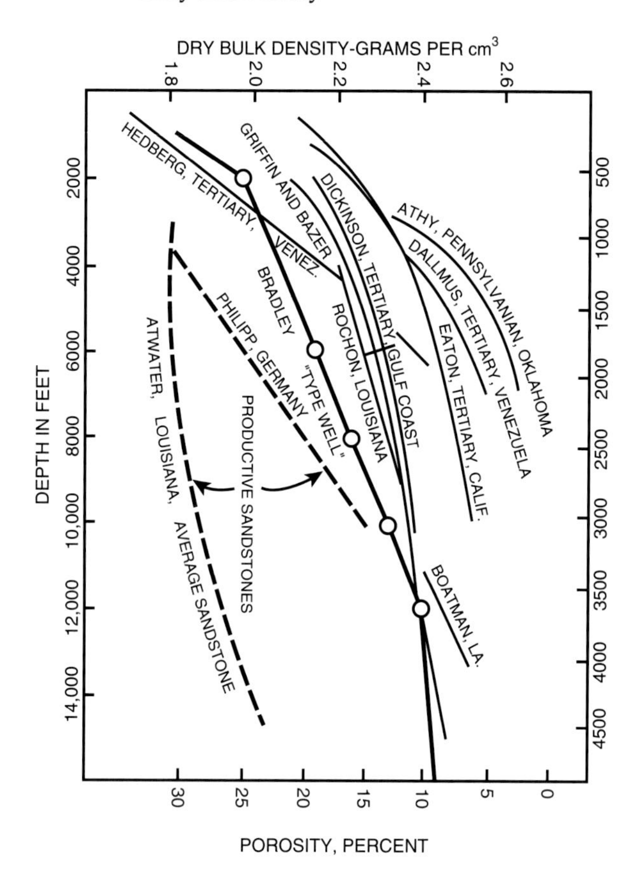

Figure 9. Porosity vs. depth—shales. After Dickey (1976).

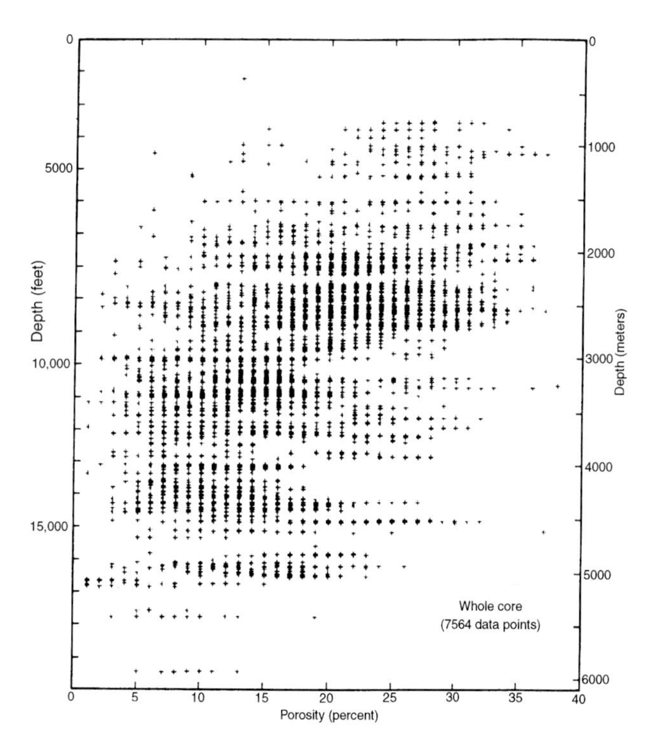

Figure 10. Sandstone porosity vs. depth from whole-core analyses for Lower Tertiary units along the Texas Gulf Coast. Sandstones with and without matrix were included. From Loucks et al. (1984).

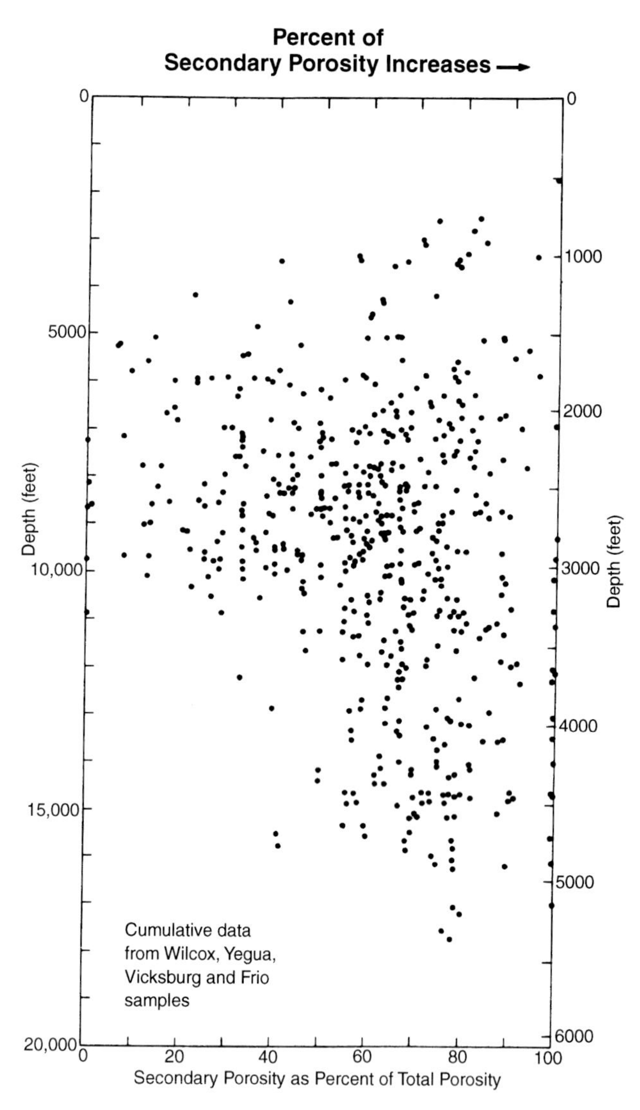

Figure 11. Secondary porosity as a percent of total porosity vs. depth for Lower Tertiary sandstones. Note that below 3000 m (10,000 ft) most pore networks are composed predominantly of secondary porosity. Each datum point represents a thin-section analysis. From Loucks et al. (1984).

through fractures (Figure 4) that may "heal" with blow down (Ghaith et al., 1990).

Whenever the internal fluid pressure across the seal exceeds the fracture pressure of the seal (P_F), the seal will fracture (Figure 4) and fluids will escape from the compartment (Bradley, 1975).

$$P_F = \sigma_H + P + T_S$$

where P_F = fracture pressure, psi (within the seal); σ_H = minimum effective horizontal stress, psi (outside the seal); P = pore pressure, psi (outside the seal); and T_S = tensile strength of seal, psi.

There has been, and continues to be, a great deal of discussion about the relative importance of the various possible origins of pressure/flow. Much of the

Table 5. Unit conversions.

145 psi = 1 MPa

1 psi = .006897 MPa

1 ft = .3048 m

1 m = 3.281 ft

$$\frac{1}{1°F} = \frac{1}{5/9°C}$$

$$\frac{1}{1°C} = \frac{1}{9/5°F}$$

1 md = $1 \times 10^{-15} m^2 = 1 \times 10^{-11}$ cm

$$1\frac{°F}{100 \text{ ft}} = 18.23\frac{°C}{\text{km}}$$

$$1\frac{\text{psi}}{\text{ft}} = 22.63\frac{\text{MPa}}{\text{km}}$$

psi	pounds per square inch
MPa	mega Pascals
ft	feet
m	meter
°F	degrees Fahrenheit
°C	degrees Celsius
md	millidarcy
cm	centimeter
km	kilometer

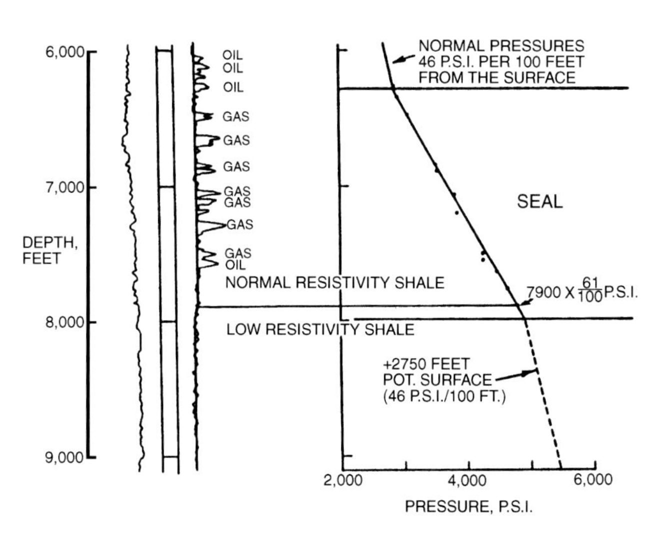

Figure 12. Pressure test well, Baram field, offshore Sarawak. After Powley (1987).

problem stems from equating a porosity-depth curve with a compaction curve (Neuzil, 1986a, b). "Compaction" is defined as the loss of porosity due to stress and, as such, is related to compressibility and

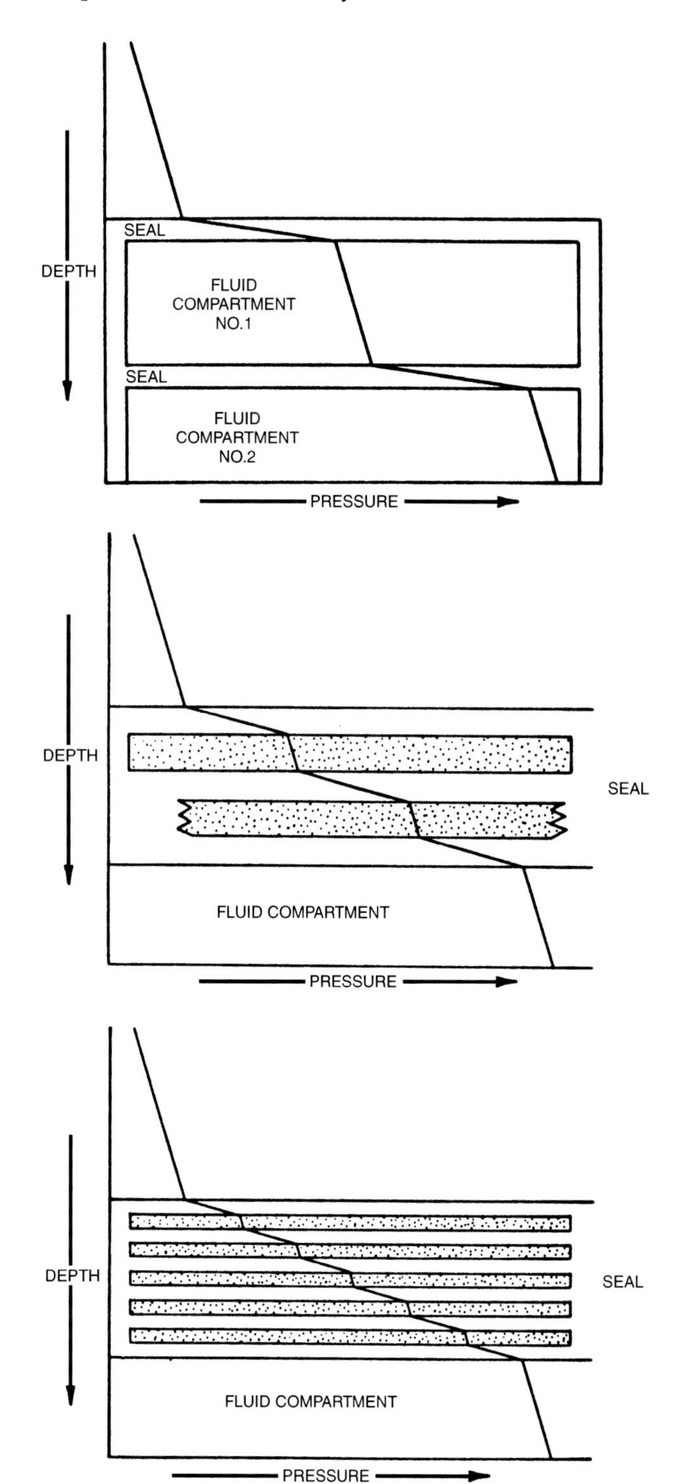

Figure 13. Seal thickness uncertainty. After Powley (1987).

pore-volume compressibility. The measured compressibilities of sedimentary rocks are two to four orders of magnitude less than the compressibilities calculated from the porosity-depth curve (Figure 5) (Neuzil, 1986a). The use of the porosity-depth compressibilities results in an orders-of-magnitude

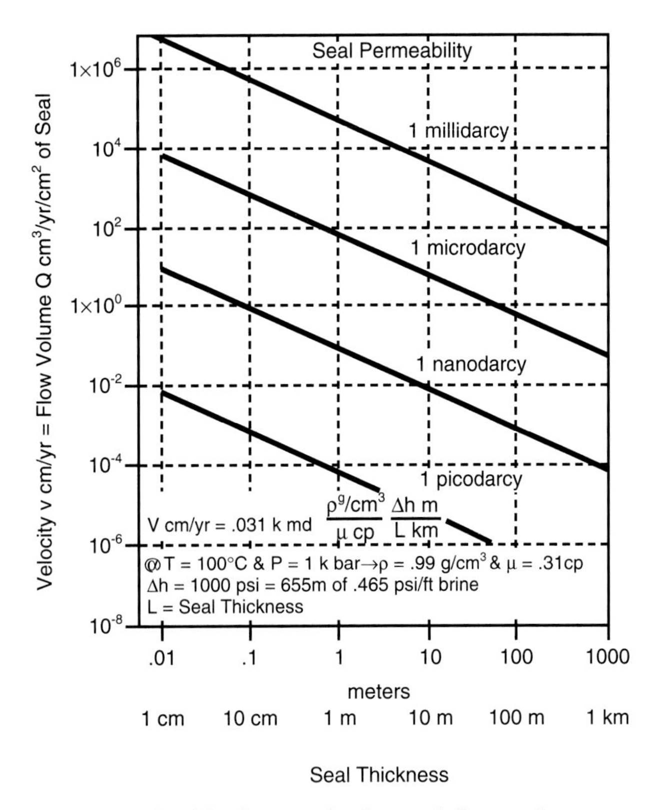

Figure 14. Seal leakage velocity and flow volume.

overemphasis on compaction disequilibrium (stress effects) relative to thermal and pore pressure effects.

The changes in pore volume and fluid volume, which are the origins of pressure/flow, are themselves the algebraic sum of several contributing factors. The total pore volume change with burial or unroofing is the algebraic sum of the bulk rock volume change, the grain volume change, the fluid volume change, and the pore volume change due to diagenetic processes. The diagenetic volume change is, in turn, the sum of precipitation effects, such as cements or authigenic clays, and of solution effects, such as stylolites or vugs. Replacement, such as silicification and pseudomorphs, and recrystallization, such as aragonite to calcite, may be thought of as microsolution and microprecipitation effects. The net result of all these volume changes is the change in porosity for a given depth increment. Note that porosity may increase or decrease with depth as a result of diagenetic effects.

Both the thermal and the stress-induced volume changes of rock and fluid proceed approximately smoothly with depth, as do some aspects of the diagenetic change. However, some diagenetic changes may occur at certain depth conditions T, P, and σ, and these changes alter parameters such as porosity, coefficient of thermal expansion, and compressibility, which, in turn,

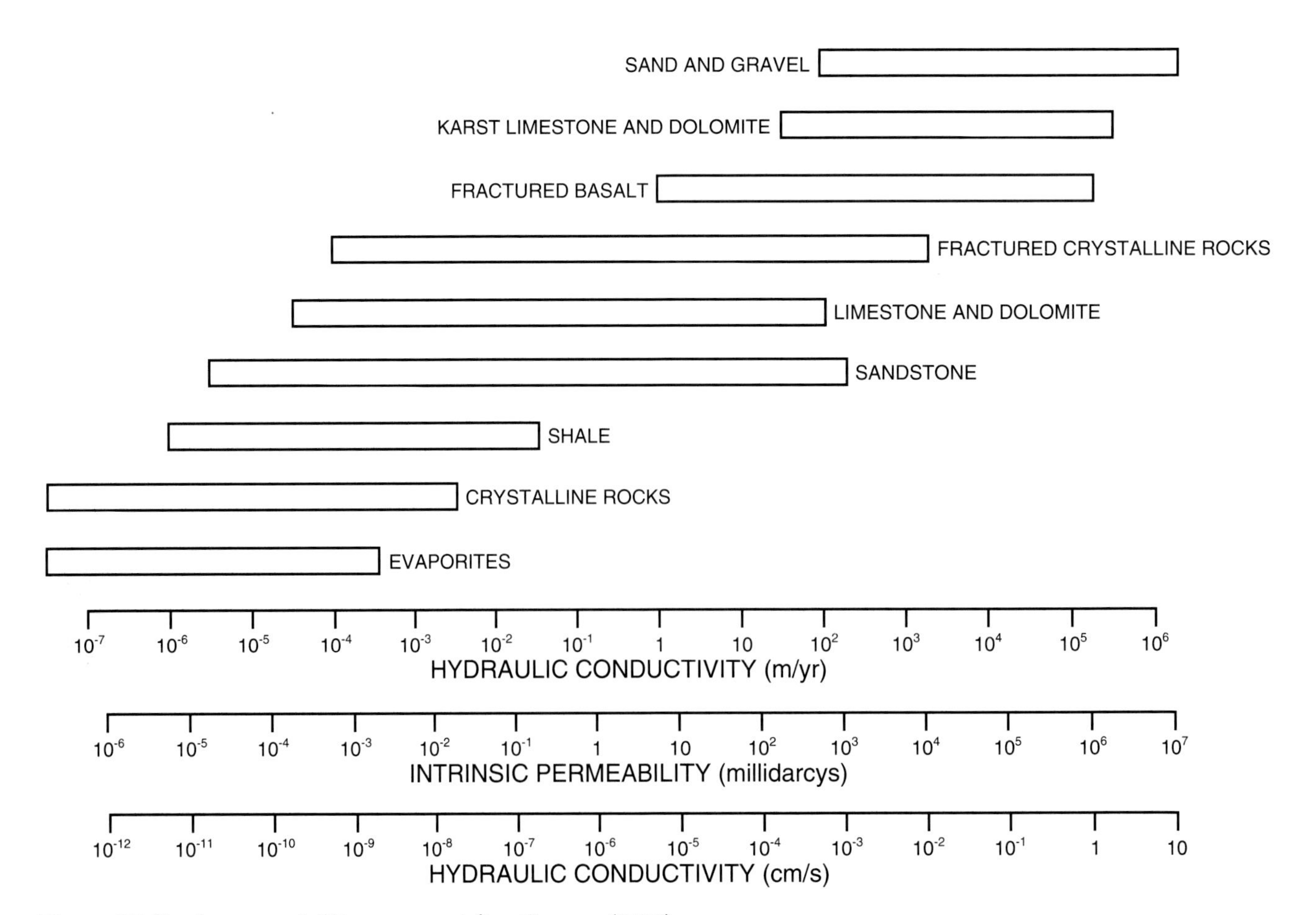

Figure 15. Rock permeability ranges. After Garven (1986).

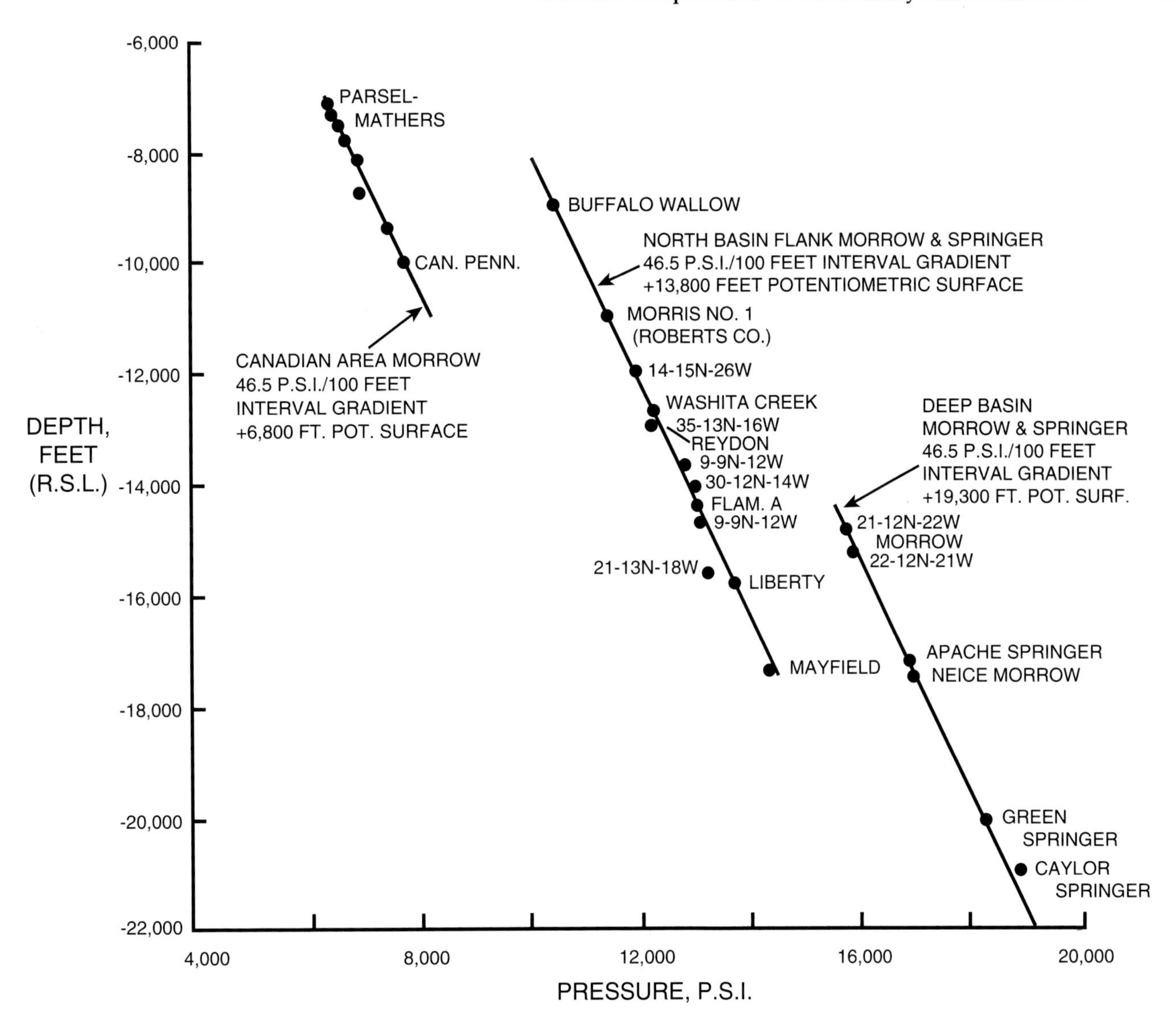

Figure 16. Pressures measured in the Morrow and Springer formations, western Anadarko basin. After Powley (1983).

alter the approximately linear-with-depth solutions for stress and temperature. The intricate complexity of the physical and chemical feedbacks in the system creates a nightmare for chemical basin modelers.

From the thermoporoelastic equations, Higgs (1991) has developed equations relating pore strain, flow, and abnormal pressures to epeirogenesis, i.e., burial with sedimentation and uplift with erosion. The flow (in an open system) is a function of the change in porosity or total pore volume strain.

The total pore strain on the pore fluid is the algebraic sum of all the contributing effects: (1) the change in bulk volume due to stress, pore pressure, and temperature, (2) the change in grain volume due to stress, pore pressure, and temperature, (3) the change in fluid volume due to pore pressure and temperature, and (4) the net change in pore volume due to diagenesis. The diagenetic effect on porosity is not yet determinable from theoretical principles but may be determined by difference. The total pore strain (change in porosity determined from a porosity-depth plot) less the bulk, grain, and fluid effects is the diagenetic effect. Notation is in Table 2.

$$\Delta\varepsilon_D = \Delta\varepsilon_T - \left(\frac{\Delta\varepsilon_B}{\phi} - \frac{1-\phi}{\phi}\Delta\varepsilon_G - \Delta\varepsilon_F \right)$$

The grain volume strain due to increments of effective stress cannot easily be calculated from theoretical principles because the grains are subjected to a heterogeneous stress field at the microscopic scale, which causes both volume changes and distortions. Three-dimensional finite element models of an idealized rock microstructure determined that for a 26% porosity rock (idealized by spherical grains in a rhombohedral packing with a grain contact diameter equal to 10% of the grain diameter), the grain compressibility should be corrected by a factor (f) of approximately 0.5 to correct for distortion (Higgs, 1991, personal communication). This assumption is made in the calculation below.

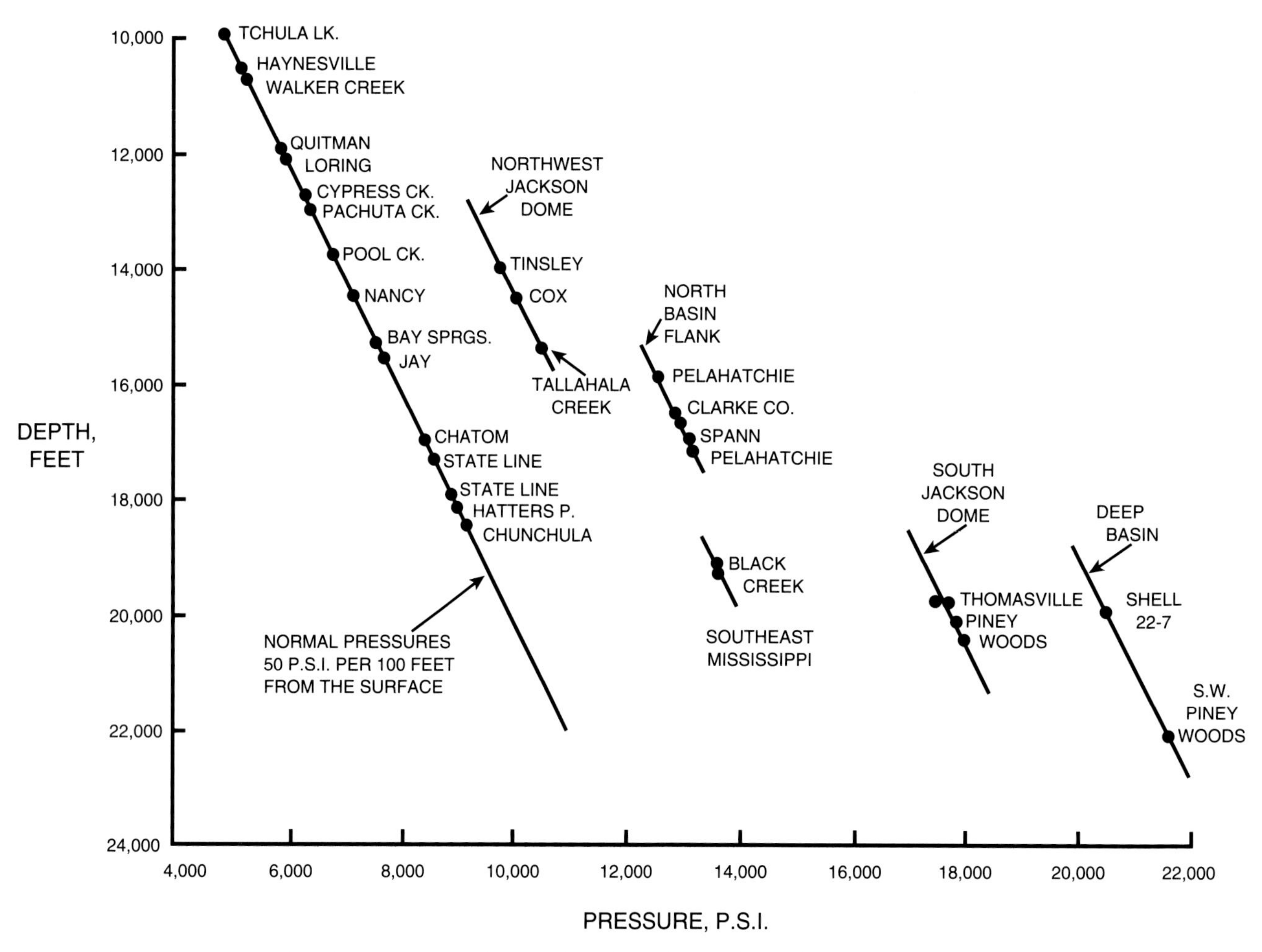

Figure 17. Pressures vs. depth, Smackover Formation, Mississippi salt basin. After Powley (1983).

To get some feel for the relative effects of changes of temperature, stress, and diagenesis with burial, a model may be examined in which a stratum subsides (with sedimentation) from 7500 to 7600 ft (zone of maximum shale porosity change, Figure 6).

$$\Delta D = 100 \text{ ft}$$

$$\Delta S @ \frac{1 \text{ psi}}{\text{ft}} = 100 \text{ psi}$$

$$\Delta T \text{ °F} @ \frac{1.2\text{°F}}{100 \text{ ft}} = 1.2\text{°F}$$

$$\varepsilon_T = \frac{\Delta V_{total}}{V} = \frac{0.002\text{cm}^3}{\text{cm}^3} \quad (2\% \text{ per 1000 ft}) \text{ (Figure 6)}$$

Measured shale parameters (Preston, 1976) are:

$$\phi = 0.067$$

$$K = 2.19 \times 10^6 \text{ psi}$$

$$\beta_B = \frac{1}{K} = \frac{0.46 \times 10^{-6}}{\text{psi}}$$

$$G = 1.78 \times 10^6 \text{ psi}$$

Based on the thermoporoelastic equations, Higgs (1991) derives expressions for changes in porosity with burial in an open system:

$$d\phi = \left[\frac{1}{M} - (1-\phi) f \beta_G\right](dS - dP) + \left\{\left[\frac{K}{M} - (1-\phi)\right]\beta_G - \phi\beta_F\right\} dP - \left\{\left[\frac{K}{M} - (1-\phi)\right]\alpha_G - \phi\alpha_F\right\} dT$$

And for changes of pressure with burial in a closed system:

$$dP = \frac{[1-(1-\phi)Mf\beta_G]dS + [\phi M\alpha_F - K\alpha_G + (1-\phi)M\alpha_G]dT}{\phi M\beta_F + 1 - (1-\phi)Mf\beta_G - K\beta_G + (1-\phi)M\beta_G}$$

These equations are used with the parameters above to calculate expected porosity and pressure changes in the model. The results of the calculations are shown in Table 3.

For the open system, the porosity would decrease 0.00314% with the 100 ft of burial. Porosity is the pore volume divided by the bulk volume, so 0.0000314 cm^3

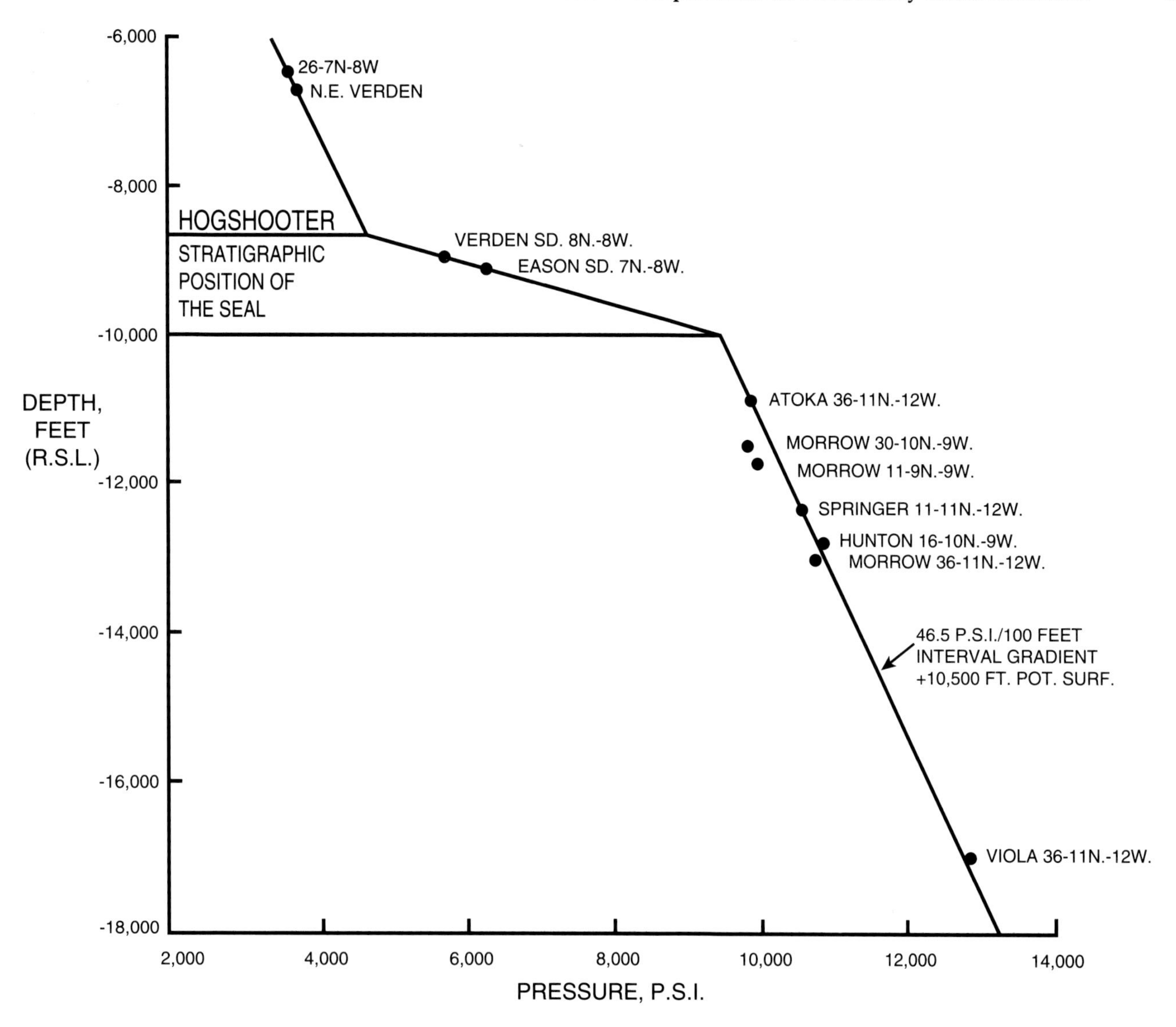

Figure 18. Pressure vs. depth, N.W. Chickasha area, Marchand sand trend, Grady County, Oklahoma. After Powley (1983).

of fluid would be expelled from 1 cm^3 of porous rock.

For the closed system, the pressure would increase by 125.32 psi, 28.70 psi due to stress effects and 96.61 psi due to thermal effects.

No diagenetic effects are calculated by these expressions. The diagenetic effects are presumed to be the remaining unaccounted-for porosity changes contained in the porosity-depth curve. These diagenetic effects are calculated.

$$\Delta\varepsilon_D = \Delta\varepsilon_T - \left(\frac{\Delta\varepsilon_B}{\phi} - \frac{1-\phi}{\phi}\Delta\varepsilon_G - \Delta\varepsilon_F \right)$$

$$\Delta\varepsilon_D = 0.002 - 0.0000314$$

$$\Delta\varepsilon_D = 0.00197$$

At greater depths, the total change in porosity of shale with burial approaches 0%/100 ft (Figure 6). So the diagenetic effect decreases essentially to zero, while the volume changes due to temperature and stress remain the same. Table 4 shows the relative distribution of effects for 100 ft of additional burial at two depths.

It is obvious that the choice of a porosity-depth curve, whether for shale, sandstone, or limestone, controls the relative effects of stress, temperature, and diagenesis. Unfortunately, there is no unanimity of thought on how porosity varies with depth in various lithologies and mixtures of lithology (Figures 6, 7, 8, and 9). Even when a brave author publishes a porosity-depth curve, it is wise to remember the origin of the line (Figures 10 and 11).

It should be noted that the volume changes due to stress, pressure, and temperature are reversible (elastic) with a change in conditions, while the volume changes due to diagenetic processes are essentially irreversible (inelastic).

Table 5 shows the unit conversions used herein.

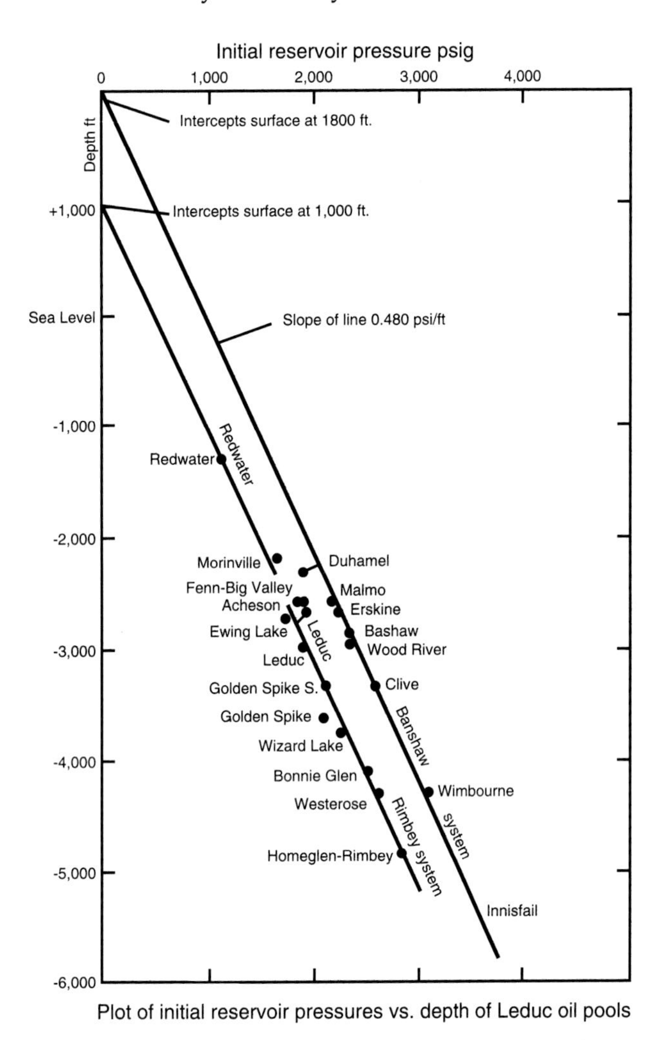

Figure 19. Pressure vs. depth—Alberta basin. After Dickey (1979).

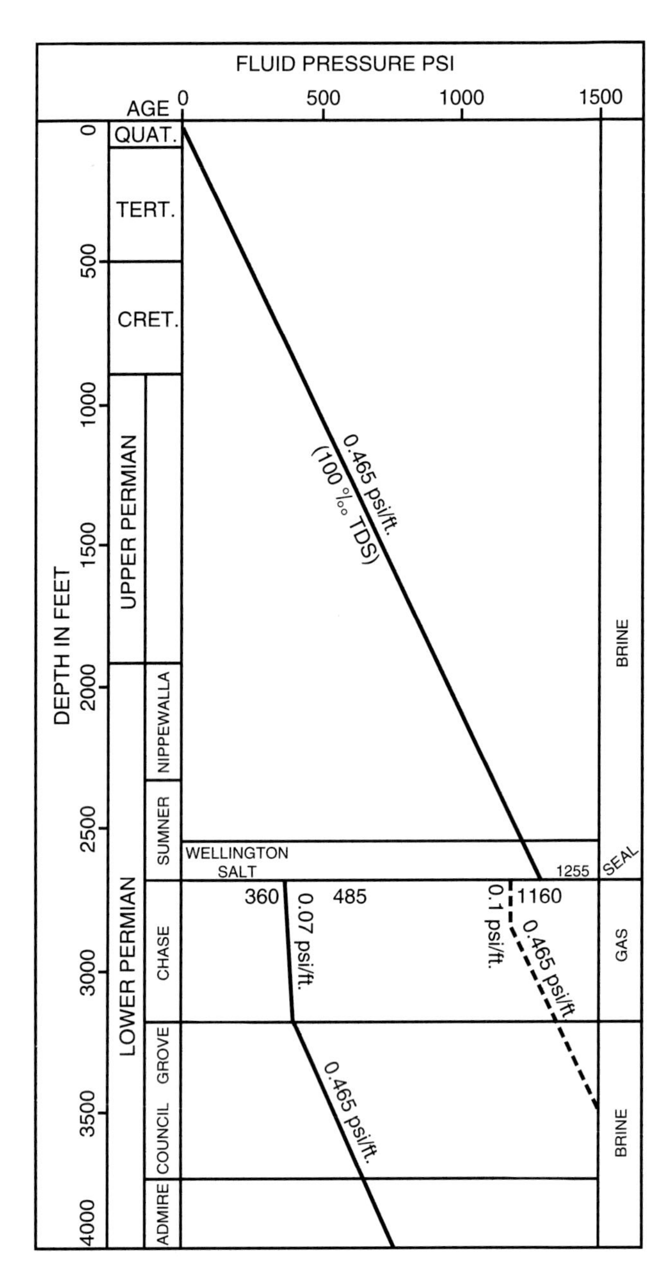

Figure 20. Pressure vs. depth—Hugoton field, Kansas. After Bradley (1985).

Seals

The existence of a common potential throughout a large volume of porous rock adjacent to similar large volumes of different potential implies the existence of a seal completely surrounding the volume. If such seals did not exist, the pressures would equilibrate. The seals appear to be relatively thin when compared with the vertical and lateral dimensions of the rock volume.

The effectively impermeable seals have been defined only by differences in hydraulic potential measured in relatively permeable reservoir beds. While seals are suspected at evaporite or clay laminae or along faults and unconformities, no example of a *directly measured, permeability-defined* seal is known, and thus seal thickness is somewhat problematical.

Theoretically, the seal could be extremely thin, a membrane one grain thick. The actual seal thickness is somewhere between the distance between a normal and abnormal pressure measurement and a membrane. Sometimes tests show intermediate pressures in reservoir zones between the normal and the fully abnormal measurements (Figure 12). Such records in a "transition zone" may indicate either a thick homogeneous seal or a series of parallel thin seals that are increasingly effective with depth (Figure 13). Inspection of bedded to laminated, highly inhomogeneous cores from the "transition zone" leads to the conclusion that multiple seals are more likely. Pending the development of shale permeameters, the number, effectiveness, and thickness of seals remain unknown.

To give some idea of the permeability of seals, a table of flow velocities and flow amounts was calculated for different seal permeabilities and thicknesses, assuming the Darcy equation is valid for such low flows. The pressure (potential) drop across the seal was kept constant at 1000 psi. The results of the calcu-

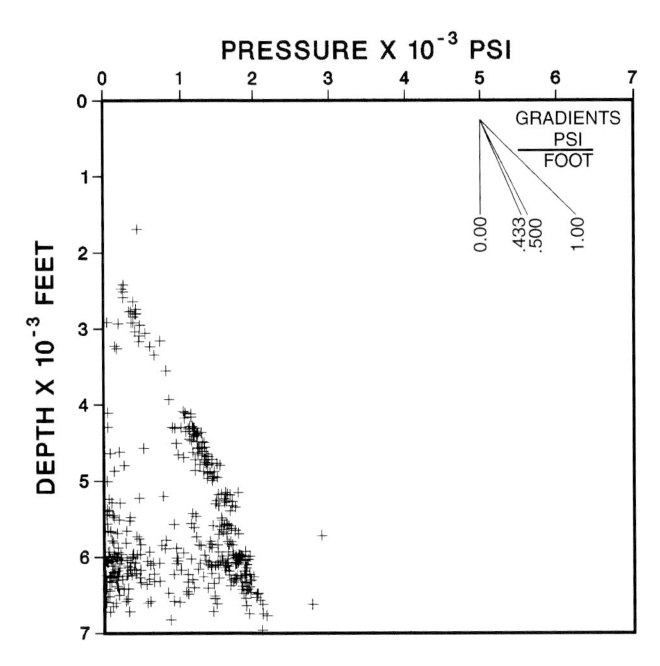

Figure 21. Pressure vs. depth—underpressures, Stevens County, Kansas.

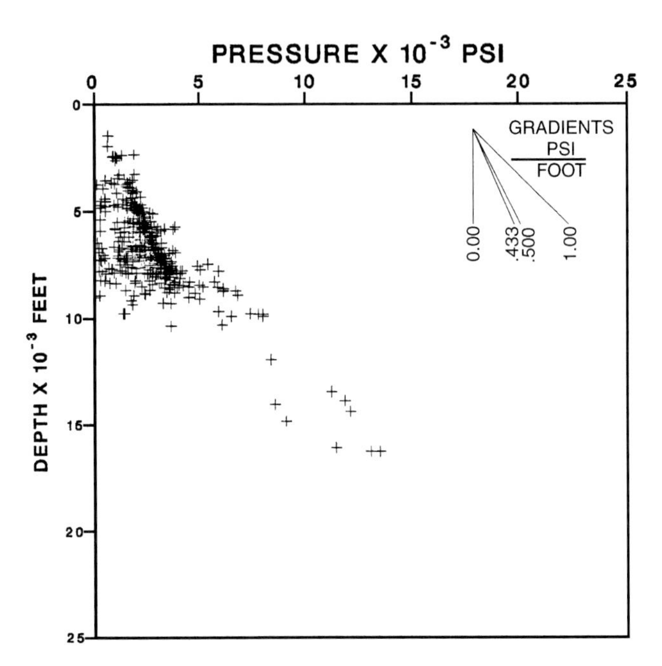

Figure 22. Pressure vs. depth—overpressures, Harris County, Texas.

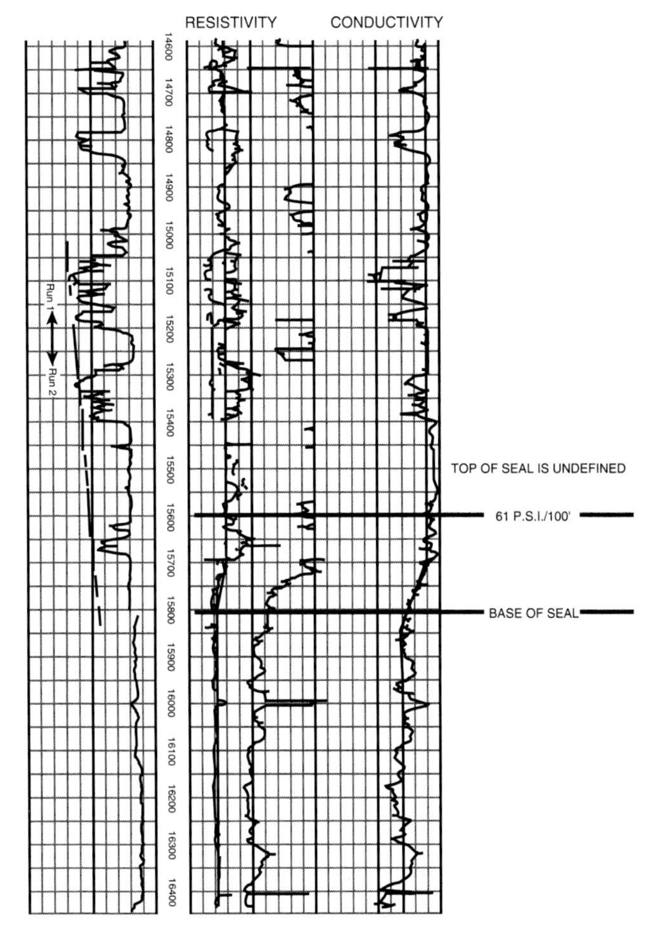

Figure 23. Resistivity log, Amoco No. 1 S.L. 4427, West Bayou Carlin field, St. Mary Parish, Louisiana. After Powley (1983).

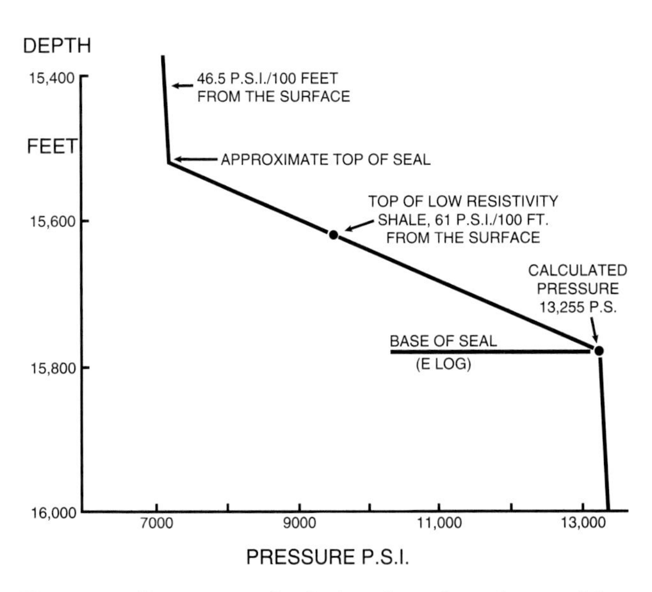

Figure 24. Pressure calculation from log, Amoco No. 1 S.L. 4427, West Bayou Carlin field, St. Mary Parish, Louisiana. After Powley (1983).

lations are plotted in Figure 14. A comparison of permeabilities for various lithologies is given in Figure 15. Because pressures can be equilibrated with very little fluid flow, the plot indicates that the seals must have extremely low permeabilities to be effective through geologic time.

One of the major problems in determining the effect of seals on fluid movement in a sedimentary basin is that only a pressure difference proves an

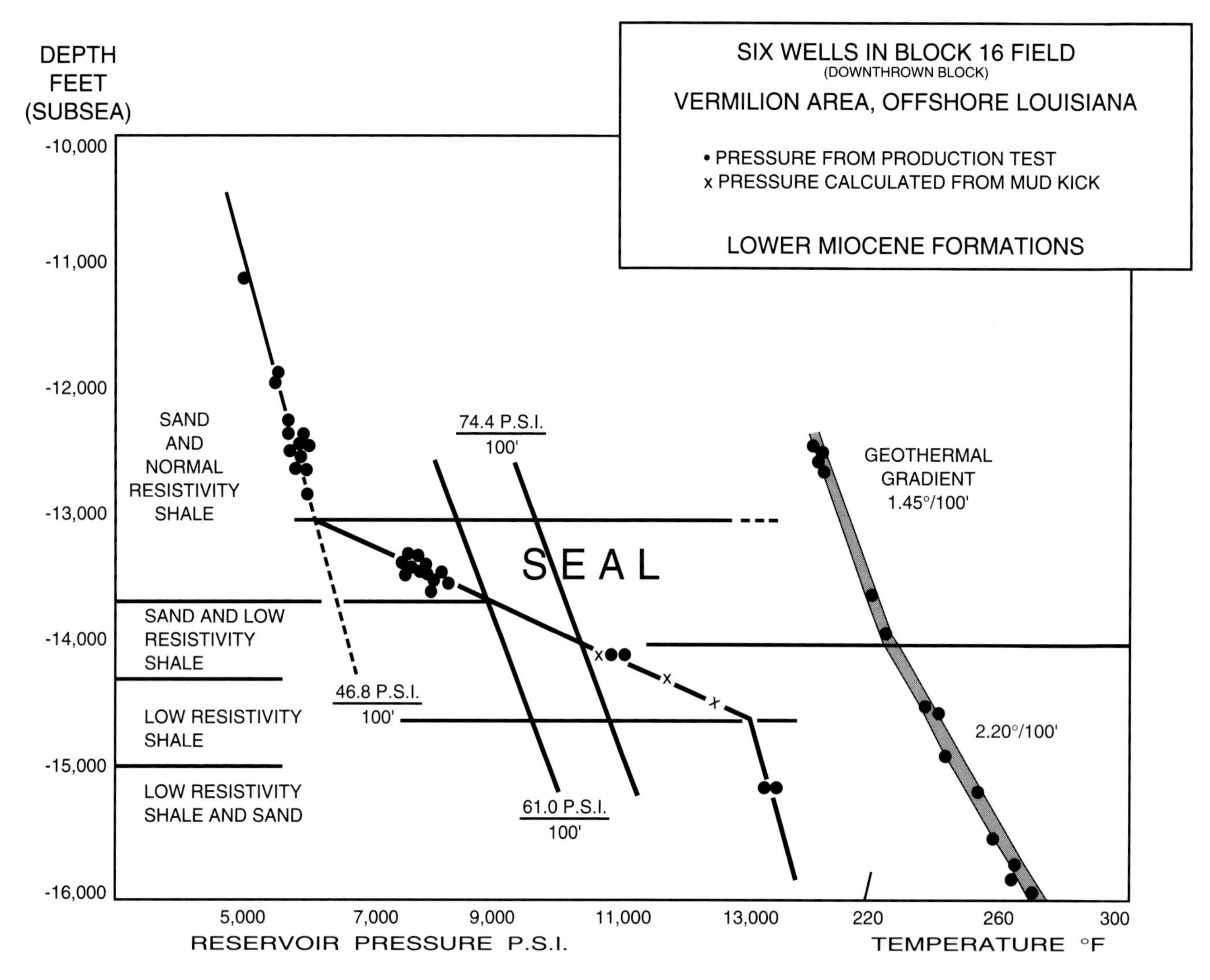

Figure 25. Comparison of pressure and temperature gradients—Vermilion area, Louisiana. After Powley (1983).

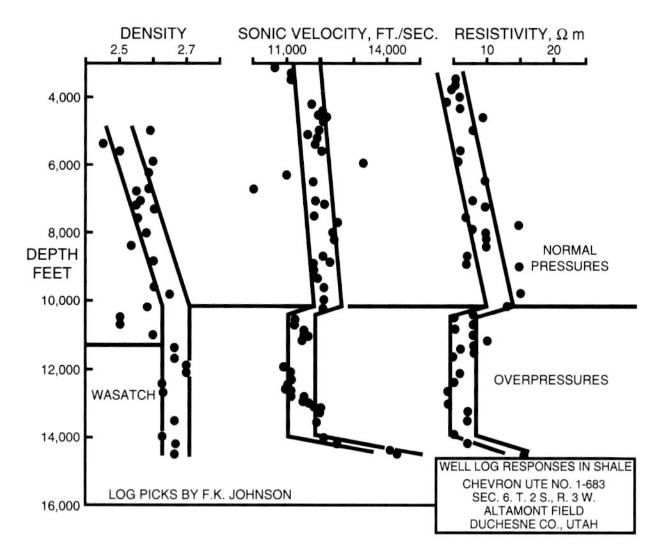

Figure 26. Log response to pressure in shale. After Powley (1983).

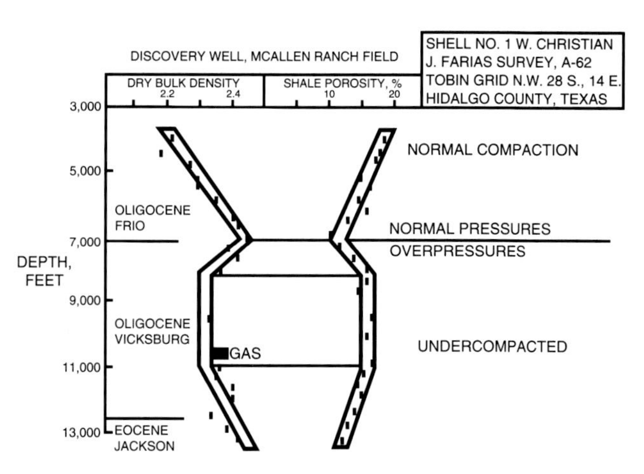

Figure 27. Shale porosities and densities—Hidalgo County, Texas. After Powley (1982).

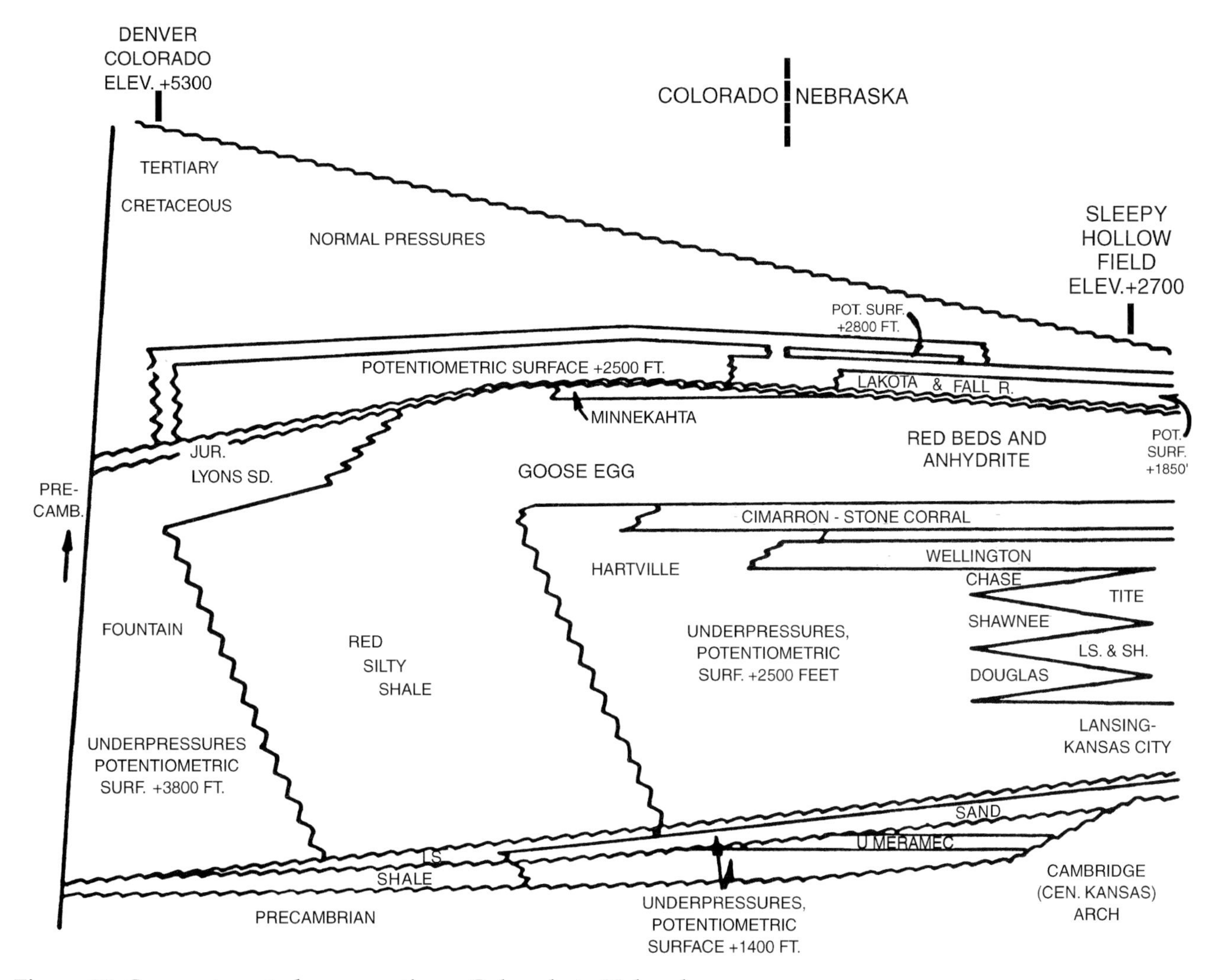

Figure 28. Compartmented cross section—Colorado to Nebraska.

effectively sealed compartment. But the seals may exist even though the pressures are now normal (Figure 1). Also, vertical seals can exist without ever having had a top seal or a change from hydrostatic pressures. In all cases, the occurrences of seals or incipient seals would tend to reduce fluid movement at basin (compartment) scales. Pressure compartment sizes can vary from one fault block in the Gulf Coast Basin (a half mile or so) to at least 100 miles in the Alberta Basin. In a compartmented basin, the flow velocities at depth would approach zero.

The origins of seals must be multiple to explain their geometric and stratigraphic occurrence. Certainly some seals appear to be sedimentary, with perhaps more or less diagenesis. Some seals, particularly those that cross stratigraphy, appear to be entirely diagenetic. The lateral seals, which appear to be subvertical to vertical, are possibly due to faulting and fracturing, or to lateral facies changes. The diagenetic effects on lateral seals are problematic because the pressure, temperature, and stress, which control the diagenetic chemistry, vary with depth along the seal.

A major investigation of seals—their origin, geometry, lifetime, and occurrence—has been undertaken by the Gas Research Institute with a multiuniversity consortium. The institutions presently involved are Indiana University, Pennsylvania State University, Oklahoma State University, Wisconsin University, Texas A&M University, Wyoming University, Woods Hole Oceanographic Institute, and Cornell University. Some of the early results of this effort are Tigert and Al-Shaieb (1990), Moore and Ortoleva (1990), Dewers and Ortoleva (1990), Ghaith et al. (1990), Bahr (1989), Engelder and Weedman (1989), Hunt (1990), and Cathles (1989).

Indicators of Compartmentation

Most indications of compartmentation are related to pore-fluid pressures, measured or implied. However, numerous chemical parameters may also infer or confirm compartmentation. Both the major ion and isotope water chemistry may (or may not) vary from compartment to compartment (Fisher,

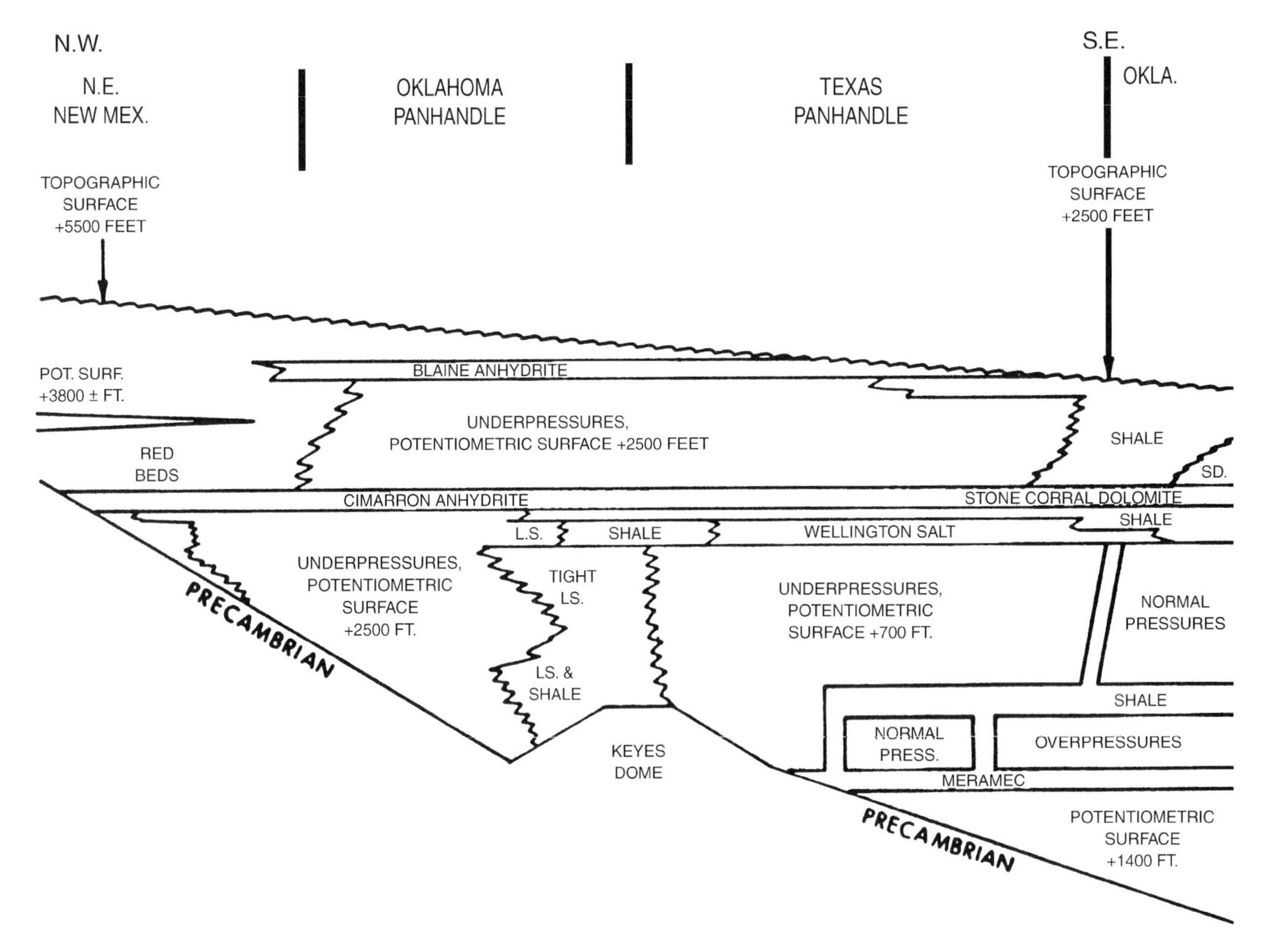

Figure 29. Compartmented cross section—New Mexico to Oklahoma.

1989, personal communication). The hydrocarbon chemistry may show similar variations. The mineralogy may also reflect the variations in water chemistry. The chemistry of fluids in fluid inclusions has been shown to differ between compartments (Smith, 1989, personal communication).

Examples of pressure compartments are shown in Figures 16, 17, and 18. The compartments are evidenced by the pressure measurements lining up on separate brine gradients. The pressure data would line up on one brine gradient if the compartments were hydraulically connected. An example from the Alberta Basin is shown in Figure 19. A Hugoton field compartment is described by Bradley (1985; Figure 20).

The most direct indication of compartmentation is pore-fluid pressure measurements from drill-stem tests (DST), repeat formation tests (RFT), and initial production tests (IP), all of which give a form of bottomhole final shut-in pressure (FSIP). Mud kicks and lost circulation may also indicate formation pressure when the formation depth and mud density are known. Much of this information is commercially available from scout ticket databases such as those from Petroleum Information and its affiliates. Perhaps the most reliable information is from the sworn records of state commissions controlling hydrocarbon production. Many of these records are available from Dwight's Petroleum Data Service. Foreign pressure information is available from CIFE, ERICO, Robertson, and Petroconsultants. Pressure-depth diagrams for counties (or smaller areas) from these data may roughly indicate pressure compartments and the areas for more detailed pressure studies (Figures 21 and 22).

Less direct, but still pressure related, is the shale resistivity (from well logs) method, which shows a reduction in resistivity with an increase in pore pressure (Figures 23 and 24). The top of the resistivity break was found empirically to be at a pressure gradient value of 0.61 psi/ft from the surface, and the bottom is assumed to be the bottom of the seal. A similar empirical indication of overpressures is the change in thermal gradient at about 0.75 psi/ft from the surface (Figure 25).

Sonic velocity logs may also indicate overpressured compartments by low velocities (Figure 26). This is probably due to the reduced effective stress on the rock as the pore pressure increases or may be due to a lower density shale.

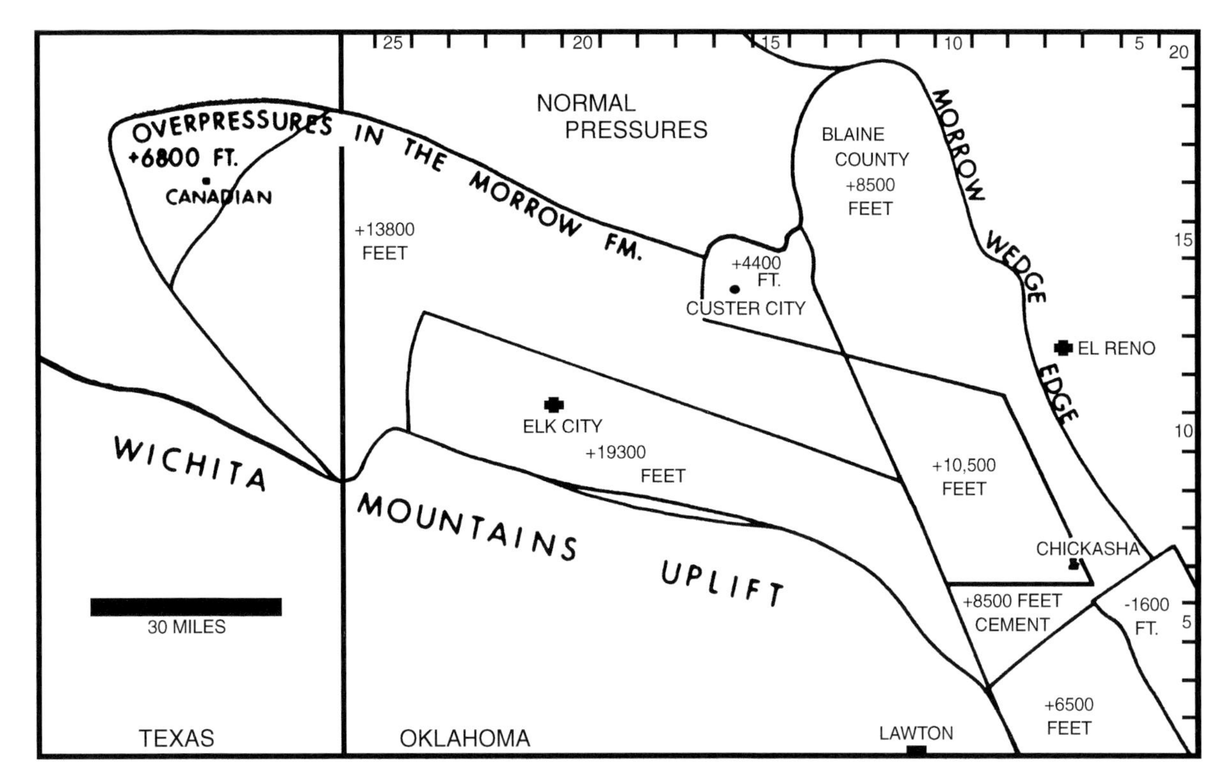

Figure 30. Compartmented map—pressure compartments and potentiometric surfaces in the Morrow Formation—Anadarko basin. After Powley (1983).

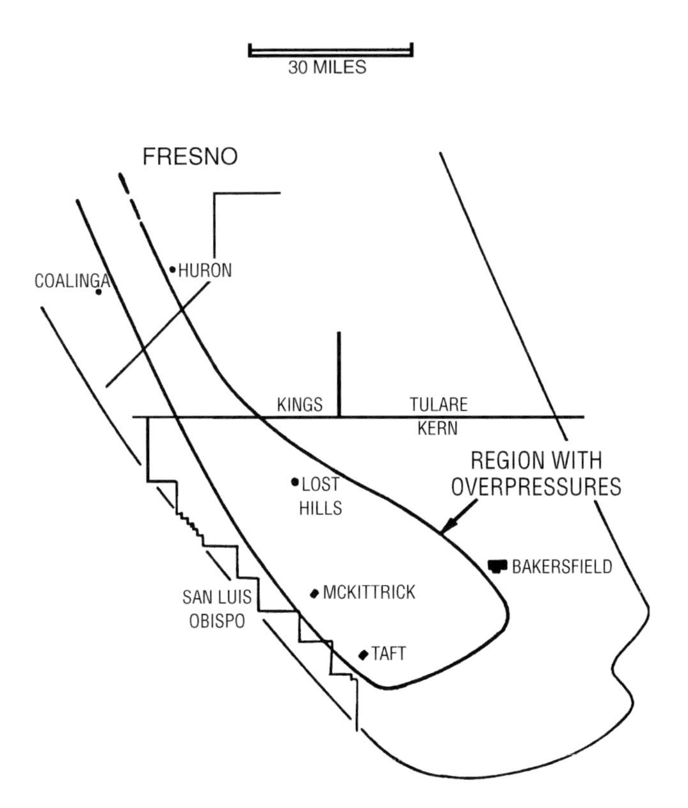

Figure 31. Overpressures—southern San Joaquin basin.

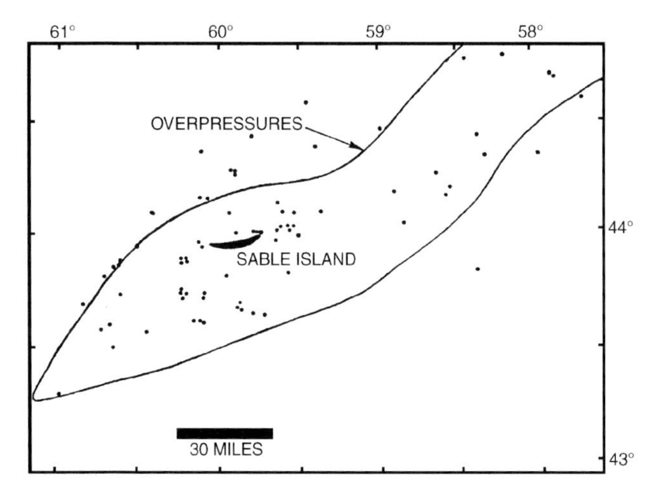

Figure 32. Overpressures—offshore Nova Scotia.

Shale densities may often, but not always, be indications of overpressures (Bradley, 1976). The shale density may be lower in the higher pressure zones (Figure 27).

Reflection seismology maps the top of overpressures as a low velocity interface. Neither seismology nor sonic logs delineate underpressures.

Since a pressure seal is characterized by extremely low permeabilities, which imply low porosities and

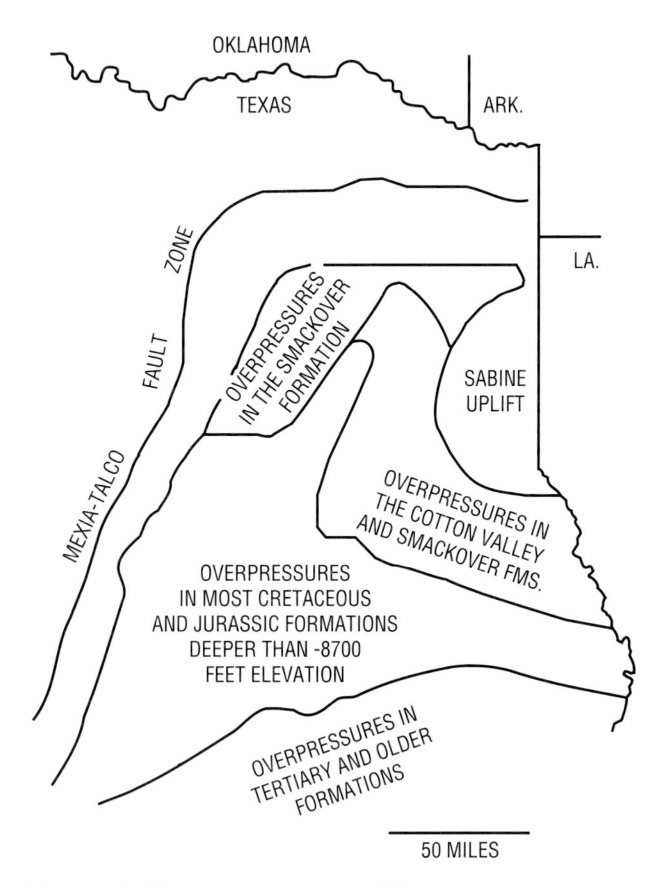

Figure 33. Overpressures—east Texas.

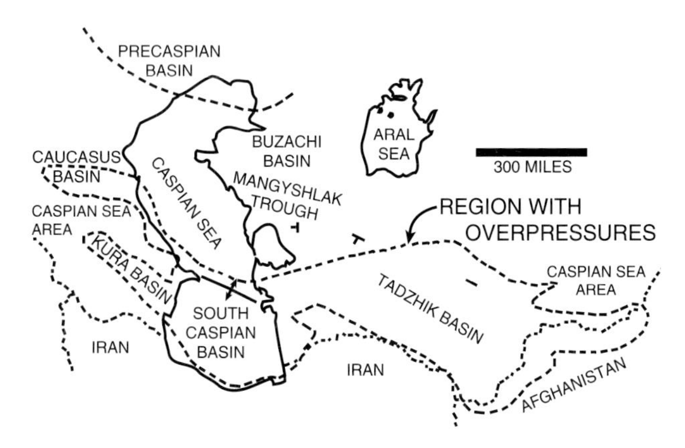

Figure 34. Overpressures—Caspian Sea area.

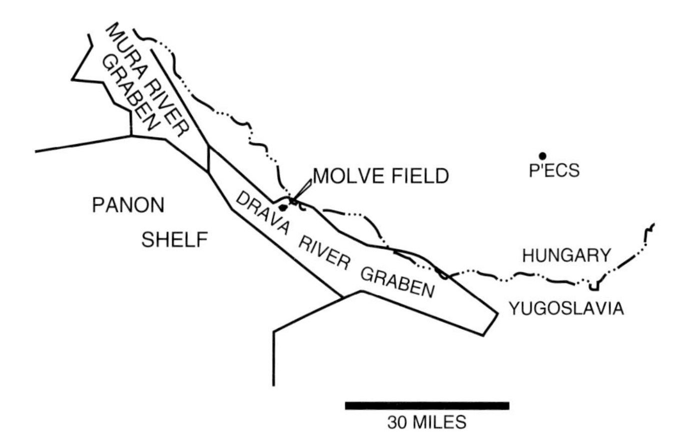

Figure 35. Overpressures—Yugoslavia.

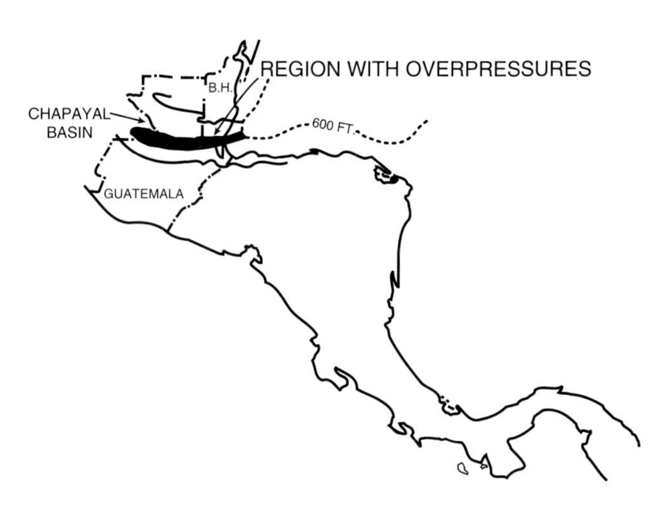

Figure 36. Overpressures—Guatemala.

high bulk densities, it should be seen on resistivity, sonic, and density logs. The fact that indications are not obvious on these logs may indicate a "thin-bed" logging problem which would be an argument for "thin" seals. Perhaps micrologs (resistivity) or microsonic logs taken through the "transition" between normal and abnormal pressures will pinpoint the seal or seals. If seals are indeed found to be relatively thick, there is the possibility that high-frequency reflection seismology could map the dense (high velocity) top seal. The high frequency seismic is limited to relatively shallow targets because of the relationship between frequency, wave length (resolution), and depth (attenuation).

Also, a seal may be indirectly implied by a sudden decrease in drilling rate caused by the high-density, low-porosity seal lithology.

By utilizing several or all of the above characteristics, but principally pressure-depth diagrams, it is possible to construct pressure compartment cross sections (Figures 28 and 29). These sections can be helpful in basin analysis and modeling to delineate flow and permeability heterogeneity. Maps of pressures in given formations may also be constructed (Figure 30).

EXAMPLES OF PRESSURE COMPARTMENTS

Pressure compartments are known on all continents except Antarctica. Many of the examples are from North America simply because the measure-

Figure 37. Overpressures—Australia and Papua.

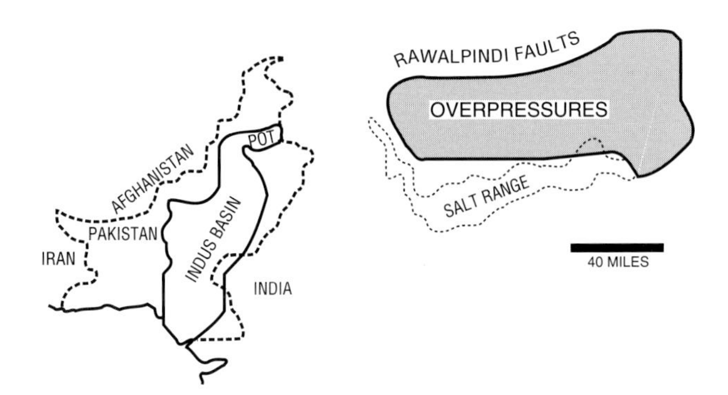

Figure 38. Overpressures—Potwar subbasin, Pakistan.

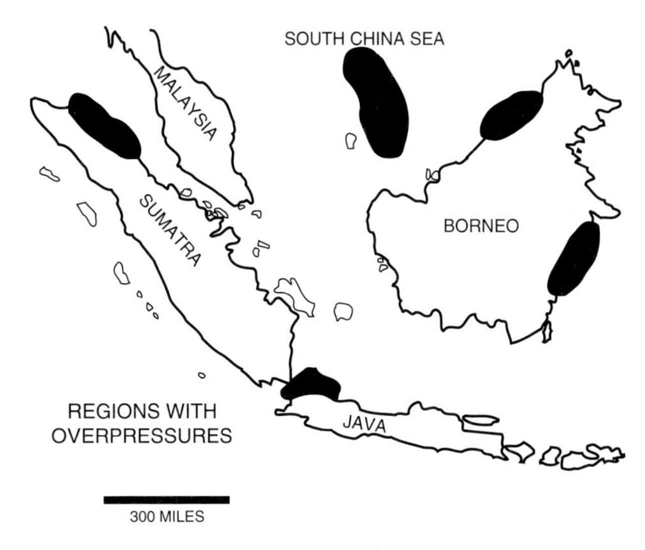

Figure 39. Overpressures—Indonesia.

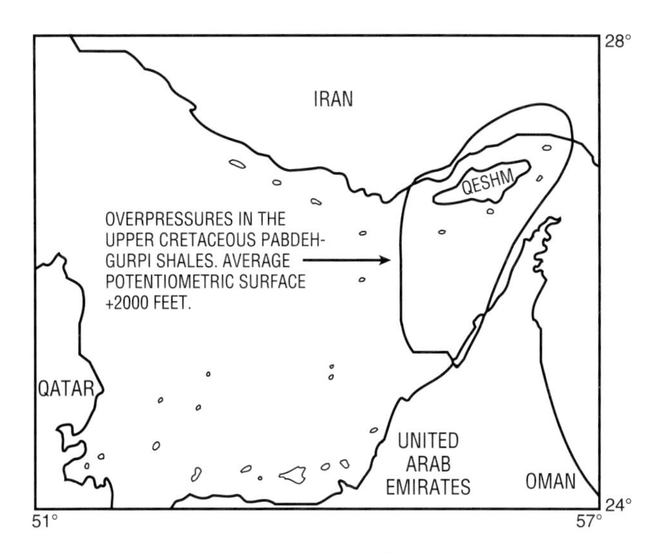

Figure 40. Overpressures—southern Persian Gulf.

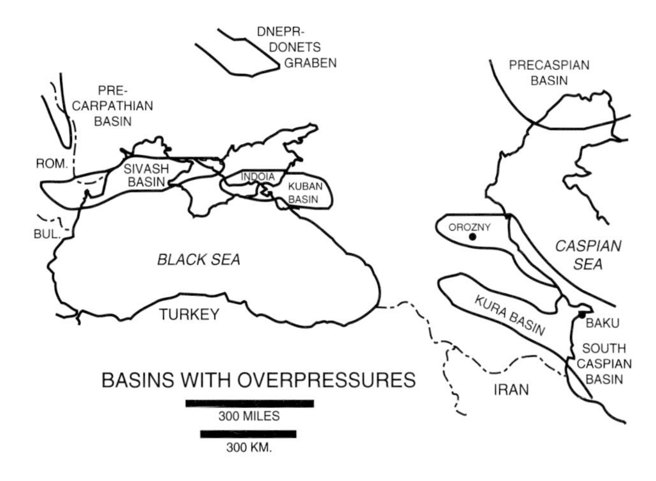

Figure 41. Overpressures—Black Sea area.

ments and other data have been more readily available there. Only a few of the available pressure-depth diagrams are provided because of space problems. Also, the availability and quality of pressure-depth information in the public domain varies widely with both the area and dates of drilling. The ubiquity of abnormal pressures is indicated by a bibliographic search of the Tulsa Data Base (University of Tulsa—Petroleum Abstracts), which listed 539 references to overpressure, 41 to underpressure, and 80 to abnormal pressure. Many of the earlier figures show examples of pressure compartments, but there are numerous other examples. Figures 31 through 48 show some additional areas where abnormal pressure

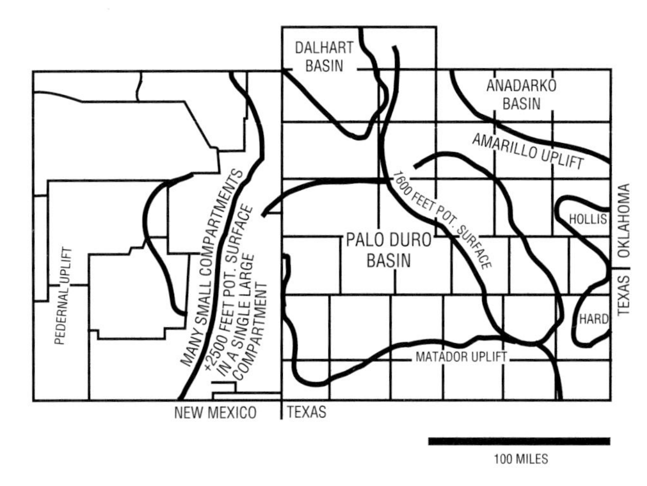

Figure 42. Fluid compartments below the seal in the Clearfork Group, North Texas and New Mexico.

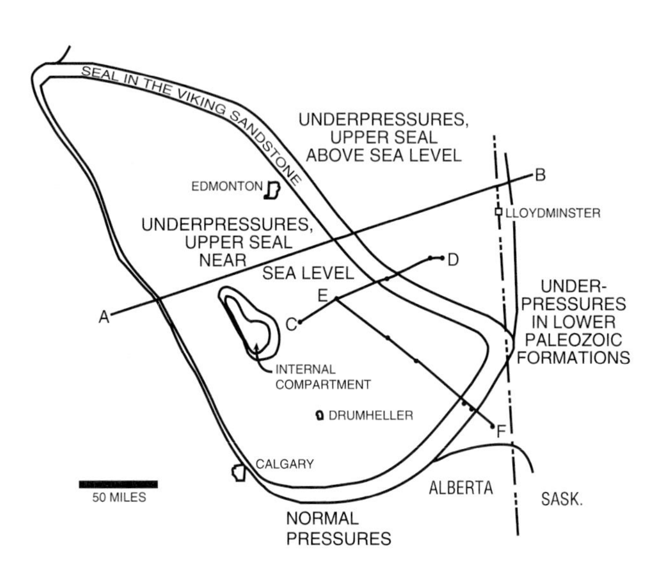

Figure 43. Pressure compartments—Alberta basin.

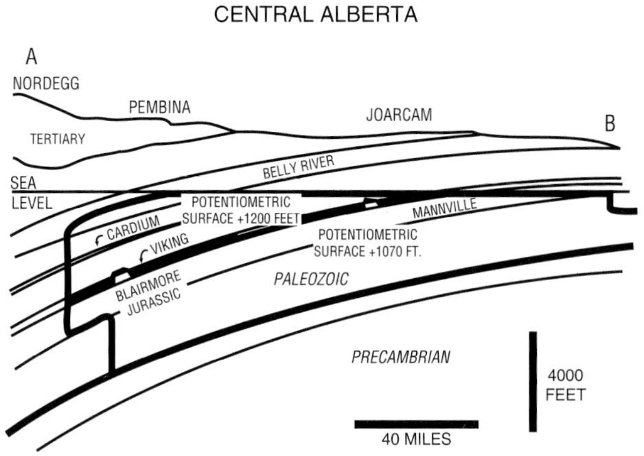

Figure 44. Pressure cross section—Alberta (A–B).

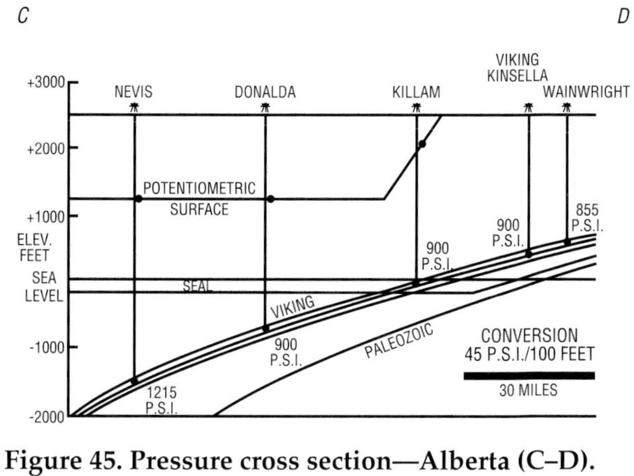

Figure 45. Pressure cross section—Alberta (C–D).

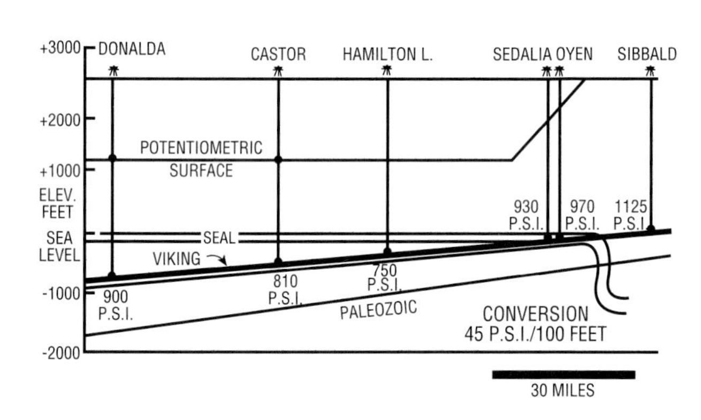

Figure 46. Pressure cross section—Alberta (E–F).

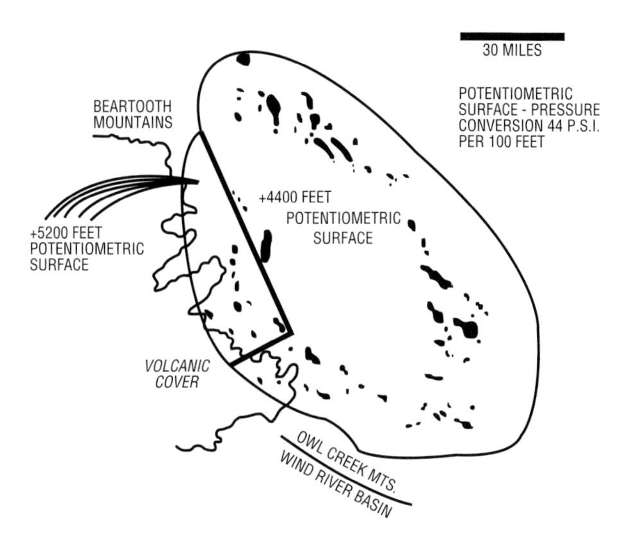

Figure 47. Compartment map—Tensleep Sandstone, Big Horn basin, Wyoming. After Powley (1983).

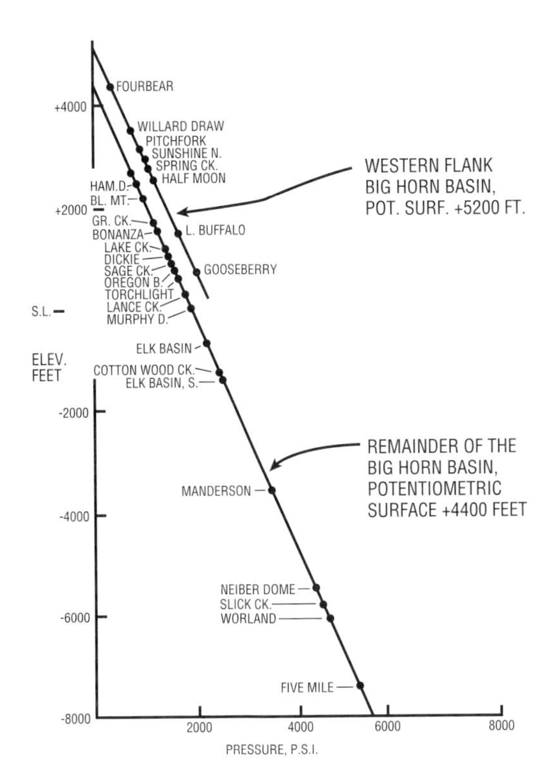

Figure 48. Pressure vs. depth—Tensleep Sandstone, Big Horn basin, Wyoming. After Powley (1983).

are known and give a few examples of pressure-depth plots and pressure cross sections.

ACKNOWLEDGMENTS

Published by permission of Amoco Production Company. This report has benefited from the scientific and editorial inputs of N. G. Higgs, D. R. Spain, R. G. Gibson, L. C. Babcock, P. J. Ortoleva, Z. Al-Shaieb, and Paul Westcott.

REFERENCES CITED

Atwater, G. I. and Miller, E. E., 1965, The effect of decrease in porosity with depth on future development of oil and gas reserves in south Louisiana: AAPG Bull., V. 49, No. 3, pt. 1, p. 334, abstract.

Atwater, G. I., Miller, E. E., and Wiggins, G. B., 1986, Effect of decreased porosity with depth on oil and gas reserves in south Louisiana sandstone reservoirs: AAPG Bull., V. 70, No. 5, p. 561, abstract. Presentation notes from R. Steinmetz, July 15, 1986, personal communication.

Bahr, J. M., 1989, Evaluation of pressure distribution in the St. Peter sandstone, Michigan Basin: Abstract, EOS, p. 1097, October 24, 1989.

Bradley, J. S., 1975, Abnormal formation pressure: AAPG Bull., V. 59, n. 6, pp. 957–973.

Bradley, J. S., 1976, Abnormal formation pressure, Reply: AAPG Bull., V. 60, n. 7, pp. 1 127–1 128.

Bradley, J. S., 1985, Safe disposal of toxic and radioactive liquid wastes: Geology, V. 13, pp. 328–329.

Bradley, J. S., 1986, Fluid movement in deep sedimentary basins: A review: Proc. 3rd Can./Am. Conf. on Hydrogeology, Banff, 1986, Nat'l. WaterWell Assoc., pp. 19–31.

Cathles, L. M., 1989, Overpressuring, episodic fluid flow, and process coupling in sedimentary basins: Abstract, EOS, p. 1097, Oct. 24, 1989.

Chapman, R. E., 1981, Geology and water, an introduction to fluid mechanics for geologists: Martinus Nijhoff/Dr. W. Junk Publishers, The Hague, 228 p.

Dewers, T., and Ortoleva, P., 1990, A coupled reaction/transport/mechanical model for intergranular pressure solution, stylolites, and differential compaction and cementation in clean sandstones: Geochemica and Cosmochemica Acta, V. 54, pp. 1609–1625.

Dickey, P. A., 1972, Migration of interstitial water in sediments and the concentration of petroleum and useful minerals: 24th World Geological Congress, Montreal, Section 5, pp. 3–16.

Dickey, P. A., 1976, Abnormal formation pressures: Discussion: AAPG Bull., V. 60, n. 7, pp. 1124–1127.

Dickey, P. A., 1979, Petroleum development geology: Petroleum Publishing Co., Tulsa, Oklahoma, 398 p.

Engelder, T., and Weedman, S., 1989, An analysis of the pressure chamber model: Abstract, EOS, p. 1097, Oct. 24, 1989.

Garven, G., 1986, The role of regional fluid flow in the genesis of the Pine Point deposit, western Canada sedimentary basin—a reply: Econ. Geol., v. 81, no. 4, pp. 1015–1020.

Ghaith, A., Chen, W. and Ortoleva, P., 1990, Oscillatory methane release from shale source rock; Earth–Science Reviews, 29, pp. 241–248.

Hunt, J. M., 1990, The generation and migration of petroleum from abnormally pressured fluid compartments: AAPG Bull., V. 74, n. 1, pp. 1–12.

Loucks, R. G., Dodge, M. M., Galloway, W. E., 1984, Regional controls on diagenesis and reservoir quality in lower Tertiary sandstones along the Texas Gulf Coast: AAPG Memoir 37, Clastic Diagenesis, pp. 15–45.

Moore, C. H., and Ortoleva, P. J., 1990, Effects of fluid and rock compositions on diagenesis: A modeling investigation: Am. Assoc. Petrol. Geol. Memoir 49, Prediction of reservoir quality through chemical modeling, I. D. Meshri and P. J. Ortoleva, eds., pp.131–146.

Neuzil, C. E., 1986a, Groundwater flow in low-permeability environments: Water Resources Research, v. 22, n. 8, pp. 1163–1195.

Neuzil, C. E., 1986b, Problems of long-term deformation of basin fill and transient flow of pore fluids: Abstract 104326, Geol. Soc. Am. 1986 Annual Meeting, San Antonio, Texas.

Powley, D. E., 1982 through 1989, Abnormal pressure classes and lectures: Sponsored by AAPG, SEG, local geological societies, and Gas Research Institute.

Powley, D. E., 1990, Pressures and hydrogeology in petroleum basins: Earth–Science Reviews, 29, pp. 215–226.

Preston, D. A., 1976, Correlation of certain physical and chemical properties in sedimentary rocks: Proc. 17th U.S. Symp., Rock, Mech., Site characterization, Snowbird, Utah, August 25–27, 1976, p. 2A8–1.

Tigert, V., and Al-Shaieb, Z., 1990, Pressure seals: their diagenetic banding patterns: Earth–Science Reviews, v. 29, pp. 227–240.

Chapter 2

Summary of Published Literature on Anomalous Pressures: Implications for the Study of Pressure Compartments

Randi S. Martinsen
University of Wyoming
Laramie, Wyoming, U.S.A.

ABSTRACT

The phenomenon of anomalous geopressure has been a focus of study for almost 40 years. Until recently, most studies were concerned with either the detection of overpressure or with the processes by which anomalous pressure is generated. Two processes, disequilibrium compaction and hydrocarbon generation/maturation, probably account for the majority of overpressured rocks observed worldwide. Because anomalous pressure is often associated with undercompacted rocks, various techniques based on detecting undercompaction (in addition to formation pressure tests) are commonly used to detect anomalous pressure indirectly. Whereas formation pressure tests can only be obtained in porous and permeable reservoir-type rocks, indirect techniques can only be applied in thick shale sequences. Therefore, data sets obtained by these two different approaches are mutually exclusive.

There has been a great deal of uncertainty with regard to the existence and nature of pressure seals. Overall, researchers studying anomalous pressure fall into one of two categories: those who accept the existence of effective seals and those who do not. Researchers who do not believe seals exist support classic hydrologic interpretations that suggest anomalous pressures are bounded by low-hydraulic-conductivity rocks and are, therefore, geologically ephemeral. Those who do believe seals exist, for the most part, do not present specific evidence of what comprises a seal. Although the sealing capacity of various rocks for hydrocarbon accumulations has been investigated, very little research has been conducted specifically concerning the seals that contain anomalous pressures. The influence of capillary forces on hydrocarbon entrapment is widely recognized, but capillary phenomena have largely been ignored in studies dealing with geopressures, even though hydrocarbon generation is commonly cited as the mechanism by which overpressure is generated. The delineation of pressure compartments, especially those of known geologic duration, is an important tool by which potential seals can be mapped and investigated. Without specific knowledge of seal characteristics, well-founded interpretations of seal effectiveness and pressure compartment longevity cannot be made.

INTRODUCTION

The beginning of widespread interest in anomalous geopressure is usually credited to George Dickinson (1953) for his study of anomalous pressure in the Gulf Coast. According to Gretener (1990), Dickinson made his initial presentation, including a preprint with several pressure-depth plots, in 1951 to the third World Petroleum Congress at The Hague. Since then, interest in anomalous pressure has grown and numerous articles have been written concerning the phenomenon. For many years, interest was centered on the causes, recognition, and prediction of anomalous pressure, because of the impact anomalous pressure has on drilling for and producing hydrocarbons. Only in recent studies (Hitchon et al., 1990; Hunt, 1990; Jansa and Urrea, 1990; Miles, 1990; Pilkington, 1988; Powley, 1982, 1990; Tigert and Al-Shaieb, 1990) has the significance of anomalous pressure in the exploration for hydrocarbons been emphasized. This paper is intended to be a review and analysis of the various observations concerning geopressure phenomena presented in the literature up to the current phase of exploration-oriented pressure analysis.

ANOMALOUS PRESSURES

Definition of Terms

Normal formation pressure is that pressure required to raise a column of fluid from the subsurface formation to the surface. Any deviation from this is considered anomalous (Bradley, 1975; Fertl, 1976; Gretener, 1990; Hunt, 1990; Pickering and Indelicata, 1985; Sahay and Fertl, 1988). Normal formation pressure is also known as *hydrostatic pressure*. Hydrostatic pressure is the product of the unit weight and vertical height of a fluid column. The size and shape of this fluid column have no effect on the magnitude of hydrostatic pressure. The unit weight is affected by a number of things, including temperature and the concentration of dissolved solids and gases. Normal *hydrostatic pressure gradients*, therefore, vary from area to area, depending upon the nature of the subsurface fluid. Typical hydrostatic gradients vary from 0.433 psi/ft for freshwater columns to 0.465 psi/ft for saltwater columns.

Standard practice for calculating the "normal" hydrostatic pressure of a zone is to multiply the accepted hydrostatic pressure gradient (0.433 psi/ft in the Rocky Mountain region) by the depth of the zone. Errors in estimates of "normal" pressures, however, can range up to several hundred psi by assuming a constant salinity gradient (Bradley, 1975). Therefore, it is probably better to determine what gradient is best to use in a specific area from available measured formation pressures, rather than use a constant salinity gradient to calculate expected normal formation pressures.

Formation pressures either less than or in excess of hydrostatic pressure in a specific geologic environment are defined as *abnormal formation pressures* (with formation pressures more than hydrostatic denoted *overpressures* and *surpressures*, formation pressures less than hydrostatic denoted *subnormal formation pressures*, *subpressures*, and *underpressures*). Some authors (e.g., Breeze, 1970) include only overpressures in the term *abnormal* and use the term *subnormal* in reference to underpressures; this convention is generally followed in the present paper.

Characteristics of and Phenomena Associated with Anomalous Pressure

1. Regionally, abnormal pressure does not exceed the lithostatic (overburden) pressure gradient (approximately 1 psi/ft); locally, several areas worldwide have been documented that exceed the 1 psi/ft gradient (Gretener, 1990; Miles, 1990; Sahay and Fertl, 1988).
2. Anomalous pressure may exist at any depth, but abnormal pressure is more common at depths greater than 10,000 ft (3050 m).
3. Anomalous pressure occurs in rocks of all ages, but is more common in younger (Cretaceous and Tertiary) rocks. Where known to occur in older rocks, the time of the formation of the anomalous pressure is usually not known. For example, a rock deposited in the Paleozoic may not have been deeply buried until the Cretaceous or later, and probably would not have developed abnormal pressure until then.
4. Anomalous pressure is a global phenomenon and is found in many of the world's deep basins (Gretener, 1990; Hunt, 1990; Powley, 1982).
5. Many but not all anomalously pressured zones (both overpressured and underpressured) are undercompacted and have higher porosities (lower densities) than lithologically similar but normally pressured sediments at similar depths. According to Hunt (1990), onshore overpressured zones in Texas and Oklahoma are typically normally compacted, while offshore Gulf Coast overpressured zones are undercompacted. Some zones, however, are characterized by normal density (Gretener and Feng, 1985) and some even by high density, suggesting greater-than-normal compaction (Bradley, 1975, 1976).
6. Anomalously pressured zones typically have higher geothermal gradients, probably due to the fact that the greater fluid-filled porosity makes the zones less thermally conductive. Lower geothermal gradients in overpressured zones may also be related to lower grain-contact stresses therein.
7. Formation water salinity in anomalously pressured zones is often lower than in nearby normally pressured zones.
8. The interval wherein the change from a normal to anomalous pressure occurs is usually referred to as the transition zone. The transition zone may be absent, thin and sharp, or thick (thousands of meters) and gradual. Pressure gradients are highest in the transition zone. Many authors

(Chapman, 1972, 1980; Daines, 1982; Gretener, 1990; Sahay and Fertl, 1988) believe that the existence of a transition zone implies a leaky or incomplete seal, and therefore incomplete isolation.

9. The geometry and distribution of anomalously pressured zones frequently do not appear to correlate with any observed structural or stratigraphic framework (Breeze, 1970, 1971; Hunt, 1990; Law, 1984; Law and Dickinson, 1985; Powley, 1982).
10. Anomalously pressured zones, both overpressured and underpressured, are commonly, but not always, associated with hydrocarbons, especially gas, and often have no free water—i.e., do not have either bottom or edge water and contain only interstitial water (Davis, 1984; Gies, 1984; Gretener, 1990; Gretener and Feng, 1985; Law, 1984; Law and Dickinson, 1985).

Mechanisms Capable of Generating Anomalous Pressure

The pertinent literature indicates that there are numerous ways to create anomalous subsurface pressure. The more commonly cited mechanisms are summarized below.

Abnormal (High) Pressure

1. *Nonequilibrium compaction.* During burial and compaction, water is physically expelled from sediments. In thick, rapidly deposited shale sections, reductions in porosity and permeability related to compaction inhibit the flow of water out of the shale and cause compaction to cease, or at least to slow down. As burial continues and the weight of the overburden increases, fluid pressure increases in response to bearing the ever-increasing weight of the overburden. Most authors (Dickinson, 1953; Hubbert and Rubey, 1959; Rubey and Hubbert, 1959; Dickey, 1976; Fertl, 1976; Chapman, 1980; Plumley, 1980; Pickering and Indelicata, 1985; Bethke et al., 1988; Sahay and Fertl, 1988; Gretener, 1990), except Bradley (1975), believe nonequilibrium compaction to be a viable mechanism for creating abnormal pressure. Disagreement centers on whether or not compaction is the only, or even the dominant, mechanism. Rocks with abnormal pressure generated by nonequilibrium compaction should be undercompacted and therefore have higher porosity than rocks that have undergone normal compaction. Because so many of the techniques used to identify and quantify anomalous pressure in actuality only identify anomalous pressure associated with undercompacted rocks, there is a distinct tendency to study anomalously pressured rocks associated with undercompacted shales. Consequently, studies of rock sequences in which anomalous pressure was generated by means other than nonequilibrium compaction are much less common than studies of sequences in which nonequilibrium compaction is believed to be the cause of overpressuring. The existence of abnormally pressured rocks that are normally compacted has suggested to several authors that mechanisms in addition to nonequilibrium compaction must be active (Barker, 1972; Plumley, 1980; Carstens and Dypvik, 1981).

2. *Tectonic compression.* Tectonic compression creates abnormal pressure similarly to the nonequilibrium compaction mechanism. Generally, horizontal compressive forces (rather than vertical forces that cause compaction), directed at a rock sequence in which zones of low porosity and permeability inhibit the escape of pore fluids, act on the pore fluids and cause pore pressure to increase (Hubbert and Rubey, 1959; Fertl, 1976; Pickering and Indelicata, 1985; Sahay and Fertl, 1988).

3. *Aquathermal pressuring.* Increasing temperature with depth of burial causes pore waters to expand at a greater rate than the rock expands. If the pore waters are prevented from escaping by a flow barrier, pore pressure will increase (Barker, 1972; Barker and Horsfield, 1982). Some authors (Chapman, 1972, 1982; Daines, 1982) dismiss aquathermal pressuring as a viable mechanism because it requires a perfect seal, and perfect seals are considered geologic rarities. Even very slight leakage across the seal would quickly dissipate any overpressuring due to aquathermal (or any other) processes, once generation of overpressure has ceased. Other authors (Barker, 1972; Magara, 1975; Plumley, 1980; Gretener, 1990) reason that in abnormally pressured, normally compacted rocks, mechanisms other than nonequilibrium compaction, such as aquathermal pressuring, must be active; therefore, perfect seals must exist. However, no description of, or mechanism for generating, a perfect seal, except for thick salt beds, has been presented.

4. *Transformation of smectite to illite.* At a temperature of about 221°F, smectite begins altering to illite and expels a large volume of structural water in the process. If the rock is sealed, the addition of this released water, combined with the thermal expansion of the pore fluids, will increase the formation pressure to above normal (Fertl, 1976; Berg and Habeck, 1982; Pickering and Indelicata, 1985; Sahay and Fertl, 1988; Freed and Peacor, 1989). This mechanism is also believed to produce barriers to fluid flow, as illite is more densely and efficiently packed than smectite (Colten-Bradley, 1987; Freed and Peacor, 1989; Bethke et al., 1988). Colten-Bradley (1987), however, does not believe that this transformation can generate significant overpressuring because the reaction is inhibited by increasing pressure and would stop in response to significant overpressuring.

5. *Hydrocarbon generation.* The reactions that convert organic matter into hydrocarbons during progressive burial also cause fluid volume increases that result in overpressuring within isolated compartments. Numerous studies (Meissner, 1981; Momper, 1978; Tissot, 1984) suggest that fracturing produced by the overpressuring associated with maturation is the mechanism by which hydrocarbons migrate out of source rocks into more porous and permeable reservoir rocks. Specifically, methane generation has been cited as the cause of overpressuring in numerous

reservoirs. Gas is typically associated with anomalously pressured zones, and anomalous pressure is characteristic of gas-saturated regions (Davis, 1984). As either organic matter in source rocks or oil trapped in reservoirs is converted to methane, there is a significant volume increase that is capable of generating extremely hard geopressure in restricted compartments (Law, 1984; Gies, 1984; Law and Dickinson, 1985; Hunt, 1990; Barker, 1990). Pressure associated with source rocks generating gas is great enough to inject gas into rocks with very high capillary pressure, expelling water in the process, such that even though barriers may exist that impede flow, few are capable of disallowing flow. Where an effective seal exists, continued generation of methane can raise the pressure over lithostatic pressure and cause fracturing and leakage through the seal. Methane generation is potentially a highly effective mechanism for generating anomalous pressure, especially in rocks closely associated with source rocks. Continuous methane generation can produce such extreme geopressure that seals will not exist indefinitely; they will either leak continuously or will periodically fracture and leak. However, even if the seal is breached (because the overpressuring exceeds either the threshold displacement pressure or the lithostatic gradient and fractures the seal) it is possible the seal will "heal" (through water imbibition or closure of the fractures) before normal pressure is obtained; the compartment will still be overpressured, just not as highly overpressured as before leakage.

6. *Osmosis.* Osmosis is the mass transfer of water through a semipermeable membrane from fresher water to saltier water. Osmotic flow may cause abnormally high pressure in an isolated zone. If the pore water within an isolated zone is saltier than the pore water surrounding the zone then osmotic flow will be inward and the pore pressure will increase in the isolated zone. Disagreement exists about how effective osmotic flow is in creating anomalous pressure. Bradley (1975) concludes that osmotic flow is not a major factor in subsurface pressure systems, as the amount of flow required to cross a seal and equalize pressure is relatively small. Using Hunt's (1990) definition of a seal, osmotic flow, which requires the movement of water across a semipermeable membrane, does not qualify as a mechanism capable of generating anomalous pressure in hydraulically sealed compartments.

7. *Reverse osmosis.* Studies have documented the occurrence of reverse osmosis, the process of water flowing from a high-pressure, high-salinity zone to a lower-pressure, lower-salinity zone (Fertl, 1976; Pickering and Indelicata, 1985; Sahay and Fertl, 1988). This results in increased pressure (above normal) in the low-salinity zone as water flows into it from a highly overpressured zone. Again, this mechanism is not applicable for generating anomalous pressure in effectively sealed compartments.

8. *Gypsum/anhydrite transformations.* Both the transformation of gypsum to anhydrite plus water and the rehydration to gypsum at depth have been cited as possible mechanisms for generating abnormal pressure in carbonates (Fertl, 1976; Pickering and Indelicata, 1985; Sahay and Fertl, 1988).

9. *Buoyancy.* Difference in density between hydrocarbons, especially gas, and water can create anomalous pressure at the tops of hydrocarbon accumulations (Hubbert and Rubey, 1959). The longer the hydrocarbon column and the greater the density contrast between the hydrocarbon and the surrounding water, the greater will be the overpressuring. Generally, buoyancy differences can cause pressure to increase on the order of hundreds of psi, not thousands. Also, anomalous pressure is confined to the upper reaches of the hydrocarbon column and is typically the result of capillary trapping. Because hydrocarbon accumulations can occur in confined but open hydraulic systems, pressure differences due to buoyancy are not necessarily indicative of sealed compartments.

10. *Irregularities in the potentiometric surface.* Artesian conditions, or the existence of a permeable conduit between a shallow formation and a deeper, more highly pressured formation, can cause pore pressure to be higher than would be expected (Fertl, 1976; Sahay and Fertl, 1988).

11. *Isolation and uplift of deep, gas-filled compartments.* As temperature within an isolated, gas-filled compartment decreases in response to uplift and removal of overburden, volumetric contraction of the gas results in pressure within the compartment decreasing at a rate lower than the hydrostatic gradient, such that the isolated compartments become overpressured upon uplift relative to adjacent hydrostatically pressured rocks (Barker, 1979).

Although many mechanisms capable of generating overpressure have been described in the literature, most studies of anomalously pressured rocks cite either disequilibrium compaction (Bethke et al., 1988; Bredehoeft and Hanshaw, 1968; Bredehoeft et al., 1988; Dickinson, 1953; Dickey, 1976; Pickering and Indelicata, 1985) or hydrocarbon generation (Hedberg, 1974; Davis, 1984; Gies, 1984; Hitchon et al., 1990; Jansa and Urrea, 1990; Law, 1984; Law and Dickinson, 1985; Meissner, 1978; Pickering and Indelicata, 1985) as responsible for the generation of overpressure.

Subnormal (Low) Pressure

1. *Decrease in temperature and stress due to uplift and removal of overburden.* When rocks isolated at depth are uplifted, formation temperature is reduced. In response to lower temperature, water volumes decrease, thereby reducing pore pressure. Any increase in porosity due to matrix expansion would cause additional pore pressure reduction (Russell, 1972; Barker, 1972; Fertl, 1976; Sahay and Fertl, 1988).

2. *Burial of gas-saturated reservoirs.* Subnormal pressure can develop in rocks that are gas saturated (probably from bacterially generated gas), noncompacting (for example, chalk), and isolated at shallow depths of burial. Low pressure develops because the thermal pressuring of the gas with depth occurs at a lower rate than the increase in hydrostatic pressure, and because the gas dissolves in any water present (Barker, 1979, 1987).

3. *Nonequilibrium flow.* In regions where a confined aquifer is separated from its recharge area by a zone of low permeability and low flow, subnormal pressure can be generated due to discharge from the region being greater than recharge into it. Accurate mapping of head within the aquifer should indicate a sloping potentiometric surface (Belitz and Bredehoeft, 1988).

4. *Leaky seals.* If the seals associated with an overpressured, gas-saturated reservoir leak, then overpressure will be maintained only as long as the rate of pressure generation exceeds the rate of loss. When the rate of loss begins to exceed the rate of generation, overpressuring will decrease until normal and then subnormal pressure develops. Because of relative permeability, water is unable to flow into gas-saturated rocks, even if the pressure in the water-saturated zones is greater than the pressure in the gas-saturated zones. This mechanism could explain the existence of subnormal pressure in undercompacted sediments (Gries, 1984; Law and Dickinson, 1985).

5. *Osmosis.* If the water in the isolated zone is fresher, water will flow outward across a semipermeable membrane, resulting in abnormally low pressure within the zone.

6. *Removal of subsurface fluids.* The production of fluids—oil, gas, or water—from a subsurface zone will result in a reduction of formation pressure.

7. *Irregularities in the potentiometric surface.* The existence of a permeable conduit between confined formations with different head will allow fluids to flow from the formation with greater head to the one with lesser head, thereby locally reducing formation pressure in the one with greater head.

8. *Low water table.* Calculations of normal hydrostatic pressure assume that the water column extends to the ground surface. If the water table is significantly below the ground surface, as in places in the Middle East, then pressure will appear subnormal.

Implications for the Study of Pressure Compartments

All of the mechanisms proposed for the generation of anomalous pressure, with the exception of miscalculating what the normal hydrostatic gradient for an area is (wrong salinity gradient, low water table), require the existence of barriers or retardants to fluid flow, but very few of them require the existence of a closed, pressure-sealed compartment. Buoyancy, nonequilibrium flow, irregularities in potentiometric surfaces, and removal of subsurface fluids, for instance, only require confined systems. The level of barrier effectiveness required to generate abnormal pressure due to nonequilibrium compaction is highly relative and probably highly variable. Osmosis and reverse osmosis require fluid flow across barriers. Hydrocarbon generation commonly results in such large fluid volume increase that short-term overpressure could develop even in systems without effective barriers. Only pressure anomalies related to small, temperature-dependent changes in pore fluid volume (i.e., aquathermal pressuring, alteration of smectite to illite, cooling due to uplift and erosion) *require* the existence of seals (which are perfect or effective by definition) and cannot develop as a result of low hydraulic conductivity of enveloping rocks. Therefore, because many mechanisms that generate anomalous pressure do not require seals, the common occurrence of anomalous pressure, without knowledge of timing and mechanism of formation, does not contribute to answering the question of whether or not seals exist. Conversely, the fact that seals are not always required to develop anomalous pressure does not indicate that seals aren't present. In order to gain better understanding of the potential longevity of pressure compartments, more research on the time and mechanism of origin of specific pressure anomalies is needed.

The two mechanisms most often cited as the underlying cause of abnormal pressure are disequilibrium compaction and hydrocarbon generation. Insofar as these processes are depth related, abnormal pressure should develop as a function of depth and not necessarily be constrained by either structure or stratigraphy. That is, in a basin in which abnormal pressure is being currently generated, the top of the overpressuring should correlate with depth and may not necessarily correlate with any single seal or set of seals capable of maintaining the pressure once generation has ceased.

TECHNIQUES FOR MEASURING SUBSURFACE PRESSURES

Techniques for evaluating subsurface pressure fall into two distinct categories. *Indicated pressures* are indirectly derived from the analysis of logs, geothermal gradients, mud weights, drilling penetration rates, etc., and are mostly limited to interpreting pressures in shales. *Measured pressures* are direct measurements of subsurface pressures that can be obtained from analysis of various pressure-time tests (production tests, drill-stem tests, repeat formation tests) that are commonly performed on wells.

Indicated pressures require two stages of interpretation. An interpretation of the quality of the data obtained must be made, and then the relationship of the characteristic measured to pressure must be interpreted. Measured pressures require only an analysis of the quality of the data and therefore may be more reliable indicators of subsurface pressures. However, good quality, interpretable measured pressure data are much less abundant than logs from which indicated pressure data can be derived. Therefore, much of the pressure data presented in the literature is indicated pressures.

Measured Pressure Data

The most commonly available measured pressures in many regions are from drill-stem tests (DSTs). Wireline devices capable of obtaining small samples of formation fluids and making pressure measurements also exist. Schlumberger markets these devices under the name Repeat Formation Tester (RFT). A similar device developed by Dresser Atlas is the Formation Multi Tester (FMT). The RFT/FMT tool works like the drill-stem test tool but can obtain multiple pressures during a single run. However, the RFT/FMT samples a much smaller interval than the DST and is strongly affected by local well-bore conditions, such as formation damage.

The most accurate measured pressure data are obtained from tests on producing wells, because they are of longer duration and are less adversely affected by local well-bore conditions. These data, however, are difficult to obtain. In addition, pressures measured during production tests within isolated reservoirs are always lower than original reservoir pressures.

Indicated Pressure Data

Most indicated pressure detection analyses are based on the observation that many overpressured rocks, and even some subnormally pressured rocks, are associated with undercompacted shales. When fluids cannot escape during burial, they support the overburden and prevent compaction of the sediments. The greater the fluid pressure, the less the compaction and the higher the porosity. Degree of compaction can be ascertained from techniques that utilize various logging and drilling data to estimate porosity. Because pores are fluid filled and fluids are poorer conductors of heat than rock, high-porosity layers also have an insulating effect and raise the geothermal gradient of underlying rocks. Anomalously pressured shales are also commonly characterized by low resistivity responses. Such low resistivity response is believed to result from the higher than "normal" porosity and higher temperature associated with overpressured zones. With the exception of mud weight analyses, all indicated pressure data are obtained from within thick shale sequences. Pressures interpreted for the shales are then extrapolated into porous and permeable reservoirs within the shales.

Types of Indicated Pressure Analyses

The most common indicated pressure profiles in the literature use acoustic travel time, formation resistivity, and bulk density data obtained from well logs. The method of constructing pore pressure profiles using acoustic and resistivity measurements was initially established by Hottman and Johnson (1965) and has been elaborated by numerous authors. Basically, these techniques require first establishing a normal response trend and then comparing the normal trend with the trends observed in various wells to determine if, where, and how much they diverge from the normal trend (see Maucione et al., this volume). Generally it is easier to establish trends of acoustic travel time versus depth than resistivity versus depth, because more variables influence resistivity measurements.

Other types of data used to determine indicated pressures are discussed below.

Weight and Viscosity of the Drilling Mud

The weight of the drilling mud is continuously measured to ensure a safe balance between the hydrostatic and formation pressures. During drilling, influx of formation fluids or gas from high-pressure reservoirs will cause a decrease in the density and viscosity of the mud coming out of the well. In order to compensate for the high pressure, the density and viscosity of the mud going into the well will be increased. Ideally, the minimum mud weight capable of balancing subsurface pressure is desired in order to minimize formation damage during drilling. Therefore, under ideal conditions, mud density and viscosity should increase in proportion to pressure increase.

Rate of Penetration

The rate of penetration of the drill bit increases in overpressured zones. Typically, the shale penetration rate decreases uniformly with depth, due to the compaction of shales in response to overburden stress and diagenesis. However, the penetration rate through transition zones of anomalously pressured sections is commonly twice the normal rate. Accurate analysis of subsurface pressures using rate of penetration requires stable drilling parameters (bit weight, rotary speed, mud weight, and hydraulics) and uniform lithologies.

Shale Cuttings Characteristics

The size, shape, volume, and density of the shale cuttings change as the underbalance between mud density and pore pressure increases during the drilling of a transition zone. The volume and size of the cuttings increase. Shape may become conchoidal, long and splintery, or sometimes brick-shaped. The density of shale cuttings should depart from the normal trend—become less dense with depth into the transition zone. When mud weight is increased and drilling resumed, cuttings size and volume decrease. Mud incompatibility with the shale and mechanical damaging of the wellbore increase the size and volume of cuttings.

Gas in Mud

Frequently, overpressured shales are rich source rocks and contain mature gas that is liberated by the bit during drilling. Therefore, overpressured zones commonly show gas kicks.

Salinity of Formation Waters

It has been shown that formation water salinity varies inversely with the porosity of adjacent shale. If the shale is normally compacted, porosity decreases with depth, and formation water salinity increases. In overpressured zones, apparent shale porosity increases and formation water salinity decreases. Calculations of formation water salinity can be made using spontaneous potential logs.

Implications for the Study of Pressure Compartments

It is significant that pressure can only be accurately measured in rocks that have at least fair porosity and permeability because pressure measurement depends upon fluids being able to flow from the formation into the measuring device (Iverson, 1990). Therefore, pressure can only be measured in reservoir-type rocks and cannot be measured in potential seals. Conversely, indirect measurement of pressure is more commonly obtained from low-permeability, potential sealing rocks such as shales. For the most part, studies assume (perhaps incorrectly; see Maucione et al., this volume) that the pressure regime in the porous and permeable reservoir rocks is similar to that in the low-permeability rocks and that pressure data obtained from these two different methods can be correlated.

Furthermore, although indirect pressure techniques are widely used, they have several significant limitations that must be addressed in order to make reasonable interpretations of subsurface pressures. Most of the limitations are concerned with obtaining good quality data and can be overcome with careful analysis. The greatest limitation, however, is that indirect techniques may be influenced by variables other than pressure or porosity, e.g., gas saturation. Furthermore, indirect techniques are mostly applicable in areas in which overpressuring is accompanied by undercompaction and so may provide a bias against detecting anomalous pressures in rocks not accompanied by undercompaction. Because rocks characterized by abnormal pressure resulting from disequilibrium compaction are always undercompacted, indirect techniques probably also predispose their recognition.

PRESSURE SEALS

Definition of Terms

According to Hunt (1990, p. 2) the term *seal* refers to a "zone of rocks capable of hydraulic sealing . . . that prevent essentially all pore fluid movement over substantial intervals of geologic time." This definition does not exclude rocks that may *periodically* fail and leak but requires that, at least under some conditions, the rocks do not allow the flow of any fluids across them.

Barriers are basically low-hydraulic-conductivity rocks that *continuously* allow the flow of fluids across them, but at a reduced rate compared with the flow rates through surrounding rocks.

Characteristics of Seals Associated with Anomalous Pressures

1. Seals typically are characterized by higher than normal hydrostatic gradients; but because seals are tight, measured pressures cannot be obtained within them, and gradients are typically an extrapolation of pressure data derived from above and below the seal. Some data are available from porous and permeable zones within thick seals, such as those found in the North Sea, and indicate that pressure gradients in seals can be high.

2. Evidence suggests seals leak at least periodically (Gies, 1984; Law, 1984; Law and Dickinson, 1985), and no evidence exists in support of a "perfect" seal that retains its integrity indefinitely.

3. Seals typically have transition zones. The pressure gradient within a transition zone is considerably higher than the hydrostatic gradient, but generally remains below the lithostatic gradient (1 psi/ft). Transition zones should occur above, below, and lateral to anomalously pressured zones. Most recognized transition zones occur above or across anomalously pressured zones because only rarely are wells drilled through anomalously pressured zones (Chapman, 1980; Daines, 1982; Gretener, 1990; Hunt, 1990; Sahay and Fertl, 1988). It is widely accepted that the existence of a transition zone above an overpressured compartment is evidence of a leaky seal (Chapman, 1972, 1980; Daines, 1982; Gretener, 1990). Thick transition zones probably are most often the result of leakage across either a fluid seal or a zone containing multiple stacked seals wherein a sequential outward rupture of the individual seals occurs within the overall thick fluid seal zone.

4. Seals are characterized by a slowdown in penetration rate. Overpressured, undercompacted sediments drill faster than normally pressured, normally compacted sediments because the high pressure reduces the mechanical strength of the rock. As pressure increases in the transition zone, therefore, the drilling rate increases. The rate slows down, however, as nonpermeable tight zones in the transition zone are penetrated (Chapman, 1972; Jordan and Shirley, 1966; Pilkington, 1988).

5. No evidence is presented in the literature of what lithologically could be considered a "perfect" seal (i.e., is impermeable, ductile, and laterally continuous), with the possible exception of salt.

6. Cores taken from seals in the North Sea indicate that the seals there consist of interbedded permeable and nonpermeable layers. Permeable layers frequently produce hydrocarbons, and nonpermeable layers typically are calcite or silica cemented (Hunt, 1990). Cores taken from seals overlying overpressured compartments in the Anadarko basin show similar alternating silica and carbonate cemented and porous intervals that give the rock a banded appearance (Tigert and Al-Shaieb, 1990).

7. Cores also indicate that whereas overpressured compartments are sometimes characterized by open, nonmineralized fractures, seals contain calcite-filled fractures (Hunt, 1990).

8. Pore pressure plots using well logs commonly show evidence of a very resistive shale "cap" that overlies anomalously pressured zones. Shale cuttings from the cap zones show them to be limey in comparison with other intervals (Breeze, 1970; Magara, 1981b).

Types and Origin of Seals

Almost nothing has been written specifically about the characteristics and origins of seals that envelop anomalously pressured regions. Exceptions include the recent studies of Hunt (1990), Jansa and Urrea (1990), and Tigert and Al-Shaieb (1990). Much more is written about the origins and characteristics of anomalous pressures than the origins and characteristics of whatever contains them. Regarding seals, researchers studying anomalous pressure generally *assume* that either (1) all barriers leak (continuously) because the concept of a perfect seal is difficult to accept geologically (Chapman, 1972, 1980; Law, 1984; Bethke et al., 1988), or (2) effective seals must exist because anomalous pressure exists (Russell, 1972; Barker, 1972; Bradley, 1975; Hunt, 1990). Those opposing the concept of "perfect" seals commonly present pressure equilibration estimates based on Darcy's law of fluid flow, which show that even the tightest shales are not impermeable and therefore cannot maintain anomalous pressure over significant periods of geologic time (Bradley, 1975; Bredehoeft and Hanshaw, 1968; Hanshaw and Bredehoeft, 1968). No one, however, in analyzing pressure seals and estimating flow rates, addresses the impact of capillary forces on rock/fluid systems, even though the significance of capillary forces in hydrocarbon trapping is well known. In fact, capillary seals have been excluded from definitions of what comprises a seal in relation to pressure compartments (Hunt, 1990).

Several excellent papers have been written about how to evaluate seal quality in terms of a seal's capacity to contain hydrocarbons (Smith, 1966, 1980; Berg, 1975; Schowalter, 1979; Downy, 1984; Watts, 1987; Jennings, 1987). These papers all conclude that it is the capillary forces within a multiple-fluid-phase system with variable permeabilities that determine the holding capacity of a seal (both in terms of pressure and length of fluid columns contained), and not some absolute characteristic of the rock. In fact, definitions of effective seals (at least for hydrocarbon accumulations) do not incorporate any absolute permeability (or nonpermeability) requirement, but describe seals as thick, laterally continuous, ductile rocks with high capillary entry pressure (Schowalter, 1979; Downy, 1984).

According to Watts (1987), two general types of capillary seals can be recognized, on the basis of the mechanism by which they will leak. The first type, called a membrane seal, periodically leaks whenever the pressure differential across the seal exceeds the threshold displacement pressure and enables fluids to enter and pass through the capillary pore system of the seal. A membrane seal leaks just enough to bring the pressure differential below the displacement pressure, and then "re-seals." The second type, called a hydraulic seal, preferentially leaks by fracturing: this type of seal has such a high displacement pressure that the pressure gradient required for fracturing is less than the pressure gradient required for fluid displacement. Failure of hydraulic seals also occurs periodically, not continuously, and leaks "re-seal" when the internal pressure is reduced due to fluid escape and the fractures close. Seal thickness is not a consideration in evaluating the quality of membrane seals but is an important factor in evaluating hydraulic seals. Neither membrane seals nor hydraulic seals require the existence of impermeable rocks, yet they are capable of sustaining relatively high pressure differentials.

Some evidence has been presented which suggests that the mechanisms that generate overpressures may also generate the formation of low-porosity and low-permeability zones. The compaction of clay minerals reduces permeability, especially vertical permeability, and creates barriers to fluid flow. The deeper the burial and the greater the compaction, the more permeability is reduced (Hubbert and Rubey, 1959; Bredehoeft and Hanshaw, 1986; Gretener, 1990). The alteration of smectite to illite also causes a reduction in permeability that may be capable of effectively inhibiting fluid flow (Colten-Bradley, 1987; Bethke et al., 1988; Freed and Peacor, 1989). Highly mineralized, low-permeability cap rocks sometimes found overlying overpressured zones may be the result of precipitation from carbonate-rich fluids expelled from source rocks during hydrocarbon maturation. Even if the cap rocks are products of and not the cause of initial overpressuring in underlying formations, as is suggested by Magara (1981b), the formation of cap rocks should improve the efficiency of the seals associated with the overpressuring.

Implications for the Study of Pressure Compartments

Whereas some pressure compartments are bounded by barriers, others appear to be bounded by seals. Both leaky seals and barriers can be effective and cause anomalous pressure to exist, provided that the anomalous pressure is being generated at a rate greater than the rate at which pressure is equilibrating across the seal or barrier. The key element, then, in the impact of seals on the potential longevity of pressure compartments is the modifier *effective*. More studies need to be done on the effectiveness of seals. We need to know: (1) What are the lithologic characteristics of effective seals? (2) Are some seals more effective than others? If so, why? (3) How effective are seals? (4) Under what conditions are seals effective? (5) For how long can seals maintain their effectiveness?

Many workers (Bredehoeft and Hanshaw, 1968; Hanshaw and Bredehoeft, 1968; Chapman, 1972; Gretener, 1990) believe that the longevity of anomalous pressure is simply a function of the magnitude of the pressure generated and the hydraulic conductivity of the barriers to flow. For the most part, their analyses of flow rates across pressure barriers assume single-fluid-phase flow conditions. This assumption is not always well founded, especially in systems wherein large quantities of hydrocarbons have been generated and therefore the subsurface fluid system is characterized by multiple fluid phases. In these systems, flow calculations based on Darcy's Law are no longer applicable (see Iverson et al., this volume). Although the qualitative aspects of capillary seals are generally presented in terms of hydrocarbon column holding capacity (Schowalter, 1979; Sneider et al., 1991), capillary seals inhibit the flow of *all fluids* present in the system, not just hydrocarbons. Typically, because most rocks in the subsurface have water-filled pore spaces, they are characterized by single-fluid-phase conditions. It is the addition of hydrocarbons to a subsurface system that converts it from a water-only system to a multi-fluid-phase system. Rocks that may only be barriers under single-fluid-phase flow conditions (see Iverson et al., this volume) become capillary seals under multi-fluid-phase conditions. Capillary seals do not allow the flow of any fluids across them until their integrity is disrupted either by achieving threshold displacement pressure or by fracturing. If abnormal pressures are dominantly a product of disequilibrium compaction of water-saturated rocks (as for many areas of the Gulf Coast), the duration of abnormal pressures may be geologically short (depending on the barriers' effectiveness in inhibiting flow). Similarly, if the subsurface system were filled completely with oil or gas, single-fluid-phase conditions would also exist; there would not be any capillary sealing, and any abnormal pressures generated probably would equilibrate shortly after generation ceased. Therefore, capillary seals should be considered in any study of seals in relation to pressure compartments. Furthermore, in that the flow of water in the subsurface is no longer thought to control the migration of hydrocarbons (except locally), emphasis on identifying seals capable of inhibiting water flow seems misplaced.

Considerable evidence exists that seals leak. Knowledge of when, where, and why seals leak is almost as important as knowledge of their existence. In our search for hydrocarbons, we must delineate the paths they have followed prior to trapping and accumulation. Knowledge of when and where seals are likely to have ruptured is necessary in order to delineate these paths. The issues then become: When did the seal form—before, during, or after hydrocarbon generation and migration? And, does the seal leak or does it disintegrate? A change in the properties of the seal with time such that its holding capacity is reduced would work against the seal being an effective barrier to hydrocarbon migration over geologic time. Pressure loss with time does not mean that the seal loses its ability to trap hydrocarbons or influence the secondary migration of hydrocarbons, especially if pressure normalization was the result of local seal rupture due to fracturing. Pressure analysis is a tool for the recognition and study of seals.

Very little material has been published on the three-dimensional aspects of pressure compartments. We need to determine if the boundaries of the zones of anomalous pressure generation coincide with the boundaries of sealed pressure compartments. What is more important to oil and gas exploration? The seals, because they influence fluid flow throughout geologic time, or the pressure, which perhaps only temporarily (geologically) influences fluid flow? Again, the relationship between anomalous pressure and seals needs to be better understood.

CONCLUSIONS

Many mechanisms capable of producing anomalous pressure in the subsurface exist. Of these many mechanisms, two appear to be dominant and probably account for the majority of overpressuring observed. These mechanisms are disequilibrium compaction and hydrocarbon generation. Both these mechanisms may be enhanced by aquathermal pressuring.

Many techniques, both direct and indirect, exist for detecting anomalous pressure. Indirect techniques are mostly applicable in areas in which overpressuring is accompanied by undercompaction—and so may provide a bias against detecting anomalous pressures in rocks not accompanied by undercompaction. Because rocks characterized by abnormal pressure resulting from disequilibrium compaction are always undercompacted, indirect techniques probably predispose their recognition. Furthermore, whereas direct measurements of pressure can only be obtained in porous and permeable rocks, indirect measurements of pressure are more commonly obtained from low-permeability rocks such as shales. For the most part, studies assume—perhaps incorrectly—that the pressure regime in the porous and permeable reservoir rocks is similar to the pressure regime in the low-permeability rocks, and that pressure data obtained from these two different methods can be correlated.

Whereas the phenomenon of anomalous geopressure and the question of whether or not there exists in nature a "perfect" seal have been the focus of numerous studies, very little research has gone into the study of seals and the quantification of the sealing capacity of various rocks in regard to pressure compartmentation.

The existence of pressure compartments indicates that barriers to fluid flow exist. Hypotheses concerning rates of flow and pressure equilibration across pressure compartment boundaries based on single-fluid-phase conditions, and without consideration of capillary forces, are probably not applicable to systems characterized by multiple fluid phases, such as systems known to include significant hydrocarbons. That at least some pressure compartments have been in existence for geologically significant periods of time (see Surdam et al., this volume) indicates that seals capable of maintaining anomalous pressures for geologically significant periods of time also exist.

The delineation of pressure compartments of known geologic duration is an important tool that allows the distribution of seals to be mapped. Once seal distributions are established, data can be collected on their characteristics and the characteristics of the systems in which they occur. With this data, analyses of seal efficiency and mechanisms of formation, as well as mechanisms or conditions under which seals leak or even disintegrate, can be made. Improved understanding of seals will greatly enhance our ability to analyze the distribution of subsurface fluids and better predict the occurrence of hydrocarbons.

ACKNOWLEDGMENTS

This material covers research performed through November 1991. The author is grateful to Dave Powley, who first introduced me to the concept of pressure compartments and their significance during an American Association of Petroleum Geologists "Petroleum Exploration School" held in 1982. Appreciation is also extended to the American Association of Petroleum Geologists for the grant that allowed me to attend that course, as well as grants to attend others of its short courses that have been beneficial to an understanding of geopressure and seals. I also want to express thanks to Tim Schowalter, Bob Berg, Robert Sneider, and Pete D'Onfro for sharing their knowledge and insights concerning seals. Discussions with Ron Surdam, Jim Steidtmann, Brian Fuller, Bill Iverson, and Hank Heasler are also appreciated. David Copeland edited and greatly improved the manuscript. This study was funded by the Gas Research Institute under contract no. 5089-260-1894.

REFERENCES CITED

Barker, C., 1972, Aquathermal pressuring—Role of temperature in development of abnormal-pressure zones: American Association of Petroleum Geologists Bulletin, v. 56, p. 2068–2071.

Barker, C., 1979, Role of temperature and burial depth in development of subnormal and abnormal pressures in gas reservoirs (abstract): American Association of Petroleum Geologists Bulletin, v. 63, p. 414–415.

Barker, C., 1987, Development of abnormal and subnormal pressures in reservoirs containing bacterially generated gas: American Association of Petroleum Geologists Bulletin, v. 71, p. 1404–1413.

Barker, C., 1990, Calculated volume and pressure changes during the thermal cracking of oil to gas in reservoirs: American Association of Petroleum Geologists Bulletin, v. 74, p. 1254–1261.

Barker, C., and B. Horsfield, 1982, Mechanical versus thermal cause of abnormally high pore pressures in shales: Discussion: American Association of Petroleum Geologists Bulletin, v. 66, p. 99–100.

Belitz, K., and J. D. Bredehoeft, 1988, Hydrodynamics of Denver Basin: Explanation of subnormal fluid pressures: American Association of Petroleum Geologists Bulletin, v. 72, p. 1334–1359.

Berg, R. R., 1975, Capillary pressure in stratigraphic traps: American Association of Petroleum Geologists Bulletin, v. 59, p. 939–956.

Berg, R. R., and M. F. Habeck, 1982, Abnormal pressures in the Lower Vicksburg, McAllen Ranch field, south Texas: Transactions of the Gulf Coast Association of Geological Societies, v. 32, p. 247–253.

Bethke, C. M., W. J. Harrison, C. Upson, and S. P. Altaner, 1988, Supercomputer analysis of sedimentary basins: Science. v. 239, p. 261–267.

Bradley, J. S., 1975, Abnormal formation pressure: American Association of Petroleum Geologists Bulletin, v. 59, p. 957–973.

Bradley, J. S., 1976, Abnormal formation pressure: Reply: American Association of Petroleum Geologists Bulletin, v. 60, p. 1127–1128.

Bredehoeft, J. D., and B. B. Hanshaw, 1968, On the maintenance of anomalous fluid pressures: I. Thick sedimentary sequences: Geological Society of America Bulletin, v. 79, p. 1097–1106.

Bredehoeft, J. D., R. S. Djevanshir, K. R. Belitz, 1988, Lateral fluid flow in a compacting sand-shale sequence: South Caspian Basin: American Association of Petroleum Geologists Bulletin, v. 72, p. 416–424.

Breeze, A. F., 1970, Abnormal-subnormal pressure relationships in the Morrow sands of northwestern Oklahoma: M.S. thesis, University of Oklahoma, Norman, 122 p.

Breeze, A. F., 1971, Abnormal-subnormal pressure relationships in the Morrow sands of northwestern Oklahoma: Shale Shaker, Oklahoma City Geological Society, p. 172–193.

Carstens, H., and H. Dypvik, 1981, Abnormal formation pressure and shale porosity: American Association of Petroleum Geologists Bulletin, v. 65, p. 346–350.

Chapman, R. E., 1972, Clays with abnormal interstitial fluid pressures: American Association of Petroleum Geologists Bulletin, v. 56, p. 790–795.

Chapman, R. E., 1980, Mechanical versus thermal cause of abnormally high pore pressures in shales: American Association of Petroleum Geologists Bulletin, v. 64, p. 2179–2183.

Chapman, R. E., 1982, Mechanical versus thermal cause of abnormally high pore pressures in shales: Reply: American Association of Petroleum Geologists Bulletin, v. 66, p. 101–102.

Colten-Bradley, V. A., 1987, Role of pressure in smectite dehydration—Effects on geopressure and smectite-to-illite transformation: American Association of Petroleum Geologists Bulletin, v. 71, p. 1414–1427.

Daines, S. R., 1982, Aquathermal pressuring and geopressure evaluation: American Association of Petroleum Geologists Bulletin, v. 66, p. 931–939.

Davis, T. B., 1984, Subsurface pressure profiles in gas-saturated basins, *in* J. Masters, ed., Elmworth—Case Study of a Deep Basin Gas Field: American Association of Petroleum Geologists, 316 p.

Dickey, P. A., 1976, Abnormal formation pressure: Discussion: American Association of Petroleum Geologists Bulletin, v. 60, p. 1124–1128.

Dickinson, G., 1953, Geological aspects of abnormal reservoir pressures in Gulf Coast Louisiana: American Association of Petroleum Geologists Bulletin, v. 37, p. 410–432.

Downy, M. W., 1984, Evaluating seals for hydrocarbon accumulations: American Association of Petroleum Geologists Bulletin, v. 68, p. 1752–1763.

Fertl, W. H., 1976, Abnormal Formation Pressures: New York, Elsevier, 382 p.

Freed, R. L., and D. R. Peacor, 1989, Geopressured shale and sealing effect of smectite to illite transition: American Association of Petroleum Geologists Bulletin, v. 73, p. 1223–1232.

Gies, R. M., 1984, Case history for a major Alberta deep basin gas trap: The Cadomin Formation, *in* J. Masters, ed., Elmworth—Case Study of a Deep Basin Gas Field: American Association of Petroleum Geologists, 316 p.

Gretener, P.E., 1990, Geomechanics in production geology and geophysics: Short Course Notes, Rocky Mountain Association of Geologists, Denver, 97 p.

Gretener, P. E., and Zeng-Mo Feng, 1985, Three decades of geopressures—Insights and enigmas: Bulletin, Vereinigung schweizerischer petroleum-geologen und -ingenieur, Basel, v. 51, no. 120, p. 1–34.

Hanshaw, B. B., and J. D. Bredehoeft, 1968, On the maintenance of anomalous fluid pressures: II. Source layer at depth: Geological Society of America Bulletin, v. 79, p. 1107–1122.

Hedberg, H. D., 1974, Relation of methane generation to undercompacted shales, shale diapirs, and mud volcanos: American Association of Petroleum Geologists Bulletin, v. 58, p. 661–673.

Hitchon, B., J. R. Underschultz, S. Bachu, and C. M. Sauveplane, 1990, Hydrogeology, geopressures and hydrocarbon occurrences, Beaufort-Mackenzie Basin: Bulletin of Petroleum Geology, v. 38, p. 215–235.

Hottman, C. E., and R. K. Johnson, 1965, Estimation of formation pressures from log-derived shale properties: SPE Paper 1110, Journal of Petroleum Technology, v. 17, p. 717–722.

Hubbert, M. K., and W. W. Rubey, 1959, Role of fluid pressure in mechanics of overthrust faulting, I: Geological Society of America Bulletin, v. 70, p. 115–166.

Hunt, J., 1990, Generation and migration of petroleum from abnormally pressured fluid compartments: American Association of Petroleum Geologists Bulletin, v. 74, p. 1–12.

Iverson, W. P., 1990, Drill stem tests: Second Quarter Report on Contract No. 5089-260-1984, Multidisciplinary Analysis of Pressure Chambers in the Powder River Basin, Wyoming and Montana, Gas Research Institute, Chicago.

Jansa, L. F., and V. H. N. Urrea, 1990, Geology and diagenetic history of overpressured sandstone reservoirs, Venture gas field, offshore Nova Scotia, Canada: American Association of Petroleum Geologists Bulletin, v. 74, p. 1640–1658.

Jennings, J. B., 1987, Capillary pressure techniques: Application to exploration and development geology: American Association of Petroleum Geologists Bulletin, v. 71, p. 1196–1209.

Jordan, J. R., and O. Shirley, 1966, Application of drilling performance data to overpressure detection: SPE Paper 1407, Journal of Petroleum Technology, v. 18, p. 1387–1394.

Law, B. E., 1984, Relationships of source-rock, thermal maturity, and overpressuring to gas generation and occurrence in low-permeability upper Cretaceous and lower Tertiary rocks, Greater Green River Basin, Wyoming, Colorado, and Utah, *in* J. Woodward, F. Meissner, and J. Clayton, eds., Hydrocarbon Source Rocks of the Greater Rocky Mountain Region: Rocky Mountain Association of Geologists, p. 469–490.

Law, B. E., and W. W. Dickinson, 1985, Conceptual model for origin of abnormally pressured gas accumulations in low-permeability reservoirs: American Association of Petroleum Geologists Bulletin, v. 69, p. 1295–1304.

Magara, K., 1975, Importance of aquathermal pressuring effect in Gulf Coast: American Association of Petroleum Geologists Bulletin, v. 59, p. 2037–2045.

Magara, K., 1981a, Mechanisms of natural fracturing in a sedimentary basin: American Association of Petroleum Geologists Bulletin, v. 65, p. 123–132.

Magara, K., 1981b, Fluid dynamics for cap-rock formation in Gulf Coast: American Association of Petroleum Geologists Bulletin, v. 65, p. 1334–1343.

Meissner, F. F., 1978, Petroleum geology of the Bakken Formation, Williston Basin, North Dakota and Montana, *in* The Economic Geology of the Williston Basin, Montana, North Dakota, South Dakota, Saskatchewan, Manitoba: Williston Basin Symposium, Montana Geological Society, p. 207–227.

Meissner, F. F., 1981, Abnormal pressures produced by hydrocarbon generation and maturation and their relation to processes of migration and accumulation: American Association of Petroleum Geologists Bulletin, v. 65, p. 2467.

Miles, J. A., 1990, Secondary migration routes in the Brent Sandstones of the Viking Graben and East Shetland Basin: Evidence from oil residues and subsurface pressure data: American Association of Petroleum Geologists Bulletin, v. 74, p. 1718–1735.

Momper, J.A., 1978, Oil migration limitations suggested by geological and geochemical considerations, *in* Physical and Chemical Constraints on Petroleum Migration: American Association of Petroleum Geologists Continuing Education Short Course Note Series 8, p. B1–B60.

Pickering, L. A., and G. J. Indelicata, 1985, Abnormal formation pressure: A review: The Mountain Geologist, v. 22, p. 78–89.

Pilkington, P. E., 1988, Uses of pressure and temperature data in exploration and new developments in overpressure detection: SPE Paper 17101, Journal of Petroleum Technology, v. 40, p. 543–549.

Plumley, W. J., 1980, Abnormally high fluid pressure: Survey of some basic principles: American Association of Petroleum Geologists Bulletin, v. 64, p. 414–430.

Powley, D., 1982, Pressures, normal and abnormal: American Association of Petroleum Geologists,

Advanced Exploration Schools unpublished lecture notes, 48 p.

Powley, D., 1990, Pressures and hydrogeology in petroleum basins: Earth-Science Reviews, v. 29, p. 215–226.

Rubey, W. W., and M. K. Hubbert, 1959, Role of fluid pressure in mechanics of overthrust faulting, II: Geological Society of America Bulletin, v. 70, p. 166–205.

Russell, W. L., 1972, Pressure-depth relations in Appalachian region: American Association of Petroleum Geologists Bulletin, v. 56, p. 528–536.

Sahay, B., and W. H. Fertl, 1988, Origin and Evaluation of Formation Pressures: Boston, Kluwer, 292 p.

Schowalter, T. T., 1979, Mechanics of secondary hydrocarbon migration and entrapment: American Association of Petroleum Geologists Bulletin, v. 63, p. 723–760.

Smith, D.A., 1966, Theoretical consideration of sealing and non-sealing faults: American Association of Petroleum Geologists Bulletin, v. 50, p. 363–374.

Smith, D.A., 1980, Sealing and non-sealing faults in Louisiana Gulf Coast basins: American Association of Petroleum Geologists Bulletin, v. 64, p. 145–172.

Sneider, R. M., K. Stolper, and J. S. Sneider, 1991, Petrophysical properties of seals: American Association of Petroleum Geologists Bulletin, v. 75, p. 75.

Tigert, V., and Z. Al-Shaieb, 1990, Pressure seals: Their diagenetic banding patterns: Earth Science-Reviews, p. 1–14.

Tissot, B. P., 1984, Recent advances in petroleum geochemistry applied to hydrocarbon exploration: American Association of Petroleum Geologists, v. 68, p. 545–563.

Watts, N. L., 1987, Theoretical aspects of cap-rock and fault seals for single- and two-phase hydrocarbon columns: Marine and Petroleum Geology, v. 4, p. 274–307.

Chapter 3

Basin Compartmentation: Definitions and Mechanisms

Peter J. Ortoleva
Indiana University
Bloomington, Indiana, U.S.A.

ABSTRACT

A classification scheme for compartments and seals is introduced and physico-chemical processes underlying their genesis and evolution are suggested. The sedimentary basin is viewed as a chemical reactor of epic scale constantly being driven out of equilibrium. As a result, it sustains a variety of important compartmentation and sealing phenomena. Practical implications of this can be obtained by building a comprehensive model accounting for operating physico-chemical processes and then developing computer codes to simulate it. We argue that this is feasible and can contribute greatly to the development of exploration, production, and resource assessment strategies.

Diagenesis deep in a sedimentary basin involves a number of strongly coupled reaction, transport, and mechanical (RTM) processes. When a coupled RTM system is driven sufficiently far-from-equilibrium, it can become organized in space or time in ways that have no direct relation to patterns imposed at the basin boundary or through sedimentary input. Rather, these patterns organize themselves. Our results to date suggest that many aspects of compartment and seal genesis and dynamics appear to be a manifestation of this far-from-equilibrium dynamic.

If sedimentation was very slow, then all fluids could escape and rock at depth would be porosity-free. But beyond some critical subsidence and burial rate, fluid can get trapped in compartments for appreciable times. This is because relatively uncompacted rock finds itself at appreciable depth. This rock is therefore far-from-equilibrium—a large free energy difference exists between the uncompacted and the compacted system due to the overburden stress.

A most dramatic manifestation of far-from-equilibrium conditions occurs when the system develops periodic or other oscillatory variations in space or time. A sequence of episodic fluid releases from an overpressurizing compartment can occur via a cycle of fracture generation and healing. Also, alternating layers of contrasting texture or mineralogy can develop to produce textural banding that has been found to be at the root of a number of pressure seals.

We set forth the general point of view that a basin is a far-from-equilibrium system capable of sustaining a variety of compartmentation phenomena. Compartments, banded seals, and episodic fluid migration and other phenomena key to petroleum exploration, production, and resource assessment are found to be consequences of the far-from-equilibrium basin dynamic.

To turn these general observations into practical strategies one must be more specific in correlating these phenomena with basin tectonic, thermal, and sedimentary history. The complexity arising from the many coupled RTM processes demands that this can only be done by the development of a basin RTM simulator, whereby we can

- predict new compartment and seal types;
- extrapolate findings from one basin to other basins;
- predict the location, internal structure and extent, and contents of compartments;
- understand episodic and other petroleum release phenomena in compartmented basins for the exploration of petroleum expelled from compartments; and
- design production strategies for compartmented reservoirs.

INTRODUCTION

The concept of Powley and Bradley is that at depth (roughly below 10,000 ft [3000 m]) a sedimentary basin can be divided into a boxwork of domains each of which is essentially in hydraulic isolation from its surroundings (Bradley, 1975; Powley, 1975, 1980, 1990; Powley and Bradley, 1987; see also Bradley and Powley, this volume). These compartments are most easily recognized by their abnormal fluid pressures; compartments may be either over- or underpressured (OP or UP) relative to hydrostatic. What distinguishes their interpretation of OP and UP domains from that of others is that they propose that compartments can exist in three-dimensional isolation; hence the OP or UP does not arise trivially through hydraulic connection with higher- or lower-lying fluid reservoirs, respectively.

To be sure, no domain of rock can be in perfect hydraulic isolation. All rocks have some residual permeability. The Powley and Bradley notion is an idealization based on their recognition of an important separation of time scales. While no domain of rock can be sealed off from its environment forever, it can be so isolated for appreciable durations of geologic time—i.e., on the million-year or greater time scale. Furthermore, by definition, within a compartment nonhydrostatic pressure gradients become dissipated on a much shorter time scale than that associated with exchange between the interior and the environment of the compartment. With these caveats, then, our goal is to characterize these compartments and the seals that bound them and, furthermore, determine how they develop and what is their structure and dynamic once in existence.

An idealized compartment is suggested in Figure 1. Its bounding seal leaves it in three-dimensional hydraulic isolation—i.e., an encasing shell of rock that is of very low permeability to the fluids within the compartment interior or in the environment. The interior of the compartment is of relatively high permeability so that there is good hydraulic communication within the compartment. Hence deviations from a hydrostatic pressure gradient within the compartment are rapidly dissipated in comparison to the rate for exchange of fluid between the compartment interior and its surroundings.

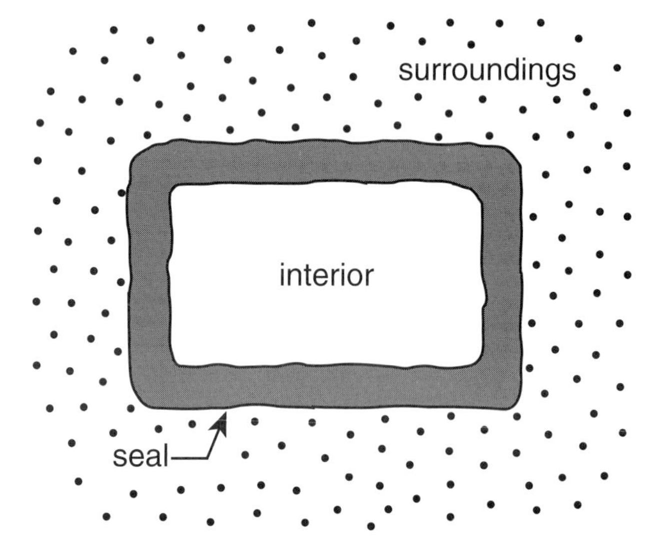

Figure 1. Generic compartment of Powley and Bradley showing an interior of relatively good hydraulic connectivity separated from its surroundings by a low-permeability shell or seal.

From this notion, a number of key questions immediately follow. How are seals formed and, in light of tectonic activity, hydrofracturing, or chemical degradation, how are they healed, once breached? How are porosity and permeability within a compartment preserved? Are there universal properties of seals that fit them into a single or a few easily identifiable classes, or are they all rather special cases? Our objective in the present work is to address these and other issues via a combination of observations and modeling studies.

COMPARTMENTS, SEALS, AND LENGTH SCALES—DEFINITIONS

In order to orient the discussion, let us first consider a number of definitions and concepts that emerge from our work and the Powley and Bradley compartment concept. In later sections we shall discuss further evidence of these phenomena and make conjectures on the mechanisms by which they develop.

Compartments

A Compartment and Its Surrounding Seal

By definition, a compartment is a domain of rock of relatively good hydraulic connectivity and porosity surrounded by a shell-like domain of rock of sufficiently low permeability that the fluids within the compartment do not have appreciable exchange with the environment for long periods of geologic time (Figure 1).

Powley-Bradley Compartment Boxwork

According to Powley and Bradley, a basin can be divided into a boxwork of compartments separated by seals (Figure 2). The Powley and Bradley boxwork may even extend into basement rock as suggested (D. Powley, 1985, personal communication).

Nested Compartments

Compartments may reside within compartments (Figure 3). This nesting can exist over a large range of spatial scales from the supra-kilometer to the sub-meter scale. Compartment nesting is not necessarily a case of fractal behavior. The latter implies that the physics of the genesis of the compartments is independent of scale over several orders of magnitude. It does not appear that this is the case for the millimeter-, centimeter-, meter-, kilometer-, and basin-scale compartmentation observed although invariance might apply between some of these scales.

Megacompartment and Its Complex Interior and Satellites

A megacompartment is a basin-scale compartment with its top seal and its (typically stratigraphically associated) basal seal (see Al-Shaieb et al., this volume) (Figure 4). The top seal is diagenetic and may traverse stratigraphy. Its interior may be subdivided

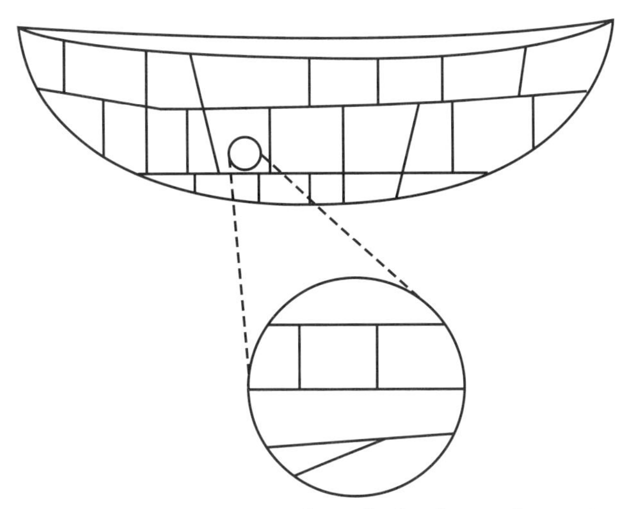

Figure 3. Compartmentation of a basin can be nested—i.e., there can be a hierarchy of compartmentation within compartmentation.

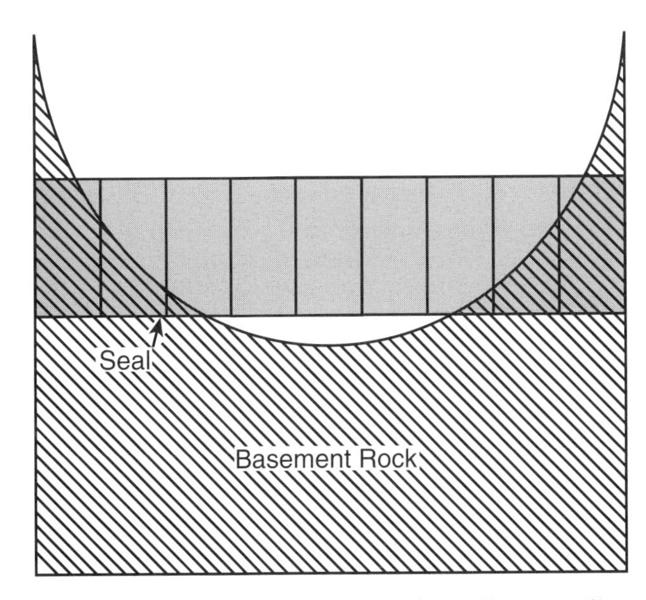

Figure 2. According to Powley and Bradley, a sedimentary basin can be divided into a boxwork of compartments at depth. The compartments may not only be side-by-side (as shown) but may also be stacked, comprising a three-dimensional boxwork.

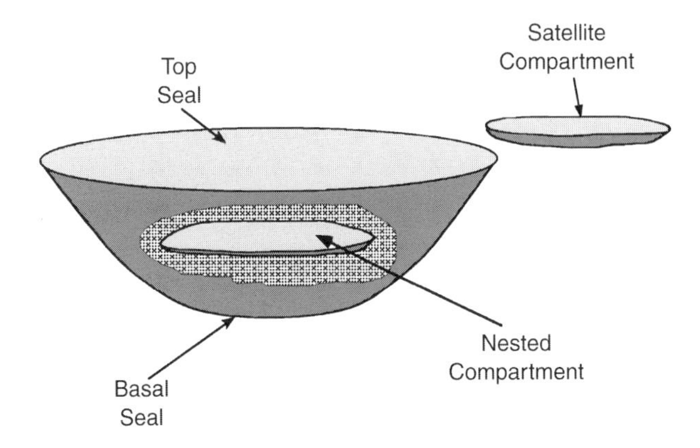

Figure 4. Compartmentation can be basinwide, forming a "megacompartment." The latter may have inner compartments forming a megacompartment complex and may have near-lying satellite compartments.

into an array of smaller compartments; the ensemble being denoted a megacompartment complex. A megacompartment may involve several levels of internal compartment nesting. Smaller scale external compartments—i.e., satellites—often reside outside a megacompartment.

Columns

A column is a relatively elongate domain that is similar to a compartment except that it has no identifiable bottom—i.e., permeability and porosity may decrease gradationally, perhaps even extending into basement rock (Figure 5).

Intrastratum Compartment

An individual stratum may constitute a compartment; its bounding seal may exist wholly within the stratum or be at or associated with the interface between it and surrounding rock (Figure 6). In the most interesting case the seal is not simply impermeable surrounding rock. Rather, both the compartment interior and the seal develop within the original sedimentary stratum.

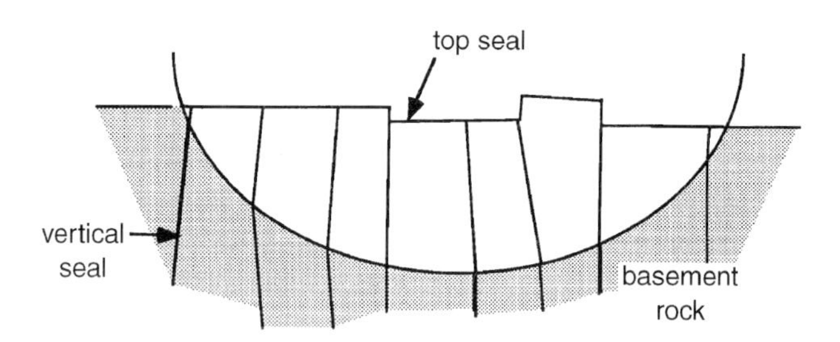

Figure 5. Columnar or other compartments without a well-defined basal seal may likely form through the interplay of fault blocking and processes of top seal genesis.

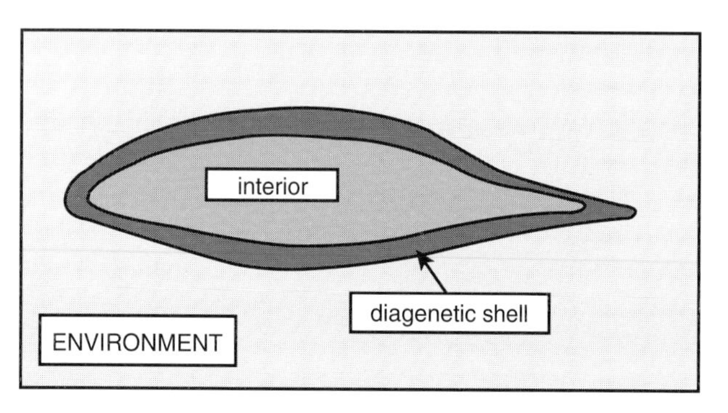

Figure 6. A compartment may exist wholly within a sedimentary stratum. This may occur when the stratum is surrounded by low-permeability strata or when some diagenetic process tends to cause augmented cementation or compaction near the perimeter of the stratum.

Microcompartment

A very thin layer of rock lying wholly within a single stratum may be hydraulically isolated (Figure 7). Such microcompartments may appear as a layered sequence.

Seals

Seal Geometry

Seals may be roughly categorized in terms of their spatial relationship to the compartment. Thus compartments are bounded above and below by top and bottom (basal) seals. Compartments may be bounded on their sides by lateral (or vertical) seals although such side seals may be absent when closure occurs by a convergence of the top and bottom seals.

Stratigraphic Seals

These seals are comprised of a single, roughly uniform lithologic unit that has apparently been compacted (or cemented) preferentially due to its original chemistry or texture. Common examples are shales and anhydrite beds.

Some horizontal seals seem to have risen during burial diagenesis. These *migrating seals* appear to have ascended through lithologic units and may cross them (D. Powley, 1988, personal communication).

Diagenetically Banded Seals

Seals may have an internally layered structure (as suggested in the horizontal seal of Figure 7) that has developed through diagenesis and that is on a spatial scale (i.e., has an interband distance) which is much shorter than the thickness of an individual lithologic unit (Dewers and Ortoleva, 1988; Tigert and Al-Shaieb, 1990). Diagenetically banded vertical seals have also been observed (P. D'Onfro, 1991, personal communication).

Repetitively Banded Seals

Seal banding can involve many alternations of the same textural repetition unit. The most commonly

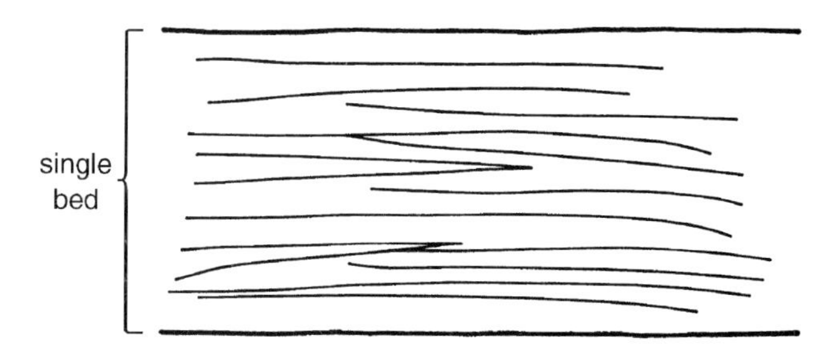

Figure 7. Seals may have a large degree of layered internal structure from mesoscopic phenomena such as stylolites, clay seams, and cementation/compaction alternations. This internal structure may also, in fact, constitute a fine-scale array of layered microcompartments.

observed are composed of (Dewers and Ortoleva, 1988):

- roughly regularly spaced arrays of stylolites, dissolution seams, and related features; and
- bands of augmented compaction and porosity alternating with bands of relatively highly cemented rock.

Examples are given in Tigert and Al-Shaieb (1990) and Shepherd et al. (this volume). For further discussion see below and Qin and Ortoleva (this volume).

Precipitated Seals

Some seals have resulted from the precipitation of cements—notably, carbonate in sandstones (Dewers and Ortoleva, 1988; Tigert and Al-Shaieb, 1990). They may have filled pores or fractures or even replaced the original grains. The cements may have been derived from closely lying beds or from afar. Mechanisms for precipitation of carbonate bands in sandstones are presented in Ortoleva et al. (1993) and Chen et al. (this volume).

Gradational Seals

Seals are not necessarily a well-localized horizontal or vertical sheet but rather may take the form of gradual textural changes culminating in very low-permeability, roughly planar domains.

Fault Associated Seals

Many vertical seals are associated with a fault. The fault may constitute (or contain) the seal or may have served as a disturbance or nucleus that developed a seal adjacent to it through an interaction of the fault and its environment during diagenesis (Ortoleva et al., in press).

Free Vertical Seals

Vertical seals have sometimes formed in the absence of a fault (Powley and Bradley, 1988, personal communication). They may intersect many lithologic units and may meander somewhat as they do so.

The Micro-, Meso-, Macro-, and Megascopic Scales: Definitions and Interrelationships

A key element of compartmentation appears to be the existence and interrelationships among phenomena on a broad range of length scales. What is, in fact, most interesting and important is that the phenomena on these wide-ranging scales strongly couple to reinforce each other in many ways. Let us define these scales and then note their interrelationships.

Microscale phenomena are defined here to be those that take place on a length scale of the order of a single or perhaps a few grain diameters. Most notable among these are grain growth/dissolution/nucleation reactions, grain comminution, grain coating, and plastic grain deformation or mechanical rearrangement. These processes underlie compaction or porosity/permeability occlusion or enhancement.

Mesoscopic phenomena, by definition, take place on scales ranging (roughly) on the ten grain diameter scale to perhaps the meter scale. A key aspect of mesoscopic phenomena is that they involve a statistically significant number of grains so that they might be described in terms of the spatial distribution of the local "texture." Texture here is taken to mean the average grain volume, shape and orientation of the various minerals present but could be more detailed—i.e., be described in terms of the probability distributions for the aforementioned quantities. Examples of mesoscopic phenomena are stylolites, banded compaction/cementation alternations in sandstones, differentiated or enhanced marl/limestone alternations, and banded carbonate cements in sandstones.

Macroscopic phenomena are exemplified by intrastratum compartments or compartments encompassing several strata. They are on the supra-meter to perhaps tens of kilometers scale.

The largest (basin-scale) phenomenon is the megacompartment. It may have a complex, nested interior structure of hydrologically isolated domains.

All these phenomena may be strongly related. The mesoscopic processes involve the spatial localization of microscopic processes. In turn, the conditions promoting or repressing the mesoscopic processes—i.e., stress or fluid pressure—are affected by the macroscopic and megascopic phenomena. In turn, the mesoscopic phenomena affect the macroscopic permeability and hence may play a key role in the development of seals defining a compartment or megacompartment. In summary, phenomena on one scale affect and are affected by phenomena on one or more other scales—compartmentation involves a complex network of cross-coupled processes on a hierarchy of scales.

While many details on the phenomena on all scales need to be clarified, we believe that this viewpoint puts these phenomena in an orderly perspective. In the following sections we present more detailed remarks on a number of the aforementioned micro-, meso-, macro-, and megascopic phenomena and review processes that likely underlie their development and evolution.

THE SEDIMENTARY BASIN: A CHEMICAL REACTOR OF GRAND PROPORTIONS

The basin is a chemical reactor of grand proportions, in which reactants (sediment and fluid) are continually being exchanged with the environment. As a result, diagenetic reactions take place, producing a variety of products (authigenic mineralization, altered pore-fluid composition, and organic phases—oil and gases). The basin reactor is continuously subjected to changing stress and thermal conditions at the boundary and may exchange fluid and sediment mass there also.

As a result of the imposed stresses and the exchange of mass and energy, the basin reactor not only can be driven out of equilibrium but, because some processes proceed slowly, can be maintained far-from-equilibrium with respect to chemical and mechanical processes. Chemical engineering reactors and biological systems driven far-from-equilibrium can display a richness of very interesting phenomena (Nicolis and Prigogine, 1977; Field and Burger, 1985; Ortoleva, 1992). Their temporal dynamics and the internal spatial distribution of matter, temperature, and stress may take on patterns that bear no direct relation to the time course of reactant injections or the spatial pattern of temperature or composition imposed at the boundary. In a real sense, far-from-equilibrium reactors may take on an "autonomous" dynamic whereby they organize themselves in space and time in a manner dictated more by their internal reaction-transport-mechanical (RTM) process network than by the detailed pattern of their interactions with the environment.

Key processes associated with the genesis and evolution of compartments and associated petroleum may be far-from-equilibrium:

- organic reactions such as kerogen → petroleum;
- the rising of fluid to the top and minerals descending to the bottom via Darcy flow and compaction;
- the rising of less-dense organic hot fluids;
- the deformation of the solid matrix in response to shear (as occurs during thrusting) or other tectonic forces applied to the basin; and
- imposition of thermal nonuniformity (the geothermal gradient).

Because these processes can take place on a geological time scale they allow the basin to be maintained far-from-equilibrium on this scale.

With this observation we expect that the grand basin reactor can also be self-organizing key aspects of its internal distribution and energy. If this is true, then the spatial distribution of traps, compartments, and water and hydrocarbon migration paths can have rather distinct features not directly related to the geometry and timing of sedimentation or applied stress and temperature at the basin boundary.

A key question is whether these rather general notions can be developed into practical strategies for the petroleum industry. At yet another level one might ask whether such phenomena are even manifest in real basins. We believe that both can be answered in the affirmative—see Figure 8.

Classical exploration strategies based on the geometry of sedimentary bedding and structures have yielded a great triumph in exploration. However, there is another, complementary part of the story that can be used to

- explore for new types of reservoirs;
- explain the genesis of already-discovered reservoirs that are paradoxical from the classical perspective; and
- delineate conditions under which classical reservoirs actually contain petroleum in light of the diagenetic processes that may have altered them.

The complementary viewpoint is that the basin reactor can organize itself in space and time when it is driven sufficiently far-from-equilibrium.

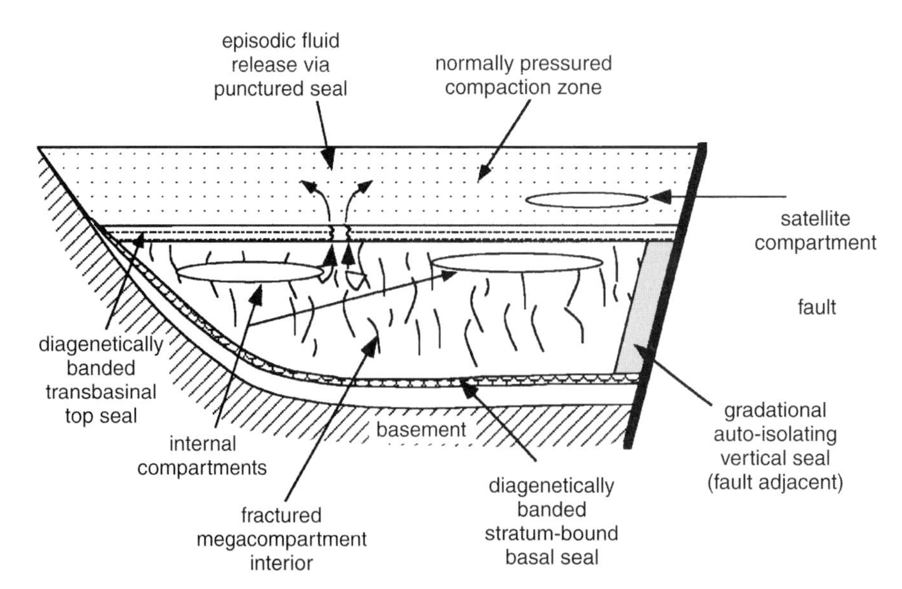

Figure 8. Schematic view of a basin-scale compartment defined by a banded transbasinal top seal, a stratum-localized bottom seal, a fault-associated autolocalizing side seal, and a fractured interior. Compartmentation is nested (as suggested by the interior compartments). There is also a satellite compartment. A rupture in the top seal with concomitant episodic fluid release is also suggested. The compartmenting basin is a complex dynamical system evolving via a number of strongly coupled chemical, transport, and mechanical processes.

Experience with far-from-equilibrium reaction-transport phenomena gained over the past two decades (Nicolis and Prigogine, 1987; Field and Burger, 1985; Ortoleva, 1992) suggests it is typically the case that the potential for a system to self-organize increases with the degree to which it is displaced from equilibrium (see also in the geological context Ortoleva et al., 1987a, b, 1990; Dewers and Ortoleva, 1990a; Ortoleva, 1994a). Increased displacement also favors an enhanced degree of complexity of the self-organized spatial or temporal patterns. Hence one might expect that the spatial scale of compartmentation and the time between episodic fluid migrations should decrease with increasing displacement from equilibrium. Furthermore, length scales of heterogeneity associated with sedimentary bedding may exceed those associated with differentiated layering (Ricken, 1986) that self-organizes during diagenesis (and forms layered seals) (Tigert and Al-Shaieb, 1990; Dewers and Ortoleva, 1990c; Ortoleva, 1994b; see also Qin and Ortoleva, this volume). Thus we might expect that exploration strategies based on the notion of the far-from-equilibrium basin reactor should be most successful when conditions favor large displacement from equilibrium.

A key challenge is then to develop RTM models with which we can delineate the specific attributes of the far-from-equilibrium phenomena that can occur and the sedimentary, thermal, and tectonic histories favorable for their existence. The development of such a basin reactor simulator is one of the central goals of the work being carried out at Indiana University for the GRI compartment project.

DISPLACEMENT FROM EQUILIBRIUM

Much of geochemistry is describable by equilibrium thermochemistry. A common view is that kinetics is of minor importance; however, even if many processes of an RTM system are fast and reversible (and hence are maintained at equilibrium), the system may be in an overall state of disequilibrium if *any* of the processes are out of equilibrium. Let us examine how the basin reactor may be maintained out of equilibrium with respect to key diagenetic processes.

Engineering reactors are of two basic types. The batch reactor is fed an initial mixture of reactants and then as it proceeds to equilibrium various products are produced. An open reactor receives a continuous influx of reactants and loses mass out the outlet.

The basin is a hybrid of the above two limiting cases. Sediment reactants are continuously being injected (during epochs of infilling) but product minerals are not being removed. Rather, the entire sediment package is subject to increasing burial.

Through thermal activation, burial-initiated reactions can take place that proceed at long (i.e., geological) time scales. A primary example is the suite of petroleum-generating processes that consume and create the kerogenic reactants.

Mineral reaction processes may take place on long time scales in at least two ways. Pressure solution or other stress-mediated processes are generally driven by weak free energy differences and thereby proceed slowly. Clay or other grain coatings or growth poisons can greatly slow reactions. In the case of pressure solution-type compaction, the driving force is indeed small for the redistribution of matter from contacts to free faces; however, the overall process (porous rock at depth → pore-free rock at depth) is far-from-equilibrium. Thus the main theme, a process driven by a large free energy force but not proceeding rapidly due to inhibited kinetics, can leave the basin far-from-equilibrium for long times.

An important aspect of stress in maintaining the basin far-from-equilibrium is that its effects can be manifest at all points within the system. Compressive stress at the lateral boundaries of the basin affects the state of stress throughout it. The effect of gravitational forces is similarly expressed at all points in the deep basin interior. As these forces drive mechano-chemical reactions (e.g., pressure solution), they can sustain the system far-from-equilibrium.

The deep penetration of stress applied at the boundaries proceeds with the speed of sound. This is in contrast to the maintenance of far-from-equilibrium conditions by bringing the system in contact with a composition bath at the boundary. Then the advance of reactants into the system proceeds by diffusion. This is too slow to be of importance in a basin. Of course, the reactor could be stirred to bring injected reactants into the reactor interior as is done in engineering contexts. However, there is negligible stirring of the rock matrix in a basin. There can be percolation flows. These bring in crustal or meteoric fluids that can react with solid matter in the basin.

Let us define stirring as flow whose pattern is dominated by the pattern of externally imposed forces. There is little such stirring in the lithified basin medium. As a result of stress-mediated processes and sediment reactant input the basin can be maintained far-from-equilibrium without having internally generated (i.e., self-organized) patterns destroyed by stirring. As we shall discuss further below, a key element of the development of spatial and temporal order—feedback (Ortoleva et al., 1987a, 1990; Ortoleva, 1994a, b) in the RTM network—is also commonplace in the sedimentary basin. We thus conclude that the basin should be rich in self-organized spatial structure.

TRANSIENTS FAR-FROM-EQUILIBRIUM AND COMPARTMENTATION

The real basin is only transiently maintained in one type of far-from-equilibrium state. If sedimentation stops and stresses and thermal effects at the basin boundary are fixed, then the basin will evolve toward a state of zero-porosity rock underlying density-sorted lighter fluids and with all reactions at equilibrium.

However, the history of sediment and fluid input, erosion, and stresses can be very complex. Within this history there can be long periods wherein the basin is sustained far-from-equilibrium. During these epochs self-organization can take place.

But the finite duration of these active epochs requires that expressed self-organization be developed on a time scale shorter than that of these epochs. Furthermore, some pattern formation phenomena could, for example, only take place in a uniform sandstone. If the characteristic length of this pattern is greater than the size of available sand bodies, then such patterns are also ruled out by length-scale constraints.

Thus, transience times and imposed characteristic lengths may eliminate certain self-organized patterns from being expressed.

FAR-FROM-EQUILIBRIUM COMPARTMENTATION

The stimulus for the GRI compartment project has been the notion of Powley and Bradley (Powley, 1975, 1980, 1990; Bradley, 1975; Powley and Bradley, 1987; see also Bradley and Powley, this volume) that the deeper part of a sedimentary basin has been divided into a few or even a complex boxwork of hydrologically isolated compartments. In this published work and in discussions, Powley and Bradley have been very convincing in forwarding the hypothesis that compartmentation must be understood in terms of some notions lying outside the realm of classical hydrology.

To address the question raised by the Powley and Bradley conjecture, let us ask whether compartmentation or some examples of compartmentation can be understood as a manifestation of far-from-equilibrium order or is compartmentation simply a consequence of imposed patterns (due to sedimentation structures or fault arrays). Let us term the latter *template compartmentation* and attempt to understand how compartments in the narrower sense as self-organized structures can come to be.

Imagine that we are running the basin reactor, varying conditions on the hundred-million-year time frame. We start with a very slow sedimentation rate. There is ample time for compaction to proceed to completion and fluid to escape upward. The result is a continued decrease of porosity with depth grading into pore-free rock below.

The combined effects of increased stress and rate of pressure solution due to the temperature dependence of rate coefficients and solubility as well cause compaction to start to proceed faster than the upward escape of fluids. The relative incompressibility of the fluid causes pore fluid pressure to rise. This tends to slow compaction and allows for the fluid to trap itself. The continuous burial traps the fluid into what one might term a compartment.

To give a more convincing scenario we must show how a few more features of observed compartments follow naturally. These include:

- seals that are thin relative to compartment size;
- the existence of lateral as well as horizontal seals;
- seals that are often diagenetically banded;
- compartment interiors that are often fractured; and
- fluid release from overpressured (or overpressuring) compartments that may be episodic or occur as a sequence of episodes.

To make our argument that compartmentation is a far-from-equilibrium process, we should show that these aspects of compartmentation follow naturally from our perspective.

The involvement of fracturing and episodic fluid release follows straightforwardly. As the aforementioned fluid pressure builds at depth because of impeded upward escape, the tensile strength plus least-principal compressive stress can be exceeded and the rock will be fractured. As this cannot only fracture the compartment interior but also seal rock, fluid can be expelled in an episode because of the permeability that is greatly enhanced by the fracturing. As fluid pressure drops, fractures may close or heal and pressurization can recommence to get the system set for a subsequent episode (see Figure 9).

The seal is a region of low-permeability rock bounded by higher-permeability rock. If the interior is characterized by preserved porosity or fluid pressure induced fracturing, then its interior boundary can be relatively sharp.

However, the thinness of the sealing rock may be the result of some type of autolocalizing feedback. Suppose there is a process whereby the fluid pressure gradient promotes a reaction that decreases perme-

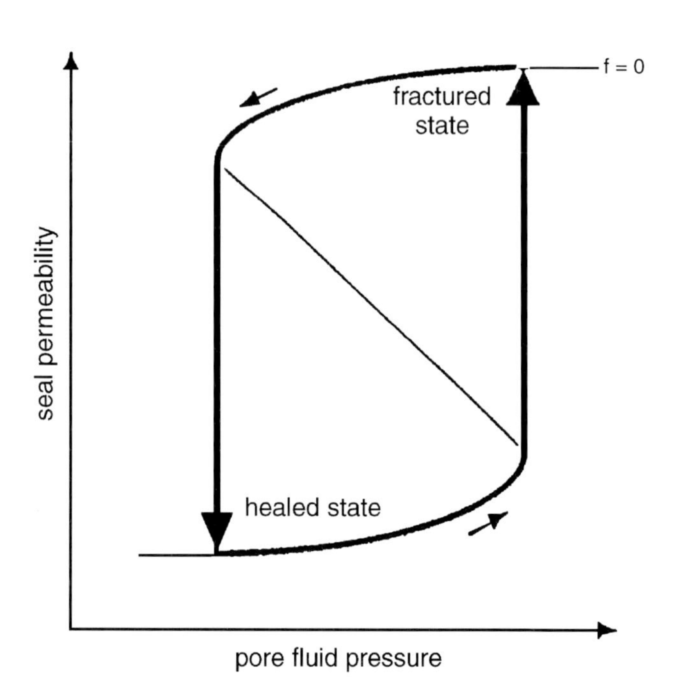

Figure 9. Repetitive fluid release from an overpressuring compartment can occur through a cycle of overpressuring with the seal in the healed (low-permeability) state → fracturing (or fracture aperture opening) → compartment depressuring with the seal in its leaky state → fracture healing (or aperture closing).

ability loss and thereby promotes a greater gradient—i.e., a type of fluid pressure gradient autocatalysis. Thus if there is one layer of relatively low permeability, then this process could make it even lower in permeability than its surroundings. Also, if within the putative region of lower permeability there is a smaller scale, low-permeability region, then that would become the lowest permeability region of all. Thus, there would be a general tendency toward the development of a seal of increasing quality and decreasing thickness. In fact, such rocks would present themselves as examples of what might be termed "smart" seals. The greater the pressure gradient applied across them, the better they develop their quality as seals.

A common pattern into which a far-from-equilibrium system can self-organize is one that consists of alternating bands or layers of one composition alternating with bands of another of contrasting character. For the porous rock compacting through pressure solution, we have shown in modeling studies of sandstones and carbonates (Merino et al., 1983; Dewers and Ortoleva, 1990b, c, 1994a) that there can be such a patterned genesis involving bands of augmented compaction and porosity alternating with ones of less compaction, higher overgrowth, and lower porosity (see Figure 10) (see Qin et al., this volume, for citations of geological examples).

There are two significant properties of self-organized mechano-chemical banding. Like traditional (i.e., stable, nonbanding) compaction, banded or stylolitized compaction is slowed at elevated pore pressure. At lower pore pressure it proceeds to create layers of very low permeability that tend to dominate (and thereby depress) the overall permeability for flow normal to the layered structure. Thus the property of a rock to have a banded compaction response makes it an ideal candidate for sealing rock (see Qin and Ortoleva, this volume, for further discussion).

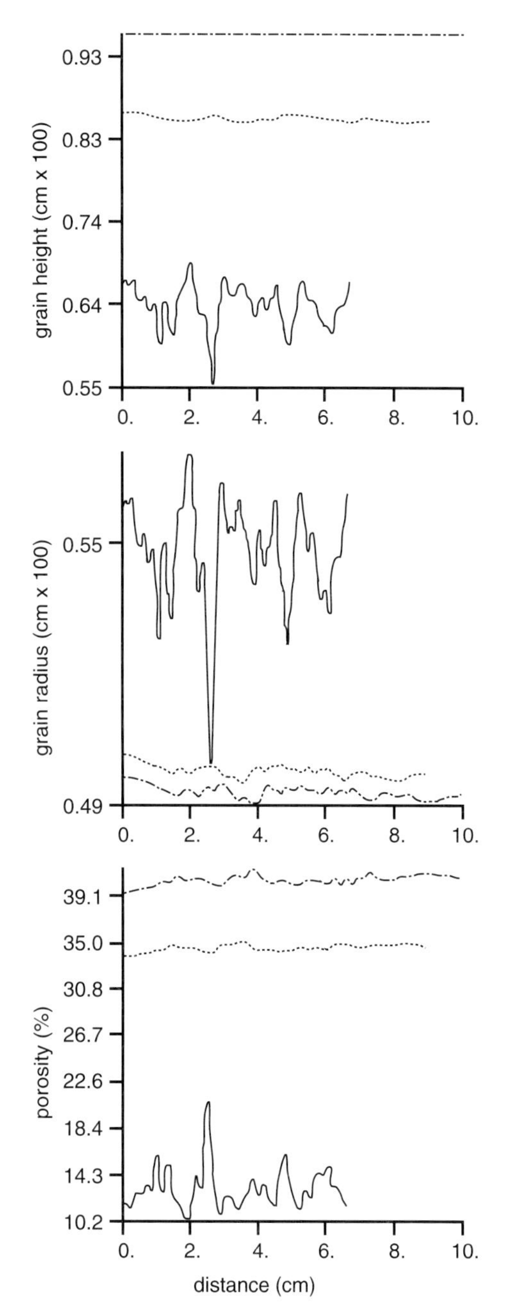

Figure 10. Predictions from Dewers and Ortoleva (1990b) model of compaction in a sandstone. Note that as compaction proceeds (and hence porosity decreases), small heterogeneities are amplified into a differentiated layering. The layering involves zones of low porosity, high overgrowth, and low compaction (i.e., larger grain height) alternating with zones of opposite character. Such bands are apparently one of the types observed in banded seals.

AUTOISOLATION: ISOLATED BEDS AND FAULT-ADJACENT SEALS

Consider the hypothesis that diagenetic compartmentation occurs when a fluid body at depth can somehow isolate itself from its surroundings. We term this property *autoisolation*. Let us now discuss several mechanisms for it and note their implications.

For autoisolation to take place, two elements must be in place:

- the fluid, in its interaction with the rock matrix containing it, must somehow preserve porosity and permeability; and
- at the periphery of the isolating fluid body, there must be some process whereby permeability is destroyed.

By such a general scheme, if it exists under diagenetic conditions, a fluid body can promote its own isolation by creating a compartment around it.

Perhaps the most straightforward autoisolation dynamic follows from the fluid pressure-dependence of pressure solution-mediated compaction. High overpressure inhibits this compaction, while low fluid pressure at the periphery of an overpressuring domain can lead to augmented compaction there that can isolate the interior.

Possible examples of this mechanism are suggested in Figures 11 and 6. A "fault-adjacent" seal is seen in Figure 11. There, high fluid pressure in the center of the overpressuring body preserves porosity. However, contact with the fault leads to some fluid leakage *into* the fault and reduction of pore pressure there as the fault here is assumed to act as a conduit in contact with hydrostatic fluid overhead. Such fault-associated seals can be several kilometers thick (as observed in the Anadarko basin—see Al-Shaieb et al., this volume). This runs contrary to the commonly held notion that a fault itself is the vertical seal (due to cementation or gouge within it).

Another example of this mechanism is an autoisolated fine-grained bed surrounded by coarser sandstone as suggested in Figure 6. Here the coarser sand does not compact much relative to the fine-grained lens (because of the grain size dependence of the rate of pressure solution). However, at the periphery of the lens fluid pressure is lower and so a "rind" of compacted rock can develop through augmented pressure solution that serves as a seal, isolating the lens.

A distinct second mechanism for the autoisolation of a lens as in Figure 6 can operate based on a type of displaced pressure solution-driven process. Assume that the quartz grains in the fine lens are coated with overgrowth inhibiting clay coatings. Then in the interior the activity of $SiO_2(aq)$ builds up to its value in equilibrium at the state of stress in the grain-grain contacts; thereby pressure solution-mediated compaction is repressed. However, if the surrounding coarse sandstone has no such overgrowth-inhibiting clay coating then $SiO_2(aq)$ in the fine lens near the periphery diffuses to the surrounding coarse sandstone and precipitates on the grains there. Thereby a surrounding layer of quartz overgrowth is produced that can serve as a permeability barrier. This process can be further augmented by the accumulation of clay on the fine-grained side of the interface between the beds; this could itself be a further barrier to flow. Also it is observed (and shown via RTM modeling) that the formation of stylolites and diagenetic banding tends to be promoted at a region of sedimentary textural contrast (Dewers and Ortoleva, 1990b, c).

It is often noted that petroleum can preserve reservoir porosity/permeability by tending to repress diagenetic reactions that would destroy reservoir quality (notably pressure solution). One may argue that this may lead to a dynamic of autoisolation of a petroleum-filled compartment and that this is a strictly far-

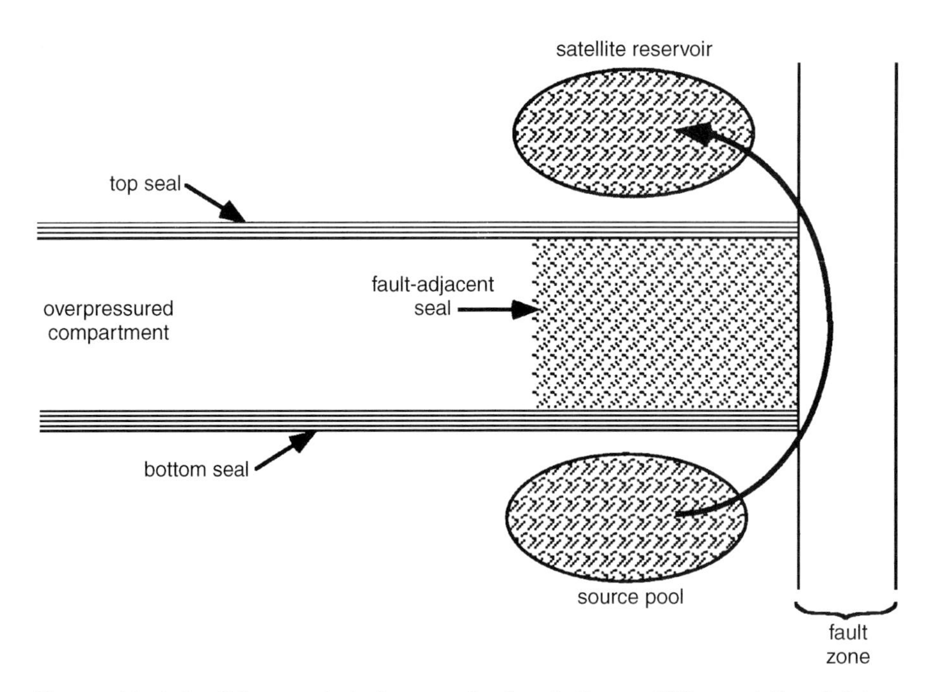

Figure 11. A fault here acts to impose hydrostatic conditions at the right-hand end of the otherwise sealed, overpressurized bed. As a result, compaction is not arrested near the fault, unlike away from the fault where it is slowed by overpressuring. The net result is that the bed isolates itself from connection to hydrostatic pressure through the fault. Thus vertical seals may develop near but not within faults that serve as fluid conduits. However, rocks below the bottom seal cannot develop such a seal (due to unfavorable lithology) and hence petroleum there could migrate up through the fault and can be sequestered in reservoirs overlying the autoisolated compartment—this likely occurred near the Wichita fault in the Anadarko basin.

from-equilibrium phenomenon. If burial is sufficiently fast then before petroleum generation and upward migration is complete, sufficient depth can be obtained for pressure solution-mediated compaction to take place. This can proceed to sufficient extent to slow down upward migration of petroleum that a barrier develops above a rising petroleum mass. Compaction at the sides of the mass can also prevent its lateral escape. The bottom of the petroleum-filled compartment could then develop as the petroleum leaves the rock in this basal region and is replaced by pressure solution-promoting aqueous fluid.

From the above we see that petroleum autoisolation involves a number of balances of the timings of various processes. Thus it can clearly take place only in some range of basin histories.

RESPONSE OF THE UNSTABLE BASIN REACTOR TO SEDIMENTARY, FAULTING, AND OTHER TEMPLATES

Having attempted to develop the case for compartmentation of a diagenetic nature we return to the problem of templated compartments and the possible difficulties discriminating between them and self-organized ones. Templated compartments are the consequence of a sedimentary and/or fault pattern that is just so as to completely yield a volume of hydraulically connected rock that is surrounded by low-permeability "sealing" rock in three dimensions. The latter is supposed to be due to its specific sedimentary origin or is a flow-inhibiting fault (i.e., from gouge).

An example of such a templated compartment of completely sedimentary origin would be a coarse-grained channel or valley fill completely encased by mudstone. For such a compartment to have survived as an overpressured system containing hydrocarbons, we must recognize the following:

- As mudstone is supposed to be a permeability barrier, not much of the petroleum produced in it can get into the sandstone.
- If closure was a completely sedimentary feature, then the sequestered petroleum must have come from the mudstone near the sand/mud interface.
- In light of the last two points, how can one account for the total mass of petroleum sequestered?
- Appreciable transport of aqueous fluids across the mudstone is prohibited, so that if the sand is petroleum filled, where did the original pore-filling water go?

If the above points imply some contradictions to the templated origin of such a compartment then we conclude that there must have been appreciable exchange between the sedimentary template protocompartment and its surroundings.

These cited "paradoxes" could be resolved if the concept of diagenetic self-organized compartments is valid. In this view the sedimentary protocompartment served as a kind of nucleus for the formation of a compartment. If its dimensions and the properties of the rock as well as its burial and thermal history would allow for self-organized compartmentation, then the sedimentary protocompartment will develop into a true compartment. Its early phase of incompleteness (i.e., gaps in the closure or protosealing rock that is too leaky to be quality sealing rock) would allow for early exchange of fluids that would ultimately be terminated as the protocompartment → compartment transformation took place via diagenesis.

This recalls two general themes from the theory of self-organization in far-from-equilibrium systems.

1. Omnipresent nonuniformity in any physical system serves as the disturbance that can be amplified into a well-defined pattern through the self-organization dynamic.
2. The further a system is driven out of equilibrium the greater the number of pattern geometries or types that it can sustain.

Property 2 implies that under very far-from-equilibrium conditions favoring compartmentation, any of a very large class of sedimentary, faulting, or other features could serve as viable nuclei for compartment genesis.

COMPARTMENTATION AND RESERVOIR HETEROGENEITY

As noted earlier, the further the basin is driven from equilibrium, the greater the degree of complexity one might expect in the development of patterns of compartmentation. Hence, far-from-equilibrium basin evolution is likely to lead to reservoirs with heterogeneity on shorter scales.

There are two opposing tendencies determining the scale of reservoir heterogeneity. The increased degree of tendency toward complexity of self-organized compartmentation promotes downscaling. Hydrofracturing due to overpressuring tends to break through barriers and thereby increases the scale of reservoir hydraulic continuity. Clearly the actual resultant scale is a complex balance of these competing effects.

Another natural implication of the tendency toward increasing compartmental complexity is the possibility of compartment nesting—i.e., compartments formed within compartments, within compartments, and so on. Such a hierarchy does not likely have a fractal character. This is because the mechanism for forming compartments on the various scales is likely rather different.

As a final note, again consider the upscaling role of overpressure derived from its ability to induce fracturing. In Figure 12 we suggest the existence of a basinwide front of hydrofracturing that can develop during active sedimentation. The front rises through the incoming sediment. When the latter exceeds a critical rate the fracture front does not rise steadily but rather rises in episodes. This was found in the simulation of Figure 13 (Dewers and Ortoleva, 1994b).

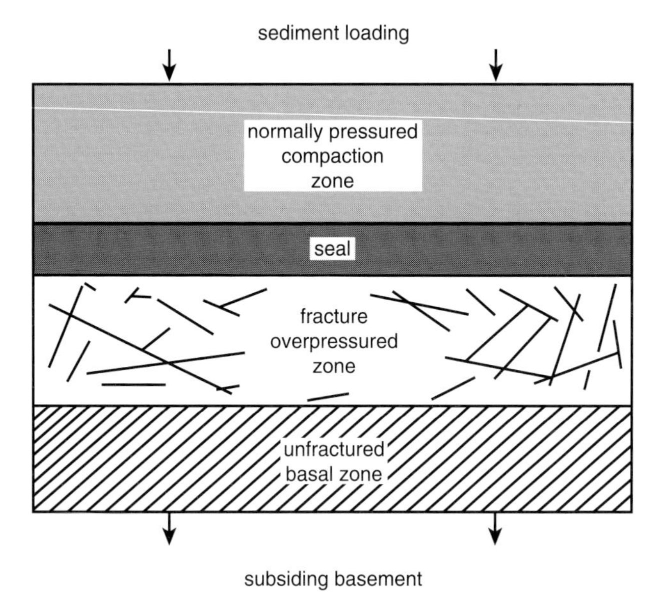

Figure 12. Configuration of a simple burial diagenesis model showing sediment infilling at the top, zone of compaction under hydrostatic fluid pressure, "seal" region where overall permeability (matrix plus fracture contributions) is a minimum, hydrofractured "compartment interior," unfractured basal zone, and subsiding basement rock. At depth, fracturing could be repressed because of increased lateral stress giving an effective lower seal.

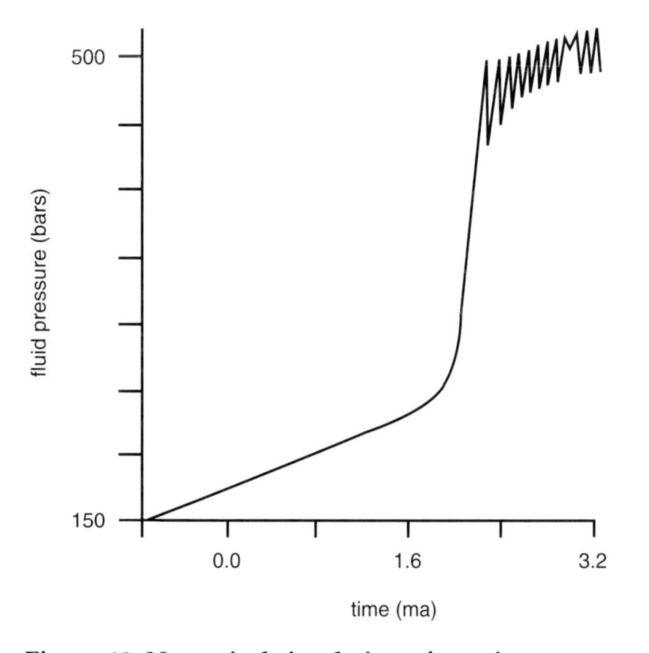

Figure 13. Numerical simulation of reaction-transport-mechanical model of the system in Figure 12 (adapted from Dewers and Ortoleva, 1994a). Shown is the temporal evolution of the fluid pressure at the sediment-basement interface.

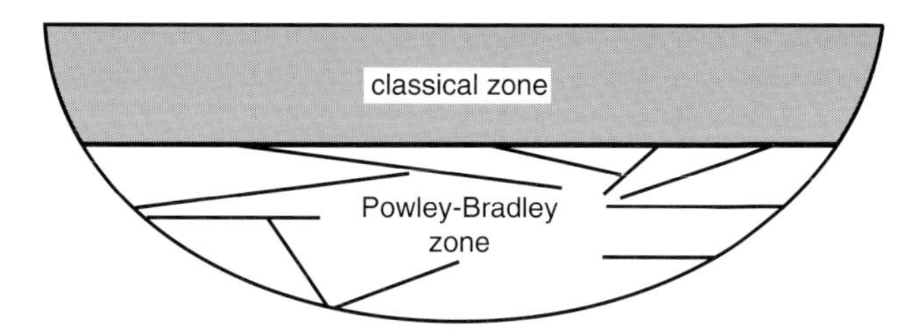

Figure 14. Schematic view of a sedimentary basin divided into a classical zone and a lower Powley and Bradley zone subdivided by seals into a boxwork of compartments.

CONCLUSIONS

The above discussion suggests the following two key points.

1. The basin may be roughly divided into an upper "classical" zone and the Powley-Bradley compartmentation (Powley, 1975, 1980, 1990; Bradley, 1975; Powley and Bradley, 1987; see also Bradley and Powley, this volume) zone (see Figure 14). The behavior of the classical zone is dominated by the pattern of sedimentology, surface topography and of pressure heads imposed from the surface. The Powley-Bradley zone is dominated by the strongly coupled RTM internal processes. Imposed patterns left over from the classical zone may be dramatically altered on many spatial scales through diagenesis.

Because of the great complexity of these systems, computer RTM simulations give the only hope of turning conclusion (2) into a practical tool for exploration and production.

2. The relatively large degree of autonomy of the Powley and Bradley regime implies that RTM modeling can predict key aspects of reservoir location in a basin and its degree of internal heterogeneity.

ACKNOWLEDGMENTS

Research supported in part by contracts from the Gas Research Institute (No. 5092-260-2443), Amoco Production Company, and Mobil Exploration and Producing Services, and grants from the Basic Energy Sciences Program of the U.S. Department of Energy (No. DEFG291ER14175), Shell Development Company, IBM, Intel, and Sun Microsystems.

REFERENCES CITED

Bradley, J.S., 1975, Abnormal formation pressure: AAPG Bulletin, v. 59, p. 957–973.

Dewers, T., and P. Ortoleva, 1988, The role of geochemical self-organization in the migration and trapping of hydrocarbons: Applied Geochemistry, v. 3, p. 287–316.

Dewers, T., and P. Ortoleva, 1990a, Geochemical self-organization III: A mean field, pressure solution model of spaced cleavage and metamorphic segregational layering: American Journal of Science, v. 290, p. 473–521.

Dewers, T., and P. Ortoleva, 1990b, A coupled reaction/transport/mechanical model for intergranular pressure solution, stylolites, and differential compaction and cementation in clean sandstones: Geochimica et Cosmochimica Acta, v. 54, p. 1609–1625.

Dewers, T., and P. Ortoleva, 1990c, The interaction of reaction, mass transport, and rock deformation during diagenesis: Mathematical modeling of intergranular pressure solution, stylolites, and differential compaction/cementation, *in* I. Meshri and P. Ortoleva, eds., Prediction of Reservoir Quality through Chemical Modeling: AAPG Memoir, v. 49, p. 147–160.

Dewers, T., and P. Ortoleva, 1994a, Formation of stylolites, marl/limestone alternations, and clay seams through unstable chemical compaction of argillaceous carbonates: in K.H. Wolf and G.V. Chilingarian, eds., Diagenesis Vol. IV: Amsterdam, Elsevier, p. 155–216.

Dewers, T., and P. Ortoleva, 1994b, Nonlinear dynamical aspects of deep basin hydrology; fluid compartment formation and episodic fluid release: American Journal of Science, v. 294, p. 513–755.

Field, R.J., and Burger, M., 1985, Oscillations and Traveling Waves in Chemical Systems: New York, John Wiley and Sons.

Merino, E., P. Strickholm, and P. Ortoleva, 1983, Generation of evenly spaced pressure-solution seams during (late) diagenesis: A kinetic theory: Contributions to Mineralogy and Petrology, v. 82, p. 360–370.

Nicolis, G., and I. Prigogine, 1977, Self-Organization in Nonequilibrium Systems: From Dissipative Structures to Order Through Fluctuations: Wiley-Interscience.

Ortoleva, P., 1992, Nonlinear Chemical Waves: New York, John Wiley and Sons.

Ortoleva, P., 1994a, Geochemical Self-Organization: New York, Oxford University Press, in press.

Ortoleva, P., 1994b, Development of diagenetic differentiated structure through reaction-transport feedback, *in* K.H. Wolf and G.V. Chilingarian, eds., Diagenesis, Vol. IV, in press.

Ortoleva, P., E. Merino, J. Chadam, and C.H. Moore, 1987a, Geochemical self-organization I: Reaction-transport feedbacks and modeling approach: American Journal of Science, v. 287, p. 979–1007.

Ortoleva, P., E. Merino, J. Chadam, and A. Sen, 1987b, Geochemical self-organization II: The reactive infiltration instability: American Journal of Science, v. 287, p. 1008–1040.

Ortoleva, P., B. Hallet, A. McBirney, I. Meshri, R. Reeder, and P. Williams, eds., 1990, Self-organization in geological systems: Proc. of a Workshop held 26–30 June 1988, University of California at Santa Barbara: Earth Science Reviews, v. 29, Amsterdam, Elsevier.

Ortoleva, P., T. Dewers, and B. Sauer, 1993, Modeling diagenetic bedding, stylolites, concretions and other mesoscopic mechano-chemical structures, *in* R. Rezak and D. Lavoie, eds., Proceedings of Carbonate Microfabrics: New York, Springer-Verlag, p. 291–300.

Ortoleva, P., Z. Al-Shaieb, and J. Puckette, in press, Genesis and dynamics of basin compartments and seals: American Journal of Science.

Powley, D.E., 1975, Course notes used in AAPG Petroleum Exploration Schools.

Powley, D., 1980, Normal and abnormal pressure. Lecture presented to AAPG Advanced Exploration Schools, 1980–1987.

Powley, D., 1990, Pressures, hydrogeology and large scale seals in petroleum basins: *in* P. Ortoleva, B. Hallet, A. McBirney, I. Meshri, R. Reeder, and P. Williams, eds., Self-Organization in Geological Systems, Proceedings of a Workshop held 26–30 June 1988, University of California at Santa Barbara, Earth Science Reviews, v. 29, p. 215–226.

Powley, D., and J.J. Bradley, 1987, Pressure compartments: Unpublished manuscript.

Ricken, W., 1986, Diagenetic Bedding: Berlin, Springer Verlag.

Tigert, V., and Al-Shaieb, Z., 1990, Pressure seals: Their diagenetic banding patterns, *in* P. Ortoleva, B. Hallet, A. McBirney, I. Meshri, R. Reeder, and P. Williams, eds., Self-Organization in Geological Systems, Proceedings of a Workshop held 26–30 June 1988, University of California at Santa Barbara, Earth Science Reviews, v. 29, p. 227–240.

II. The Anadarko Basin

Chapter 4

Megacompartment Complex in the Anadarko Basin: A Completely Sealed Overpressured Phenomenon

Zuhair Al-Shaieb
James O. Puckette
Azhari A. Abdalla
Patrick B. Ely
Oklahoma State University
Stillwater, Oklahoma, U.S.A.

ABSTRACT

Integrated pore pressure, potentiometric, and geologic data in the Anadarko basin demonstrate the existence of a basinwide, completely sealed overpressured compartment, called the megacompartment complex. All reservoirs within this complex exhibit pressure gradients that exceed the normal gradient of 10.515 kPa/m (0.465 psi/ft). These reservoirs have produced large quantities of natural gas, particularly from the Pennsylvanian Red Fork and Morrowan sandstones.

This megacompartment complex is enclosed by top, basal, and lateral seals. The top seal zone, which is located between 2290 m and 3050 m (7500 and 10,000 ft) below the surface, is relatively horizontal, dips slightly to the southwest, and appears to cut across stratigraphy. However, the diagenetically enhanced basal seal is stratigraphically controlled and seems to coincide with the Devonian Woodford Shale. The complex is laterally sealed to the south by a vertical cementation zone associated with the frontal fault zone of the Wichita Mountain uplift and by the convergence of the top and basal seals along the eastern, northern, and western boundaries.

The interior of the complex is subdivided into a myriad of smaller compartments with distinct pressure gradients. In addition, local overpressured compartments are present outside the megacompartment complex in normally and near-normally pressured regions.

INTRODUCTION

Basin compartmentation, introduced by Bradley (1975) and Powley (1987), is an important concept in the exploration and production of hydrocarbons in deep sedimentary basins. A compartment is a two-component system that consists of a porous internal rock volume and a surrounding low-permeability seal. Compartmentation can occur beyond a certain depth due to the interplay of a number of geological processes: in particular, subsidence and sedimentation rates, lithologies, and diagenetic modifications. These compartments exist as abnormally pressured rock volumes that exhibit distinctly different pressure regimes in comparison with their immediate surroundings. They are most easily recognized on pressure-depth profiles (PDPs) by their departure from the normal hydrostatic gradient to the surface. This chapter focuses primarily on abnormally overpressured compartments in the Anadarko basin, their geometrical configuration, distribution within the basin, and their associated seals. Integrated pore-pressure and subsurface geological data indicate the presence of a basinwide overpressured and completely sealed compartment in the Anadarko basin. The term *megacompartment complex* (MCC) was introduced by Al-Shaieb (1991) to describe this phenomenon. The MCC in the Anadarko basin encompasses rock intervals representing the Devonian, Mississippian, and the Pennsylvanian systems. Its internal volume consists of a network of totally isolated smaller, nested compartments.

The isolation of the MCC is maintained through considerably long geological times (early Missourian to present) via encasement by top, basal, and lateral seals (Al-Shaieb et al., 1990). Compartments nested within the MCC are isolated from each other by an intricately complex framework of seals. Seal rocks display unique diagenetic banding structures that formed as a result of the mechano-chemical processes of compaction, dissolution, and precipitation.

GEOLOGICAL SETTING

Tectonic History of the Anadarko Basin

The Anadarko basin is considered one of the deepest foreland Paleozoic basins on the North American craton. Located in western Oklahoma and the northern Texas Panhandle, this basin covers an area of approximately 90,639 km^2 (35,000 mi^2) (Figure 1). It is bounded to the east by the Nemaha Ridge and by the ancient eroded Amarillo-Wichita Mountain Front to the south. To the west and north, the Anadarko basin is flanked by shallow platform areas.

The basin is elongated, west–northwest-trending, and exhibits a marked cross-sectional asymmetry due to a complex fault zone on its southern margin that separates the basin from the Wichita Mountain uplift. Maximum structural displacement between the uplift and the basin floor exceeds 9144 m (30,000 ft).

The Anadarko basin is one of several west–northwest-trending basins and uplifts that extend from the Ouachita foldbelt in southern Oklahoma to the Texas Panhandle. Other basins include the Marietta and Ardmore. Positive structural features include the Wichita-Criner uplift, the Muenster-Waurika arch, and the Arbuckle-Tishomingo uplift (Figure 1).

Southern Oklahoma was initially described as an aulacogen by Schatski (1946). Burial history curves for the Anadarko basin show a relatively rapid rate of subsidence in Cambro-Ordovician times. This rapid rate was followed by relatively slow subsidence rates in the Silurian, Devonian, and Early Mississippian times. Extremely rapid rates of subsidence for the Late Mississippian to Pennsylvanian age coincide with the tectonic development of the Anadarko basin and the Wichita Mountain uplift. By Early Pennsylvanian, 3050 m (10,000 ft) of Springer-Morrowan and Atokan sediments were deposited in the Anadarko basin. Pennsylvanian-age deformation within the aulacogen was dominated by displacements along major high-angle fault zones. As time progressed, flooding of the craton occurred, followed by a slowing in subsidence rate, regression of the sea, and filling of the basin from east to west (Rascoe and Adler, 1983).

Stratigraphy and Sedimentology of Key Intervals in the Anadarko Basin

While the middle to late Paleozoic sequences are primarily of clastic composition, the lower Paleozoic stratigraphy of the Anadarko basin is mostly dominated by carbonates (Figure 2). The lower Paleozoic carbonate reservoirs along with the Pennsylvanian sandstones are credited for most of the hydrocarbon production in the basin. During the course of this study, all available pressure data in the basin were examined. To show general changes in pressure regimes from the shelf to basinal reservoirs within the MCC and in the pressure domain of normally pressured rocks below the MCC, maps of selected stratigraphic intervals were used. These intervals are characterized as having adequate data to illustrate these changes. The Missourian/Virgilian interval data include pressure measurements from eight different sandstone units. Completions are limited in any one of these reservoirs, but collectively they provide basinwide coverage. The Desmoinesian Red Fork Formation is used to represent the Middle Pennsylvanian since it contains an abundance of completions. The Morrowan interval includes both upper and lower reservoir data since together they provide the most complete Lower Pennsylvanian data across the basin. The paucity of Mississippian pressure data prevents their use. The eroded karsted Hunton reservoirs are often indistinguishable at the formation level, so measurements for the total group were used. In the following section, a brief description of the stratigraphy and sedimentology of each interval will be presented.

Missourian/Virgilian

During the Late Pennsylvanian, the Ouachita Mountains were a major sediment source for the Missourian/Virgilian sequence. Extensive clastic sediments were deposited during early Missourian time. During periods of low clastic influx, carbonates were deposited in the shelf areas. These carbonates are relatively thin units sandwiched within thicker clastic intervals. Carbonates were more prevalent on the northern shelf of the basin.

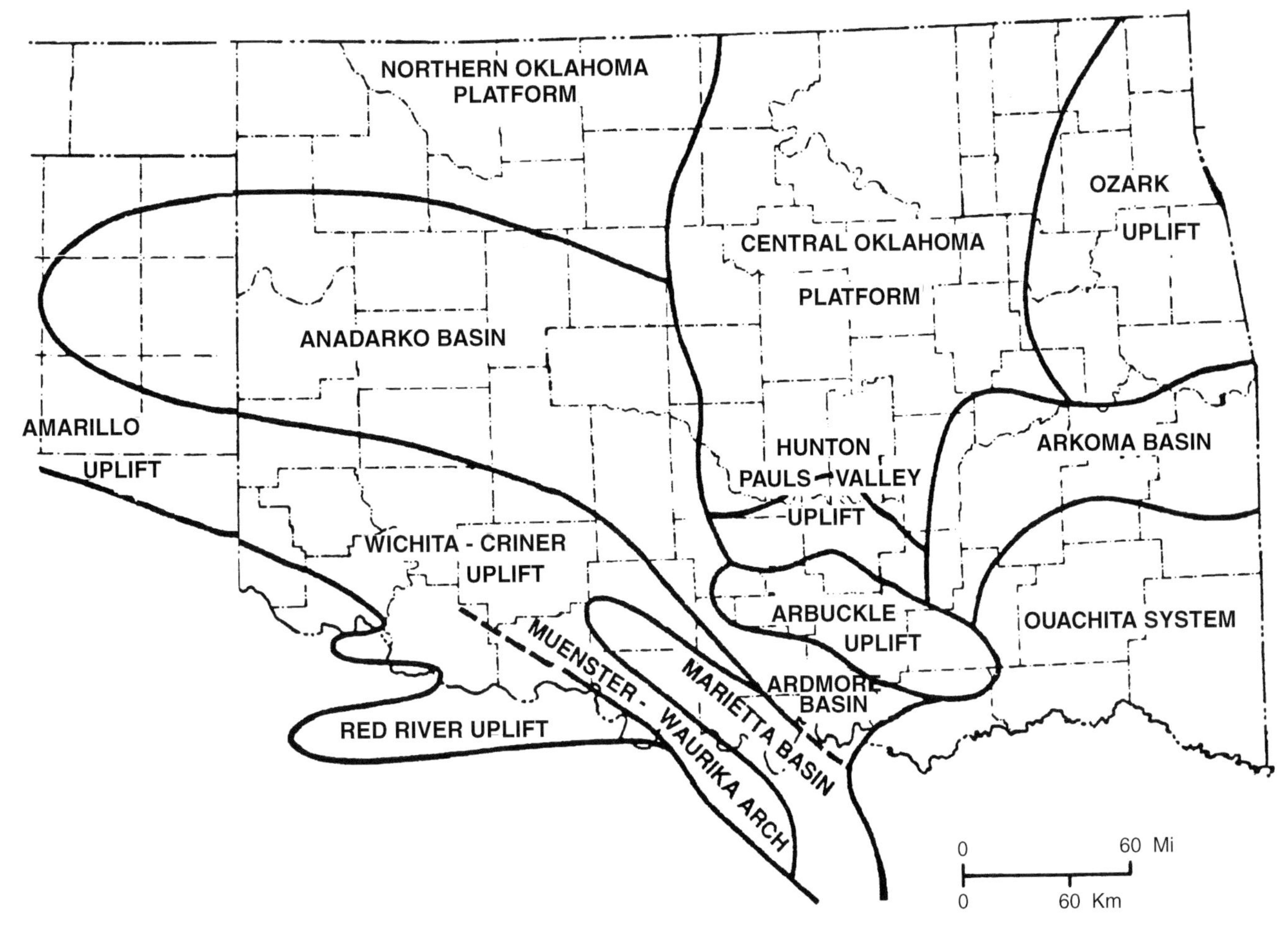

Figure 1. Tectonic map showing location of the Anadarko basin and the major structural features of Oklahoma. After Al-Shaieb and Shelton (1977); Arbenz (1956).

A second sediment source during the Missourian/Virgilian is associated with the Wichita-Amarillo uplift. Granites and, less important, carbonates were eroded from the Wichita Mountains and deposited in the Anadarko basin as granite wash and minor carbonate wash (Wade, 1987; Rascoe and Adler, 1983). The southerly source provided clastics throughout the Missourian and reached its peak in early Virgilian. Deposition of carbonate units within the thicker intervals of clastic sediments was characteristic of these sequences.

An exception to the normal clastic/carbonate sequences is the Heebner Shale, which marks a period of unusual quiescence in Virgilian time. This shale is of deep marine origin and has considerable regional continuity.

Interpretation of depositional environments indicates that sediments of the Missourian and Virgilian series were deposited in deltaic, fan-delta, shallow marine, and deep marine settings. These settings provided conditions favorable for the generation, migration, and trapping of petroleum. Several zones within the Missourian/Virgilian are considered prolific oil and gas producers in the Anadarko basin, especially the Granite Wash, Marchand, and Cottage Grove sandstones.

Red Fork (Desmoinesian)

The Middle Pennsylvanian (Desmoinesian-age) rocks in the Anadarko basin are represented primarily by clastic sequences. The Red Fork Sandstone is the most important petroleum-producing reservoir in this sequence. Sources for the Red Fork sediments were primarily to the north and east and less significantly to the south. Alluvial plain, deltaic, and incised channels flowed from the north and east southward and westward into the basin. During lowstand conditions the shelf edge was exposed, the streams were incised during extension, and submarine fans were deposited on the basin floor (Anderson, 1991). As sea level rose, deposition occurred in the channels as valley fill. During temporary stillstands, deltaic conditions developed in the lower stretches of the streams, and relatively thin delta-fringe sands formed sheet-like units. Further rise of sea level inundated the stream systems, and estuarine settings preceded widespread shallow marine conditions (Al-Shaieb et al., 1989b). The Red Fork channel-fill sandstones are commonly

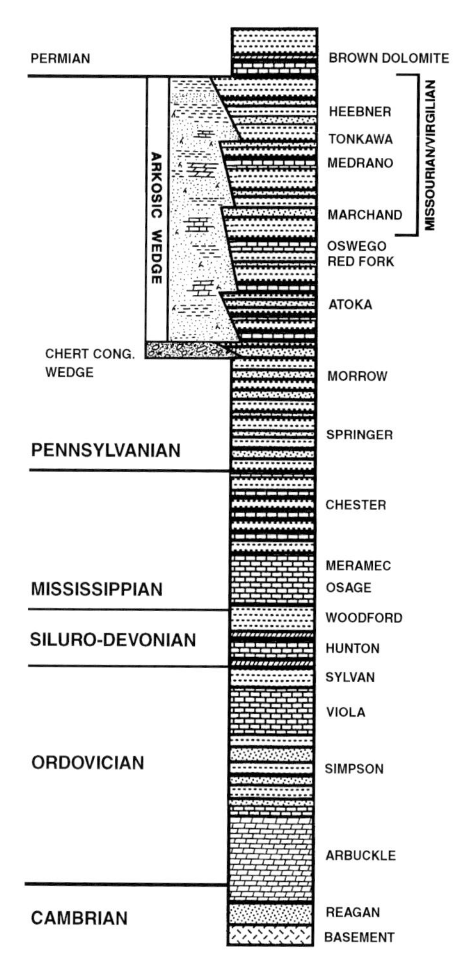

Figure 2. Generalized stratigraphic column of the Anadarko basin. After Evans (1979).

multistoried and laterally restricted. These sandstones generally have higher reservoir potential than the delta-fringe sandstones because they are coarser grained and thicker.

Morrowan Series

The Morrow Formation is a predominantly clastic, prolific hydrocarbon-producing unit in the Lower Pennsylvanian. The sediments of the Morrow interval are dominantly shales with discontinuous and erratic sandstones and limestones. Distribution of coarse terrigenous clastics within the Morrow interval indicates several sources for these clastics. One source is to the west-northwest of the Oklahoma and Texas panhandles. A second source is to the north-northeast of the present shelf area. The third significant source is the Wichita Mountain uplift to the south-southwest, which provided clastics to form the Upper Morrowan chert conglomerate sequences. A cross section of the basin shows a wedge-shaped accumulation of the Upper Morrowan near the Wichita Mountain uplift. This wedge thins to the north and northeast toward the basin shelf.

Several depositional models are proposed for the Morrowan clastics. Generally, lower Morrow sandstones are characterized by features indicative of marine environment. These sandstones have been interpreted as shallow marine sandstones deposited parallel to the ancient shoreline (Al-Shaieb and Walker, 1986; Swanson, 1979). The Upper Morrow sandstones are generally considered as deltaic facies that prograded to the west-northwest from the Wichita Mountain uplift. Morrow deltas are believed to be small and tidal dominated.

Regressive fluvial/deltaic systems are also recognized in the Upper Morrow interval. Transgressive valley-fill sandstones in incised channels are documented in the Morrow interval in the northern shelf area (Krystinik, 1989; Harrison, 1990). The Upper Morrowan chert conglomerate in the southwestern Anadarko basin represents a braid and fan-delta system that prograded northeastward into the shallow marine basin adjacent to the Wichita Mountain uplift (Puckette et al., 1993; Al-Shaieb et al., 1989a; Shelby, 1979).

Hunton Group (Ordovician–Silurian–Devonian)

The Hunton Group is a series of carbonates that were deposited during the Upper Ordovician, Silurian, and Lower Devonian times. In contrast to older Paleozoic carbonates, the Hunton represents extremely slow rates of deposition.

The Chimneyhill, Henryhouse-Haragan, and Bois D'Arc intervals of the Hunton are considered peritidal carbonate sequences. They are represented by several distinct facies belts, including open shelf subtidal mudstones and wackestones, intertidal shoal packstones and wackestones, peloidal lagoonal mudstones and wackestones, and supratidal-tidal flat algal boundstones and mudstones (Al-Shaieb et al., 1985; Medlock, 1982; Beardall, 1983). The Frisco interval was deposited on an unconformity surface of the older eroded Henryhouse-Haragan–Bois D'Arc units. This eroded paleotopography was conducive to the development of crinoidal bioherms in the Frisco interval (Medlock, 1982).

The alteration of original rock textures in the Hunton Group has created both fabric-selective and nonfabric-selective reservoirs. Dolomitization and karsting account for most porosity observed in the Hunton reservoirs.

PROCEDURE

Pressure compartments can be identified on pressure-depth profiles (PDPs) as deviations from a "normal" fluid-pressure gradient of 10.515 kPa/m (0.465 psi/ft) for a standard 80,000 ppm formation brine. Depths to sealing intervals associated with these compartments can be estimated on PDPs at the transition zones between portions of the PDP curve representing normally pressured and overpressured rocks (Powley, 1986).

Several sources were utilized to collect pressure data for various reservoirs in the Anadarko basin. The data sources include drill-stem tests, recorded bottom-hole pressures in production records, and calculated bottom-hole pressures from static initial well-head shut-in pressures. During the course of the study, 28,452 pressure-data values were analyzed and screened with respect to the time elapsed since the specific well was spudded. From this group, which covered various producing intervals in the Ordovician, Silurian, Devonian, Mississippian, and Pennsylvanian systems, 4439 bottom-hole (reservoir) pressure-data points were entered into a computer database. These consisted of 2579 reservoir-pressure-data points calculated from static initial well-head shut-in pressures (Dwight's, 1990), 1787 data points from the shut-in bottom-hole pressures of drill-stem tests (Amoco, 1989), and 73 recorded bottom-hole pressures from P/Z plots in production records (Dwight's, 1990).

Pressure-depth profiles were then constructed for various geographical areas across the Anadarko basin (Figures 3, 4, 5, 6, 7, and 8; locations are shown on pressure-gradient maps). Observations from these profiles integrated with the stratigraphy of the Anadarko basin established the following general relationships:

1. Normal to near-normal gradients are observed in all stratigraphic horizons down to approximately 2290 to 3050 m (7500 to 10,000 ft) below the surface.
2. Overpressuring is observed in all reservoirs between 2290 and 3050 m (7500 to 10,000 ft) deep and the Mississippian or Woodford intervals.
3. A return to normal and near-normal pressure gradients is observed in the Hunton and older Paleozoic reservoirs. This pattern in PDP curves is observed in all areas of the Anadarko basin that lie within a finite overpressured volume of rocks referred to as the megacompartment complex.

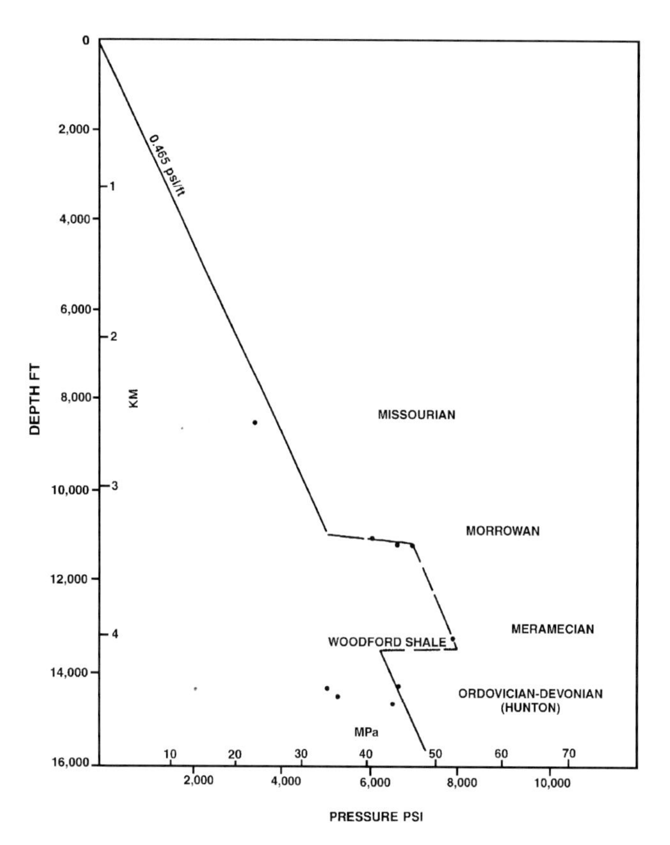

Figure 4. Pressure-depth profile from northeastern Custer County, Oklahoma (Thomas field area), showing part of the northern portion of the MCC.

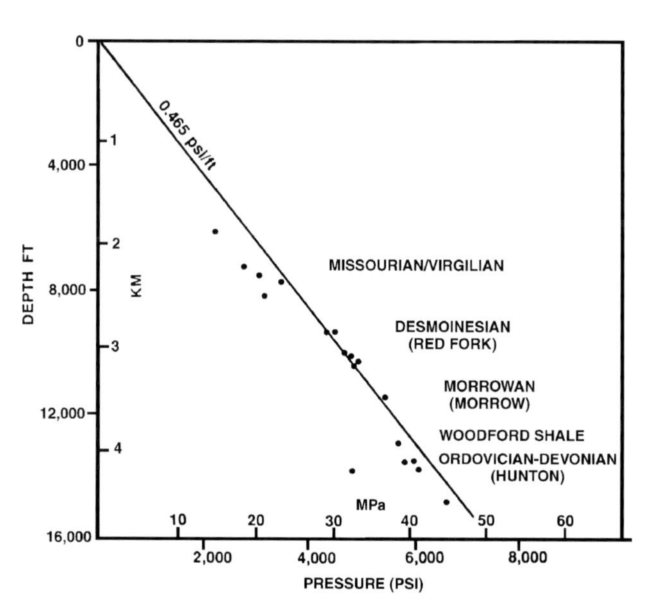

Figure 3. Pressure-depth profile from northern Dewey County, Oklahoma, indicating near-normal pressure gradients throughout the rock column outside the boundaries of the MCC. (See pressure-gradient maps figures for location.)

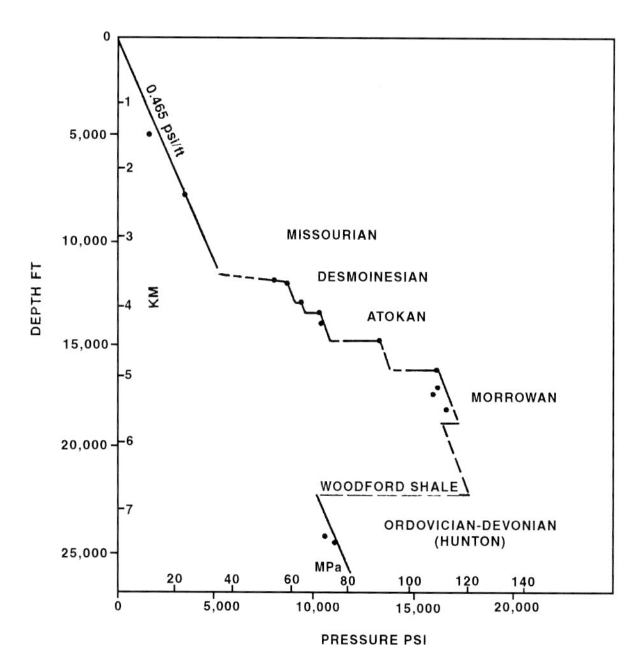

Figure 5. Pressure-depth profile from western Roger Mills County, Oklahoma. Overpressuring begins in the lower Missourian and continues in the Desmoinesian Red Fork and Morrowan intervals. Exiting the MCC into the Hunton, the pressure returns to normal and subnormal.

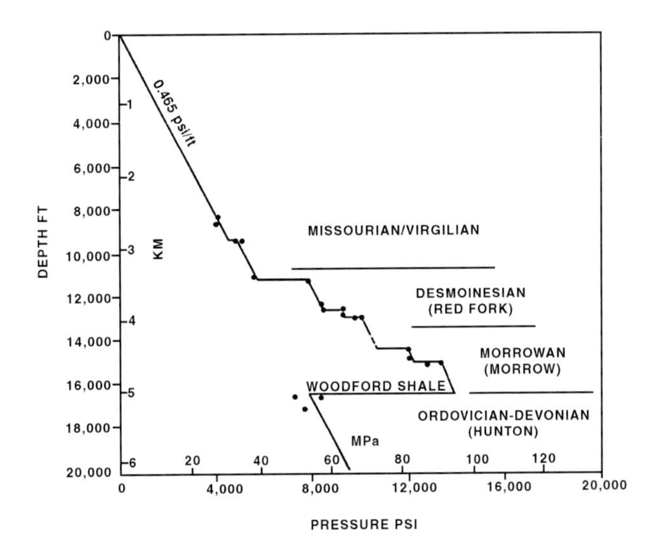

Figure 6. Pressure-depth profile from Beckham and Roger Mills Counties, Oklahoma, identifying the overpressured reservoirs within the MCC and normal pressure gradients in the underlying Hunton Group.

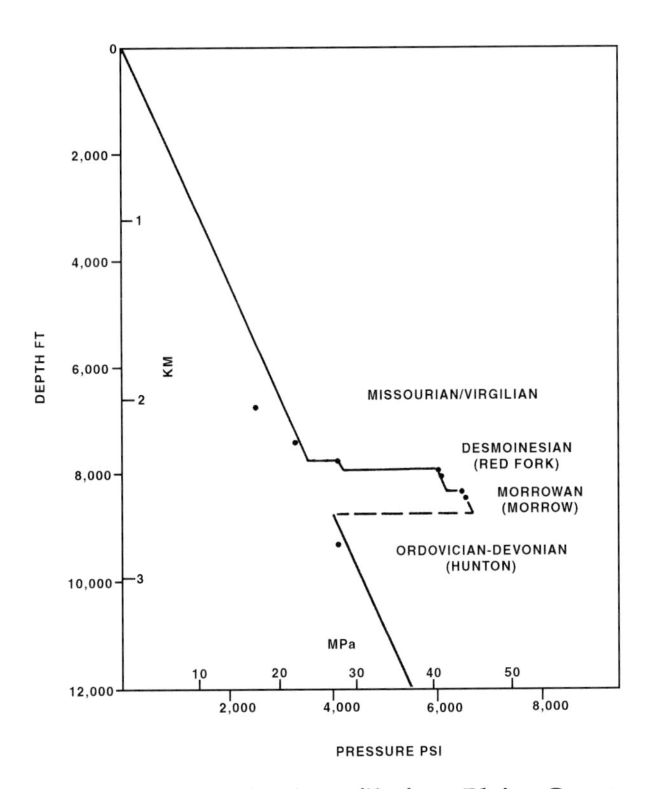

Figure 7. Pressure-depth profile from Blaine County, Oklahoma, showing overpressuring in the Watonga Trend in northeastern Anadarko basin.

Once this pattern of overpressuring was recognized, a systematic approach to mapping pressure gradients in the basin was formulated. Gradient maps were constructed for the Missourian/Virgilian, Red

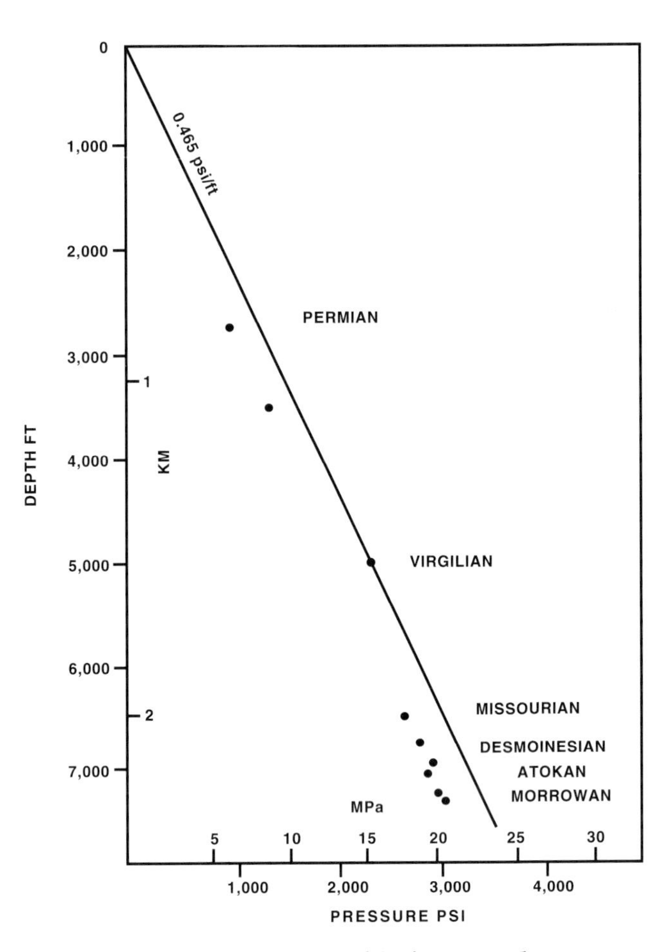

Figure 8. Pressure-depth profile from southern Beckham County, Oklahoma. This profile covers an area south of the Wichita frontal fault zone, outside the MCC. Notice all intervals are predominantly subnormally and normally pressured.

Fork (Desmoinesian), Morrow (Morrowan), and Hunton (Ordovician–Silurian–Devonian) intervals. Three-dimensional diagrams of potentiometric values were computer generated using the bottom-hole pressure data for the same intervals in accordance with the procedure discussed in Dahlberg (1982).

GEOMETRICAL CONFIGURATION OF THE MCC

Three-dimensional potentiometric diagrams and fluid-pressure-gradient maps were used to study the vertical and lateral configuration of the overpressured domain in the Anadarko basin. The mapped horizons were selected on the basis of stratigraphic position, pressure regime, and availability of data. Figure 9 is a cross section illustrating the magnitude of potentiometric head values of the mapped horizons relative to the average topographic elevation across the Anadarko basin.

The MCC is an elongated body of overpressured rocks that is approximately 241 km (150 mi) long and

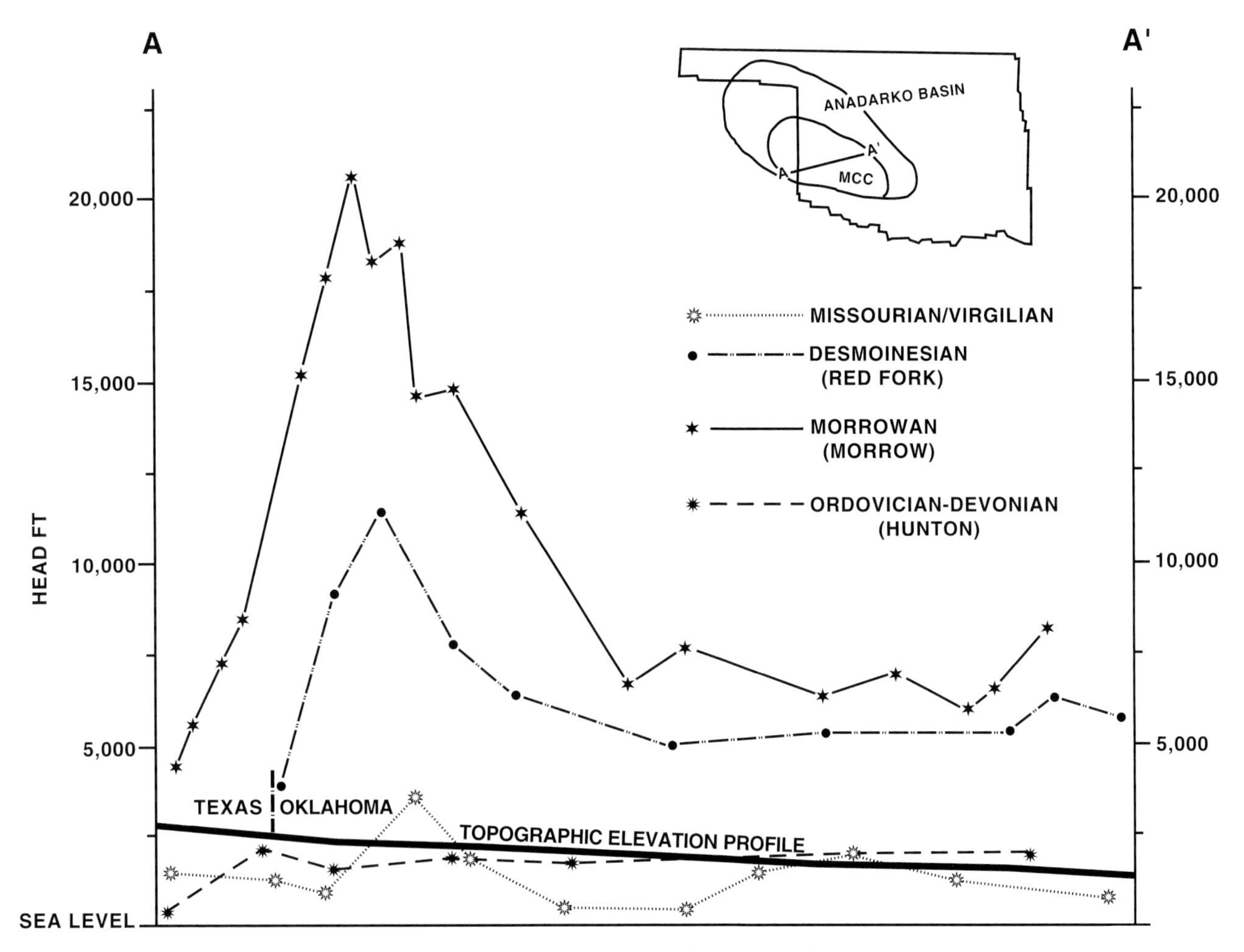

Figure 9. Profile across the megacompartment complex (southwest-northeast) illustrating variation in potentiometric surface values of the Missourian/Virgilian, Red Fork, Morrow, and Hunton intervals relative to the topographic elevation of the Anadarko basin.

113 km (70 mi) wide and has a maximum thickness of 4877 m (16,000 ft) (Figure 10). Internally, the MCC is subdivided into a myriad of nested compartments that display varied overpressured domains. These nested compartments are recognized as closed and isolated features on pressure-gradient contour maps, as anomalous peaks on potentiometric 3-D diagrams, or can be indicated by the "stairstep" pattern on pressure-depth profiles.

Our basinwide study indicates the presence of two overpressured compartments that are located outside the MCC. One is in the Hunton Group in Custer County, Oklahoma, exhibiting a fluid pressure gradient of 0.74 psi/ft, and the other, in the Ordovician Simpson Group in McClain County, Oklahoma, with an average fluid-pressure gradient of 0.5 psi/ft (Figure 10).

Pressure-Gradient Maps and Potentiometric 3-D Diagrams

Missourian/Virgilian

The Missourian/Virgilian pressure-gradient map (Figure 11) exhibits relatively near-normal values across the Anadarko basin (6.78–10.62 kPa/m, or 0.3–0.47 psi/ft). Maximum gradients of 11.31–14.70 kPa/m (0.5–0.65 psi/ft) are observed in the Marchand sandstone and Granite Wash sequences, buried below 3050 m (10,000 ft). Figure 12, a computer-generated three-dimensional diagram of potentiometric values, illustrates the same pressure pattern for the Missourian/Virgilian sequences as discussed above.

Red Fork (Desmoinesian)

The Red Fork reservoirs, which are stratigraphically below the Missourian/Virgilian strata, show a distinct change in pressure-gradient and potentiometric values (see Figures 13 and 14). On the shelf area to the north, these reservoirs are underpressured and/or normally pressured (9.05–11.31 kPa/m)(0.4–0.5 psi/ft) while they are overpressured (13.57–18.09 kPa/m)(0.6–0.8 psi/ft) in the deeper parts of the basin to the south and southwest. The transition from underpressured to overpressured rocks is generally observed at a depth of 3050 m (10,000 ft). However, along the eastern fringe of the basin in the proximity of the Nemaha Ridge, this transition is observed at 2286 m (7500 ft) deep.

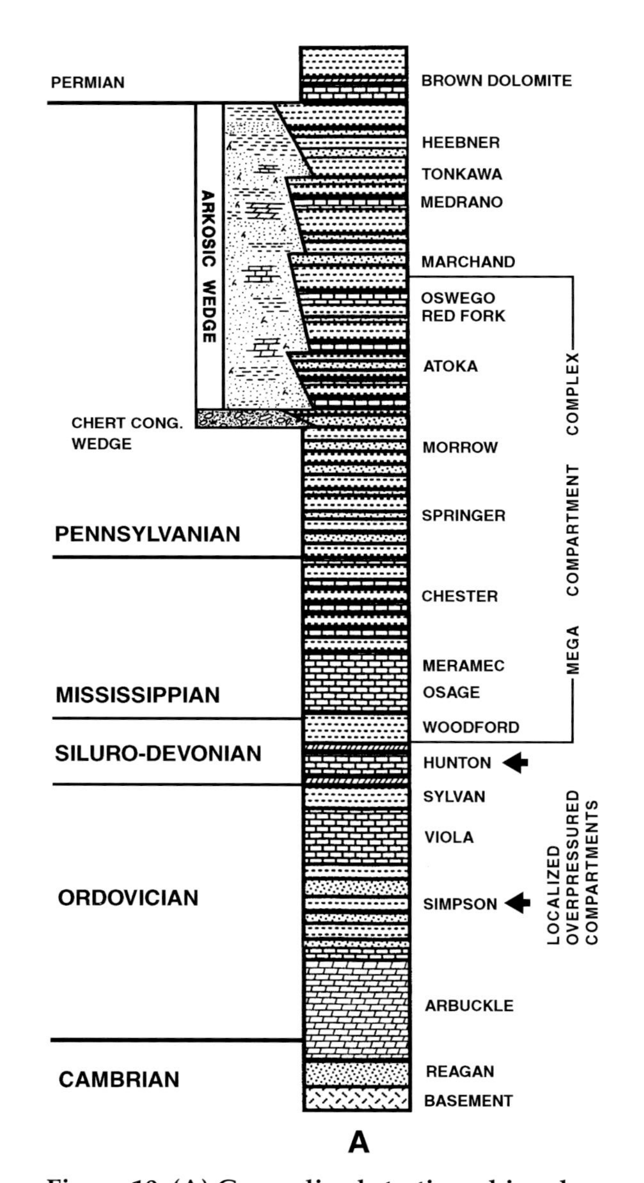

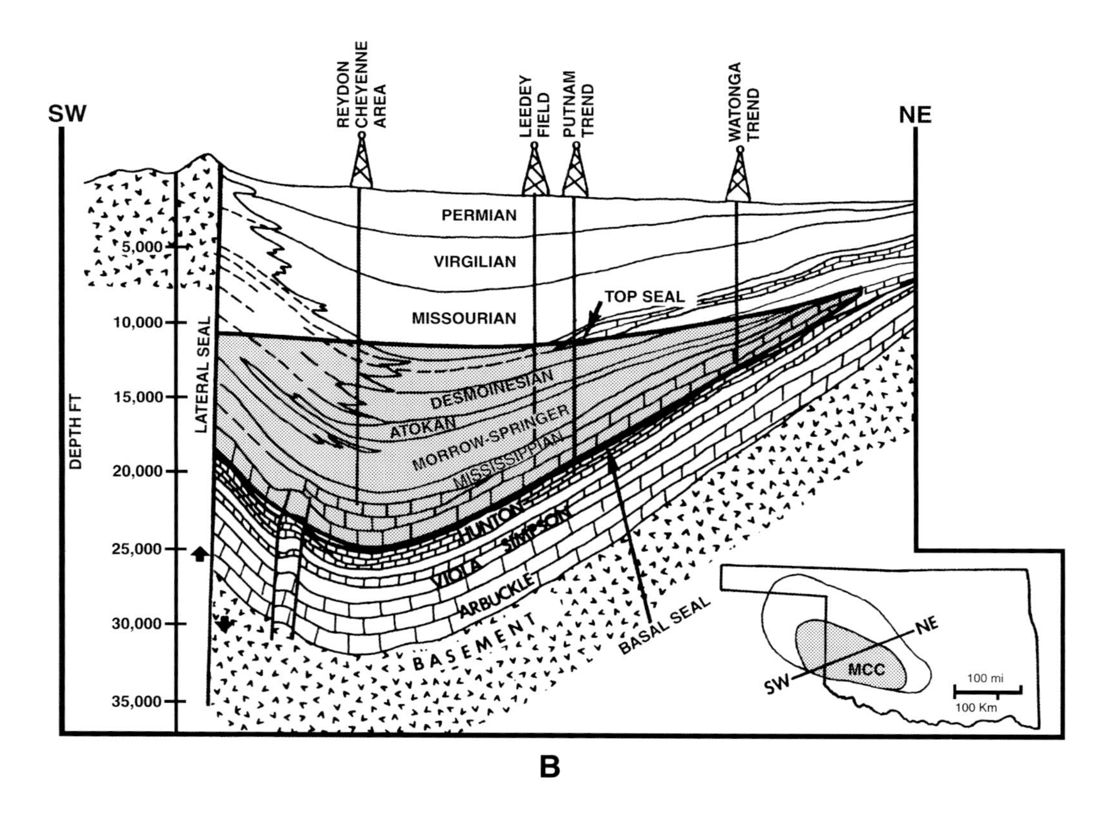

Figure 10. (A) Generalized stratigraphic column of the Anadarko basin showing the intervals contained within the MCC; also the stratigraphic position of two localized overpressured compartments (arrows) outside the megacompartment complex (after Evans, 1979). (B) Generalized cross section of the Anadarko basin showing the spatial position of the MCC within the basin. Geopressures within the MCC are maintained by top, lateral, and basal seals.

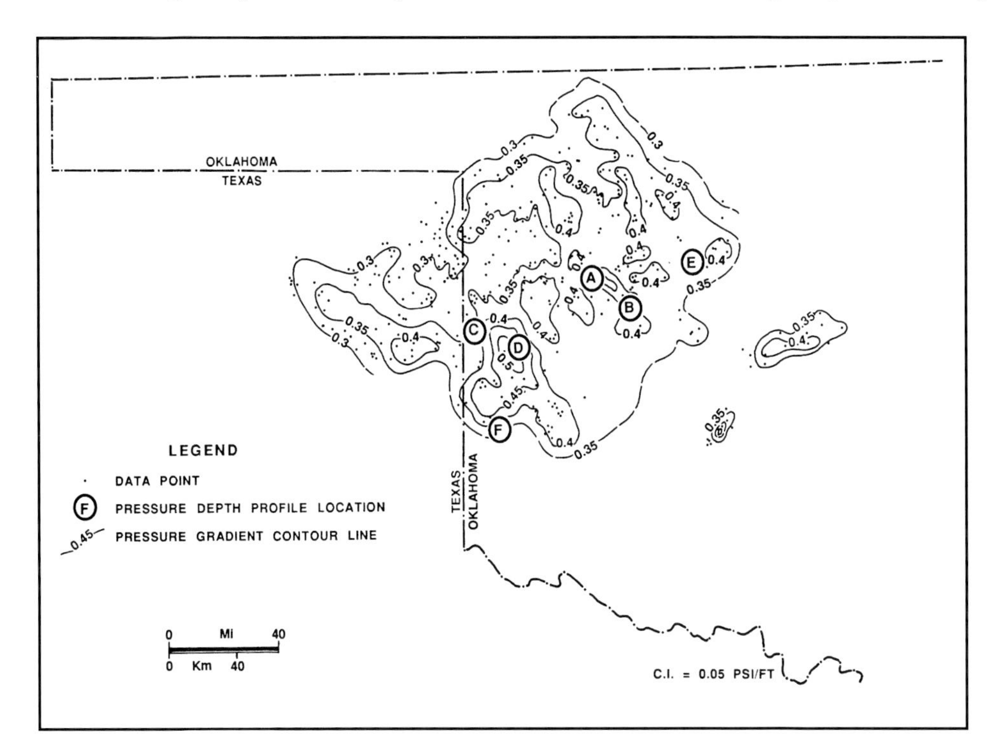

Figure 11. Pressure-gradient map of the Missourian/ Virgilian sequences. Contour lines indicate a predominantly normal pressure regime.

Morrow (Morrowan)

The Morrowan reservoirs exhibit a similar pressure pattern to the overlying Red Fork sandstones. However, the magnitude of the pressure gradients (Figure 15) and the potentiometric head values (Figure 16) are considerably higher. The transition from normal pressures to overpressures occurs at approximately 3050 m (10,000 ft) deep, except in eastern Anadarko basin where the change occurs at 2438 m (8000 ft). In the deep basin, Morrowan rocks show pressure gradients as high as 22.16 kPa/m (0.98 psi/ft) (more than 6096 m [20,000 ft]) of potentiometric head. The 3-D diagram of potentiometric heads (Figure 16) delineates two anomalous regions of overpressuring within the Morrowan interval. The positive ridge along the eastern edge of the diagram is the Watonga trend while the area of high peaks in the southwestern part of the diagram is the Morrowan chert conglomerate. These represent two major nested overpressured compartments within the MCC.

Hunton Group (Ordovician–Silurian–Devonian)

Crossing the Mississippian/Devonian Woodford Shale, which constitutes the basal seal, into the Hunton Group, the pressure regime returns to a normal domain. Pressure gradients within these reservoirs (Figure 17) generally range from 6.78 to 11.31 kPa/m (0.3 to 0.5 psi/ft). The potentiometric diagram (Figure 18) also indicates normal to subnormal pressures. An exception is indicated by an abnormally overpressured value of 16.73 kPa/m (0.74 psi/ft) in Custer County, Oklahoma. This anomaly is interpreted as an isolated Hunton compartment outside the MCC and is shown by a star on Figure 17.

The integration of stratigraphic, pressure, and hydrodynamic data illustrates very clearly the geometrical configuration of the MCC. Pressure-depth profiles over the northern platform and outside the MCC indicate normal to subnormal pressures prevailing in Permian through lower Paleozoic sequences. However, basinward the PDPs show deviations from a normal gradient of 10.515 kPa/m (0.465 psi/ft) within the lower Missourian rocks. Maximal deviation is attained in Middle Pennsylvanian through Upper Mississippian sequences. The pressure gradients return to near normal in the Hunton Group.

Seals of the Megacompartment Complex

In the preceding section, we illustrated the extent and geometry of the overpressured MCC. The confinement of geopressures within the MCC for considerable periods of geological time is maintained by low-permeability pressure seals that compose the nonreservoir portion of the MCC. These are top, lateral, and basal seals.

Top Seal

The top seal zone of the MCC is encountered around 3050 m (10,000 ft) deep in the western and central parts of the basin. However, the extension of this seal along the eastern fringe of the basin dips slightly to the southwest and hence the top seal there is approximately 2290 m (7500 ft) below the surface. The top seal is located in the argillaceous Mississippian carbonate and Pennsylvanian shale/sandstone sequences along the eastern fringe of the basin. Proceeding westward, the seal is flat-lying and crosses stratigraphic boundaries into the younger Missourian–Virgilian rocks.

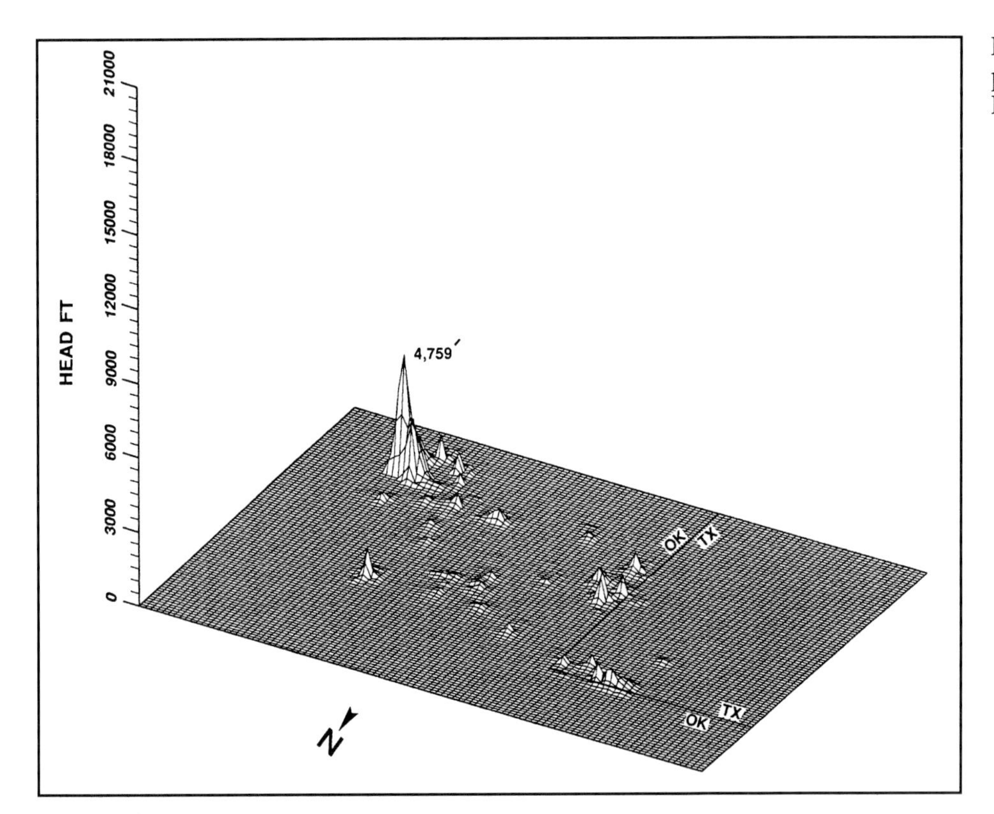

Figure 12. 3-D diagram of potentiometric values of the Missourian/Virgilian.

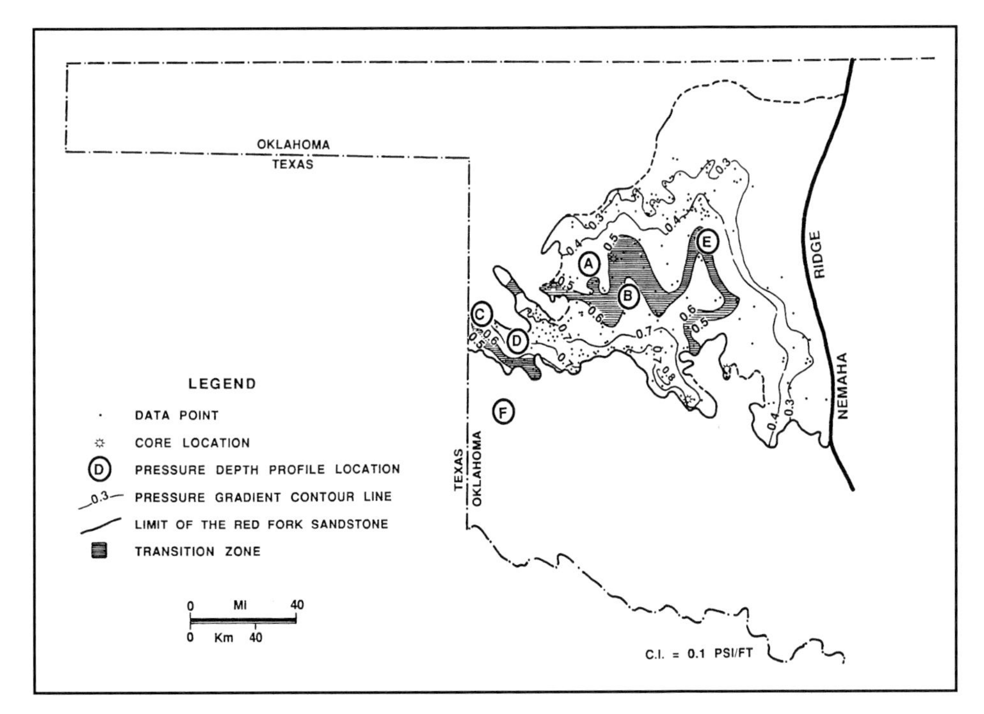

Figure 13. Pressure-gradient map of the Red Fork Sandstone. Shaded area indicates transition zone from normally pressured rocks to overpressured intervals within the MCC.

Basal Seal

The basal seal of the MCC appears to coincide with the Woodford Shale in the northern and western parts of the basin. In the southern part of the basin, the basal seal may coincide with the Mississippian Caney and/or Woodford shales. Delineating the precise location of the basal seal is difficult due to the paucity of Mississippian pressure data. However, overpressured Meramec and Chester reservoirs in the northern part of the basin suggest that the Mississippian Limestone is overpressured and sealed within the MCC by the Woodford Shale.

Lateral Seals

The lateral seals of the MCC are controlled by the structural and stratigraphic elements of the basin. The southern lateral seal of the MCC is associated with the frontal fault zone of the Wichita Mountain uplift. This

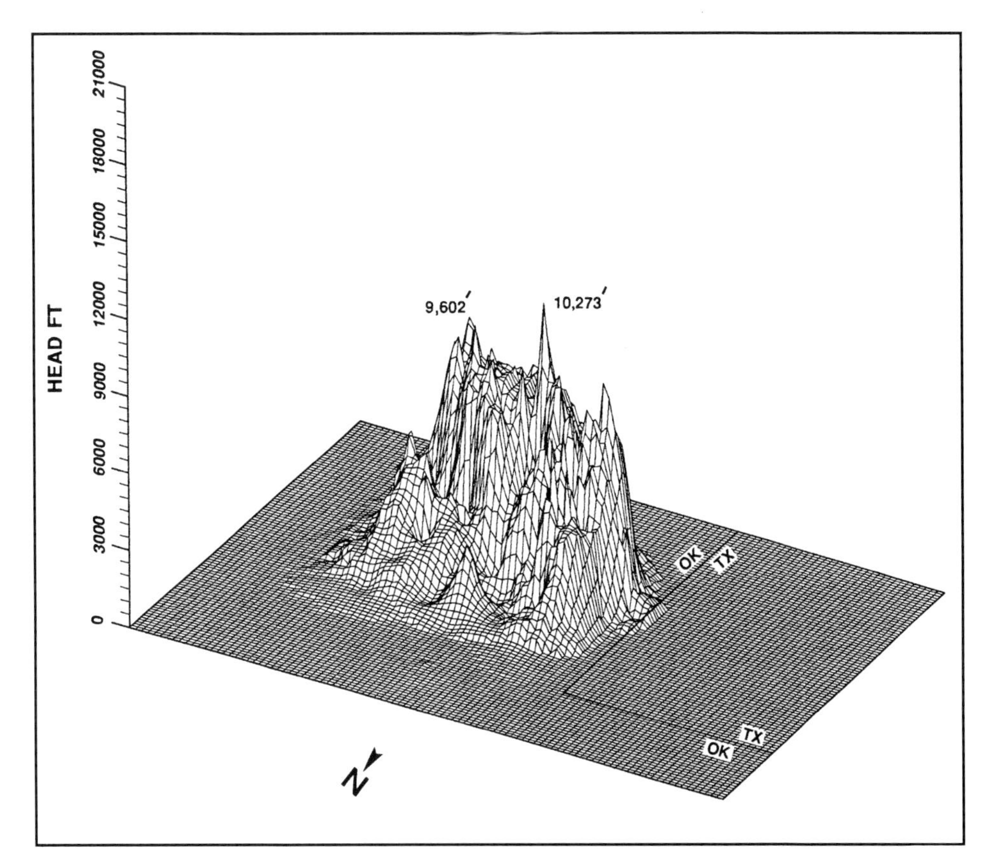

Figure 14. 3-D diagram of the potentiometric data of the Red Fork Sandstone. The plane of zero ft "head" coincides with the surface elevation of the Anadarko basin; therefore, all peaks represent overpressured Red Fork rocks within the MCC.

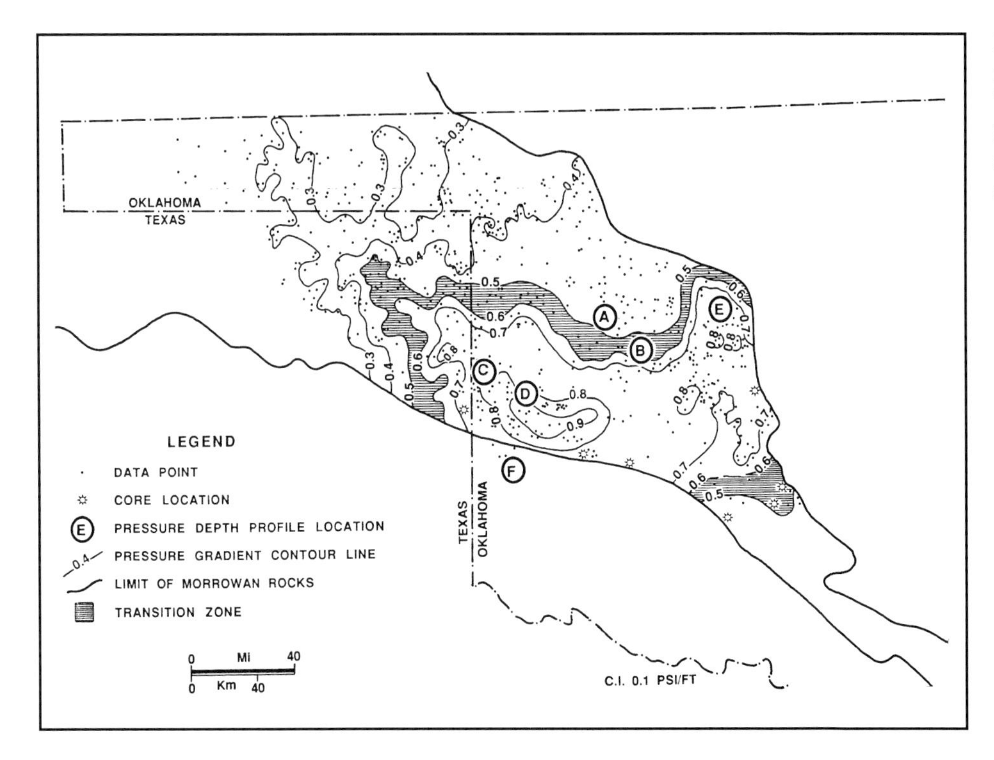

Figure 15. Pressure-gradient map of the Morrowan series. The transition zone between normally pressured and overpressured rocks is shaded.

seal is nearly vertical and includes the fault zone and the highly cemented rocks adjacent to that zone.

The lateral seal along the western, northern, and eastern boundaries of the MCC is formed by the convergence of the top and basal seals along the shelf margin of the basin.

Diagenetic Banding of Seals

Diagenetic alterations of rocks within the seal intervals of the MCC have generated a series of bands that are an integral part of the sealing mechanism. These bands have unique mineralogical and morphological

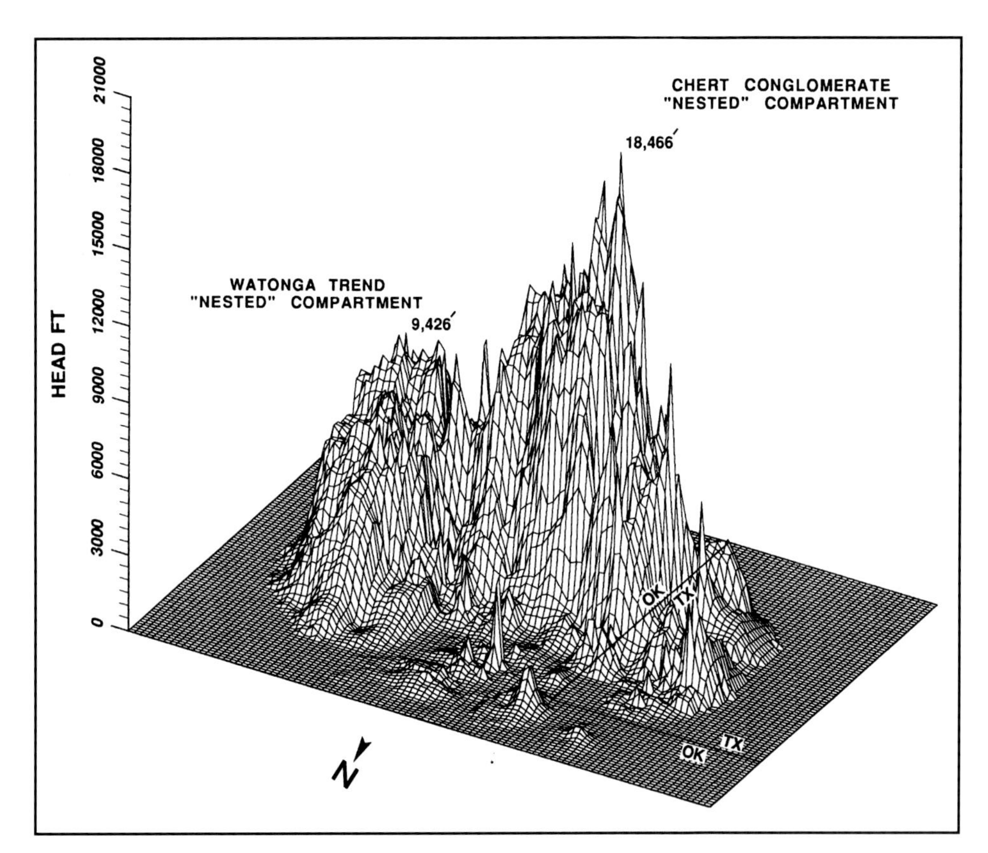

Figure 16. 3-D diagram of potentiometric surface values of the Morrowan Series that constitute the middle and part of the lower portion of the megacompartment complex. All peaks represent overpressured Morrowan rocks within the MCC (the plane of zero ft coincides with the surface elevation of the Anadarko basin).

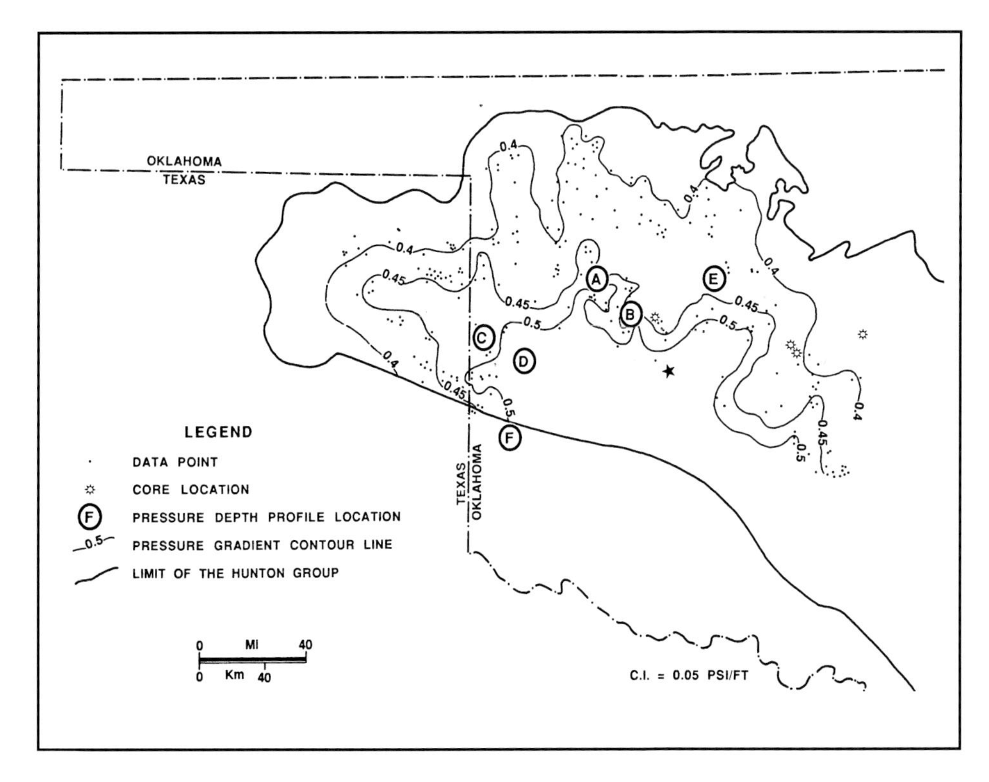

Figure 17. Pressure-gradient map of the Hunton Group indicating the return to normal pressures below the Woodford basal seal. Solid star represents the isolated Hunton compartment.

characteristics that reflect the host rock composition. Banded seal structures are prevalent in rocks that have been buried below 3050 m (10,000 ft) and are noticeably absent in rocks that have not entered "the seal window." (These banded structures are described in detail in Chapter 23, *The Banded Character of Pressure Seals* by Al-Shaieb et al., in this volume.)

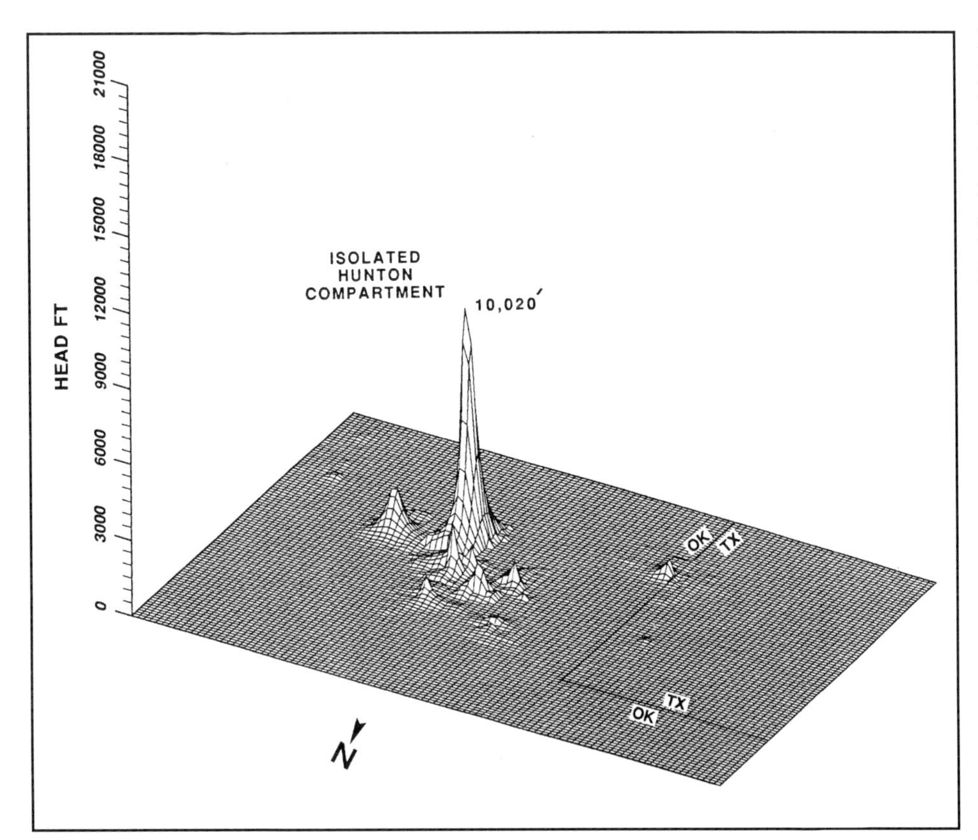

Figure 18. 3-D diagram of potentiometric surface values of the Hunton Group, which is stratigraphically below the basal seal of the megacompartment complex. An isolated Hunton compartment with a potentiometric head value of 3552 m (10,020 ft) above the ground elevation is indicated.

SUMMARY

The Anadarko basin contains an overpressured domain of rocks called the megacompartment complex. The interior of the complex is subdivided into smaller nested compartments with distinct pressure regimes. These compartments are isolated by seals that contain cementation bands whose mineralogical and morphological characteristics reflect the host lithologies. The overpressured MCC includes Upper Devonian (Chattanoogan) to Pennsylvanian (Missourian) age strata. The top seal zone of the MCC is found approximately 2290 m to 3050 m (7500 to 10,000 ft) deep and crosses stratigraphic boundaries. The base of the MCC appears to coincide with the Woodford Shale since the pressure regime changes to a normal domain below that interval. The shelfward lateral boundaries of the MCC are formed by the convergence of the top and basal seal zones. The southern boundary of the MCC is formed by intense cementation along the Wichita Mountain frontal fault zone.

Due to their hydraulically isolated nature, nested pressure compartments may provide drilling prospects that are not constrained by structural position or proximity to existing reservoirs. Predicting the compartment and seal geometries and internal reservoir quality should improve drilling success ratios and diminish hazards associated with drilling abnormally pressured rock sequences.

ACKNOWLEDGMENTS

The authors gratefully acknowledge the Gas Research Institute for funding this research through Contract No. 5089-261805. The authors also thank David Powley and John Bradley, who gave helpful suggestions concerning this study. Stimulating discussions with Peter Ortoleva have contributed to the refinement of the concept. Amoco, Dwight's Energydata Inc., and KOPCO Inc. provided drilling, completion, and production data essential to completion of this study.

REFERENCES CITED

Al-Shaieb, Z., 1991, Compartmentation, fluid pressure important in Anadarko exploration: Oil and Gas Journal, v. 89, no. 27, p. 52–56.

Al-Shaieb, Z., and J. W. Shelton, 1977, Evaluation of uranium potential in selected Pennsylvanian and Permian units and igneous rocks in southwestern and southern Oklahoma: U.S. Department of Energy, Open File Report GJBX-35 (78), 248 p.

Al-Shaieb, Z., and P. Walker, 1986, Evolution of secondary porosity in Pennsylvanian Morrow Sandstones, Anadarko basin, Oklahoma: American Association of Petroleum Geologists Studies in Geology, no. 24, p. 45–67.

Al-Shaieb, Z., G. B. Beardall, Jr., and G. F. Stewart, 1985, Depositional facies, dolomitization, and porosity of the Henryhouse Formation (Silurian), Anadarko basin, in Ordovician and Silurian rocks of the Michigan basin and its margins, *in* K. R. Cercone and J. M. Budai, eds., Michigan Basin Geological Society Special Paper no. 4, p. 173–196.

Al-Shaieb, Z., P. L. Alberta, and M. Gaskins, 1989a, Role of diagenesis in development of Upper

Chapter 5

Three Levels of Compartmentation within the Overpressured Interval of the Anadarko Basin

Zuhair Al-Shaieb
James O. Puckette
Azhari A. Abdalla
Patrick B. Ely
Oklahoma State University
Stillwater, Oklahoma, U.S.A.

ABSTRACT

Deep basin pressure compartments can be classified on the basis of their size, stratigraphy, and pressure regimes. Detailed investigations of the geologic setting and pressure gradients of numerous reservoirs in the Anadarko basin reveal the presence of three distinct levels of compartmentation.

Level 1 is a basinwide feature known as the megacompartment complex (MCC). This complex is an overpressured volume of rocks that is completely enclosed by seals. It is approximately 241 km (150 mi) long, 113 km (70 mi) wide, and has a maximum thickness in excess of 4877 m (16,000 ft). Gas reserves of the overpressured reservoirs within the MCC are speculated to be approximately 20 tcf. The other two compartmentation levels are further subdivisions of the internal volume of the MCC. Level 2 compartmentation consists of multiple, district-, or field-sized configurations within a particular stratigraphic interval. These compartments are 32 to 49 km (20 to 30 mi) long, 19 to 32 km (12 to 20 mi) wide, and 122 to 183 m (400 to 600 ft) thick. Their reserve estimates can exceed 2 tcf. Examples of this type are the upper Morrowan Chert Conglomerate reservoirs in the Cheyenne/Reydon field area. Level 3 consists of a single, small field or a particular reservoir nested within Level 2. These compartments are generally 3 to 6 km (2 to 4 mi) long, <1.6 to 5 km (<1 to 3 mi) wide, and 3 to 30 m (10 to 100 ft) thick. Reserve estimates range from <1 to several hundred bcf. Individual channel-fill reservoirs of the "Pierce" chert conglomerate represent adequate examples of Level 3 compartments.

The hierarchical classification of compartments has important implications for petroleum exploration and development. Recognizing the upper and lower boundaries of the MCC is essential to safe drilling practices. The similar pressures within Level 2 compartments allow the prediction of approximate reservoir pressures across trend-size areas of the basin (tens to hundreds of square kilometers). Delineating Level 3 compartments improves reservoir boundary prediction and exploration and development strategies.

The integration of tectonic history, stratigraphic relationships, facies distribution, thermal history, and diagenetic patterns of seal zones suggests that the three levels of compartmentation evolved during the Pennsylvanian orogenic episode. This occurred during the rapid subsidence phase of the orogeny over a period of approximately 30 million years.

INTRODUCTION

The overpressured domain in the Anadarko basin is composed of a multitude of individual pressure compartments that collectively form the megacompartment complex (MCC). These compartments within the MCC are identified on the basis of their distinct pressure regimes. Pressure-gradient (also called pressure-depth ratio or surface gradient) and potentiometric maps were constructed to establish the geometric configuration of these compartments. Pressure gradient is the slope of a line connecting pressure-depth plot to the surface (0 depth, 0 pressure) on a pressure-depth profile. As a result, a hierarchical classification of compartments based on their size, stratigraphy, and pressure regimes was developed.

Three distinct levels of compartmentation were recognized. Level 1 is a regional feature that transects stratigraphic boundaries and includes most of the overpressured rocks in the basin. Level 2 is a field- or district-sized feature with relatively uniform pressure gradients that occurs within a stratigraphic interval. Level 3 is the smallest size and consists of small field- to reservoir-size compartments within a particular stratigraphic unit. The size and geometry of these compartments are strongly linked to their depositional setting and facies.

LEVELS OF COMPARTMENTATION

Level 1

The first level of compartmentation is a basinwide overpressured volume called the megacompartment complex (MCC). The MCC is an elongate body of overpressured rocks approximately 240 km (150 mi) long and 113 km (70 mi) wide that has a maximum thickness of around 4880 m (16,000 ft) (Figure 1). Pressure data indicate that the top of the MCC is located between 2290 m (7500 ft) and 3050 m (10,000 ft) below the surface. The top is shallower in the proximity of the Nemaha Ridge in central Oklahoma and dips gently toward the basin axis until it is found around 3050 m (10,000 ft) deep in western Oklahoma. The MCC contains all rock sequences from its top in the Upper Pennsylvanian at 2290 m to 3050 m to its base at the Woodford Shale. All reservoirs within this complex appear to be overpressured. They exhibit a wide range of gradients (from slightly to extremely overpressured) and are totally isolated by seal zones from the realm of normally pressured rocks outside the complex.

Total gas reserves in the MCC portion of the Anadarko basin are estimated to be 20 tcf. A large portion of the proven reserves is found in the Pennsylvanian Red Fork and Morrowan stratigraphic intervals whose fields/districts and reservoirs compose the second and third levels of compartmentation.

Level 2

Within the MCC, certain anomalous field- or district-sized areas of overpressuring are easily recognized by their similar pressure regimes on pressure-gradient and potentiometric maps. These anomalies are designated as Level 2 compartments and they are typically restricted to a specific stratigraphic horizon. Pressure measurements within each horizon are relatively uniform and predictable across the trend. Examples of these areas, such as the Watonga Trend or Chert Conglomerate, may be seen on the Morrowan Series pressure maps and identified by their distinct pressure regimes. The three-dimensional diagram of Morrowan potentiometric head values (Figure 2) delineates the Watonga trend as a "ridge-like" feature along the eastern edge of the basin; whereas peaks of the Chert Conglomerate potentiometric values are identified as the "castle-like" feature in the southwestern corner of the basin. The individual peaks within these Level 2 compartments represent the third compartmentation level.

Level 3

Level 3 compartments are single, small field- or reservoir-size subdivisions usually nested within Level 2. The geometry of these compartments is often closely linked to the depositional facies of the associated reservoirs. Level 3 compartments are generated within the various reservoirs that compose a stratigraphic interval (Al-Shaieb et al., 1990b). Examples of this type include the Southwest Leedey Red Fork sandstone reservoir (Al-Shaieb et al., 1991), the "Old Woman" channel-fill reservoir in the North Geary

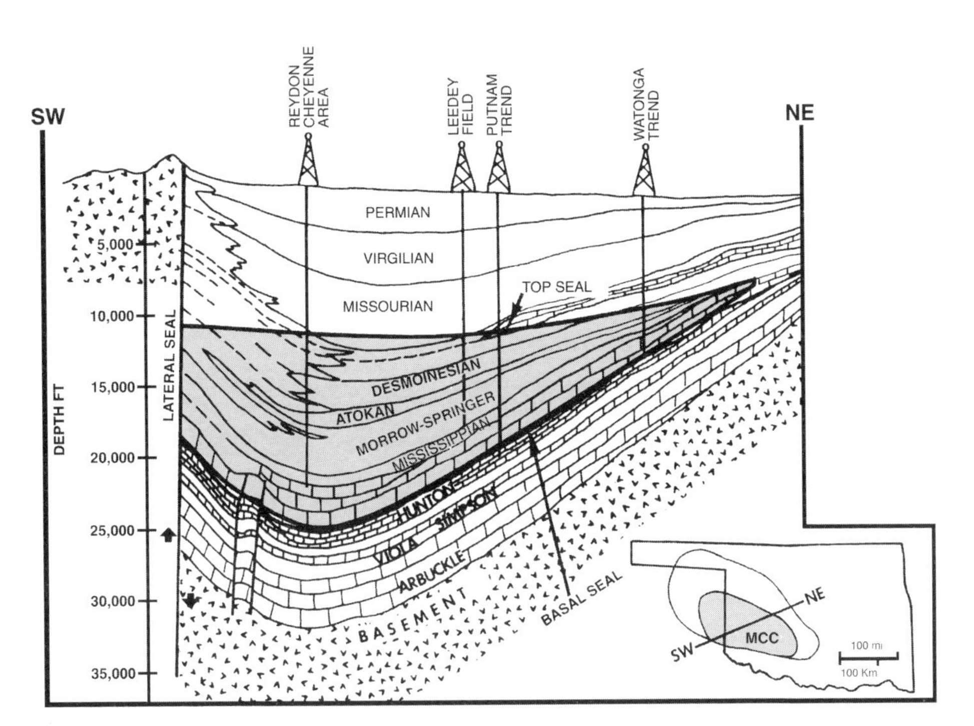

Figure 1. Generalized cross section of the Anadarko basin, showing the spatial geometry of the megacompartment complex (MCC), the first level of compartmentation in the basin. Arrows in the lower left corner indicate relative fault movement. Inset map shows extent of the MCC with respect to the basin's dimensions.

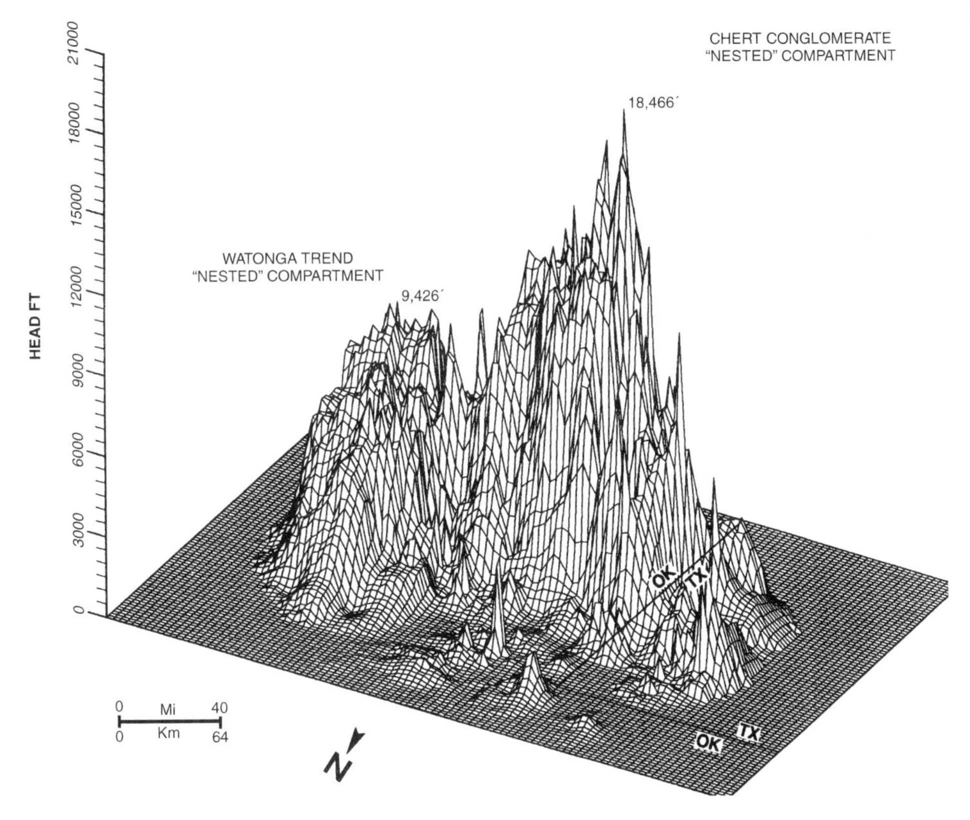

Figure 2. Three-dimensional diagram of Morrowan potentiometric head values in the Anadarko basin (measured from ground level). The Watonga Trend and the Chert Conglomerate —Level 2 compartments— are major producing areas where reservoirs exhibit similar pressures.

area of the Watonga Trend (Abdalla et al., 1992), and individual reservoirs within the chert conglomerate producing area. These Level 3 compartments can be identified and delineated by their distinct pressure gradients, fluid types (gas/oil and water ratios), and/or pressure-decline curves that indicate isolation from nearby reservoirs. Figure 3 is a schematic diagram that depicts the relationship between the three levels of compartmentation in the Anadarko basin. Figure 4 is a pressure-depth profile that graphically

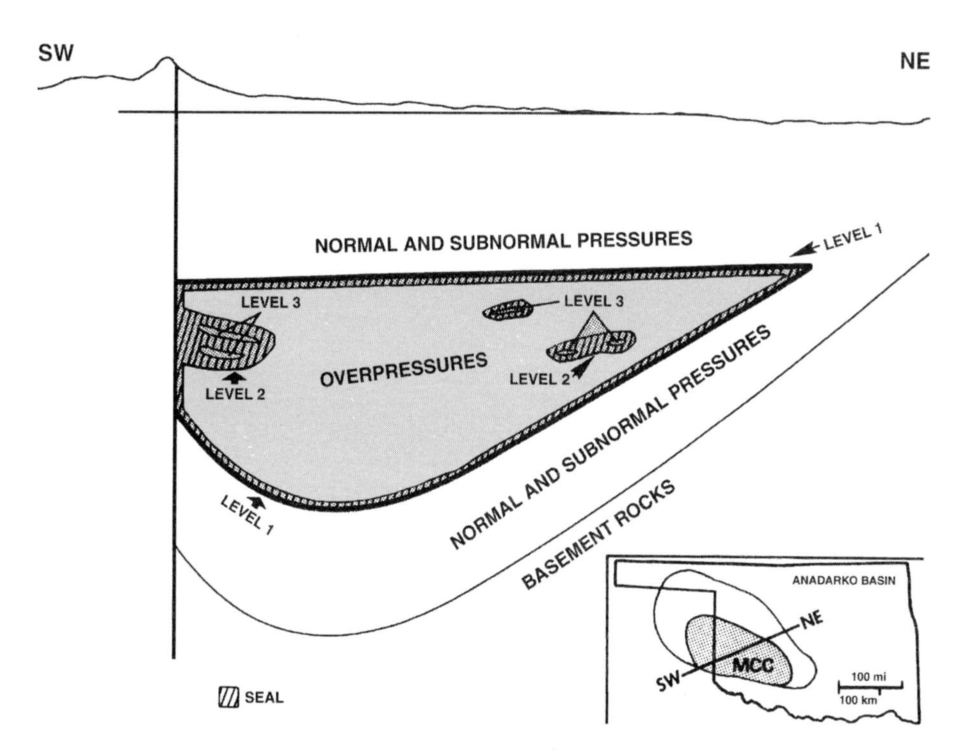

Figure 3. Schematic diagram illustrating the spatial relationship of the three levels of compartmentation in the Anadarko basin. Inset map shows the areal extent of the MCC within the basin.

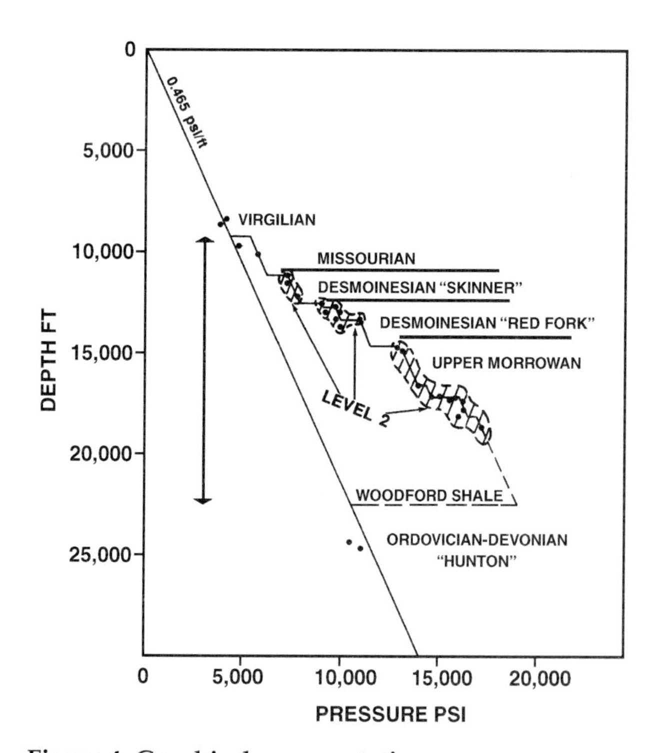

Figure 4. Graphical representation on a pressure-depth profile illustrating the relationship among Levels 1, 2, and 3. Note Level 2 compartments are essentially clusters of isolated Level 3 compartments. The pressure-depth profile was constructed in the Reydon-Cheyenne area in western Oklahoma.

portrays the relationship among Level 1, 2, and 3 compartments.

LEVEL 2 AND 3 COMPARTMENTS' CASE HISTORY: THE UPPER MORROWAN CHERT CONGLOMERATE

The upper Morrowan Chert Conglomerate provides a classic case history regarding the spatial distribution and geometry of Level 2 and 3 compartments in the Anadarko basin. The entire conglomeratic braid- and fan-delta sequence in the Cheyenne-Reydon field (CRF) area represents a Level 2 compartment within the Level 1 MCC. Furthermore, this sequence is itself subdivided into individual, completely sealed, Level 3, "nested compartments" that correspond to the different reservoirs within the interval. Pressure-gradient and potentiometric surface maps were used to delineate these individual chert compartments and to demonstrate their unique pressure regimes.

In order to understand the development of these compartmentalized systems, it is necessary to integrate the stratigraphy and depositional facies to the geopressure regime of these reservoirs.

Stratigraphy

The upper Morrowan Series in the Reydon-Cheyenne study area (Figure 5) is defined as all strata

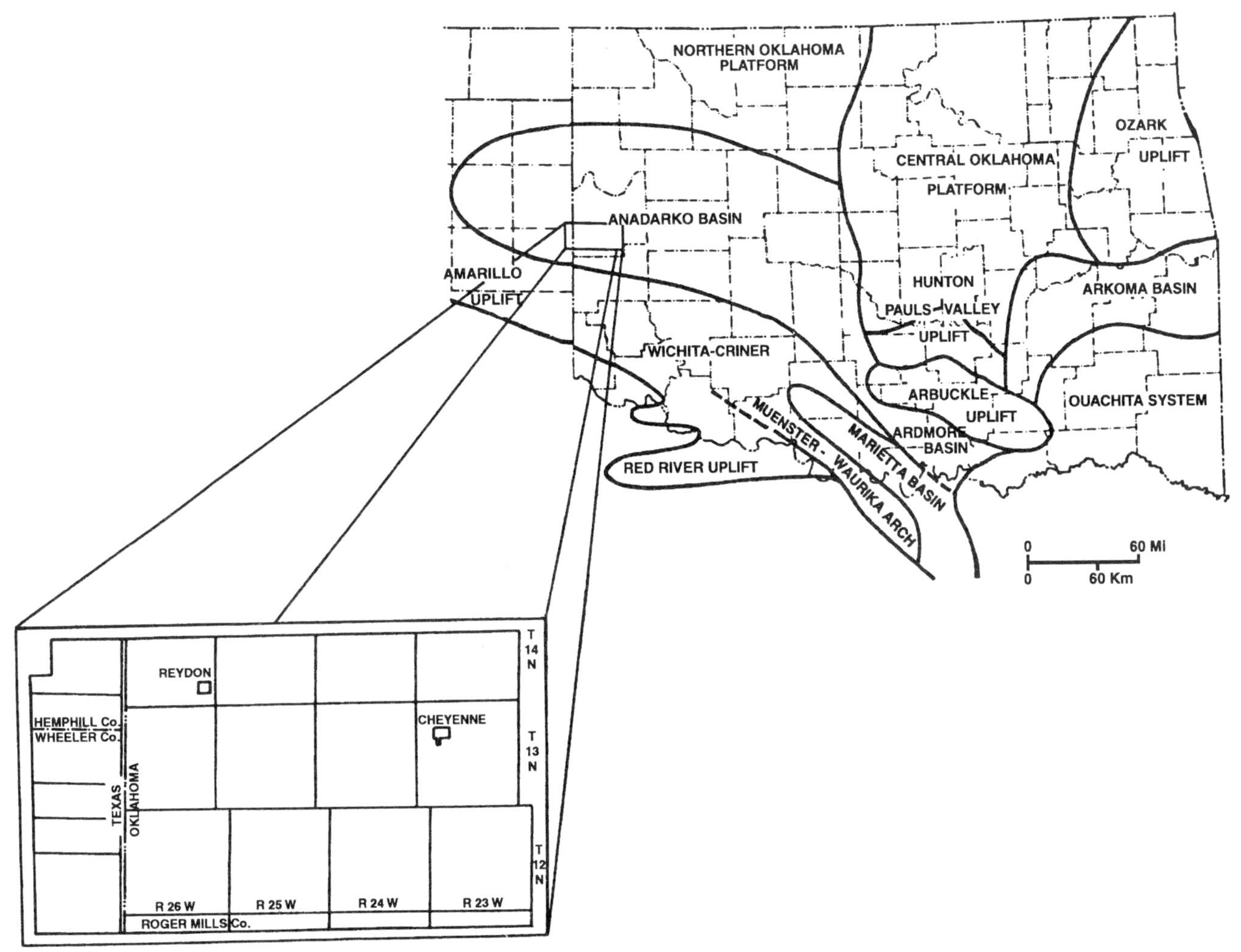

Figure 5. Area of study of the Chert Conglomerate as an example of Level 2 and 3 compartments. Major geological provinces of Oklahoma are also shown (Al-Shaieb and Shelton, 1977; Arbenz, 1956).

between the base of the Atokan "13 Finger Limestone" and the top of the underlying "Squaw Belly Limestone" (Figure 6). Within the northern part of the study area, the upper Morrowan Series is a shale-dominated sequence that contains thinner chert conglomerate and chert sand reservoirs. The conglomerate-to-shale ratio increases to the south, and the upper Morrowan Series becomes a conglomerate-dominated sequence in the vicinity of the Wichita Mountain frontal fault zone.

The individual reservoirs are named locally and called "members" of the chert sequence. In descending order these are the Coffey, Purvis (Allison and Key in Texas), Puryear, Hollis, Pierce, Bradstreet, and Armstrong (Petroleum Information, 1990; Dwight's, 1989). The more commonly recognized reservoirs are shown in Figure 6.

Depositional Environment and Facies

The upper Morrowan chert conglomerate sequences have been interpreted as fan-delta complexes (Alberta, 1987; Al-Shaieb et al., 1990a), coastal alluvial fans (Johnson, 1990), and braid- and fan-delta complexes (Puckette et al., 1993) that formed from Mississippian-age chert detritus eroded off the rising Wichita Mountain uplift. Alberta (1987) and Al-Shaieb et al. (1990a) conducted an integrated study that examined the sedimentary features, geometry, and wireline log characteristics of the upper Morrowan chert interval. They recognized various depositional facies including main and minor distributary channels, crevasse splays, overbank deposits, delta plain, lignites, swamp, delta front, and shallow marine/prodelta (Figure 7).

Reservoir distribution and geometries suggest that most chert conglomerate deposition occurred in northward-prograding braid- and fan-delta systems and incised valley fills. The fan-delta systems are composed of stacked and amalgamated braided channels that form thick conglomeratic sequences adjacent to the uplift. The braid-delta systems are composed of laterally extensive distributary channels that formed as braided fluvial streams prograded

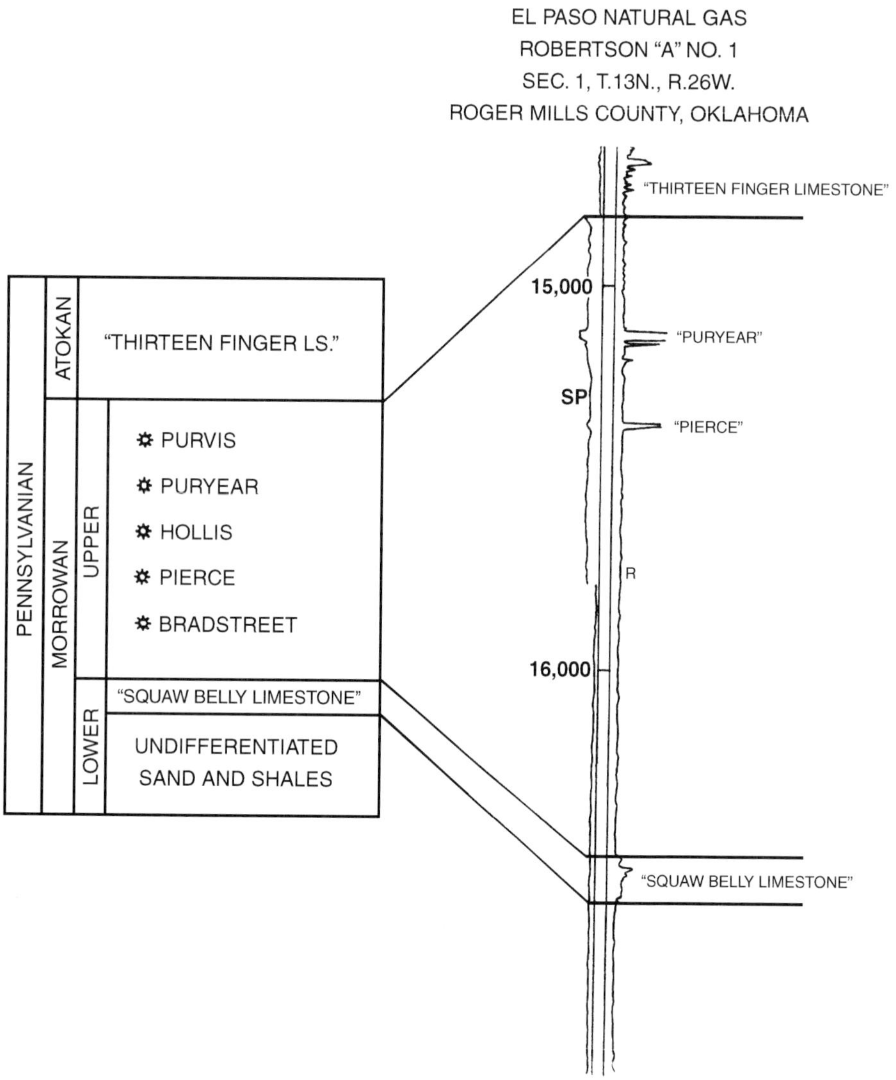

EL PASO NATURAL GAS
ROBERTSON "A" NO. 1
SEC. 1, T.13N., R.26W.
ROGER MILLS COUNTY, OKLAHOMA

PENNSYLVANIAN | ATOKAN | "THIRTEEN FINGER LS."
MORROWAN | UPPER | PURVIS, PURYEAR, HOLLIS, PIERCE, BRADSTREET
LOWER | "SQUAW BELLY LIMESTONE" | UNDIFFERENTIATED SAND AND SHALES

"THIRTEEN FINGER LIMESTONE"
15,000
"PURYEAR"
SP
"PIERCE"
R
16,000
"SQUAW BELLY LIMESTONE"

Figure 6. Stratigraphic column of an upper Morrowan type log, Cheyenne and Reydon fields study area (Hawthorne, 1984). Depth in feet.

across a coastal plain into the Morrowan sea. Incised channel systems apparently developed in response to sea level fluctuations.

The most pertinent facies to this study are the distributary channels, crevasse splays, and overbank deposits. These facies represent the most productive reservoirs found in the chert conglomerate depositional system. Their encasement in shale facilitated sealing and compartmentation.

Thickness and Potentiometric Maps

Thickness maps for the chert conglomerate reservoirs were constructed using wireline logs. The mapped facies all exhibit clean gamma-ray log signatures characteristic of conglomerate and sandstone lithologies.

Channel-fill facies were recognized by blocky to bell-shaped gamma-ray log signatures (Figure 7). These signatures often exhibit a fining-upward nature that is characteristic of waning flow and abandonment. The geometry of some channel-fill reservoirs suggests that the streams were rather confined and formed a distributary channel system. The thick nature of some reservoirs (greater than 30 m) indicates that channelization or incision of older sediments occurred during sea level lowstands.

Sand-sized overbank deposits are characterized by a thin (less than 3 m thick) gamma-ray log signature that is spike-like in appearance (Figure 7). These deposits are located in interchannel areas between the more prominent braid- and fan-delta systems.

Along the southern boundary, the chert conglomerate sequence is dominated by stacked upper mid- to proximal-fan braided-stream deposits.

Potentiometric and pressure-gradient maps were constructed from available reservoir pressure data. Static bottom-hole (reservoir) pressures were calculated from reported initial well-head shut-in pressures using software developed by Echometer (1986). Pressure gradients were calculated by dividing reservoir pressure by the perforation depth. Only single zone completions were utilized in the study.

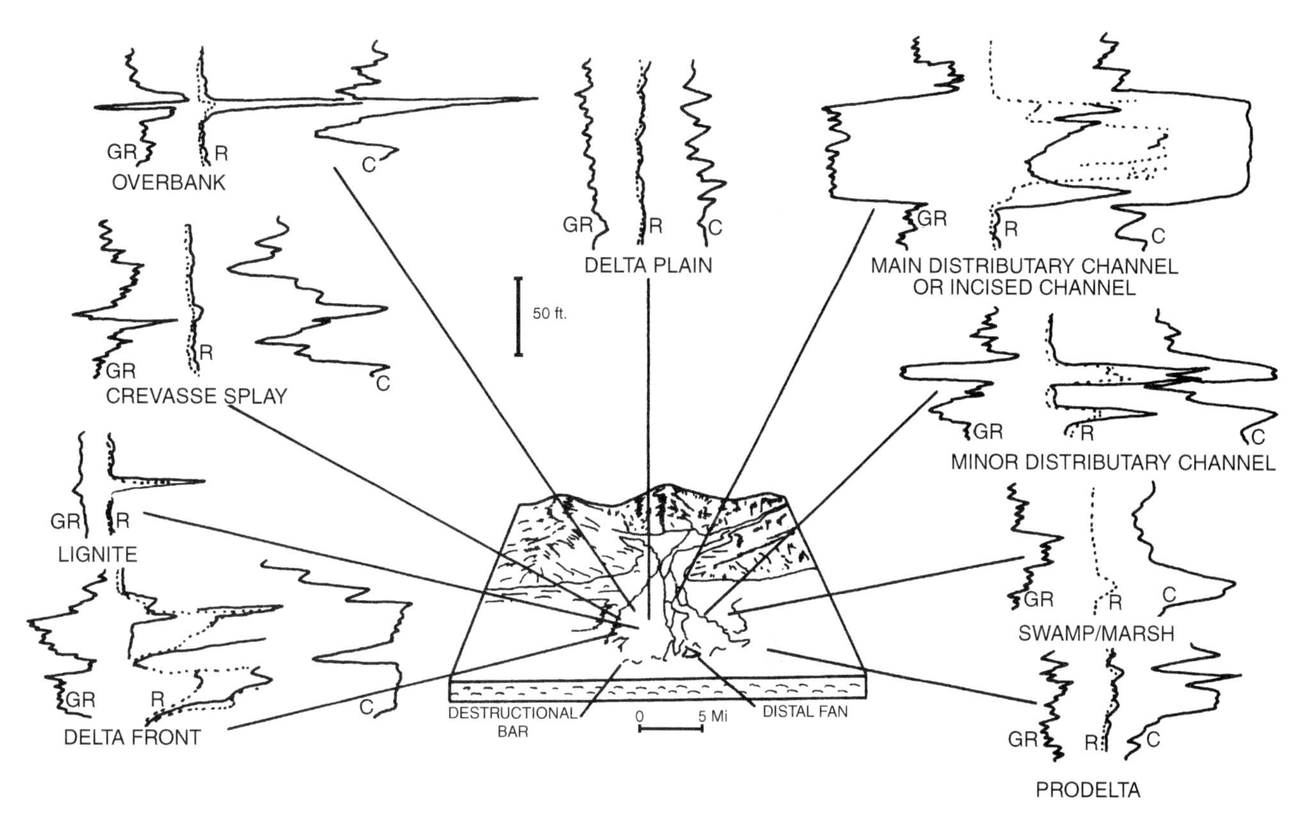

Figure 7. Characteristic log signatures of various depositional facies proposed within the study area (Alberta, 1987).

Potentiometric surface values (*hw*) were calculated using Dahlberg's formula (1982):

$$hw = Z + \frac{P}{\text{grad } p}$$

where P = reservoir pressure, hw = hydraulic head (potentiometric surface), Z = elevation in feet above or below an arbitrary datum, and grad p = static pressure gradient (0.465 psi/ft). A constant fluid gradient of 10.515 KPa/m (0.465 psi/ft) was used in the conversion of bottom-hole (reservoir) pressures to potentiometric head values. All head values are standardized to sea level.

The thickness and potentiometric maps of three chert conglomerate members are used to demonstrate the existence of Level 3 compartmentation within the chert conglomerate interval. These conglomerate sequences represent stratigraphic intervals where the most pressure data were available. In ascending order, the reservoirs are the "Pierce," "Puryear," and "Purvis" chert conglomerates.

Pierce Chert Conglomerate

Thickness Map

The thickness map of the Pierce (Figure 8) indicates two prominent northeast-trending braid-delta distributary systems. The western lobate deposit is larger and trends northeast from the area of stacked upper-mid-fan sequences in the southwest corner of the mapped area. Within T13N, R25W, and T13N, R26W, areally widespread deposition resulted from avulsion and channel migration.

The eastern Pierce deposit has a dominant northeast trend. In this area, thickening within the Pierce interval indicates that braided streams incised into the delta-plain/shallow marine silts and muds in response to a sea level drop.

A small isolated chert deposit in Sections 1 and 2, T12N, R25W, and Sections 35 and 36, T13N, R25W, is located between the two prominent Pierce braid-delta systems and is therefore believed to be an overbank deposit.

Pressure Regimes

Reservoir pressure measurements for the Pierce (Figure 9) indicate that the different Pierce facies display significantly diverse pressure domains. The eastern braided distributary system has pressure gradients that range from 20.58 to 21.03 KPa/m (0.91 to 0.93 psi/ft) and potentiometric surface values in excess of 5970 m (19,600 ft). The western lobe has a maximum pressure gradient of 18.54 KPa/m (0.82 psi/ft) and a potentiometric surface value of approximately 4328 m (14,200 ft). The thin overbank Pierce reservoir has a pressure gradient of 22.16 KPa/m (0.98 psi/ft) and a potentiometric surface value in excess of 6280 m (20,600 ft).

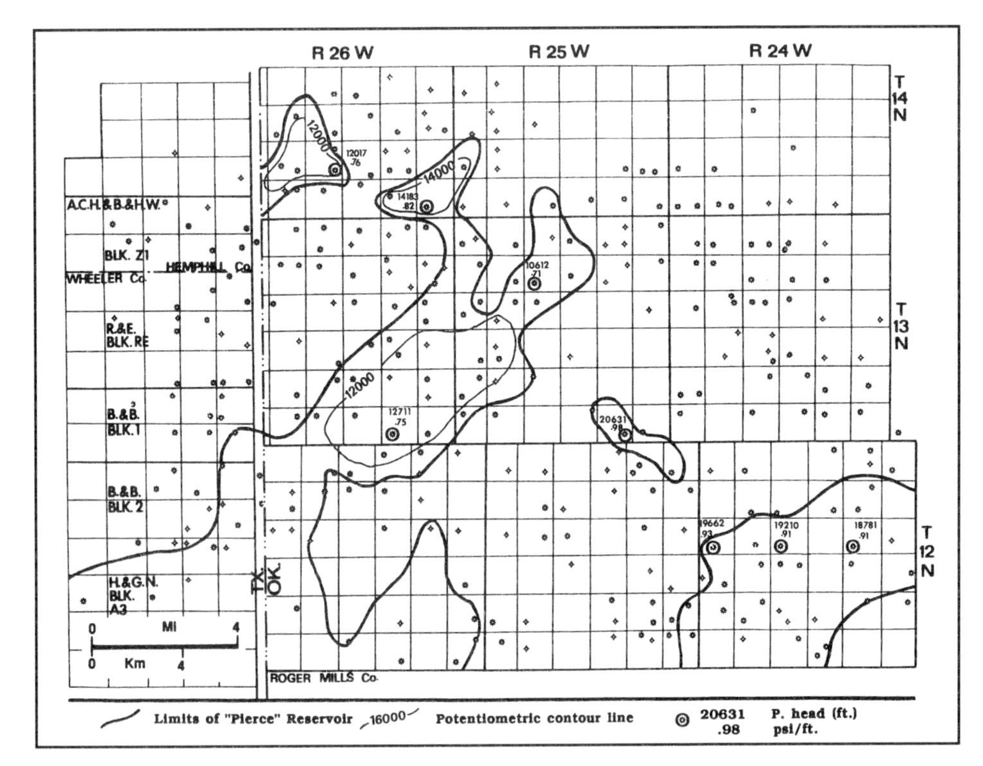

Figure 8. Thickness map of the Pierce chert conglomerate. (Contour interval = 20 ft.)

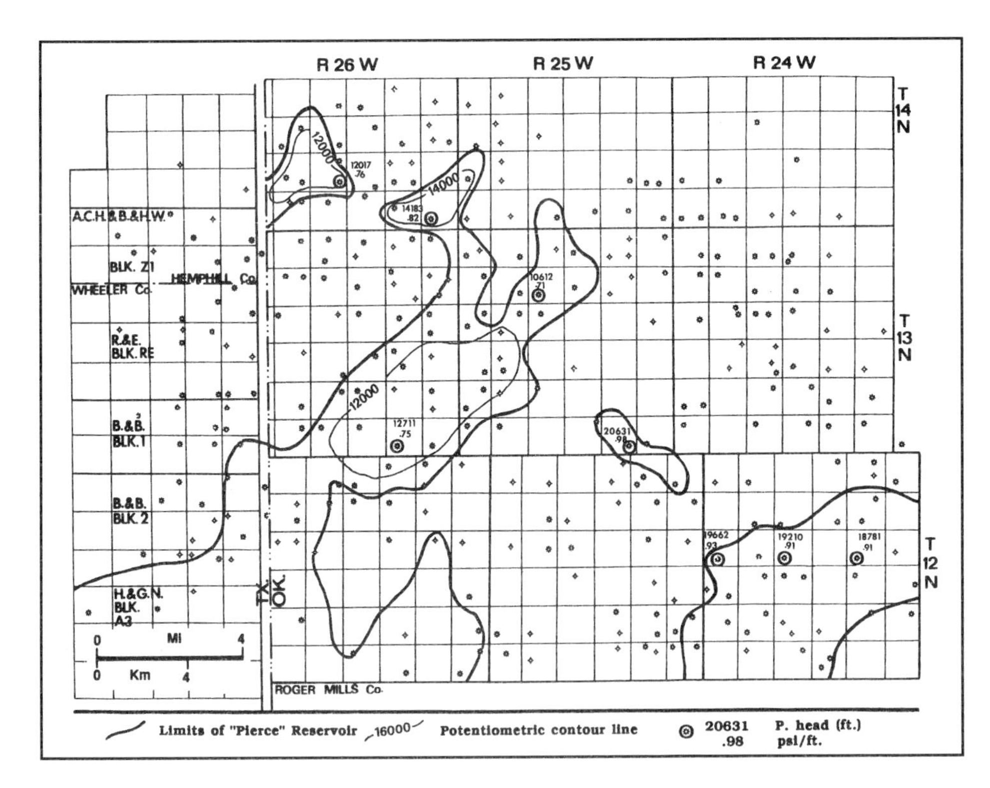

Figure 9. Potentiometric map (ft) of the Pierce chert conglomerate. Corresponding pressure-gradient values (psi/ft) are also shown.

The substantial difference in pressure regimes for these three separate Pierce reservoirs indicates complete isolation of the reservoirs through compartmentation of the Pierce stratigraphic interval. The extremely high pressure of the areally small overbank reservoir may result from early autoisolation of the reservoir through mechano-chemical processes (Ortoleva and Al-Shaieb, 1991).

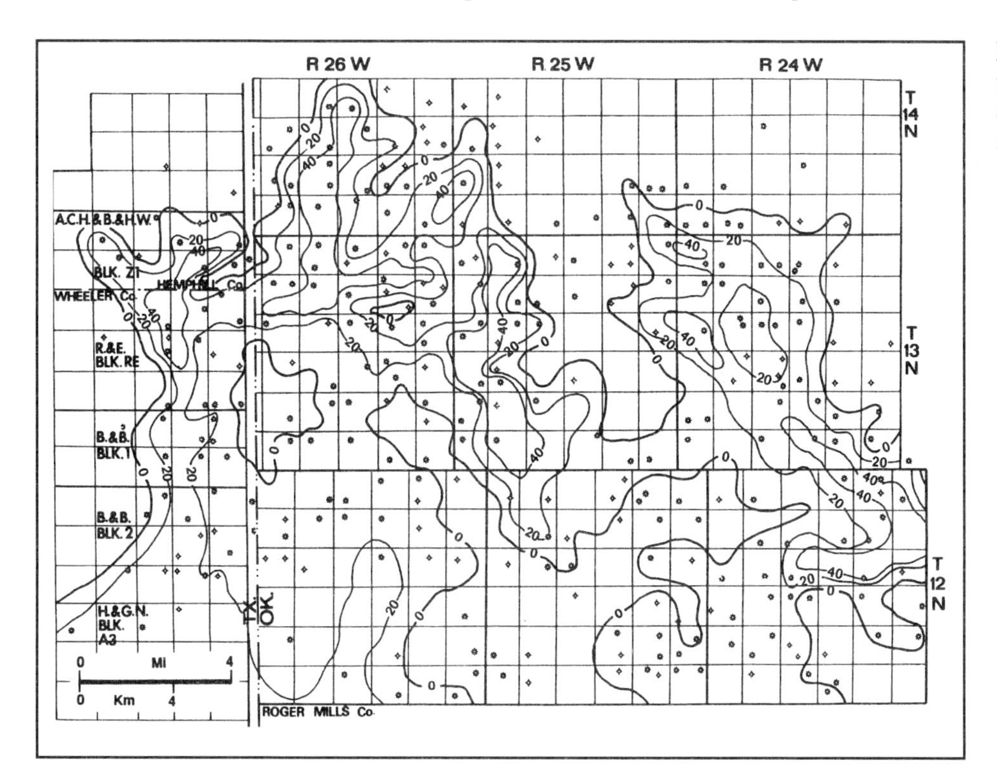

Figure 10. Thickness map of the Puryear chert conglomerate. (Contour interval = 20 ft.)

Puryear Chert Conglomerate

Thickness Map

The Puryear is the most extensive chert conglomerate. This is clearly shown on the thickness distribution map (Figure 10). It is possible that the abundant chert detritus may have been triggered by a significant pulse of tectonism in the Wichita Mountain area during Puryear time. These extensive deposits are represented by a series of northerly trending lobes that are part of the fan- and braid-delta system.

Two dominant Puryear trends are seen in the study area. The western lobe trends north approximately in the same direction as the underlying (Pierce) lobe. The channel within this western trend area bifurcated in northwesterly and northeasterly directions. The eastern lobe trends northwestward and also exhibits bifurcation and merging. Thick channel-fill reservoirs (greater than 30 m) suggest deep incision of the underlying deltaic muds and silts by Puryear channels.

Pressure Regimes

Pressure data for the Puryear reservoirs (Figure 11) suggest that the two primary lobes are not in communication. Pressure gradients in the eastern lobe range from 19.90 to 20.80 KPa/m (0.88 to 0.92 psi/ft) while gradients for the western lobe have a maximum value of 18.77 KPa/m (0.83 psi/ft). Potentiometric surface values for the eastern lobe exceed 5270 m (17,300 ft) while the maximum potentiometric surface values of the western lobe are approximately 4270 m (14,000 ft). This differential in reservoir pressures between the two lobes suggests isolation and compartmentation of the Puryear interval. Therefore, reservoirs in the western and eastern Puryear lobes comprise separate Level 3 compartments. It is also important to note that the pressure and lithologic data for the Puryear suggest complete isolation from the underlying Pierce reservoirs.

Purvis Chert Conglomerate

Thickness Map

The Purvis is the uppermost chert conglomerate. It has a distribution pattern (Figure 12) that is similar to those of the older conglomerates and apparently was derived from the same source. The dominant site of Purvis deposition is a northward-trending braided-stream channel fill along the Texas/Oklahoma border. The primary channel bifurcates to form northerly and east-northeasterly trends. The sedimentary structures (crude stratification, imbrication, and grading) observed in a core located in Section 35, T13N, R26W (Figure 13) support the braided-stream interpretation. A relatively minor, isolated trend develops in T12N, R25W. This may be a splay or overbank deposit.

In the southernmost part of the study area, the Purvis thickens and is widespread and sheet-like. It is interpreted to be stacked and amalgamated braided channels, which are typical of alluvial fan settings.

Pressure Regimes

The Purvis interval contains completely isolated reservoirs with their own distinct pressure values (Figure 14). Pressure gradients for the main Purvis reservoirs are 19.22–19.45 KPa/m (0.85–0.86 psi/ft).

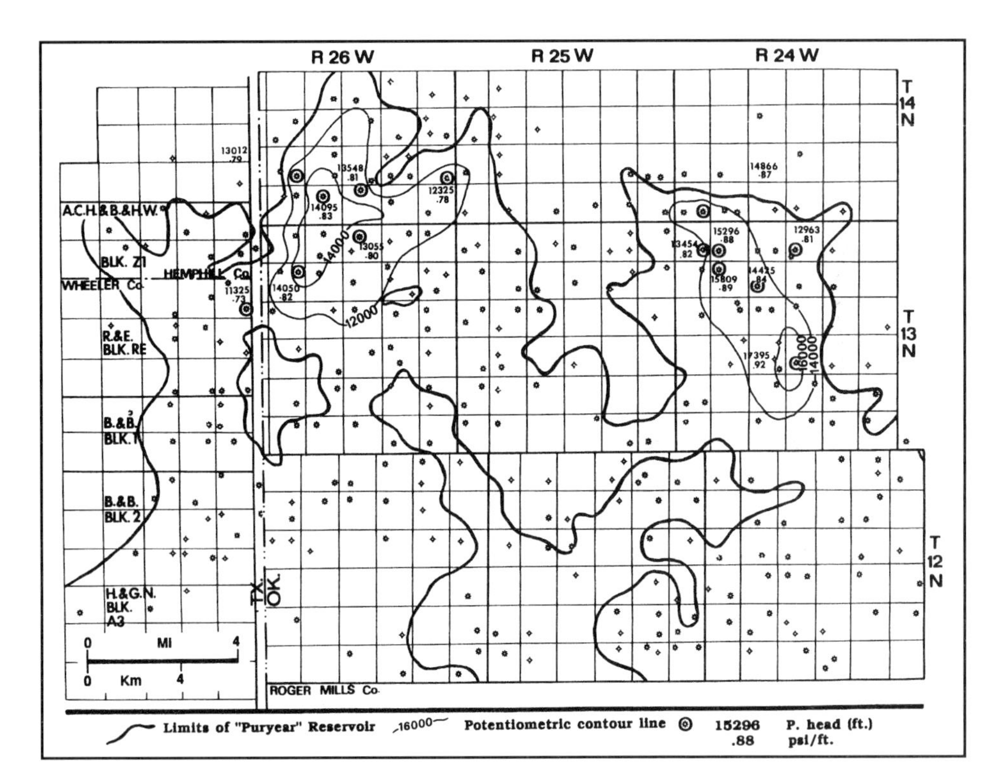

Figure 11. Potentiometric map (ft) of the Puryear chert conglomerate. Corresponding pressure gradient values (psi/ft) are also shown.

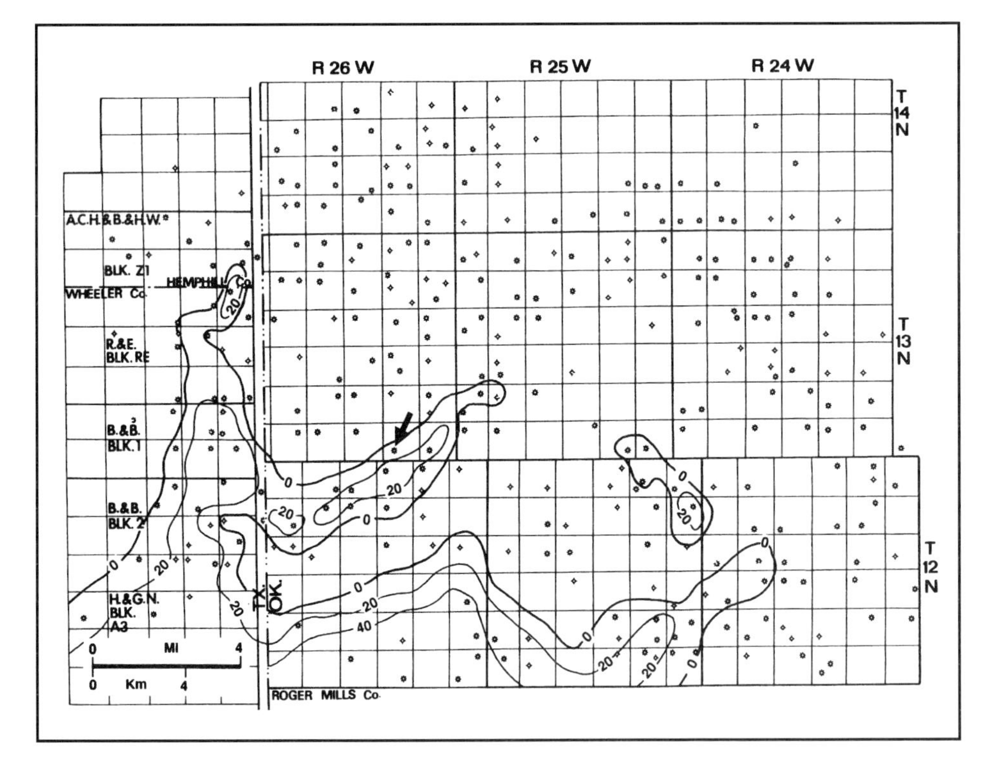

Figure 12. Thickness map of the Purvis chert conglomerate. Location of core studied in Section 35, T13N, R26W is shown by arrow. (Contour interval = 20 ft.)

The maximum potentiometric surface values range from 4880 to 5060 m (16,000 to 16,600 ft) above sea level. However, the interchannel splay or overbank deposit in T12N, R25W has a pressure gradient of 22.16 KPa/m (0.98 psi/ft) and a potentiometric surface value of 6293 m (20,645 ft). The marked difference in pressure gradients and potentiometric heads between the channel-fill and floodplain deposits is strong evidence of compartmentation of the Purvis chert interval. Pressure and lithologic data for the Purvis indicate complete isolation from the underlying Puryear reservoirs.

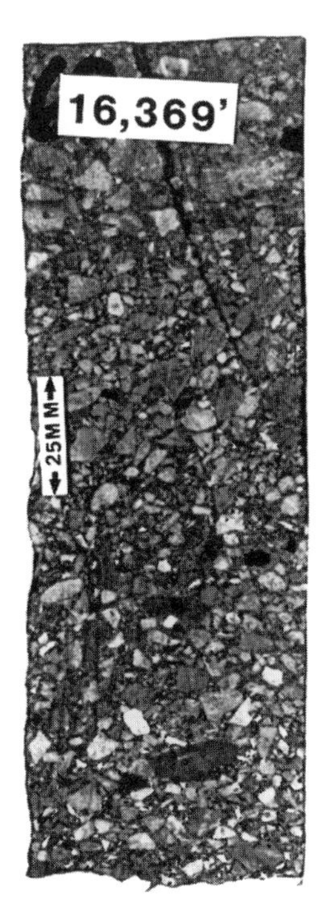

Figure 13. Clast-supported, poorly sorted, and crudely stratified conglomerate and sandstone indicative of braided-stream channel fill. Purvis chert conglomerate, Section 35, T13N, R26W.

The pressure data for the chert conglomerate demonstrate that the members are isolated from each other to form Level 3 compartments within the Level 2 chert interval. Additional Level 3 compartments are formed where cementation and facies changes serve to isolate reservoirs within the members.

SEALING AND COMPARTMENTATION TIMING

Seals

The entire Chert Conglomerate compartment is sealed laterally due to changes in lithofacies from conglomerate and sandstone to dark gray shale along the distal northern boundary of the chert clastic wedge. The encasement of individual members within this shale isolates these reservoirs to form Level 3 compartments.

Stacked conglomerate sequences in the southern proximal part of the complex are sealed by cementation associated with the frontal fault zone of the Wichita Mountain uplift. Figure 15, a cross section extending from the frontal fault zone in Texas to the Reydon field, depicts this relationship. Cementation in the proximity of the fault zone formed a wide band of highly indurated conglomerate that serves as a southern lateral seal for the entire chert interval. This extensive cementation is observed in the cored interval of the Hunt, Bryant 1-57 well in Wheeler County, Texas (Figure 16). This core contains a segment of the stacked conglomerate sequences that are believed to correlate to the Puryear/Pierce interval. Chemical diagenesis was manifested by the extensive silica cementation that obliterated the porosity in these rocks (Figure 17).

This cementation forms a diagenetic seal that divides the thick conglomerate sequences in this area

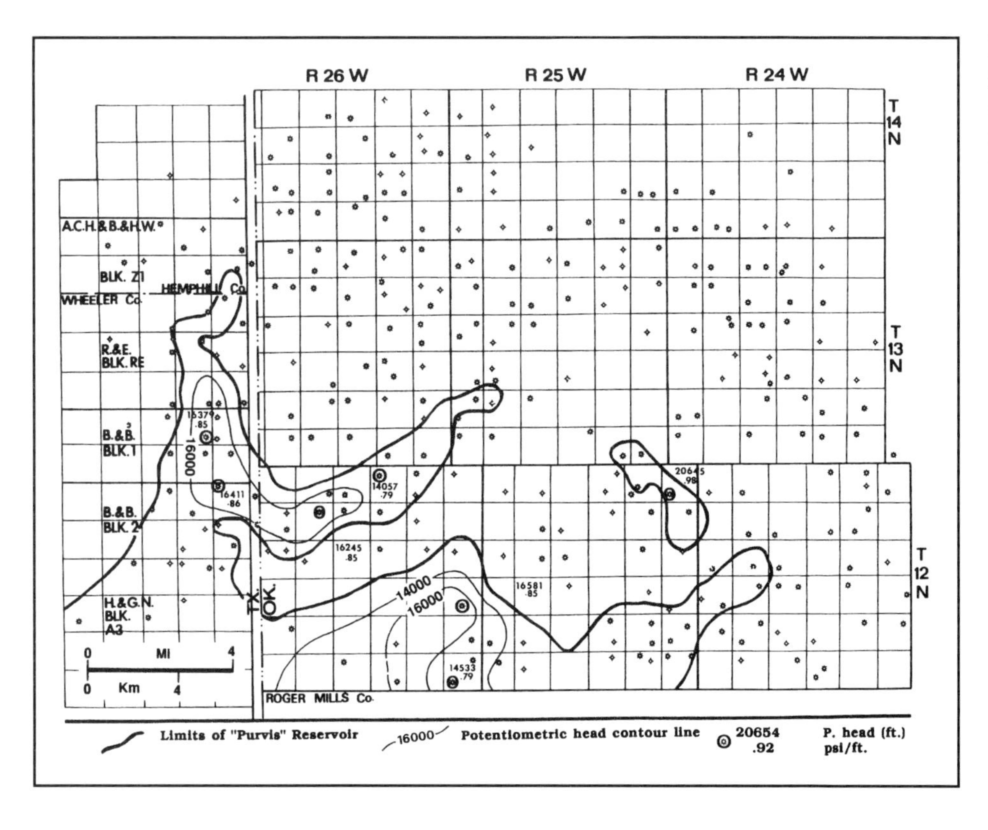

Figure 14. Potentiometric map (ft) of the Purvis chert conglomerate. Corresponding pressure gradient values (psi/ft) are also shown.

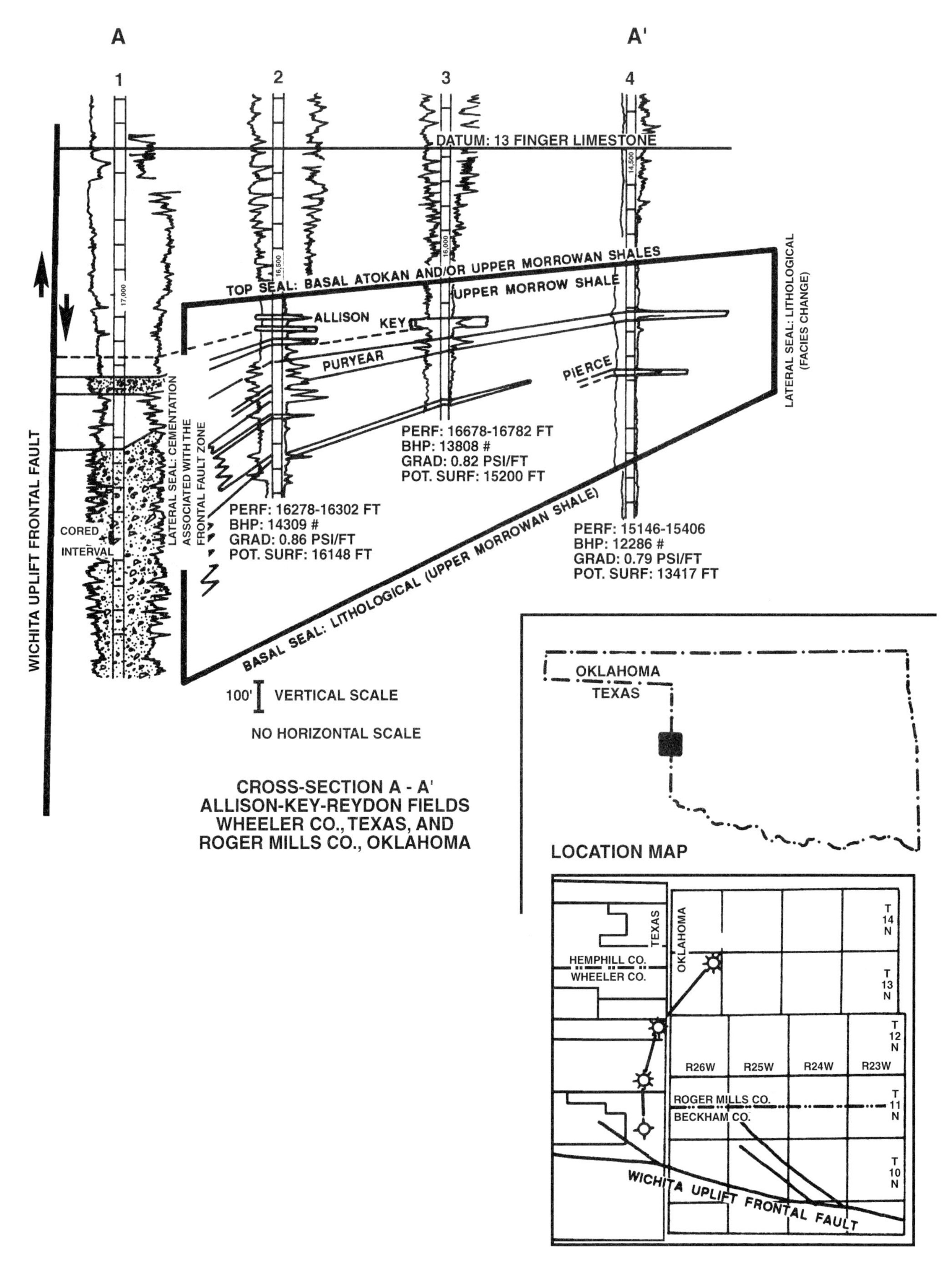

Figure 15. Cross section extending from the Wichita Mountain frontal fault zone through the Allison, Key, and Reydon fields depicting the location of pressure seals and isolated nature of the reservoirs. Cored interval in the Hunt, Bryant 1-57 well (no. 1 on cross section) is indicated by a heavy black bar.

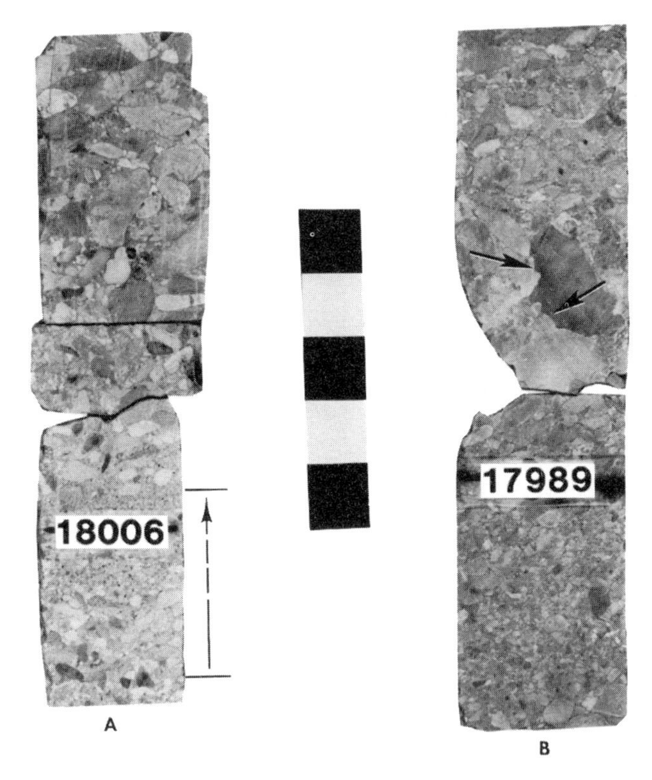

Figure 16. (A) Extensively cemented, poorly sorted, interstratified conglomerate and coarse sandstone. Vertical arrow identifies fining-upward sequence. Hunt, Bryant 1-57, Wheeler County, Texas. (B) Compaction dissolution represented by sutured-seam grain contacts between larger chert pebbles (arrows). Hunt, Bryant 1-57. Scale bar has 1 in. divisions.

Figure 17. Cements occluding intergranular porosity between chert grains (Ct). (A) Coarse, equant quartz cement (Eq). (B) Radial-fibrous chalcedonic quartz cement (Cq).

into small low-porosity/permeability reservoirs. Thus, wells completed in the vicinity of the fault zone are generally low-volume producers. However, reservoir quality and productivity show significant improvement away from the highly cemented fault zone to the south. Highly productive braided-stream channel-fill and overbank reservoirs occur in the Cheyenne and Reydon fields that are located along the northern margin of the chert clastic wedge. These reservoirs are completely sealed by the diagenetically modified shale that encloses them (Figure 17).

Timing of Compartmentation

Compartmentation is intricately related to basin evolution. Overpressuring requires the development of pressure seals prior to the generation of extremely high reservoir pressures via thermal expansion and cracking of reservoir hydrocarbons. Modeling of the Anadarko basin suggests an early protoseal formed prior to source rock maturation and petroleum migration. Rapid subsidence associated with the Pennsylvanian orogeny (Figure 18) created rapidly changing rock/fluid geochemistries that facilitated precipitation of various cements and strengthened seal rocks. Sealed reservoirs were heated and overpressured, which tended to slow compaction. Overpressured reservoirs adjacent to faults vented pressure into the faults, encouraging fluid movement toward the faults and precipitation in the adjacent depressurized regions of the reservoirs. Consequently, reservoirs in the proximity of the frontal fault zone became highly cemented while more distant reservoirs retained higher pressures and were less compacted and cemented (Ortoleva and Al-Shaieb, 1991). Diagenetic alteration of shales strengthened their confining capabilities. Probable episodic ruptures of some seals occurred, but most of these breaches were cemented by the solute-laden fluids escaping the overpressured reservoirs. The seals and compartments presently observed in the Anadarko basin were likely in place by the end of the Pennsylvanian orogeny. Additional slow burial and heating occurred until the Laramide Orogeny, when uplift and slight cooling ensued.

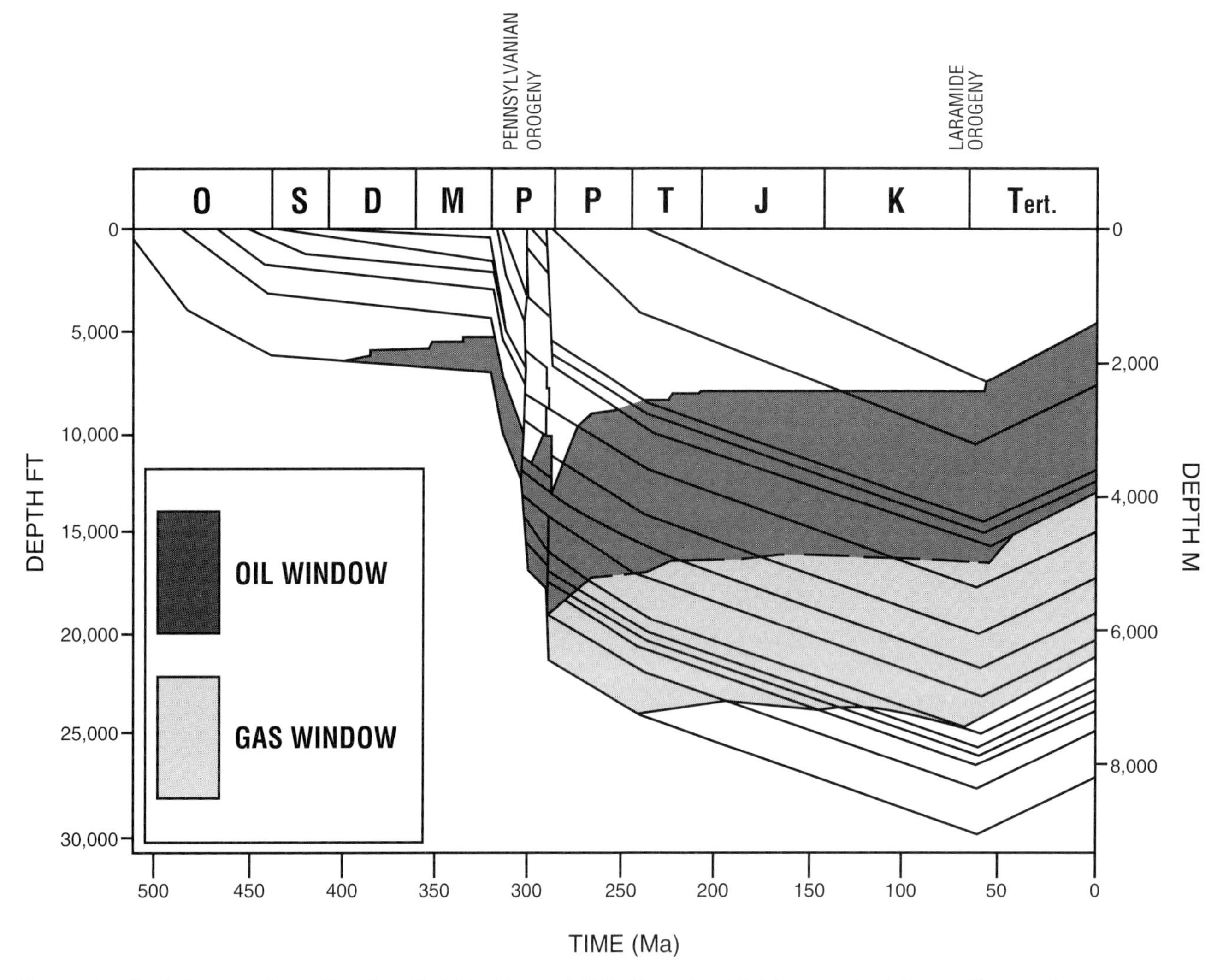

Figure 18. Burial curve from Reydon field in Roger Mills County, Oklahoma. Seal formation and compartmentation occurred during rapid subsidence of the Pennsylvanian orogeny.

SUMMARY

In the Anadarko basin, all reservoirs between the top seal zone at approximately 2290 m to 3050 m (7500 to 10,000 ft) deep and the Woodford Shale are overpressured and comprise the Level 1 compartment called the megacompartment complex. In a given area of the basin, such as the Watonga Trend or the Cheyenne-Reydon fields, all reservoirs within the same stratigraphic interval were subjected to similar burial and heating scenarios. Consequently, these reservoirs have similar pressure gradients and comprise Level 2 compartments. Detailed scrutiny of pressure and production data including reservoir pressures, decline rates, and gas/liquid ratios allows the subdivision of these Level 2 compartments into smaller reservoir-size compartments. Each reservoir with a distinct pressure regime represents a sealed Level 3 compartment.

The hierarchical classification of compartments has many implications for exploration and development in deep sedimentary basins. Knowing the boundaries of the Level 1 MCC is essential to safe drilling practices and the prevention of lost circulation, blowouts, or reservoir damage by unbalanced mud systems. Comparable pressures within Level 2 compartments allow gradient prediction across entire trends or fields. This information is useful in designing properly balanced drilling fluid systems and recognizing depletion due to production or leaking. Level 3 compartments may be identified by integrating subsurface, initial reservoir pressure, fluid type (gas/oil water ratios), and pressure decline data. Recognizing compartment boundaries is essential to the development of drilling strategies. Within field boundaries, nonproducing compartments may exist in close proximity to producing ones. These undrilled or underdeveloped compartments offer additional targets for deep, overpressured gas exploration.

ACKNOWLEDGMENTS

The authors gratefully acknowledge the Gas Research Institute for funding this research through Contract No. 5089-26-1805.

REFERENCES CITED

Abdalla, A., Z. Al-Shaieb, J. Puckette, and A. Rice, 1992, Levels of compartmentation: Implications for exploration and production in the Anadarko basin: Session Abstracts of Second Symposium on Deep Basin Compartments and Seals, Gas Research Institute and Oklahoma State University, Stillwater, Oklahoma.

Alberta, P. L., 1987, Depositional facies analysis and porosity development of the (Pennsylvanian) upper Morrow Chert Conglomerate "Puryear" Member, Roger Mills and Beckham counties, Oklahoma: M. S. Thesis, Oklahoma State University, 135 p.

Al-Shaieb, Z., and J. W. Shelton, 1977, Evaluation of uranium potential in selected Pennsylvanian and Permian units and igneous rocks in southwestern and southern Oklahoma: U.S. Department of Energy, Open File Report GJBX-35 (78), 248 p.

Al-Shaieb, Z., et al., 1990a, Annual Report prepared for the Gas Research Institute, Chicago, Illinois.

Al-Shaieb, Z., P. L. Alberta, and M. Gaskins, 1990b, Depositional environment, petrology, diagenesis, and porosity of the upper Morrow Chert Conglomerate in Oklahoma: Transaction Volume of the 1989 AAPG Mid-Continent Section Meeting, September 24–26, 1989, Oklahoma City, Oklahoma.

Al-Shaieb, Z., et al., 1991, Annual Report prepared for the Gas Research Institute, Chicago, Illinois.

Arbenz, J. K., 1956, Tectonic map of Oklahoma: Oklahoma Geol. Sur. Map GM-3.

Dahlberg, E. C., 1982, Applied hydrodynamics in petroleum exploration: Springer-Verlag, New York, 161 p.

Dwight's Energy Data Inc., 1989, Natural gas well production histories: Richardson, Texas.

Echometer Company, 1986, Analyzing well performance, Wichita Falls, Texas.

Hawthorne, H. W., 1984, Upper Morrow (Pennsylvanian) chert conglomerates and sandstones of the Reydon and Cheyenne fields, Roger Mills County, Oklahoma: Unpublished M.S. Thesis, Baylor University, 117 p.

Johnson, B., 1990, Regional geology of the Pierce member of the Upper Morrow Formation in the Anadarko basin, with a detailed look at South Dempsey field in Roger Mills County, Oklahoma: Transaction Volume of the 1989 AAPG Mid-Continent Section Meeting, September 24–26, 1989, Oklahoma City, Oklahoma.

Ortoleva, P., and Z. Al-Shaieb, 1991, Supplemental Report prepared for the Gas Research Institute, Chicago, Illinois.

Petroleum Information, 1990, Oil and gas well completion data: Oklahoma City, Oklahoma.

Puckette, J., A. Abdalla, Z. Al-Shaieb, and A. Rice, 1993, The Upper Morrowan reservoirs: Complex fluvio-deltaic depositional systems: Abstract volume of fluvial-dominated deltaic reservoirs in the southern Mid-continent, Okla. Geological Survey and U.S. Dept. of Energy, p. 5.

III. Gulf Coast

Chapter 6

Deep Pressure Seal in the Lower Tuscaloosa Formation, Louisiana Gulf Coast

Suzanne D. Weedman
Albert L. Guber
Terry Engelder
Penn State University
University Park, Pennsylvania, U.S.A.

ABSTRACT

Repeat formation tester (RFT) pore pressure measurements spanning a depth range of 5500–6060 m in the lower Tuscaloosa Formation (Upper Cretaceous) document a pressure discontinuity of >20 MPa at ~5680 m forming a pressure seal in two natural gas fields in the Tuscaloosa trend, Louisiana. In the Morganza field the depth to the top of overpressure varies by less than 30 m across two adjacent fault blocks, though equivalent strata are downthrown by 100 to 120 m. In contrast, the depth to the top of overpressure in the nearby Moore-Sams field rises slightly across the same fault. Therefore, the nearly horizontal top of overpressure does not appear to coincide with time- or lithostratigraphic boundaries.

The overpressures in all of the Moore-Sams and some of the Morganza fields wells follow a local hydrostatic gradient with increasing depth indicating that pore fluids below the pressure seal are in communication, and demonstrating that sandstone connectivity occurs below the pressure seal as well as above. In the remaining Morganza wells, overpressure increases with depth in a stair-step manner that may comprise offset local hydrostatic gradients, to magnitudes of 117 MPa at depths of 5.9 km. The occurrence of the pressure seal within interbedded sandstones and shales, where high sandstone connectivity is expected, suggests that the sandstones of the seal zone are unusually tight.

The above observations coupled with a petrographic study of sandstones from the vicinity of the pressure seal suggest that extreme compaction of the sandstones after dissolution of carbonate cements may have contributed to the low permeability indicated by the pressure data, and that the seal formed a kilometer or more shallower than it is today.

INTRODUCTION

The occurrence and maintenance of abnormally high fluid pressures in sedimentary basins is explained by processes that account for pore size reduction, pore fluid volume increase, and/or seal formation (see Gretener and Feng, 1985, for a review). Several pressure-generating mechanisms have been proposed for overpressures in sandstones such as thermal cracking of oil to gas (Hedberg, 1974), the migration of overpressured fluid into sandstones from undercompacted shales (Dickinson, 1953), the addition of pore water produced by the smectite to illite transformation (Powers, 1967; Perry and Hower, 1970, 1972), or the thermal expansion of pore water with burial (Barker, 1972). Additionally, the maintenance of abnormal pressure over geologic time requires either seals of extremely low permeability, the recharging of pressures by burial compaction and heating, or continuous addition of fluids by hydrocarbon generation or by topographically driven flow (Bradley, 1975; Gretener and Feng, 1985; Hunt, 1990).

Fluid flow models have reproduced observed pore pressures in sedimentary basins (England et al., 1987; Ungerer et al., 1987; Bethke et al., 1988; Mann and Mackenzie, 1990; Harrison and Summa, 1991). The pressure-generating mechanism in these models is shale compaction disequilibrium (Dickinson, 1953; Magara, 1971, 1978). Permeability barriers required to maintain overpressure in vertically flowing pore water require permeabilities in the nanodarcy range (Ungerer et al., 1987; Harrison and Summa, 1991; L. Cathles, 1991, personal communication), which for shales may require burial depths of at least 3 km (Gretener and Feng, 1985) and would exclude most siltstones and sandstones.

In the Tertiary section of the Gulf Coast basin the top of overpressure typically occurs near the base of massive deltaic sandstones at the contact with underlying marine shales (Dickinson, 1953; Wallace et al., 1979; Bruce, 1984). However, there remain some unusual hydrocarbon reservoirs where the present-day transition to overpressure occurs at unexplained permeability barriers that have either stopped migrating overpressured pore fluids or have sealed fluids that have subsequently expanded with heat or diagenesis (Bradley, 1975; Hunt, 1990). Very little is known about pressure seals within sandstones or sandstone-dominated sections (Jansa and Noguera Urrea, 1990; Tigert and Al-Shaieb, 1990; Moline et al., 1991; Weedman et al., 1992a, b), especially of the type that seal overpressured fluids with local hydrostatic gradients characteristic of the pressure compartments described by Powley (1990) and Hunt (1990).

The very porous sandstones (>25% porosity) of the deltaic portion (Smith, 1985) of the lower Tuscaloosa Formation (Upper Cretaceous) in Louisiana produce gas from 5400 to 6400 m depth. A transition to overpressure occurs within the formation in the study area including the Moore-Sams and Morganza fields (Figure 1). Less than 10 km up-dip from these fields, the formation is normally pressured, and less than 10 km down-dip it is entirely overpressured (McCulloh and Purcell, 1983). In this paper, we present closely spaced repeat formation tester (RFT) pore pressure measurements that constrain the geometry of the top of a pressure seal in the study area and use the data to assess pressure generation and seal formation mechanisms.

METHODOLOGY

RFT pressure measurements used in this study were taken during the course of drilling wells in the early 1980s and have been graded by the authors using criteria described in Smolen (1977). Although the accuracy of the pressure gauge on the tool is reported as ±0.2 MPa (Smolen, 1977), pressure measurements in this study at the same depths vary as much as 2 MPa. Sand/shale ratios were estimated from gamma-ray logs by defining the 100% sandstone and 100% shale lines, and considering any unit registering 30% to the left of the shale line on the gamma-ray trace to be sandstone. Ratios were calculated every 30.5 m and averaged over the entire formation in the producing wells. The top of the lower Tuscaloosa Formation is identified and correlated by a distinctive, high-resistivity calcareous shale bed known as the Pilot, or Bain marker (Billingsley, 1980).

The term *overpressure* is used here to denote pore pressures above normal, i.e., greater than the hydrostatic pressure indicated by ρgh, where ρ is the density of the pore fluid, g is gravity, and h is depth. Pressure gradients are calculated from the surface by dividing pore pressure by depth. Reference to local hydrostatic gradient means that pore pressures increase with depth at a rate that is parallel to the hydrostatic gradient to the ground surface and is interpreted to mean that the fluids in that interval are in hydraulic communication despite being overpressured.

Sandstones were sampled from available cores from these fields. The wells were cored above the pressure seal in the Ravenswood B and Butler wells from depths of 5483 m to 5613 m, and below the seal from the Fontaine well from depths of 5726 m to 5745 m. Discussion of these samples is given in more detail in Weedman et al. (1992a). A persistent problem in the study of pressure seals is the unavailability of core samples through the seal zone because of the hazardous nature of coring through a very high pressure gradient. Therefore, to infer the nature of the rocks in the seal zone we use indirect evidence such as pore pressure changes, differences in diagenesis of sandstones above and below the seal, and geophysical log characteristics.

RESULTS

A plot of all pressure data versus depth for both fields is shown in Figure 2. A transition from normal to overpressure occurs at about 5680 m near the top of the lower Tuscaloosa Formation. Other studies have documented a shallower transition zone, as well, in these and other Tuscaloosa trend fields (Matheny,

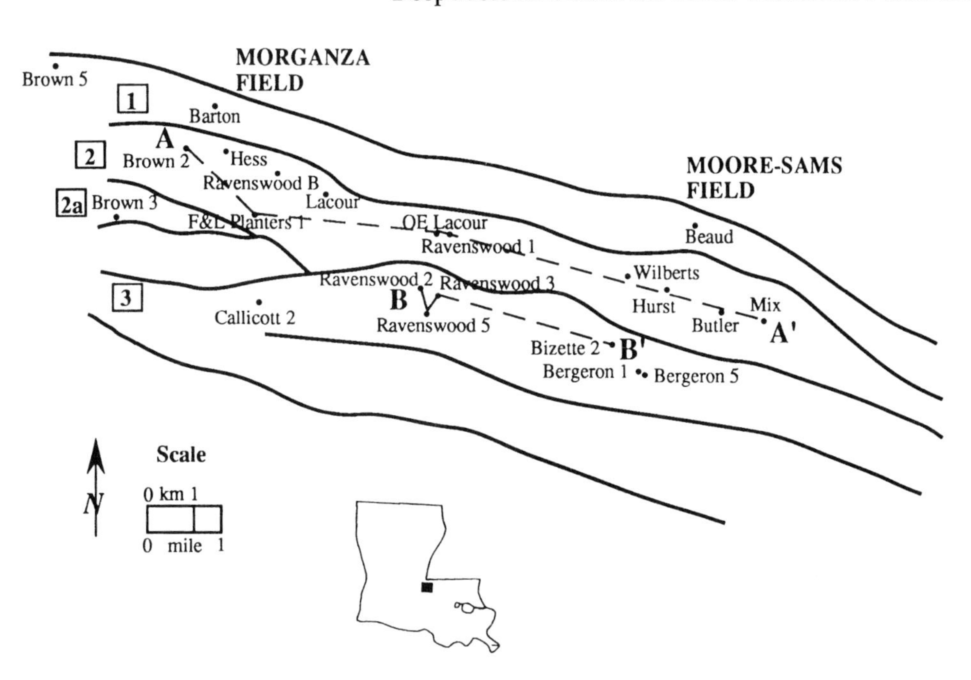

Figure 1. Map of the Morganza and Moore-Sams fields showing the locations of wells and the listric normal faults that subdivide the fields into fault blocks. Location in Louisiana shown in inset. Lines of cross section, A–A′ and B–B′, are shown for Figures 4 and 5 (after AMOCO map).

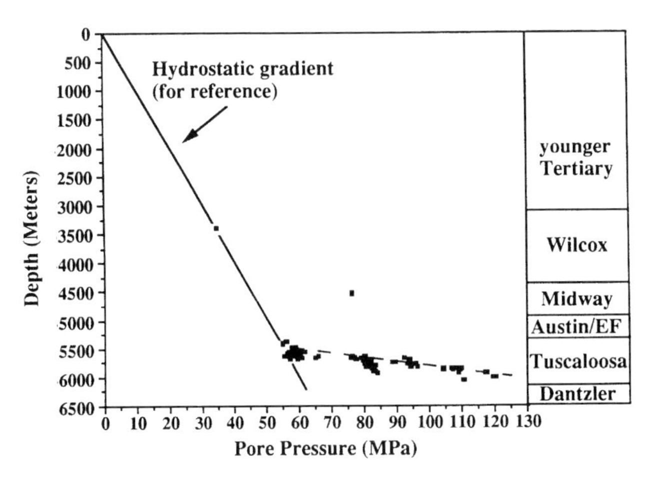

Figure 2. Plot of pressure vs. depth of all successful RFT data from the Morganza and Moore-Sams fields. Data indicate perhaps two overpressured zones separated by ~300 m of normally pressured fluids at 5350 m in the upper sandstones of the lower Tuscaloosa Formation. Approximate depths to formation tops shown at right; EF = Eagle Ford. N = 203 total data points, n = 5 at 3440 m, n = 3 at 4530 m. Data tables are given in Weedman et al. (1992b).

1979; Pankonien, 1979; Gill, 1980; McCulloh and Purcell, 1983; McCulloh, 1985) and an interval of normal pressure in between the two overpressured zones.

Moore-Sams Field

RFT pore pressure data from within the lower Tuscaloosa Formation in the Moore-Sams field are shown for three fault blocks (Figure 3A). All pore pressures in fault block 1 are normal, while below the transition zone pressures in blocks 2 and 3 follow a local hydrostatic gradient down to depths of 5850 m, reaching magnitudes of 82.6 MPa. In block 2, the transition zone from normal to overpressure occurs at 5728 ± 45 m, while in block 3 it occurs at 5674 ± 15 m; the top of the formation is displaced down-to-the-south across the fault separating blocks 2 and 3 by 105 m. Therefore, the transition zone is higher stratigraphically in fault block 3 than in 2, and the variation in the depth to the top of overpressure between fault blocks 2 and 3 is less than the displacement of the strata across the fault. Additionally, fluid communication in the overpressured zone below the pressure seal and across the fault is indicated by the nearly equal magnitude of pore pressures at the same depths on both sides of the fault.

Morganza Field

RFT pore pressure data from four fault blocks of the Morganza field are shown in Figure 3B. The pore pressures in the overpressured intervals in the Morganza field follow apparent local hydrostatic gradients, but increase as much as 26 MPa across certain shaly intervals. As for the Moore-Sams field, all pore pressure data from fault block 1 of the Morganza field are normal, while the overpressures encountered in fault blocks 2 and 3 reach a greater magnitude than in the Moore-Sams field, up to 117 MPa at depths of 5975 m. The variation in pore pressure over the depth range of 5675 to 5850 m suggests that several pressure seals may exist within this field, not only along faults (for example, between blocks 2 and 2A), but within fault blocks as well (blocks 2 and 3).

Pressure Seal Geometry

Both normal and overpressure RFT measurements are available for eight wells that constrain the depth and thickness of the pressure seal in the two fields.

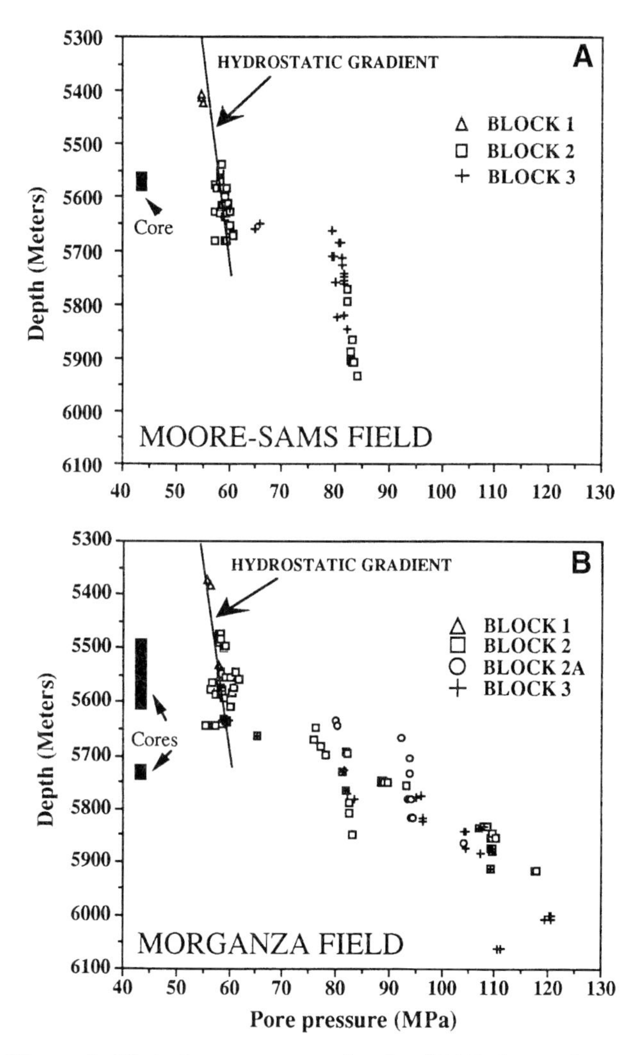

Figure 3. Plot of pressure vs. depth of RFT data. The hydrostatic gradient to the surface is shown for reference. (A) Pressure vs. depth in the Moore-Sams field from the fault blocks listed at upper right. Core taken from interval indicated. (B) Pressure vs. depth in the Morganza field from fault blocks listed at upper right. Cores taken from the intervals indicated.

Cross sections of those wells from fault blocks 2 and 3 are shown in Figures 4 and 5, respectively. The shaded pressure seal interval, constrained by the deepest normal pressure and the shallowest RFT overpressure measurements, is characterized by interbedded sandstone and shale.

Comparison of the Bizette 2 with the Mix well (Figures 4 and 5) shows that the pressure seal is in the middle part of the lower Tuscaloosa Formation in fault block 2 and in the upper part of the formation in fault block 3. The pressure seal in the Mix well occurs within a thick shale interval that may be correlated from across both fields. However, the pressure data show that this apparently laterally extensive shale, indicated in Figures 4 and 5 by a dashed line, is not the pressure seal in the other wells.

The maximum possible thickness of the pressure seal in block 2 varies from 28 m (OE Lacour) to 137 m (F&L Planters), and in block 3 varies from 67 m (Bizette 2) to 131 m (Ravenswood 5). In fault block 3 of both fields, the depth to the top of the pressure seal is consistent from well to well, varying from ~5620 m (Bizette 2) to ~5640 m (Ravenswood 5). It is possible that the seal is less than 28 m thick throughout the fields. The upper part of the pressure seal zone in nearly all wells in Figures 4 and 5 is characterized by high resistivity in both the sandstones and shales. High resistivity is generally attributed to the presence of hydrocarbons or extremely low porosity, or both. Below the resistivity maximum, but in the seal zone, there is a sharp decline in resistivity, indicating either increased porosity or more conductive pore fluids, or both. This resistivity signature of the top of overpressure, due to the assumed undercompacted state of the sediments, was first described by Hottman and Johnson (1965) and is used in the Tuscaloosa trend, as well as in most of the Gulf Coast, to anticipate the onset of overpressure (Gill, 1980).

DISCUSSION

The pressure transition zone in the study area is unusual in that it is nearly horizontal, contains sandstones and shales, is thin (28 m in one well), and appears to cross-cut stratigraphic boundaries. This horizontality could reflect the greatest depth at which interconnected, normally pressured sandstones are juxtaposed at faults of the type of situation described by Mann and Mackenzie (1990)—an interpretation that requires a laterally extensive, low-permeability shale that reaches from fault to fault to vertically isolate normal from overpressured sandstones. The lithology at the transition zone is not a thick shale but a zone of interbedded (~3 m thick) sandstone and shale, typically upward-coarsening. Evidence from gamma-ray logs suggests that the only thick shale that might extend from fault to fault, the shale break at 5700 m in block 2 and at 5900 m in block 3, clearly is not the seal in most of the wells (Figure 4). We think that correlation across these distances in deltaic intervals is difficult and perhaps unreliable with log data alone. If the thick shale shown in the Mix well with shading (Figure 4) is continuous to the west as indicated, it clearly does not form the pressure seal in wells to the west. However, if the thick shale at the Mix well is not laterally continuous as indicated, then it must pinch out to the west. In either case, a thick shale does not form the pressure seal in any well but perhaps the Mix well, where the shale coincides with the pressure transition as indicated by the RFT measurements.

Weber's (1982) investigation of shale length (lateral extent) as a function of depositional environment suggests that in the delta front environment laterally continuous shales typically have a length that is smaller than the distance from fault to fault in the Moore-Sams and Morganza fields (2–5 km), suggesting further that a shale bed is an unlikely candidate for a pressure seal in the study area.

A study of sandstone diagenesis from above and below the pressure seal of cores from the Ravens-

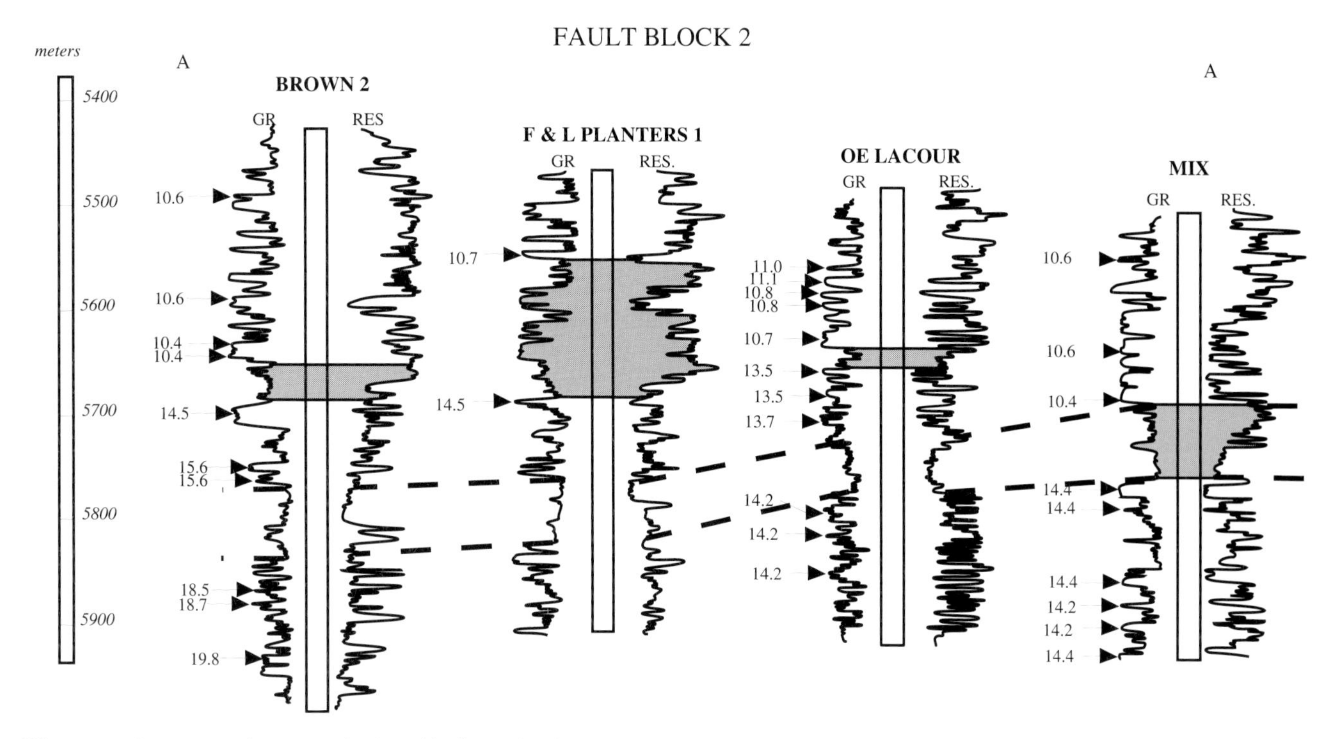

Figure 4. Cross section A–A′ of wells for which normal and abnormal pressures were measured in fault block 2. Line of cross section is shown in Figure 1. Arrows indicate measurement location, numbers are pressure gradients in MPa/km or KPa/m. The logs are gamma ray on the left and resistivity on the right. The logs begin at the top of the lower Tuscaloosa Formation at the Bain Marker bed and extend to the bottom of the well. The depths between the deepest normal pressure measurement and the shallowest overpressure measurement are shaded. A thick shale is correlated (dashed line) and may be the only laterally extensive shale in the two fields.

wood B, Butler, and Fontaine wells demonstrates that the pressure transition zone separates sandstone strata of unusually high secondary porosity of up to 26% (Weedman et al., 1992a). In addition, pressure data show that there is sufficient sandstone connectivity below the seal to maintain a local hydrostatic gradient in the overpressured zone, a characteristic of a pressure compartment (Hunt, 1990; Powley, 1990). The rocks that form the seal and maintain a pressure discontinuity of >20 MPa over a depth range of 28 to 137 m are interbedded thin (~3 m) sandstones and shales. King (1990) has shown by three-dimensional modeling of hypothetical random networks of sandbodies in shale that within an interval where the net to gross ratio (sand/sand + shale) is 0.8, the connected sand fraction approaches 100%. The producing wells of the lower Tuscaloosa Formation have sand/shale ratios of at least 4:1 or a net to gross value of 0.8. This observation suggests to us that while the thin sandstones in the seal zone were probably interconnected with sandstones above and below when deposited, they are now very tight with sufficiently low permeability to act with the shales as a pressure seal that maintains the pressure anomaly.

Compaction parameters of packing density and packing proximity were measured on three sandstone populations in the vicinity of the pressure seal: normally pressured with >18% cement, normally pressured with <10% cement, and overpressured with <10% cement. Packing proximity is the percentage of grain contacts along a traverse that are in grain-to-grain contact; packing density is the percentage of a traverse that is occupied by framework grains and not cements or pore spaces. We have presented evidence elsewhere that the samples with less than 10% cement had lost a previous carbonate cement by dissolution (Weedman et al., 1992a). Results show that the packing of framework grains is similar between normally pressured, cemented sandstones and overpressured, low-cement sandstones, while normally pressured, low-cement sandstones exhibit as much as 20% greater compaction compared to the other two populations (Figure 6). These results suggest to us that the compaction of framework grains resumed after decementation in the normally pressured zone but was inhibited below the pressure seal because of high fluid pressures. Therefore, the pressure seal became effective in isolating porous and interconnected sandstones soon after the dissolution of grain-supporting calcite cement, and a process analogous to shale undercompaction can exist in overpressured sandstones. In addition, small samples of cuttings taken from within the seal zone, where cores are unavailable, show extensive pressure solution and fitted textures that would, if as extensive as suspected, provide permeability barriers within the thin sandstones of the seal zone (Albrecht, 1992; Weedman et al., 1992a). If observed in only one well, the highly compacted textures may be interpreted as fault gouge; however, cuttings were examined from three

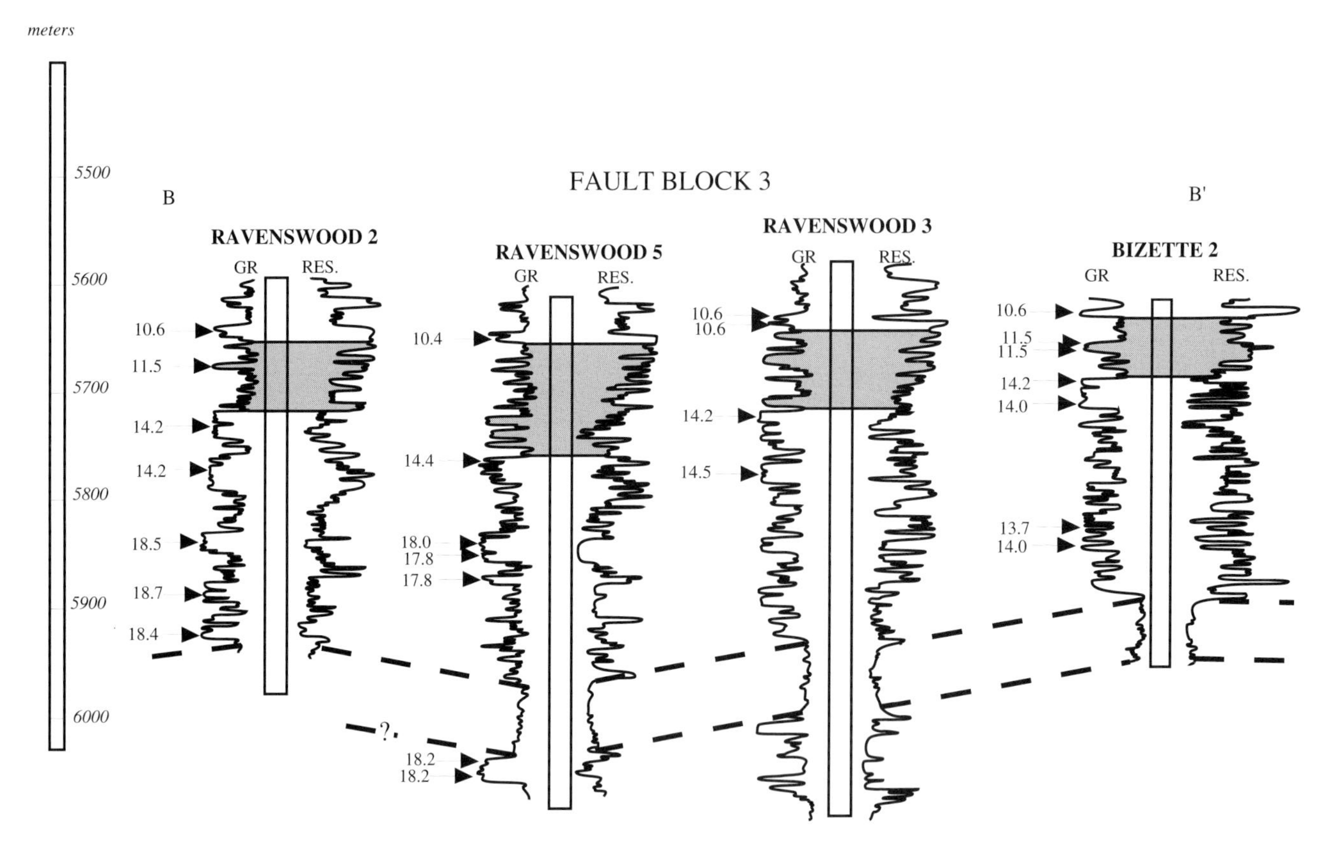

Figure 5. Cross section B–B′ of wells for which normal and abnormal pressures were measured in fault block 3. Line of cross section is shown in Figure 2. Arrows indicate measurement location, numbers are pressure gradients in MPa/km. The logs are gamma ray on the left and resistivity on the right. The logs begin at the top of the lower Tuscaloosa Formation at the Bain Marker bed and extend to the bottom of the well. The depths between the deepest normal pressure measurement and the shallowest overpressure measurement are shaded. The shale that is correlated may be the only laterally extensive shale in the two fields.

widely spaced wells (Brown 2, Ravenswood 5, and V.J. Hurst; see Figure 1) and the pressure solution texture was observed only within the seal zone as defined by RFT pressure data (Albrecht, 1992). If this compaction texture is the consequence of faulting, the fault must be nearly horizontal across the two fields.

In several studies of sandstones from the Gulf Coast Tertiary, calcite cements are thought to have been dissolved in reservoirs at temperatures of 75° to 125°C (Franks and Forester, 1984). Assuming a geothermal gradient of 25°C for the study area, that depth today would be between 2 and 4 km. Superimposing that depth interval on a simple burial history curve for a well in the Moore-Sams field, Figure 7, suggests that the pressure seal within the lower Tuscaloosa Formation could be as old as 30 million years. Bethke (1989) and Harrison and Summa (1991) calculate that the onset of overpressures in Late Cretaceous rocks at the depths and approximate strike location of the study area commenced in about Oligocene time, which is consistent with the above estimate based on petrography.

There may be multiple sources for the overpressured fluids in these fields. Organic-rich shales in the lower Tuscaloosa Formation, thought to be the source rocks for these reservoirs (Sassen, 1990), are deep enough to produce gas (Hunt, 1979). In addition, the shale resistivity declines across the pressure seal zone suggest that the overpressured shales are still undercompacted and could be a source of overpressured fluids to the sandstones. If the pressure seal formed at a kilometer or more shallower depth than it is today, as suggested by compaction differences of sandstones in the vicinity of the seal, some of the overpressure could be attributed to aquathermal pressuring or to trapped pore fluids generated by clay diagenesis at shallower depths.

None of the popular pressure-generation mechanisms can explain the maintenance of such a high pressure discontinuity (>20 MPa) for the amount of time indicated from petrography. We think that, in the absence of a laterally extensive thick shale, a sealing mechanism is required to explain the pressure anomaly and observed grain packing, and propose the process of secondary compaction of high porosity sandstones.

CONCLUSIONS

Repeat formation pressure data across a pressure transition zone in the deep Tuscaloosa trend have been evaluated to document the geometry and lithology of

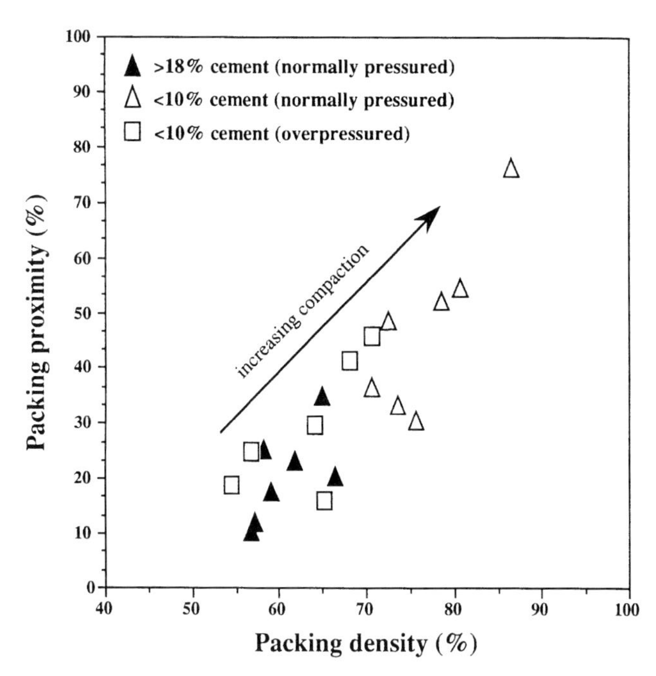

Figure 6. Packing proximity vs. packing density for 20 thin sections cut perpendicular to bedding. The sandstones that have suffered the greatest compaction (Δ) are normally pressured and have <10% cement, due to dissolution of calcite cement. Normally pressured sandstones with >18% cement (▲) are compacted to a degree similar to overpressured sandstones with <10% cement (❑). This plot documents secondary compaction after secondary porosity and the inhibition of compaction by overpressured pore fluids below the pressure seal.

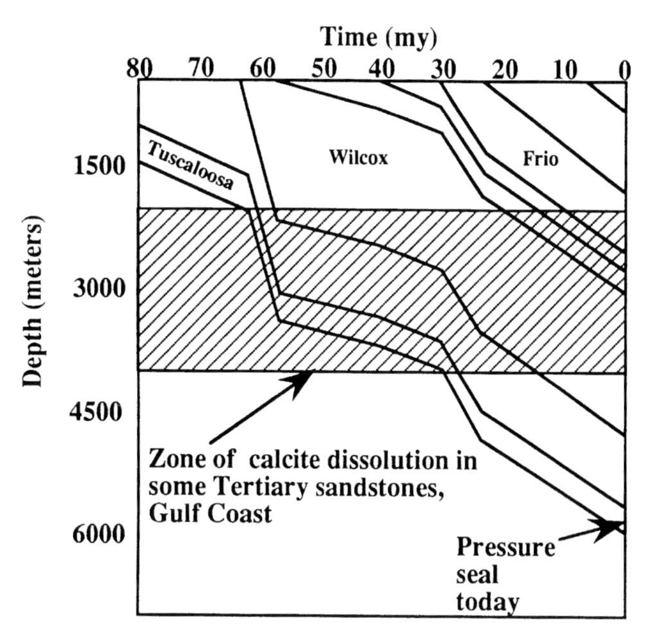

Figure 7. Simple burial history curve for the V.J. Hurst well in the Moore-Sams field from the present day to approximately 80 m.y. ago (compaction has been neglected). Lower Tuscaloosa (Upper Cretaceous), Wilcox (Eocene), and Frio (Oligocene) clastic wedges are identified. The depths at which temperatures range from 75° to 125°C, the temperature range of calcite dissolution in the Tertiary of the Gulf Coast according to data from Franks and Forester (1984), are approximated with diagonal lines; these depths are derived using a geothermal gradient of 25°C/km. The lower Tuscaloosa Formation may have been in this temperature range from approximately 30 to 60 million years ago.

the seal zone. The pressure seal is nearly horizontal and does not follow a laterally continuous lithologic horizon but is characterized by thinly bedded sandstones and shales. While horizontality may be controlled by the juxtaposition of normally pressured sandstones across growth faults, the problem of characterizing a pressure seal remains. The only potentially laterally extensive shale in the formation forms the pressure seal only in one well. Without high resolution pressure data, the pressure seal may have been attributed to that shale; pressure data shown in Figures 4 and 5 demonstrate that it is not.

An alternative interpretation is that the high sandstone secondary porosity has collapsed in places producing zones of extensive pressure solution and extremely low permeability. This compaction has been documented above the seal but can only be inferred from cuttings within the seal zone. Where those collapsed sandstones are interbedded with shales on a fine scale, there may be sufficiently low permeability to maintain a pressure anomaly of >20 MPa. Diagenetic study of core and cuttings from the vicinity of the pressure seal coupled with burial history suggests that the seal may have formed as long as 30 million years ago and subsided to the present depth of ~5.6 km.

ACKNOWLEDGMENTS

Funding for this research is provided by the Gas Research Institute contract number 5088-260-1746. We thank AMOCO Production Company for access to pressure data, cores, cuttings, logs, and maps of the study area. Access to data was facilitated by E. Batchelder and B. Ward. Also, we have benefited greatly from discussions with D. E. Powley and J. S. Bradley, both of AMOCO, and S.L. Brantley.

REFERENCES CITED

Albrecht, W., 1992, Geochemistry of diagenetic pressure seal formation: Unpublished Ph.D. dissertation, Penn State University, University Park, 265p.

Barker, C., 1972, Aquathermal pressuring—role of temperature in development of abnormal-pressure zones: American Association of Petroleum Geologists Bulletin, v. 56, p. 2068–2071.

Bethke, C. M., 1989, Modeling subsurface flow in sedimentary basins: Geologische Rundschau, v. 78, n. 1, p. 129–154.

Bethke, C. M., W. J. Harrison, C. Upson, and S. P. Altaner, 1988, Supercomputer analysis of sedimen-

tary basins: Science, v. 239, p. 261–267.
Billingsley, A. L., 1980, Unconventional energy resources in Tuscaloosa sediments of the "Tuscaloosa trend," South Louisiana: Unpublished Master's Thesis, University of New Orleans.
Bradley, J.S., 1975, Abnormal formation pressure: American Association Petroleum Geologists Bulletin, v. 59, p. 957–973.
Bruce, C. H., 1984, Smectite dehydration—its relation to structural development and hydrocarbon migration in northern Gulf Coast basin: American Association of Petroleum Geologists Bulletin, v. 68, p. 673–683.
Dickinson, G., 1953, Geological aspects of abnormal reservoir pressures in Gulf Coast Louisiana: American Association of Petroleum Geologists Bulletin, v. 37, p. 410–432.
England, W.A., A. S. Mackenzie, D. M. Mann, and T. M. Quigley, 1987, The movement and entrapment of petroleum fluids in the subsurface: Journal of the Geologic Society of London, v. 44, p. 327–347.
Franks, S.G., and R. W. Forester, 1984, Relationships among secondary porosity, pore-fluid chemistry, and carbon dioxide, Texas Gulf Coast, *in* McDonald, D.A., and R. C. Surdam, eds., Clastic Diagenesis: American Association of Petroleum Geologists Memoir 37, p. 63–79.
Gill, J.A., 1980, Multiparameter log tracks Tuscaloosa/Woodbine pressures: Oil and Gas Journal, v. 78, n. 44, p. 55–64.
Gretener, P. E., and Zeng-Mo Feng, 1985, Three decades of geopressures—insights and enigmas: Bull. Ver. schweiz. Petroleum-Geol. u. -Ing., 51, Nr. 120, p. 1–34.
Harrison, W.J., and L. Summa, 1991, Paleohydrology of the Gulf of Mexico Basin: American Journal of Science, 291, p. 109–176.
Hedberg, H.D., 1974, Relation of methane generation to under-compacted shales, shale diapirs, and mud volcanoes: American Association of Petroleum Geologists Bulletin, v. 53, p. 661–673.
Hottman, C.E., and R. K. Johnson, 1965, Estimation of formation pressures from log-derived shale properties: Journal of Petroleum Technology (June 1965), p. 717–722.
Hunt, J.M., 1979, Petroleum Geochemistry and Geology: San Francisco, W.H. Freeman, 617 p.
Hunt, J.M., 1990, The generation and migration of petroleum from abnormally pressured fluid compartments: American Association of Petroleum Geologists Bulletin, v. 74, p. 1–12.
Jansa, L.F., and V. H. Noguera Urrea, 1990, Geology and diagenetic history of overpressured sandstone reservoirs, Venture gas field, offshore Nova Scotia, Canada: American Association of Petroleum Geologists Bulletin, v. 74, p. 1640–1658.
King, P.R., 1990, The connectivity and conductivity of overlapping sand bodies: *in* Buller, A.T., E. W. Berg, O. Hjelmeland, J. Kleppe, O. Torsæter, and J. O. Aasen, 1990, North Sea Oil and Gas Reservoirs—II: Proceedings of the 2nd North Sea Oil and Gas Reservoirs Conference of Norwegian Institute of Technology, Trondheim, Norway, May 8–11, 1989, p. 353–362.
Magara, K., 1971, Permeability considerations in generation of abnormal pressures: Society of Petroleum Engineers Journal, v. 11, p. 236–242.
Magara, K., 1978, Compaction and Fluid Migration—Practical Petroleum Geology: New York, Elsevier Publishing Company, 319 p.
Mann, D. M., and A. S. Mackenzie, 1990, Prediction of pore fluid pressures in sedimentary basins: Marine and Petroleum Geology, v. 7, p. 55–65.
Matheny, S. L., 1979, AMOCO develops lower Tuscaloosa: Oil and Gas Journal, Apr. 9, 1979, p. 122–126.
McCulloh, R. P., 1985, Patterns of fluid flow in the central Tuscaloosa trend, Louisiana: Transactions of the Gulf Coast Association of Geological Societies, v. 35, p. 209–214.
McCulloh, R. P. and M. D. Purcell, 1983, Hydropressure tongues within regionally geopressured lower Tuscaloosa Sandstone, Tuscaloosa trend, Louisiana: Transactions of the Gulf Coast Association of Geological Societies, v. 33, p. 153–159.
Moline, G.R., P. A. Drzewieki, and J. M. Bahr, 1991, Identification and characterization of pressure seals through the use of wireline logs: a multivariate statistical approach: American Association of Petroleum Geologists Bulletin, v. 75, p. 638.
Pankonien, L. J., 1979, Operators scramble to tap deep gas in South Louisiana: World Oil, Sept. 1979, p. 55–62.
Perry, E., and J. Hower, 1970, Burial diagenesis in the Gulf Coast pelitic sediments: Clays and Clay Minerals, v. 8, p. 165–177.
Perry, E., and J. Hower, 1972, Late stage dehydration in deeply buried pelitic sediments: American Association of Petroleum Geologists Bulletin, v. 56, p. 2013–2021.
Powers, M. L., 1967, Fluid release mechanisms in compacting marine mud-rocks and their importance in oil exploration: American Association of Petroleum Geologists Bulletin, v. 51, p. 1240–1254.
Powley, D. E., 1990, Pressures and hydrogeology in petroleum basins, Earth Science Reviews, v. 29, p. 215–226.
Sassen, R., 1990, Lower Tertiary and Upper Cretaceous source rocks in Louisiana and Mississippi; implications to Gulf of Mexico crude oil: American Association of Petroleum Geologists Bulletin, v. 74, p. 857–878.
Smith, G.W., 1985, Geology of the deep Tuscaloosa, Upper Cretaceous, gas trend in Louisiana, *in* Habitat of Oil and Gas in the Gulf Coast, Gulf Coast Section of the Society of Economic and Paleontologic Mineralogists, Foundation Fourth Annual Research Conference Proceedings, June 1985, p. 153–190.
Smolen, J. J., 1977, RFT Pressure Interpretation: Schlumberger Well Services, 50 p.
Tigert, V., and Z. Al-Shaieb, 1990, Pressure seals: their diagenetic banding patterns: Earth Science Reviews, v. 29, p. 227–240.

Ungerer, P., B. Doligez, P. Y. Chenet, J. Burrus, F. Bessis, E. Lafargue, G. Giror, O. Heum, and S. Eggen, 1987, A 2–D model of basin-scale petroleum migration by two-phase flow application to some case studies, *in* Doligez, B. (ed.), Migration of hydrocarbons in sedimentary basins, 2nd Institut Francais du Petrole Exploration Research Conference, Carcans, France, June 15–19, 1987.

Wallace, R.H., Jr., T. F. Kraemer, R. E. Taylor, and J. B. Wesselman, 1979, Assessment of geopressured–geothermal resources in the northern Gulf of Mexico basin: U.S. Geological Survey, Circular 790, p. 132–155.

Weber, K.J., 1982, Influence of common sedimentary structures on fluid flow in reservoir models: Journal of Petroleum Technology (March 1982), p. 665–672.

Weedman, S.D., S. L. Brantley, and W. Albrecht, 1992a, Secondary compaction after secondary porosity: Can it form a pressure seal?, v. 20, p. 303–306.

Weedman, S.D., A. L. Guber, and T. Engelder, 1992b, Pore pressure variation in the Morganza and Moore-Sams fields, Louisiana Gulf Coast: Journal of Geophysical Research, v. 97, no. B5, p. 7193–7202.

Chapter 7

Pressure Seals—Interactions with Organic Matter, Experimental Observations, and Relation to a "Hydrocarbon Plugging" Hypothesis for Pressure Seal Formation

Jean K. Whelan
Lorraine Buxton Eglinton
Woods Hole Oceanographic Institution
Woods Hole, Massachusetts, U.S.A.

Lawrence M. Cathles III
Cornell University
Ithaca, New York, U.S.A.

ABSTRACT

Organic geochemical characteristics diagnostic of pressure seals have been determined for two wells in the Moore-Sams field of the Tuscaloosa trend, Louisiana Gulf Coast (Mix and Bizette wells) and one well penetrating a much weaker pressure transition zone of the Anadarko basin, Oklahoma (Weaver well). Preliminary data suggest these characteristics of organic matter in zones of pressure seals: a rapid increase in vitrinite reflectance near the top of the pressure seal; fractionation of bitumens through the pressure seal with a gradual change from lighter to heavier n-alkanes with increasing depth in the pressure seal; a buildup of hydrocarbons just beneath the pressure seal; and an enhancement of asphalt (or asphaltene) throughout the general zone of the pressure seal. For all three wells, very tight associations of carbonate cements, fine pyrite, asphaltenes, and micrinite (generally considered to be a residual product of hydrocarbon generation) were observed in the general zone of pressure seals, suggesting that interactions of organic and inorganic materials may be required for pressure seal formation and maintenance, even in fairly organic lean wells such as Weaver. A sharp jump in thermal maturity, as measured by vitrinite reflectance, occurs at the top of the Mix pressure transition zone. Maturity levels below the seal reach gas thermal window levels, suggesting that gas formation within and below the (seal) zone is contributing both to overpressuring and sealing of pressure seals investigated here. It is proposed that all these observations can be accommodated if the pressure drop across the seal pressure transition zone causes separation of oil and gas and deposition of asphalt from the upward-streaming oil/gas that carried them from sources at greater depths. Permeability is reduced through a combination of asphalt hydrocarbon plugging, inorganic alteration, and (most important) gas-water capillary effects.

BACKGROUND AND INTRODUCTION

Recent observations and measurements have shown that large cells, or fluid compartments, persist in the subsurface that remain out of hydrostatic equilibrium with adjacent sedimentary rocks over geologic time (Bradley, 1975; Powley, 1980, 1990; Hunt, 1990; Weedman et al., 1992). Furthermore, some mechanisms of pressure seal formation and maintenance appear to have a significant geochemical component as described in several contributions in the current volume.

Throughout this paper, the term *pressure seal* is used to mean a zone in the subsurface of inferred extremely low permeability that maintains a measurable water pressure anomaly and is used synonymously with *pressure transition zone*.

The purpose of this research was to clarify the role of sedimentary rock organic matter, oil, and gas with respect to three aspects of pressure cells and their associated seals:

1. To determine if fluid compartments and their associated pressure seals showed unique organic geochemical fingerprints that might help in their future detection, particularly in shales where the repeat formation tester (RFT), currently the most reliable means of obtaining accurate downhole pressures (Hunt, 1990), gives no data. The problem arises because the RFT operates by punching a hole into the adjacent sedimentary rock in the well, collecting about one-half cup of water, and measuring its pressure. Because shales, for all practical purposes, extrude no water over the time period of the RFT measurement, the RFT is not useful for measuring shale pressures.
2. To gain preliminary information on the effects of pressure compartments and their associated seals on oil and gas distributions in sedimentary rocks.
3. To carry out preliminary research on how organic maturation and migration processes might contribute to pressuring and sealing mechanisms (versus occurring in parallel to independent inorganic sealing processes caused by the same time-temperature regime).

SAMPLES

Samples from two different geographic areas were examined in this work: drill cuttings samples from the Mix M J well (API 17-077-20241; lat. 30.67753; long. 91.47746) and Bizette R2 well (API 17-077-20290; lat. 30.67170; long. 91.51810) from the Moore-Sams field in the Louisiana Gulf Coast Tuscaloosa trend as described in Weedman et al. (1992) (Figure 1A), and from the Weaver No. 1 well in the N.W. Lindsay field of McClain County, Oklahoma of the Anadarko basin (Figure 1B, API No. 35-087-35445-00, location Section 17, T5N, R4W) as described in Tigert and Al-Shaieb (1990).

Samples from the two Gulf Coast wells represent intervals above, within, and below pressure transition zones as defined by RFT measurements (Weedman et al., 1992). The RFT is a downhole tool that operates by measuring the pressure in fluids extracted from sandstone intervals. In the case of the Mix well, water pressures of 8726 and 12,128 psi (60 and 84 MPa) were measured at the top and bottom of the pressure transition zone at depths of 18,586 ft (5665 m) and 18,881 ft (5755 m), respectively. The pressure gradients at the bottom and top of the same zone were 0.469 and 0.649 psi/ft (10.4 and 14.4 kPa/m, respectively) compared to a normal hydrostatic gradient of 0.43 to 0.53 psi/ft (9.54 to 11.8 kPa/m, depending on the salinity of the pore waters). For the Bizette-2 well pressure transition zone, pressures of 8700 and 11,759 psi (60.0 and 79.5 MPa) and gradients 0.469 and 0.649 psi/ft (10.4 and 14.4 kPa/m) were measured at depths of 18,445 ft (5622 m) and 18,734 ft (5710 m), respectively.

Only cuttings were available from the pressure transition zones of Mix and Bizette 2 wells. Because of potential problems with downhole slumping and drilling contamination, these were examined carefully using organic petrographic microscopy to ensure that drilling contamination was minimal. There was no mention of drilling additives in the drilling reports.

Data are also presented for conventional cores from the Weaver well of the Oklahoma Anadarko basin (Figure 1B) (Tigert and Al-Shaieb, 1990), which penetrates a much weaker pressure transition zone than the one described above for the Gulf Coast. In this zone, a transition from underpressured to slightly overpressured conditions, as defined regionally by final shut-in pressure from drill-stem test (DST) measurements, was found in the depth range of 11,000 to 12,000 ft (3353 to 3658 m). However, the highest average gradient measured in the overpressured zone was only about 0.51 psi/ft (11.3 kPa/m) at the bottom of the pressure transition zone, or seal zone.

METHODS

Pyrolysis, pyrolysis gas chromatography (Py-GC), and pyrolysis gas chromatography mass spectroscopy

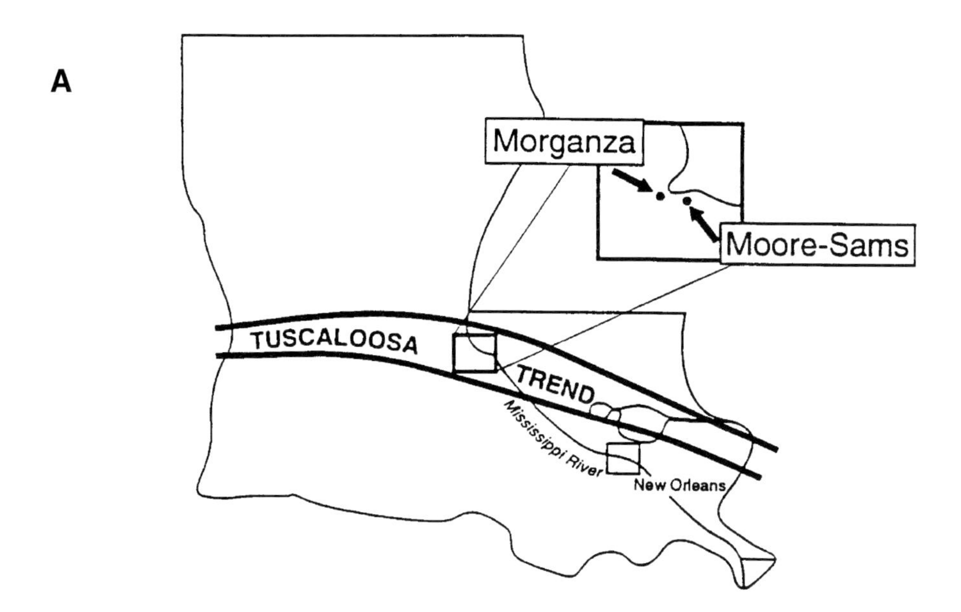

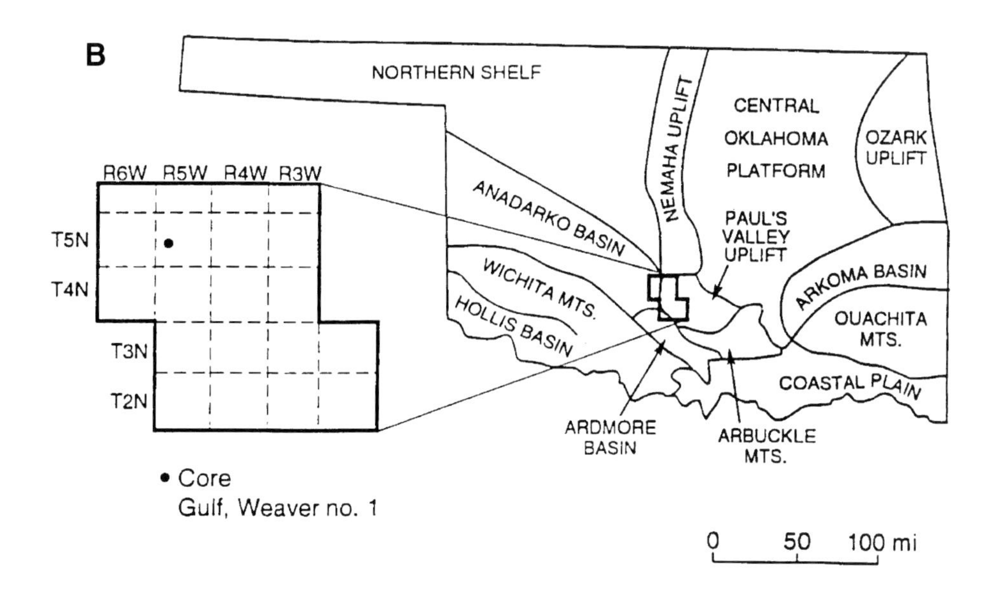

Figure 1. Location of (A) Mix and Bizette wells, Louisiana Gulf Coast Tuscaloosa Trend and (B) Weaver well, Anadarko basin, Oklahoma.

(Py-GCMS) were performed with a Chemical Data Systems (CDS) Pyroprobe used in the desorption mode as described previously (Whelan et al., 1990; Tarafa et al., 1988). The procedures and data are similar, but not identical to, those obtained from Rock Eval pyrolysis (Espitalié et al., 1984; Espitalié, 1986). Typically, a 10 to 50 mg sample of either whole rock or kerogen is heated at a constant rate of 30°C/min in a helium flow. The total volatilized hydrocarbons that evolve as a function of temperature are measured with a flame ionization detector to produce a pyrogram, such as those in Figure 3, where P_1 represents the volatile lower molecular weight constituents ($<C_{25}$) of generated petroleum and P_2 represents the sum of the residual petroleum generating capacity of the sedimentary rock. P_2 is derived from the sum of the heavier, less volatile ($>C_{25}$) cracking of more polar constituents of the in situ generated petroleum and petroleum generated by pyrolytic cracking of kerogen (Tarafa et al., 1988 and references cited). Cracking products of asphaltenes (the most polar component of petroleum) often elute along with the less volatile oil components as a low temperature shoulder on the front of the P_2 peak (Tarafa et al., 1988).

T_{max} is the temperature at which maximum P_2 evolution occurs. If P_2 is only a product of kerogen cracking with no contribution from residual or migrated heavy oil or asphaltene cracking, T_{max} generally increases with the thermal maturity, or time-temperature history, of the kerogen in the sedimentary rock (Espitalié, 1986; Peters, 1986). The T_{max} values reported here are the actual temperatures of the sample measured by a thermocouple placed adjacent to the sample. These are typically 50 to 60°C higher than values obtained from the Rock Eval pyrolysis instrument (Espitalié et al., 1977, 1984; Whelan et al., 1986).

The production index (P.I.) is the ratio of P_1 to the sum of P_1 plus P_2 and represents the ratio of the $<C_{25}$

fraction of generated in situ petroleum to total generation potential. In this paper, the low and high temperature pyrolysis peaks are referred to as P_1 and P_2, respectively, rather than S_1 and S_2 as used for Rock Eval pyrolysis, because of the differences in separation characteristics, responses, and T_{max} temperatures in the CDS Pyroprobe and Rock Eval pyrolysis instruments.

Total carbon and total organic carbon (TOC) measurements were performed using a Coulemetrics CO_2 Coulometer by procedures previously described (Whelan et al., 1990).

Iatroscan analysis was performed on an Iatroscan TH-10, Mk IV (Iatron Labs, Inc., Tokyo), equipped with a flame ionization detector (FID), interfaced with an electronic integrator (Perkin-Elmer LCI-100) by the method of Karlsen and Larter (1991).

Vitrinite determinations were made on whole rock cuttings samples that were coarsely ground to a particle size of approximately 2 mm and cold set in an epoxy resin. The samples were then ground and polished according to standard procedure (ICCP, 1963, 1971, 1975). Vitrinite determinations were performed on a Zeiss Universal microscope-photometer system. The system and determinations were standardized according to Stach et al. (1982). Determinations represent the mean reflectance population of indigenous, randomly oriented vitrinite particles (% R_0 average), i.e., not those deemed as reworked from older sediments or sedimentary rocks. Fluorescence blue-light examination was performed on the same microscope fitted with a 100w HBO mercury arc bulb and using an LP520, BP450-490 filter set.

The term *matrix bitumen stain* refers to bitumen that impregnates the mineral matrix, does not readily polish, and often exhibits weak dark-brown fluorescence in blue light. Solid bitumen is that portion of the bitumen which takes a polish and can be identified in incident white light and fluorescence light (Jacob, 1989). These terms are distinct from matrix bituminite as characterized by Creaney (1980), which refers to very finely dispersed indistinguishable amorphous material within the mineral matrix and is analogous to the amorphous and perhaps some fraction of the liptodetrinitic material referred to in kerogen concentrates.

RESULTS

Summary of Observations from the Bizette 2 and Mix Wells, Louisiana Gulf Coast

Pyrolysis Data, Mix Well

Pyrolysis was carried out on samples from within as well as above and below the pressure seal zone, as defined by the pressure transition zone. P_1, P_2, TC, TOC, P_1/TOC, P_2/TOC, P.I., and T_{max} are given in Table 1, and downhole profiles are plotted in Figure 2. The solid horizontal lines show the pressure transition or pressure seal zone, as defined by the RFT measurements. Distinct changes in all pyrolysis parameters can be observed just above as well as within and below the pressure seal zone in comparison to shallower and deeper intervals. P_1, P_1/TOC, and P.I. maximize at the top and just under the seal for the Mix well (Figure 2) reflecting a maxima in light (<C_{25}) generated petroleum in this zone. A gradual decrease in both P_1 and P.I. is also observed with increasing depth in the seal zone. Based on P.I. values, all of the Mix samples are influenced by either generated or migrated petroleum hydrocarbons: P.I. values of less than 0.1 are typical of immature or organic lean intervals; 0.1 to 0.4 is typical of source rocks within the oil generation window; values of 0.4 to 1 are typical of sedimentary rocks containing migrated hydrocarbons (Peters, 1986; Vandenbroucke and Durand, 1983; Whelan et al., 1986). P_2 shows an increase approaching the top of the pressure seal zone and remains fairly high through the pressure transition zone and just below.

The lithology and shapes of the P_1 and P_2 peaks as a function of depth for the Mix well are shown in Figure 3. The low temperature shoulder, indicative of asphaltene and C_{25+} components of generated oil (Clementz, 1979; Tarafa et al., 1988), is commonly observed on the front of P_2 peak (e.g., indicated by circles for 17,690 and 18,570 ft [5392 and 5660 m] samples in Figure 3). This low-temperature P_2 shoulder is present in most of the Mix well samples shown in Figure 3, including those above and below as well as within the seal zone. An enhancement of the area of the shoulder in relation to the rest of the pyrogram is most evident through and below the seal zone.

T_{max} shows an overall decrease with depth in the Mix well (Figure 2F), contrary to the more normal case where T_{max} increases with increasing depth and maturity. This decreasing T_{max} trend is particularly pronounced within the seal zone and is due to the presence of heavy and polar components of generated or migrated petroleum within this zone (discussed below). T_{max} values start at 490°C at the top of the seal zone and decrease to 480°C at the bottom (equivalent to a decrease from 440 to 430°C on the Rock Eval pyrolysis T_{max} scale) and below the seal (Figure 2F and Table 1).

The partial hydrocarbon compositions of P_1 for the Mix well samples for the samples numbered 7, 11, 14, 17, and 21 (Table 1 and Figure 2A) are shown in Figure 4. Only the lighter hydrocarbon compositions are shown because they are the most diagnostic of various migration processes (Thompson, 1979, 1988; Leythaeuser et al., 1980; Hunt, 1985; Whelan et al., 1984). Mix #7, just above the seal zone at 18,580 ft (5663 m), shows a predominance of the aromatic hydrocarbon xylene, a hydrocarbon that tends to migrate easily with the aqueous phase (Thompson, 1979, 1988; Whelan et al., 1984). Samples crossing from the top to the bottom of the seal zone, Mix #11 (18,700 ft, 5700 m), Mix #14 (18,790 ft, 5727 m), and Mix #17 (18,880 ft, 5756 m), show a general progression from a predominance of the lowest molecular weight alkane (nC_8) toward the heavier nC_{14} at the bottom of the seal in Mix #17. In contrast, Mix #21 at 19,000 ft (5791 m), just below the seal, shows no

Table 1. Summary of pyrolysis, TC, TOC, and vitrinite reflectance results, Mix, Bizette, and Weaver wells.

Depth (ft)	Sample no.	P_1 (mg HC/g rock)	P_2 (mg HC/g rock)	P.I.*	T_{max} (°C) (CDS)	T_{max} Rock Eval calculated**	TC (%)	TOC (%)	P_1/TOC (mg HC/g TOC)	P_2/TOC (mg HC/g TOC)	Iatroscan yields (Polar fraction) (mg/g rock)	% R_o #	Number of points
Mix Well													
17,440												0.84±0.11	11
17,470	1	4.29	7.71	0.36	492	442	3.93	1.9	226	406			
17,660												0.97±0.10	12
17,690	2	0.07	0.38	0.15	508	458	1	0.43	16	89			
18,060												0.94±0.10	6
18,090	3	0.35	0.74	0.32	489	439	1.05	1.06	33	70			
18,150												0.94±0.02	5
18,180	4	0.20	1.33	0.13	488	438	1.69	1.2	17	111			
18,430												0.93±0.04	4
18,450	5	0.65	3.09	0.17	488	438	2.97	1.61	40	192			
18,560												0.93±0.08	6
18,570	6	3.19	4.30	0.43	495	445	2.62	2.29	139	188	2.71		
18,580	7	2.69	3.08	0.47	490	440	1.71	0.83	324	371		0.91±0.09	8
18,610	8	6.93	5.16	0.57	492	442	2.11	1.21	572	426		0.90±0.09	4
18,640	9	3.62	3.63	0.50	490	440	2.53	1.76	206	206		1.05±0.09	12
18,670	10	6.79	6.51	0.51	490	440	2.38	1.3	522	500	3.6	1.20±0.08	6
18,700	11	2.98	3.39	0.47	483	433	2.6	1.63	183	203			
18,730	12	2.65	6.10	0.30	498	448	2.41	1.7	156	359		1.29±0.09	24
18,760	13	2.85	3.43	0.45	482	432	2.15	1.61	177	213	5.17	1.29±0.06	12
18,790	14	3.33	3.86	0.46	481	431	2.67	1.59	210	242		1.24±0.08	5
18,820	15	4.09	5.51	0.43	479	429	2.61	1.58	259	349		1.28±0.09	18
18,850	16	2.90	4.38	0.40	478	428	2.35	1.59	183	276		1.29±0.08	9
18,880	17	2.24	4.57	0.33	488	438	1.73	1.27	176	360			
18,900	18	1.68	4.08	0.29	482	432	2.82	2.02	83	202	10.3	1.26±0.10	5
18,930	19	1.77	1.36	0.57	481	431	1.74	0.95	187	144		1.23±0.11	7
18,960	20	3.82	4.83	0.44	486	436	1.93	1.33	287	363	5.26		
19,000	21	5.13	5.59	0.48	478	428	2.78	1.95	263	286		1.31±0.07	4
19,010	22	2.56	1.16	0.69	488	438	1.56	1.25	205	93			
Bizette Well													
17,775	1	0.05	0.32	0.12	522	472	1.12	0.74	6.1	44		1.08±0.09	10
17,875	2	0.06	0.36	0.14	562	512	1.35	0.81	7.0	45			
18,025	3	0.07	0.37	0.16	523	473	2.31	0.89	8.2	42			
18,150	4	0.07	0.85	0.078	510	460	2.56	2.15	3.3	39			
18,300	5	0.14	0.46	0.24	528	478	1.22	0.93	15.4	4.9			

* P.I. = $P_1/(P_1 + P_2)$.

** Calculated Rock Eval T_{max} = CDS T_{max} – 50°C.

Average vitrinite reflectance % in oil (ne1.517@23°C) @ 546 nm.

Underlined values designate values within the pressure transition zone.

Continued on next page

Table 1. Summary of pyrolysis, TC, TOC, and vitrinite reflectance results, Mix, Bizette, and Weaver wells (continued).

Depth (ft)	Sample no.	P_1 (mg HC/g rock)	P_2 (mg HC/g rock)	P.I.*	T_{max} (°C) (CDS)	T_{max} Rock Eval calculated**	TC (%)	TOC (%)	P_1/TOC (mg HC/g TOC)	P_2/TOC (mg HC/g TOC)	Iatroscan yields (Polar fraction) (mg/g rock)	% R_o #	Number of points
18,325											8.83		
18,350	7	0.14	0.44	0.24	507	457	2.18	1.21	11	36			
18,412	8	0.01	0.96	0.006	441	391	2.45	1.19	0.53	81			
18,450	9	0.62	2.88	0.18	487	437	4.44	4.32	14	67		0.90±0.09	12
18,475	10	0.73	0.86	0.46	477	427	1.93	1.93	38	45		0.98±0.08	15
18,500	11	0.61	0.61	0.50	559	509	1.74	1.92	32	32	20.04	1.10±0.07	20
18,525	12	6.88	6.52	0.51	495	445	3.09	3.5	197	186	25.2	1.02±0.09	20
18,550	13	3.01	3.97	0.43	502	452	2.63	2.44	123	163	4.36		
18,575	14	7.91	8.50	0.48	489	439	3.71	3.4	233	250			
18,625	15	13.47	5.42	0.71	482	432	4.46	4.1	329	132	6.78		
18,650	16	1.40	3.04	0.31	491	441	2.57	2.54	55	120			
18,675	17	4.16	6.43	0.39	504	454	3.64	3.1	134	207			
Weaver Well													
10,190	1	0.0031	0.136	0.02	504	454	0.14	0.14	2.21	97			
10,537	2	0.0012	0.07	0.02	490	440	8.05	0.25	0.48	28		0.50±0.09	4
11,014	3	0.014	0.181	0.07	514	464	1.14	0.13	10.8	139		0.46±0.02	4
11,057	4	0.0064	0.0978	0.06	512	462	0.24	0.08	8.00	122			
11,085	5	0.0028	0.151	0.02	495	445	0.08	0.08	3.50	189		0.97±0.05	4
11,171	6	0.0052	0.114	0.04	488	438	1.83	0.18	2.89	63		0.75±0.09	6
11,204	7	0.016	0.758	0.02	504	454	0.84	0.63	3	120		0.60±0.10	10
11,274	8	0.013	0.654	0.02	507	457	1.68	0.53	2.45	123		0.82±0.02	4
11,344	9	0.294	1.033	0.22	519	469	0.4	0.36	82	287		0.90±0.06	8
11,384	10	0.0011	0.08	0.01	500	450	0.27	0.15	1	53		0.86±0.05	10
11,446	11	0.01	0.333	0.03	498	448	0.62	0.33	3.03	101		0.95±0.06	20
11,472	12	0.002	0.087	0.02	495	445	0.17	0.1	2.00	87			
11,521	13	0.0016	0.157	0.01	500	450	4.85	0.3	0.53	52			
11,543	14	0.0025	0.143	0.02	495	445	1.55	0.18	1.39	79			
11,556	15	0.003	0.464	0.01	498	448	2.47	0.43	0.70	108			
11,650	16	0.0044	0.318	0.01	488	438	0.46	0.43	1.02	74		0.89±0.05	4
11,720	17	0.0068	0.146	0.04	493	443	0.43	0.3	2.27	49			
11,961	18	0.017	0.229	0.07	502	452	3.22	0.43	3.95	53		1.02±0.04	8
11,992	19	0.003	0.059	0.05	486	436	0.4	0.14	2.14	42			
12,068	20	0.016	0.08	0.17	492	442	8.42	0.66	2.42	12			

* P.I. = $P_1/(P_1 + P_2)$.
** Calculated Rock Eval T_{max} = CDS T_{max} − 50°C.
\# Average vitrinite reflectance % in oil (ne1.517@23°C) @ 546 nm.
Underlined values designate values within the pressure transition zone.

Continued on next page

Table 1. (Continued)

Source Rock Classification (from Peters, 1986)
Typical values of the parameters above:

Source rock quality	P_1 (mg HC/g rock)	P_2 (mg HC/g rock)	TOC (%)
Poor	0 to 0.5	0 to 2.5	0 to 0.5
Fair	0.5 to 1	2.5 to 5	0.5 to 1
Good	1 to 2	5 to 10	1 to 2
Very good	2 plus	10 plus	2 plus

Typical source rock maturation values (from Tissot and Welte, 1984; Peters, 1986; Mukhopadhyay, 1992)

	P.I.*	T_{max} Rock Eval	$\%R_o$# (Kerogen Type II/III)
Immature		<435	<0.45%
Marginally mature	~0.1	435 to 445	~0.45–0.5%
Total oil window			0.55–1.1%
Maximum oil generation zone	~0.1 to 0.4	435 to 470	~0.7–0.9%
Wet gas zone			~1.1–2%
Dry gas zone		>470	>2.0%

* P.I. = $P_1/(P_1 + P_2)$.
** Calculated Rock Eval T_{max} = CDS T_{max} – 50°C.
\# Average vitrinite reflectance % in oil (ne1.517@23°C) @ 546 nm.

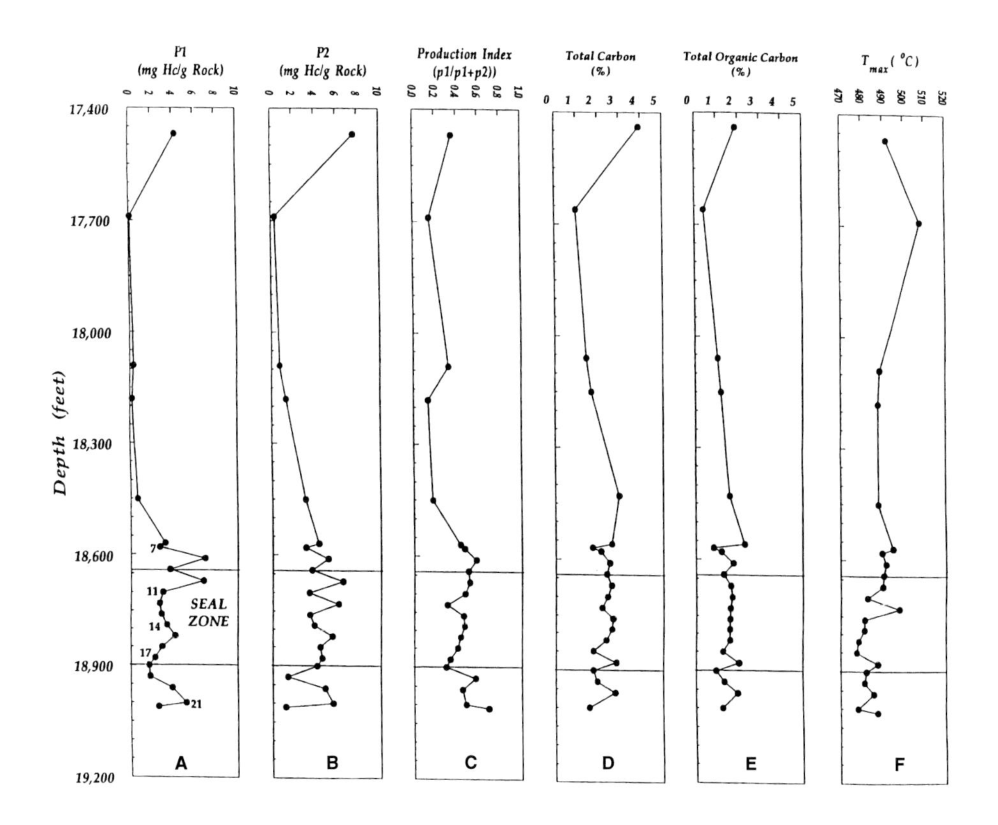

Figure 2. Pyrolysis data across pressure seal, Mix well—Gulf Coast Tuscaloosa trend.

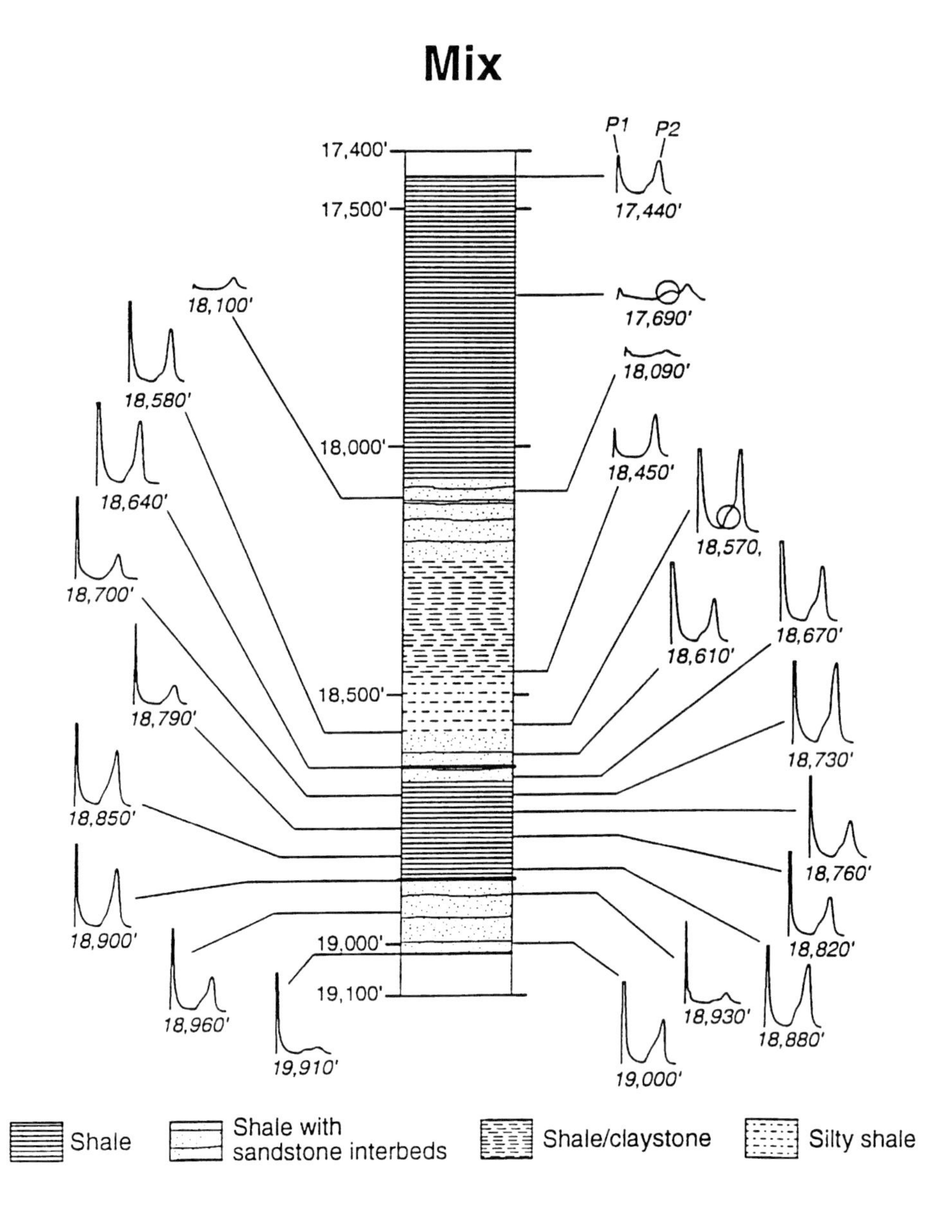

Figure 3. Downhole pyrograms, Mix well. Pressure seal is located between the heavy lines.

particular predominant alkane—all of the nC_7 through nC_{14} hydrocarbons, including the aromatic compounds toluene and xylene, are present.

Unlike the P_1 products, the compositions of P_2 products (Figure 5) are similar, except for depletion of the heavier hydrocarbons nC_{14} and nC_{16} in Mix #14 and Mix #17. The similarity in P_2 products is consistent with the initial cracking of a high molecular weight organic matrix (i.e., asphaltene or kerogen), which is similar for all of the Mix samples (Larter, 1984; Whelan et al., 1980, 1986; Dembicki et al., 1983; Horsfield, 1984).

Pyrolysis Data, Bizette Well

Downhole profiles of TOC and pyrolysis data are shown in Table 1. TOC is 0.7 to 1.2% above the pressure seal zone and increases across and below the seal zone with a maximum value of 4.3% at the top of the seal (18,450 ft, 5624 m). Increases are apparent for P_1, P_2, P_1/TOC, P_2/TOC, and P.I. at the bottom and beneath the seal. P_2 and P_2/TOC also show maxima at the top of the seal zone. P.I. increases to values above 0.4 within and below the seal, probably due to petroleum that has migrated into this silty and sandy shale (Figure 6).

The difference between the generally low amounts of petroleum above the seal and the much larger amounts just above, within, and below is striking in the downhole pyrograms (Figure 6). Increasing light oil (strong P_1) as well as polar and/or heavy oil constituents (shoulder on the front of the P_2 peak) throughout deeper intervals of the Bizette well can be observed in the downhole pyrograms (Figure 6).

The C_7 to C_{16} hydrocarbon compositions of P_1 and P_2 for several Bizette samples from above, below, and within the seal zone were also examined for this well. Unfortunately, sample coverage is not extensive enough to discern whether or not the hydrocarbon patterns observed previously in the Mix well also apply here.

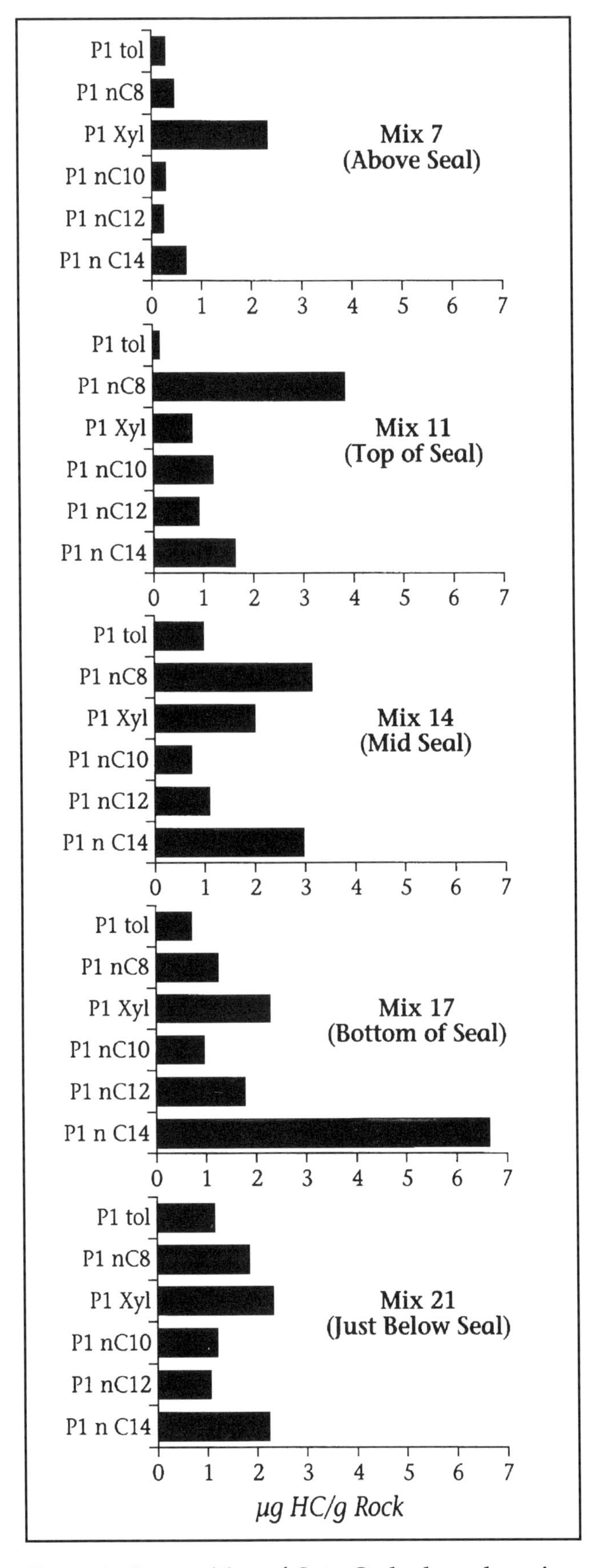

Figure 4. Composition of C_7 to C_{14} hydrocarbons in P_1, Mix well.

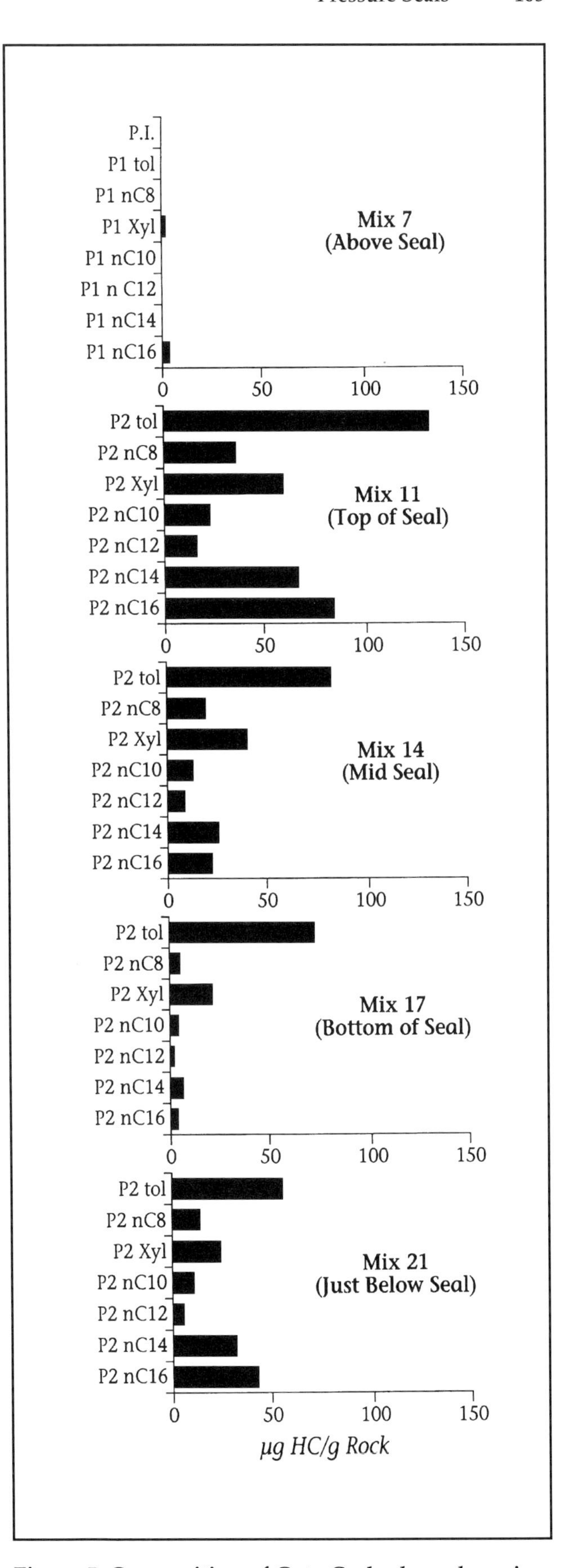

Figure 5. Composition of C_7 to C_{14} hydrocarbons in P_2, Mix well.

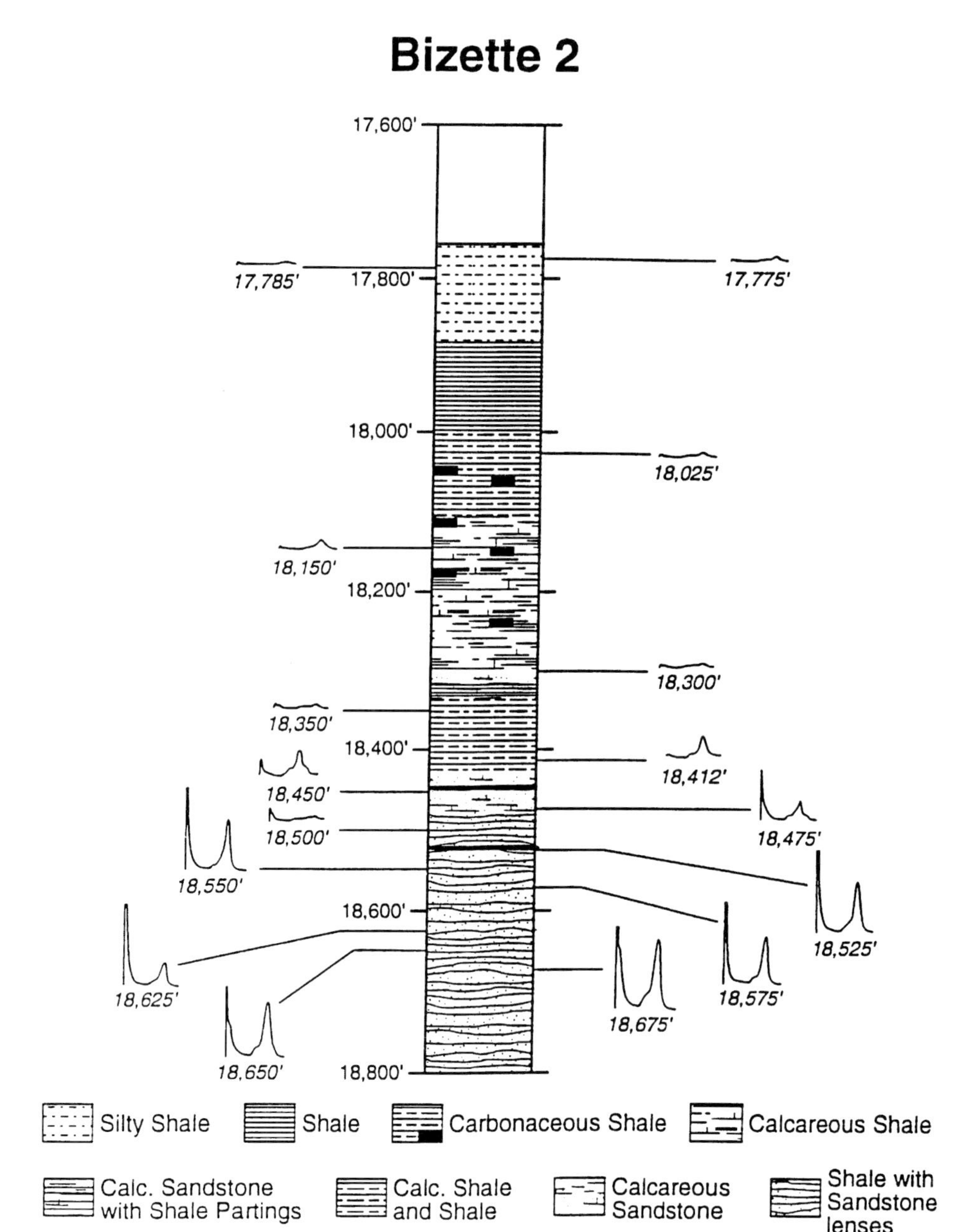

Figure 6. Downhole pyrograms, Bizette 2 well—Gulf Coast Tuscaloosa trend. Pressure seal is located between the heavy lines.

Microscope Observations

Mix Well

Sample Description

Eighteen drill cuttings samples between 17,445 and 19,000 ft (5317 and 5791 m) were prepared for thermal maturation determinations by vitrinite reflectance and for simultaneous petrographic observation. Samples above, within, and below the seal show an intimate association of quartz, carbonate, pyrite, bitumen, asphaltenes, and micrinite (a granular, <1 mm residual product from hydrocarbon generation from bituminite).

The sampled section spans a depth interval of 1560 ft (475 m). The section begins in Cretaceous sedimentary rocks at the top of the upper Tuscaloosa Formation (17,440 ft, 5317 m), as defined by the Bain marker bed from gamma-ray and resistivity logs (Weedman et al. 1992). Repeat formation tester measurements (RFT, trademark of Schlumberger and described in detail in Weedman et al., 1992) show an increase from normally pressured to overpressured strata within the Lower Tuscaloosa Formation as shown in Figure 7 along with the well log data. This pressure transition zone between 18,649 and 18,910 ft (5684 and 5764 m) is interpreted in this paper as a zone containing a permeability barrier to escaping overpressured fluids. This barrier is referred to as a pressure seal.

Within the Upper Tuscaloosa Formation, two microlithotypes are characterized as interbedded thin laminated shales (≈0.5–1.0% TOC) and silty shales (≈<0.5% TOC). Both microlithologies contain terrestri-

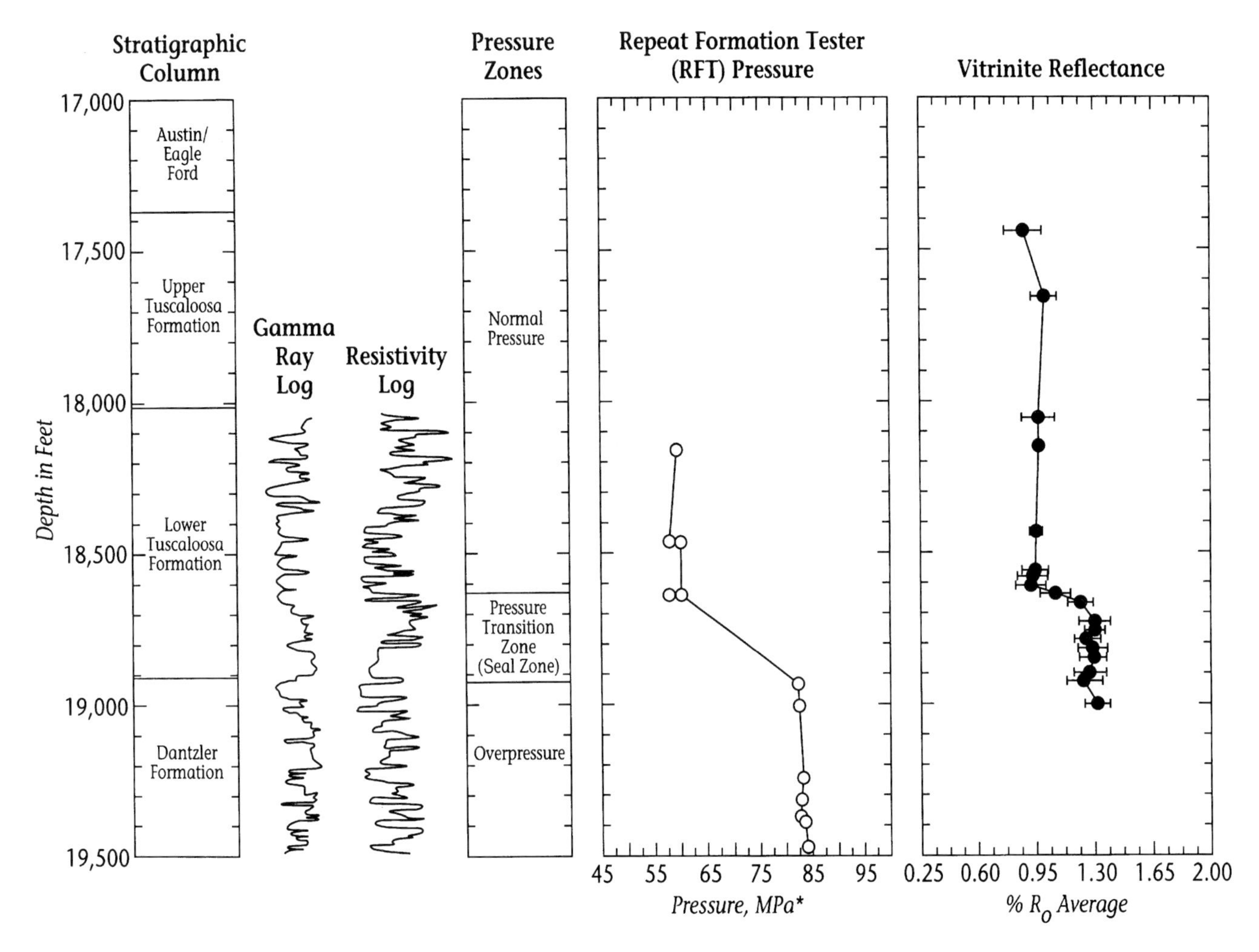

Figure 7. Vitrinite reflectance trend across pressure seal, Gulf Coast Mix well. Typical vitrinite reflectance values for the peak oil generation zone are 0.8 to 1.2%; those for the gas generation window are typically 1.1 to 1.8%. Also shown are pressure and well log data (gamma-ray and resistivity logs).

ally derived organic matter characterized as Type III/IV kerogen (Tissot and Welte, 1984; Mukhopadhyay and Wade, 1990).

Fifteen samples from the lower Tuscaloosa Formation were analyzed spanning a depth interval from 18,060 to 18,930 ft (5505 to 5770 m). These are characterized by interbedded silty shales, siltstones, and laminated organic-rich shales (≈1–2% TOC). This shale is rich in matrix bitumen staining and framboidal pyrite and shows an increase in fine rutile (TiO_2) needles in the mineral matrix with depth. Quartz grains in the silty shales are generally dispersed but occasionally form more tightly packed domains. The organic material in the shales is dominated by terrestrially derived Type III and III/IV solid organic particles. A sandstone at 18,150 ft (5532 m) contains high abundance of dark brown matrix bitumen surrounding subrounded quartz grains. This bitumen fluoresces dark orange-brown in blue light.

Between 18,580 and 18,640 ft (5663 and 5681 m), a well-laminated shale containing high abundances of terrestrially derived Type III organic material (≈2.5% TOC) is interbedded with a lean (≈<0.5% TOC) silty shale containing well-rounded quartz grains that are probably dissociated foraminifera chambers. This shale is heavily stained with dark-brown matrix bitumen and contains abundant fine grains of pyrite evenly dispersed throughout the matrix. The cuttings sample below 18,610 ft (5672 m) is dominated by a fine-grained sandstone. Pyrite infills pit on the surface of many subangular quartz grains. There is an enrichment of coarser phyllosilicate laths of rutile in the matrix of some silty shale cuttings at this depth. Acicular radiating groups (rutile) occur in the interior of quartz grains but are more abundant in the shaly matrix where multiple knee-shaped twins occasionally form wheels. This may be indicative of an igneous source of the grains, although it may be representative of low-grade metamorphism. Between 18,670 and 18,790 ft (5691 and 5727 m), there is little variation in organic facies from predominantly terrestrial Type III woody material. The microlithologies show a similar distribution of laminated shales, silty shales, and fine sandstones. In different cuttings of the silty shale the quartz grain shapes range from subangular to subrounded and in general these grains are poorly sorted and are persistently matrix supported, although a minor siltstone lithology at 18,670 ft (5691 m) contains tightly packed elongated grains (grain supported). Cracks are apparent in the larger quartz grains at 18,730 ft (5709 m) and in a siltstone at 18,760 ft (5718 m) where they are surrounded by smaller quartz grains that are stained with light-brown bitumen that fluoresces a dull orange-brown in blue light.

Three samples between 18,820 and 18,900 ft (5736 and 5761 m) contain cuttings of an unusual coarse-grained silty shale in which the matrix minerals exhibit a very heavy dark-brown–black nonfluorescing bitumen stain. The quartz grains associated with this lithology are cracked. It is difficult to assess whether these cracks are a feature of pressure solution at grain contacts due to the limited information available from a two-dimensional viewing plane. An important microlithologic observation noted from this sample is the presence of cuttings of a tightly packed siltstone in which quartz grains are elongate. This grain packing produces very fine sutures at grain margins that yield an extremely low porosity and could conceivably be considered as a permeability barrier. Although the extent of this lithology is difficult to determine from cuttings samples alone, it may be represented by a finger-shaped drop in the gamma-ray log at ~18,000 ft (5486 m) (Figure 7). An increase in matrix bitumen staining occurs at 18,850 ft (5745 m), and by 18,900 ft (5761 m) there is an abundance of disseminated fine pyrite, micrinite, and heavy bitumen staining of the shaly mineral ground mass.

A poorly laminated shale containing a low to moderate (≈0.5–1% TOC) abundance of terrestrial Type III and Type III/IV organic matter occurs at 18,930 ft (5770 m). There is only a faint matrix bitumen staining in this microlithology. A second microlithology consists of a silty shale that contains subangular grains of quartz dispersed in a shaly matrix. These quartz grains are not cracked.

One sample collected from the Dantzler Formation (19,000 ft, 5791 m) contains a tightly packed silty shale in which the matrix minerals are heavily stained with dark-brown–black bitumen. Small quartz grains (≈10 μm ϕ) also appear to be cracked, but it is impossible to determine, using incident light techniques, whether these cracks have been formed by pressure solution or are due to the provenance of the sedimentary rocks. The organic facies remains terrestrial Type III in this sample.

Thermal Maturity

Figure 7 shows a vitrinite reflectance depth plot for the Mix well and data are shown in Table 1. Vitrinite reflectance values remain fairly constant between 17,440 and 18,640 ft (5317 and 5681 m) at about 0.94% R_o. A maturation jump to 1.20% R_o occurs at 18,730 ft (5709 m). This jump is coincidental with an initial increase in formation pressures at the top of a pressure transition zone as shown in Figure 7. The maturity remains constant at about 1.28% R_o throughout this zone. These maturity determinations together with the organic matter type are consistent with a capacity for the generation of gaseous hydrocarbons.

Summary

The organic facies remains virtually constant throughout the depth interval and can be characterized as terrestrial-derived organic material comprised of woody Type III kerogen and inertinitic Type IV kerogen with a capacity at this thermal maturity for gas generation. The sudden occurrence of rutile and peculiarly cracked quartz surrounded by heavy bitumen staining is coincidental with the pressure transition zone. This observation seems to be consistent with those reported in Weedman et al. (1992) where they describe unusually fractured quartz grains in a quartz-rich matrix with closely spaced pressure-solution seams that are seen only within the pressure transition zone. Further work is in progress to determine if these fractures might be caused by the overpressuring or whether they are an artifact of drilling. In any case, the present work suggests that these fractures, whatever their origin, are fairly characteristic of the pressure transition zone.

Bizette Well

Samples from this well showed lithologies and organic material (Type III/IV) similar to that observed in Mix (Figure 6). Reliable vitrinite data were more difficult to obtain because of the oxidation of the vitrinite. However, available data are shown in Table 1. Solid bitumen asphaltic material was present together with bitumen staining and framboidal pyrite above the seal in this well. Just above the seal, heavy organic staining, a micrinitic residue, indicative that oil generation has occurred, and quartz grains surrounded by calcite and pyrobitumen were present. Kerogen Type III persists into the middle and bottom of the seal, along with occurrences of pyrite, micrinite, secondary cementation, and etching of quartz grains. Asphalt that has passed through the oil window (termed here as pyrobitumen; Hunt, 1979) was evident as 25% of the organic matter in samples from 18,500 to 18,525 ft (5639 to 5646 m) at the bottom of the seal, giving an indication of the higher maturity in comparison to samples above the seal. Thermal maturity is in the range of 0.90 to 1.10% R_o, which is at the end of the oil window.

Iatroscan Results

Mix and Bizette Wells

Both organic petrographic and pyrolysis data showed the presence of asphaltic material in the vicinity of the seal zones. The approximate relative concentrations of this fraction were determined by Iatroscan (Table 1).

Relatively high amounts of asphaltenes are present in all four of the Mix samples examined, one from above, two from within, and one from below the seal. These results are in good agreement with microscopic and pyrolysis results described above which show the presence of asphaltenes as major organic phases in these intervals.

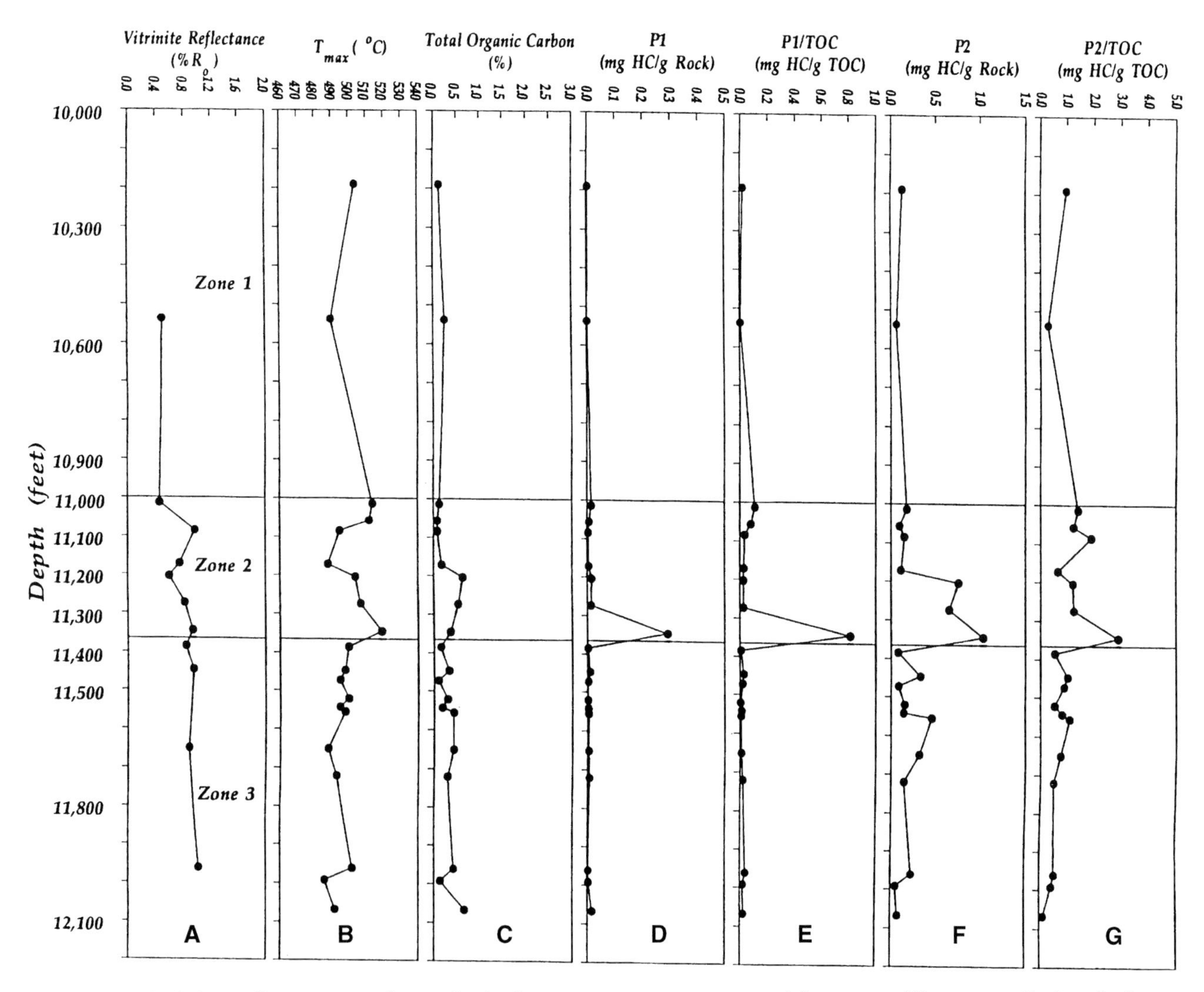

Figure 8. Vitrinite reflectance and pyrolysis data across pressure transition zone, Weaver well, Anadarko basin, Oklahoma.

Six samples from the Bizette well were also analyzed. High amounts of asphaltenes were detected within the seal zone, with lower amounts detected above and below.

Summary of Observations for Weaver Well, Anadarko Basin

Pyrolysis and petrographic microscopic studies were also carried out on much more organic lean core samples from the Weaver well, Anadarko basin, Oklahoma (Figure 1). The cores came from a much weaker pressure transition zone than that described above for the Gulf Coast wells. The downhole profiles of vitrinite reflectance and pyrolysis data (T_{max}, P_1, P_2, P_1/TOC, P_2/TOC, and P.I.) are shown in Table 1 and Figure 8, with zone 2 being the pressure transition zone and approximate position of the postulated upper seal (Tigert and Al-Shaieb, 1990). The values of TOC, P_1, P_2, and P.I. are all much lower than for Mix and Bizette. Several of these parameters show very erratic behavior, particularly within the pressure transition zone. All parameters related to P_1, P_2, or TOC show maxima within this interval.

Downhole vitrinite reflectances for Weaver are erratic because of changing organic facies and reworked particles. However, thermal maturity generally increases (from R_o 0.46 to 0.82%) through zone 2. These values cover the beginning to the middle of the oil window (see data from Peters, 1986, at the bottom of Table 1) over a relatively short depth interval of 400 ft. T_{max} goes through a minimum at ≈11,200 ft, as was also the case with the Mix well within the pressure transition zone, consistent with the presence of asphaltenes within the zone of maximum maturity within the seal zone. The vitrinite population examined at Weaver was more heterogeneous than was the case for the Mix well, so there is the possibility that the R_o changes were due to differences in organic matter type rather than maturation. The organic leanness coupled with the generally poorer quality of the vitrinite and the higher abundance of reworked phyto-

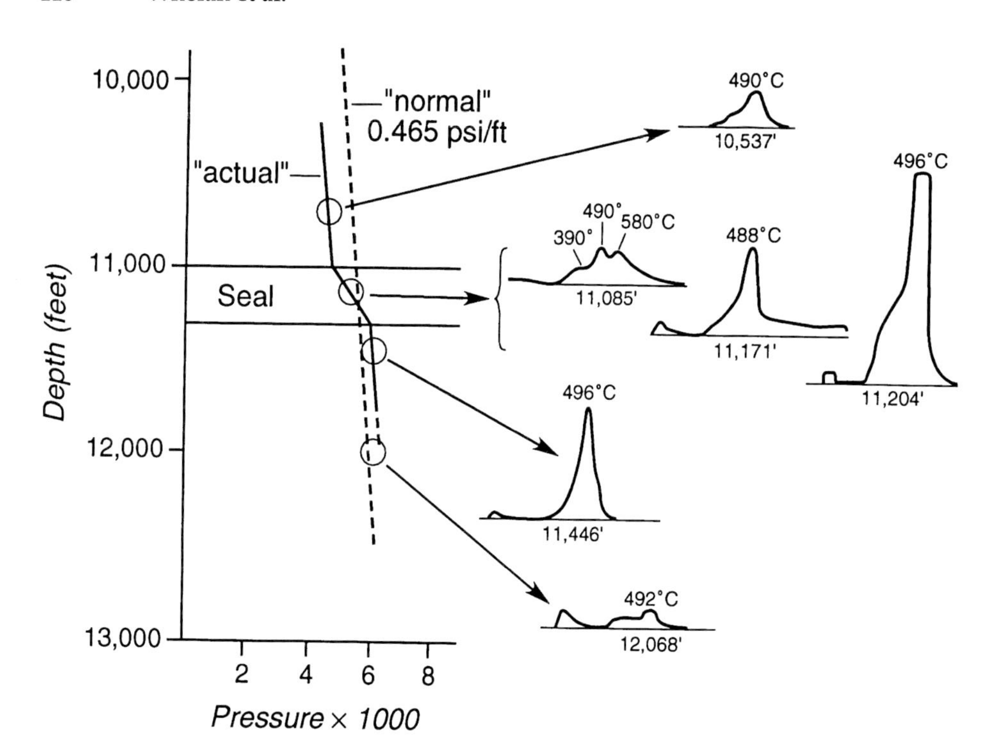

Figure 9. Representative pyrograms from across pressure transition zone, Weaver well, Anadarko basin, Oklahoma.

clasts have all affected the vitrinite measurements, producing greater scatter in the vitrinite determinations.

In spite of the much leaner nature of the Weaver samples, several pyrolysis parameters reflect zones containing asphaltene or heavy oil within the seal zone: the low values of T_{max} within the seal in zones of increasing R_o values (Figure 8) and the low-temperature lobe on many of the P_2 peaks (Figure 9). The lithology and organic petrography also show the close association between organic and inorganic phases, similar to those observed for the seal zones of Mix and Bizette wells. For example, in some intervals in the lower part of the pressure transition zone, pyrite that is heavily bitumen stained represents about 70% of the matrix mineral ground mass which surrounds carbonate grains and completely occupies the porosity in other clasts. At 11,204 ft (3415 m) in this well a fine sandstone contains rounded quartz grains that show an interesting inorganic and organic relationship. An organic bitumen directly coats the quartz grain margin that was encased by a subsequent phase of secondary silica overgrowths which trapped the bitumen next to the grain boundary. This relationship indicates hydrocarbons had entered these intervals prior to secondary cementation. These observations are similar to those made in deeper parts of the pressure seal in the Mix well.

Within the pressure transition zone (zone 2 in Figure 8), samples from 11,014 to 11,274 ft (3357 to 3436 m) are similar, have low to moderate organic contents (0.1 to 0.6% TOC), and have a shale lithology that becomes progressively more silty and calcite cemented with depth. There is also an increase in silicate dissolution at grain boundaries and precipitation of calcite as a cement with depth in fine sand to silty shale, but especially at 11,344 ft (3458 m). This sample contains considerable amounts of fine hydrocarbon wisps surrounding quartz grains in the calcite-cemented zones. Fine pyrite and matrix bitumen staining are closely associated and increase slightly in abundance with depth in this well. Pyrolysis P_1 and P.I. values are all low for this interval, however, increasing to somewhat higher values at the bottom of the pressure transition zone at 11,344 ft (3458 m). These changes occur within the zone showing some vitrinite reflectance change (Table 1 and Figure 8). Based on the microscopic and pyrolysis data, it seems probable that most of the relatively low amounts of TOC in zone 2 is present as heavy asphaltic bitumen.

Zone 3, below the pressure transition zone, is characterized as an organic lean interval from about 11,384 to 11,543 ft (3470 to 3519 m). Abundant fine pyrite is observed throughout this zone but is particularly abundant in samples from 11,446 to 11,650 ft (3489 to 3551 m). There is an increase in organic content by 11,720 ft (3572 m) that is dominated by pyrobitumen stringers closely associated with pyrite in a shaly claystone. Sedimentary rocks then become progressively more pyrobitumen dominated with depth to 11,921 ft (3634 m). This sample shows a moderate degree and constant intensity of bitumen staining and fine organic matter coatings of grains in a calcareous silty sand. In deeper intervals below 11,921 ft (3634 m), samples contain fine-grained calcite-cemented quartz sandstones but are organically lean with very low traces of matrix bitumen.

DISCUSSION

The most important observation that must be accommodated by any mechanism explaining the formation and maintenance of pressure seals over geologic time is that the pressure seal zones must have extremely low permeability. Most estimations, including those of Iverson et al., this volume, show that only the most impermeable known rock types, such as unfractured metamorphic and igneous rocks or shales, even approach the required permeabilities. Therefore, it appears that some agent in addition to the rock must be contributing to these low permeabilities. We propose that sediment organic matter, particularly asphaltic materials and gas, are important components in forming such an impermeable pressure seal via the processes described below. We also propose that the observations about the nature of seals and the organic materials associated with them are giving important clues about the nature of the pressure sealing mechanism. These are discussed below.

1. Zoning of hydrocarbons has been observed within the pressure transition zone as discussed by Vandenbroucke and Durand (1983) for the Mahakam Delta and shown in Figure 4 for the Mix well. In both cases, lighter hydrocarbons tend to focus at the top of the seal, while heavier components predominate at the bottom of the seal. In addition, the GC patterns of the Mix P_2 pyrolysis products in comparison to P_1 show that migration phenomena, rather than changes in in situ organic source, are responsible for the P_1 C_8 to C_{14} hydrocarbon distributions. Within and below the seal, the P_2 GC patterns for all of the Mix samples are very similar. If the P_1 hydrocarbons were derived from a localized in situ source of kerogen or asphaltene, then the P_1 GC compositions should also be similar. In fact, the small changes in P_2 GC patterns are contrary to the P_1 patterns. Figure 5 shows enhancement of P_2 C_{14} and C_{16} products at the top of the seal (e.g., Mix 11) and some depletion at the bottom of the seal, just the opposite of the trends observed for the P_1 aliphatic hydrocarbon compositions (Figure 4). Therefore, migration processes must be invoked as the cause of the P_1 compositional gradients across the seal.

2. A sharp increase in thermal maturity gradient has been observed in pressure transition zones from the Gulf Coast Mix well (Figure 7). A weaker maturity increase also occurs for the Oklahoma Anadarko Weaver well (Figure 8) and possibly also in a Nova Scotia Sable Island Venture Field well (Well H-22 in Jansa and Norguera, 1990; Eglinton and Whelan, unpublished results). In addition, other microscopic differences between the two samples at the top and bottom of the Mix well seal zone are also consistent with a large maturity difference occurring over this very short depth interval. Thus, the 18,640 ft (5681 m) sample appears to contain bitumens typical of the oil window, while the sample at 18,850 ft (5745 m) shows higher maturity pyrobitumen, or bitumens that have passed through the oil window. In the Gulf Coast Bizette well, minimal fluorescence of organics is observed above the seal at 17,750 ft (5410 m), indicative of maturity around 0.9% R_o. However, at 18,425 ft (5616 m) and below, pyrobitumen is observed, diagnostic of higher maturities. Therefore, as with Mix, it appears that fairly large changes in maturity have occurred across the relatively short distance of seal zone. These maturational changes are accompanied by increasing amounts of carbonate cementation within and below the seal zone.

3. Precipitated asphalts and pyrobitumens are commonly an important part of the material in the pressure transition zone as shown by pyrolysis, petrographic microscopic results, and Iatroscan, as discussed above and shown in Figures 2 and 6 and in Table 1. This observation applies to the strong pressure transition zone in two Gulf Coast wells as well as to the much weaker pressure transition zone in Weaver well of the Anadarko basin, which has generally low TOC (Table 1). For example, one sample from the Weaver core (11,274 ft, 3436 m) shows extremely fine "sutures" infilled with bitumen giving the appearance of a grain coating or filling (Eglinton and Whelan, in preparation). Samples from Weaver differ from the Mix well in that their hydrocarbon phases are discontinuous. However, in all three wells, these organic precipitates are closely associated with secondary inorganic phases including carbonate cements and pyrite.

4. In both the Alaskan North Slope and in the Gulf Coast, some relatively low TOC rocks (1.5% TOC or less) appear to be producing and expelling oil (Huc and Hunt, 1980; Whelan et al., 1986), contrary to current wisdom that petroleum source rocks must contain enough TOC to form a separate oil phase before primary petroleum expulsion and migration can occur (Lewan, 1987; Cooles et al., 1986; Durand, 1988).

We propose that these observations can be explained as the consequence of: (1) the generation and migration of a separate gas phase in the deep subsurface; (2) the high solubility of oil in gas at high pressure; (3) the pressure dependence of the solubility of oil in gas; and (4) the low thermal conductivity of oil and gas.

Figure 10 shows an estimate of types and amounts of breakdown products of a typical Type III kerogen as it is heated with increasing burial along the gradient in

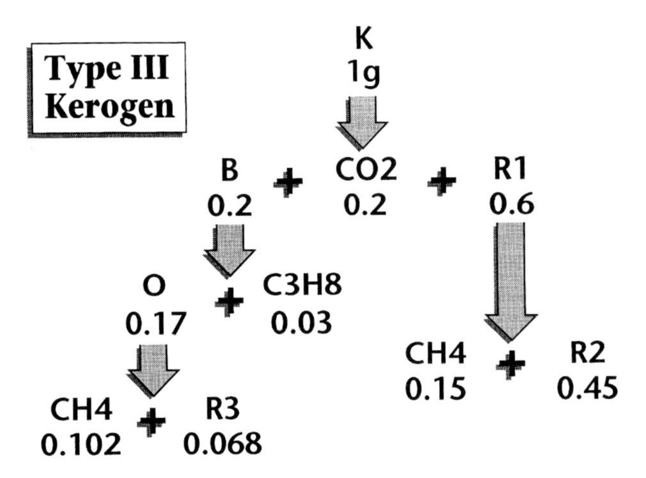

Figure 10. Initial estimation of breakdown of 1 g of Type III kerogen.

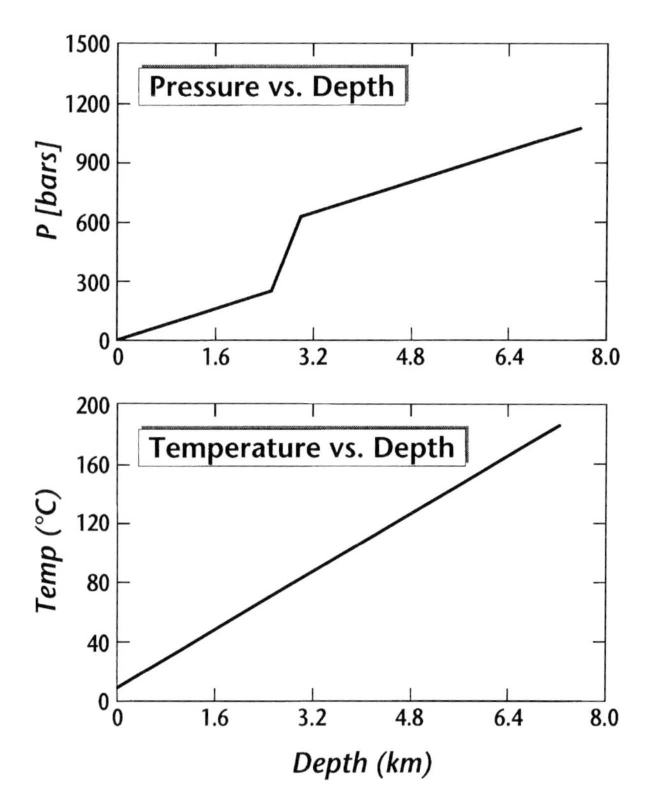

Figure 11. Pressure vs. depth and temperature vs. depth curves used in Figure 12.

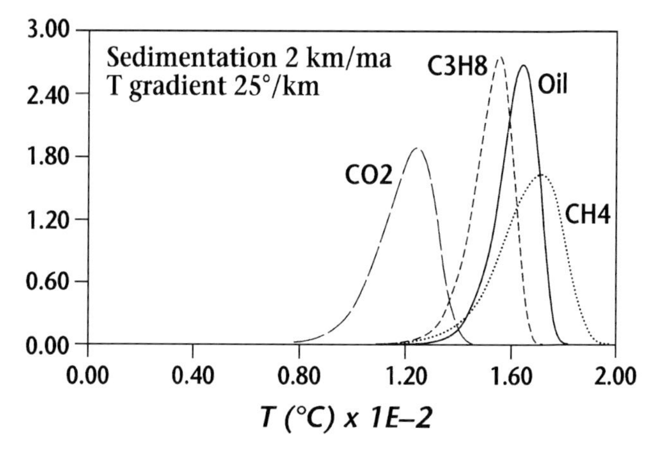

Figure 12. Calculated evolution of gas and oil as a function of depth, using breakdown scheme in Figure 12 and activation energies from Hunt et al. (1991) and Hunt and Hennet (1992). The Y axis represents the fraction of reaction that has gone to completion, $(dm_{ii}/m_i)/dt$, where m_{ii} is the ith product and t is time.

Figure 11. The kerogen maturation scheme in Figures 10 and 12 is based on experimental data from Ungerer et al. (1981, 1988) and data summarized in Hunt and Hennet (1992). The products were confined to dry gas, methane; one typical wet gas component, propane; and oil (Figure 10). It was further assumed that with further burial, residue R_1 can undergo further cracking to 0.15 g methane and 0.068 mg of a pyrobitumen residue, R_3. Reasonable activation energies and frequency factors for each of these reactions were estimated from the literature, as summarized in Hunt et al. (1991) and Hunt and Hennet (1992).

The first point we would make from Figures 10 and 12 is that even very low TOC rocks produce enough methane to saturate basin pore waters and produce a separate gas phase. Bonham (1978) shows that the solubility of methane in the overpressured parts of basins is ≈3000 to 14,000 ppm. But 0.2% TOC generates over 10,000 ppm methane in 10% porosity sediments of density 2 g/cc; 0.4 weight percent TOC sediments would generate 20,000 ppm methane. Methane that is generated in excess of that which can be dissolved in pore water will form a separate gas phase. The portion of the TOC value assumed to contribute separate phase reaction gas in Figure 13 is 0.2%. The total TOC of the sediments would be about 0.4%. The approximate average TOC value for all open-ocean sediments worldwide is about 0.2% (summarized in Calvert and Pedersen, 1992), a value about ten times lower than typically found for even a marginal quality petroleum source rock. But 0.4% TOC is typical of many intervals of the Weaver well (Figure 8 and Table 1). This value of TOC thus represents the minimum that could typically be expected from most rocks in most sedimentary basins, even those which are too organically lean to be considered as petroleum source rocks.

If we assume for the moment that all the gases in Figure 12 go into a separate gas phase and escape vertically, the vertical flux of gases out of the basin due to maturation of 0.2% kerogen sediments buried at 2 km/m.y. is shown in Figure 13. The flux was calculated for CH_4 and oil at the pressures and temperatures at the top of overpressure using the Behar et al. (1985) equation of state. Figure 13 shows that a flux of >1.4 liter/cm^2/ma of methane could be expected into the base of the oil window in a basin with 0.2 wt% TOC subsiding at 2 km/m.y.

The reason that this flux of CH_4 is interesting is that the solubility of oil-in-methane at pressures greater than 200 bars is very large. Figure 14 shows some of the solubility of oil in methane data found in experiments reported by Price et al. (1983). Referring to the solubility of oil in methane is misleading in one sense because oil and methane form a single mixed phase at the pressures and temperatures encountered in the deep parts of a basin. This single phase crosses a bubble curve at which a separate gas phase separates at about the depths at which pressure seals form. This process and its consequences will be discussed in a separate paper. Deep kerogen maturation produces a separate gas phase that becomes a gas/oil mixture as it dissolves oil generated higher in the basin. In the sense that this oil is dissolved in gas streaming from greater depths, the oil can be considered to be soluble in the gas.

The amount of oil that can be dissolved by methane along the P-T path of Figure 11 was measured directly by Price et al. (1983). Those data were

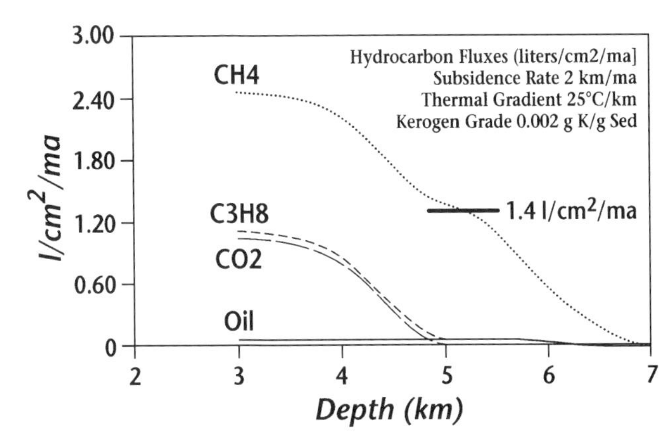

Figure 13. Calculated fluxes of gas and oil as a function of depth using kerogen breakdown scheme in Figure 11 and activation energies from Hunt et al. (1991) and Hunt and Hennet (1992). Volumes calculated at 87.5°C and 290 bars. ρCH4 = ρCO2 = 0.16, ρC3H8 = 0.49.

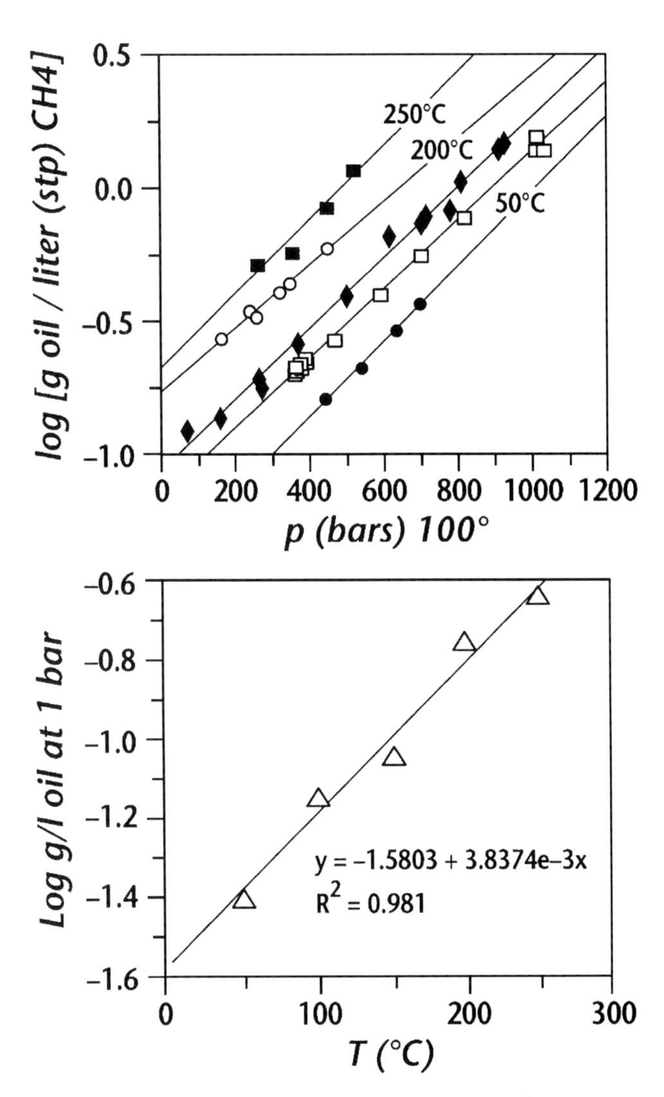

Figure 14. Solubility of oil in gaseous methane at P > 200 bar. Data from Price et al. (1983). Log Coil = –1.58 + 3.837 × 103 T (°C) + 1.3 × 10–3 P (bars).

used to obtain the oil-in-methane "solubility" equation in Figure 14 and the curve shown in Figure 15. The curve is dashed below the depth at which the methane generated from a sediment of any particular kerogen grade (i.e., type and richness) can dissolve all of the oil generated in sediments of the same grade. Since the generative ratio is 0.67 g oil/g CH_4 (Figure 10), and CH_4 has a density of 0.66 g CH_4/liter at 25°C and 1 bar, the curve is dashed to: (0.67)(0.66) = 0.44 g oil/liter methane at STP. Price's experiments were carried out using a Spindle Top oil that is generally similar to oils from many other basins.

Figure 15 shows that gas streaming through a basin can mobilize oil upward but that there is a steep drop in the solubility of oil across the pressure transition zone where gas exsolves from the gas/oil mixture to produce separated oil and gas phases. Oil transported by the gas will be dumped out of the gas/oil phase going through the pressure transition at the top of the seal zone and become available to aid in the plugging process. Such a precipitation process is consistent with our observation of heavy oil, asphalt, or pyrobitumen in and around every seal zone observed to date. Note that in our calculations, the generative ratio of oil to gas is 0.44 g oil/liter CH_4, STP, so that at a uniform weight percent kerogen content in the sediments, all oil could be dissolved and moved upward to the end of the dashed line in Figure 15. The transport capabilities depend on the generative ratio and are the same for all kerogen concentrations. Thus, in either very organic lean rock, such as assumed in these calculations, or in very organic rich rocks, enough methane would be produced to carry oil generated below the seal zone up into the pressure transition zone.

The data of Price (1983) can also be used to estimate the composition of oil precipitated across selected 0.2 km depth intervals along the P-T curves of Figure 11, as shown in Figure 16. The average composition of oil dropped out at shallower depth above the seal, e.g. 2.2 km, has a maxima in average carbon numbers in the range of C_{12} to C_{18}, while the oils precipitated between 3 and 6 km (in and below the seal) tend to have more of the heavier carbon numbers. This change corresponds to observations in the Mix well (Figure 4) as well as in the pressure transition zone of a Mahakam Delta well (Vandenbroucke and Durand, 1981), where lighter hydrocarbon components predominate at the top of the seal while heavier components predominate at the bottom.

The oil precipitated within the pressure transition zone will also reduce the thermal conductivity of sediments, as shown in Figure 17. The points represent experimental thermal conductivities measured for quartz sand filled with air, oil, or brine with the lines representing the theoretical curves calculated by Cathles (unpublished results) based on series and parallel combinations of pore spaces filled with oil and/or air using fabric theory (Jowett et al., 1993). Both the experimental results (points on Figure 17) and theoretical calculations (lines and large circles)

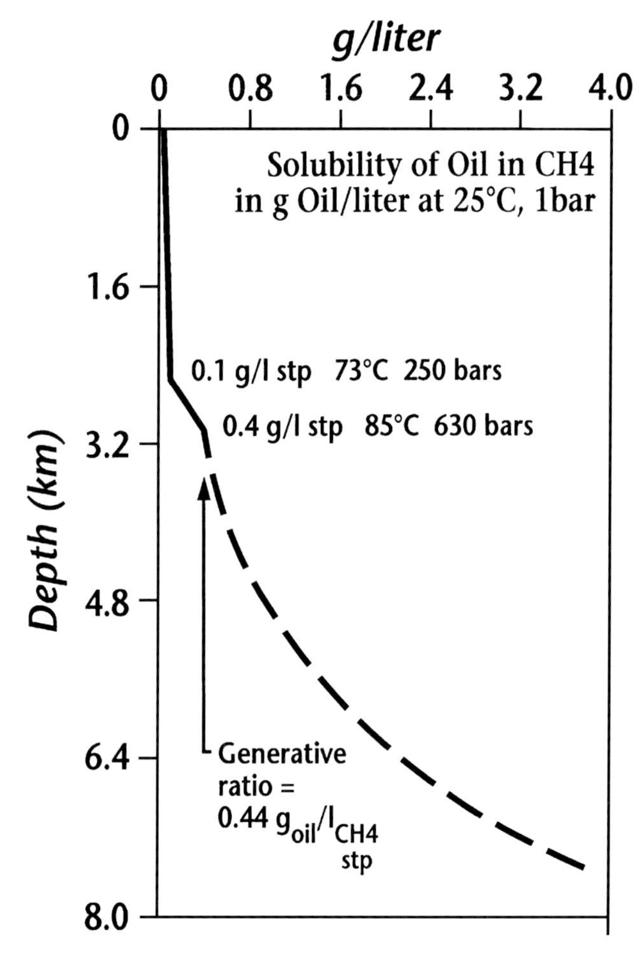

Figure 15. Methane in gas phase will mobilize oil to near top of geopressure.

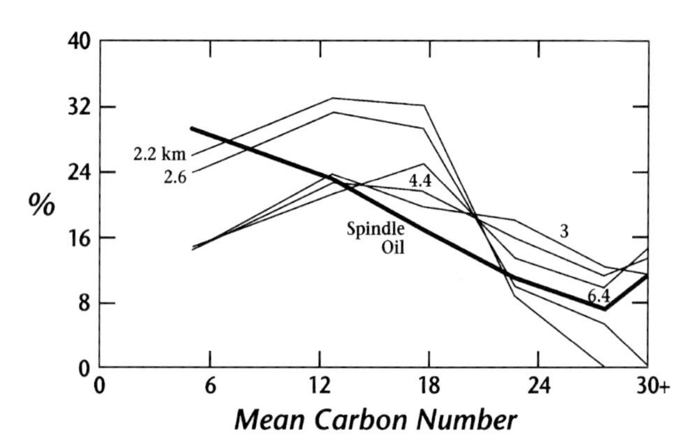

Figure 16. Expected composition of condensate oil dropped out of gas phase in passing depth indicated. Composition of start in Spindle Oil is shown by heavy line. Note predominance of lower carbon numbers in going from shallower depths. (Based on data from Price et al., 1983.)

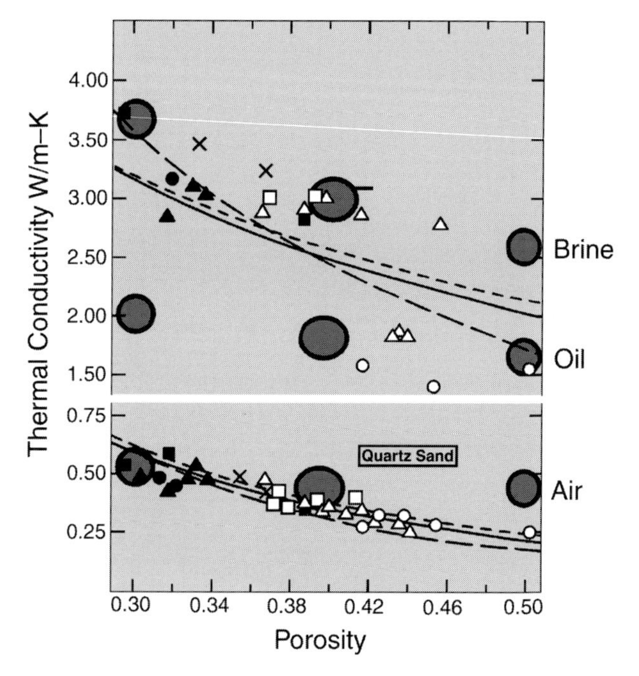

Figure 17. Calculated and experimentally observed effect of oil vs water in sediment pores on thermal conductivity.

show that filling pores with oil can halve the thermal conductivity observed when pores are filled with brine. Note that the presence of any free gas would decrease the thermal conductivity even further. The decrease in thermal conductivity across the pressure transition zone is consistent with the higher thermal gradient in the pressure transition zone observed both by Jones (1975) and Engelder et al. (unpublished results) as part of this project, as well as the sharp change in vitrinite maturity gradient observed at the top of the pressure transition zones of the Gulf Coast Mix well (Figure 7), a weaker gradient for the Anadarko basin Weaver well (Figure 9), and possibly also for the Nova Scotia Venture Field H-22 well (Whelan and Eglinton, unpublished results).

The precipitation of polar and hydrocarbon constituents can plug intervals that already have low permeability due to their lithology, inorganic cementation, extreme compaction, or other factors. The plugging is a combination of heavy oil condensation and capillary effects that will be discussed in a subsequent publication. Inorganic mineral alteration constituents have a very low solubility in water (<0.1 wt.%) compared to the solubility of hydrocarbons in methane (66 wt.%) so that this oil condensation provides a comparatively effective mechanism for healing ruptured seals. The composition of the condensed oil through the pressure transition zone may record the status of gas and oil maturation and migration deeper in the basin.

CONCLUSIONS

The purpose of this research was to develop specific organic analyses that could aid in finding and characterizing pressure seals, particularly in shales where RFT measurements are not possible. In addition, we wished to gain some preliminary information on how sedimentary rock organic matter might be involved in the formation and maintenance of pressure seals. Summarized below are the findings of this work with respect to each of these.

1. Determination of organic geochemical signatures diagnostic of pressure compartments and seals, as distinguished from those of similar normally pressured sedimentary rocks: (a) A zone of increase in vitrinite reflectance at the top of the pressure seal zone; organic petrographic evidence of bitumen maturation within the same zone. (b) Characteristic patterns of P_1 C_7–C_{16} hydrocarbon compositions above, within, and below the seal zone, including a progression of aliphatic n-alkanes from a predominance of the lighter nC_8 at the top of the seal to the heavier nC_{14} at the bottom; a piling up of all C_7 to C_{16} hydrocarbons just below the seal; and possibly a predominance of the aromatic compounds xylene and toluene above the seal. (c) A predominance of asphaltene and pyrobitumen over less-polar aliphatic and aromatic hydrocarbons in the vicinity of the seal zone. (d) Close association of asphaltene/micrinite with carbonate cementation, silicate dissolution and precipitation, and, sometimes, epigenetic pyrite phases in the vicinity of the seal zone. (e) Some preliminary evidence of microfractured quartz within and below seal zones.
2. Preliminary information on the effects pressure compartments and seals might have on oil and gas distributions in sedimentary rocks: see (b) and (c) above.
3. Preliminary research on how organic maturation and migration processes might be contributing to pressuring and sealing mechanisms (versus being parallel inorganic processes affected by the same time-temperature regime).

A plugging hypothesis is presented in which oil dissolved in upward-streaming gas is precipitated by the pressure drop across the seal zone. The precipitated bitumens plus the associated gas acting together with inorganic precipitation/cementation are postulated to produce a seal. This hypothesis accounts for the very low permeabilities observed in seals, the characteristic hydrocarbon distributions across the seals, the presence of a high proportion of heavy residual oil (i.e., asphalt and pyrobitumen) within and adjacent to the seal zone, and the increases in maturity observed within the seal. The hypothesis also takes into account the ability of the upward-streaming gas to aid in sealing. In cases of seal rupture, precipitation of heavy oil components in the seal would aid in subsequently healing the ruptures. The very tight association of carbonate cements, pyrite, asphaltene, and micrinite is strongly suggestive that the interaction of organic and inorganic reactions are important in seal formation and maintenance. Organic bitumen was observed to coat the quartz grains that were encased by a subsequent phase of secondary silica overgrowths in some samples from both the Mix and Weaver wells. This relationship indicates hydrocarbons had entered these intervals prior to secondary cementation. The presence of both mineral cements (calcite and silica) and organic asphalt-micrinite phases in the vicinity of seals suggests that inorganic cemented intervals interspersed with the organic matter might produce a seal more impermeable to passage of both aqueous and hydrocarbon fluids than either acting alone.

At the Mix well, the bottom of the seal zone occurs within the gas maturation window so that gas generation and pressure buildup could be contributing to the overpressuring of these sedimentary rocks, as proposed by Hedberg (1974) and discussed more recently by Barker (1987 and 1990) with the sealing being aided by the capillary gas pressure as proposed by Iverson et al., this volume.

Preliminary calculations show that very low TOC rocks (possibly as low as 0.2 to 0.5% TOC) may be capable of generating sufficient methane to form a separate hydrocarbon phase at depth. Because almost all sedimentary rocks, including nonsource rock open-ocean sedimentary rocks, contain at least these low amounts of TOC, the "hydrocarbon plugging" mechanism of seal formation could occur very widely both in petroleum and nonpetroleum source rock. An example of such a "low TOC" seal studied in this research may be a weak pressure transition zone in the Anadarko basin Weaver well.

ACKNOWLEDGMENTS

We thank Jeff Seewald, John Hunt, and Suzanne Weedman for careful reading of the manuscript and helpful discussions; Mary Zawoysky for editorial help; Carl Johnson for running GCMS analyses; Bob Nelson and Mary Zawoysky for carbon analyses; and our collaborators at Pennsylvania State University and Oklahoma State University, particularly Susan Weedman, Terry Engelder, and Vennessa Tigert for providing samples and background information as well as helpful discussion and encouragement. Help from Amoco, particularly John Bradley and Dave Powley, in initiating this research and in providing samples and background information is gratefully acknowledged. We thank the editor, Peter Ortoleva, for his help and patience during the preparation of this manuscript, and three anonymous reviewers for their careful reading of the manuscript and helpful suggestions. This work was supported by the Gas Research Institute, Contract Nos. 5088-260-1746 and 5091-260-2298. Woods Hole Oceanographic Institution Contribution No. 7984.

REFERENCES CITED

Barker, C., 1987, Development of abnormal and subnormal pressures in reservoirs containing bacterially generated gas: American Association of Petroleum Geologists Bulletin, v. 71, p. 1404–1413.

Barker, C., 1990, Calculated volume and pressure changes during the thermal cracking of oil to gas in reservoirs: American Association of Petroleum Geologists Bulletin, v. 74, p. 1254–1261.

Behar, E., R. Simonet, and E. Rauzy, 1985, A new non-

cubic equation of state: Fluid Phase Equilibria, v. 21, p. 237–255.
Bonham, L.C., 1978, Solubility of methane in water at elevated temperatures and pressures: American Association of Petroleum Geologists Bulletin, v. 62, p. 2478–2488.
Bradley, J.S., 1975, Abnormal formation pressure: American Association of Petroleum Geologists Bulletin, v. 59, p. 957–973.
Calvert, S.E. and T.F. Pedersen, 1992, Organic carbon accumulation and preservation in marine sediments: how is anoxia important? *in* J.K. Whelan and J.W. Farrington, eds., Organic Matter: Productivity, Accumulation, and Preservation in Recent and Ancient Sediments: New York, Columbia University Press, p. 231–263.
Clementz, D.M., 1979, Effect of oil and bitumen saturation on source-rock pyrolysis: American Association of Petroleum Geologists Bulletin, v. 63, p. 2227–2232.
Cooles, G.P., A.S. Mackenzie, and T.M. Quigley, 1986, Calculation of petroleum masses generated and expelled from source rocks: Organic Geochemistry, v. 10, p 235–245.
Creaney, S., 1980, The organic petrology of the Upper Cretaceous Boundary Creek formation, Beauford-Mackenzie Basin: Bulletin of Canadian Petroleum Geology, v. 28, p.112–119.
Dembicki, H., Jr., B. Horsfield, and T.T.Y. Ho, 1983. Source rock evaluation by pyrolysis-gas chromatography: American Association of Petroleum Geologists Bulletin, v. 67, p 1094–1103.
Durand, B., 1988, Understanding of hydrocarbon migration in sedimentary basins (present state of knowledge): Organic Geochemistry, v. 13, p 445–459.
Espitalié, J., 1986, Use of T_{max} as a maturation index for different types of organic matter. Comparison with vitrinite reflectance, *in* J. Burrus, ed., Thermal Modeling in Sedimentary Basins: Paris, Editions Technip, p. 475–496.
Espitalié, J., F. Marquis, and I. Barsony, 1984, Geochemical logging, *in* K.J. Voorhees, ed., Analytical Pyrolysis: Boston, Butterworths, p. 276–304.
Espitalié, J., J.L. LaPorte, M., Madec, F. Marquis, P. Leplat, J. Paulet, and A. Boutefeu, 1977, Méthode rapide de caractérisation des roches méres de leur potentiel pétrolier et de leur degré d'evolution: Reviews de l'Institute Francais du Pétrol., v. 32, p. 23–42.
Hedberg, H., 1974, Relation of methane generation to undercompacted shales, shale diapirs, and mud volcanoes: American Association of Petroleum Geologists Bulletin, v. 58, p. 661–673.
Horsfield, B., 1984, Pyrolysis studies and petroleum exploration, *in* J. Brooks and D. Welte, eds., Advances in Petroleum Geochemistry, v. 1: London, Academic Press, p. 247–298.
Huc, A.Y., and J.M. Hunt, 1980, Generation and migration of hydrocarbons in offshore South Texas Gulf Coast sediments: Geochimica et Cosmochimica Acta, v. 44, p. 1081–1089.
Hunt, J.M., 1979, Petroleum Geochemistry and Geology: San Francisco, Freeman, 617 pp.
Hunt, J.M., 1985, Generation and migration of light hydrocarbons: Science, v. 226, p. 1265–1270.
Hunt, J.M., 1990, Generation and migration of petroleum from abnormally pressured fluid compartments: American Association of Petroleum Geologists Bulletin, v. 74, p. 1–12.
Hunt, J.M., and R. J-C Hennet, 1992, Modeling petroleum generation in sedimentary basins, *in* J.K. Whelan and J.W. Farrington, eds., Organic Matter: Productivity, Accumulation, and Preservation in Recent and Ancient Sediments: New York, Columbia University Press, p. 20–51.
Hunt, J.M., M.D. Lewan, and R. J-C. Hennet, 1991, Modeling oil generation with time-temperature index graphs based on the Arrhenius equation: American Association of Petroleum Geologists Bulletin, v. 75, p. 795–807.
ICCP (International Committee for Coal Petrology), 1963, 1971, 1975, International Handbook of Coal Petrography, Paris, CNRS.
Jacob, H., 1989, Classification, structure, genesis and practical importance of natural solid oil bitumen ("migrabitumen"): International Journal of Coal Geology, v. 11, p. 65–79.
Jansa, L.F., and Norguera, V., 1990, Geology and diagenetic history of overpressured sandstone reservoirs, Venture Gas Field, offshore Nova Scotia, Canada: American Association of Petroleum Geologists Bulletin, v. 74, p. 1640–1658.
Jones, P.H., 1975, Geothermal and hydrocarbon regions, Northern Gulf of Mexico Basin, *in* M.H. Dorfrom and R.U. Deller, eds., Proceedings First Geopressured Energy Conference, Center for Energy Studies, University of Texas at Austin, p. 15–97.
Jowett, E.C., L.M. Cathles, and B.W. Davis, 1993, Predicting depths of gypsum dehydration in evaporitic sedimentary basins: AAPG Bulletin, v. 77, p. 402–413.
Karlsen, D., and S.R. Larter, 1991, Analysis of petroleum fractions by TLC-FID: applications to petroleum reservoir description: Organic Geochemistry, v. 17, p. 603–617.
Larter, S.R., 1984, Application of analytical pyrolysis techniques to kerogen characterization and fossil fuel exploration/exploitation, *in* K.J. Voorhees, ed., Analytical Pyrolysis Techniques and Explorations: London, Butterworths, p. 212–275.
Lewan, M.D., 1987, Petrographic study of primary petroleum migration in the Woodford Shale and related rock units, *in* B. Doligez, ed., Migration of Hydrocarbons in Sedimentary Basins: Paris, Edition Technip, p. 113–130.
Leythaeuser, D., H.W. Hagemann, A. Hollerbach, and R.G. Schaefer, 1980, Hydrocarbon generation in source beds as a function of type and maturation of their organic matter: a mass balance approach: Tenth World Petroleum Congress Proc., v. 2, p. 31–41.
Mukhopadhyay, P.K., 1992, Maturation of organic

matter as revealed by microscopic methods: applications and limitations of vitrinite reflectance, and continuous spectral and pulsed laser fluorescence spectroscopy, *in* K.H. Wolf and G.V. Chilingarian, eds., Diagenesis, III. Developments in Sedimentology: Elsevier Science Publishers, New York, v. 47, p. 435–510.

Mukhopadhyay, P.K., and J.A. Wade, 1990, Organic facies maturation of sediments from three Scotian Shelf wells: Bulletin of Canadian Petroleum Geology, v. 38, p. 407–425.

Peters, K.E., 1986, Guidelines for evaluating petroleum source rock using programmed pyrolysis: American Association of Petroleum Geologists Bulletin, v. 70, p. 318–329.

Powley, D.E., 1980, Pressures, normal and abnormal: AAPG Advanced Exploration Schools Unpublished Lecture Notes, Tulsa, OK, American Association of Petroleum Geologists, 38 p.

Powley, D.E., 1990, Pressures and hydrogeology in petroleum basins: Earth-Science Reviews, v. 29, p. 215–226.

Price, L.C., L.M. Wenger, T. Ging, and C.W. Blount, 1983, Solubility of crude oil in methane as a function of pressure and temperature: Organic Geochemistry, v. 4, p. 201–221.

Stach, E., Mackowsky, M Th., Teichmüller, M., Taylor, G.H., Chandra, D., and Teichmuller, R., 1982, Textbook of Coal Petrology: 3rd ed.: Gebrüder Borntraeger Berlin, Stuttgart, 535 pp.

Tarafa, M.E., J.K. Whelan, and J.W. Farrington, 1988, Investigation on the effects of organic solvent extraction on whole-rock pyrolysis: Multiple-lobed and symmetrical P2 peaks: Organic Geochemistry, v. 12, p. 137–149.

Thompson, K.F.M., 1979, Light hydrocarbons in subsurface sediments: Geochimica et Cosmochimica Acta, v. 43, p. 657–672.

Thompson, K.F.M., 1988, Gas-condensate migration and oil fractionation in deltaic systems: Marine and Petroleum Geology, v. 5, p. 237–246.

Tigert, V., and Z. Al-Shaieb, 1990, Pressure seals: their diagenetic banding patterns: Earth-Science Reviews, v. 29, p. 227–240.

Tissot, B. P., and D.H. Welte, 1984: Petroleum Formation and Occurrence: Berlin, Springer-Verlag, 609 p.

Ungerer, P., E. Behar, and D. Discamps, 1981, Tentative calculation of the overall volume expansion of organic matter during hydrocarbon genesis from geochemistry data. Implication for primary migration: Advances in Organic Geochemistry: New York, Wiley, p. 129–135.

Ungerer, P., E. Behar, M. Villalba, O.R. Heum, and A. Audibert, 1988, Kinetic modelling of oil cracking: Organic Geochemistry, v. 13, p. 857–868.

Vandenbroucke, M., and B. Durand, 1983, Detecting migration phenomena in a geological series by means of C1–C35 hydrocarbon amounts and distributions: Advances in Organic Geochemistry: New York, Wiley, p. 147–155.

Weedman, S.D., A.L. Gruber, and T. Engelder, 1992, Pore pressure variation within the Tuscaloosa Trend: Morganza and Moore-Sams Fields, Louisiana Gulf Coast: Journal of Geophysical Research, v. 97, p. 7193–7202.

Whelan, J.K., J.M. Hunt, and A.Y. Huc, 1980, Applications of thermal distillation-pyrolysis to petroleum source rock studies and marine pollution: Journal of Analytical and Applied Pyrolysis, v. 2, p. 79–96.

Whelan, J.K., J.M. Hunt, J. Jasper, and A. Huc, 1984, Migration of C_1–C_8 hydrocarbons in marine sediments: Organic Geochemistry, v. 6, p. 683–694.

Whelan, J.K., J.W. Farrington, and M.E. Tarafa, 1986, Maturity of organic matter and migration of hydrocarbons in two Alaskan North Slope wells: Organic Geochemistry, v. 10, p. 207–219.

Whelan, J.K., Z. Kanyo, M. Tarafa, and M.A. McCaffrey, 1990, Organic matter in Peru Upwelling sediments-analysis by pyrolysis, pyrolysis-gas chromatography, and pyrolysis-gas chromatography–mass spectrometry, *in* E. Suess, R. von Huene et al., eds., Proceedings of the Ocean Drilling Program, Scientific Results, College Station, TX, v. 112, p. 573–590.

Chapter 8

The Characteristics of Geopressure Profiles in the Gulf of Mexico Basin

John T. Leftwich, Jr.
Terry Engelder
The Pennsylvania State University
University Park, Pennsylvania, U.S.A.

ABSTRACT

This paper is a summary of our work on the relationship between undercompacted shale and abnormal pressure in the Tertiary portion of the Gulf of Mexico Basin. A major objective of this study is to map the depth to the top of the undercompacted shale, as located using shale density (cuttings samples), conductivity, sonic, plus bulk density logs relative to the depth to the top of abnormal pore pressure as determined using bottom-hole pressure (BHP) and repeat formation tester (RFT) data. Geopressure profiles (formation pressure versus depth curves) were most useful for such mapping because the geopressure profiles showed linear segments in which pressure gradients were constant. Although the top of abnormal pressure and the top of undercompaction sometimes occur at the same depth in a given field area, often these boundaries are separated by hundreds of feet (tens of meters) and in some cases by vertical distances of over 2000 ft (600 m). Two types of field areas in the Tertiary portion of the Gulf of Mexico are distinguished based on whether or not the top of undercompaction and the top of abnormal pressure correspond. Tertiary fields have geopressure profiles characterized by pressure gradients showing either three (the Alazan-type field) or four (the Ann Mag-type field) linear segments. One consequence of our work is that electropressure methods which assume that the top of abnormal pressure is always coincident with the top of the zone of undercompaction are unreliable when used for a quantitative estimate of geopressure.

INTRODUCTION

Upon the discovery that shale properties correlate, at least indirectly, with formation pressures, researchers proposed techniques for estimating virgin reservoir pressure in sandstone beds by the analysis of electric logs of adjacent shale beds (Wallace, 1964; Hottman and Johnson, 1965; MacGregor, 1965; Foster and Whalen, 1966; Ham, 1966; Mathews and Kelley, 1967; Fertl, 1976). Electric logs detect changes in shale properties (i.e., conductivity and sonic velocity) that vary as a function of shale porosity. In general, porosity of a shale decreases with depth as water is displaced by cementation of pores and squeezed out by mechanical compaction in response to overburden weight (Bradley, 1986). If water is unable to escape, further compaction is impeded, and with addition of overburden, the shale is said to become undercompacted. The zone of low-density shale is known as the zone of undercompaction and consists of shales having anomalously high porosity and concomitant low density associated with a high water content. The word *undercompaction* is used because shale density in the undercompacted zone is lower than it would be for normal compaction at the depth in question. The basis for estimating formation pressure using logs, electropressure techniques, is that shale compacted under hydrostatic conditions exhibits a characteristic density, acoustic travel time, or conductivity which changes as a function of depth of burial, depending on the age of the shale. A common expectation in the literature is that a divergence from those electric log signals expected for normally compacted shale indicates undercompaction and presumably marks a concomitant buildup of abnormal fluid pressure (Hottman and Johnson, 1965).

The thesis of this paper is that undercompaction as detected by electric logs does not necessarily signal the top of abnormal formation pressure. To show this we map the relationship between depth to the top boundary of abnormal pressure and depth to the shale density reversal as signaled by electric logs. At the same time we examine the characteristics of formation pressure as a function of depth in the Tertiary portion of the Gulf of Mexico.

THE GEOLOGY OF THE NORTHERN GULF OF MEXICO

Our study in the northern Gulf of Mexico Basin focuses on 20 oil and gas fields divided into two groups: onshore South Texas fields and offshore Texas and Louisiana fields (Figure 1). Fifteen of the 20 fields are located in the Tertiary oil- and gas-producing trends of South Texas, geologically a part of the Rio Grande Embayment. Onshore South Texas is characterized by rapid Eocene and Oligocene deltaic sedimentation and a complex system of coastward-dipping syndepositional growth faults that become progressively younger to the east and toward the Gulf. The thickest regressive sequences of the Eocene–Oligocene deltaic sedimentation are thick progradational wedges of sand deposited during the late Paleocene–early Eocene (Wilcox) and the mid- to late Oligocene (Frio–Vicksburg). During Frio–Vicksburg time, the Rio Grande Embayment was filled by the Norias delta system that deposited thick sections of sand throughout the area.

The second group of Gulf Coast fields is situated in the heart of the Plio-Pleistocene producing trend (offshore Texas and Louisiana). Fields in this area are generally larger and situated on salt domes or involved tectonically with salt. Some of these fields are located close to the axis of the Mississippi River delta and thus are in areas of rapid and thick late Tertiary sedimentation. Rapid deltaic sedimentation, growth faults, and salt domes characterize the depositional and structural style of the offshore Texas and Louisiana area. Fields in the Plio-Pleistocene trend allow analysis of abnormal pressure and compaction in areas of extremely high recent sedimentation rates with active salt tectonism.

METHOD OF DATA COLLECTION

Compaction and formation pressure data were collected from 20 oil and gas fields to determine the depths to the top boundaries of the undercompacted shale zone and the abnormal pressure. The data are then compiled in the form of pressure-depth profiles for further analysis.

Determination of Top of Undercompaction

The depth to the top boundary of the undercompacted zone was identified using a combination of conductivity, sonic, and density logs. The top boundary is defined as an abrupt increase in electrical conductivity, an increase in sonic travel time, and a drop in density, all of which are associated with a retention of pore water in the undercompacted shales (Fertl, 1976; Magara, 1978; Leftwich, 1993).

An initial electric-log database of compaction plots consisting of 471 conductivity plots, 232 sonic log plots, 119 shale density plots, and 46 density log plots was provided by various oil companies. Additional plots were prepared from logs purchased from Petroleum Information Service (PI). A mean depth to the top of undercompaction was determined for each field by using the compaction plots on several wells in each field area. A well-by-well compilation is given in Leftwich (1993).

Determination of Depth to Top of Abnormal Pressure

Geopressure profiles (pressure-depth curves) were constructed largely from original shut-in bottom-hole pressure (BHP) measurements recorded in numerous wells within each field. Such data were recorded by the various operators during the completion and production phases of each well and are available through a Houston-based company, Petroleum Information Service (PI). In some cases repeat formation tester (RFT) and other wireline (WLT) pressure data supplemented the BHP data.

Geopressure profiles for many fields show several segments having linear trends. As a consequence, we computed a best fit or we hand-fitted a line to linear segments of the pressure-depth profiles. A pressure gradient for each segment was then calculated and recorded on each plot. The depth of the top of abnormal pressure, a break in slope of the geopressure profiles, was determined for each field by visual inspection to ±100 ft (±30.48 m).

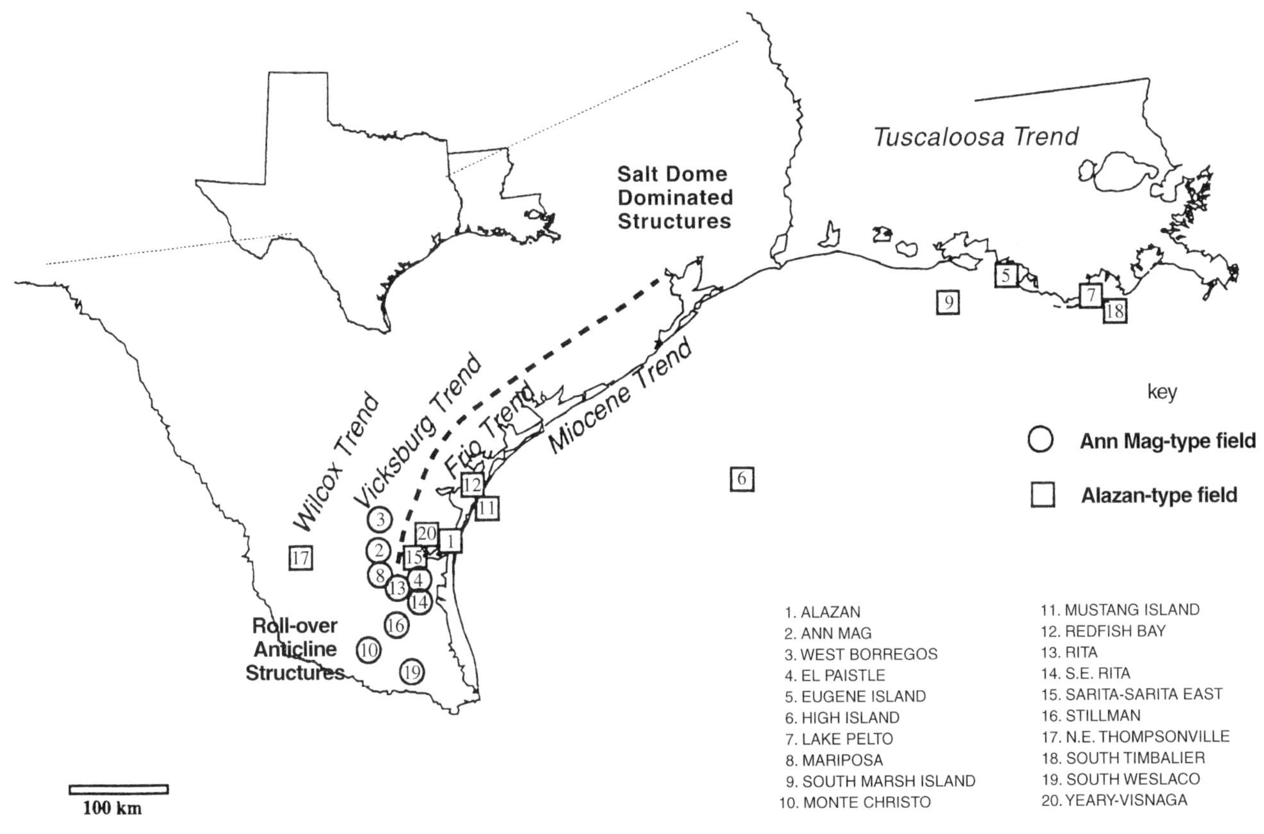

Figure 1. Map of Gulf Coast oil and gas fields for which geopressure profiles have been compiled in this paper. Ann Mag-type and Alazan-type fields are distinguished on this map.

OBSERVATIONS

A Case Study: The Ann Mag Field

The Ann Mag field is operated by Maguire Oil Company of Dallas, Texas, and is located in Brooks County, Texas. Geologically, the field is situated in the Frio–Vicksburg trend of the South Texas Rio Grande Embayment (Figure 1). This field exhibits a typical example of the structure, stratigraphy, and geopressure development observed in many onshore fields of the northwestern Gulf of Mexico Basin. In addition, the Ann Mag field is fairly well developed and contains many abnormally pressured wells in which the operator has made numerous wireline (WLT) and bottom-hole pressure (BHP) measurements. Furthermore, Maguire Oil Company graciously granted complete access to all available logs, detailed well files, cuttings samples, and cores on all wells in the field.

Top of Undercompaction

The top of undercompaction was identified in Ann Mag field wells from electric log conductivity plots (Leftwich, 1993). Using these tops of undercompaction the mean top of the zone of undercompaction in the Ann Mag field was established at 8980 ft (2737 m). The conductivity anomalies (tops of undercompaction) identified on plots of Ann Mag field wells illustrate that there is a relationship between the sand-shale stratigraphy and the top of undercompaction in the Ann Mag field. These logs show that the top of the undercompacted zone starts at depths where sands become less abundant above long shale sections (Leftwich, 1993).

Top of Abnormal Pressure

A geopressure profile constructed for the Ann Mag field allows us to identify the top of abnormal pressure at a depth of approximately 7500 ft (2286 m). The most striking feature of the Ann Mag pressure-depth profile is that it is divided into four linear segments with characteristic pressure gradients (Figure 2). The top segment is the normal pressure gradient of 0.445 psi/ft (10.1 MPa/km). Immediately below the top of abnormal pressure is a normally compacted section with a pressure gradient of 1.094 psi/ft (24.7 MPa/km). Below the top of undercompaction is a section with a high gradient of 2.391 psi/ft (54.1 Mpa/km) down to a depth of approximately 10,500 ft (3200 m). The deepest section has a pressure gradient of about 0.84 psi/ft (18.9 MPa/km) that is poorly constrained because of the small number of data points. When encountering these linear pressure-depth segments in other fields, they are labeled as segments ONE through FOUR

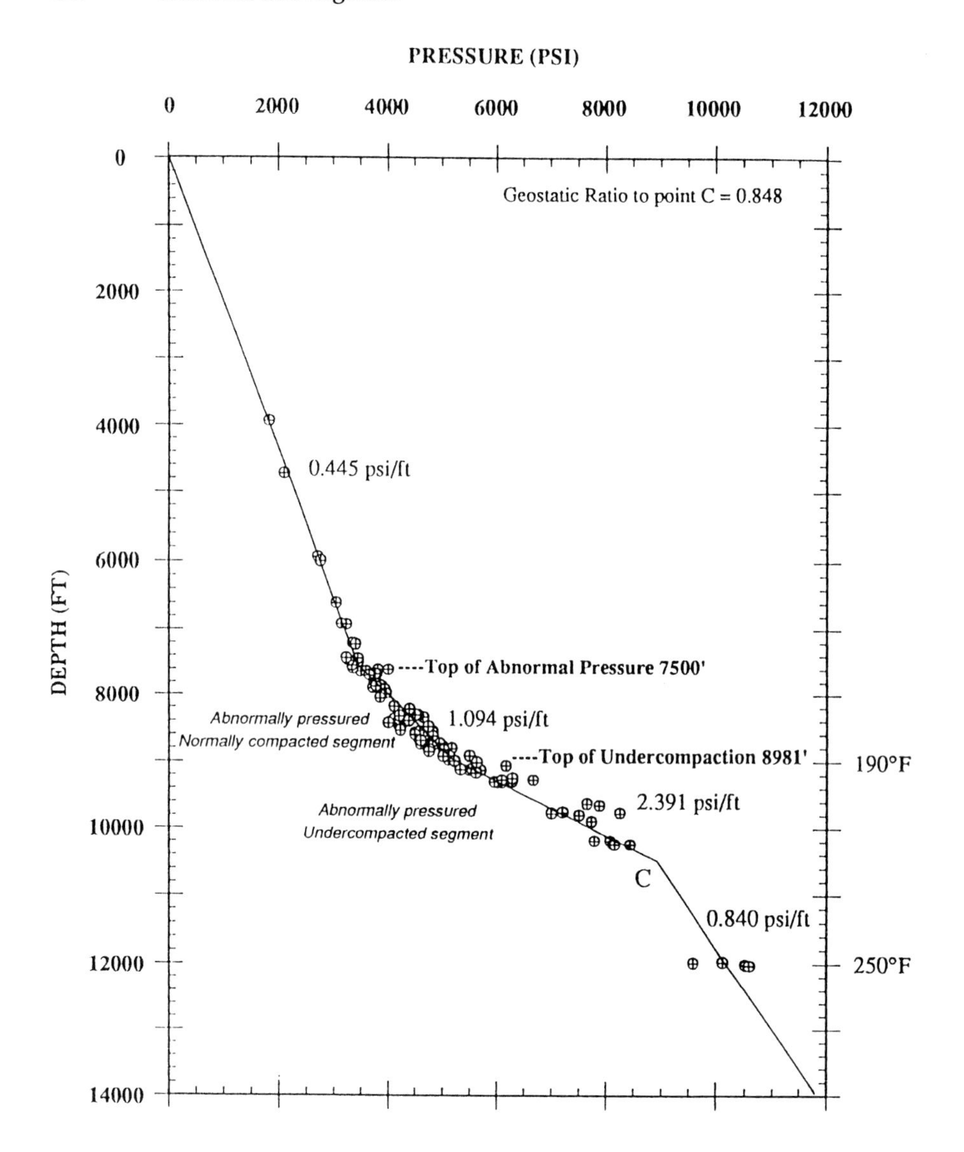

Figure 2. The geopressure profile for the Ann Mag field, Brooks County, Texas. Database includes BHP and RFT tests (5 data points excluded).

moving downward in a field. Segment FOUR is better defined in other fields where more data exist. Note that the mean top of the zone of undercompaction in the Ann Mag field at a depth of 8980 ft (2737 m) is 1480 ft (451 m) below the mean top of abnormal pressure in this field.

The skeptic will undoubtedly argue that the raw data in Figure 2 follow a curve more like the "lazy S" of Chapman (1980). However, Figure 2 is the compilation over an entire field. Pressure data from single wells serve as the best evidence for linear segments in pressure-depth profiles. Pressure segment TWO for the Rupp 1 well is particularly well defined (Figure 3). A computer-calculated curve fit to the data points comprising this segment indicates a linear curve fit with a regression coefficient of 0.993. An abrupt change in pressure gradient that occurs in going from the normally pressured section to the abnormally pressured section is also consistent with linear pressure-depth segments. Furthermore, the change from normal compaction to undercompaction occurs abruptly, rather than gradually. This abrupt change is also accompanied by an abrupt change in pressure gradient in going from pressure segment TWO to pressure segment THREE in the Ann Mag field.

Geopressure profiles were constructed from a number of wells at various locations on the structure that define the field. Because the depth to sandstone beds within each field is controlled by structure, the local structure must affect the geopressure profiles as is seen in two wells from the Ann Mag field, the Rancho Neal Rupp 1 and the Maguire Oil Sullivan 9 (these wells are 2200 ft apart) (Figure 4). The two wells are structurally the same down to approximately 8200 ft with their pressure versus depth curves essentially coincident (point A, at a depth of 8225 ft). Below point A, the Sullivan 9 well gradually gains structure relative to the Rupp 1 well, and a log correlation of the two wells at a mean depth of 8565 ft reveals that the Rupp 1 is 50 ft high to the Sullivan 9 well. At a depth of 8565 ft the geopressure profile for the Sullivan 9 well is offset approximately 90 ft high relative to the Rupp 1 geopressure profile. The difference in structure between the two wells accounts for the offset in the pressure curves. As the depth increases, the structural difference between the two wells

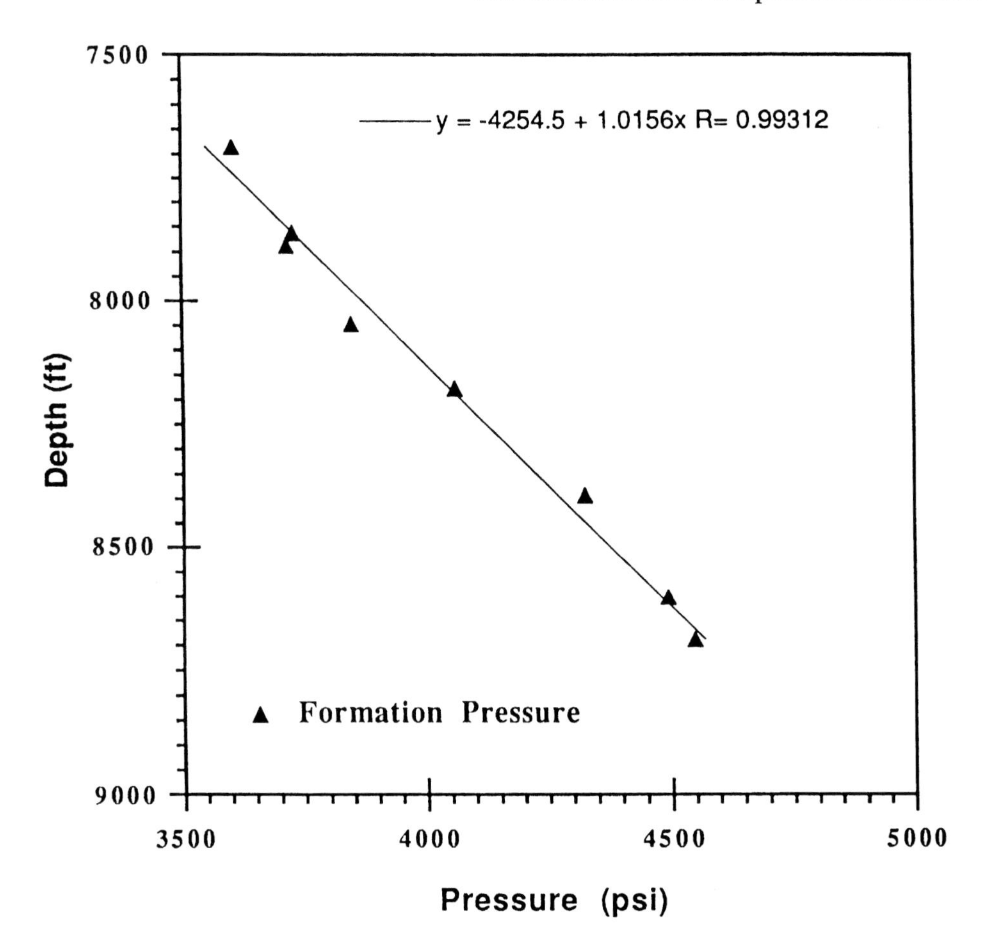

Figure 3. Geopressure profile segment TWO for the Rancho Neal Rupp 1 well in the Ann Mag field, Brooks County, Texas. Database includes RFT tests.

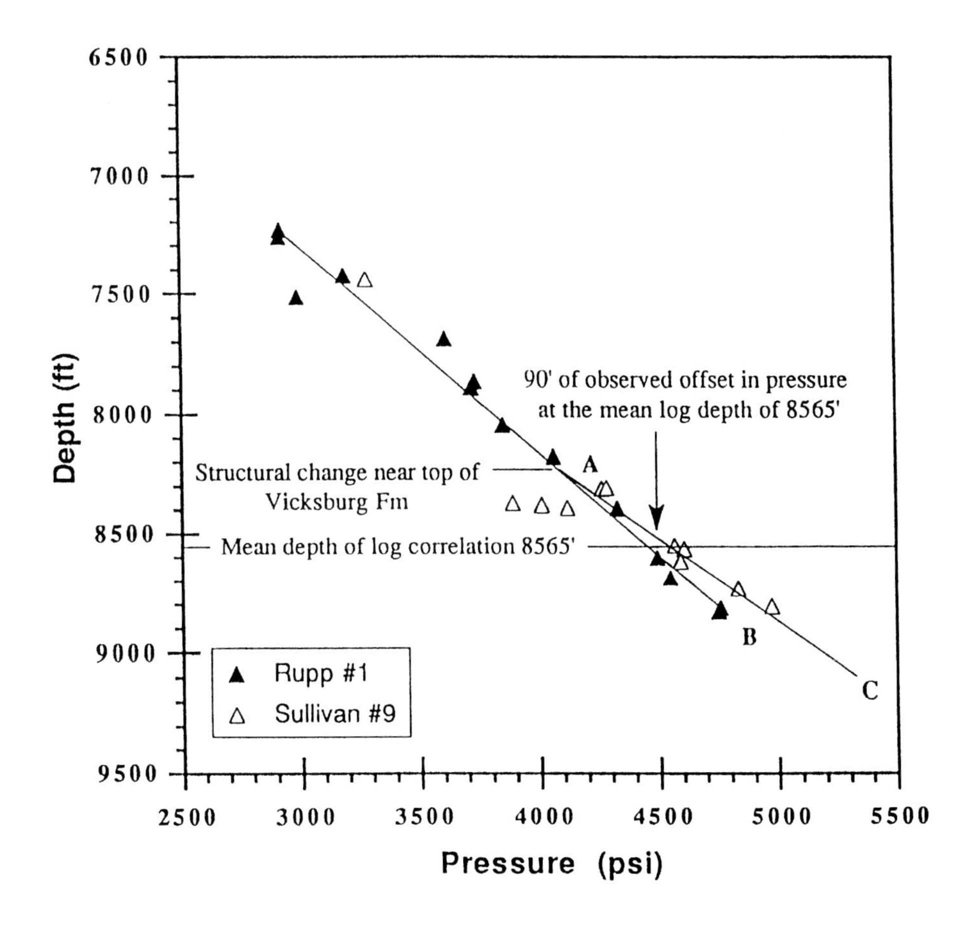

Figure 4. Geopressure profiles for the Rancho Neal Rupp 1 and the Maguire Oil Sullivan 9 wells in the Ann Mag field, Brooks County, Texas. This plot demonstrates the effects of structure on geopressure profiles.

Morrow Fan-Delta reservoirs in Anadarko basin: American Association of Petroleum Geologists Annual Convention, San Antonio, Texas, v. 73/3, p. 326.

Al-Shaieb, Z., J. W. Shelton, and J. O. Puckette, 1989b, Sandstone reservoirs of the mid-continent; syllabus for short course, Mid-continent Section of the American Association of Petroleum Geologists, 1989 meeting: Oklahoma City Geological Society, 125 p.

Al-Shaieb, Z., J. Puckette, P. Ely, and A. Abdalla, 1990, Annual Report prepared for the Gas Research Institute, Chicago, Illinois.

Amoco Production Company, 1989, Drill-stem test data for various counties in Oklahoma.

Anderson, C. J., 1991, Distribution of submarine fan facies of the Upper Red Fork interval in the Anadarko basin, western Oklahoma: Unpublished M.S. Thesis, Oklahoma State University, 210 p.

Arbenz, J. K., 1956, Tectonic map of Oklahoma: Oklahoma Geological Survey Map GM-3.

Beardall, G. F., Jr., 1983, Depositional environment, diagenesis, and dolomitization of the Henryhouse Formation, in the western Anadarko basin and northern shelf, Oklahoma: Unpublished M.S. Thesis, Oklahoma State University, 127 p.

Bradley, J. S., 1975, Abnormal formation pressure: American Association of Petroleum Geologists Bulletin, v. 59, p. 957–973.

Dahlberg, E. C., 1982, Applied hydrodynamics in petroleum exploration: Springer-Verlag, New York, 161 p.

Dwight's Energy Data Inc., 1990, Natural gas well production histories: Richardson, Texas.

Evans, J. L., 1979, Major structural and stratigraphic features of the Anadarko basin, *in* N. J. Hyne, ed., Pennsylvanian sandstones of the Mid-Continent: Tulsa Geological Society Special Publication no. 1, p. 97–113.

Harrison, J. C., 1990, Upper Morrow Purdy sandstones in parts of Texas and Cimarron counties, Oklahoma: Unpublished M.S. Thesis, Oklahoma State University, 95 p.

Krystinik, L. F., 1989, Depositional and diagenetic controls on production in Morrow valley fills, central state line area, Colorado/Kansas, Abstract: American Association of Petroleum Geologists Bulletin, v. 73, no. 8, p. 1048.

Medlock, P. L., 1982, Depositional environment and diagenetic history of the Frisco and Henryhouse formations in central Oklahoma: Unpublished M.S. Thesis, Oklahoma State University, 146 p.

Powley, D. E., 1986, Pressures, normal and abnormal: Lectures presented at American Association of Petroleum Geologists Advanced Exploration Schools.

Powley, D. E., 1987, Abnormal pressure seals, Gas Research Institute; Gas Sands Workshop, Chicago.

Puckette, J., A. Abdalla, Z. Al-Shaieb, and A. Rice, 1993, The Upper Morrowan reservoirs: Complex fluvio-deltaic depositional systems: Abstract Volume of Fluvial-dominated deltaic reservoirs in the southern Midcontinent; Okla. Geol. Survey and U.S. Dept. of Energy, p. 5.

Rascoe, B., Jr., and F. J. Adler, 1983, Permo-Carboniferous hydrocarbon accumulations, Mid-continent USA: American Association of Petroleum Geologists Bulletin, v. 67, no. 6, p. 979–1001.

Schatski, N. S., 1946, The Great Donets basin and Wichita system. Comparative tectonics of ancient platforms: Akademiya Nauk S.S.S.R. Izvestiya Seriya Geologicheskaya, no. 1, p. 5–62.

Shelby, J. M., 1979, Upper Morrow fan-delta deposits of Anadarko basin, Abstract: American Association of Petroleum Geologists Bulletin, v. 63, p. 2119.

Swanson, D. C., 1979, Deltaic deposits in the Pennsylvanian Upper Morrow Formation of the Anadarko basin, *in* Pennsylvanian sandstones of the Mid-Continent: Tulsa Geological Society Special Publication no. 1, p. 115–168.

Wade, B. J., 1987, The petrology, diagenesis and depositional setting of the Pennsylvanian Cottage Grove Sandstone in Dewey, Ellis, Roger Mills, and Woodward counties, Oklahoma: Unpublished M.S. Thesis, Oklahoma State University, 174 p.

increases and the separation between the pressure profiles increases. The structural relief in the 20 field areas of this study was on the order of a few hundred feet and this added an apparent scatter to the pressure data in a given geopressure profile.

In summary, the Ann Mag field shows four linear pressure segments with the top of undercompaction defining the boundary between segments THREE and FOUR. Although the fourth segment is poorly constrained in the Ann Mag field due to lack of data, a linear trend is well constrained in other Tertiary fields.

Undercompaction in the Gulf of Mexico Basin

Log Correlations

Various logs from the same well may indicate slightly different depths to top of the zone of undercompaction. Hence, it is important to establish that the depth to undercompaction obtained by any one of the four logs (i.e., conductivity, shale density [from cuttings], sonic, and bulk density) is, within a given tolerance, the same as that obtained from the other logs. To establish the reliability of individual logs as a measure of the top of undercompaction, a large industry database was searched and evaluated. The study confirms that the depths to the top of undercompaction as indicated by the different logs are consistent within the same wellbore to ±500 ft (Leftwich, 1993).

Correlation of undercompaction across a field is another problem. The top of undercompaction does not correlate with any particular time-stratigraphic boundary across a field, but rather correlates very well to the top boundary of certain thick marine shale lithostratigraphic units. Variations in the depth to the top of undercompaction in a given field are largely associated with variations in structure through the field. One reason for this is that the growth faults and their associated folds in a given field cause offsets and variations in the depth to the controlling lithostratigraphic unit, which in turn causes variations in the depth to the top of undercompaction. These differences in elevation of the top boundary of undercompaction in a given field lead to correlations across fields of no better than 500 ft (154 m).

A Data Compilation

We compiled data on undercompaction and formation pressure from 20 fields in the Gulf of Mexico Basin (Leftwich, 1993). In some fields (e.g., El Paistle and Alazan) the top of undercompaction varies as much as 2000 ft (610 m) from well to well, whereas in other fields (e.g., Ann Mag) the well-to-well variation is less than 500 ft (154 m). In a large number of cases the top of abnormal pressure and the top of the zone of undercompaction do not occur at the same depth in a given well or field. The top of abnormal pressure may occur above the top of undercompaction, may be coincident with the top of undercompaction, or may be below the top of undercompaction. The next question concerns the nature of the geopressure profile associated with these three cases.

Geopressure Profiles in the Gulf of Mexico Basin

Ann Mag-Type Fields

The quantity and depth distribution of pressure data vary from field to field. However, enough data are available to show that the Ann Mag field geopressure profile is typical of fields in which the top of undercompaction is below the top of abnormal pressure. Another example of a field with the top boundary of abnormal pressure above the top boundary of undercompaction is the South Weslaco field in Hidalgo County, Texas (Figure 5). A pressure-versus-depth plot for this field was constructed using RFT and bottom-hole pressure (BHP) data. The field is normally pressured (0.458 psi/ft. or 10.4 kPa/m gradient) down to a depth of approximately 7500 ft (2300 m). The top of the undercompacted zone was established using electrical conductivity methods and was found to occur in the main portion of the field at a depth of 9000 ft (2770 m). Again a separation is observed between the top of abnormal pressure and the top of undercompaction, which in this case is approximately 1500 ft (462 m). A straight line fit to the abnormally pressured data shows a gradient of 1.512 psi/ft (34.2 kPa/m). In the Weslaco field the available pressure data define two linear segments (i.e., segments ONE and TWO). Only two of the four linear segments appear because there are no pressure data below the top of undercompaction.

The West Borregos field (Figure 6) reveals another example suggesting that pore pressure increases in linear segments with depth. The stratigraphic section containing hydrostatic pore pressures has a pressure gradient of 0.450 psi/ft (10.2 kPa/m) down to 6050 ft (1844 m). From 6050 ft (1844 m) down to 7900 ft (2408 m), the uppermost abnormally pressured segment consists of normally compacted sediments but has a pore pressure gradient of 0.944 psi/ft (21.4 kPa/m). Below 7900 ft (2408 m) in the abnormally pressured and undercompacted section of the West Borregos field, the pore-pressure gradient is much higher, 3.650 psi/ft (82.6 kPa/m). At the West Borregos field three linear segments are defined (i.e., segments ONE, TWO, and THREE). The fourth linear segment may exist at depth but wells in the field are not deep enough to measure it.

In Monte Christo (Figure 7), where a large number of pressure measurements exist, four pressure-depth segments (segments ONE, TWO, THREE, and FOUR, respectively) are defined. Pressure segment FOUR (0.706 psi/ft) is better constrained than in the Ann Mag field. In the three fields mentioned above—South Weslaco, West Borregos, and Monte Christo—the gradient of pressure segment TWO is 1.512, 0.994, and 1.231 psi/ft, respectively. These gradients, which fall between the top of abnormal pressure and the top of undercompaction, are lower than the gradients of pressure segment THREE of the Ann Mag and Monte Christo fields. Four-part segmentation is apparent where enough pressure data exist to delineate the pressure-depth profiles and where the top of abnor-

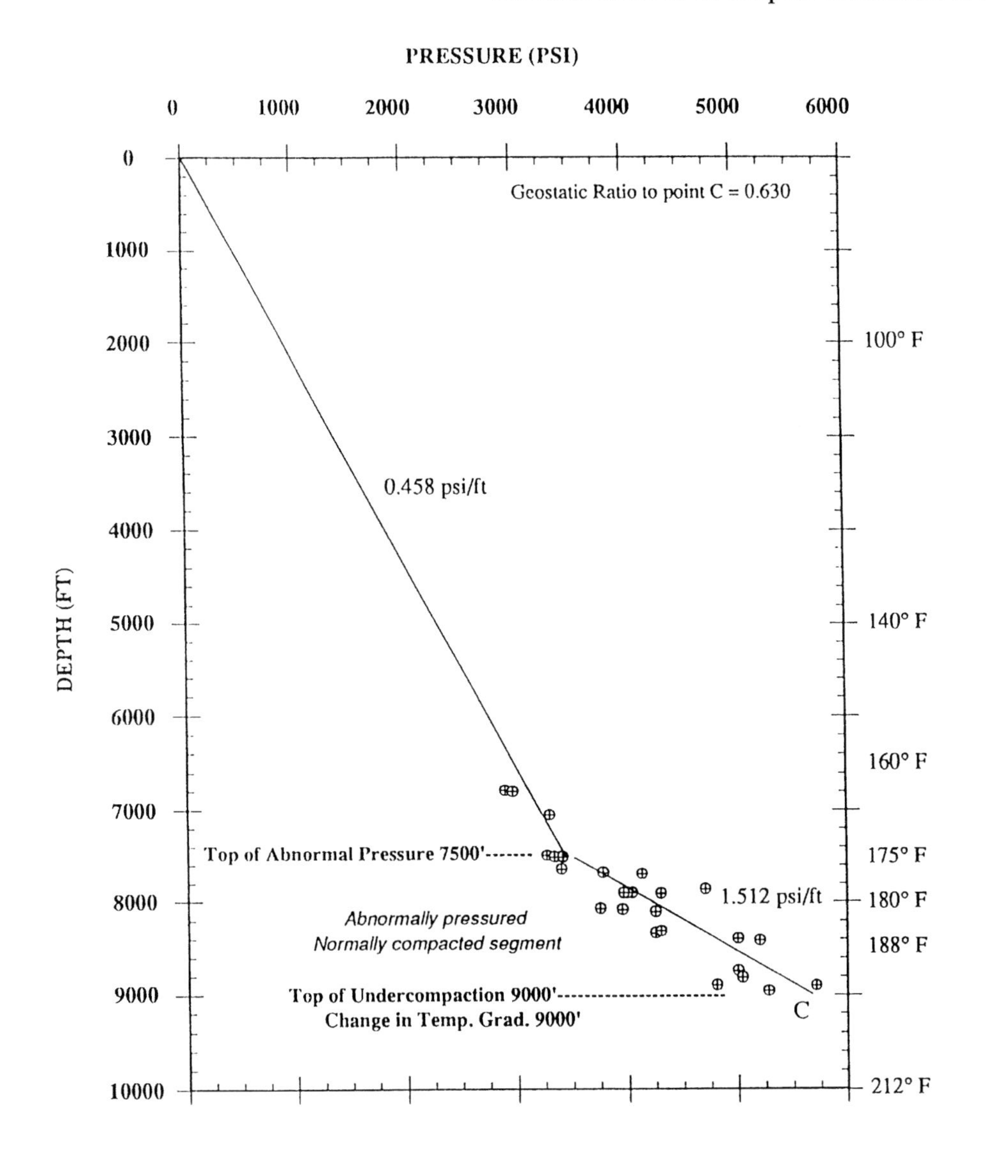

Figure 5. Geopressure profile for the South Weslaco field, Hidalgo County, Texas.

mal pressure occurs above the top of undercompaction. We have labeled these segments as ONE, TWO, THREE, and FOUR from top to bottom. Gulf Coast fields having a pressure segment TWO are classified as Ann Mag-type fields.

Alazan-Type Fields

In the Alazan field (Figure 8) the top of undercompaction occurs at essentially the same depth as the top of abnormal pressure. Three segments exist with the middle segment having a pressure gradient as high as that of segment THREE in the Ann Mag and Monte Christo fields. Segments in the Alazan field are equivalent to segments ONE, THREE, and FOUR in the Ann Mag and Monte Christo fields. Segment TWO is missing because the top of abnormal pressure and the top of undercompaction are coincident. Fields with segment TWO missing are classified as Alazan-type fields.

The distribution of Ann Mag-type and Alazan-type fields is shown in Figure 1. The preliminary results from the South Texas region suggest that Ann Mag-type fields (generally Vicksburg sandstones, i.e., older) are indeed found further west and down section from Alazan-type fields (generally Frio sandstones, i.e., younger). The dashed line in Figure 1 divides the Frio and Vicksburg trends. While this line divides Ann Mag-type and Alazan-type fields in South Texas, such a division to the northeast is yet to be tested.

DISCUSSION

Electropressure techniques to quantify subsurface pressures led to the common notion that the top of abnormal pressure and the top of the undercompacted zone occur at the same depth. Such a notion was reinforced by the assumption that all of the sediments on the normal compaction trend were always normally or hydrostatically pressured. We have now shown using wireline-test pressure data that this assumption is incorrect. In some fields the top of abnormal pressure

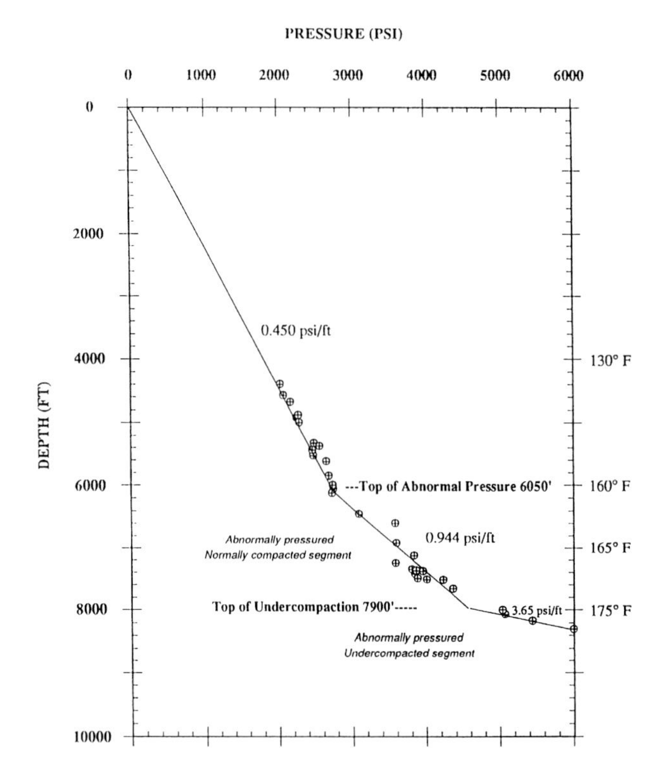

Figure 6. Geopressure profile for the West Borregos field, Kleberg County, Texas.

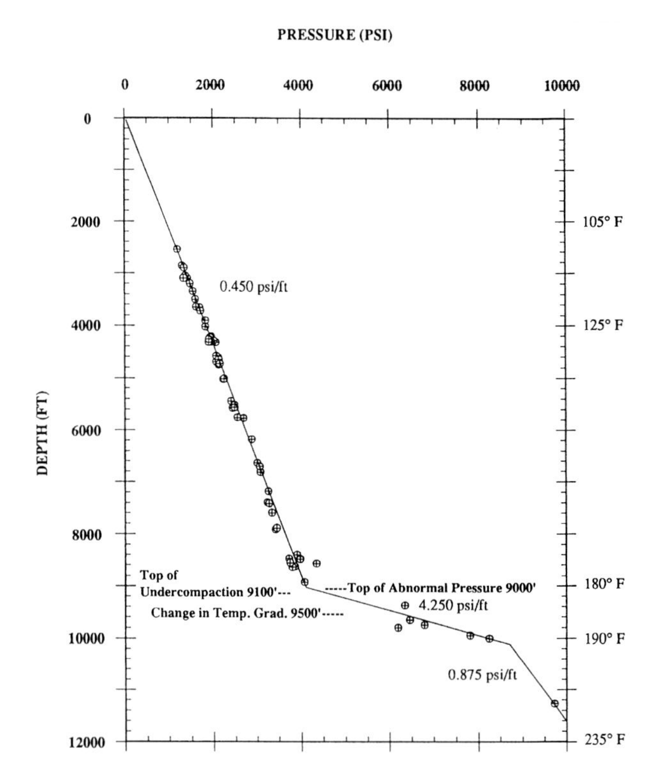

Figure 8. Geopressure profile for the Alazan field, Kleberg County, Texas.

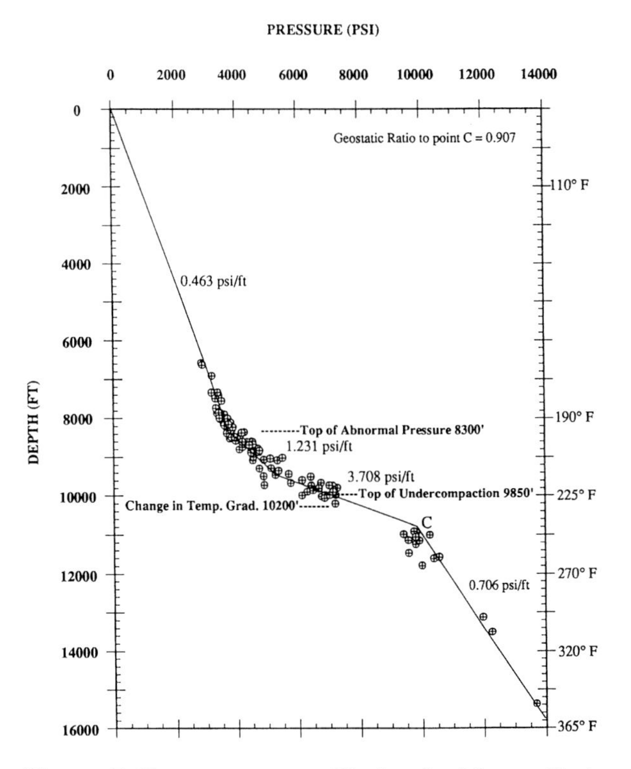

Figure 7. Geopressure profile for the Monte Christo field, Hidalgo County, Texas. (100 BHP measurements from 87 wells with 11 data points excluded.)

and the top of undercompaction occur at the same depth; in most cases they are separated. The separation varies from field to field and usually averages about 1000 ft (308 m), but separations as great as 2500 ft (770 m) are observed.

Pore-Pressure Gradients

Calculated pressure gradients were determined for the top three pressure segments of 13 of the 20 field areas studied (Leftwich, 1993). Based on the pressure gradients, four pressure-compaction states can be identified: normally pressured, normally compacted; abnormally pressured, normally compacted; abnormally pressured, undercompacted; and normally pressured, undercompacted sediments. The gradients for pressure segments TWO and/or THREE of each field are plotted against the separation between the top of abnormal pressure and depth to the top of undercompaction for each area (Figure 9). Abnormally pressured, undercompacted sections invariably contain pressure gradients (segment THREE) that are higher than pressure gradients found in normally compacted but abnormally pressured sections (segment TWO) (>2 psi/ft [45.3 kP/m] versus ≈ 1 psi/ft [22.6 kPa/m]). In fields where undercompaction and the top of abnormal pressure are nearly coincident, pressure segment TWO is missing and the pressure gradient jumps immediately to values in excess of 2.0 psi/ft (45.3 kP/m) (i.e., Alazan and Yeary fields).

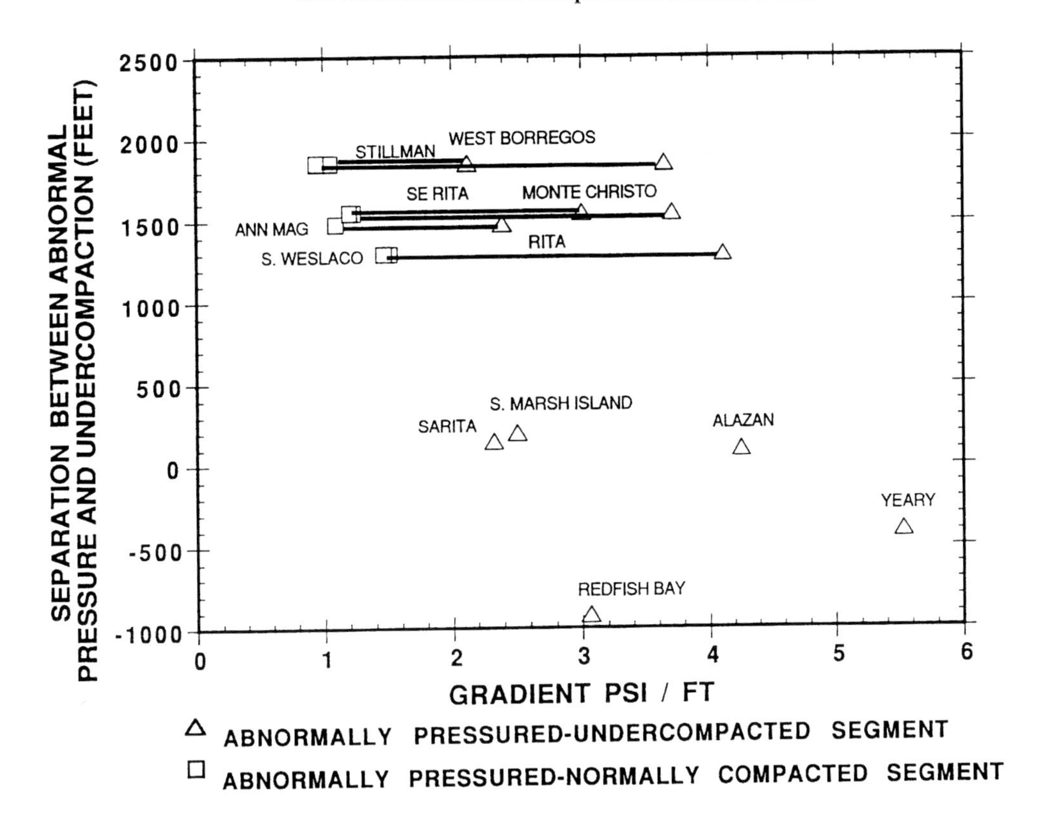

Figure 9. Gradients of segments versus the separation between the top of abnormal pressure and the top of undercompaction. The square symbols on the plot represent the gradients of the abnormally pressured, normally compacted sections plotted as a function of the corresponding separations between the top of abnormal pressure and undercompaction. The triangular symbols represent the gradients of the abnormally pressured, undercompacted sediments plotted as a function of the corresponding separations. Fields showing both an abnormally pressured, normally compacted section and the abnormally pressured, undercompacted section are plotted with both gradient symbols connected by a line. Other fields plotted with only one symbol are areas where there are data from either the abnormally pressured, normally compacted segment or the abnormally pressured, undercompacted section but not both.

The incremental increases in pressure gradients that were observed in the Gulf of Mexico can be modeled using layers of different hydraulic conductivity. Variations in hydraulic conductivity in the overburden section are in turn principally a function of the sand-shale ratio and the distribution of sand and shale in the overburden section (Leftwich, 1993). Generally, the more sands in the section the higher the hydraulic conductivity. The hydraulic conductivity of the normally pressured, normally compacted section is higher than that of the abnormally pressured, normally compacted section. Further, the hydraulic conductivity of the abnormally pressured, normally compacted section is higher than that of abnormally pressured, undercompacted section. The three different pressure-compaction regimes indicated on the Ann Mag-type curve shown in Figure 10 might be thought of as layered with respect to hydraulic conductivity, where $K_1 > K_2 > K_3$. This three-layer model is similar to that presented by Wallace et al. (1979) for the correlation between gross lithology and fluid pressure.

The hydraulic conductivity of the undercompacted zone (K_3) has the lowest value, and this fits well with observation because undercompaction usually occurs in a geologic sequence where rocks of lowest permeability predominate, i.e., marine shale sections. In the West Borregos field the top of the Jackson marine shale occurs at approximately 7900 ft (2408 m), which is the same depth as the occurrence of the top of undercompaction (Leftwich, 1993). Likewise, in the Ann Mag field (Figure 2) the top of undercompaction that occurs in this area at 8980 ft (2737 m) is coincident with the top of a thick marine shale section which occurs in the field at that depth.

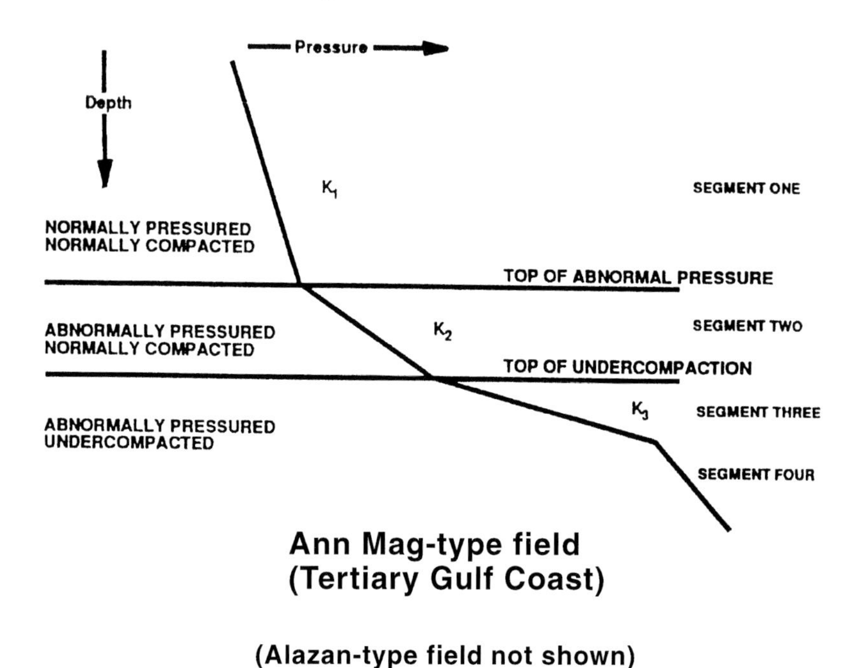

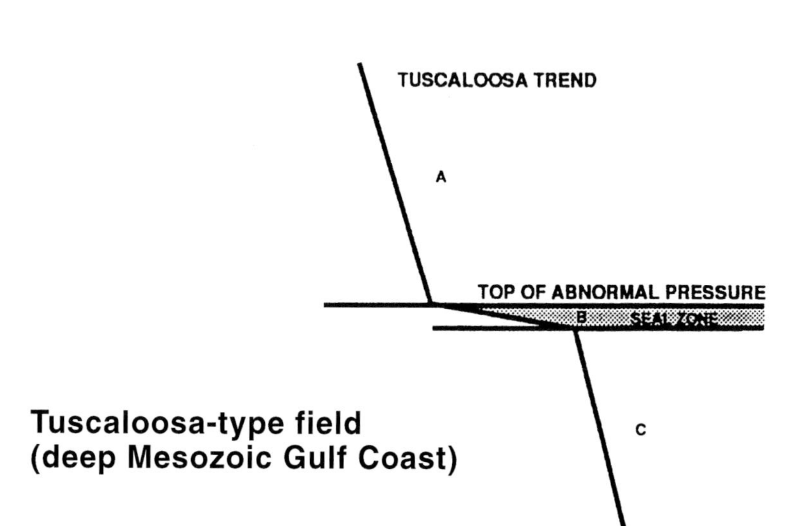

Figure 10. Ann Mag-type (Gulf Coast Tertiary) and Tuscaloosa-type (Louisiana Cretaceous) (Weedman et al., 1992) geopressure profiles. Also shown are three possible pressure-compaction states and hydraulic conductivities where $K_1 > K_2 > K_3$.

Gulf Coast Tertiary Versus Cretaceous Tuscaloosa Geopressure Profiles

Both types of Tertiary geopressure profiles are distinct from those encountered in older rocks of the Gulf of Mexico where a pressure seal cuts through thick Cretaceous sand bodies within the deep lower Tuscaloosa trend of Louisiana (Weedman et al., 1992a, b). The distinction in geopressure profiles is attributed to differences in lithology, stratigraphy, and seals. Geopressure profiles in the deep Tuscaloosa sandstones are characterized by a much thinner pressure transition (≈ 60 ft) (i.e., a pressure seal à la Powley, 1990) where the local pressure gradient is as much as 19 psi/ft (0.43 MPa/m) (pressure segment B of Figure 10). We know less about the shallower Tertiary geopressure profile above the Tuscaloosa trend, but we know that it is overpressured at depths of ~14,000 ft, with a regression to normal pressure at the top of the lower Tuscaloosa Formation. Tuscaloosa-type geopressure profiles are characteristic of pressure compartments (Hunt, 1990) with pressure segments A and C having a gradient of about 0.45 psi/ft, which is characteristic of freely communicating pore fluid in relatively permeable rocks. Above pressure segment A of the Tuscaloosa-type geopressure zone, there is a shallower abnormal pressure zone that could be characterized by either an Ann Mag or Alazan geopressure profile. The present data are too sparse to distinguish between the two.

In comparing Tuscaloosa-type and the two Tertiary geopressure profiles, pressure segments A and ONE are the same. While pressure segment C has a hydrostatic gradient, it is distinguished from pressure segment A by abnormal fluid pressures and it does not have the same pressure gradient as pressure segment FOUR. Likewise, pressure segments B, TWO, and THREE are all distinct from each other (Figure 10).

Table 8-1. Ranges for pressure gradients for each pressure segment.

Segment:	ONE & A	TWO	THREE	FOUR	B	C
Gradient :	0.46–0.48	0.9–1.5	2.0–4.5	0.85–0.90	20.00 +	0.46–0.48

Thus, we have recognized that geopressures within the Gulf of Mexico are characterized by at least six different pressure segments. Table 8-1 gives ranges for pressure gradients for each of the six pressure segments in terms of psi/ft.

CONCLUSIONS

The top of abnormal pressure and the top of the zone undercompaction as indicated by shale density, sonic, and conductivity logs usually do not occur at the same depth at a given location. Rather, these tops in most wells are separated by hundreds of feet (tens of meters) and in some wells by vertical distances of over 2000 ft (600 m). Furthermore, the top of abnormal pressure is usually above or coincident with the zone of undercompaction, and therefore the zone of undercompaction in many cases is abnormally pressured. Since the top of abnormal pressure and the top of the zone of undercompaction are usually not coincident, the use of electropressure methods to determine geopressures quantitatively yields results that are in many cases inaccurate and unreliable. The Tertiary section of the Gulf of Mexico Basin is characterized by two geopressure profiles consisting of three or four linear segments. These are the Ann Mag-type profile with four segments and Alazan-type with three segments.

ACKNOWLEDGMENTS

A number of oil companies, including Amoco, Exxon Company, U.S.A., Maguire Oil Company, Maxus Energy, Mobil Oil, Texaco, and Texas Oil and Gas Corporation, have contributed significantly to this study by providing logs, samples, and other subsurface information without which this study would not have been possible. The writers are especially indebted to Maguire Oil Company for allowing us to log and cut cores in their Sullivan M-1 well located in Brooks County, Texas. Mr. William Potthoff of Maguire Oil Company was most helpful in this regard.

We wish to extend thanks and gratitude to the Gas Research Institute (GRI contract 5088-260-1746), Amoco, and The Pennsylvania State University for providing financial support for this project.

REFERENCES CITED

Bradley, J. S., 1986, Fluid movement in deep sedimentary basins: a review, *in* Hichon, B., Bachu, S., and Saveplane, C.M.: Proceedings Third Canadian/American Conference on Hydrogeology: Hydrogeology of sedimentary basins: Application to Exploration and Exploitation, p. 19–31.

Chapman, R.E., 1980, Mechanical versus thermal cause of abnormally high pore pressures in shales: AAPG Bulletin, v. 64, p. 2179–2183.

Fertl, W. H., 1976, Abnormal Formation Pressures: New York, Elsevier Scientific Publishing Co., 382 p.

Foster, J. B., and J. E. Whalen, 1966, Estimation of formation pressures from electrical survey—offshore Louisiana: Journal of Petroleum Technology, v. 18, p. 165–171.

Ham, H. H., 1966, A method of estimating formation pressures from Gulf Coast well logs: Gulf Coast Association of Geological Societies Transactions, v. 16, p. 185–197.

Hottman, C. E., and R. K. Johnson, 1965, Estimation of formation pressures from log derived shale properties: Journal of Petroleum Technology, v. 17, p. 717–722.

Hunt, J. M., 1990, The generation and migration of petroleum from abnormally pressured fluid compartments: AAPG Bulletin, v. 74, p. 1–12.

Leftwich, J. T., 1993, The development of zones of undercompacted shale relative to abnormal subsurface pressures in sedimentary basins: Ph.D. dissertation, The Pennsylvania State University.

MacGregor, J. R., 1965, Quantitative determination of reservoir pressures from conductivity log: AAPG Bulletin, v. 49, no. 9, p. 1502–1511.

Magara, K., 1978, Compaction and Fluid Migration Practical Petroleum Geology: New York, Elsevier Scientific Publishing Co., 319 p.

Matthews, W R., and J. Kelly, 1967, How to predict formation pressure and fracture gradient: Oil and Gas Journal, v. 65, no. 8, p. 92–106.

Powley, D., 1990, Pressures and hydrogeology in petroleum basins: Earth-Science Reviews, v. 29, p. 215–226.

Wallace, R. H., T. F. Kraemer, R. E. Taylor, and J. B. Wesselman, 1979, Assessment of geopressured–geothermal resources in the northern Gulf of Mexico Basin, *in* Muffler, L. J. P., Jr., ed., Assessment of geothermal resources in the United States, 1978: U.S. Geological Survey Circular 790, p. 132–155.

Wallace, W.E., 1964, Will induction log yield pressure data?: Oil and Gas Journal, v. 62, no. 37, p. 124–126.

Weedman, S. D., S. L. Brantley, and W. Albrecht, 1992a, Secondary compaction after secondary porosity: Can it form a pressure seal?: Geology, v. 20, no. 4, p. 303–306.

Weedman, S. D., A. L. Guber, and T. Engelder, 1992b, Pore pressure variation within the Tuscaloosa Trend: Morganza and Moore-Sams Fields, Louisiana Gulf Coast: Journal of Geophysical Research, v. 97, no. B5, p. 7193–7202.

Chapter 9

Seismic Amplitude Versus Offset (AVO) Character of Geopressured Transition Zones

John N. Louie
Abu M. Asad
The University of Nevada
Reno, Nevada, U.S.A.

ABSTRACT

The presence of diagenetically cemented seals distinguishes between possible mechanisms for the maintenance of abnormal fluid pressures in sedimentary basins. The velocity gradients and Poisson's ratio variations surrounding a cemented seal affect the pre-stack amplitude versus offset behavior of seismic reflections from the seal. Acoustic synthetic seismograms based on well logs from sealed transitions demonstrate their unusual AVO character. A long-offset COCORP reflection line near Port Lavaca, Texas, shows similar effects from a seal at 6000 ft (1800 m) depth. Local conductivity logs and regional drilling mud weight compilations establish the presence of this pressure transition zone. It is associated with a strong, low-frequency reflector near its base and displays linear AVO trends markedly stronger than from deeper reflections within the overpressured compartment. The fact that a seal exhibits such prominent physical property characteristics suggests that AVO analysis techniques may locate other seals in basins worldwide, where proper conditions for diagenetic sealing rather than shaliness and subsidence may be needed to maintain geopressures.

INTRODUCTION

Bradley (1975) and Powley (1975) proposed the existence within sedimentary basins worldwide of closed but internally conductive "compartments" containing abnormal pore fluid pressures. They and their co-workers compiled accurate downhole pressure measurements into profiles showing dramatic changes in pressure gradient with depth. Based on these profiles Bradley and Powley divided abnormally pressured regions into high-gradient transition zones or "seals" surrounding hydrostatic but abnormally pressured "compartments." Hunt (1990) reviewed this hypothesis and some of the data from which it originated. This paper briefly explores how seismic reflection surveys may identify such pressure seals and transition zones. We studied both models of wave propagation in seals and reflection data from abnormally pressured oil and gas fields. Our objective is to suggest seismic acquisition and analysis strategies that should increase the likelihood of identifying overpressured formations in advance of drilling.

In the oil and gas industry, work on the seismic signature of abnormally pressured formations has concentrated on the analysis of normal-incidence, stacked reflection data. For many years, the high-porosity, reduced-velocity rocks within large overpressured "compartments" have been identified by stacking velocity analyses (Bellotti and Giacca, 1978; Bilgeri and Ademeno, 1982). More recently, industry specialists have investigated how the presence of abnormal pressures affects the instantaneous characteristics of stacked reflection wavelets.

To complement the velocity analysis work, we investigate the seismic properties of the pressure seals or transition zones themselves rather than those of the abnormally pressured compartment interiors. The seismic character of pressure seals may be similar for both over- and underpressured compartments and may allow the accurate location of vertical as well as horizontal seals. Characterization methods based on compartment velocities, on the other hand, whether based on stacking or tomographic velocity analysis, suffer from poor lateral resolution and the inability to identify underpressured formations.

Bethke's (1986) modeling of the hydrodynamics of abnormally pressured sections only requires a basin to subside and be relatively shaly to develop large pressure gradients. Bradley (1975), Powley (1975), Dewers and Ortoleva (1988), and Tigert and Al-Shaieb (1990) suggested alternatively that bands of calcite mineralization observed within transition zones, not necessarily parallel to lithologic layering, provide the sealing mechanism to maintain the overpressures below. The concept of a diagenetically cemented seal would explain how the many basins listed by Hunt (1990) exhibiting slower subsidence and less shaliness than required by Bethke (1986) could develop abnormal pressures.

We base our modeling and data analysis on the results of geological and geochemical efforts to characterize sealing mechanisms (this volume). The diagenetic silica or carbonate cementation mechanisms may yield transition zones that form seismic reflectors having identifiable properties. Analysis of reflections before stacking may be a reliable way to detect and measure such properties.

METHODS

Our assessment of pre-stack pressure seal reflection properties took two parallel tracks. One effort modeled the properties of pressure seal reflections with synthetic seismograms to search for distinguishing seismic characteristics. The other effort analyzed seismic reflection and well data to see if their characteristics matched those of the synthetics.

Modeling

In a sand or shale, the amount of secondary cementation required to retain overpressures also drastically alters the physical properties of the seal rocks. The presence of porosity can affect rock rigidity more than the compressional modulus (Christensen, 1985). Thus, as porosity is reduced in seals by cementation, rigidity rises faster than incompressibility, leading to distinguishable alterations in the Poisson's ratio property of the seal. Pre-stack seismic analysis can often detect reflectors with unusual Poisson's ratios (Ostrander, 1984).

The higher total incompressibility of the seal may also have other effects on offset seismograms, due to velocity gradients above and below the seal. Figure 1 shows the generalized geometry of near- and far-offset reflection rays from a horizontal seal. At larger incidence angles i, the presence of a strong velocity gradient with depth at the seal can cause rays to bend in directions not predicted by simple reflection theory. While at some angles the rays may bend up, throwing extra energy at some receiver interval (Figure 1, "turning ray"), at other angles rays may

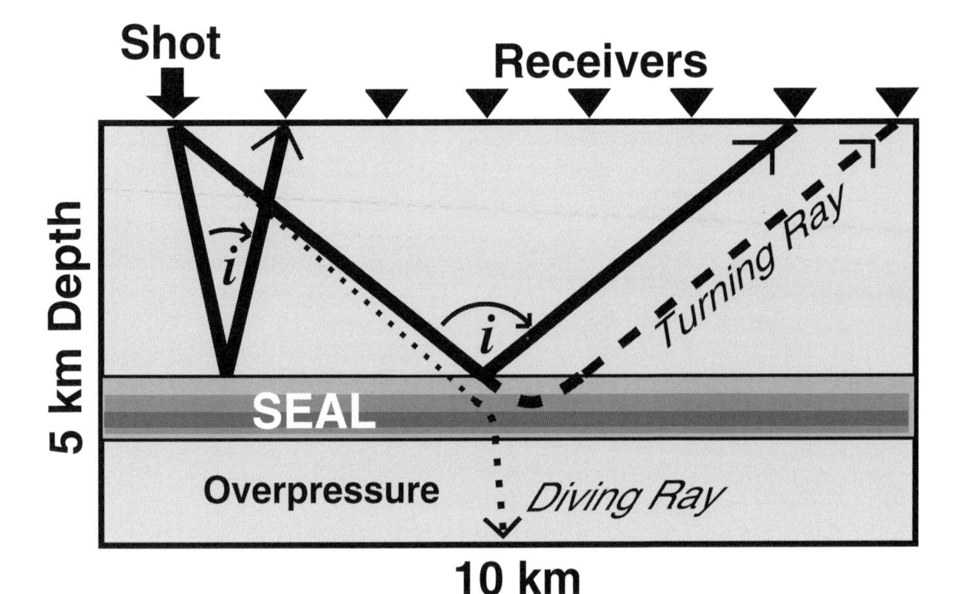

Figure 1. Geometry of multi-offset reflection rays from a horizontal seal at a range of incidence angles *i*. Positive velocity gradients with depth in the top of the seal lead to turning rays at some angles of incidence. At other angles, the negative velocity gradients at the bottom of the seal bend diving rays down and away from the receivers.

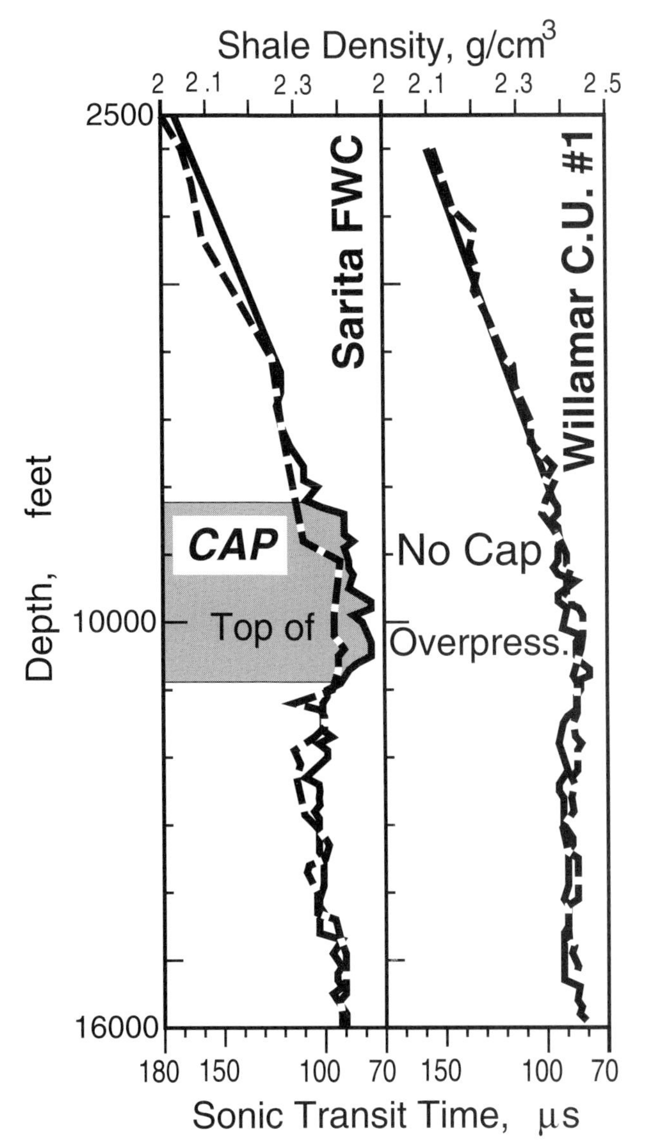

Figure 2. Shale density (solid curves) and sonic transit time (dashed curves) measurements from two typical wells on the south Texas coast that penetrate pressure transition zones. The Sarita well exhibits a high-velocity, high-density "cap" at the top of the overpressured section. The Willamar well shows no cap but exhibits constant velocity and density within the overpressured section.

bend down, depriving other receivers of returned energy (Figure 1, "diving ray"). We can model this ray-bending effect with purely acoustic synthetics.

Velocity measurements in wells confirm that sufficient gradients exist to produce such ray bending. We generated synthetic seismograms using velocity and density profiles from wells penetrating seals and abnormally pressured compartments. Two fields in the south Texas coast provided a suite of well logs with complete shale density and sonic transit time information. The completeness of the logs allowed acoustic seismogram modeling with a minimum of assumptions.

Figure 2 shows the log information from two of these wells, which we label the Sarita and the Willamar. In the Sarita well, shale density (solid curve on Figure 2) rises to high values between 8000 and 10,000 ft (2400 and 3000 m) depths, just above the onset of overpressure. Sonic transit times (dashed curve) similarly fall to a minimum in this depth range. Thus sonic velocities would reach a maximum.

We call this high-density, high-velocity region within the pressure transition zone a "cap." Several wells from our south Texas suite exhibited a cap. Some did not, despite the presence of overpressure. One example, the Willamar well, is also shown in Figure 2. Instead of a distinct cap, densities and velocities simply increase to the top of the overpressured zone. Below, the high compartment pressures preserve porosity and velocity at nearly constant values in both wells. The absence of a cap suggests that the pressure seal is not particularly well cemented.

The velocity gradients above and below such a cap may give the reflections arising from it distinguishing characteristics. These effects will be strongest in the comparison of narrow-angle (*i*; Figure 1) seismograms against wider-angle seismograms, in terms of changes in reflection amplitude versus source-receiver offset (AVO). The velocity gradients may not reflect strongly at short wavelengths (relative to the breadth of the gradients) and at normal incidence. However, at wider angles reflection rays may turn in positive velocity gradients, producing strong amplitudes. At certain angles, the reflections may lose energy downwards through the negative gradients at the bottom of the cap.

To investigate how the cap affects reflection amplitudes relative to incidence angle we generated multi-offset, pre-stack synthetic seismic gathers with acoustic finite-difference methods. Figure 3 shows two of the gathers, from the Sarita and Willamar profiles, respectively, at relatively low frequencies. We also made sets of higher-frequency synthetics. Our synthetics include all multiples, reverberation, interference, and other full-wave acoustic effects.

Data Analysis

We were able to obtain a public-domain seismic reflection data set from an abnormally pressured region. The Consortium for Continental Reflection Profiling (COCORP) program at Cornell University provided three lines from the Port Lavaca area of Texas, extending about 100 km inland. The COCORP data have several advantages for our study. First, Line 4 of the data set is close to the coast near Port Lavaca (Figure 4), so the expected depth to the onset of abnormal pressure is much shallower at 6000 ft (1800 m) than for other regions further inland or offshore (R. Spiller of Maxxus Energy, 1989, personal communication). Second, the COCORP line was recorded using offsets to 12 km, so it provides wide incidence-angle coverage. Moreover, COCORP acquisition parameters ensure good coverage of relatively low reflection frequencies between 10 and 35 Hz. Frequencies above 35 Hz are not included in the data,

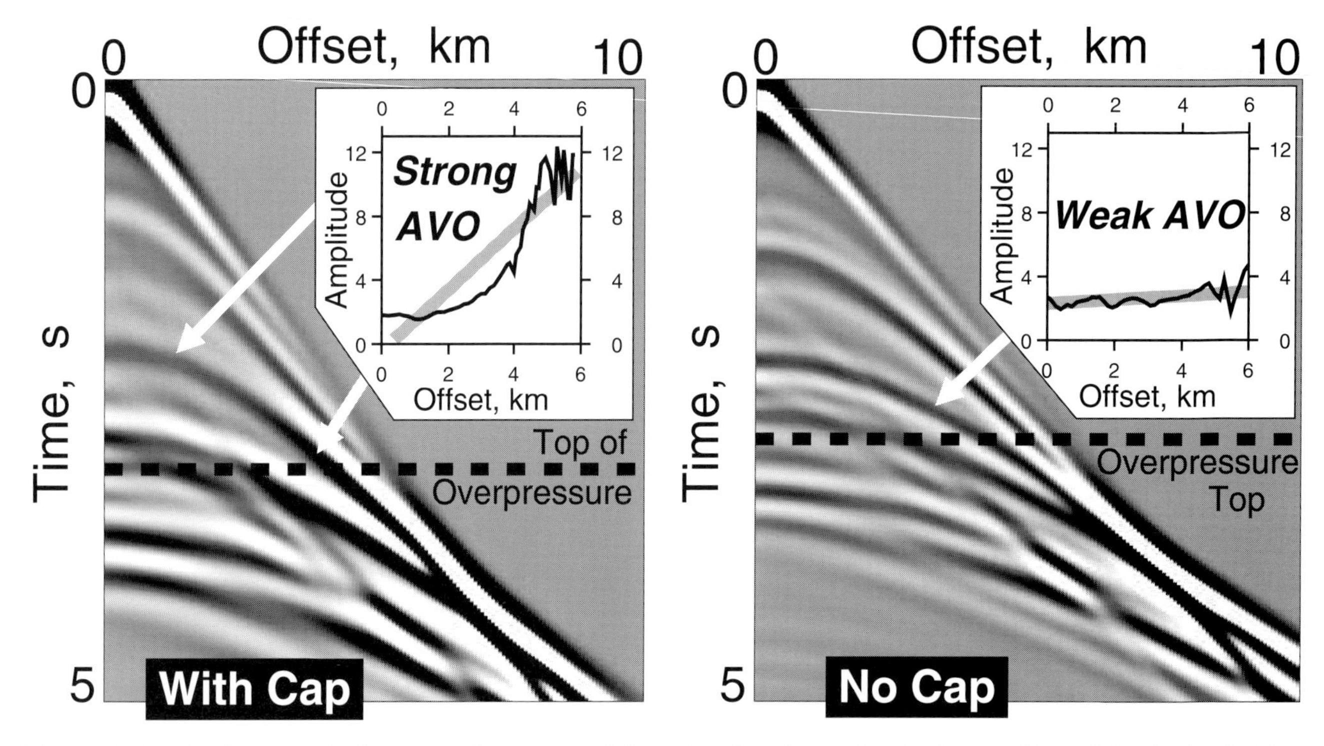

Figure 3. Synthetic acoustic shot records computed from the density and velocity profiles of the Sarita (with cap, left) and Willamar (no cap, right) wells of Figure 2. The arrows point to a reflection near the top of the seal in each model. The insets show the amplitude of each reflection versus offset. The amplitude of the reflection from the velocity gradient at the top of the cap in the Sarita model (left inset) increases at larger offsets because of turning rays (Figure 1), producing a strong linear AVO trend. Reflection amplitudes from the Willamar model without any cap (right inset) show little AVO trend.

however, limiting resolution to features comparable in size to the ~80 m seismic wavelength.

Figure 4 shows a map of COCORP Texas Line 4 near Port Lavaca. The numbered dots indicate the locations of wells for which we obtained all available geophysical logs. Our search revealed principally conductivity logs. Figure 5 displays COCORP's stacked seismic section for Line 4, converted to a depth scale using velocities from regional type logs and COCORP velocity spectra. The thick lines superimposed on the section show shale conductivities we picked from eight of the nine wells. Figure 5 suggests that a relatively low-frequency reflector at about 6000 ft (1800 m) depth across the section is at the base of a gradual rise in shale conductivity. The presence of a shale diapir below the southeast end of the section provides a mechanism for a relatively shallow transition to fluid overpressures, as shale diapirism is often a result of the buoyancy of an undercompacted, less-dense shale section. These observations suggest that the shale conductivities in Figure 5 are increasing through a pressure transition zone 1000 to 2000 ft (300–600 m) thick.

Offshore information on the depth to overpressure from reported drilling mud weights (R. Spiller, 1989, personal communication) supports the presence of a transition zone near the depth of the low-frequency reflector. The wells in this compilation closest to Line 4 reported overpressures as shallow as 6100 to 6300 ft (1860–1920 m). Further offshore, depths to the onset of geopressure in this region range from 4700 ft (1400 m) to 15,000 ft (4600 m), clustered around 8000 ft (2400 m). The local shale conductivities together with the offshore drilling information establish the existence of a geopressured section beginning ~6000 ft (1800 m) below COCORP Texas Line 4.

The low-frequency reflection may arise at the top of the seal or at its base, where it meets the overpressured compartment. Assuming that velocity gradients over 300 ft (100 m) depth intervals, as in Figure 2, control the reflectivity of the seal, the seal boundaries could appear as reflections having a generally lower frequency content than the reflections from sharper stratigraphic intervals. The low-frequency reflection we observe in COCORP Texas Line 4 at the 6000 ft (1800 m) depth may be a direct seismic characteristic of a fluid-pressure transition zone.

After identifying the reflections associated with the top of overpressure, we analyzed the pre-stack data for amplitude changes with incidence angle. We used the methods of Louie (1990), which most effectively find AVO changes for reflections that are strong and flat-lying in a stacked section. Figure 6 shows our own stack of the southern portion of Line 4 (box on Figure 5), emphasizing the near-horizontal structures. To develop this section we began with COCORP-

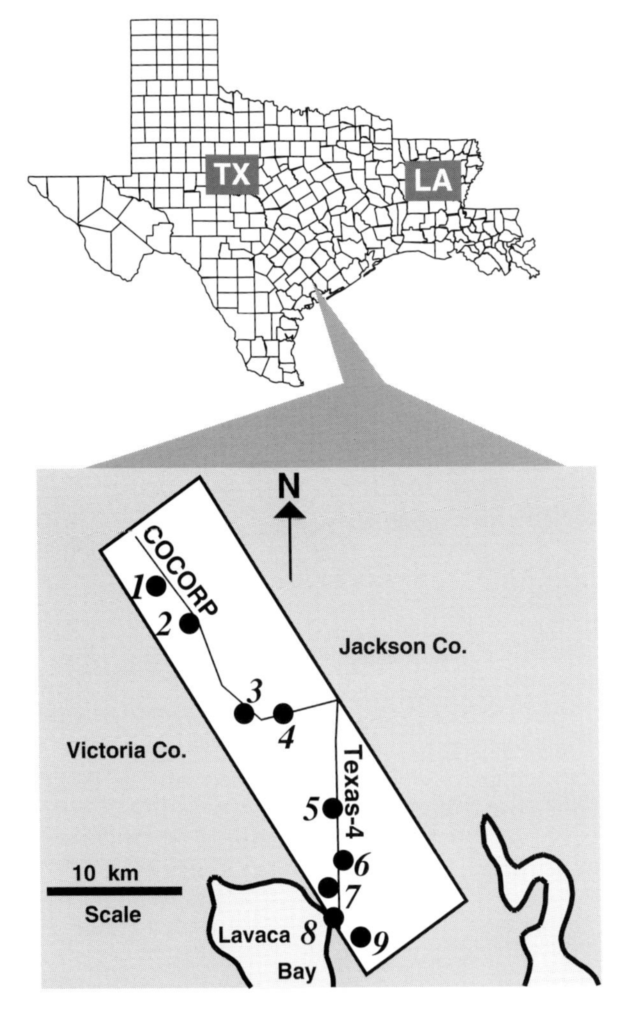

Figure 4. Location map of COCORP Texas Line 4 near Port Lavaca, Texas. The numbered circles locate wells providing shale conductivity measurements. The rectangle shows the range of stack midpoints near the line of section included in Figure 5.

correlated field tapes and applied quantile equalization, spherical divergence correction, mutes, and crooked-line common-midpoint sorting before normal-moveout correction and stack. Stacking velocities were picked from a set of constant-velocity stacks. Bayesian statistics applied after stack enhance the reflections most continuous before stack. We applied similar techniques also to the northwestern portion of Line 4 (not shown).

In the manner of Louie (1990), we developed linear AVO statistics for our stacked sections, using Figure 6 from the southern portion of Line 4. We also developed the AVO statistics for a northwestern portion of Line 4, which we do not show here. The true-amplitude stack of Figure 6 identifies the strong, continuous reflections above the shale diapir (Figure 4). Louie (1990) showed from COCORP data across southern California that such events are the only ones amenable to analysis with AVO trend stacks.

Figure 7 overlays the wiggle traces of the amplitude trend stack on top of the variable-density traces of the portion of the stack boxed on Figure 6. Wiggles to the right represent increasing reflection amplitude with offset, whereas wiggles to the left are on top of reflections having decreasing amplitude with offset. In examining reflections that are strong and laterally continuous in the underlying variable-density traces, the wiggle traces show the corresponding linear trend of amplitude versus offset. Where seismic velocity gradients in the seal produce turning or diving rays, we expect to see larger absolute values for the AVO trend, from the resulting rapid changes in reflection amplitude with offset. Inconsistent AVO trends appear in low-fold "wedges" near the top of the section (Figure 7). High-fold reflections above 1.5 s, near the bottom of the pressure transition zone, show strong, laterally consistent AVO changes. Reflections below 1.5 s, within the overpressured compartment, do not show such strong variations in AVO.

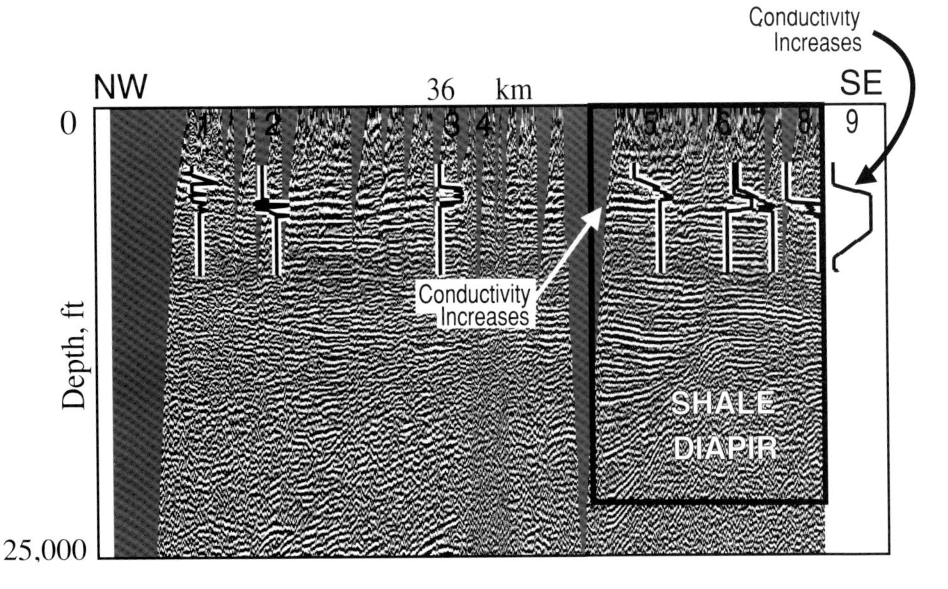

Figure 5. Approximately depth-converted COCORP stacked section for Texas Line 4. The superimposed curves give shale conductivities picked from the numbered wells. Greater conductivities are to the right. Where the curves are constant no shale conductivities could be picked. The box identifies the area shown in Figure 6. A thick section of shale with increasing conductivity overlies a relatively low-frequency reflector at ~6000 ft (1800 m) depth.

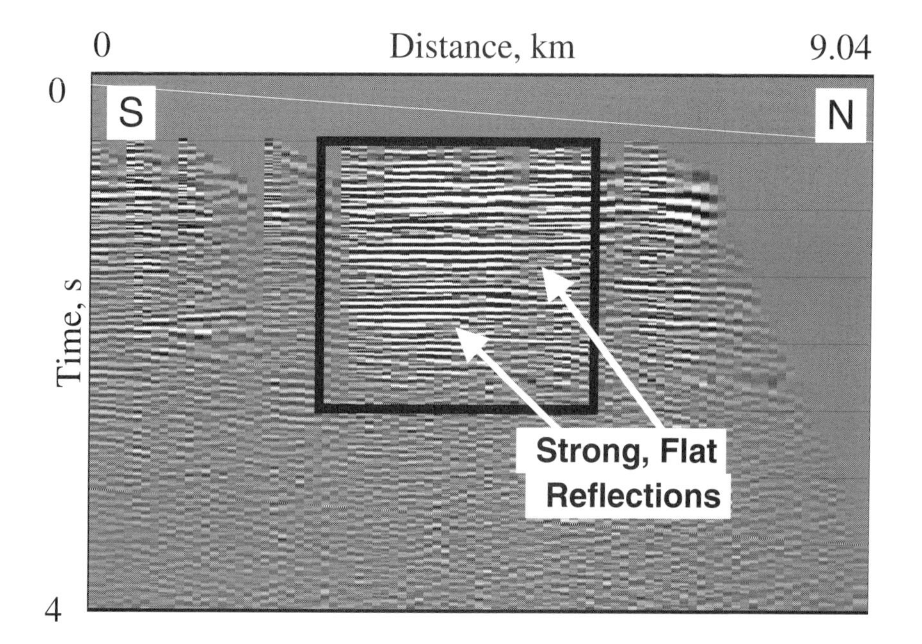

Figure 6. Stacked section derived from the southern portion of the pre-stack COCORP Texas Line 4 data. Note that the north-south sense of this section is reversed relative to that of Figure 5. This image emphasizes strong, flat reflectors having the most predictable normal moveout before stack. The box identifies the higher-fold region analyzed for reflection amplitude versus offset trends.

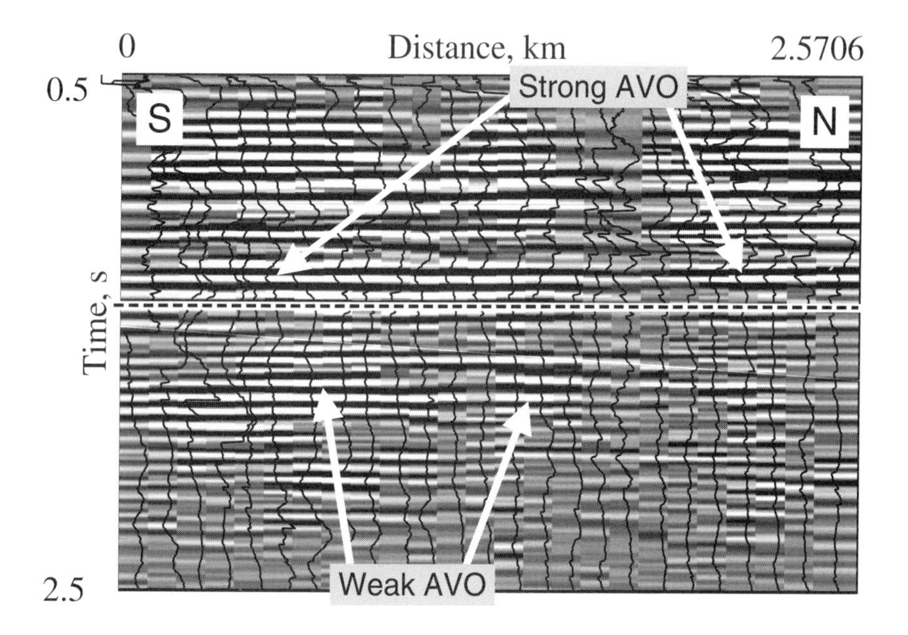

Figure 7. Linear amplitude versus offset trends (wiggle traces) superimposed on our COCORP Texas Line 4 reflection section (variable density traces). Inconsistent AVO trends appear in low-fold "wedges" near the top of the section. High-fold reflections above ~1.4 s, near the bottom of the pressure transition zone (dashed line), show strong, laterally consistent AVO trends. Reflections below ~1.4 s, within the overpressured compartment, show only weak AVO trends.

RESULTS

For our modeling study, Figure 3 shows that the presence of a high-density, high-velocity cap at the pressure transition zone will result in strong AVO variations. The AVO is complex in that it cannot be explained in terms of reflection coefficients from a single interface. Instead, it is the product of reflections turning in the positive velocity versus depth gradients and leaking energy down through the negative gradients. The left panel of Figure 3 shows how as offset increases, strong reflections within the cap will have sudden increases and decreases in amplitude. The amplitude of the reflection from the velocity gradient at the top of the cap in the Sarita model increases at larger offsets because of turning rays (Figure 1), producing a strong linear AVO trend (Figure 3, left insert). In contrast, reflections within a transition zone that is not cemented well enough to have developed a cap (right panel of Figure 3) have little AVO variation. Reflection amplitudes from the Willamar model, lacking a cap (Figure 3, right inset), show only the weakest AVO trend.

We have identified similar AVO phenomena in the COCORP Texas Line 4 data set, which passes over a relatively shallow seal. Shale conductivities and regional mud weight reports show the seal to be associated with a relatively low-frequency reflection near 1 s time and 6000 ft (1800 m) depth on the stacked sections of Figures 5, 6, and 7. Figure 7 and stacks of other parts of Line 4 (not shown) indicate that this reflection has large AVO variations. Below it, however, the AVO variations become much smaller. The lack of strong AVO trends suggests that these deeper reflections, within the overpressured compartment, do not exhibit the degree of Poisson's ratio variation or the velocity gradients associated with the almost complete cementation of the pressure transition zone above.

CONCLUSIONS

Our work leads to several recommendations for identifying pressure seals with seismic reflection work. Essentially, high-quality reflections are needed over a large range of incidence angle. In addition, relatively low frequencies may be more diagnostic than the high frequencies that must be used for development and stratigraphic work. To acquire a survey to locate seals, we suggest the use of frequencies below 50 Hz and maximum offsets at least as large as the target depths. In processing such data, true-amplitude techniques (such as described by Louie, 1990) need to be maintained. Finally, careful interpretation must compare the AVO trend section with the strength and quality of the stacked reflections and against other geophysical and geochemical information suggesting the depth of the transition zone.

This work has examined only a top seal from a small area of the U.S. Gulf Coast. However, our verification that strong reflection AVO effects are associated with the pressure transition zone suggests that it embodies steep velocity gradients or unusual variations in Poisson's ratio. Coincident with an observed pressure transition zone or seal, either observation would point to the presence of a diagenetically cemented seal as proposed generally by Bradley (1975) and Powley (1975). While regional shaliness and subsidence as advanced by Bethke (1986) could assist in the formation of geopressures in the Gulf Coast, diagenetic sealing would allow many other basins worldwide to develop abnormal fluid pressures. Seismic reflection techniques can locate and characterize such seals.

ACKNOWLEDGMENTS

We thank the Industrial Associates of the Gas Research Institute for generously providing seismic and borehole data and Cornell University for providing COCORP seismic sections. This study was supported by the Gas Research Institute under Contract No. 5088-260-1746, while the authors resided at the Department of Geosciences of The Pennsylvania State University at University Park. Deborah Dann interpreted the conductivity logs, and John Leftwich supplied the south Texas logs. Shelton Alexander and Terry Engelder of Penn State provided valuable insights and guidance to this work.

REFERENCES CITED

Bellotti, P., and Giacca, D., 1978, Seismic data can detect overpressures in deep drilling: Oil and Gas Jour., v. 76, no. 34 (Aug. 21), p. 47–52.

Bethke, C., 1986, Inverse hydrologic analysis of the distribution and origin of Gulf Coast-type geopressured zones: J. Geophys. Res., v. 91, p. 6535–6546.

Bilgeri, D., and Ademeno, E. B., 1982, Predicting abnormally pressured sedimentary rocks: Geophysical Prospecting, v. 30, p. 608–621.

Bradley, J. S., 1975, Abnormal formation pressure: AAPG Bull., v. 59, p. 957–973.

Christensen, N., 1985, The influence of pore pressure and confining pressure on dynamic elastic properties of Berea sandstone: Geophysics, v. 50, p. 207–213.

Dewers, T., and Ortoleva, P., 1988, The role of geochemical self-organization in the migration and trapping of hydrocarbons: Appl. Geochemistry, v. 3, p. 287–316.

Hunt, J. M., 1990, Generation and migration of petroleum from abnormally pressured fluid compartments: AAPG Bull., v. 74, p. 1–12.

Louie, J. N., 1990, Physical properties of deep crustal reflectors in southern California from multioffset amplitude analysis: Geophysics, v. 55, p. 670–681.

Ostrander, W. J., 1984, Plane-wave reflection coefficients for gas sands at nonnormal angles of incidence: Geophysics, v. 49, p. 1637–1648.

Powley, D. E., 1975, Course notes used in Am. Assoc. Petrol. Geol. Petroleum Exploration Schools.

Tigert, V., and Al-Shaieb, Z., 1990, Pressure seals: their diagenetic banding patterns, *in* Ortoleva, P., Hallet, B., McBirney, A., Meshri, I., Reeder, R., and Williams, P., eds., Self-Organization of Geological Systems, proc. of workshop 26–30 June 1988 at Univ. Calif., Santa Barbara; Earth-Science Rev., v. 29, p. 227–240.

Chapter 10

Pore Fluid Chemistry of a Pressure Seal Zone, Moore-Sams–Morganza Gas Field, Tuscaloosa Trend, Louisiana

Thomas P. Ross
Arthur W. Rose
Simon R. Poulson
Pennsylvania State University
University Park, Pennsylvania, U.S.A.

ABSTRACT

A set of water and gas samples from 17 wells in the Moore-Sams and Morganza gas fields, producing from 17,800 to 19,100 ft (5400 to 5800 m) depths on the deep Tuscaloosa trend, have been chemically analyzed in order to investigate possible mechanisms for forming the pressure seal separating overpressured from normally pressured fluids in these fields. Calculated corrections for condensation of water from the gas phase for these wells indicate that hydration of CO_2 in the gas phase of these high-CO_2 gases is significant.

Two main types of water are present in the reservoirs. Type 1, with about 20,000 mg/l Cl, appears to be modified seawater that is leaking from the overpressured zone into the normally pressured zone. Type 2 has about 33,500 mg/l Cl, was probably derived by moderate evaporation of seawater, and occurs mainly near the northwest corner and in the upper reservoirs of the Morganza field. In general, pore waters in these fields and the lower Tuscaloosa are heterogeneous, indicating complex hydrology. Median concentrations of dissolved SiO_2 are 340 mg/l, greatly supersaturated relative to quartz at reservoir temperatures of 160–175°C. The high supersaturation suggests active silicate breakdown, combined with inhibition of precipitation by chlorite coats on quartz grains. P_{CO_2} decreases from overpressured horizons to normally pressured horizons. Exsolution of CO_2 into the gas phase on leakage of overpressured fluid to normally pressured conditions should cause carbonate precipitation and also act to seal the overpressured zone.

INTRODUCTION

A current model for some pressure seals separating overpressured and normally pressured regimes proposes that sealing is diagenetic, and that the seal is maintained against cracks and other leakage by ongoing mineral precipitation or similar processes (Hunt, 1990; Powley, 1990; Tigert and Al-Shaieb, 1990; Dewers and Ortoleva, 1988). If seals are being maintained currently, then the chemistry of the pore fluid in and near the seal should furnish insight into the processes involved, which may include precipitation caused by cooling of fluids, pressure release possibly accompanied by gas exsolution, and mixing of fluids. Extreme dissolution followed by secondary compaction and cementation is a version of these processes that has been proposed for the study area (Weedman et al., 1992a). In addition to their relevance for current processes, the fluids may be representative of past fluids that originally formed the seal.

The Moore-Sams and Morganza gas fields in Point Coupee Parish, Louisiana, produce gas at depths of 17,800 to 19,100 ft (5400 to 5800 m) from the Cretaceous lower Tuscaloosa sandstones (Figure 1). Most wells are operated by AMOCO Production Co. and a few by Chevron U.S.A. Formation pressures within the gas fields range from hydrostatic to varying degrees of overpressure (Weedman et al., this volume). The boundary or seal between pressure regimes is not restricted to a specific stratigraphic horizon and is proposed by Weedman et al. (1992b) to be diagenetic, caused by extensive removal of carbonate cement followed by crushing of framework grains (secondary compaction) and induration to form a seal within the sandstones.

In order to examine fluid chemistry in and near this seal zone, gas and brine have been carefully collected and analyzed from 17 wells in the two fields, and several additional samples have been studied from nearby fields in the Tuscaloosa trend.

GEOLOGY OF THE MOORE-SAMS AND MORGANZA FIELDS

The geology of these fields has been described by Weedman et al. (1992a, 1992b, this volume) and the characteristics of the Tuscaloosa trend by Funkhouser et al. (1980) and Smith (1985).

At the time of this study, a total of 23 wells in the Moore-Sams–Morganza (MSM) fields were producing an average of about 4 mmcf/day of gas each and 75 bbl/day of petroleum condensate each from the lower Tuscaloosa Formation at depths of 17,800 to 19,140 ft (5400 to 5800 m). The lower Tuscaloosa Formation in the producing area consists of interbedded sandstones and shales, with an average sandstone/shale ratio of about 4/1 (Weedman et al., 1992b). The formation is dissected into several blocks by a number of east–west-trending growth faults.

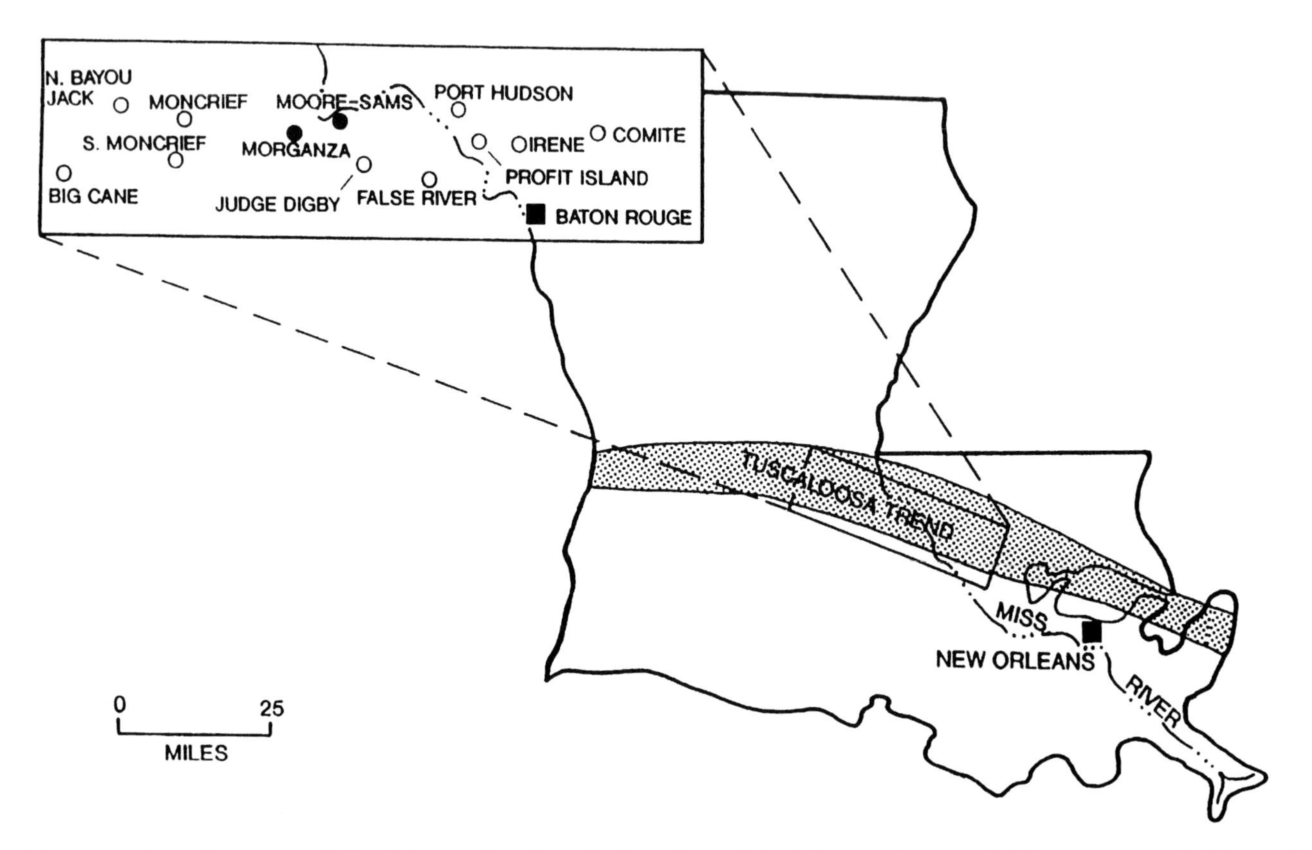

Figure 1. Tuscaloosa trend, Louisiana, showing the location of the Moore-Sams and Morganza gas fields and other nearby fields of the trend (modified after Funkhouser et al., 1980).

Sandstones are well sorted with an average framework grain composition of 88% quartz, 7% volcanic rock fragments, 3% chert, and 2% others. Porosity averages 10% and is inferred by Weedman et al. (1992a) to be dominantly secondary, by dissolution of carbonate cement. The diagenetic sequence is interpreted as (1) quartz cementation, (2) calcite cementation, (3) dissolution of calcite and rock fragments, and precipitation of ferroan dolomite, accompanied by compaction and initiation of seal formation, (4) precipitation of grain-rimming chlorite, and (5) precipitation of kaolinite and late quartz. Shales contain illite and chlorite along with the above components, and minor illite may be present in the sandstones.

In general, most wells in fault block 3 produce from overpressured zones, and most wells in blocks 1 and 2 from normally pressured zones, but some wells produce from zones overlapping the seal.

SAMPLING AND ANALYTICAL METHODS

Gas and entrained water from the AMOCO wells are piped to two central facilities where they are passed through a high-pressure separator (P = 900–1200 psi, T = 90–185°F) to separate gas from liquid, and then into a low-pressure separator where petroleum condensate is separated from water (P = approx. 60 psi, T = 90–180°F). Water samples were collected from the low-pressure separator and gas samples from the high-pressure separator, while the wells were actively flowing. For the Chevron wells, both water and gas were collected from high-pressure separators at the wellhead.

Analytical methods generally followed the procedures of Lico et al. (1982). Water from the separator was run into a 5 gallon carboy to separate petroleum condensate. Temperature, specific conductivity, pH, and Eh were measured immediately, and alkalinity titrated with 0.05N H_2SO_4 to an inflection point at pH 3 to 4.5. Water was then filtered through a 0.45 μm millipore filter under N_2 pressure. A 500 ml filtered sample was acidified with 3 ml conc. HNO_3 for cation analyses, another 500 ml was collected for anion analyses, 1:1 and 1:4 dilutions with deionized water were prepared for silica determination, and 50 ml was collected for oxygen isotope analyses.

Gas samples were collected by flowing gas through high-pressure samplers connected to the exit line of the high-pressure separator. They were analyzed at Southern Petroleum Laboratories, Lafayette, Louisiana, by gas chromatography.

Inorganic anions (Cl, Br, SO_4) were determined by ion chromatography after appropriate dilutions (Ross, unpublished master's thesis). Alkali metals (Na, K, Li) were determined by flame emission spectrometry, and other cations (Ca, Mg, Sr, Ba, Fe, Mn) by DC plasma emission spectrometry after appropriate dilutions. Silica was determined by the molybdate blue colorimentric method (Lico et al., 1982). Total organic acids were measured by titration following chromatographic separation (Lico et al., 1982), and volatile organic acids were further investigated by gas chromatography.

The analyses and well characteristics are listed in Tables 1, 2, and 3. Charge balance is within 3% for all but four samples, indicating that major constituents

Table 1. Characteristics of wells in the Moore-Sams and Morganza fields, Tuscaloosa trend, Louisiana.

	Reservoir			Production			
Well	Press. (psi)	Temp. (°C*)	Overp.	Gas (mmcf/d)	Water (bbl/d)	Pet. Cond. (bbl/d)	Depth (ft)
Bergeron-5	11,780	169	O†	7.27	236	255	18,572–98
Bizette-1	11,760	169	O	1.90	1729	70	18,544–790 (6)‡
Butler-1	8610	163	N	4.22	29	116	18,268–320
Curet-1	11,830	173	O	0.59	705	11	18,602–19314 (5)
Deville-1	11,790?	170	O?	5.46	368	139	18,555–950 (3)
Mix-1	8500	163	N	2.75	18	89	18,246–276 (2)
Brown-2	8510	166	N	9.47	405	88	18,422–464
Brown-4	8410	161	N	7.68	182	87	18,032–80 (2)
Debetaz-1	8560	165	N-O	2.95	194	54	18,094–702 (4)
G.G. Lacour-1	8300	163	N	4.11	1025	15	18,174–359 (3)
Hess-1	8480	161	N	9.58	71	97	18,050–130 (2)
Howard-1	8410	161	N	1.33	680	11	18,074–94
O.E. Lacour-1	8910	163	N	3.19	699	114	18,189–241 (2)
Planters-2	8480	167	N	9.54	208	211	18,472–532 (4)
Ravenswood-B1		163	N	9.70	153	94	18,094–298 (5)
Ravenswood-3	11,810	163	O	0.76	2298	18	18,720–68
Ravenswood-5	11,920	174	O	0.96	1057	34	18,900–19,140 (2)

* Estimated by method of Kehle (1972); see also Deming (1989).
† O = overpressured, N = normally pressured.
‡ Number in parentheses indicates number of intervals perforated within the total interval listed, if more than one.

Table 2. Analyses of waters from the Moore-Sams and Morganza fields.

Well ID	Cl	Br	SO_4	HCO_3	Ac	Na	K	Li	Ca	Mg	Sr	Ba	Fe
Bergeron-5	16,800	62	11	323	159	10,000	75	7	420	28	55	22	3
Bizette-1	20,300	70	10	379	221	12,500	110	10	550	45	67	37	3
Butler-1	210	0.8	0.9	77	77	14	0.7	0.1	27	0.1	0.6	0.2	3
Curet-1	18,800	68	10	361	269	11,600	105	9	570	39	63	41	5
Deville-1	15,900	56	15	359	240	9800	80	8	380	22	44	26	2
Mix-1	30	0.5	0.9	81	41	4	0.2	0.1	10	0.1	0.3	0.02	4
Brown-2	15,700	55	22	250	180	9700	95	8	540	40	80	16	7
Brown-4	19,500	66	18	205	123	11,500	110	10	1100	100	260	42	14
Debetaz-1	26,700	95	14	162	249	16,300	115	11	940	71	88	38	8
G.G. Lacour-1	19,600	54	20	266	132	11,600	120	10	860	73	120	26	5
Hess-1	8200	25	13	104	125	4400	55	4	440	32	46	14	20
Howard-1	33,500	105	16	362	240	18,700	180	19	2350	270	680	74	16
O.E. Lacour-1	16,500	52	16	231	286	10,400	85	8	470	38	62	20	3
Planters-2	8200	29	30	423	190	5300	40	3	130	8.5	18	4.2	3
Ravenswood-B	14,200	45	17	232	127	8400	95	7	730	60	89	29	1
Ravenswood-3	21,100	68	10	384	252	13,300	110	11	710	52	94	40	2
Ravenswood-5	13,200	50	10	366	216	8400	70	8	300	23	42	15	16

Well ID	Mn	SiO_2	$\delta^{18}O$*	$\delta^{13}C$†	TDS	Bal§	pH	Alkalinity	Eh	T	Conduct.	Factor	Date
Bergeron-5	1.4	260	3.67	−6.5	28,226	−4.6	6.08	5.94‡	0.143**	42††	45,500§§	1.28‡‡	6/90
Bizette-1	1.0	450	8.12	−7.7	34,753	−0.7	6.21	9.99	0.053	65	46,500	1.01	6/90
Butler-1	0.1	140	−1.22	−10.0	552	−260.9	5.49	1.26	0.222	46	330		6/90
Curet-1	0.8	226	7.80	−4.3	32,167	−0.2	6.05	10.4	0.200	35	8000	1.00	2/91
Deville-1	0.5	204	7.04	−3.2	27,137	−2.0	6.60	9.89	0.101	69	27,000	1.12	2/91
Mix-1	0.1	20	−2.22	−10.2	192	−217.1	5.50	1.33	0.234	44	400		6/90
Brown-2	1.9	270	6.24	−8.9	26,965	1.2	5.93	7.10	0.120	49	45,000	1.25	6/90
Brown-4	1.8	220	2.01	−8.3	33,270	2.8	5.65	5.40	0.145	45	53,000	1.48	6/90
Debetaz-1	1.3	340	5.42	−8.2	45,132	0.8	5.94	6.78	0.123	56	55,100	1.14	6/90
G.G. Lacour-1	1.0	260	7.32	−8.3	33,147	−0.2	5.99	6.59	0.089	54	51,500	1.03	6/90
Hess-1	0.9	130	0.86	−9.6	13,609	−7.8	5.82	3.78	0.200	45	33,500	127	6/90
Howard-1	5.0	420	7.12	—	56,937	1.8	5.32	5.94	0.138	35	93,000	1.02	6/90
O.E. Lacour-1	0.9	340	8.10	−10.6	28,512	1.6	5.77	8.55	0.143	45	33,500	1.03	6/90
Planters-2	0.05	250	3.05	−7.0	14,629	−1.3	6.32	10.20	0.248	57	23,500	1.61	6/90
Ravenswood-B	0.1	250	2.50	−7.8	24,282	1.0	5.85	5.94	0.147	45	42,500	1.59	6/90
Ravenswood-3	0.5	330	8.65	−8.6	36,464	2.7	6.44	10.49	0.082	82	48,500	1.00	6/90
Ravenswood-5	1.4	460	8.23	−9.8	23,177	0.6	5.99	9.56	0.118	58	27,000	1.01	6/90

* Per mil (SMOW).
† Per mil (PDB).
§ Charge balance (%).
‡ Total alkalinity to endpoint defined by inflection at pH 3.8–4.5, meq/l.
** Volts.
†† Temperature of field measurements, °C.
§§ Conductivity corrected to 25°C.
‡‡ Multiplier to correct analyses for dilution.

Table 3. Gas analyses (mole %).

Well ID	Nitrogen	CO_2	Methane	Ethane	Propane	I-butane	N-butane	I-pentane	N-pentane	Hexanes	Heptanes+
Bergeron-5	0.12	6.09	85.41	4.85	1.37	0.47	0.32	0.27	0.13	0.27	0.70
Bizette-1	0.14	6.22	87.09	4.11	0.99	0.38	0.22	0.21	0.10	0.20	0.34
Bizette-2	0.12	5.75	78.97	6.66	2.81	1.08	1.00	0.85	0.45	0.83	1.48
Curet-1	0.17	7.33	90.00	1.66	0.25	0.08	0.05	0.04	0.02	0.04	0.36
Deville-1	0.12	5.97	83.88	5.58	1.66	0.58	0.43	0.36	0.19	0.44	0.79
Mix-1	0.13	6.01	87.32	3.71	1.04	0.34	0.28	0.21	0.10	0.18	0.68
V.J. Hurst	0.13	6.00	86.78	4.17	1.15	0.42	0.27	0.23	0.11	0.23	0.51
Brown-2	0.13	5.93	89.28	2.84	0.55	0.20	0.12	0.12	0.06	0.17	0.60
Brown-4	0.12	5.90	86.82	2.99	0.62	0.25	0.17	0.26	0.15	0.52	2.20
Brown-7	0.14	6.25	84.99	5.50	1.51	0.32	0.33	0.18	0.10	0.17	0.51
Debetaz-1	0.14	6.16	88.39	3.27	0.68	0.23	0.16	0.16	0.08	0.21	0.52
G.G. Lacour-1	0.13	6.04	88.84	2.88	0.55	0.20	0.12	0.14	0.07	0.16	0.87
Gustin-1	0.00	6.09	89.09	3.25	0.64	0.22	0.14	0.13	0.06	0.12	0.26
Hess-1	0.15	5.74	88.79	3.28	0.72	0.21	0.16	0.12	0.07	0.12	0.64
Howard-1	0.17	6.08	89.67	2.85	0.51	0.17	0.10	0.09	0.04	0.10	0.22
O.E. Lacour-1	0.13	6.00	87.67	3.74	1.04	0.34	0.27	0.20	0.10	0.16	0.35
Planters-1	0.14	6.15	88.47	3.27	0.69	0.24	0.16	0.15	0.08	0.19	0.46
Planters-2	0.13	5.98	88.85	3.15	0.64	0.22	0.14	0.13	0.07	0.13	0.56
Ravenswood-B	0.14	5.87	89.49	2.92	0.57	0.20	0.12	0.11	0.05	0.11	0.42
Ravenswood-2	0.13	5.81	87.58	4.04	0.99	0.34	0.23	0.19	0.10	0.18	0.41
Ravenswood-3	0.13	7.40	88.72	2.22	0.35	0.15	0.07	0.09	0.04	0.14	0.69
Ravenswood-5	0.13	5.90	90.29	2.22	0.34	0.15	0.08	0.10	0.05	0.14	0.60

are relatively accurate. The two very dilute waters (Butler-1, Mix-1) have significant unbalance, probably because the analytical methods used for brines were not applicable to these waters.

Oxygen isotopes of waters were determined by CO_2–water equilibration at 25°C, using a fractionation factor of 1.0412 (Friedman and O'Neil, 1977). Correction for dissolved salts was negligible.

CORRECTION FOR WATER CONDENSATION DURING PRODUCTION

In general, a mixture of gas, aqueous liquid, and possibly petroleum liquid enters these wells from the reservoir sandstones, and during decrease in temperature and pressure as the fluids flow to the separators, water condenses out of the gas phase and dilutes the solutes in the aqueous phase (Kharaka et al., 1977). For true compositions at depth, a correction must be made for this effect. For wells producing large ratios of water/gas, the correction is probably minimal, but if water/gas is small, a large dilution may have occurred.

Initially an attempt was made to calculate the proportion of condensed water using the T and P of the reservoir and high-P separator, and the methods of McKetta and Wehe (1958) and Kobayashi et al. (1987, Fig 25.21 and eq. 12). However, this procedure accounted for only about 30% of the very low salinity water from the Mix-1 and Butler-1 wells, which is reasonably considered to be mainly condensate water. Also, the Cl concentration in water from many wells with high water/gas production ratios (>100 bbl/mmcf) was approximately 20,000 mg/l, but wells with low water/gas ratios remained much lower in Cl after the correction. The likely reason for this discrepancy is that the correction procedure underestimates the true amount of condensed water, probably because of the relatively high CO_2 content of these gases (about 6 mole %), and perhaps the presence of significant amounts of heavy petroleum condensates in the gas phase at P and T. Coan and King (1971) show that H_2O interacts with gaseous CO_2 to form gaseous H_2CO_3 or a similar molecule. Experimental data to estimate the extent of the CO_2–H_2O interaction in the gas phase at reservoir conditions have not been found in the literature. However, extrapolation of the lower temperature data of Coan and King (1971) suggests that amounts of H_2O associated with CO_2 would be large enough to explain the water in the two low-water wells.

In view of these relations, condensed water has been estimated by assuming that all water from Mix-1 and Butler-1 is condensed from the gas phase (average of 6.75 bbl/mmcf), and that T and P effects in these CO_2-rich gases are proportional to those in the methane system. The condensation correction has therefore been calculated as

$$V_{i,corr} = 6.75 G_i V_{i,h} / V_{m,h} \quad (1)$$

where $V_{i,corr}$ is the estimated water condensed from the gas phase (bbl/day) of well i, G_i is the gas production of well i (mmcf/day),and $V_{i,h}$ and $V_{m,h}$ are the calculated water production of well i and Mix-1, using equation 25-12 of Kobayashi et al. (1987), to allow for differences in T and P of reservoir and high-pressure separator among wells. The measured concentrations can then be corrected to reservoir concentrations by

$$F_i = V_{i,p} / (V_{i,p} - V_{i,corr}) \quad (2)$$

where F_i is the factor for correction for condensate dilution, and $V_{i,p}$ is the observed volume of water produced by well i. Evidence that this correction is approximately correct is that good correlations are found among $\delta^{18}O$, Cl content, and estimated proportion of reservoir water in the sample ($V_{i,p} - V_{i,corr}$), as discussed below.

PRESENCE OF MULTIPLE WATER TYPES

The ^{18}O of water in the reservoirs is expected to depend on temperature and ^{18}O of minerals in the reservoir. Because both of these vary only slightly among wells, a limited range of $\delta^{18}O$ is expected for reservoir water. Similarly, a homogeneous Cl concentration is a reasonable first assumption. Condensate water would have different $\delta^{18}O$ and very low Cl, thereby changing both these parameters in proportion to the amount of dilution. For the raw data, a plot of $\delta^{18}O$ vs. chloride clearly shows two trends (Figure 2). One trend extends from the Butler-Mix wells (30–210 mg/l Cl, $\delta^{18}O$ = –1.6‰) to about 20,000 mg/l Cl and 8‰ $\delta^{18}O$. The upper end of this trend is defined by five wells with water/gas ratios exceeding 100 bbl/mmcf (Curet-1, Bizette-1, Ravenswood-3, G.G. Lacour-1, O.E. Lacour-1), which must represent formation water. Several wells lie between this cluster and the Butler-1

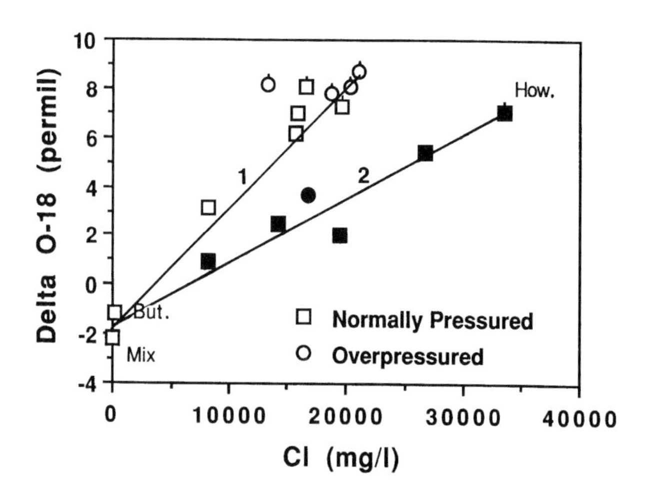

Figure 2. Chloride vs. $\delta^{18}O$ for waters from the Moore-Sams and Morganza fields, showing waters of type 1 (open symbols) and type 2 (filled; symbols). How = Howard-1 well; But = Butler-1 well; Mix = Mix-1 well. Tick mark above symbol indicates high water/gas ratio (>100 bbl/mmcf).

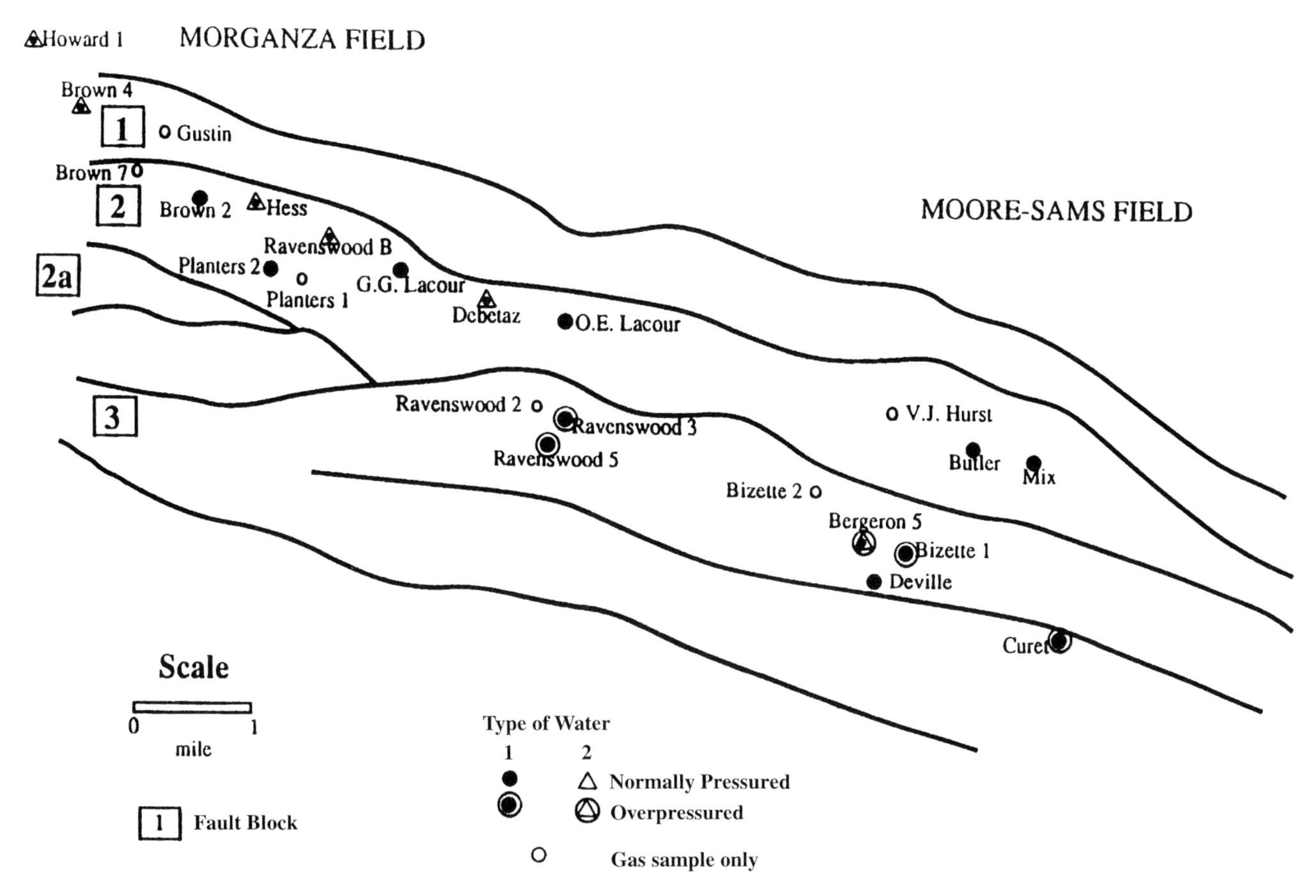

Figure 3. Location of wells, Moore-Sams and Morganza fields, showing distribution of water types, overpressured wells, and fault blocks. Modified after Weedman et al. (1992b).

and Mix-1 composition at 30–210 mg/l Cl and –1.6‰ $\delta^{18}O$. Estimates of dilution assuming that all water from Mix-1 and Butler-1 is condensed from the vapor (i.e., 6.75 bbl water/mmcf gas) also produce linear relations against Cl and ^{18}O, supporting the hypothesis that waters tapped by all of this group of wells are similar, with Cl of about 20,000 mg/l and $\delta^{18}O$ of about 8‰. This type of water is termed type 1 or lower-Cl water. Two wells (Ravenswood-5, O.E. Lacour-1) deviate to lower Cl content from the main trend, suggesting an additional minor component.

An end member for a second type of water is indicated by the Howard-1 well, with 33,500 mg/l Cl, $\delta^{18}O$ of 7.1%, and a high ratio of water to gas (511 bbl/mmcf). On the Cl vs. $\delta^{18}O$ plot, the Debetaz-1, Bergeron-1, Ravenswood B-1, Brown-4 and possibly Hess-1 wells plot close to a line between Howard-1 and the Butler-Mix compositions.

The above relations indicate that two main types of water exist in the gas reservoirs of the MSM fields. The areal distribution of these waters is illustrated on Figure 3. The type 2 (Howard-type) waters occur mainly in the northwestern end of the Morganza field in fault block 1 and along the north edge of fault block 2. The Hess-1, Ravenswood-B1, and Debetaz-1 wells produce from relatively shallow parts of the lower Tuscaloosa Formation, whereas the Brown-2, Planters-2, and probably the G.G. Lacour-1 wells, which produce the lower-Cl type 1 water, produce from deeper zones of the lower Tuscaloosa Formation. Type 2 waters therefore appear characteristic of block 1 and the shallow horizons of block 2, in this northwestern part of the field.

Of five wells definitely producing from the overpressured zone, four (Ravenswood-3, Ravenswood-5, Bizette-1, and Curet-1) have type 1 water. The overpressured Bergeron-5 well, producing type 2 water, is perforated over a 230 ft (70 m) interval that was measured by the RFT method to be overpressured at several depths, but sandstones in the upper part of this interval appear to correlate with a normally pressured zone in the nearby Bizette-2 well, so the relation of Bergeron-5 remains ambiguous. Similar ambiguities from production spanning large intervals apply to Deville-1, which produces type 1 water, but for which no RFT pressure data are available; it is inferred to be overpressured by correlation to nearby holes. The fact that the clearly overpressured wells produce type 1 water suggests that this water represents the overpressured zone.

Type 1 water also is present in several normally pressured wells (O.E. Lacour-1, G.G. Lacour-1, Planters-2, Brown-2, Deville-1). This water is tentatively attributed to leakage of overpressured water into the normally pressured zone. Posey et al. (1985) found evidence for similar leakage from overpressured to normally pressured zones in southern Louisiana.

Nearby fields, mainly normally pressured, show several different types of water, based on water analyses obtained from oil company files. At the normally pressured Port Hudson field, associated with a salt dome, formation water is much more saline (125,000–198,000 mg/l Cl). Relations on a plot of Cl vs. MCl_2 (Carpenter, 1978) suggest it contains a component of residual brine from evaporites, perhaps leaking out of the salt dome. Of three wells sampled for this study from the nearby Judge Digby field (Figure 1), one appears to be our type 2 water, one appears to be type 1, and one is intermediate. A third type of water with lower Cl/MCl_2 also is represented among company data for the lower Tuscaloosa fields of the area.

The data clearly indicate that several types of water are present in the lower Tuscaloosa Formation. They suggest that the normally pressured zone is commonly occupied by more saline waters (type 2 and the very saline Port Hudson type mentioned above) than the overpressured zone, but that the overpressured type (type 1) has leaked into the normally pressured zone at the Moore-Sams–Morganza field. Clearly the lower Tuscaloosa is characterized by a relatively complex set of formation waters, implying a complex flow history through the formation.

ORIGIN OF WATER

The amounts of Cl and Br in water are conservative in most processes affecting water (Carpenter, 1978). The Cl/Br ratio of seawater is about 295. On evaporation of seawater, this ratio is maintained until halite precipitates, when the ratio in the residual brine decreases significantly. Dilution of seawater with fresh water does not change the ratio, but mixing of the residual brine (after halite precipitation) with fresh water or seawater can decrease the ratio. In contrast, dissolution of halite can increase it.

Ratios of Cl/Br for the Moore-Sams–Morganza samples range from 260 to 330 and average approximately 295, indicating that seawater is the probable major source, and that only very small proportions of residual evaporite brine or water that has dissolved halite can be a component. The type 1 and 2 waters do not clearly differ in Cl/Br, although most samples from overpressured reservoirs tend to have slightly lower Cl/Br than normally pressured waters. This relation suggests that the overpressured waters may contain a small component of residual evaporite brine and that type 2 waters may contain a small component from halite dissolution.

The concentration of Cl in ocean water (35‰ salinity) is about 19,300 mg/l (Holland, 1978). The type 1 waters, averaging 20,000 mg/l, are therefore very similar to seawater with respect to Cl, Br, and Cl/Br, and undoubtedly were derived mainly from seawater. A minor component of residual evaporite brine may raise the Cl concentration to 20,000 mg/l. Minor dilution with water from clay mineral dehydration may be responsible for samples with $\delta^{18}O$ near 8‰ and concentrations of Cl lower than 19,000 mg/l, as in the samples from the Ravenswood-5 and O.E. Lacour wells.

The type 2 waters can be explained as dominantly formed by moderate evaporation of seawater, short of halite precipitation, in order to reach Cl values of 33,500 mg/l in the Howard well. Some halite dissolution may also be involved, as noted above. The salt dome present at the Port Hudson field about 5 mi (8 km) east of the Moore-Sams–Morganza fields is a possible source (Funkhouser et al., 1980), and Jurassic salt is presumably present beneath the region. Migration of a more saline water into the Moore-Sams–Morganza area is not surprising.

SILICA SATURATION

At 160–175°C, the solubility of quartz is 170 to 220 mg/l (Holland, 1979). Even before correction for condensation, most samples exceed the quartz saturation level. After correction of waters using equations (1) and (2), SiO_2 ranges from 204 to 462 mg/l, with a median value of 345 mg/l (Figure 4). Samples from wells with high water production span most of the range and include the highest values. Therefore, the reservoir waters are considerably supersaturated in quartz. Supersaturation does not correlate with overpressuring or with type of water; all types of water are supersaturated.

These high silica concentrations are unusual if not unique in the Gulf Coast region. Values this high have not been reported in the literature.

A possible explanation for supersaturation is the presence of organic complexing (Surdam and MacGowan, 1987; Bennett et al., 1988). However, the total organic alkalinity of the MSM samples is equivalent to only 41 to 286 mg/l acetate, which in comparison with the much higher values used in experiments of Surdam and MacGowan (1987) and Bennett et al. (1988) does not appear adequate to account for appreciable dissolved silica.

The marked supersaturation of quartz clearly indicates that it is capable of precipitating to seal fractures and pores. The rate of precipitation will vary with temperature and surface area of silica phases available (Rimstidt and Barnes, 1980). For conditions of

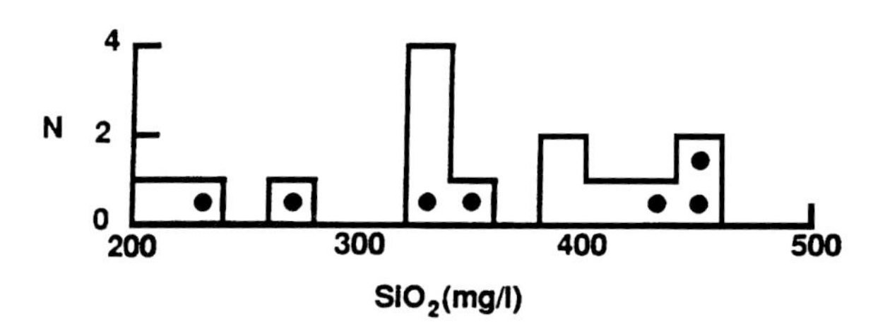

Figure 4. Histogram of SiO_2 contents after correction for condensate. Dot indicates high water/gas ratio (>100 bbl/mmcf) with negligible correction.

closely packed sand grains, 1 mm in diameter, at 160–175°C, the time required to approach within 63% of equilibrium is about 1 month, and equilibrium would be expected to be effectively reached in about 6 months (Rimstidt and Barnes, 1980, Figures 5 and 6). Thus, any fracturing or secondary compaction that exposes quartz surfaces to these solutions will quickly lead to quartz precipitation and reduction of dissolved silica concentration until saturation is reached. If additional solution of similar supersaturation flows into the zone, the fractures and pore spaces will rapidly become completely sealed, assuming no inhibition of precipitation.

The fact that MSM samples are so highly supersaturated indicates either that they have only recently become supersaturated, or that quartz precipitation is inhibited in some way, or both. The presence of chlorite coatings on the quartz grains must occlude the surface and inhibit precipitation, and probably serves to maintain supersaturation. A corollary is that the chlorite coatings may be a key factor in maintaining quartz supersaturation until some event exposes fresh surfaces, after which the supersaturated solution can reseal the zone of disturbance.

Two sources of the supersaturation can be envisioned. One possibility is that the reservoir fluid has been recently derived from deeper in the basin where temperatures are higher. Quartz solubility reaches 350 mg/l, the median value, at about 220°C. At a geothermal gradient of 30°C/km, typical of shallower depths in the area, the fluids would have to come from about 2 km (6500 ft) deeper in the basin to account for 350 mg/l SiO_2, and still deeper to explain the highest uncorrected values of 420–460 mg/l. Because the lower Tuscaloosa sandstones pinch out within a few miles downdip, and because maintenance of this supersaturation during the time necessary for fluids to migrate from depth is difficult to accept, this source of silica seems unlikely.

The alternative is that some silicate-decomposition reaction, such as feldspar, mica, or smectite breakdown, is releasing silica within or very close to the reservoirs. Such reactions commonly produce silica concentrations exceeding quartz solubility, up to amorphous silica saturation. Such a silica-releasing reaction would have to be active either currently or very recently (probably within tens of years) in order to account for the observed supersaturation, even if chlorite coatings are inhibiting quartz precipitation. The existence of supersaturation in both type 1 and type 2 water and in both overpressured and normally pressured waters argues for currently active silica release in the vicinity of the gas reservoirs.

CO_2 PRESSURE AND CARBONATE SATURATION

The CO_2 concentrations in gas from the MSM wells are uniformly about 6 mole % (Table 3). Wells in nearby fields of the Tuscaloosa trend have up to 12% CO_2 (Rougan-1 well, Profit Island field, based on data collected in this study). A CO_2 level of even 6% is relatively high for natural gases, at least for those from shallower depths, and implies that some reaction has occurred to produce CO_2 at or deeper than the MSM production. A high P_{CO_2} in the aqueous phase of the MSM reservoirs is also implied. The reservoir P_{CO_2} can be estimated by

$$P_{CO_2} = P_t X_{CO_2}$$

where P_t is the total pressure, obtained from the RFT measurements, and X_{CO_2} is the mole fraction of CO_2 in the gas phase, as shown by the analyses.

Calculations of saturation state of calcite and dolomite in the reservoir waters, after corrections for dilution, have been made using the program SOLMINEQ.88 PC/SHELL (Perkins et al., 1991; Kharaka et al., 1988). The saturation index is defined as

$$SI = \log(AP/K_{sp}) \tag{3}$$

where AP is the observed activity product in the water sample, and K_{sp} is the corresponding solubility product.

Using SOLMINEQ.88, if CO_2 is added to the waters (after correction for dilution) until the calculated P_{CO_2} is reached, dolomite is found to be slightly undersaturated (SI = –0.12 to –0.67, median = –0.35 for wells producing significant water). Calcite is slightly more undersaturated. The slight calculated undersaturation of dolomite is small enough that it may be caused by some combination of impurities in the dolomite, decrease of reservoir pressures from predrilling values with resultant dolomite precipitation, underestimation of reservoir temperature or degree of dilution, or errors in the thermodynamic data or activity coefficients. If temperature is increased 15°C, a change that is entirely possible, most samples are slightly supersaturated with dolomite, though still undersaturated with calcite. Since dolomite is common as a cement in the sandstones, it is likely that the waters are actually saturated with dolomite.

Because of the decrease in total pressure from the overpressured conditions to normally pressured reservoirs, the P_{CO_2} of the latter is significantly lower. Under conditions where CO_2 is present in a gas phase, dissolution of calcite may occur by two different reactions. If carbonate provides the main pH buffer,

$$CaCO_3(s) + CO_2(g) + H_2O = Ca^{2+} + 2HCO_3^- \tag{4}$$

Conversely, if pH is buffered by other species, such as organic acids or mineral reactions (kaolinite–illite or kaolinite–chlorite), then the effective reaction may be

$$CaCO_3(s) + 2H^+ = Ca^{2+} + CO_2(g) + H_2O \tag{5}$$

Analogous equations apply for dolomite.

The samples from normally pressured and overpressured wells show similar SI values for carbonate minerals and have similar calculated pH (4.40 ± 0.05) after CO_2 addition. Since P_{CO_2} is higher in overpressured reservoirs, some other constituent must adjust to maintain saturation. Both the room temperature analyses and the SOLMINEQ calculations at elevated temperature and high P_{CO_2} show that HCO_3^- is generally higher in the overpressured waters. This pattern indicates that reaction (4) occurs in the reservoirs, and that release of CO_2 pressure will cause carbonate to precipitate. The precipitation of carbonate on leakage of overpressured waters furnishes an additional method of sealing any leaks in the overpressured zone.

A tendency for Ca to be higher in the type 2 waters is also evident, but these waters are also near saturation in dolomite.

Weedman et al. (1992a) show that carbonate cement in the seal zone was dissolved, followed by secondary compaction. A possible scenario is ingress of high P_{CO_2}-gases and accompanying waters that were able to dissolve carbonate, resulting in collapse to form a seal. On formation of the seal and near stagnation, the waters became saturated in dolomite. Any further leakage of overpressured fluid with high P_{CO2} into a normally pressured regime containing a gas phase would result in CO_2 exsolution, carbonate precipitation, and extension of the seal.

CONCLUSIONS

Waters associated with the pressure seal at the Moore-Sams–Morganza field are characterized by several unusual features. Two main types of water are present in the reservoirs of the area. One type approximates seawater in salinity and in Cl/Br, and characterizes the overpressured zone and some of the normally pressured zone. A second, more saline water, apparently formed mainly by slight evaporation of seawater, is found in the northwest corner of the field and perhaps elsewhere. Minor amounts of mixing with other types of water are also suggested by the data, and highly saline waters are associated with a nearby salt dome. In general, the waters of this zone are very heterogeneous, suggesting that flow in relatively complex patterns has characterized the lower Tuscaloosa zone.

Both main types of water are markedly supersaturated with quartz, to levels not documented in other waters of the Gulf Coast. The high degree of supersaturation and its occurrence in both types of waters implies an active silicate-decomposition reaction in waters both above and below the seal. The precipitation of quartz (or other forms of silica) would reestablish the sealed condition if any fractures should occur. Maintenance of silica supersaturation probably depends on chlorite coatings on the quartz grains.

The gases of the fields contain high levels of CO_2, which are identical in the normal and overpressured regimes. Owing to the higher total pressure in the overpressured zone, the activity of CO_2 in the aqueous phase differs in the different pressure regimes. The data indicate that leakage of overpressured water to a lower pressure zone with a gas phase would be accompanied by CO_2 exsolution and precipitation of carbonate.

This pressure seal is relatively deep compared with most such seals, which occur in the vicinity of 10,000 ft (3000 m) (Powley, 1990). However, at least some of the phenomena and processes indicated for this area may be operative in forming shallower seals.

ACKNOWLEDGMENTS

We are indebted to Suzanne Weedman and Susan Brantley for valuable discussion and review of the manuscript and to Terry Engelder for encouragement and assistance throughout the project. AMOCO was very helpful in collecting the samples and providing information on the wells. The research was supported by the Gas Research Institute under contract 5088-260-1746.

REFERENCES CITED

Bennett, P.C., Melcer, M.E., Siegel, D. I., and Hassett, J. P., 1988, The dissolution of quartz in dilute aqueous solutions of organic acids at 25°C: Geochimica et Cosmochimica Acta, v. 52, p. 1521–1530.

Carpenter, A. B., 1978, Origin and chemical evolution of brines in sedimentary basins: Oklahoma Geological Survey, Circular 79, p. 60–77.

Coan, C. R., and King, A. D., Jr., 1971, Solubility of water in compressed carbon dioxide, nitrous oxide and ethane. Evidence for hydration of carbon dioxide and nitrous oxide in the gas phase: J. Amer. Chem. Soc., v. 93, p. 1857–1862.

Deming, D., 1989, Application of bottom-hole temperature corrections in geothermal studies: Geothermics, v. 18, p. 775–786.

Dewers, T., and Ortoleva, P., 1988, The role of geochemical self-organization in the migration and trapping of hydrocarbons: Applied Geochemistry, v. 3, p. 287–316.

Friedman, I., and O'Neil, J. R., 1977, Compilation of stable isotope fractionation factors of geochemical interest: U.S. Geological Survey Professional Paper 440-KK.

Funkhouser, L. W., Bland, F. X., and Humphris, C. C., 1980, The deep Tuscaloosa gas trend of southern Louisiana: Oil and Gas J., Sept. 8, p. 96–101.

Holland, H. D., 1978, The chemistry of the atmosphere and oceans: Wiley–Interscience, New York, p. 154.

Holland, H. D., 1979, The solubility and occurrence of non-ore minerals: *in* Geochemistry of Hydrothermal Ore Deposits, ed. by H. L. Barnes, Wiley–Interscience, New York, p. 469.

Hunt, J. M., 1990, Generation and migration of petroleum from abnormally pressured fluid compartments: AAPG Bull., v. 74, p. 1–12.

Kehle, R. O., 1972, Geothermal energy of North America (Annual Progress Report): Amer. Assoc. Petroleum Geol. Tulsa.

Kharaka, Y. K., Callender, E., and Wallace, R. H., 1977, Geochemistry of geopressured geothermal waters from the Frio Clay in the Gulf Coast region of Texas: Geology, v. 5, p. 241–244.

Kharaka, Y. K., Gunter, W. D., Aggarwal, P. K., Perkins, E. H., and Degraal, J. D., 1988, SOLMINEQ.88: A computer program for geochemical modeling of water–rock interactions: U.S. Geol. Survey Water Res. Inv. Rept. 88-4227, 420 p.

Kobayashi, R., Song, K. Y., and Sloan, E. D, 1987. Phase behavior of water/hydrocarbon systems, *in* Petroleum Engineering Handbook, ed. by H. B. Bradley, Soc. of Petroleum Engineers, Richardson, TX, p. 25-11 to 25-15.

Lico, M. S., Kharaka, Y. K., Carothers, W. W., and Wright, V. A., 1982, Methods for collection and analysis of geopressured geothermal and oil field waters: U.S. Geological Survey Water Supply Paper 2194, 21 pp.

McKetta, J. J., and Wehe, A. H., 1958, How to determine the water content of natural gases: World Oil, July, p. 122–123.

Perkins, E. H., Kharaka, Y. K., Gunter, W. D., and Debraal, J. D., 1991, Geochemical modeling of water–rock interactions using SOLMINEQ.88, *in* Chemical modeling of aqueous systems II, ed. by D. C. Melchior and R. L. Bassett, Symposium Series No. 416, American Chemical Soc., p. 117–127.

Posey, H. H., Workman, A. L., Hanor, J. S., and Hurst, S. D., 1985, Isotopic characteristics of brines from three oil and gas fields, southern Louisiana: Trans. Gulf Coast Assoc. of Geol. Soc., 35th Ann. Meeting, p. 261–267.

Powley, D., 1990, Pressure and hydrogeology in petroleum basins: Earth-Science Reviews, v. 29, p. 215–226.

Rimstidt, J. D., and Barnes, H. L., 1980, The kinetics of silica–water reactions: Geochimica et Cosmochimica Acta, v. 44, p. 1683–1699.

Smith, G. W., 1985, Geology of the deep Tuscaloosa (Upper Cretaceous) gas trend in Louisiana: *in* Habitat of oil and gas in the Gulf Coast: Gulf Coast Section, Soc. Econ. and Pal. Min., Proc. of 4th Ann. Research Conf., p 153–190.

Surdam, R. C., and MacGowan, D. M., 1987, Oilfield waters and sandstone diagenesis: Applied Geochemistry, v. 2, p. 127–149.

Tigert, V., and Al-Shaieb, Z., 1990, Pressure seals: their diagenetic banding patterns: Earth–Science Reviews, v. 29, p. 227–240.

Weedman, S. D., Brantley, S. L., and Albrecht, W., 1992a, Secondary compaction after secondary porosity: Can it form a pressure seal?: Geology, v. 20, p. 303–306.

Weedman, S. D., Guber, A. L., and Engelder, T., 1992b, Pore pressure variation within the Tuscaloosa trend: Morganza and Moore-Sams fields, Louisiana Gulf Coast: J. of Geophysical Res., v. 97B, p. 7193–7202.

IV. The Michigan Basin

Chapter 11

Anomalous Pressures in the Deep Michigan Basin

Jean M. Bahr
Gerilynn R. Moline
Gregory C. Nadon
University of Wisconsin—Madison
Madison, Wisconsin, U.S.A.

ABSTRACT

In this chapter we examine pressures in the St. Peter Sandstone and associated formations of the deep Michigan basin. A comparison of computed brine heads to surface elevations reveals a large area of overpressures within the St. Peter Sandstone and the Glenwood Formation to the west and north of Saginaw Bay. Contrary to the patterns expected for a steady-state, topographically driven flow system, heads in these formations are highest in the regional discharge area. Vertical gradients between the Glenwood and the St. Peter and between the St. Peter and the Prairie du Chien Group would generate downward flow in the regional discharge area, which is also inconsistent with a steady-state flow system. Low-permeability zones must exist within the Glenwood and St. Peter to inhibit equilibration with normal pressures in the underlying and overlying units. Overpressures appear to be dissipating by both upward and downward leakage through high-permeability zones located in anticlinal structures and possibly related to basement faults. Within the larger area of anomalous pressures, repeat formation test data from selected wells reveal even greater overpressures locally within the St. Peter. These vertical variations in pressures are associated with vertical variations in permeability, suggesting a stacked system of compartments separated by low-permeability zones of diagenetic origin.

INTRODUCTION

Abrupt offsets in plots of fluid pressure versus depth or elevation have been used to infer the existence of abnormally pressured "compartments" that are isolated from the surrounding hydrodynamic regime by low-permeability seals (Hunt, 1990; Powley, 1990). However, as is well known in the hydrogeologic literature (e.g., Freeze and Cherry, 1979), deviations from a hydrostatic pressure profile can be expected in an open system as a result of topographically driven regional flow. Downward flow, which occurs in recharge areas of a basin, is associated with underpressures (relative to hydrostatic) at depth. Conversely, regional discharge areas are characterized by overpressures at depth. Steady-state hydrodynamic models have been used in recent years to explain regional underpressuring in the Denver (Belitz and Bredehoeft, 1988) and Palo Duro (Senger and Fogg, 1987) basins. Furthermore, Darcy's Law requires that vertical flow across a low-permeability confining unit must be accompanied by a steep hydraulic gradient. Toth et al. (1991) provide additional discussion of these points. The conclusion that can be reached from these considerations is that although abrupt offsets in a pressure profile generally do result from a large permeability contrast, they do not necessarily indicate limited hydraulic communication with a surrounding hydrodynamic system.

In a hydrodynamic sense, overpressures and underpressures resulting from topographically driven flow are not truly anomalous. However, there are other mechanisms besides regional flow that could give rise to "anomalous" overpressures and underpressures. Such anomalous pressures could be the consequence of currently active pressure-generating mechanisms, such as compaction or hydrocarbon maturation. They could also be manifestations of relatively slow equilibration to a steady-state system following a change in boundary conditions or the elimination of a pressure source. In cases of slow equilibration, the anomalous pressure zones could be considered to be in limited hydraulic communication with the surrounding system and would thus at least partially fulfill the definition of a compartment as proposed by Powley (1990). However, it should be recognized that because no formation is completely impermeable, such nonequilibrium conditions and the zones of anomalous pressures themselves are inherently transient. Nevertheless, as demonstrated by Toth and Millar (1983) for a one-dimensional example, significant permeability contrasts are required to allow anomalous pressures to persist over geologic time scales.

In this chapter we examine evidence for anomalous pressures in the deep Michigan basin. The primary formation of interest is the Middle Ordovician St. Peter Sandstone, which became an active gas play in the 1980s with production extending to depths of over 3500 m. Related chapters in this volume (Drzewiecki et al.; Shepherd et al.; and Wang et al.) describe the sedimentologic, diagenetic, and thermal processes that have affected this formation. Figure 1 shows structure contours on the top of the St. Peter, in meters relative to sea level (masl), along with locations of major gas fields. Over most of the Lower Peninsula of Michigan, the top of the St. Peter occurs at elevations of less than −1700 masl. Producing zones are concentrated in the deep areas of the basin to the west and north of Saginaw Bay and in a shallower area of the west-central basin.

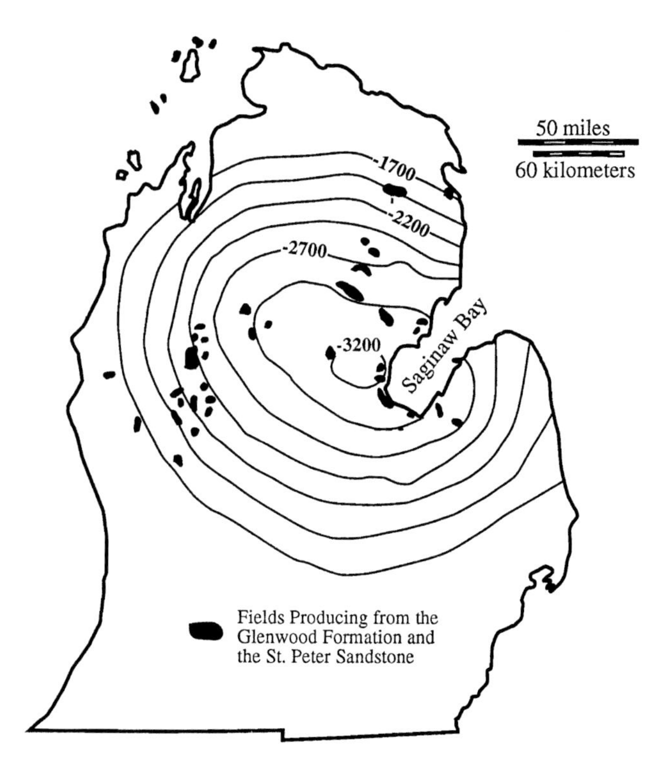

Figure 1. Structure contours on the top of the St. Peter in the Michigan basin and locations of major producing fields in the Glenwood and St. Peter. Producing fields after Catacosinos et al. (1990). Contour interval = 250 m.

REGIONAL SCALE PRESSURE ANOMALIES

Areal Distribution of Heads

Identification of anomalous pressures requires a three-dimensional approach that considers not only vertical pressure profiles but also the areal distribution of pressure across a basin. Comparison of the observed pressure distribution to the general pattern expected for a topographically driven flow system should allow one to distinguish between zones of overpressures or underpressures that are a consequence of regional flow and zones in which pressures are anomalous with respect to basin hydrodynamics. Regional flow paths in a steady-state, topographically driven system extend from recharge areas in topographic highs to discharge areas in topographic lows.

The distribution of regional recharge and discharge areas for the Michigan basin can be inferred from the generalized topography of the Lower Peninsula of Michigan illustrated in Figure 2. Recharge areas are located in the north-central portion and at the southern margins of the Lower Peninsula. Discharge occurs to the Great Lakes to the east and west. The broad lowland region surrounding Saginaw Bay would be expected to focus discharge of groundwater that enters the system in both the north-central and southern recharge areas. The presence of saline groundwater at shallow depths in this region (Long et al., 1988), which is likely the result of upward flow of basin brines, suggests that the topographically driven, regional flow system extends to considerable depth. Modeling results reported by Mandle and Westjohn (1989) support the existence of a topographically driven flow system extending to the base of the Mississippian Marshall Sandstone.

Fluid flow in porous media, including regional-scale flow in sedimentary basins, is governed by

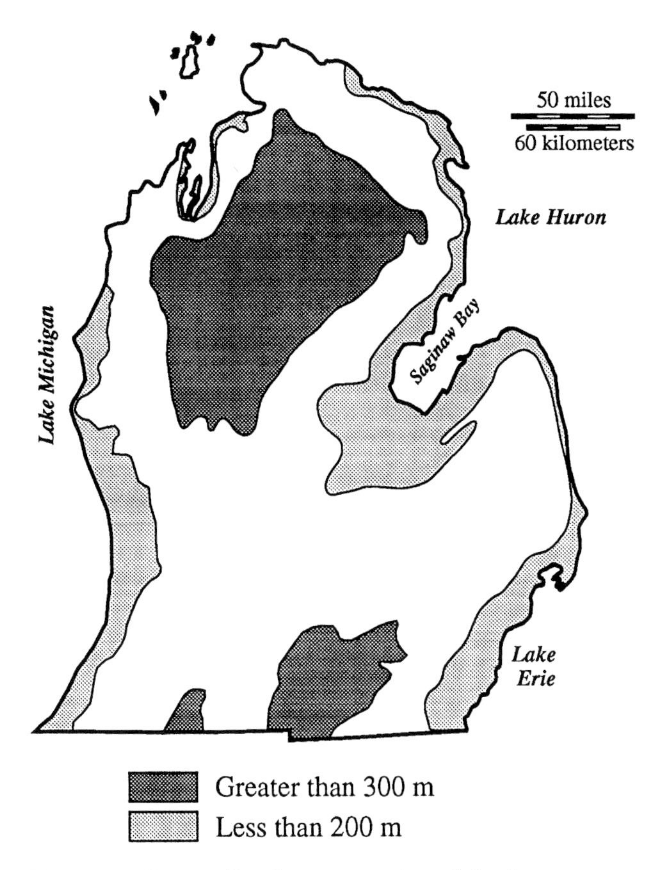

Figure 2. Generalized topography of the Lower Peninsula of Michigan, showing regions of surface elevation exceeding 300 m above sea level and regions of surface elevation less than 200 m above sea level.

Darcy's Law. In the most general form, this law can be expressed as

$$q = \frac{k}{\mu}\left(\nabla p + \rho g \nabla z\right) \tag{1}$$

where q is the specific discharge, k the permeability, μ the fluid viscosity, p the fluid pressure, ρ the fluid density, g the gravitational constant, and z the elevation relative to a datum. This form of Darcy's Law holds for fluids in which density can vary as a function of pressure or solute concentration. For a fluid of a constant density, Darcy's Law can also be written as

$$q = \frac{k\rho g}{\mu}\nabla h \tag{2}$$

where h, the fluid head, is a potential corresponding to the mechanical energy per unit weight of the fluid, measured relative to the fluid in a standard state (Hubbert, 1940). Head is computed by

$$h = \frac{p}{\rho g} + z \tag{3}$$

When fluid flow can be described by equation 2, the distribution and gradients of head can be used directly to determine directions of flow.

Within the Michigan basin, fluids vary from fresh water in most shallow portions of the system to dense brines in the deep basin. This precludes the use of head as the dependent variable for basin-scale flow modeling that incorporates both the shallow and deep portions of the system. However, salinities in the Michigan basin appear to reach an asymptotic value at depths of approximately 2 km (Hanor, 1979). This means that it should be possible to use heads to characterize the potential gradient responsible for fluid flow in deeper parts of the basin. In contrast to the typical groundwater flow problem, in which heads are computed using a freshwater density, heads in this case should be computed using a brine density that corresponds to an integrated average of the fluid density from the water table to the depth at which the pressure measurement is made. Comparison of the resulting brine head to the elevation of the water table then provides a direct measure of the pressure deviation from hydrostatic.

Data from the Michigan Department of Natural Resources (MDNR) indicate that fluids from depths greater than 1700 m, while showing significant variations in cation composition between some formations, have relatively constant specific gravities between 1.27 and 1.30 when measured at surface conditions. The higher in situ temperatures and pressures will modify the fluid density from that measured at the surface, complicating the estimation of an integrated average density from surface-specific gravity measurements. An alternate approach to obtaining estimates of average in situ density for brine head computations is to examine profiles of pressure versus elevation constructed using data from throughout the basin. Extrapolated pressures from drill-stem test (DST) results were obtained from the MDNR and supplemented with additional DST pressures included in a database purchased from Petroleum Information, Inc. (PI). The DST pressures from the PI database were edited to exclude tests yielding extremely low pressures. In most cases, these low pressures could be directly attributed to short shut-in times. Data were grouped by formation after cross-checking formations listed in the PI database with formation picks from wireline logs of nearby wells. The data for each formation were used to generate the plots shown in Figure 3 for the Trenton/Black River formations (combined), the Glenwood Formation, the St. Peter Sandstone, and the Prairie du Chien Group. Lines corresponding to the gradient predicted for an average fluid density of 1.16 g/cm^3 (gradient of 0.50 psi/ft), projected downward from the surface elevation of Lake Huron, are shown on these figures for reference.

The lower pressure points for each formation, particularly the extrapolated DSTs from MDNR, cluster along the 1.16 g/cm^3 line. This suggests that this is a reasonable estimate of the average in situ density. Use of a higher density or surface elevation would shift the reference line to the right toward higher pressures at a given elevation. For all units except the Prairie du

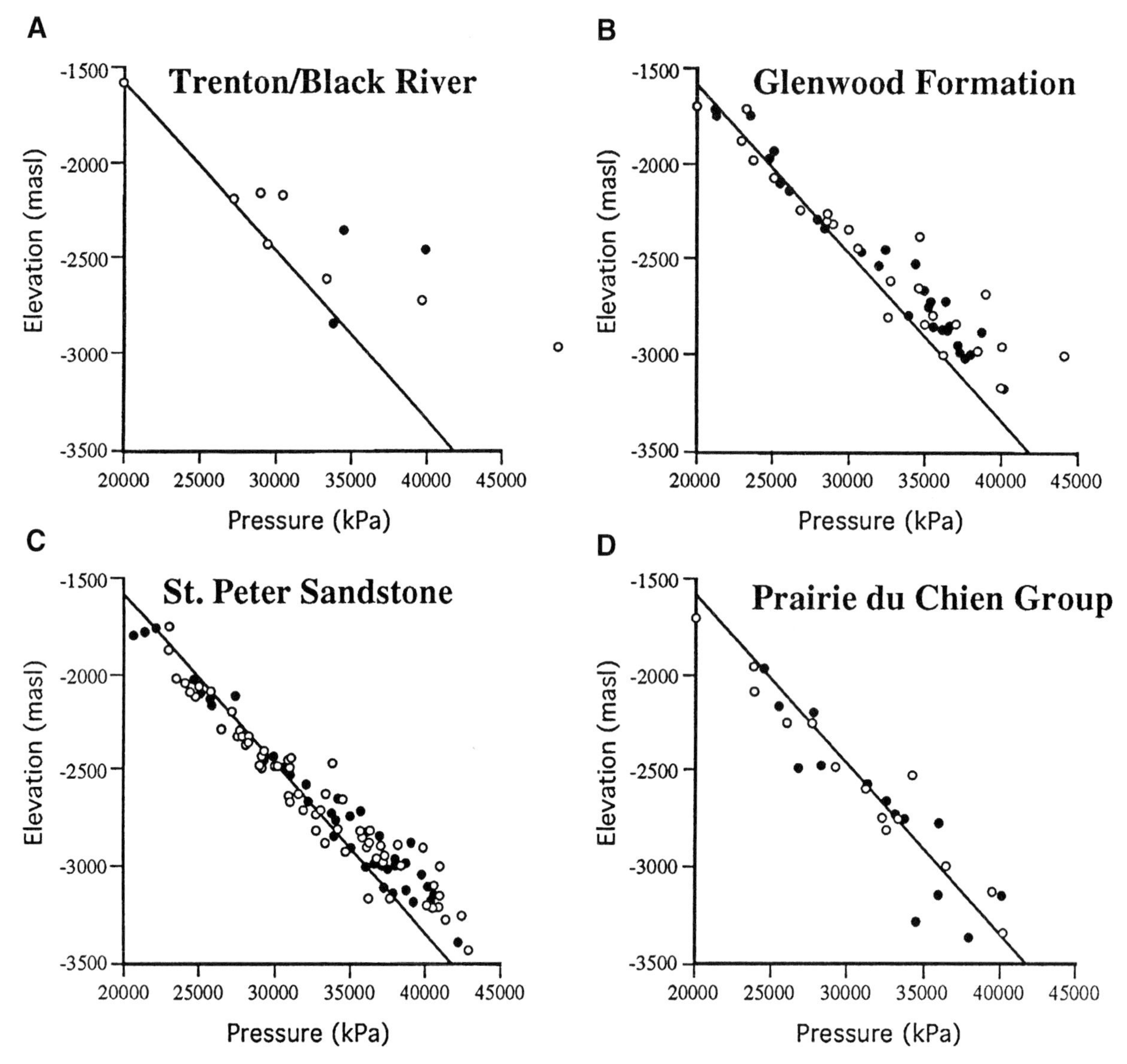

Figure 3. Plots of pressure versus elevation for (A) Trenton and Black River formations, (B) Glenwood Formation, (C) St. Peter Sandstone, and (D) Prairie du Chien Group. Solid symbols are extrapolated drill-stem test pressures obtained from the Michigan Department of Natural Resources; open symbols are pressures from the Petroleum Information, Inc., database. The solid line in each figure is the theoretical pressure profile for a column of fluid with a density of 1.16 g/cm^3 extending to the surface of Lake Huron.

Chien, a majority of the points fall to the high pressure side of the reference line, suggesting the existence of overpressures relative to hydrostatic, particularly at elevations lower than –2500 masl. The combined data for the Trenton/Black River show the largest deviations from the reference line. All the high-pressure points for the Trenton/Black River come from a relatively restricted area of producing fields in the north-central portion of the Lower Peninsula (see Figure 4). The Prairie du Chien Group, also a relatively limited data set compared to the St. Peter and Glenwood, shows approximately equal numbers of points falling above and below the reference line.

Brine heads were computed using measured pressures, elevations relative to sea level, and assuming a brine density of 1.16 g/cm^3. Contours of brine heads are shown in Figure 4 for the Prairie du Chien, the St. Peter, and the Glenwood. The limited number of data points for the Trenton and Black River formations does not allow reasonable contouring of heads over the basin.

All three formations exhibit high heads in the north-central portion of the basin, which corresponds to the general vicinity of the recharge area. The St. Peter and Glenwood head maps also show areas of high heads to the west and north of Saginaw Bay, in the regional discharge area. The high heads in this area would generate lateral flow through these formations in a direction away from the discharge area, opposite to the flow path expected for a steady-state

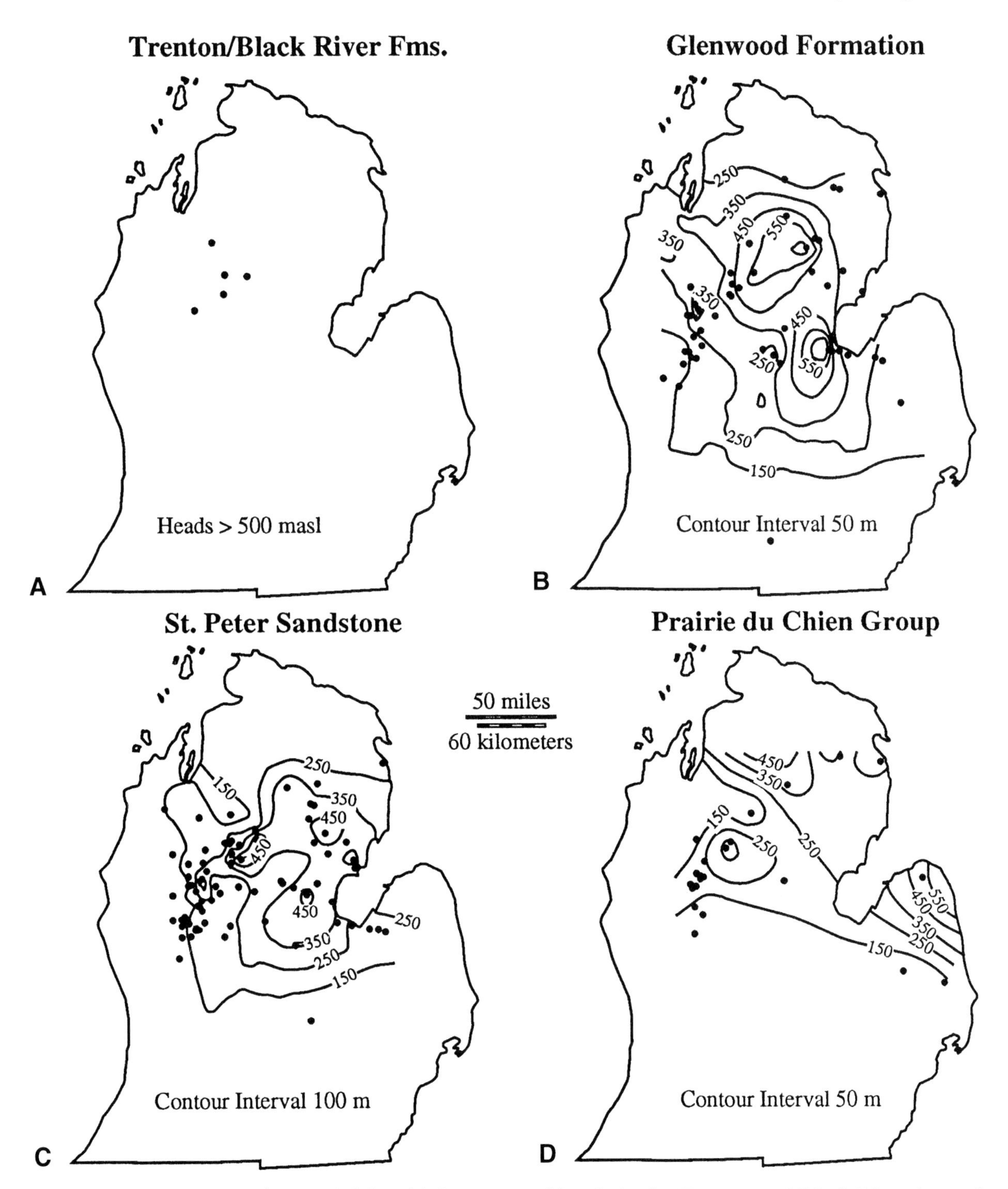

Figure 4. (A) Locations of tests yielding high computed heads in the Trenton and Black River formations. The remaining figures show computed brine heads, in m above sea level, for (B) the Glenwood Formation, (C) the St. Peter Sandstone, and (D) the Prairie du Chien Group. Symbols represent locations of pressure measurement used to compute heads and construct contours.

regional flow system. Contour maps of head computed using a higher brine density show essentially the same patterns as those illustrated in Figure 5 but with a lower absolute magnitude of the computed head. Similar patterns were also obtained assuming a variation in average density with depth of the formation. In that case the highest average densities were assigned to the area around Saginaw Bay, which is the deepest area of the basin as well as the locus of shallow saline groundwater.

Another way to examine the head distribution is to compare computed heads to surface elevation as a

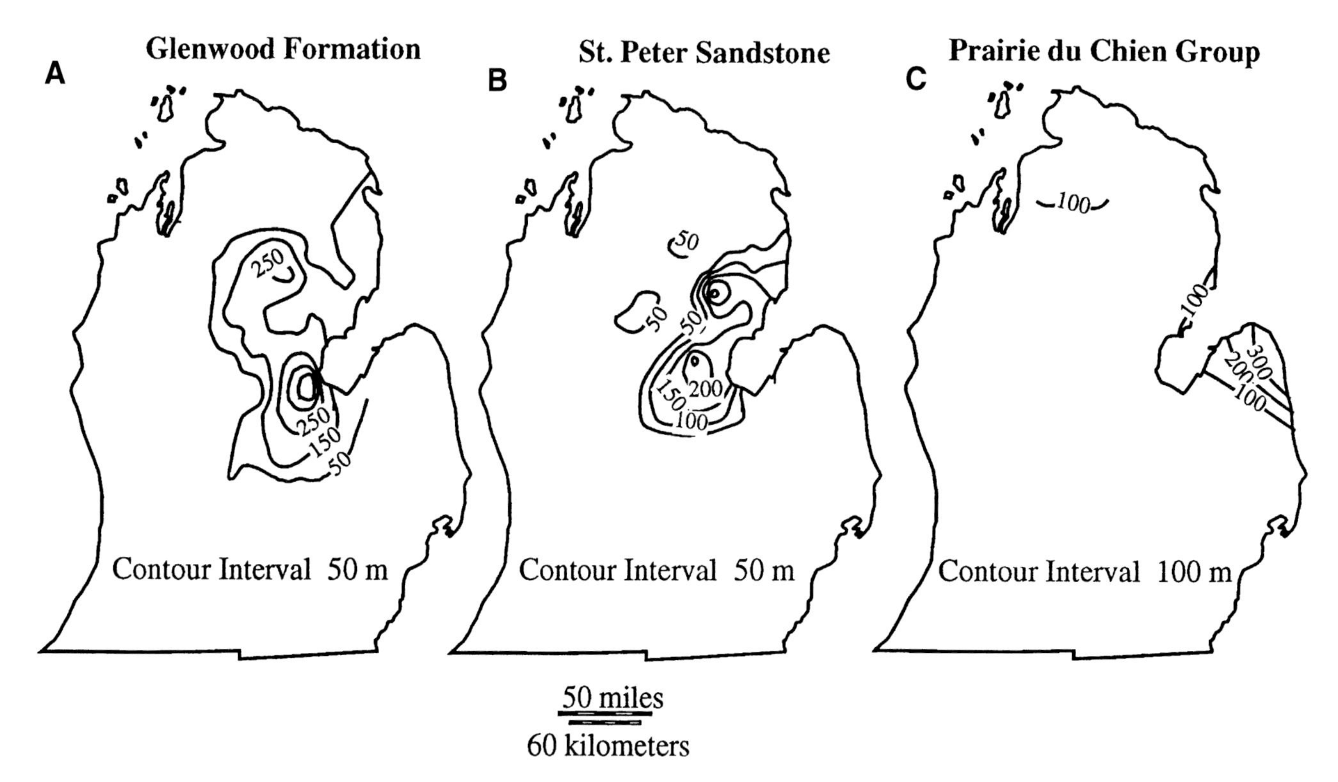

Figure 5. Excess heads (in m above sea level), determined by subtracting surface elevation from computed brine head, for (A) the Glenwood Formation, (B) the St. Peter Sandstone, and (C) the Prairie du Chien Group.

measure of deviation from hydrostatic pressure. For a steady-state flow system, heads at depth should be less than the surface elevation in the recharge area and greater than surface elevation in the discharge area. Figure 5 shows contours of computed head minus surface topography for the Prairie du Chien, the St. Peter, and the Glenwood. The pattern of deviations from hydrostatic conditions in the Prairie du Chien is similar to that expected for a steady-state flow system, with heads below the land surface in the north-central basin and near or slightly exceeding surface elevation near Saginaw Bay. The St. Peter, however, exhibits heads exceeding surface elevations by more than 100 m within the recharge area and by more than 200 m in the discharge area near Saginaw Bay. The Glenwood also shows high heads relative to hydrostatic near Saginaw Bay. Although heads exceeding the land surface elevation are not inconsistent with a regional flow system, the relative magnitudes of the deviations from hydrostatic for the three formations are contrary to those expected in a discharge area. Near Saginaw Bay, heads are highest in the Glenwood, which would generate downward flow to the St. Peter rather than the upward flow typical of a discharge area. Heads greater in the St. Peter than in the Prairie du Chien would also generate downward flow between these two units. Thus, pressures in the St. Peter and Glenwood appear to be anomalous over most of the Saginaw Bay lowland.

Cross Sections of Head

Three cross sections through the zones of excess head in the St. Peter were chosen for further comparison of heads and surface topography as well as to attempt to relate anomalous pressures to subsurface structure. Section A–A′ goes from the regional recharge area across the Saginaw Bay lowland, intersecting zones of maximum head in the recharge area and to the west of Saginaw Bay. Section B–B′ is oriented along the axis of the large anomalous pressure region that extends west and north of Saginaw Bay. Section C–C′, oriented subparallel to section A–A′, originates in a smaller zone of high heads in the northern portion of the recharge area and crosses the zone of maximum excess heads to the north of Saginaw Bay. For each section, head profiles in the Prairie du Chien, St. Peter, and Glenwood were generated from the contoured head maps and then plotted along with surface topography. Values of head for the Trenton or Black River were also plotted on the cross sections at locations corresponding to the projection of nearby wells. Three pressure measurements were available from Trenton/Black River wells located near section A–A′ and one measurement was available from a well near section C–C′. Structure profiles for the Prairie du Chien, St. Peter, Glenwood, Trenton, and Utica formations were also generated along these three sections using contours obtained from formation picks of 163 wireline logs. Zones of

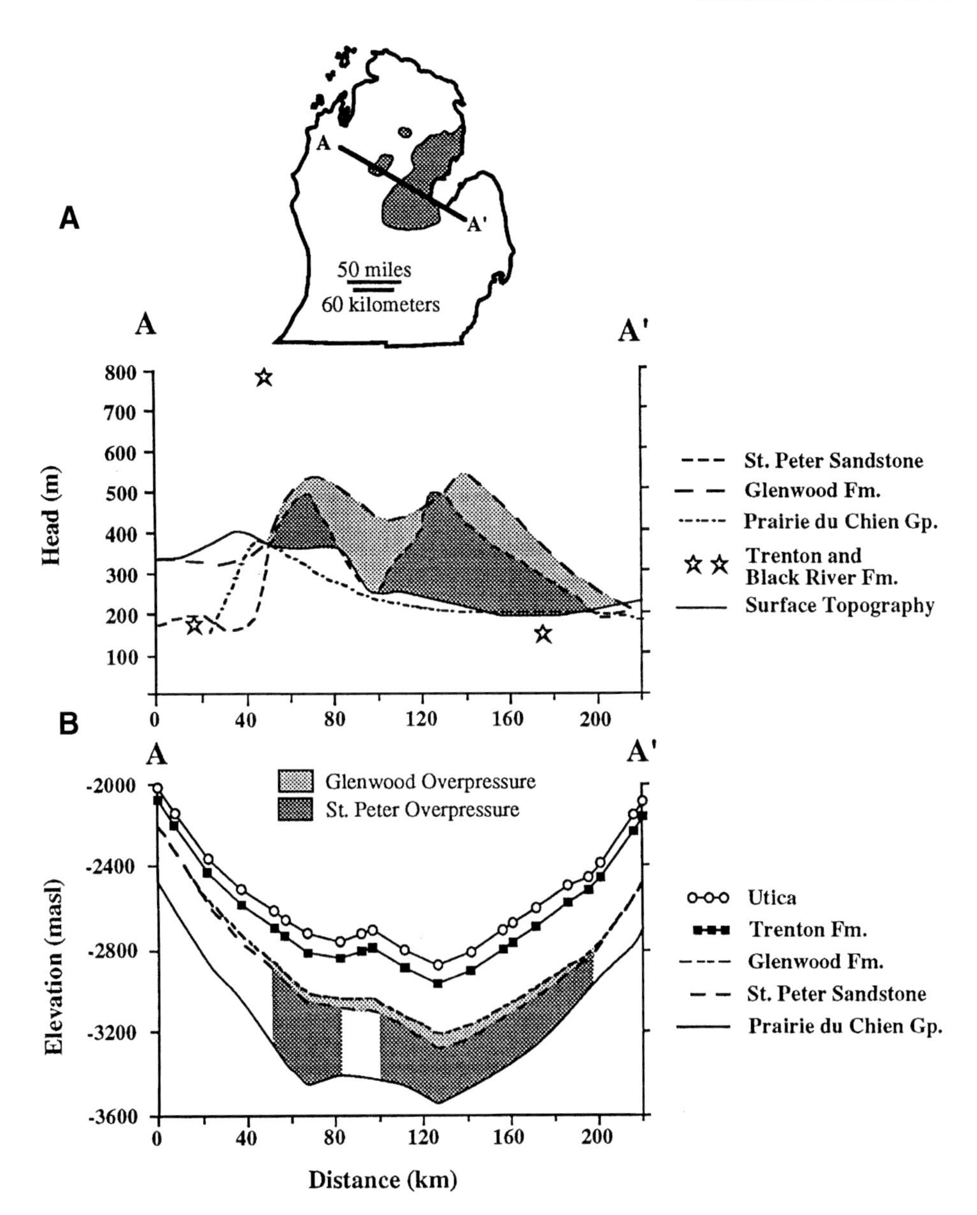

Figure 6. (A) Heads and surface elevation along section A–A′; (B) structure section and zones of overpressure along A–A′ showing top of the Utica, Trenton, Glenwood, St. Peter, and Prairie du Chien.

overpressure are indicated by shading on the head and structure profiles. The head and structure cross sections are shown on Figures 6, 7, and 8.

Along section A–A′ (Figure 6A), heads in the Prairie du Chien are fairly close to surface elevations and decrease steadily from the recharge area in the west to the discharge area in the eastern portion of the section. Heads in the St. Peter and Glenwood show two peaks, one in the recharge area and one in the discharge area. For both formations, these peaks are 150 m or more above the land surface. However, near the center of the cross section, St. Peter heads approximately coincide with surface elevations. The structure cross section (Figure 6B) shows a significant anticlinal structure in all formations at this same location. According to Fisher and Barratt (1985), structures of this type in the Michigan basin are related to basement faults. Faults extending through the base of the St. Peter at this location could create a high-permeability drain, allowing pressures in the St. Peter to equilibrate with normal pressures in the underlying Prairie du Chien. The lower heads in the St. Peter at this location would also induce downward flow from the Glenwood. Partial adjustment of heads in the Glenwood is suggested by the decrease in heads in this formation at a similar location along the cross section. Low-permeability zones in the upper St. Peter or within the Glenwood would inhibit complete equilibration. One computed head value in the Trenton/Black River along this cross section indicates significant overpressures above those in the St. Peter and Glenwood. Overpressuring in the Trenton/Black River cannot be used to explain the regional-scale overpressures in the Glenwood and St. Peter, however, since the pressures in these formations are approximately hydrostatic at the cross section location where overpressures have been measured in the Trenton/Black River.

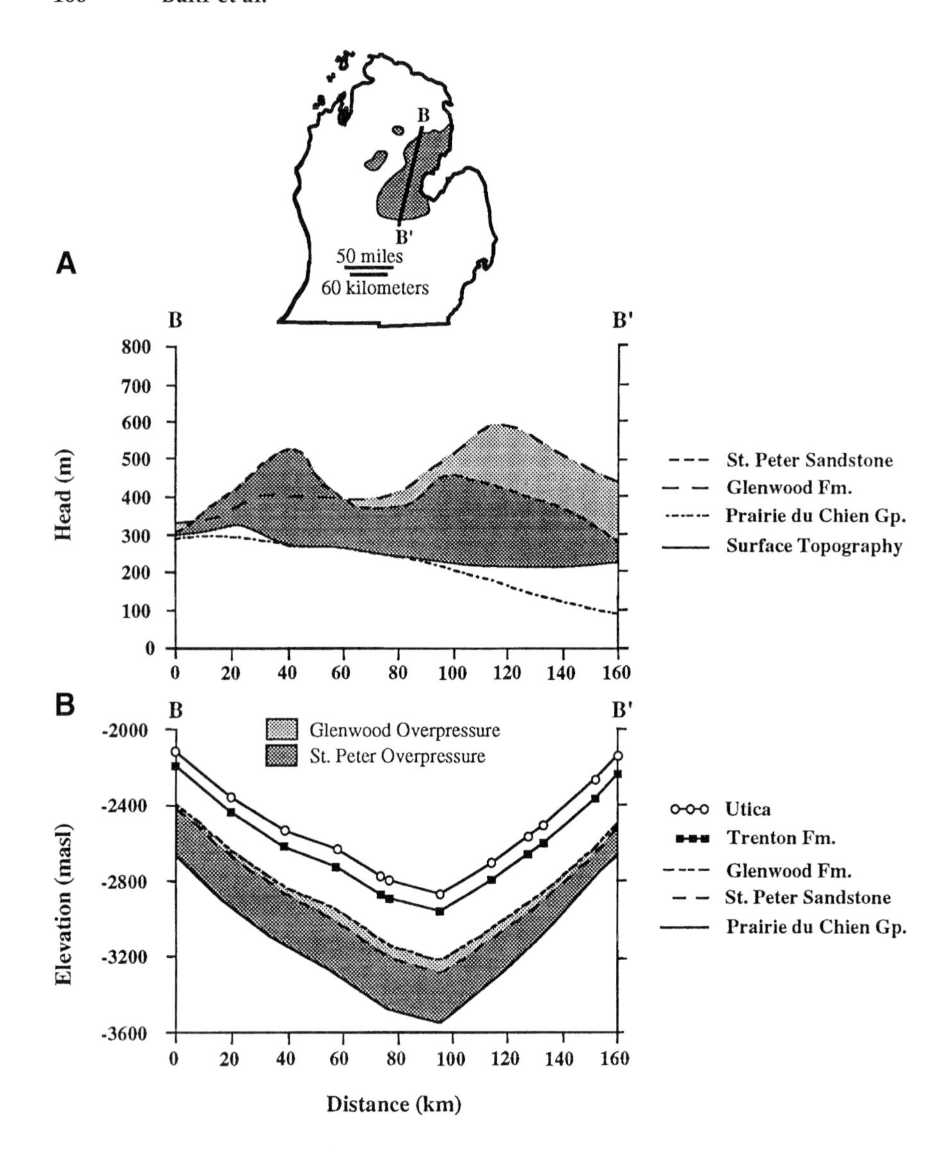

Figure 7. (A) Heads and surface elevation along section B–B′; (B) structure section and zones of overpressure along B–B′ showing top of the Utica, Trenton, Glenwood, St. Peter, and Prairie du Chien.

Section B–B′ shows a similar pattern of nearly hydrostatic heads in the Prairie du Chien and heads significantly above surface elevations in the St. Peter and Glenwood (Figure 7A). As in section A–A′, a decrease in head within the St. Peter occurs near the middle of the section. In this case, however, heads in the St. Peter remain significantly above surface elevations and the effect on heads in the Glenwood is less clear. A subtle bulge in the structure profiles (Figure 7B) at this location could correspond to another anticlinal structure related to basement faulting. Even more intriguing is the fact that at the location of this apparent drain, the gradients between the Glenwood and St. Peter reverse. In the northern portion of the section, heads in the St. Peter are higher than those in the Glenwood, which would induce upward flow, while at the southern end of the section, heads in the Glenwood are higher, which would induce downward flow. These head relations are the opposite of those expected for a topographically driven flow system since surface elevation along this section decreases from north to south.

Section C–C′ also shows a reversal in gradients between the St. Peter and Glenwood (Figure 8A), but in this case the reversal is somewhat more consistent with topography. Near the topographic high, only slight overpressures occur in the St. Peter, but the Glenwood exhibits heads significantly above the land surface. Farther east along the section, heads in the Glenwood decrease as heads in the St. Peter increase. The structure cross section (Figure 8B) shows no obvious features in the area of rapid head changes. However, faulting along anticlinal structures in the West Branch field, which is located to the south of this cross section at the approximate location of the gradient reversal, has been documented by other workers (Prouty, 1988).

Interpretation of Regional-Scale Pressure Anomalies

The areas of excess heads in the St. Peter and Glenwood encompass most of the deep producing fields in these formations. The geometry of the zone of excess heads in the St. Peter, which appears to par-

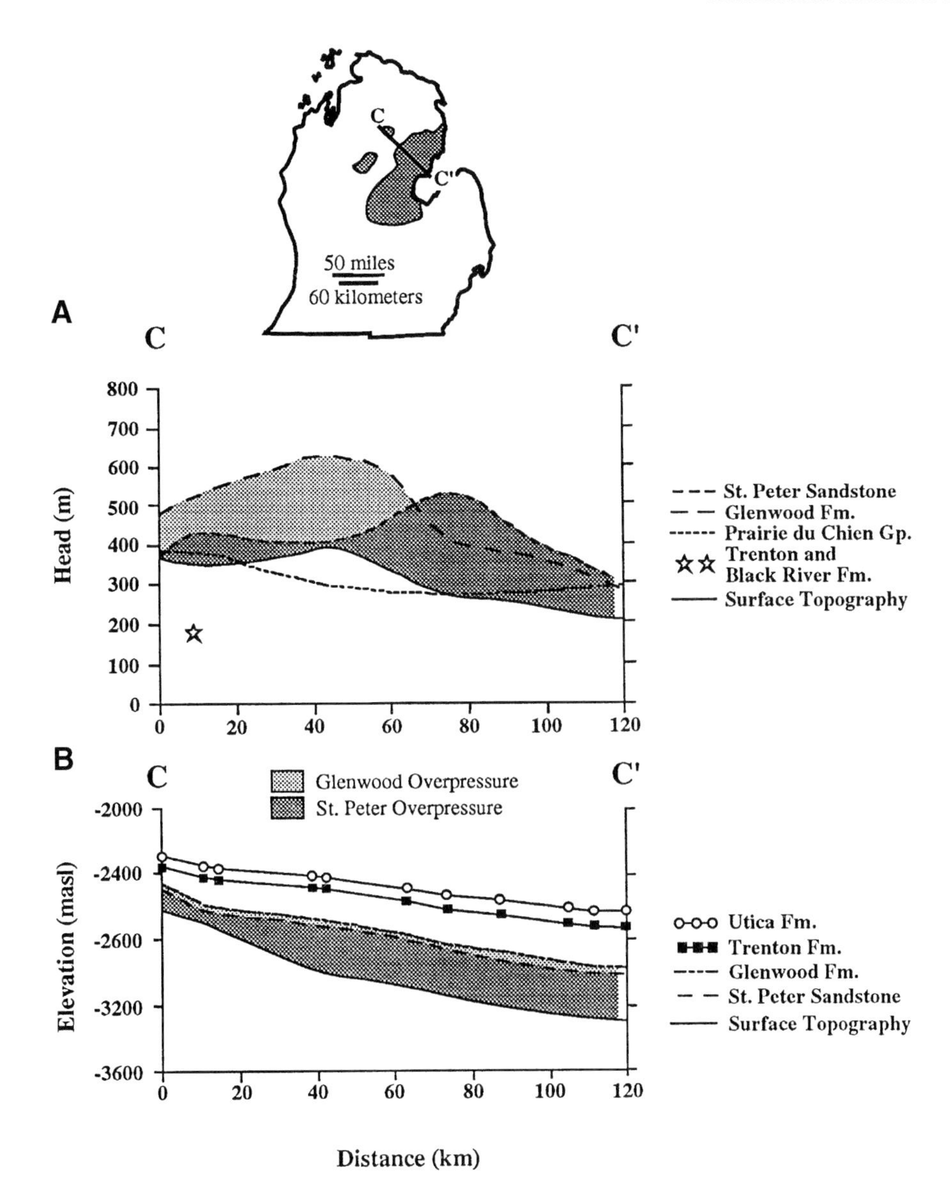

Figure 8. (A) Heads and surface elevation along section C–C′; (B) structure section and zones of overpressure along C–C′ showing top of the Utica, Trenton, Glenwood, St. Peter, and Prairie du Chien.

allel the major glacial moraines surrounding Saginaw Bay as shown in the map of Farrand and Bell (1982), suggests that glacial loading during the advance of ice through the Saginaw Bay lowland may have generated the current overpressures. Given the absence of current conditions promoting hydrocarbon maturation or active compaction, the existing overpressures must have persisted at least since the end of the last glaciation. The region extending west and north of Saginaw Bay may have at one time formed a relatively continuous zone of anomalous pressures in the St. Peter, the Glenwood, and possibly the Trenton and Black River formations. In the terminology of Al-Shaieb et al. (1991), this broad region could be considered a megacompartment. The upper seal probably consists of low-permeability zones in the Glenwood that in some areas extend into the St. Peter or the Trenton. The basal seal appears to be located in the lower St. Peter or in the Prairie du Chien.

Overpressures appear to be dissipating by leakage through high-permeability zones located in anticlinal structures and possibly related to basement faults. In some areas of this region, maximum overpressures are found in the Glenwood, while in others the St. Peter contains the highest overpressures. Thus, both upward and downward drainage appear to be occurring. Over much of the region, low-permeability zones within the Glenwood and within the St. Peter must be present to inhibit equilibration with normal pressures in the underlying Prairie du Chien and overlying Black River.

LOCAL-SCALE ANOMALIES

Evidence for the importance of low-permeability zones within the St. Peter comes from a more detailed examination of pressure profiles in individual wells. Repeat formation test (RFT) pressure and permeability data for seven wells were obtained from MDNR files. Three of these wells are discussed below.

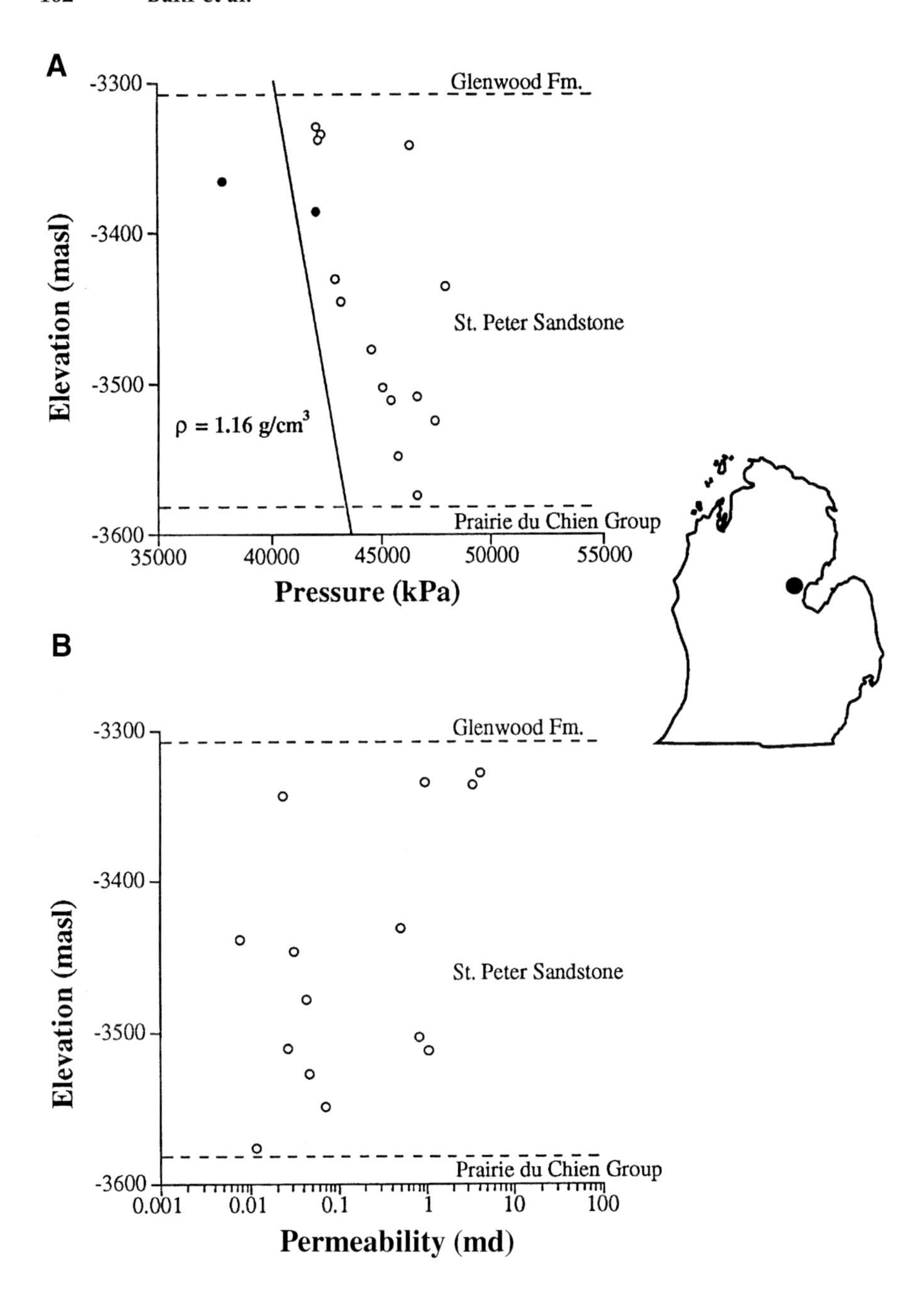

Figure 9. Pressure and permeability data determined from repeat formation tests in the Kielpinski well, Bay County. (A) Pressures versus elevation; open symbols are RFT pressures; solid symbols are DST pressures from other wells in the region; the solid line is the theoretical pressure profile for a column of fluid with a density of 1.16 g/cm^3 extending to the land surface. (B) Permeability values determined from RFT tests, in millidarcys.

Pressure and permeability data for the Kielpinski well in Bay County are plotted on Figure 9. This well is located near the middle of section B–B′, in the region where gradients between the Glenwood and St. Peter reverse. The pressure data show that the St. Peter is consistently overpressured relative to the line corresponding to a density of 1.16 g/cm^3 projected from the surface. Also plotted for reference on Figure 9A are two DST measurements from other wells in Bay County. One of these shows an overpressure of magnitude similar to that of the majority of the RFT measurements. Several of the RFT pressures are significantly higher than the others. Such abrupt vertical variations in pressure within a single well could not be the result of local fluid density variations but, instead, require variations in permeability to maintain a large pressure gradient. The permeability data in Figure 9B show variations of about three orders of magnitude within the St. Peter, with high-permeability contrasts occurring in the depth intervals corresponding to the greatest overpressures. The correspondence of zones of permeability contrasts to large vertical variations in pressure indicates that permeability variations within the St. Peter have led to a hierarchical system of stacked subcompartments. These subcompartments are intervals that are adjusting even more slowly than the St. Peter as a whole to normal pressures in the regional hydrodynamic system.

Similar patterns of variable permeability and vertical variations in pressure are found in the State Foster 1-19 well (Figure 10). This well is located in the Rose City field in Ogemaw County, near section C–C′, in an area where heads computed from DST measurements in the St. Peter are nearly hydrostatic while the

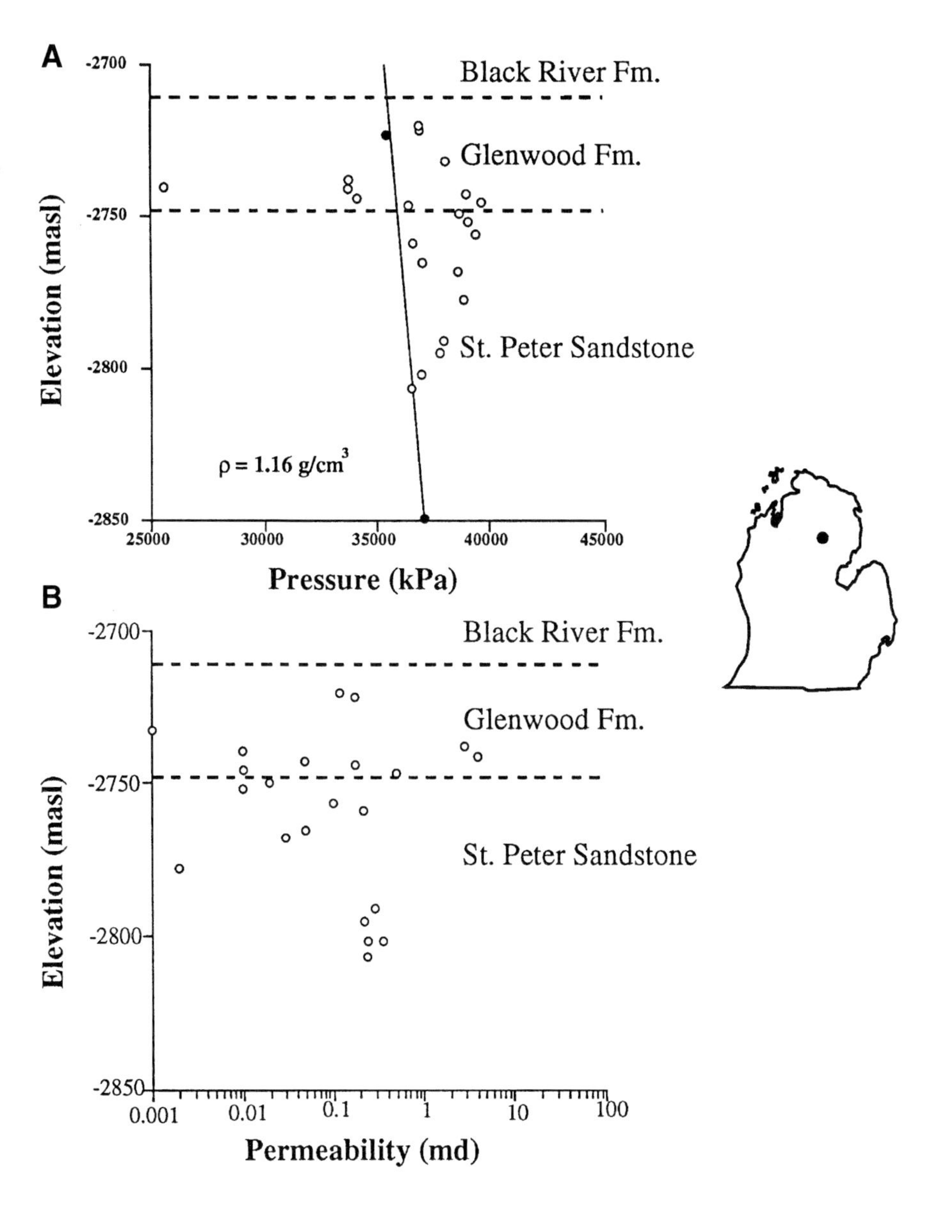

Figure 10. Pressure and permeability data determined from repeat formation tests in the State Foster 1-19 well, Ogemaw County. (A) Pressures versus elevation; open symbols are RFT pressures; solid symbols are DST pressures from other wells in the region; the solid line is the theoretical pressure profile for a column of fluid with a density of 1.16 g/cm^3 extending to the land surface. (B) Permeability values determined from RFT tests, in millidarcys.

Glenwood is overpressured. DST measurements from two other wells in Ogemaw County plot along the hydrostatic line shown in Figure 10A, corresponding to a density of 1.16 g/cm^3 projected from the surface. It should be noted that the DST measurement plotted within the elevation interval of the Glenwood was actually obtained from an interval corresponding to the St. Peter in the tested well. The apparent discrepancy between formations results from the fact that the DST well is located updip from the State Foster 1-19 well. The RFT measurements show that significant overpressures exist in the upper portion of the St. Peter as well as in the Glenwood and the transition zone between the two formations. Underpressures obtained for some RFTs in the transition zone are the result of production elsewhere in the field. Permeability of the St. Peter in this well, shown in Figure 10B, varies by almost four orders of magnitude.

A final example of the variations of pressure and permeability within the St. Peter is illustrated in Figure 11 for the State Foster 1-12 well, located in the same area as the State Foster 1-19 but outside the Rose City field. Only 5 RFT measurements were performed in this well. DST measurements from other wells located updip from the State Foster 1-12 but in the same general area of the basin are also shown on Figure 11. Although several of these appear to fall within the Black River, within the wells tested these elevations actually correspond to the Glenwood and St. Peter. DST measurements from several intervals in the upper St. Peter and Glenwood show normal pressures. Two RFT measurements near the base of the St. Peter yield normal pressures relative to the local hydrostatic gradient, while measurements at a higher elevation within the St. Peter and within the transition zone indicate significant overpressures. Two DST measurements within the middle and lower St. Peter from other wells also show overpressures, although these are less pronounced than those from the RFT measurements. The change from overpressure to normal pressure at the base of the St. Peter requires a basal seal within the lower St. Peter. A low-permeability diagenetically

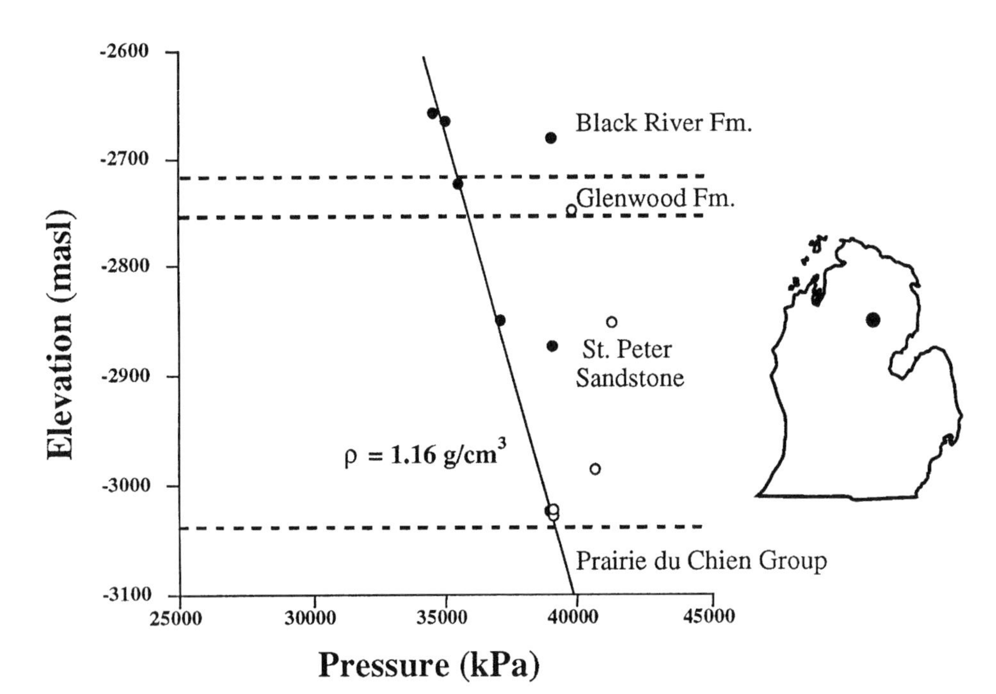

Figure 11. Pressure vs. elevation in the State Foster 1-12 well, Ogemaw County. Open symbols are RFT pressures; solid symbols are DST pressures from other wells in the region; the solid line is the theoretical pressure profile for a column of fluid with a density of 1.16 g/cm^3 extending to the land surface.

banded zone in core from near the base of the St. Peter in State Foster 1-12 could provide such a seal. Hydrologic and diagenetic characteristics of this banded zone are described in related chapters in this volume (Shepherd et al.; Drzewiecki et al.; Moline et al.).

SUMMARY AND CONCLUSIONS

Anomalous pressures within the deep Michigan basin, although subtle compared to overpressures observed in regions of active subsidence such as the Gulf Coast, are nevertheless identifiable through a careful evaluation of heads and gradients between units. A large area of overpressures, which can be considered a megacompartment, exists within the St. Peter Sandstone and the Glenwood Formation to the west and north of Saginaw Bay. The overpressures must be considered inconsistent with a steady-state, topographically driven flow system because heads are highest in the regional discharge area and because vertical gradients are reversed compared to those expected for the current regional flow system. The megacompartment includes most of the producing fields in the deep areas of the St. Peter and Glenwood. Abrupt vertical variations in pressure within individual wells suggest the presence of a series of smaller, stacked compartments within the megacompartment.

It seems likely that the existing overpressures have persisted at least since the last glaciation. Overpressures within the megacompartment appear to be slowly equilibrating with normal pressures within the surrounding stratigraphic units, primarily by leakage through anticlinal structures that likely contain high-permeability fault and fracture zones. The large vertical variations in pressure revealed by the RFT tests require low-permeability zones within the St. Peter to inhibit flow and adjustment to the surrounding pressures.

The primary significance of the pressure patterns revealed by this study lies not in the existence of the overpressures themselves, but in what these overpressures reveal about the controlling permeability distribution within the basin. When permeability contrasts are the result of lithologic variations such as those between carbonate rocks and sandstone, they are relatively easy to infer from standard wireline log signatures. However, in a sedimentologically homogeneous sandstone such as the St. Peter, significant permeability contrasts can only be explained as the result of diagenetic features, which may be difficult to identify using conventional log analysis. The pattern of overpressures observed in the St. Peter, particularly the vertical variations that suggest stacked compartments, has led to a closer examination of diagenetic banding within the formation, described in the chapters by Drzewiecki et al. and Shepherd et al. in this volume. It has also prompted development of statistical techniques suitable for detecting subtle changes in wireline log signatures that can be correlated to porosity and permeability, as Moline et al. describe in this volume. Both the diagenetic and log signature studies should enhance understanding of the distribution of low-permeability zones within the basin and provide tools for improved identification of potential reservoirs and traps.

ACKNOWLEDGMENTS

This research was supported by the Gas Research Institute. We gratefully acknowledge the assistance of Raymond Vugrinovich of the Michigan Department of Natural Resources who provided extrapolated DST pressures and other data from his files.

REFERENCES CITED

Al-Shaieb, Z., J. O. Puckett, P. Ely, and A. Abdalla, 1991, Megacompartment complex in the Anadarko basin: a completely sealed overpressured phenomenon, *in* Session Abstracts, Symposium on Deep Basin Compartments and Seals: Gas Research Institute and Oklahoma State University, Stillwater, Oklahoma, May 15–18, 1991.

Belitz, K., and J. D. Bredehoeft, 1988, Hydrodynamics of the Denver Basin: Explanation of subnormal fluid pressures: AAPG Bulletin, v. 72, p. 1334–1359.

Catacosinos, P. A., P. A. Daniels, and W. B. Harrison III, 1990, Structure, stratigraphy and petroleum geology of the Michigan Basin, *in* M. W. Leighton et al., eds., Interior Cratonic Basins: AAPG Memoir 51, p. 561–601.

Farrand, W. R., and D. L. Bell, 1982, Quaternary Geology of Southern Michigan (map): Ann Arbor, Michigan, Department of Geological Sciences, University of Michigan, 1 sheet.

Fisher, J. H., and M. W. Barratt, 1985, Exploration in Ordovician of central Michigan Basin: AAPG Bulletin, v. 12, p. 2065–2076.

Freeze, R. A., and J. A. Cherry, 1979, Groundwater: Englewood Cliffs, NJ, Prentice Hall, 604 p.

Hanor, J. S., 1979, The sedimentary genesis of hydrothermal fluids, *in* H.L. Barnes, ed., Geochemistry of Hydrothermal Ore Deposits (2nd edition): New York, Wiley, p. 137–168.

Hubbert, M. K., 1940, The theory of ground-water motion: Journal of Geology, v. 48, p. 785–944.

Hunt, J. M., 1990, Generation and migration of petroleum from abnormally pressured fluid compartments: AAPG Bulletin, v. 74, p. 1–12.

Long, D. T., T. P. Wilson, M. J. Takacs, and D. H. Rezabek, 1988, Stable-isotope geochemistry of saline near-surface ground water: East-central Michigan Basin: GSA Bulletin, v. 100, p. 1568–1577.

Mandle, R. J., and D. B. Westjohn, 1989, Geohydrologic framework and groundwater flow in the Michigan basin; *in* L. A. Swain and A. I. Johnson, eds., Aquifers of the Midwestern Area: Am. Wat. Res. Assoc. Monograph 13, p. 83–109.

Powley, D. E., 1990, Pressures and hydrogeology in petroleum basins: Earth-Science Reviews, v. 29, p. 215–226.

Prouty, C. E., 1988, Trenton exploration and wrenching tectonics—Michigan basin and environs, *in* B. D. Keith, ed., The Trenton Group (Upper Ordovician Series of Eastern North America—deposition, diagenesis, and petroleum: AAPG Studies in Geology 29, p. 207–236.

Senger, R. K., and G. E. Fogg, 1987, Regional underpressuring in deep brine aquifers, Palo Duro Basin, Texas: 1. Effects of hydrostratigraphy and topography: Water Resources Research, v. 23, p. 1481–1493.

Toth, J., and R. F. Millar, 1983, Possible effects of erosional changes of the topographic relief on pore pressures at depth: Water Resources Research, v. 19, p. 1585–1597.

Toth, J., M. D. Maccagno, C. J. Otto, and B. J. Rostron, 1991, Generation and migration of petroleum from abnormally pressured fluid compartments: Discussion: AAPG Bulletin, v. 75, p. 331–335.

Chapter 12

Thermal History of the Michigan Basin from Apatite Fission-Track Analysis and Vitrinite Reflectance

Herbert F. Wang
University of Wisconsin—Madison
Madison, Wisconsin, U.S.A.

Kevin D. Crowley
Miami University
Oxford, Ohio, U.S.A.

Gregory C. Nadon
University of Wisconsin—Madison
Madison, Wisconsin, U.S.A.

ABSTRACT

Kinetic models for apatite fission-track annealing and vitrinite maturation were used to examine hypotheses for the burial and thermal history of the Michigan basin. Fission-track ages between 160 and 200 Ma were measured for Carboniferous outcrop samples (>300 Ma) near Saginaw Bay. Published vitrinite reflectance and conodont alteration data from the Michigan basin are higher than predicted from current depths and temperatures for the samples. Both sets of data are broadly consistent with elevated temperatures due to additional burial at present geothermal gradients. The depth of additional burial varies systematically from less than 1 km in the basin center to more than 2 km near the adjacent arches. The additional burial could explain the occurrence of diagenetic banding in portions of the St. Peter Sandstone that are currently at depths shallower than the critical window for this phenomenon.

INTRODUCTION

A region of overpressures has been mapped in the Ordovician St. Peter Sandstone and Glenwood Formation northwest of Saginaw Bay in the Michigan basin (Bahr et al., this volume). The overpressures are considered to be paleopressures maintained within a megacompartment. The basal seal is composed in part of a diagenetically banded sandstone (Drzewiecki et al., this volume), which can be traced laterally over large distances in the basin from well logs and core (Moline, 1992). The top of the St. Peter is at depths of 3 km west of the Saginaw Bay area but is much shallower toward the edges of the basin. Because diagenesis is a depth- and temperature-controlled chemical process, the occurrence and timing of the seal formation will depend on the fluid flow, subsidence, and temperature history of the basin.

The thermal history of the Michigan basin has been approached in two ways. Geodynamical models, which concentrate on the mechanism for basin-centered subsidence, yield thermal histories that are apparently too cool to produce organic maturities observed in the basin (Nunn et al., 1984). The second approach is to use data from thermally activated processes, vitrinite reflectance and apatite fission-track annealing, to construct thermal models consistent with the observations. Cercone (1984) and Cercone and Pollack (1991) have used the time-temperature integral (TTI) approach of Lopatin (Waples, 1980) to make the case for deeper burial in the Carboniferous and Permian. Crowley (1991) examined apatite fission-track data in and around the Michigan basin and concluded that several kilometers of basin sediments had been eroded in the early Mesozoic. In this paper we re-examine vitrinite reflectance data from the Mobil-Jelinek well with both conventional plots (e.g., Dow, 1977) and a kinetic model by Sweeney and Burnham (1990). We compare the inferred thermal histories with those derived from new apatite fission-track measurements on Carboniferous-age sediments in south-central and east-central Michigan (Figure 1) and previously published conodont alteration indices from Silurian and Ordovician strata within the basin.

VITRINITE REFLECTANCE MODELING

Cercone and Pollack (1991) used conventional vitrinite reflectance and thermal alteration data as constraints for burial and thermal histories of the Michigan basin. We have replotted the data from the Mobil-Jelinek well compiled in their paper on a more conventional semilogarithmic plot (Figure 2). Vitrinite reflectances from a single well tend to be more consistent than measurements compiled from several wells, because they are analyzed by a single laboratory. The Mobil-Jelinek data include numerous vitrinite reflectances measured in Carboniferous and Devonian strata between 0 and 1100 m (present depths). Silurian and Ordovician reflectances (>1100 m) were made on macrinitic or chitinous material (Cercone and Pollack, 1991).

The data in Figure 2 are separated into pre- and post-Devonian strata to show which of the samples probably contain measurements of true vitrinite and which contain reflectance measurements from vitrinite-like materials (Price and Barker, 1985; Hunt et al., 1991). There is a general increase in $\%R_o$ with depth, although the data form four smaller straight-line segments as is commonly the case for vitrinite values (e.g., Dow, 1977). A best-fit, least-squares regression line was fitted to each segment to minimize the cumulative error. The offset directions between the various line segments change with depth and seldom correspond to sequence boundaries.

The shallowest line segment was fitted by two lines, one through all the post-Devonian data (line 4a) and one through all the data except the upper two points (line 4b) that may be anomalously high. Both lines were extrapolated to intersect the $0.2\%R_o$ value. The depths of the intersection points, –2200 m and –1770 m, respectively, represent the amount of material eroded from the section at the well site. This represents a conservative estimate because vitrinite attains a reflectance value of $0.2\%R_o$ "in the first few hundred meters" (Dow, 1977, p. 87).

The vitrinite data plotted in Figure 2 clearly show that despite the numerous problems inherent in the use of vitrinite reflectance (Dow, 1977; Price and Barker, 1985), approximately 2 km of section have been removed at the well site location. Only one offset along the line segments, between lines 2 and 3, corresponds to a sequence boundary and a major basin-wide unconformity (Fisher et al., 1988). Extrapolation of line segments, as suggested by Dow (1977), indicates that the Tippecanoe II sequence is offset by $0.1\%R_o$. This could indicate the removal of approximately 120 m of section prior to deposition of the Kaskaskia I sequence, although the offset is more likely to be the result of a combination of measurement error and the different maturation path of the organics within the pre-Devonian sediments (Price and Barker, 1985).

Kinetic Modeling

Sweeney and Burnham (1990) have developed a spreadsheet version, EASY$\%R_o$, of a kinetic reaction model for vitrinite reflectance. The model incorporates a weighted distribution of reactants with activation energies between 34 and 70 kcal/mole (116 and 239 kJ/mole). The kinetic model represents a more rigorous approach to maturation modeling than the Lopatin time-temperature approach, provided the distribution of different activation energies is appropriate for the initial kerogen. The model calculates the fraction, F, of reactant converted as a weighted average of the individual first-order reactions. The vitrinite reflectance is determined from the equation $\%R_o = \exp(-1.6 + 3.7F)$. Sweeney and Burnham show that their results are quite similar to Waples' (1980) correlation between $\%R_o$ and TTI (time-temperature index) within the oil window for linear heating rates between 1° and 10°C/m.y.

The model results are dependent on the spectrum of activation energies present within the organic materials in the section, especially in pre-Devonian sediments in which true vitrinite is absent. Price and Barker (1985) argued that pre-Devonian $\%R_o$ values could not be used to determine the thermal history of a basin; however, Buchardt and Lewan (1990) showed that maturation of marine organic material in pre-Devonian sediments is similar to that in post-Devonian strata and can be used as an indicator of thermal alteration. The model input chosen here was the weighted distribution of activation energies of Sweeney and Burnham (1990) in the post-Devonian and a single activation energy of 48 kcal/mole (201 kJ/mole) for the pre-Devonian strata. The latter was

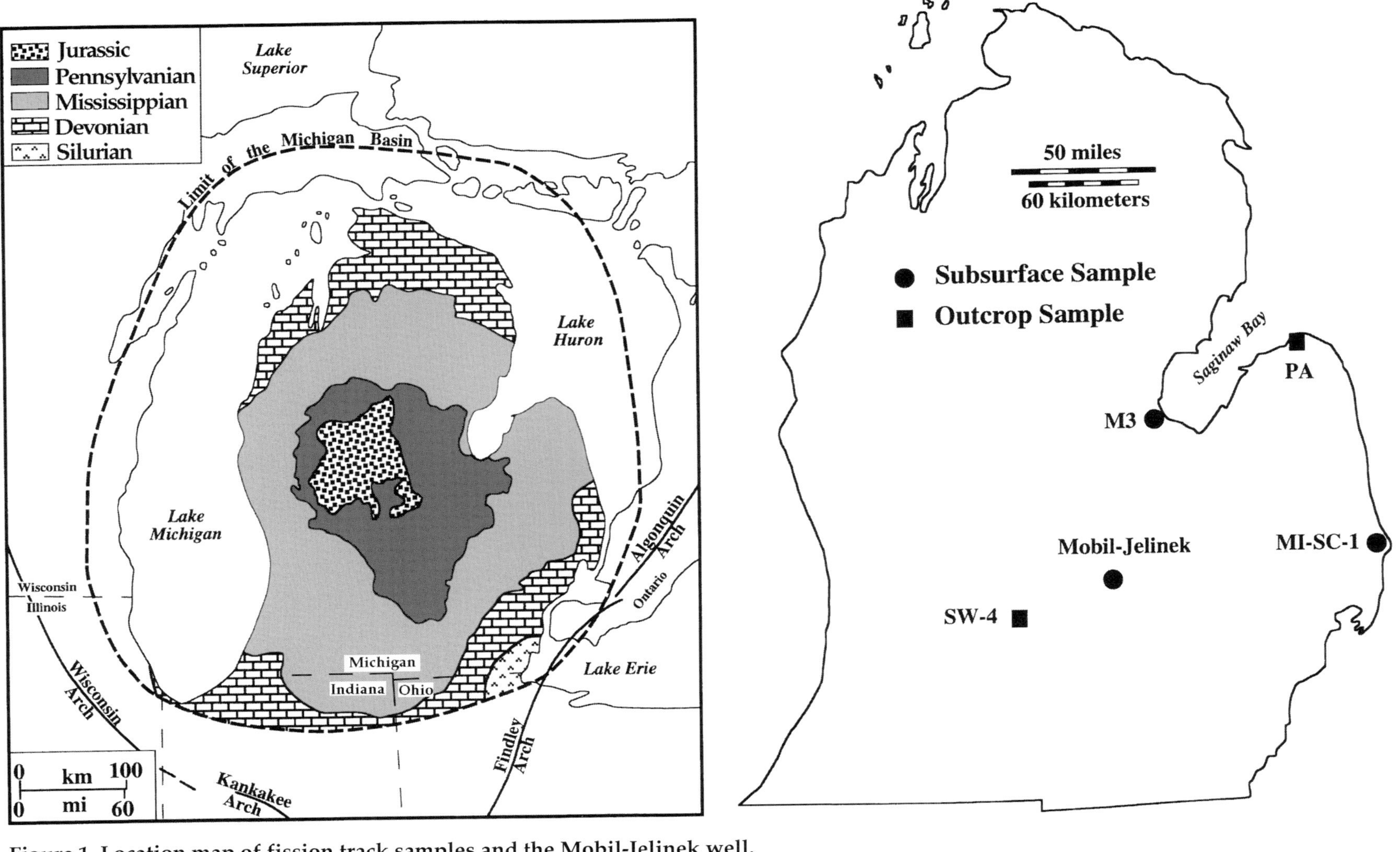

Figure 1. Location map of fission track samples and the Mobil-Jelinek well.

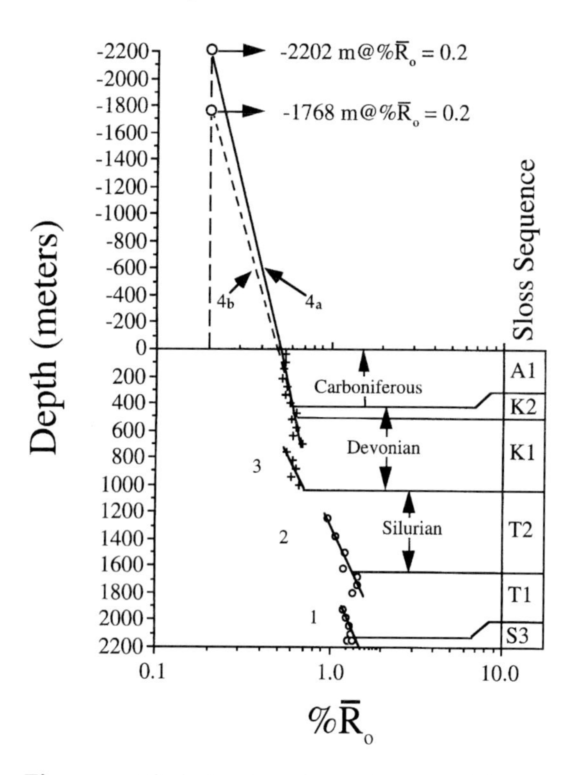

Figure 2. Vitrinite data from the Mobil Jelinek well, Shiawassee County, compiled by Cercone and Pollack (1991) and plotted on semilogarithmic coordinates. The data can be divided into four straight-line segments, each of which represents a variation in either thermal history or activation energy or both. Different symbols are used for pre- and post-Devonian reflectance values.

chosen based on data from the Alum Shale of Scandinavia (Hunt et al., 1991) as representative of Ordovician marine organic material.

Cercone and Pollack (1991) suggested that the Permian–Carboniferous sediments increased temperatures in older sediments both by increasing their depth of burial and by a "blanketing effect," whereby the thermal gradients in the Permian–Carboniferous are steepened by their low thermal conductivities. They emphasized that any combination of blanket thickness and conductivity that produces the same temperature at its base leads to the same thermal history for the older layers. They also incorporated a 10°C decrease between the Permian and the present, because global temperatures have generally cooled. The model is a fairly simple burial and thermal history, but it contains an important tradeoff—burial depth versus temperature gradient in the eroded section. As an example, Figure 3A shows Carboniferous and older sediments buried an additional 1.8 km with subsequent erosion to present levels during the Permian and Triassic. The thermal histories (Figure 3B) are based on the geologic scenario described above with a Permian temperature of 30°C and present-day thermal gradients in the different age sediments (Cercone and Pollack, 1991). The latter assumption takes into account the different thermal conductivities of the sediments, especially the high thermal conductivity of the Silurian evaporites.

The kinetic model calculations were run for the cases of no additional burial, for 1 km of additional burial with a geothermal gradient of 40°C in the eroded section, and for 2 km of additional burial with a gradient of 26.2°C/km (Figure 4). Finally, burial depth was adjusted until the calculated pre- and post-Devonian fell within the limits of the Mobil-Jelinek data set. The Permian surface temperature was set to 30°C and cooled to the present value of 10°C.

Results

The post-Devonian vitrinite reflectances calculated from Sweeney and Burnham's (1990) kinetic model are similar to those obtained by Cercone and Pollack (1991) using the Lopatin model for maturation. Thus, we find reasonable agreement with the shallow reflectances (<1100 m present depth) for 1.8 km of erosion combined with a geothermal gradient of 26.2°C/km in the eroded Permian–Carboniferous strata (Figure 4). Reflectance values in the Utica Shale (Upper Ordovician) can be matched with a 26.2°C/km geothermal gradient in 1.8 km of eroded strata (Figure 4).

The three curves presented in Figure 4 illustrate the sensitivity of reflectance modeling to temperature and organic constituents. Curve A, no additional burial and a 26.2°C/km geothermal gradient, shows that the pre-Devonian marine organic material would undergo no thermal maturation. Curve B, 1 km of additional burial and a geothermal gradient of 40°C/km, produces a better fit to the post-Devonian data although the pre-Devonian values are greatly underestimated. Curve C, 1.8 km and 26.2°C/km, was found by varying the burial depth until the calculated and observed reflectances from the middle of the Ordovician section (the base of the Utica Formation) matched. The model produces reflectance values greater than 2.0 when the temperature exceeds 111°C at maximum burial. This sensitive dependence on temperature matches experimental data on Cambrian–Ordovician organic material (Buchardt and Lewan, 1990).

The model was further tested by calculating additional burial depth to produce the equivalent vitrinite reflectances of conodont alteration indices (CAI) compiled by Cercone and Pollack (1991) and from Bowers (1989). The predicted additional depths vary from less than 1.0 km in the center of the basin to >2.0 km on the basin margins (Figure 5).

APATITE FISSION-TRACK ANALYSIS

Sample Locations and Methods

Apatite fission-track analyses were performed on four outcrop and one subsurface sample in south-central and east-central Michigan (Figure 1). Location data and drill-hole information are given in Table 1.

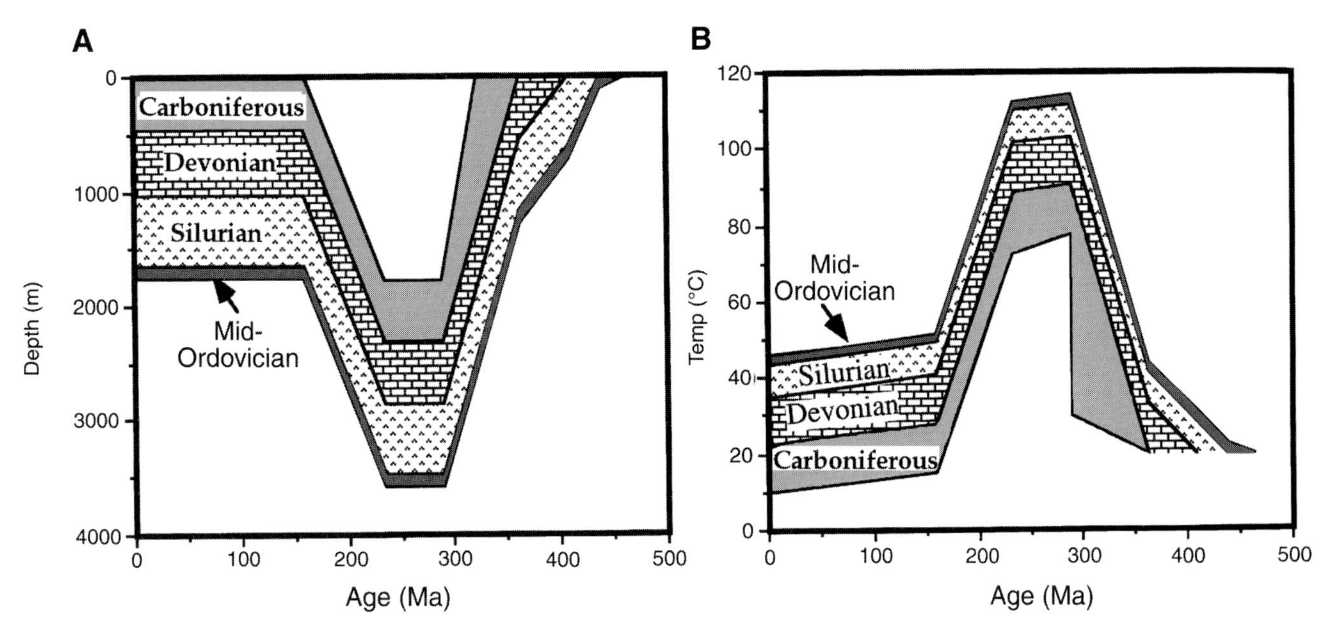

Figure 3. Burial and thermal histories for the Michigan basin (after Cercone and Pollack, 1991). (A) Depth history for additional 2 km burial and erosion between Carboniferous and Jurassic. (B) Thermal history for 2 km burial and erosion with a 26°C/km geothermal gradient in the eroded section.

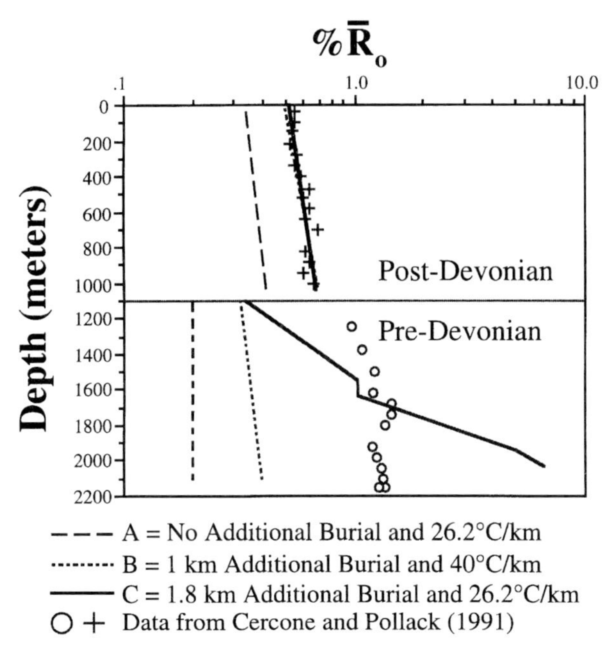

Figure 4. Calculated vitrinite reflectance profiles from Sweeney and Burnham (1990) kinetic model for several assumptions of burial/erosion and geothermal gradient in the eroded section. The dashed lines represent data for no erosion (A) and 1 km erosion (B). Solid line C is for 1.8 km of additional sediment with a geothermal gradient of 26.2°C/km. The data points from Mobil-Jelinek well, as compiled by Cercone and Pollack (1991), with separate regression lines for pre- and post-Devonian strata are also shown.

Apatite was obtained from whole rock (outcrop samples) or split core (subsurface samples) using standard separation techniques. Age and confined track-length measurements were made following the methods given in Crowley (1991, p. 697–699).

Results

Measured apatite fission-track ages, associated counting data and track-length data are given in Table 1. Measured ages range from 150 Ma ± 42 m.y. for drill-hole sample M3 to 265 ± 65 m.y. for outcrop sample MR (all uncertainties represent the 95% confidence interval). The large uncertainties in some of the estimates (particularly samples MR and ET) are a consequence of low track densities (owing to low U concentrations) or small numbers of countable grains.

Mean confined track lengths range from 12.7 ± 0.44 µm for sample M3 to 13.6 ± 0.38 µm for sample SW (uncertainties represent the 95% confidence interval). Length distributions (Figure 6) are broad, symmetrical to negatively skewed with modes at 13 or 14 µm and dispersions between 1.4 and 1.6 µm.

Interpretation

The measured fission-track ages shown in Table 1 are younger than the Mississippian and Pennsylvanian depositional ages (>290 Ma) of the host rocks, indicating that resetting, presumably by heating, has occurred since deposition. Indeed, the ages and lengths reported in Table 1 are similar to those from basement drill core in southern and southeastern Michigan and northern Indiana reported by Crowley (1991), which were interpreted to be the result of Mesozoic and/or Cenozoic heating.

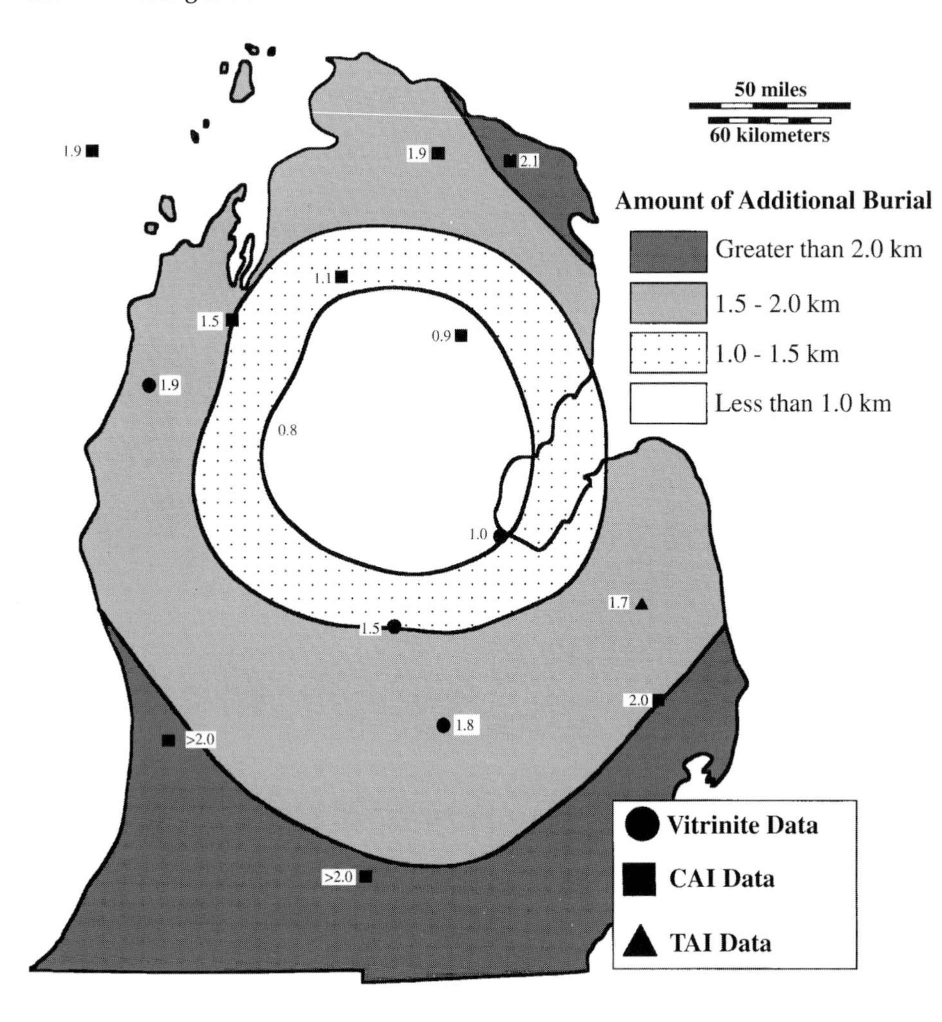

Figure 5. The distribution of additional burial depth in the Lower Peninsula of Michigan based on vitrinite reflectance, conodont alteration indices (CAI), and thermal alteration indices (TAI) data from Hogarth and Sibley (1985), Bowers (1989), and Cercone and Pollack (1991). The CAI and TAI data were converted to equivalent reflectance values using Epstein et al. (1977) and Marshall (1990). The increased thickness of strata required to match vitrinite reflectances in the northeast consists of Carboniferous as well as post-Carboniferous sediments.

Thermal histories of the samples can be elucidated by examining the measured length distributions, which are shown in Figure 6. Also shown in Figure 6 for comparison purposes is the length distribution for Durango apatite, a standard used in the age determinations (see footnotes in Table 1), as well as the length distributions for basement drill core samples from the southern part of the Michigan basin reported by Crowley (1991).

The length distribution for Durango apatite is unimodal with a slight negative skewness and has a large mean (14.6 μm) and a small dispersion (1.1 μm). The shape of this distribution is indicative of rapid cooling followed by long-term residence at ambient or near-ambient temperatures (e.g., Gleadow et al., 1986). The length distributions for the outcrop and drill core samples from the Michigan basin (Figure 6) are characterized by shorter mean lengths and larger dispersions than those for Durango apatite. Fission-track ages from the Michigan basin are partially reset; they cannot be interpreted directly using closure-temperature concepts.

A plot of measured age against mean confined track length (Figure 7) for the Michigan basin samples from this study and from Crowley (1991) indicates a pattern of resetting. This plot shows that younger ages rather than older ages are generally associated with shorter mean track lengths. Given that track length and track density (or age) decrease with increasing temperature or heating time (e.g., Crowley et al., 1991), the patterns shown in Figure 7 suggest that the younger fission-track ages represent higher degrees of thermal overprinting than do older ages. The thermally overprinted ages observed here and by Crowley (1991) indicate that sample temperatures (and basin temperatures) were elevated from at least the Early Triassic through the Early Cretaceous.

The pre-Mesozoic thermal history of the basin cannot be established with the present samples. However, this history can be constrained using fission-track ages from the Canadian Shield adjacent to the basin in Ontario published by Crowley (1991). Crowley (1991) observed that apatite fission-track ages on the shield range from Early Cambrian through Jurassic. Using confined-length distributions, he interpreted these ages to be the product of an episode of heating that peaked sometime between 200 and 250 Ma, with peak temperatures around 90°C. Thus, the temperature history prior to 200 Ma would be largely erased by fission-track annealing. Temperatures at the upper end of the partial annealing zone would require approximately 1.6 to 2.7 km additional burial for surface temperatures of 30 and 20°C, respectively, and a temperature gradient of

Table 1. Location, fission-track age, and length for outcrop and drillcore samples from the Michigan basin.

Sample Code [Interval*] Formation	Latitude Longitude	Grains Counted	$\rho_s \times 10^6$ $\rho_i \times 10^6$ $\rho_d \times 10^6$ (tracks cm^{-2})	N_s N_i N_d (tracks)	t_w [95% C.I.]** t_u [95% C.I.]** (Ma[m.y.])	Lengths Measured	Mean Length [s.d.] (μm)
M3† [714–737 ft] Berea Sandstone	N43°3818″ W84°02′30″	13	0.950 1.268 1.007	215 287 4558	143 [27]‡ 150 [42]‡	50	12.7 [1.6]
PA Port Austin	N44°02′ W82°59′	19	2.507 2.722	581 631	183 [24]‡ 204 [34]‡	51	13.6 [1.4]
SW Saginaw	N42°45′ W84°45′	16	1.529 1.730	356 403	165 [26]‡ 178 [18]‡	35	13.0 [1.5]
MR Marshall	N42°15′ W84°22′30″	6	2.086 1.541 1.034	134 99 4558	263 [71]‡ 265 [65]‡		
ET Eaton	N42°45′38″ W84°45′18″	2	1.604 1.537 0.996	24 23 4629	196 [113]‡ 206 [86]‡		

Symbols: ρ = track density; N = number of tracks counted; subscripts s, i, and d denote spontaneous, induced, and dosimeter tracks, respectively; t_w = weighted age; t_u = unweighted age; C.I. is confidence interval; s.d. is standard deviation.

* Collection interval in meters below surface for drill core samples.

** All ages were calculated using Corning Glass CN6, $\lambda_f = 7.03 \times 10^{-17}$, z = 3839 ± 250 (95% C.I.) determined using Fish Canyon and Durango apatites. Uncertainties in the ages represent the 95% C.I. Double daggers shown on the unweighted age indicate that the weighted age failed the X^2 test at the 5% level of significance.

† Mieske 3-6; Bay County, Michigan; Sec. 6, T14N, R3E.

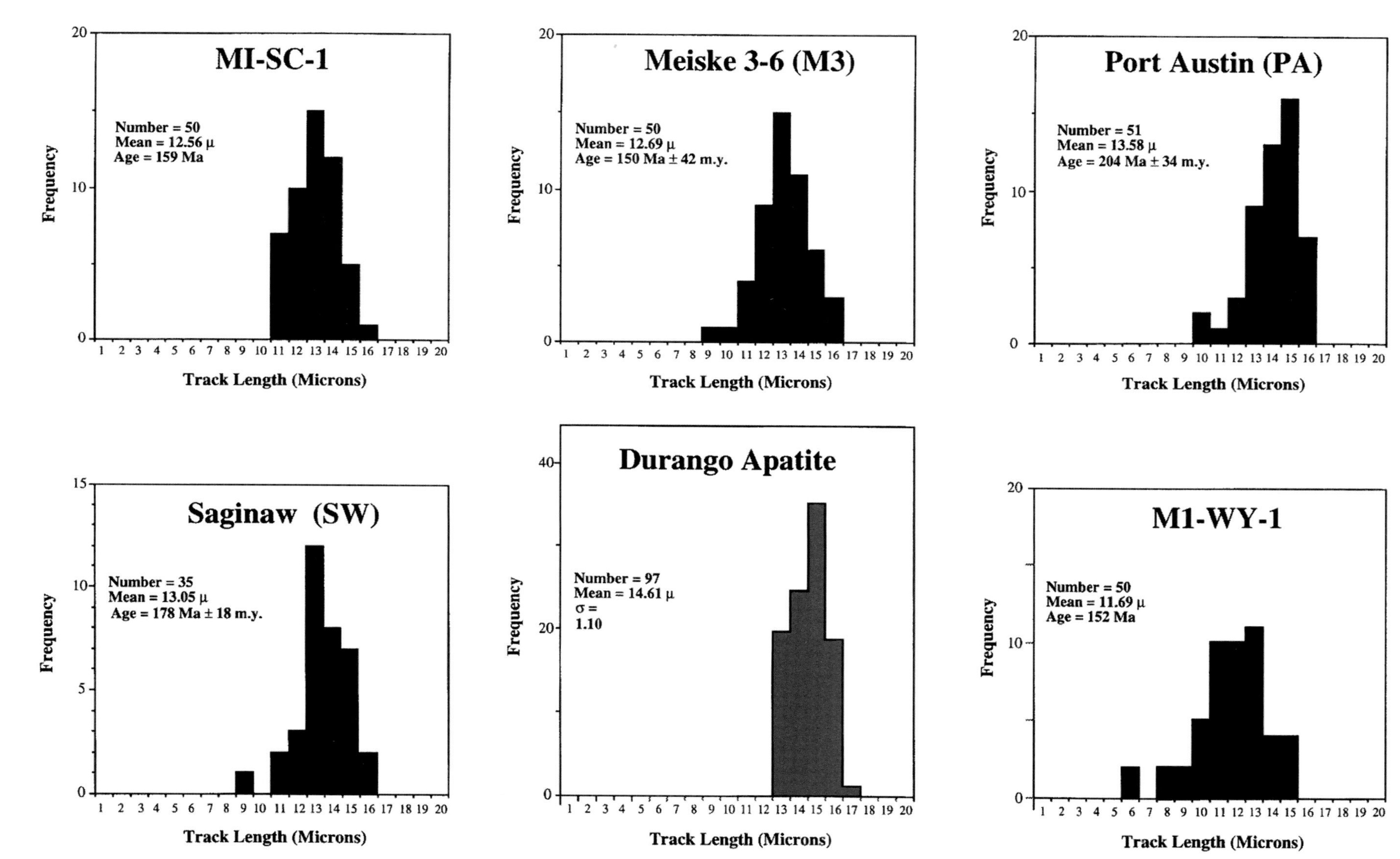

Figure 6. Fission-track length distributions for Michigan basin samples compared with the Durango apatite standard. The shorter mean lengths and larger dispersions in the Michigan basin samples indicate the F-T ages are partially reset.

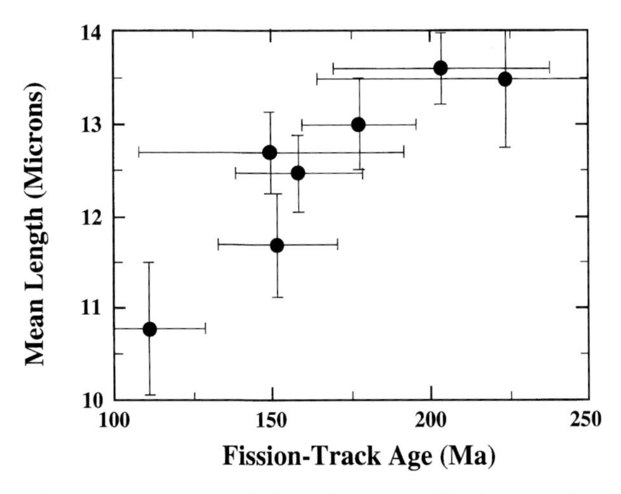

Figure 7. Mean track length versus fission-track age for the samples from the Michigan basin. The pattern shows that younger fission-track ages represent higher degrees of overprinting than older ages.

26.2°C/km in the eroded section. In summary, the fission-track data presented here and in Crowley (1991) suggest the following temperature history for the Michigan basin—increasing temperature in the late Paleozoic, peaking in the Triassic (between about 200 and 250 Ma), followed by slow cooling until the Early Cretaceous.

DISCUSSION

Our analysis of both the apatite fission-track and vitrinite reflectance data shows approximately 0.8 to >2.0 km of additional Permian–Carboniferous burial in southeastern Michigan for a normal paleogeothermal gradient of 26.2°C/km in the eroded section. Neglecting variations in paleogeothermal gradient within the basin, the kinetic model can provide constraints on the distribution of the thickness of eroded section (Figure 5), provided the selection of the activation energy of the organic material is appropriate. The general pattern of estimated additional burial matches the distribution of CAI values obtained by Hogarth and Sibley (1985) for the Ordovician Trenton Group. The values of 0.7 to 0.8 km of eroded section in the center of the basin are also similar to those determined by Vugrinovich (1988) from shale compaction. The distribution of estimated additional burial (Figure 5), which suggests more pronounced erosion on the arches than in the basin, is complementary to the present distribution of strata within the basin.

The distribution of eroded section in the basin center requires the addition of approximately 50,000 km^3 of sediments in the Permian and removal by the end of the Triassic (Jurassic sedimentation events have been ignored). This translates to 20 to 35 mm/k.y., values well within the range of long-term continental sedimentation rates (Blatt et al., 1980). Presumably fluvial and eolian processes transported the bulk of the material to the western or southern margins of North America (Peterson, 1988; Dott and Batten, 1988).

IMPLICATIONS

Powley (1990) has suggested that diagenetic sealing often appears to occur at approximately 3 km depth. The diagenetic banding within the St. Peter Sandstone, electrofacies ef4 of Moline et al. (this volume), occurs throughout the basin. This electrofacies has extremely low porosity and permeability and may be a bottom seal for the overpressures in the St. Peter Sandstone mapped to the west and north of Saginaw Bay (Bahr et al., this volume). The current depth of the St. Peter varies laterally from 1.5 km toward the northern and western edges of the basin to just over 3 km in the center of the basin. Thus, additional burial along the basin margins would have placed the flanking portions of the St. Peter into the critical window.

The timing for a Permian burial could be related to flexural coupling with Alleghenian tectonics (Beaumont et al., 1988), as sedimentation patterns in the Michigan basin have been influenced in the late Ordovician by the Taconic orogeny (Coakley et al., in press). Appalachian thrusting is the flexural load that opens space for Permian–Carboniferous sedimentation. Subsequent uplift would result from Appalachian erosion, which removes the flexural load.

Indications that high temperature fluids (~160°C) have affected Ordovician-age strata come from MVT-type mineralization in the Trenton and Black River formations (Hogarth and Sibley, 1985; Budai and Wilson, 1991) and from fluid inclusion studies in the St. Peter Sandstone and the Glenwood Formation (Drzewiecki et al., this volume). Strontium and oxygen isotopic data on the replacive dolomites within the St. Peter Sandstone indicate that the source for the fluid flux might be deeper Cambrian sediments (Winter et al., in preparation). Mechanisms such as topographically driven flow (Bethke, 1986), tectonic compression (Oliver, 1986), or deep basement convection (Nunn, 1992) have been invoked to elevate basin temperatures. However, since the wells from which the core data were obtained are almost universally drilled on geophysical anomalies arising from movement on deep-seated faults (e.g., Catacosinos et al., 1990), it is probable that the high heat flow values obtained from cores result from local, fault-controlled movement of deep-seated fluids. The concordance between the CAI isograd values and the structure of the top of the Trenton noted by Hogarth and Sibley (1985), as well as the vitrinite and fission-track data and the thermal modeling presented here, indicate that the basin as a whole did not experience an influx of high-temperature fluids.

ACKNOWLEDGMENTS

This research has been sponsored by the Gas Research Institute under Contract 5089-260-1810 (Wang and Nadon) and NSF-EAR 8916290 (Crowley). We wish to thank W.B. Harrison, III, for the samples of Berea Sandstone core, and Karen Cercone for discussions and providing timely access to the reflectance data. Discussions with colleagues Jean Bahr and Bernie Coakley have been extremely helpful.

Jerry Sweeney and Alan Burnham at Lawrence Livermore National Laboratory generously furnished the spreadsheet model for vitrinite reflectances and greatly aided in focusing the paper by bringing relevant vitrinite reflectance literature to our attention.

REFERENCES CITED

Beaumont, C., G. Quinlan, and J. Hamilton, 1988, Orogeny and stratigraphy: Numerical models of the Paleozoic in the Eastern interior of North America: Tectonics, v. 7, p. 389–416.

Bethke, C. M., 1986, Hydrologic constraints on the genesis of the upper Mississippi Valley mineral district from Illinois Basin brines: Economic Geology, v. 81, p. 233–249.

Blatt, H., G. Middleton, and R. Murray, 1980, Origin of Sedimentary Rocks: Prentice-Hall, New York, 782 p.

Bowers, T.L., 1989, Upper Niagara–Lower Salina (Mid-Silurian) sedimentology, and conodont-based biostratigraphy and thermal maturation of the southeast Michigan basin: Master's Thesis, Wayne State University, Detroit, Michigan, 119 p.

Buchardt, B., and M.D. Lewan, 1990, Reflectance of vitrinite-like macerals as a thermal maturity index for Cambrian–Ordovician Alum Shale, southern Scandinavia: AAPG Bulletin, v. 74, p. 394–406.

Budai, J. M., and J. L. Wilson, 1991, Diagenetic history of the Trenton and Black River Formations in the Michigan Basin, *in* Early Sedimentary Evolution of the Michigan Basin, ed. by P. A. Catacosinos and P. A. Daniels, Jr.: GSA Special Paper 256, p. 73–88.

Catacosinos, P.A., P.A. Daniels, Jr., and W.B. Harrison III, 1990, Structure, stratigraphy, and petroleum geology of the Michigan Basin, *in* Interior Cratonic Basins, ed. by M.W. Leighton, D. R. Kolata, D.F. Oltz and J.J. Eidel: AAPG Memoir 51, p. 561–601.

Cercone, K. R., 1984, Thermal history of Michigan Basin: AAPG Bulletin, v. 68, p. 130–136.

Cercone, K. R., and H. N. Pollack, 1991, Thermal maturity of the Michigan Basin, *in* Early Sedimentary Evolution of the Michigan Basin, ed. by P. A. Catacosinos and P. A. Daniels, Jr.: GSA Special Paper 256, p. 1–10.

Coakley, B. J., G.C. Nadon, and H. F. Wang, in press, Spatial variations in tectonic subsidence during Tippecanoe I in the Michigan basin: Basin Research, v. 6.

Crowley, K. D., 1991, Thermal history of Michigan Basin and southern Canadian Shield from apatite fission-track analysis: J. Geophysical Research, v. 68, p. 697–711.

Crowley, K. D., M. Cameron, and R. L. Schaefer, 1991, Experimental studies of annealing of etched fission tracks in fluorapatite: Geochimica et Cosmochimica Acta, v. 55, p. 1449–1465.

Dott, R.H., Jr., and R.L. Batten, 1988, Evolution of the Earth, Fourth edition: McGraw-Hill, New York, 643 p.

Dow, W.G., 1977, Kerogen studies and geological interpretations: J. Geochem. Explor., v. 7, p. 79–99.

Epstein, A.G., J.B. Epstein, and L.D. Harris, 1977, Conodont color alteration—an index to organic metamorphism: United States Geological Survey Professional Paper 995, 27 p.

Fisher, J.H., M.W. Barratt, J.B. Droste, and R.H. Shaver, 1988, Michigan Basin: in Sedimentary Cover—North American Craton: U.S., ed. by L. L. Sloss: Geological Society of America, The Geology of North America , v. D-2, p. 361–382.

Gleadow, A. J. W., I. R. Duddy, P. F. Green, and J. F. Lovering, 1986, Confined fission track lengths in apatite: a diagnostic tool for thermal history analysis: Contributions Mineralogy Petrology, v. 94, p. 405–415.

Hogarth, C.G., and D.F. Sibley, 1985, Thermal history of the Michigan Basin: evidence from conodont color alteration indices, *in* Ordovician and Silurian Rocks of the Michigan Basin and its Margins, ed. by K.R. Cercone and J.M. Budai: Michigan Basin Geological Society Special Paper 4, p. 45–57.

Hunt, J.M., M.D. Lewan, and R. J-C. Hennet, 1991, Modeling oil generation with time-temperature index graphs based on the Arrhenius equation: AAPG Bulletin, v. 75, p. 795–807.

Marshall, J.E.A., 1990, Determination of thermal maturity, *in* Palaeobiology—A Synthesis, ed. by E.G. Briggs and P.R. Crowther: Blackwell Scientific Publications, Oxford, p. 511–515.

Moline, G. R., 1992, A multivariate statistical approach to mapping spatial heterogeneity with an application to the investigation of pressure compartments within the St. Peter sandstone: Ph.D. thesis, U. of Wisconsin-Madison, 237 p.

Nunn, J. A., 1992, Free thermal convection beneath the Michigan basin: thermal and subsidence effects: Trans. American Geophysical Union, v. 73, p. 547.

Nunn, J. A., N. H. Sleep, and W. E., Moore, 1984, Thermal subsidence and generation of hydrocarbons in Michigan Basin: AAPG Bulletin, v. 68, p. 296–315.

Oliver, J., 1986, Fluids expelled tectonically from orogenic belts: their role in hydrocarbon migration and other geologic phenomena: Geology, v. 14, p. 99–102.

Peterson, J.A., 1988, Phanerozoic stratigraphy of the northern Rocky Mountain region, *in* Sedimentary Cover—North American Craton: U.S., ed. by L. L. Sloss: Geological Society of America, The Geology of North America , v. D-2, p. 83–107.

Powley, D. E., 1990, Pressures and hydrogeology in petroleum basins: Earth-Science Reviews, v. 29, p. 215–226.

Price, L.C., and C.E. Barker, 1985, Suppression of vitrinite reflectance in amorphous rich kerogen—a major unrecognized problem: J. of Petrol. Geol., v. 8, p. 59–84.

Sweeney, J. J., and A. K. Burnham, 1990, Evaluation of a simple model of vitrinite reflectance based on chemical kinetics: AAPG Bulletin, v. 74, p. 1559–1570.

Vugrinovich, R., 1988, Shale compaction in the

Michigan Basin: estimates of former depth of burial and implications for paleogeothermal gradients: Bull. Canadian Petroleum Geology, v. 36, p. 1–8.
Waples, D. W., 1980, Time and temperature in petroleum formation; application of Lopatin's method to petroleum exploration: AAPG Bulletin, v. 64, p. 916–926.

Chapter 13

Diagenesis, Diagenetic Banding, and Porosity Evolution of the Middle Ordovician St. Peter Sandstone and Glenwood Formation in the Michigan Basin

Peter A. Drzewiecki
J. Antonio Simo
P. E. Brown
E. Castrogiovanni
Gregory C. Nadon
Lisa D. Shepherd
J. W. Valley
M. R. Vandrey
B. L. Winter
University of Wisconsin
Madison, Wisconsin, U.S.A.

D. A. Barnes
Western Michigan University
Kalamazoo, Michigan, U.S.A.

ABSTRACT

The Middle Ordovician St. Peter Sandstone and Glenwood Formation of the Michigan basin are composed of alternating intervals of quartz sandstone, micritic carbonate, and an occasional thin shale. They contain abnormally pressured compartments in the deepest portion of the basin. Some of these pressure compartments are gas reservoirs and are bounded, above and below, by diagenetically banded sandstone and/or carbonate sedimentary units.

Diagenetically banded sandstones are dominated by bands of quartz cement that formed as a result of chemical compaction, quartz dissolution, and quartz precipitation during burial. Petrographic and geochemical data suggest that quartz overgrowths precipitated from fluids with a slight meteoric component. Dolomite, the second most abundant authigenic mineral in diagenetic bands and elsewhere in the St. Peter and Glenwood, postdated quartz overgrowths and precipitated from hypersaline fluids at high temperatures. Values of $\delta^{13}C$ from the dolomite indicate that the carbon was partially derived from the maturation of organic matter, and the carbon isotopic composition appears to be stratigraphically controlled. Bands of dolomite and quartz cement occur in horizontal and cross-bedded siliciclastic lithofacies that contain planar depositional discontinuities, such as grain size laminations.

Original porosity in the St. Peter Sandstone and Glenwood Formation was relatively homogeneous within a given lithofacies. Diagenesis, especially the development of diagenetic bands, resulted in heterogeneous porosity distribution. The porosity was reduced to 0 to 3% in the tightly cemented bands (quartz and dolomite) and in intervals that experienced intense intergranular pressure solution. The most significant porosity modification occurred in the deep burial environment. Depositional controls on diagenesis include bedding style, degree of bioturbation, and original mineralogical composition of the sediments. The correlation between depositional facies and diagenetic banding may allow regions of low porosity and permeability to be predicted within a sequence stratigraphic framework in deeply buried sandstones.

INTRODUCTION

The Middle Ordovician St. Peter Sandstone and overlying Glenwood Formation of the Michigan basin are composed of alternating intervals of fine- to medium-grained sandstone and dark, micritic carbonates with occasional thin shale units (Figure 1). These alternating lithologies provide an abundance of potential hydrocarbon reservoirs (porous sandstone) and seals (nonporous dolomicrites), and consequently sparked exploration interest in the Michigan basin during the 1980s (Harrison, 1987). Thirty-one gas and gas/condensate fields have produced from reservoirs in the St. Peter Sandstone and Glenwood Formation of the Michigan basin as of 1988, and the Falmouth field alone produced 5.1 billion cubic feet of gas as of 1987 (Catacosinos et al., 1990).

The identification and discussion of abnormally pressured compartments in the St. Peter Sandstone and Glenwood Formation are found in Bahr (1989), Bahr and Moline (1992), and Bahr et al. (this volume). Abnormally pressured compartments have been identified as hydrocarbon reservoirs in basins worldwide (Powley, 1990; Hunt, 1990) and contain a particular type of cement distribution characterized by bands of different cements and porosities (Hunt, 1990; Tigert and Al-Shaieb, 1990). Diagenetic banding can be defined as the alternation of cement types and porosity abundances that results from the diagenetic processes of cementation, dissolution, intergranular pressure solution, and stylolitization. It occurs on all scales from millimeters (Shepherd et al., this volume) to meters or even tens of meters (Tigert and Al-Shaieb, 1990). Five types of millimeter-scale bands have been identified in the St. Peter Sandstone and Glenwood Formation: (1) porous, (2) quartz-cemented, (3) dolomite-cemented, (4) clay-cemented, and (5) pressure solution-dominated. On a larger scale (i.e., meter-scale), bands may occur as intervals that contain high porosities, are dominated by one major cement type, or have abundant millimeter-scale banding. Banded sandstones have been identified as a seal-forming lithology for an abnormal pressure compartment in the St. Peter Sandstone (Shepherd et al., this volume).

The purpose of this paper is to compare and contrast the petrographic and geochemical characteristics of the different diagenetic events that affected the St. Peter Sandstone and Glenwood Formation in relationship to depositional textures and basin stratigraphy. This will include a description of the diagenetic sequence, diagenetic banding, stylolitization, and porosity evolution, and an examination of the depositional and diagenetic controls on band formation and distribution. These interpretations will provide information that will assist in hydrocarbon exploration in and production from deeply buried sandstone reservoirs.

REGIONAL SETTING AND MIDDLE ORDOVICIAN STRATIGRAPHY

Regional Setting

The Michigan basin is an intracratonic basin that began to subside in the Late Cambrian (Catacosinos and Daniels, 1991) and was active as late as the Jurassic (Nunn et al., 1984). Although the present Michigan basin is circular in outline and bowl-shaped in cross section, it was less simplistic in the past (Howell and van der Pluijm, 1990). Middle Ordovician sediments extend across the entire basin and reach a maximum thickness of over 400 m in north-central Michigan (Figure 2). The St. Peter Sandstone and Glenwood Formation thin toward the basin margins, especially in southern Michigan (Figure 2).

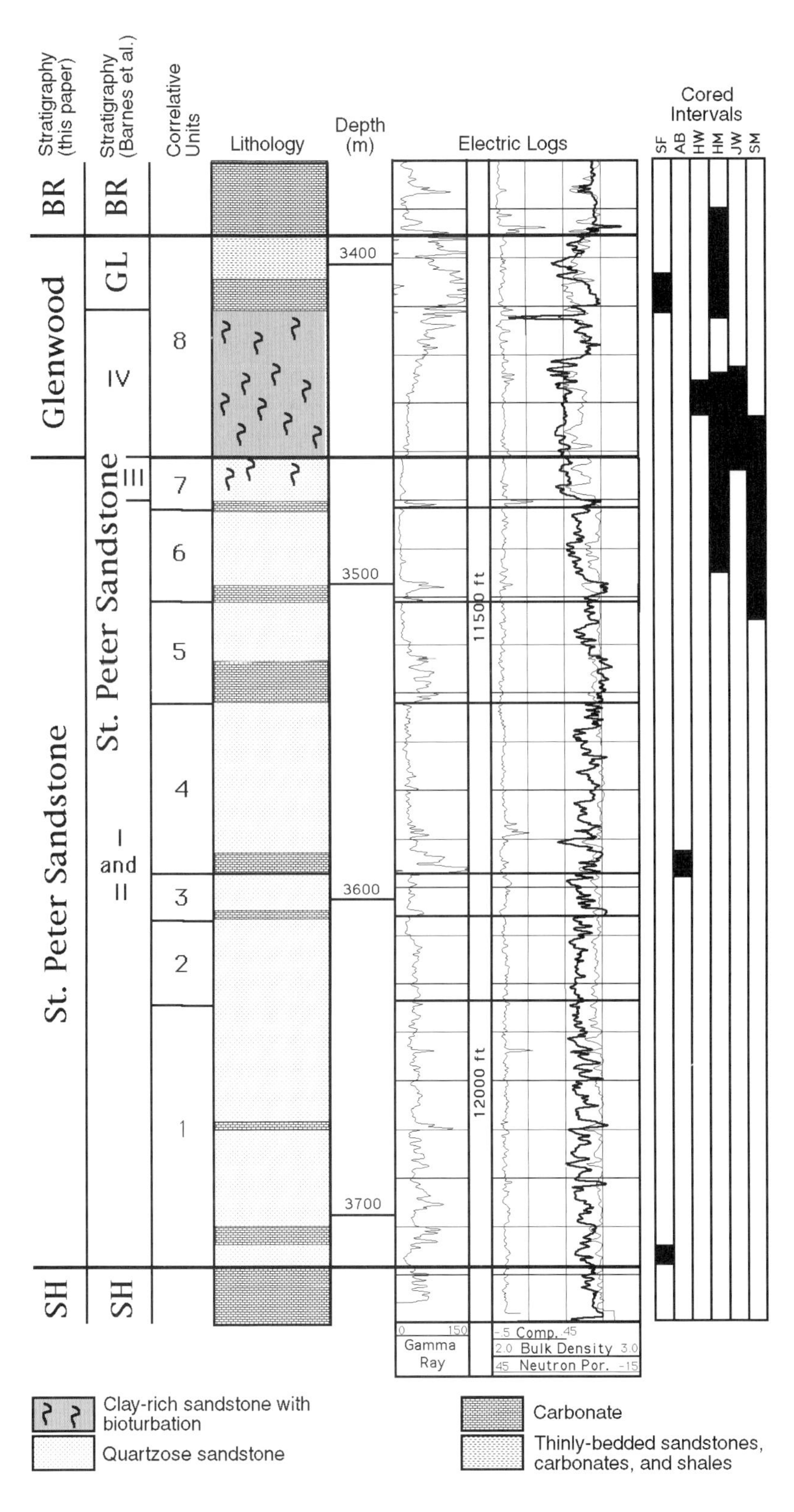

Figure 1. Core locations and log responses for the Amoco Ballentine 1-35 well, Gladwin County, in the Michigan basin. The formational boundaries used in this study are compared to those of Barnes et al. (1992a). The Roman numerals refer to the lithofacies of Barnes et al. (1992a). Note that the clay-rich bioturbated sands of Lithofacies IV (Barnes et al., 1992a) are placed in the Glenwood Formation for this study. Correlative units are based on the correlation of carbonate intervals across the basin. The relative stratigraphic position of various cores used in this study are shown next to the log responses. AB = Amoco Ballentine 1-35; HM = Hunt Martin 1-15; HW = Hunt Winterfield A-1; JW = JEM Weingartz 1-7; SF = Mobil St. Foster 1-12; SM = Sun Mentor C 1-29. SH = Shakopee Fm; GL = Glenwood Fm; and BR= Black River Fm. Locations of wells are shown in Figure 2.

Stratigraphy, Lithologies, and Sedimentology

The St. Peter Sandstone and the overlying Glenwood Formation are Middle Ordovician (Whiterockian to Mohawkian) in age (Catacosinos and Daniels, 1991; Barnes et al., 1992a). The St. Peter Sandstone lies unconformably above the Lower Ordovician dolomites of the Shakopee Formation (Figure 1) (Smith et al., 1993). The Glenwood Formation is interpreted to lie conformably above the St. Peter in the basin center and unconformably above older rocks in the southern portion of the basin where the St. Peter Sandstone is absent (Catacosinos and Daniels, 1991). Limestones of the Black River Formation conformably overlie the Glenwood, except in the southeastern and possibly the northern portions of the basin where the contact is unconformable.

Table 1. Lithofacies in the St. Peter Sandstone and Glenwood Formation: sedimentologic and diagenetic characteristics.

Lithofacies Name	Sedimentology	Occurrence	Bioturbation	Depositional Interpretation	Diagenetic Characteristics	Electro-facies
Clay-rich sandstone	Fine- to medium-grained, poorly to well-sorted quartz and K-feldspar sandstone with abundant detrital clay. Rare bedding.	Occurs at the base of the Glenwood Fm and as thin beds in the upper Glenwood Fm.	Densely bioturbated (*Skolithos, Planolites*).	Shallow subtidal, relatively quiet waters.	Dolomite and quartz cement is uniform to patchy. Pyrite is common. Solution seams present.	ef1, ef2
Shales	Mudstone to very fine grained quartz and K-feldspar sandstone. Fissile.	Occurs as thin layers in the middle to upper Glenwood Fm.	Rare burrows.	Relatively deeper subtidal, low-energy environment.	Pyrite is common. Solution seams present.	ef5, ef6
Carbonate	Mudstones, wackestones, grainstones, and algal bindstones. Mudstones are thinly laminated. Occasional mud intraclasts. Fossils and ooids are common.	Usually occurs in beds up to 10 m. Most common in the lower St. Peter Ss and upper Glenwood Fm.	Burrows common to rare.	Subtidal, low-energy environment.	Contains significant nonbanded dolomite. Solution seams common. Early marine cements are present.	ef5, ef6
Sandy carbonate	Originally carbonate mud with up to 60% quartz and K-feldspar grains (very fine to medium-grained). Rare ooids and fossils. Some horizontal bedding.	Found in intervals up to 1 m thick. Most common in the lower St. Peter Ss and upper Glenwood Fm, near carbonates.	Densely bioturbated.	Relatively deeper subtidal, low-energy environment.	Sands are cemented by nonbanded early marine replacive and late burial dolomite. Pressure solution features are rare.	ef5, ef6
Massive sandstone	Nonbedded, well- to very well sorted, fine- to coarse-grained sandstone. Composed of quartz, minor K-feldspar, and rare heavy minerals.	Found in layers decimeters to tens of meters thick. Concentrated in the upper St. Peter Ss.	Densely bioturbated (*Skolithos*).	Shallow subtidal, relatively quiet environment.	Nonbanded, rare stylolites. Clay and dolomite cement are distributed in isolated patches and/or meter-scale zones.	ef1, ef2
Horizontally bedded sandstone	Well- to very well sorted, fine- to medium-grained sandstone. Composed of quartz, minor K-feldspar, and rare heavy minerals. Parallel, horizontal, mm-scale laminations.	Found in intervals on the centimeter- to meter-scale. Most common in the lower St. Peter Ss.	Rare *Skolithos* burrows.	Shallow subtidal, high-energy environment.	Abundant intergranular pressure solution and stylolites. Silica (and some dolomite) occurs as bands of cement.	ef3, ef4, ef5
Cross-bedded sandstone	Well- to very well sorted, fine- to medium-grained sandstone. Composed of quartz, minor K-feldspar, and rare heavy minerals. Planar and trough cross-bedding is present. Occasional intraclasts.	Found in intervals on the centimeter- to meter-scale. Most common in the lower St. Peter Ss.	Rare *Skolithos* burrows.	Shallow subtidal, high-energy environment.	Abundant intergranular pressure solution and stylolites. Silica (and some dolomite) occurs as bands of cement.	ef3, ef4, ef5

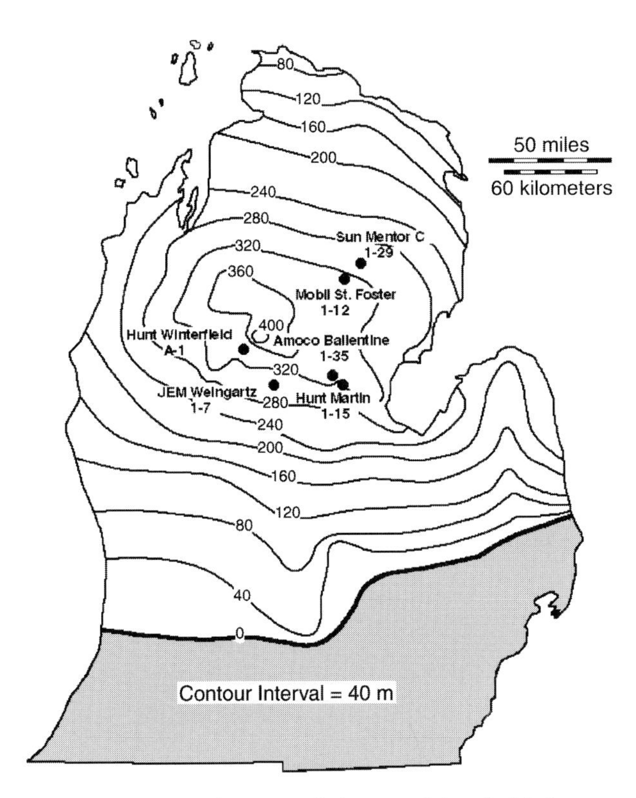

Figure 2. Isopach map of the combined thickness of the Middle Ordovician St. Peter Sandstone and Glenwood Formation in the Michigan basin. The maximum thickness occurs in the north-central part of the basin, and it thins in all directions. The locations of cores discussed in this paper are shown on the map.

A consensus of Middle Ordovician stratigraphy in the Michigan basin became important as a result of deep drilling during the 1980s. Early stratigraphic work by Ells (1967) was improved by Fisher and Barratt (1985), Brady and DeHaas (1988), Fisher et al. (1988), and Vandrey (1991). Harrison (1987) and Barnes et al. (1988) correlated the thick Middle Ordovician sandstone in the Michigan basin with St. Peter Sandstone outcrops elsewhere on the craton. Barnes et al. (1992a) studied the sedimentology and diagenesis of the St. Peter Sandstone in the Michigan basin and divided the St. Peter into four lithofacies based on core and wireline log characteristics. The formational boundary between the St. Peter Sandstone and the Glenwood Formation used in this study is described in Vandrey (1991) and differs from that of Barnes et al. (1988, 1992a) and Catacosinos and Daniels (1991) (Figure 1).

The major lithofacies that characterize the St. Peter Sandstone and the Glenwood Formation are: (1) massive sandstone, (2) clay-rich sandstone, (3) horizontally bedded sandstone, (4) cross-bedded sandstone, (5) carbonate, (6) sandy carbonate, and (7) shale (Table 1; Vandrey, 1991; Drzewiecki, 1992). The lithologies of the Glenwood Formation in the Michigan basin are similar to those exposed in outcrop in Wisconsin, Minnesota, and Illinois (Long, 1988), whereas the St. Peter differs significantly. St. Peter outcrops are characterized by burrowed and cross-bedded sandstones and lack carbonate and shale lithofacies (Mai and Dott, 1985; Dott et al., 1986).

All of the lithofacies in the St. Peter Sandstone of the Michigan basin are interpreted as shallow marine deposits (Barnes et al., 1988; Lundgren and Barnes, 1989; Nadon et al., 1992). The carbonates in the St. Peter Sandstone form widely correlative horizons that were used to subdivide the St. Peter into seven informal units (Figure 1) (Nadon et al., 1992). The difference in the relative proportions of sandstone and carbonate reflects variations in the rate of formation of accommodation space and the proximity of sand sources. There are no preserved carbonates in the northwestern portion of the Michigan basin, the inferred sand source area. The carbonate units become thinner toward the top of the St. Peter Sandstone and are typically absent in the upper one-third.

The vertical and lateral distribution of carbonate and sandstone lithologies of the Glenwood are different from those of the St. Peter. Clay-rich sandstones at the base of the Glenwood Formation probably reflect a shift from shallow marine deposits of the St. Peter Sandstone to a more distal shelf setting during Glenwood deposition and drowning of the siliciclastic source. Carbonates and phosphatic shales of the Glenwood were deposited mainly in the central portion of the basin (Vandrey, 1991), farthest from siliciclastic sources. Open marine carbonates dominated during the deposition of the Upper Ordovician Black River Formation.

DIAGENETIC SEQUENCE

The depositional framework of the St. Peter Sandstone and Glenwood Formation had a profound effect on the diagenetic processes that occurred within each lithofacies (Table 1). The sandstone lithofacies have undergone a series of diagenetic events that significantly altered the texture, porosity, and permeability of the rock (Figure 3). These events are broken into three distinct stages, each of which is characterized by a different set of diagenetic processes, pore fluid compositions, temperatures, burial depths, and cement types. The distribution and type of cements depends on lithology. Argillaceous sandstones of the Glenwood Formation contain dolomite and clay cements that have a uniform to patchy distribution. The massive sandstone lithofacies is dominated by quartz overgrowths distributed uniformly throughout the interval, and layered diagenetic features (such as diagenetic bands and stylolites) are uncommon. In contrast, the cross-bedded and horizontally bedded lithofacies contain layers with sharp grain size contrasts that promoted the development of stylolites and millimeter- to centimeter-scale diagenetic bands (Drzewiecki et al., 1991). The result is a unit characterized by alternating bands of different cement types

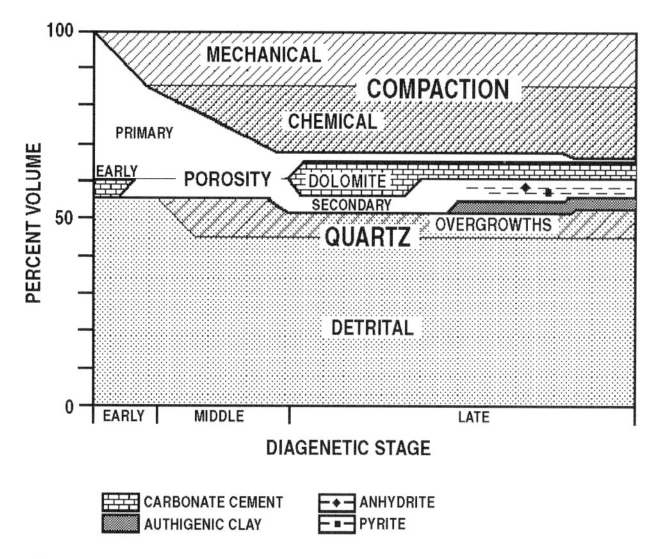

Figure 3. Paragenesis of the St. Peter and Glenwood sandstones. The vertical scale is the amount of the different authigenic and detrital components relative to the original volume of the rock. The horizontal scale is relative time divided into three diagenetic stages. Each stage is characterized by a different set of cementation and dissolution events.

and porosities (Figure 4). The carbonate facies are dominated by early carbonate cementation and late burial dolomitization. Table 2 describes the characteristics, distribution, and geochemistry of the authigenic phases in the St. Peter Sandstone and Glenwood Formation.

A comprehensive study of the St. Peter diagenesis by Barnes et al. (1992a) proposed the following diagenetic sequence: (1) early carbonate cement, (2) early dolomitization of this cement, (3) quartz overgrowths, (4) late burial dolomite, (5) dissolution of some carbonate cements and framework grains, (6) illite cement, and (7) pressure solution, minor quartz cementation, and conversion of some illite to chlorite. The diagenetic sequence proposed in this paper is generally consistent with that of Barnes et al. (1992a) and concentrates on quartz and dolomite cementation, pressure solution, and the banded characteristics of cement distribution.

EARLY DIAGENESIS

Description

Early diagenesis occurred after sediment deposition and before shallow burial. It is characterized by carbonate cementation, porosity reduction by mechanical compaction in sandstone lithofacies (Table 2, Figure 3), and micritization of skeletal components in carbonate lithofacies. Overall, early pore-lining carbonate cement (now dolomite) is uncommon (<1%) in the sandstone and carbonate lithofacies of the St. Peter Sandstone and Glenwood Formation and has a patchy distribution. It is most common in carbonates, sandy carbonates, sandstones near carbonates, and clay-rich sandstones. These early carbonate

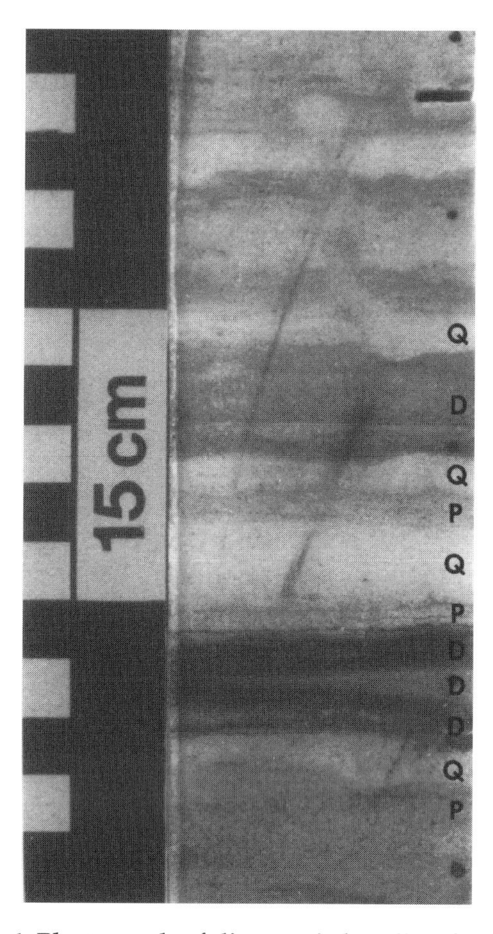

Figure 4. Photograph of diagenetic banding from the Mobil St. Foster 1-12 core (depth of 3026 m below sea level). This interval is characterized by millimeter-scale quartz-cemented (Q), dolomite-cemented (D), and porous (P) bands, and abundant stylolites. Scale is in centimeters.

cements form either dark isopachous rims (0.05 to 0.3 mm thick) that line pores, or fibrous, botryoidal pore-filling cements. Botryoidal cements are composed of acicular crystals ranging in length from 0.8 to 1.0 mm and less than 0.05 mm in width (Vandrey, 1991). Rare meniscus cements occur in oolitic grainstones of the Glenwood Formation (Vandrey, 1991).

Mechanical compaction in the sandstone lithofacies predated any quartz overgrowth cementation and resulted in reorientation of the sand grains and ductile deformation of detrital clay. There is no evidence of quartz grain fracturing. Assuming an initial porosity of 40%, Shepherd et al. (this volume) estimated that approximately 10 to 15% of the original porosity in the St. Peter Sandstone was lost by mechanical compaction (Figure 3). Fractured rim cements and the preservation of a high percentage of intergranular volume in sandstones with isopachous and botryoidal cement suggest that these cements predate significant mechanical compaction.

Interpretation

The very early diagenetic origin of the pore-lining and botryoidal carbonate cements suggests that they

Table 2. Characteristics, geochemistry, and distribution of major authigenic phases in the St. Peter Sandstone and Glenwood Formation.

Authigenic Phase	Color and Appearance	Shape	Size	CL Color	Stable Isotopes*	Fluid Inclusions**	Interpreted Pore Fluid	Occurrence	Distribution
Calcite cements	Clear to cloudy (inclusion-rich), often displaying a drusy fabric	Equant, anhedral to euhedral	<.05–1.3 mm	Zoned or mottled	O = 17 to 24‰ C = -4 to -1‰ (Vandrey, 1991)	—	Modified seawater	Rare in the carbonate facies of the Glenwood Fm, typically filling moldic porosity	Thin beds preserved between dolomite beds
Quartz overgrowths	Clear, unstrained syntaxial overgrowth cement with uniform extinction	Euhedral overgrowths with planar boundaries	Up to 0.4 mm thick	Nonlumin.	O = 13 to 18‰ (Winter and Valley, unpub. data)	81–153°C n = 90	Modified seawater	Sandstone facies, especially within burrows, and in the horizontally and cross-bedded facies. Fills primary porosity	Uniform in the Glenwood Fm and upper St. Peter Ss. Uniform or banded in lower and middle St. Peter Ss.
Potassium feldspar overgrowths	Clear to cloudy non-syntaxial overgrowth cement with uniform extinction	Euhedral overgrowths with planar boundaries	Up to 0.2 mm thick	Nonlumin.	—	—	Modified seawater	Sandstone facies, especially in the horizontally and cross-bedded lithofacies. Fills primary porosity	Uniform throughout the intervals in which it is found
Coarse replacive dolomites	Light brown, inclusion-rich core and clear rim, undulose extinction	Curved subhedral to euhedral rhombs	0.2–2.0 mm	Nonlumin., dull red, or zoned	O = 15 to 18‰ C = -2 to -1‰ n = 6	—	Hypersaline brines	Carbonate facies in the Glenwood Fm and St. Peter Ss, especially in the shaley dolomite lithofacies	Uniform to patchy in the Glenwood Fm carbonate lithofacies
Fine replacive dolomites	Light brown, inclusion-rich dolomite	Planar subhedral to anhedral rhombs	0.06–0.1 mm	Orange, nonlumin.	O = 18 to 26‰ C = -12 to -1‰ n = 36	—	Hypersaline brines	Carbonate facies in the Glenwood Fm, especially in the shaley dolomite	Uniform to patchy in the Glenwood Fm carbonate lithofacies
Coarse dolomite cements	White to tan, inclusion-rich core and clear rim, undulose extinction	Curved subhedral to euhedral rhombs	0.2–3.5 mm	Nonlumin.	O = 16 to 23‰ C = -4 to -1‰ n = 16	98–189°C n = 69; most data between 120 and 150° (n = 46)	Hypersaline brines	Found in fractures and oversized porosity in Glenwood Fm and St. Peter Ss carbonate lithofacies. Fills primary and secondary porosity	Found in fractures and oversized porosity in carbonate lithofacies
Fine dolomite cements	Light brown, inclusion-rich dolomite	Curved to planar euhedral to subhedral rhombs	0.05–1.5 mm	Nonlumin., dull red, or zoned	O = 13 to 23‰ C = -9 to -3‰ n = 50	87–155°C n = 39	Hypersaline brines	Sandstone lithofacies of the St. Peter Ss and Glenwood Fm. Fills primary and secondary porosity	Uniform to patchy in the Glenwood Fm, patchy to banded in the St. Peter Ss.
Authigenic illite cements	Light brownish green, platy to fibrous masses	Thin platy to fibrous rim coating	3–5 mm	Nonlumin.	—	—	Hypersaline brines	Sandstone facies, especially in the massive sandstone lithofacies. Fills primary and secondary porosity	Uniform throughout the intervals in which it is found. Occasionally banded.
Pyrite	Metallic gold, opaque in thin section	Euhedral crystals, some anhedral masses	Up to 1 mm	Nonlumin.	S = 12 to 27‰ n = 5	—	Hypersaline brines	Coarse, porous, clay-rich and massive sandstones facies of the Glenwood and St. Peter Ss.	Isolated crystals
Anhydrite	Clear masses (birefringent under cross-polarized light)	Blocky or bladed pore-filling cement	Up to 1.5 mm	Nonlumin.	—	—	Hypersaline brines	Sandstone facies. Fills primary and secondary porosity	Patchy

* C = $\delta^{13}C$ (PDB), O = $\delta^{18}O$ (SMOW), S = $\delta^{34}S$ (CDT).

** Not corrected for pressure.

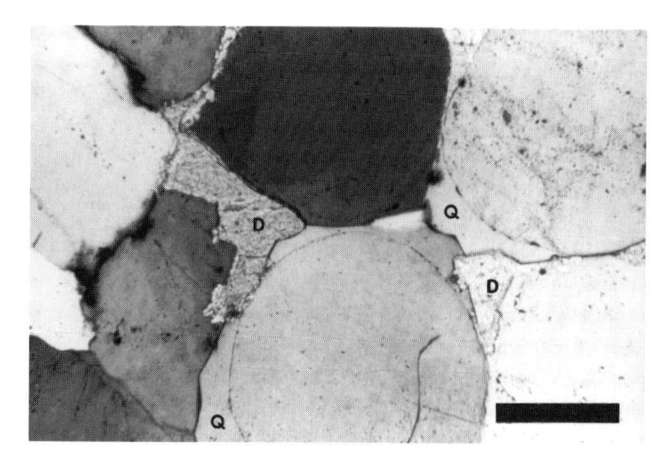

Figure 5. Typical quartz overgrowths (Q) and fine dolomite cement (D) from the St. Peter Sandstone. The dolomite postdates the quartz overgrowths. Notice the planar crystal terminations of the quartz overgrowths, which can be distinguished from the quartz grains by the presence of a fine "dust rim" separating them. This photomicrograph also shows the pore-filling nature of the fine dolomite cement. Note the well-developed dolomite crystal faces in contact with the quartz overgrowths. This is evidence of quartz replacement by dolomite. This photomicrograph was taken under crossed nicols. Scale bar is 0.5 mm. St. Peter Sandstone, Mobil St. Foster 1-12 core.

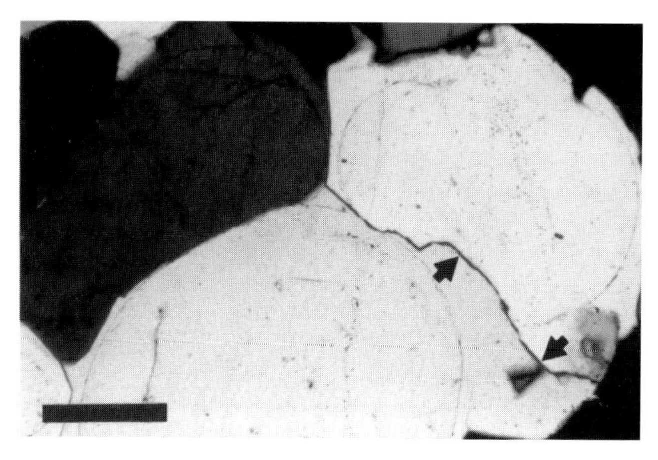

Figure 6. Relationships among quartz overgrowths that suggest they predate intergranular pressure solution. The overgrowths are partially dissolved by chemical compaction (arrows). Other samples show overgrowths postdating intergranular pressure solution. This photomicrograph was taken under crossed nicols. Scale bar is 0.5 mm. St. Peter Sandstone, Mobil St. Foster 1-12 core.

precipitated from connate marine water, presumably either syndepositionally or during shallow burial conditions. The majority of porosity loss during this stage resulted from mechanical compaction, with only a minor amount resulting from cementation.

MIDDLE DIAGENESIS

Description

This episode of diagenesis was dominated by intergranular pressure solution and the formation of quartz and potassium feldspar overgrowths in the sandstone lithofacies (Table 2, Figure 3). The carbonate lithofacies of the Glenwood Formation is characterized by the development of minor aggrading, equant, neomorphic calcite (0.01 to 0.1 mm), precipitation of ferroan to nonferroan equant calcite spar (up to 0.05 mm) in vugs, and minor secondary dissolution of skeletal components (Vandrey, 1991).

In the sandstone lithofacies, syntaxial quartz overgrowths having uniform extinction (Figures 5 and 6) are found in nearly every thin section examined and comprise about 6% of the total rock volume (it can be as high as 15% in individual samples) in the Mobil St. Foster 1-12 core (Shepherd et al., this volume). They are most abundant in the horizontal and cross-bedded lithofacies of the St. Peter Sandstone, where they typically occur in bands and form tightly cemented intervals. In the upper St. Peter Sandstone and Glenwood Formation, quartz cement is less common and does not occur as bands. Quartz overgrowths are 0.05 to 0.2 mm thick in the St. Peter Sandstone, but are typically < 0.05 mm thick in the sandstones of the Glenwood Formation. Detrital clay and early carbonate cement appear to have inhibited the formation of quartz overgrowths in parts of the Glenwood Formation and St. Peter Sandstone (Table 1). Quartz overgrowths have planar crystal terminations or compromise boundaries where they come in contact with each other (Figures 5 and 6). A dust-rim composed of inclusions or clay is often present at the grain/overgrowth boundary and is frequently the site of later secondary dissolution. Nonsyntaxial potassium feldspar overgrowths with uniform extinctions are a minor component of the St. Peter Sandstone and Glenwood Formation, making up less than 1% of the rock volume. They are usually thinner than quartz overgrowths (<0.05 to 0.1 mm).

Petrographic evidence suggests that extensive intergranular pressure solution predated and postdated quartz overgrowth precipitation (Figure 6). Intergranular pressure solution is responsible for a major decrease in porosity, by dissolution and collapse of quartz grains (Figure 3). In some intervals, nearly all porosity was destroyed by intergranular pressure solution, resulting in a "fitted fabric" texture.

Interpretation

The occurrence of equant calcite spar, aggrading neomorphic calcites, and minor dissolution of aragonite bioclastic debris suggests that the pore fluids during this stage of diagenesis were, in part, meteoric. Neomorphic calcite and calcite spar have $\delta^{18}O$ values and trace element concentrations that support a low-temperature, meteoric water origin (Vandrey, 1991).

The petrography of quartz and feldspar cement indicates that they precipitated as overgrowths in intergranular pore space where there was limited detrital clay or early carbonate cement to inhibit them. Quartz cement and intergranular pressure solution appear to have been localized by horizontal depositional heterogeneities, resulting in a banded distribution. The quartz cement may have been derived either: (1) locally by pressure solution and mineral transformations, (2) from an external source, or (3) from a combination of both. Because the timing of intergranular pressure solution is coincident with that of quartz cementation, intergranular pressure solution appears to have been the dominant source of silica. Shepherd et al. (this volume) found that in some localities in the St. Peter Sandstone sufficient quartz was dissolved by intergranular pressure solution to account for all quartz overgrowths.

The timing and depth of the middle diagenetic stage is difficult to determine. Tada and Siever (1989) suggest that a minimum depth of 1 to 2 km is required for intergranular pressure solution. This is likely the minimum depth for this diagenetic stage, which is supported by the high homogenization temperatures of fluid inclusions in the quartz cement (discussed below).

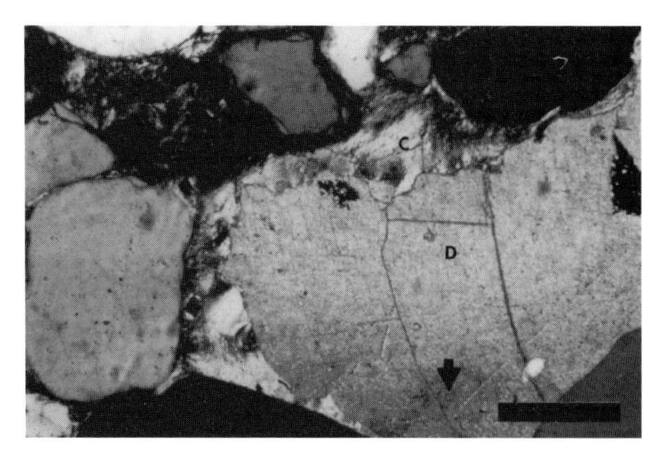

Figure 7. Coarse dolomite cement (D) from the Glenwood Formation showing the characteristic curved crystal faces and undulose extinction typical of this type of dolomite. The lower portion of the crystal (arrow) is going extinct, while the upper portion is not. Fibrous clay (C) surrounds the dolomite crystal and fills all post-dolomite porosity. The ragged edge of this dolomite implies that it has undergone dissolution before clay formation. This photomicrograph was taken under crossed nicols. Scale bar is 0.5 mm. Glenwood Formation, Hunt Winterfield A-1 core.

LATE DIAGENESIS

Description

The diagenetic sequence during late diagenesis includes quartz dissolution, dolomitization, dolomite dissolution, clay precipitation, minor anhydrite, pyrite, and late quartz cementation, stylolitization, and partial replacement of illite by chlorite.

Four types of dolomite formed in the St. Peter Sandstone and Glenwood Formation during late diagenesis (Table 2): (1) coarse (0.2 to 2.0 mm) replacive dolomite (Glenwood carbonates), (2) fine (0.06 to 0.1 mm) replacive dolomite (St. Peter and Glenwood carbonates), (3) coarse (0.2 to 3.5 mm) dolomite cement (St. Peter and Glenwood carbonates), and (4) fine (0.05 to 1.5 mm) dolomite cement (St. Peter and Glenwood sandstones). This last type forms diagenetic bands in sandstone lithofacies. The crystal size of the replacive dolomites is controlled by the depositional fabric of the original limestones (Vandrey, 1991). Coarse replacive dolomites and coarse dolomite cements commonly have curved crystal boundaries and undulose extinctions characteristic of saddle dolomite (Figure 7). Fine replacive dolomites and fine dolomite cements are usually subhedral to euhedral and have planar or curved boundaries but may be anhedral where growth space is limited.

Fine dolomite cement (Figure 5) postdates the quartz cement and comprises an average of 5% of the total sandstone volume (individual samples can have up to 20%). This cement is most common near the carbonate facies, but it also occurs as isolated diagenetic bands and patches of cement throughout the various sandstone lithofacies. Coarse dolomite cement precipitated in intergranular, moldic, and fracture porosity in the carbonate lithofacies. Fine replacive dolomites occur in the carbonate lithofacies of both the Glenwood and St. Peter, but coarse replacive dolomites are restricted to the Glenwood Formation.

Secondary porosity in the sandstone lithologies appears to have formed by the dissolution of dolomite that replaced quartz overgrowths and grains. Petrographic evidence suggesting that the fine dolomite cement partially replaced quartz as it precipitated includes embayments in the edges of quartz grains filled with dolomite cement, fingers of dolomite between quartz grains and their overgrowths, planar dolomite crystal facies in contact with quartz overgrowths (Figures 5 and 8), and dolomite rhomb-shaped pores.

Partial dolomite dissolution is evident from the presence of ragged and embayed dolomite crystals as well as rhomb-shaped pores. The amount of dissolution that occurred is unknown, but it appears to have been several percent of the total rock volume (Figure 3). Most of the secondary porosity created by dolomite dissolution remains open, but some is filled with clay cement. The relationship between clay precipitation and dolomite dissolution is unclear. In some cases, dolomite appears to have been replaced by clay (Barnes et al., 1992a), but often dolomite dissolution occurred where clay is absent.

Clay, identified as illite partially transformed to chlorite (Barnes et al., 1992a, b), is the third most

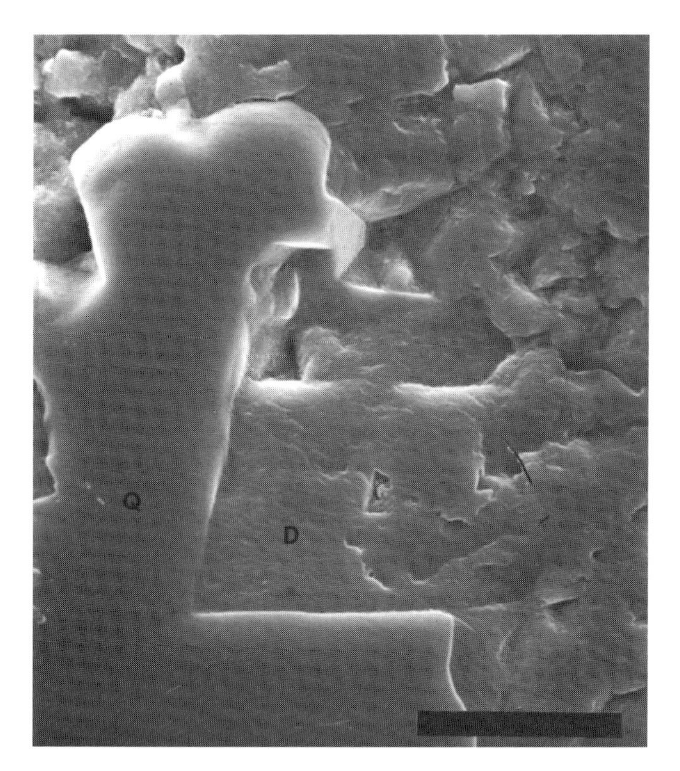

Figure 8. Scanning electron microscope image showing evidence of partial replacement of quartz (Q) by dolomite cement (D). A dolomite rhomb with planar crystal faces is growing into a quartz overgrowth. Scale bar is 10 μm. St. Peter Sandstone, Mobil St. Foster 1-12.

abundant cement in the St. Peter Sandstone and Glenwood Formation. It comprises an average of 1 to 2% of the rock volume in St. Peter Sandstone and is somewhat more common in the Glenwood sandstones. Although it is often difficult to distinguish between depositional and diagenetic clay, they can usually be differentiated on the bases of their degree of compaction and their relative position in the diagenetic sequence. Diagenetic clay is most commonly found as randomly arranged, uncompacted pore-lining cement (Figure 9) but may be pore-filling (Figure 7).

Other late diagenetic phases include anhydrite, pyrite, and quartz. Anhydrite is a minor cement phase that postdates clay cement and makes up only about 1% of some sandstone samples but is absent in most. Where present, it occurs as patches of blocky, pore-filling crystals that fill both primary and secondary porosity. Pyrite forms a minor late cement in the sands of both the St. Peter and the Glenwood. Pyrite postdates anhydrite precipitation in fractures in the Glenwood carbonates (Vandrey, 1991), and it is most abundant in the clay-rich sands of the Glenwood Formation. Late stage quartz cementation was identified in the St. Peter (Barnes et al., 1992a) and in fractures in the Glenwood Formation (Vandrey, 1991). Stylolitization (Figures 4 and 10) was a major late diagenetic process and is discussed in a later section.

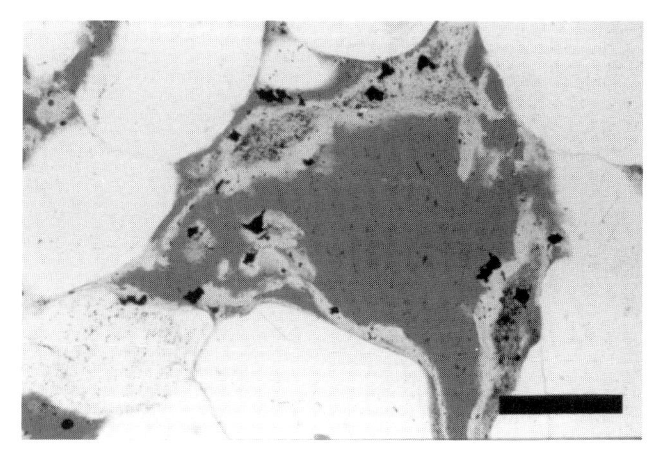

Figure 9. Photomicrograph of clay lining an intergranular pore in a porous interval of the St. Peter Sandstone. The uncompacted nature of the clay and its absence between adjacent quartz grains are evidence that the clay is authigenic. Scale bar is 0.5 mm. St. Peter Sandstone, Mobil St. Foster 1-12 core.

Interpretation

Dissolution of quartz overgrowths and grains is associated with precipitation of fine dolomite cement in the St. Peter Sandstone and Glenwood Formation (Figure 8). The solubility of silica is controlled by temperature, pH (Boggs, 1987; Brownlow, 1979), and hydrostatic and nonhydrostatic stresses (Weyl, 1959; Maliva and Siever, 1988). Silica dissolution in the St. Peter and Glenwood sandstones may be due to one or a combination of these factors. Petrographic evidence suggests that nonhydrostatic stresses created between quartz and the precipitating dolomite cement resulted in a significant amount of quartz dissolution in the St. Peter and Glenwood sandstones, but the other processes may also have been important.

Petrographic relationships suggest that all four dolomite types belong to the same general diagenetic stage, but there may have been a slight difference in the temperature, fluid composition, and amount of space available for crystal growth between the coarse dolomites and fine dolomites (see *Geochemistry* section). Replacive dolomites characterize the carbonate lithofacies, fine dolomite cement precipitated in the open pore space of the sandstone lithofacies, and coarse dolomite cement precipitated in fractures in the carbonate lithofacies. The coarse replacive dolomites and coarse dolomite cement have characteristics of saddle dolomites, which are believed to form at deep burial depths (i.e., 1.5 to 4.5 km) and elevated temperatures (i.e., 50 to 160°C) (Radke and Mathis, 1980; Choquette and James, 1987). The deep burial fluids from which saddle dolomites precipitate are typically hypersaline, have near-neutral pH, and are weakly to strongly reducing (Beales, 1971). This interpretation for the St. Peter and Glenwood dolomites is consistent with fluid inclusion and geochemical data (see below). The precipitation of anhydrite is consistent with a

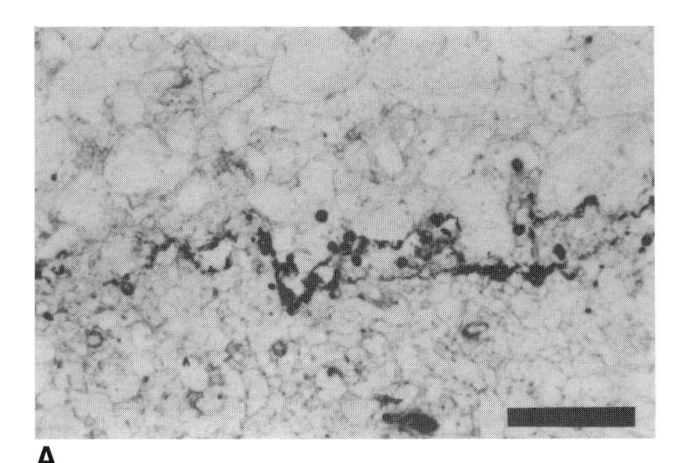

A

B

Figure 10. (A) Photomicrograph of a stylolite from the Mobil St. Foster core showing that it is localized at a textural heterogeneity. In this case, the sandstone above the stylolite is coarser than the sandstone below. The small opaque crystals are pyrite. Scale bar is 1.0 mm. St. Peter Sandstone, Hunt Winterfield A-1 core. (B) Core from the Hunt Winterfield A-1 well showing the pervasive stylolitization (dark lines) that is common in the diagenetically banded portions of the St. Peter Sandstone. Scale is in centimeters.

hypersaline pore fluid interpretation. Pyrite samples have $\delta^{34}S$ values of 12 to 27‰ CDT (Table 2), indicating that the sulfur was derived from either anhydrites or sea water and not from biologically induced sulfate reduction. This is consistent with a late burial origin for pyrite. Fine replacive dolomite and fine dolomite cement appear to be contemporaneous with their coarse counterparts in thin section but do not exhibit curved crystal faces and cleavage like the coarse dolomite cement.

Barnes et al. (1992b) determined the K/Ar age of diagenetic illite in the St. Peter Sandstone of the Michigan basin to be Late Devonian to Mississippian (367 to 327 Ma, with an average date of 346 Ma). They concluded that the illite formed at depths of 3.1 to 3.6 km and at temperatures of 150°C or more in the central portion of the Michigan basin. According to the illite ages of Barnes et al. (1992b) and burial history curves of the Michigan basin (Cercone, 1984), late diagenesis occurred during the time of maximum burial depth of the St. Peter and Glenwood. The dates obtained by Barnes et al. (1992b) have important implications for the absolute timing of all major diagenetic events. They imply that both quartz and dolomite cementation were confined to the first 100 million years of burial history and that there has been no significant diagenetic alteration or porosity modification of the St. Peter Sandstone and Glenwood Formation during the 350 million years since clay precipitation.

POROSITY EVOLUTION

The porosity in the St. Peter and Glenwood sandstone lithofacies has been strongly modified by a number of diagenetic processes. On the other hand, the carbonate lithofacies were relatively nonporous when deposited and have remained such throughout their diagenetic history. Any secondary porosity created in the carbonate lithologies by dissolution and fracturing has been almost completely occluded by dolomite cement.

Although the original porosity in the sandstones of the St. Peter and Glenwood was most likely relatively uniform throughout individual lithofacies, in the samples that exhibit well-developed diagenetic banding, porosity is now highly variable on the millimeter scale (Shepherd et al., this volume). Moline et al. (1992, this volume) identified banded sandstones as low-porosity, low-permeability intervals, and Shepherd et al. (this volume) show that they occur at the depth and location of a known abnormal pressure compartment seal in the Mobil St. Foster 1-12 core from the Michigan basin.

Figure 3 shows the evolution of sandstone porosity during the various diagenetic stages. Table 3 lists the porosity-modifying events as well as the amount of porosity added or destroyed at two well sites. Primary porosity of the Middle Ordovician sandstones was almost exclusively intergranular and is estimated at about 40% (Shepherd et al., this volume). Primary porosity distribution was controlled by depo-

Table 3. Porosity evolution of the St. Peter Sandstone from the Mobil St. Foster 1-12 and Sun Mentor 1-29 C.

	Mobil St. Foster 1-12		Sun Mentor 1-29 C	
Diagenetic Event	Porosity Change (%)	New Porosity (%)	Porosity Change (%)	New Porosity (%)
Initial porosity	—	40	—	40
Mechanical compaction	–15	25	–15	25
Pressure solution and quartz cementation	–17	8	–15	10
Quartz dissolution	+1	9	+1	11
Carbonate cementation	–6	3	minor	11
Carbonate dissolution	+3	6	minor	11
Authigenic clay cementation	minor	6	-7	4
Final porosity	6		4	

sitional characteristics and appears to have been uniform within each lithofacies. All the sandstone lithofacies of the St. Peter contain very little clay, and the only major textural differences among these facies involves sedimentary structures and burrows (Table 1). The clay-rich sandstones of the Glenwood Formation contain pore-filling clay that is in part detrital. Intergranular clay would have slightly reduced the initial porosity of this facies.

Most primary porosity in the St. Peter and Glenwood sandstones has been eliminated by compaction and cementation during burial diagenesis. Occasionally, up to 15% is still preserved, particularly in the clay-rich sandstones where detrital clay prohibited quartz cementation. Such porous, clay-rich sandstones produce gas in the Glenwood Formation. The average porosity of St. Peter and Glenwood reservoirs is 11.4%, and the average permeability is 5 md (Catacosinos et al., 1990).

Early diagenesis was dominated by major porosity loss, primarily through mechanical compaction. Middle and late diagenesis resulted in further loss of porosity by intergranular pressure solution and cementation. Porosity enhancement by dissolution of authigenic phases became significant during late diagenesis.

Secondary porosity formed at the expense of detrital quartz grains and overgrowths and consists of elongated pores between grains, embayed edges on grains and overgrowths, and pores between grains and their overgrowths. Most likely, dissolution enhanced primary porosity in porous zones, increasing it by only a few percent. The dissolution of quartz can be accounted for by both direct dissolution by silica-undersaturated pore fluids and by nonhydrostatic pressures resulting from dolomite precipitation. Occasionally, dolomite is partially or completely dissolved, resulting in isolated molds of dolomite rhombs and ragged dolomite crystal boundaries (Figure 7).

Shepherd et al. (this volume) report that, overall, compaction was more important than cementation in reducing porosity in the St. Peter Sandstone of the Mobil St. Foster core. They conclude that more quartz was dissolved by intergranular pressure solution than is present as quartz cement. Intergranular pressure solution reduced porosity through dissolution and collapse of quartz grains and provided silica for cement that further reduced porosity.

GEOCHEMISTRY

Fluid inclusion and stable isotope data were collected for the following authigenic phases: (1) quartz overgrowths, (2) fine replacive dolomite, (3) coarse replacive dolomite, (4) fine dolomite cement, and (5) coarse dolomite cement (Tables 2 and 4). These data were used to interpret fluid types and the relative timing of quartz and dolomite cementation and their significance to diagenetic band formation.

Fluid Inclusions

Samples from quartz and dolomite cements containing primary, two-phase fluid inclusions (less than 10 mm in the longest dimension) were heated until the liquid and vapor phases were homogenized. Fluid inclusions from quartz overgrowths have homogenization temperatures that range from 81°C to 153°C (n = 90). The distribution is bimodal, with a distinct mode centered at 135°C and a smaller one at 105°C (Figure 11). The smaller mode may represent an early episode of quartz cementation postdated by a second episode at 135°C. This indicates that quartz-cemented band formation may have occurred during two distinct episodes.

The homogenization temperatures of fluid inclusions from fine dolomite cement in St. Peter Sandstone range from 87°C to 155°C (n = 39), with most between 110°C and 130°C (Figure 11). Coarse fracture-filling dolomite cements in the Glenwood carbonates have homogenization temperatures ranging from 98 to 189°C (Figure 11), although most fall between 120 and 150°. The small difference in the mode of the fine and coarse cements may be an artifact of sample distribution (most of these analyses

Table 4. Stable isotope data for the St. Peter Sandstone and Glenwood Formation.

Sample	Well	Depth (m)	Depth (ft)	Electrofacies	$\delta^{13}C$ (PDB)	$\delta^{18}O$ (SMOW)	$\delta^{18}O$ (PDB)	Comments
Glenwood Formation—Fine dolomite cement in sandstone								
HM-20	Hunt Martin	3445.8	11305.0	—	–4.1	17.6	–12.8	Dolo-cemented Ss
SM-1	Sun Mentor	3074.5	10087.0	—	–6.6	19.7	–10.8	Dolo-cemented Ss
JW-2	JEM Wiengartz	3272.6	10736.9	—	–3.4	19.4	–11.1	Dolo-cemented Ss
JW-5	JEM Wiengartz	3272.9	10737.8	—	–3.4	19.7	–10.9	Dolo-cemented Ss
HW-9	Hunt Winterfield	3221.0	10567.6	—	–4.4	16.6	–13.9	Dolo-cemented Ss
HW-1	Hunt Winterfield	3221.2	10568.4	—	–4.5	16.2	–14.2	Dolo-cemented Ss
HW-2	Hunt Winterfield	3221.2	10568.1	—	–5.5	12.8	–17.5	Dolo-cemented Ss
HW-11	Hunt Winterfield	3221.2	10568.3	—	–5.1	13.9	–16.5	Dolo-cemented Ss
HW-20	Hunt Winterfield	3221.3	10568.6	—	–4.7	14.2	–16.1	Dolo-cemented Ss
HW-8	Hunt Winterfield	3221.4	10569.0	—	–4.1	17.4	–13.0	Dolo-cemented Ss
HW-5	Hunt Winterfield	3222.4	10572.2	—	–4.5	15.8	–14.6	Dolo-cemented Ss
HW-18	Hunt Winterfield	3222.5	10572.6	—	–4.2	14.1	–16.3	Dolo-cemented Ss
HW-10	Hunt Winterfield	3222.9	10573.9	—	–3.9	19.0	–11.5	Dolo-cemented Ss
HW-3	Hunt Winterfield	3223.2	10574.8	—	–4.5	16.4	–14.0	Dolo-cemented Ss
HW-4	Hunt Winterfield	3223.2	10574.9	—	–4.0	18.4	–12.1	Dolo-cemented Ss
Glenwood Formation—Coarse dolomite cement in carbonate								
AR-4	Amoco Roscomm.	3437.2	11277.0	—	–3.1	17.9	–12.5	Fracture fill
AR-5	Amoco Roscomm.	3437.2	11277.0	—	–3.8	17.5	–12.9	Fracture fill
JB-3	Joutel Brinks	3270.1	10728.7	—	–1.4	18.5	–12.0	Fracture fill
JB-4	Joutel Brinks	3270.1	10728.7	—	–1.3	19.8	–10.8	Fracture fill
JB-5	Joutel Brinks	3270.1	10728.7	—	–2.2	18.9	–11.6	Fracture fill
JB-6	Joutel Brinks	3270.1	10728.7	—	–1.6	18.5	–12.0	Fracture fill
HM-17	Hunt Martin	3411.2	11191.6	—	–2.8	22.6	–8.0	Fracture fill
HM-18	Hunt Martin	3411.3	11191.8	—	–4.4	18.2	–12.3	Fracture fill
SF-16	St. Foster	3137.5	10293.5	—	–2.9	16.1	–14.3	Fracture fill
SF-17	St. Foster	3137.5	10293.5	—	–1.6	19.6	–10.9	Fracture fill
SF-6	St. Foster	3138.0	10295.3	—	–3.0	17.7	–12.7	Fracture fill
SF-11	St. Foster	3138.1	10295.5	—	–2.3	18.7	–11.8	Fracture fill
SF-45	St. Foster	3139.0	10298.5	—	–2.4	18.0	–12.5	Fracture fill
SF-2	St. Foster	3143.7	10314.0	—	–3.0	19.9	–10.7	Shelter porosity dolomite
WM-2	JEM Workman	3259.0	10692.3	—	–1.7	17.7	–12.8	Fracture fill
WM-3	JEM Workman	3259.0	10692.3	—	–1.6	17.6	–12.9	Fracture fill
Glenwood Formation—Fine and coarse replacive dolomite in carbonate								
AR-3	Amoco Roscomm.	3437.2	11277.0	—	–4.2	19.2	–11.3	Dolomicrite
JB-2	Joutel Brinks	3270.1	10728.7	—	–1.9	18.9	–11.6	Dolomicrite
HM-15	Hunt Martin	3411.1	11191.2	—	–4.5	19.5	–11.1	Dolomicrite
HM-16	Hunt Martin	3411.1	11191.4	—	–2.6	24.0	–6.7	Dolomicrite

Continued on next page

Table 4. Stable isotope data for the St. Peter Sandstone and Glenwood Formation (continued).

Sample	Well	Depth (m)	Depth (ft)	Electrofacies	$\delta^{13}C$ (PDB)	$\delta^{18}O$ (SMOW)	$\delta^{18}O$ (PDB)	Comments
HM-19	Hunt Martin	3413.8	11200.0	—	–3.8	19.7	–10.8	Dolomicrite
SF-18	St. Foster	3137.5	10293.5	—	–1.7	19.6	–10.9	Dolomicrite
SF-7	St. Foster	3138.0	10295.3	—	–2.0	19.5	–11.1	Dolomicrite
SF-12	St. Foster	3138.1	10295.5	—	–1.9	19.4	–11.1	Dolomicrite
SF-13	St. Foster	3138.1	10295.5	—	–1.9	19.4	–11.1	Dolomicrite
SF-44	St. Foster	3139.0	10298.5	—	–1.6	19.7	–10.8	Dolomicrite
SF-19	St. Foster	3139.6	10300.6	—	–1.4	19.8	–10.7	Dolomicrite
SF-20	St. Foster	3139.6	10300.6	—	–1.5	19.7	–10.9	Rip-up clasts
SF-21	St. Foster	3139.6	10300.6	—	–1.5	19.8	–10.7	Algal mat dolomite
SF-8	St. Foster	3140.2	10302.5	—	–1.3	19.6	–10.9	Algal mat dolomite
SF-9	St. Foster	3140.2	10302.5	—	–1.4	19.6	–11.0	Dolomicrite
SF-10	St. Foster	3140.2	10302.5	—	–1.3	19.5	–11.0	Dolomicrite
SF-14	St. Foster	3140.7	10304.2	—	–1.4	19.5	–11.0	Algal mat dolomite
SF-15	St. Foster	3140.7	10304.2	—	–1.4	19.4	–11.1	Dolomicrite
SF-4	St. Foster	3141.4	10306.4	—	–1.5	19.4	–11.1	Oncolites (algal)
SF-5	St. Foster	3141.4	10306.4	—	–1.5	19.2	–11.3	Dolomicrite
WM-9	JEM Workman	3261.7	10701.1	—	–1.1	17.5	–13.0	Dolomicrite
WM-10	JEM Workman	3259.3	10693.2	—	–1.6	15.4	–15.0	Coarse replacive dolomite
WM-8	JEM Workman	3261.7	10701.1	—	–1.0	17.6	–12.9	Coarse replacive dolomite
NG-4	Nomeco Gernaat	3156.6	10356.2	—	–1.6	14.9	–15.4	Coarse replacive dolomite
NG-8	Nomeco Gernaat	3156.6	10356.2	—	–1.6	15.0	–15.4	Coarse replacive dolomite
NG-9	Nomeco Gernaat	3156.6	10356.2	—	–1.7	14.8	–15.6	Coarse replacive dolomite
NG-10	Nomeco Gernaat	3156.6	10356.2	—	–1.7	14.8	–15.6	Coarse replacive dolomite
St. Peter Sandstone—Fine dolomite cement in sandstone								
HM-1	Hunt Martin	3704.5	12154.0	—	–7.5	21.5	–9.1	Dolo-cemented Ss
HR-1	Hunt Robinson	3154.9	10350.7	—	–4.2	18.9	–11.6	Dolo-cemented Ss
HR-10348.1	Hunt Robinson	3155.3	10352.1	—	–4.9	14.9	–15.5	Dolo-cemented Ss
JW-1	JEM Wiengartz	3278.7	10757.0	—	–3.7	20.2	–10.4	Dolo-cemented Ss
JW-1A	JEM Wiengartz	3339.4	10956.2	5	–3.9	19.0	–11.5	Dolo-cemented Ss
JW-3	JEM Wiengartz	3339.4	10956.0	5	–4.6	16.5	–13.9	Dolo-cemented Ss
JW-6	JEM Wiengartz	3339.6	10956.6	5	–4.3	18.8	–11.7	Dolo-cemented Ss
SF-25	St. Foster	3433.8	11265.7	3	–7.6	18.1	–12.4	Dolo-cemented Ss (patch), W*
SF-26	St. Foster	3433.9	11266.1	3	–7.4	21.2	–9.4	Dolo-cemented Ss (bottom band), W
SF-27	St. Foster	3433.9	11266.1	3	–7.4	21.3	–9.3	Dolo-cemented Ss (middle band), W
SF-28	St. Foster	3433.9	11266.1	3	–7.6	19.8	–10.7	Dolo-cemented Ss (top band), W
SF-29	St. Foster	3433.9	11266.2	3	–7.5	20.9	–9.7	Dolo-cemented Ss (bottom band), W
SF-30	St. Foster	3433.9	11266.2	3	–7.8	17.3	–13.2	Dolo-cemented Ss (middle band), W
SF-31	St. Foster	3433.9	11266.2	3	–7.6	20.8	–9.8	Dolo-cemented Ss (top band), W

Continued on next page

Table 4. (Continued)

Sample	Well	Depth (m)	Depth (ft)	Electrofacies	$\delta^{13}C$ (PDB)	$\delta^{18}O$ (SMOW)	$\delta^{18}O$ (PDB)	Comments
SF-32	St. Foster	3437.1	11276.5	4	–7.4	18.1	–12.4	Dolo-cemented Ss (lense), W
SF-33	St. Foster	3440.3	11287.0	4	–7.5	17.7	–12.8	Dolo-cemented Ss (band), W
SF-34	St. Foster	3443.1	11296.3	4	–7.5	17.8	–12.7	Dolo-cemented Ss (lense), W
SF-35	St. Foster	3443.2	11296.6	4	–7.6	17.2	–13.3	Dolo-cemented Ss (lense), W
SF-22	St. Foster	3443.8	11298.5	4	–7.4	19.4	–11.1	Dolo-cemented Ss
SF-41	St. Foster	3443.8	11298.5	4	–7.4	19.3	–11.2	Dolo-cemented Ss (middle band), W
SF-42	St. Foster	3443.8	11298.5	4	–7.5	18.6	–11.9	Dolo-cemented Ss (top band), W
SF-43	St. Foster	3443.8	11298.5	4	–7.3	20.8	–9.8	Dolo-cemented Ss (band), W
SF-36	St. Foster	3444.0	11299.2	4	–7.6	17.5	–13.0	Dolo-cemented Ss (lense), W
SF-37	St. Foster	3444.3	11300.3	4	–7.4	19.3	–11.2	Dolo-cemented Ss (patch), W
SF-38	St. Foster	3444.4	11300.5	4	–7.3	20.1	–10.4	Dolo-cemented Ss (patch), W
SF-23	St. Foster	3444.9	11302.3	4	–7.3	20.1	–10.4	Dolo-cemented Ss
SF-39	St. Foster	3444.9	11302.3	4	–7.1	23.1	–7.5	Dolo-cemented Ss (burrow), W
SF-40	St. Foster	3445.0	11302.4	4	–7.6	18.3	–12.2	Dolo-cemented Ss (bottom band), W
AB-4	Amoco Ballentine	3588.0	11771.5	6	–9.2	22.0	–8.6	Dolo-cemented Ss, W
AB-13	Amoco Ballentine	3590.8	11780.7	5	–7.9	20.3	–10.2	Dolo-cemented Ss, W
AB-15	Amoco Ballentine	3590.8	11780.7	5	–8.0	19.6	–10.9	Dolo-cemented Ss, W
AB-16	Amoco Ballentine	3590.8	11780.7	5	–8.0	19.5	–11.0	Dolo-cemented Ss, W
AB-18	Amoco Ballentine	3590.8	11781.0	5	–4.9	15.6	–14.8	Dolo-cemented Ss, W
AB-19	Amoco Ballentine	3592.4	11786.0	6	–7.7	18.8	–11.7	Dolo-cemented Ss, W
AB-20	Amoco Ballentine	3592.4	11786.0	6	–8.5	15.1	–15.3	Dolo-cemented Ss, W
St. Peter Sandstone—Fine replacive dolomite in carbonate								
HM-21	Hunt Martin	3482.2	11424.5	5	–3.4	25.3	–5.4	Dolomicrite intraclast
SF-24	St. Foster	3444.5	11300.8	—	–7.1	21.7	–8.9	Dolomicrite intraclast
AB-3	Amoco Ballentine	3588.0	11771.5	6	–11.1	25.5	–5.2	Dolomicrite (dark mud), W
AB-5	Amoco Ballentine	3588.1	11771.9	6	–6.1	26.1	–4.6	Dolomicrite (dark mud), W
AB-9	Amoco Ballentine	3588.3	11772.6	6	–10.4	25.4	–5.3	Dolomicrite (light mud), W
AB-10	Amoco Ballentine	3588.3	11772.6	6	–10.7	26.0	–4.7	Dolomicrite (dark mud), W
AB-11	Amoco Ballentine	3588.3	11772.6	6	–10.8	25.7	–5.0	Dolomicrite (dark mud), W
AB-12	Amoco Ballentine	3588.3	11772.6	6	–10.6	25.4	–5.3	Dolomicrite (light mud), W
AB-8	Amoco Ballentine	3588.7	11773.9	6	–10.1	25.8	–4.9	Dolomicrite (dark mud), W
AB-1	Amoco Ballentine	3588.9	11774.5	5	–11.3	25.2	–5.5	Dolomicrite (mud)
AB-2	Amoco Ballentine	3588.9	11774.5	5	–11.4	25.7	–5.0	Dolomicrite (mud)
AB-6	Amoco Ballentine	3588.9	11774.5	5	–12.0	24.8	–5.9	Dolomicrite (mud/intracl.), W
AB-7	Amoco Ballentine	3588.9	11774.5	5	–11.1	24.9	–5.8	Dolomicrite (mud), W
AB-14	Amoco Ballentine	3590.8	11780.7	5	–7.8	22.7	–7.9	Dolomicrite (mud), W
AB-17	Amoco Ballentine	3590.8	11781.0	5	–8.0	23.7	–6.9	Dolomicrite, W

* W = Sample analyzed by Dr. B. Winter of the University of Wisconsin. All others analyzed by P. A. Drzewiecki of the University of Wisconsin.

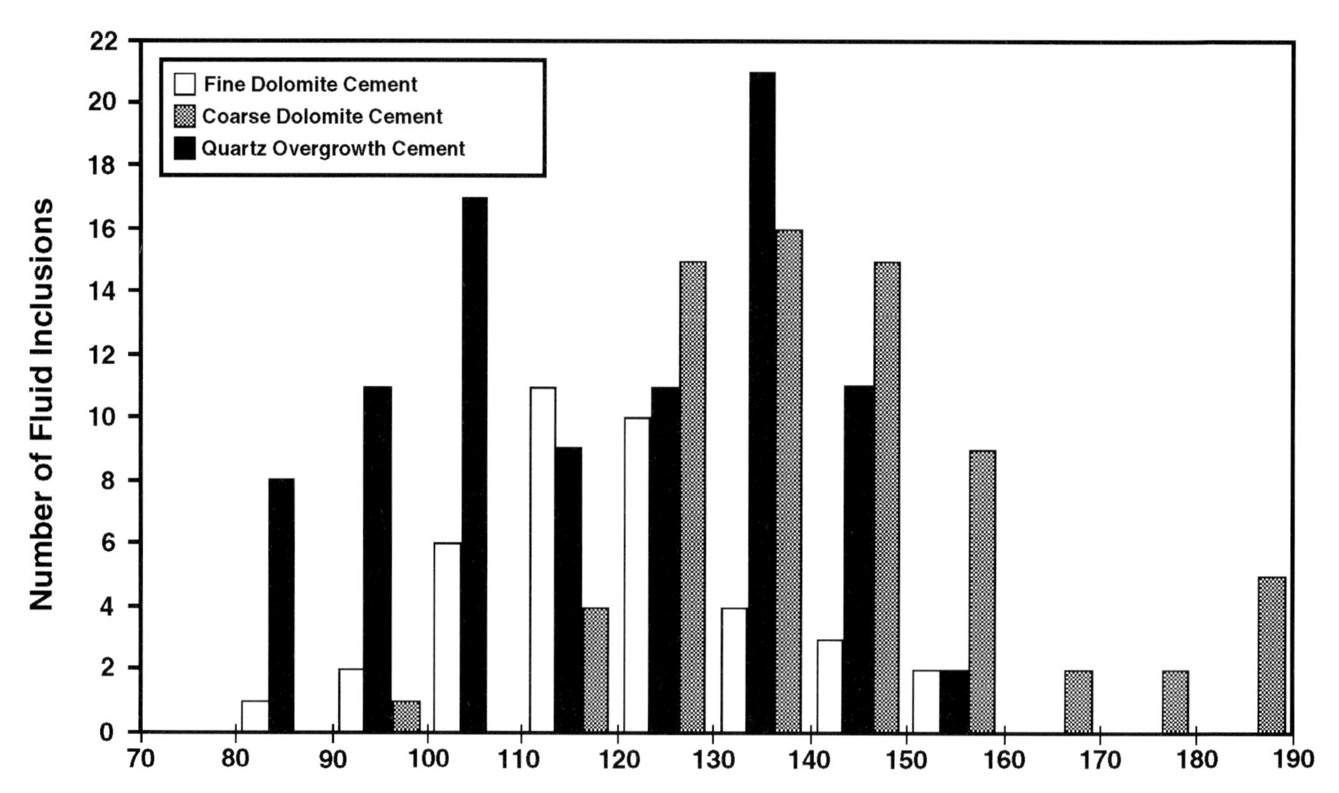

Figure 11. Homogenization temperatures of fluid inclusions from quartz overgrowths and fine dolomite cement in the St. Peter Sandstone, and from coarse dolomite cement in the Glenwood Formation. The quartz data have a bimodal distribution, with modes at 105 and 135°C. Both dolomite cement types have normal distributions. These data are not corrected for pressure.

were from a limited number of samples) and may not reflect an actual difference in crystallization temperatures. If all the fluid inclusions are primary, and using a pressure correction of 50°C for 0.5 to 0.6 kilobar (about 5 to 6 km) for both quartz and dolomite inclusions (Potter, 1977), crystallization temperatures would be around 155° and 185° ± 20°C for quartz, 185° ± 20°C for coarse dolomite cement, and 165° ± 20°C for fine dolomite cement.

Some coarse and fine dolomite cement fluid inclusions contain a cubic halite daughter product, but the small size of these inclusions did not permit us to directly measure fluid salinities. The halite and probable elevated crystallization temperatures support the interpretation that the dolomite precipitated in a deep burial environment from hypersaline fluids. Given the maximum burial depth of the St. Peter and Glenwood (about 4 km; Cercone, 1984) and a geothermal gradient of 26°C/km, the crystallization temperatures of 155 to 185°C for quartz overgrowths and dolomites are unusually high. This indicates that quartz and dolomite precipitated from hot brines or the fluid inclusions were modified during a later event.

Stable Isotopes

Values of $\delta^{13}C$ and $\delta^{18}O$ from 108 dolomite samples from the St. Peter Sandstone and Glenwood Formation were measured (Tables 2 and 4 and Figures 12 and 13). The locations of the wells from which these samples came are shown on Figure 2. Samples were extracted by drilling specific features with a 0.5 mm dental burr or by crushing whole-rock samples. Samples (1 to 50 milligrams) were reacted overnight with concentrated phosphoric acid (density = 1.92) at 50°C (McCrea, 1950; Sharma and Clayton, 1965). The CO_2 was then cryogenically purified. Values of $\delta^{18}O$ are reported in standard per mil notation relative to SMOW and $\delta^{13}C$ values are reported relative to PDB. The analytical error is ±0.1‰ for carbon and ±0.2‰ for oxygen.

Winter and Valley (unpublished data) obtained $\delta^{18}O$ values of 14.0 to 15.7‰ SMOW for quartz overgrowths in the St. Peter Sandstone of the Michigan basin. These values are significantly lower than samples from the Wisconsin Arch (16.0 to 21.6‰), consistent with higher crystallization temperatures in the Michigan basin.

Oxygen and carbon isotope data were collected from all four St. Peter and Glenwood dolomite types (Table 4 and Figures 12 and 13). Fine replacive dolomite (open and closed squares, Figure 12) ranges from −12.0 to −1.0‰ in $\delta^{13}C$ and 17.5 to 26.1‰ in $\delta^{18}O$. Samples from individual wells are characterized by comparable ranges in $\delta^{18}O$ values (about 5‰) and $\delta^{13}C$ values (up to 6‰) (Figure 12). Coarse replacive dolomite (diamonds, Figure 12) differs

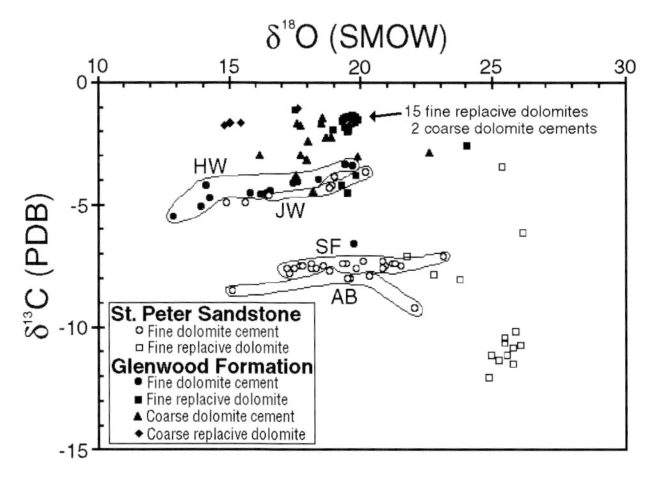

Figure 12. Oxygen and carbon isotope data from all four dolomite types of the St. Peter Sandstone and Glenwood Formation. The $\delta^{18}O$ values are reported with respect to SMOW and $\delta^{13}C$ values are reported with respect to PDB. The fine dolomite cement from each core listed below appears to plot in a unique field, characterized by a narrow range of $\delta^{13}C$ values and a wide range of $\delta^{18}O$ values (AB, HW, JW, SF). Fine replacive dolomites plot in fields characterized by comparable $\delta^{18}O$ and $\delta^{13}C$ ranges. The isotope data are provided in Table 4. HW = Hunt Winterfield; JW = JEM Weingartz; SF = Mobil St. Foster; AB = Amoco Ballentine.

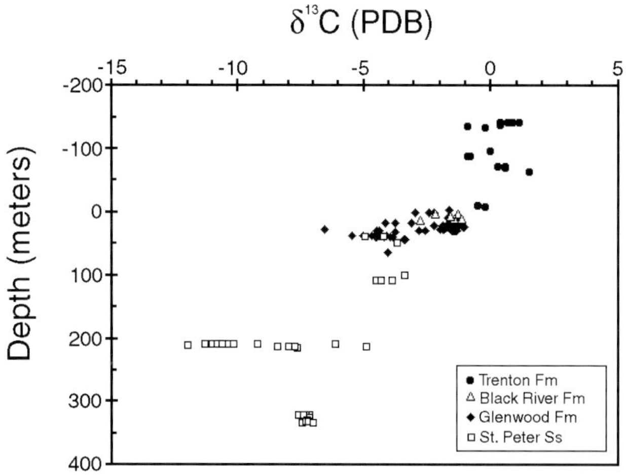

Figure 13. Carbon isotope ratios for carbonates of the St. Peter Sandstone, Glenwood, Black River, and Trenton formations in the Michigan basin plotted versus depth below the first bentonite above the base of the Black River Formation. Note the decreasing $\delta^{13}C$ values with depth. Stable isotope data for the St. Peter Sandstone and Glenwood Formation are listed in Table 4, and the Trenton Formation data are from Taylor and Sibley (1986; their locations 9 and 10).

from the fine replacive dolomite and has a very narrow range in $\delta^{18}O$ (14.8 to 15.4‰) and $\delta^{13}C$ (–1.7 to –1.6‰), with the exception of one point. These data suggest that replacive dolomite may have incorporated oxygen from its precursor calcite or that it formed from fluids with a uniform $\delta^{18}O$ composition at a constant temperature.

Fine dolomite cement (open and closed circles, Figure 12) has $\delta^{13}C$ values that range from –9.2 to –3.4‰ and $\delta^{18}O$ values that range from 12.8 to 23.1‰. Data from individual wells are characterized by a very narrow range in $\delta^{13}C$ (1 to 2‰) and a wide range in $\delta^{18}O$ (up to 6‰) (Figure 12). This wide range in $\delta^{18}O$ can be interpreted as the result of either precipitation from fluids with different $\delta^{18}O$ values or precipitation from a fluid with a uniform $\delta^{18}O$ value but at a range of temperatures. If dolomite precipitated from a fluid with a uniform $\delta^{18}O$ value of 7‰, temperatures would have to vary from 120 to 195°C. If precipitation was at a constant temperature of about 165°C, then the fluids would have a $\delta^{18}O$ range of 4 to 10‰. Coarse dolomite cement (triangles, Figure 12) has a 16.1 to 22.6‰ range in $\delta^{18}O$ and a –4.4 to –1.3‰ range in $\delta^{13}C$ with the exception of one point and may have the same horizontal trend in each well observed in the fine dolomite cement. More data are required to verify this interpretation.

Measured $\delta^{18}O$ values and pressure-corrected crystallization temperatures on the same or adjacent samples suggest the fine and coarse dolomite cement precipitated from a fluid with an average $\delta^{18}O$ value of 7‰ (fractionation equation of Matthews and Katz, 1977). These high $\delta^{18}O$ values suggest a hypersaline dolomitizing fluid. The hypersaline fluids may have been derived from the evaporative brines that occupied the Michigan basin during the Silurian and moved downward into the St. Peter Sandstone and Glenwood Formation by compaction and under the influence of density gradients. The $^{87}Sr/^{86}Sr$ values of the coarse dolomite cement (0.7083 to 0.7086; Vandrey, 1991) are consistent with this interpretation.

When plotted with data from the overlying Black River and Trenton formations, $\delta^{13}C$ values of all St. Peter Sandstone and Glenwood Formation dolomite types decrease with depth (Figure 13). This indicates that there may be a stratigraphic control on the source of the carbon. The low $\delta^{13}C$ values of many carbon analyses (Table 4) suggest a major contribution from organic matter.

DIAGENETIC BANDING AND STYLOLITIZATION

One of the most interesting diagenetic characteristics of the St. Peter Sandstone and Glenwood Formation is the presence of alternating, millimeter- to centimeter-scale poorly cemented and well-cemented bands (Figure 4). Intergranular pressure solution and stylolites are typically associated with these banded zones. Banding tends to be localized and enhanced by the planar depositional features (bedding, porosity variations) of the horizontally bedded and cross-bedded lithofacies. Diagenetic banding appears to form by differential distribution of cement types and porosity that follows these planar discontinuities. Different diagenetic band types were formed at different times in the diagenetic history of the St. Peter and Glenwood, but the majority of them formed during middle and late burial diagenesis. Quartz-cemented bands and pressure solution-dominated bands were the earliest to develop and form some of the most impermeable sandstone intervals. Silica budgets of several cores from the Michigan basin indicate that enough silica was dissolved by intergranular pressure solution to account for the cement present in the quartz-cemented bands (Shepherd et al., this volume). Mechanical and chemical compaction appear to be more important as porosity-reducing mechanisms than quartz cementation (Shepherd et al., this volume). Extreme intergranular pressure solution resulted in the formation of pressure solution-dominated bands.

Dolomite postdates quartz cement and often forms low-porosity and low-permeability bands in the lower and middle St. Peter Sandstone. The quartz-cemented bands may have provided a planar template that assisted in the localization of dolomite-cemented bands. Dolomite-cemented bands are best developed in sandstones immediately adjacent to carbonate facies, suggesting that the source of the dolomite was local primary carbonates. Partial dissolution of dolomite cement is responsible for the formation of many of the porous bands in the St. Peter and Glenwood sandstones.

Clay-cemented bands are the most poorly developed of all band types. They rarely have sharp contacts with adjacent bands. Clay-cemented bands were the last type to form and are distributed throughout the St. Peter Sandstone and Glenwood Formation but are least abundant in the lower St. Peter. They have a wide range in porosities as a result of varying amounts of intergranular pressure solution. Porous bands are defined as those that have porosities which exceed 3%. They did not form at any particular time during the diagenetic sequence. Instead, they were continually modified through time, by different cementation and dissolution episodes.

Stylolites (Figure 10) are common throughout the entire St. Peter and Glenwood but are particularly abundant in the diagenetically banded intervals. The three main types of stylolites observed in the St. Peter are sutured stylolites, solution seams (after Wanless, 1977), and microstylolites. Sutured stylolites and solution seams occur as both individual features and as groups that result from intense intergranular pressure solution. Sutured stylolites are those that occur as a single planar feature with an interdigitate appearance. In the St. Peter Sandstone and Glenwood Formation, this type of feature is restricted to the sandstones that contain very little or no clay and is rarely found in the carbonate facies. Solution seams are smooth to undulose pressure solution features. They are found in clay-rich and clay-poor sandstones, as well as in the carbonate lithofacies. The deformable nature of clay inhibits the interpenetration of the two sides of the seam, resulting in its smooth to undulose appearance. Microstylolites are only observed in thin section and are generally only a few millimeters in length. They show interdigitation like sutured stylolites and are best-developed between quartz grains in clay-poor sandstones. The distribution of stylolites is not uniform. They often occur in groups, separated by regions characterized by a sparser distribution of stylolites.

Stylolites are typically localized along roughly horizontal textural heterogeneities in the rock, such as laminae of different grain sizes (Figure 10A), cement types, porosities, and lithologies. For this reason, they are abundant in zones with well-developed diagenetic banding, where textural variations abound. Although many stylolites appear to have originated during banding formation, others are located at the contact between bands and appear to postdate them. Where no textural variations exist, stylolites are usually more common in finer-grained sandstones and in tightly cemented, low-porosity intervals. Tada and Siever (1989) report similar observations in most studies.

The amount of dissolution at each seam is difficult to determine in homogeneous sandstones, but the concentration of potassium feldspar, pyrite, and insoluble residue along well-developed stylolites indicates that stylolitization may have been a significant source of silica cement. The age of the beginning of stylolitization cannot be determined, but stylolites can form in sandstones at depths as shallow as 1 to 2 km (Tada and Siever, 1989). Petrographic evidence suggests that stylolites continued to form late into the diagenetic history of the St. Peter and Glenwood, but they may have been initiated as early as the Silurian Period.

In summary, cement and porosity banding formed throughout the history of the St. Peter Sandstone and Glenwood Formation, at different depths, and while different fluids occupied the pore space. Band formation cannot be linked to one specific set of diagenetic conditions or processes, however. The depositional texture appears to be an important control on the localization and initiation of banding by acting as a template. Initial petrographic observations are for the most part consistent with a geochemical self-organization model for banding genesis by Merino et al. (1983), Ortoleva et al. (1987), and Dewers and Ortoleva (1990), but more research is required before we are able to properly understand all of the controls on banding formation. Stylolites appear to have formed throughout the diagenetic history of the St. Peter Sandstone and Glenwood Formation and are related to the formation of diagenetic bands.

PREDICTING THE OCCURRENCE OF BANDED SANDSTONES

Moline et al. (1992, this volume) divided the St. Peter Sandstone into six electrofacies types based on wireline log signatures. They noted that each electrofacies type has a unique set of lithologic characteristics that are both depositional and diagenetic in origin (Table 1). Understanding the controls on the lithologic characteristics of electrofacies is important for several reasons. First, electrofacies analysis of wells will allow geologists to predict the lithologic and reservoir characteristics of uncored portions of these wells. Second, electrofacies analysis will allow correlation of lithologically similar intervals from well to well across a field (Moline, 1992) and provide lateral and vertical dimensions for reservoirs and seals. Finally, they will allow geologists to examine any cross-cutting relationships between seals and stratigraphy.

The lithologic and diagenetic characteristics of the six electrofacies are described in Moline et al. (this volume). Electrofacies ef4, ef5, and ef6 have the lowest porosities and permeabilities. Ef6 corresponds to fine-grained carbonates and shales, ef5 is typically composed of sandy carbonates or sandstones with abundant intergranular dolomite cement, and ef4 is characterized by quartz banding. These three electrofacies dominate the lower and middle St. Peter Sandstone, across the entire Michigan basin (Moline, 1992). This suggests that: (1) banded cement and porosity distribution may be characteristic of deeply buried quartzose sandstone, and (2) there is the potential for seal formation at many stratigraphic intervals in the St. Peter, with or without carbonate intervals. The more porous electrofacies (ef1, ef2, and ef3) are restricted to the uppermost St. Peter Sandstone and often produce gas that is trapped by the overlying Glenwood carbonates.

The electrofacies distribution correlates well with lithofacies distribution. The middle and lower St. Peter (electrofacies ef4, ef5, ef6) is characterized by alternating carbonates and cross-bedded or horizontally bedded sandstones, which usually contain well-developed diagenetic banding. The upper St. Peter (electrofacies ef1, ef2, ef3) is composed of more porous, massive, and clay-rich sandstones. This implies that the occurrence of seal-forming diagenetically banded sandstones can be predicted based on a sequence stratigraphic interpretation of the formation of interest.

CONCLUSIONS

The deeply buried St. Peter Sandstone and Glenwood Formation from the center of the Michigan basin have undergone a sequence of diagenetic events that resulted in banded cement and porosity distribution. Burial diagenetic processes (intergranular pressure solution and quartz and dolomite cementation) were the most important controls on band formation. Descriptions of diagenetically banded sandstones are rare in the scientific literature, and banding may actually be a more important control on porosity and permeability in deeply buried clastic rocks than previously thought.

Petrographic and geochemical data suggest that several different fluids were present in the St. Peter Sandstone and Glenwood Formation during their diagenetic history. Connate marine water occupied pore space during early diagenesis. Middle diagenesis was characterized by sea water with a slight meteoric component. Finally, a hypersaline, dolomite-precipitating brine was present during late diagenesis. Values of $\delta^{18}O$ and homogenization temperatures of fluid inclusions from dolomite cement in sandstones indicate that the fluid may have had a fairly uniform temperature and isotopic composition. The $\delta^{13}C$ values of fine dolomite cements suggest that carbon was stratigraphically controlled and partially derived from maturation of organic material.

Clay in the St. Peter Sandstone and Glenwood Formation is both detrital and authigenic. Detrital clay was an important control on the preservation of porosity by inhibiting quartz overgrowth and diagenetic band development. Dates obtained from authigenic clay (346 ± 20 Ma; Barnes et al., 1992b) suggest that most major diagenetic changes of the St. Peter and the Glenwood took place during the early part of their burial history.

Quartz-dominated, diagenetically banded intervals have been identified as low-porosity, low-permeability zones and may function as seals for pressure compartments in the Ordovician rocks of the Michigan basin (Shepherd et al., this volume). Evidence suggests that the distribution of silica cement may have been a function of differential compaction (mechanical and chemical) and cementation during middle diagenesis.

The electrofacies described by Moline et al. (1992, this volume) are characterized by a unique set of lithologic characteristics. These characteristics appear to be both depositionally and diagenetically controlled. The least porous and permeable electrofacies (ef5 and ef6) are found in the carbonate lithofacies or in neighboring sandstones with abundant carbonate matrix. Electrofacies ef4 is a low-porosity, low-permeability electrofacies that is characterized by diagenetic banding resulting from intergranular pressure solution and quartz cementation. Understanding the relationships between electrofacies, banding, and depositional texture will assist in predicting the distribution and size of reservoirs and seals in deeply buried sandstones.

ACKNOWLEDGMENTS

The authors of this manuscript would like to thank the professors and graduate students at the core laboratory of Western Michigan University for the opportunity to examine cores of the St. Peter Sandstone and Glenwood Formation and for their assistance in doing so. Mobil Exploration and Producing, Unocal Production, and Amoco Production provided cores and samples that were used in this investigation. This paper has benefited immensely from reviews by Samuel Savin and Eric Oswald and from valuable discussions with Jean Bahr and Geri Moline about electrofacies and Michigan basin overpressures. Mike Spicuzza and Kevin Baker provided valuable assistance in stable isotope analysis and interpretation, and

W. C. Shanks and D. E. Crowe provided sulfur isotopic data. This work was supported by a contract with the Gas Research Institute (5089-260-1810).

REFERENCES CITED

Bahr, J. M., 1989, Evaluation of pressure distribution in the St. Peter Sandstone, Michigan basin (abs.): EOS, Transactions, American Geophysical Union, v. 70, no. 43, p. 1097.

Bahr, J. M., and G. R. Moline, 1992, Modeling heterogeneity and associated anomalous pressures in the St. Peter Sandstone, Michigan basin (abs.): EOS, Transactions, American Geophysical Union, v. 73, no. 43, p. 184.

Barnes, D. A., W. B. Harrison, C. E. Lundgren, and L. M. Wieczorek, 1988, The Lower Paleozoic of the Michigan basin, a core workshop: Kalamazoo, Western Michigan University Core Research Laboratory Short Course, 65 pp.

Barnes, D. A., C. E. Lundgren, and M. W. Longman, 1992a, Sedimentology and diagenesis of the St. Peter Sandstone, central Michigan basin, United States: American Association of Petroleum Geologists Bulletin, v. 76, p. 1507–1532.

Barnes, D. A., J. -P. Girard, and J. L. Aronson, 1992b, K/Ar dating of illitic diagenesis in the St. Peter Sandstone, central Michigan basin, USA: Implications for thermal history, *in* D. W. Houseknecht and E. D. Pitman, eds., Origin, diagenesis, and petrophysics of clay minerals in sandstones: SEPM Special Publication 47, p. 35–48.

Beales, F. W., 1971, Cementation by white sparry dolomite, *in* O. P. Bricker, ed., Carbonate Cements: The Johns Hopkins University Press, p. 330–338.

Boggs, S. Jr., 1987, Principles of Sedimentology and Stratigraphy: Merrill Publishing Co., Columbus, Ohio, 784 pp.

Brady, R. B., and R. DeHaas, 1988, The "deep" (pre-Glenwood) formations of the Michigan basin: part 2, the St. Peter Sandstone: Michigan's Oil and Gas News, v. 94, p. 36–44.

Brownlow, A. H., 1979, Geochemistry: Prentice-Hall, Inc., Englewood Cliffs, New Jersey, 498 pp.

Catacosinos, P. A., and P. A. Daniels, Jr., 1991, Stratigraphy of Middle Proterozoic to Middle Ordovician formations of the Michigan Basin: Geological Society of America Special Paper 256, p. 53–71.

Catacosinos, P. A., P. A. Daniels, Jr., and W. B. Harrison III, 1990, Structure, stratigraphy, and petroleum geology of the Michigan basin, *in* M. W. Leighton, D. R. Kolata, D. F. Oltz, and J. J. Eidel, eds., Interior Cratonic Basins: American Association of Petroleum Geologists Memoir 51, p. 561–601.

Cercone, K. R., 1984, Thermal history of the Michigan Basin: American Association of Petroleum Geologists Bulletin, v. 68, p. 130–136.

Choquette, P. W., and N. P. James, 1987, Diagenesis # 12: Diagenesis in Limestones. 3. The deep burial environment: Geoscience Canada, v. 14, p. 3–35.

Dewers, T., and P. Ortoleva, 1990, A coupled reaction/transport/mechanical model for intergranular pressure solution, stylolites, and differential compaction and cementation in clean sandstones: Geochimica et Cosmochimica Acta, v. 54, p. 1609–1625.

Dott, R. H., Jr., C. W. Byers, G. W. Fielder, S. R. Stenzel, and K. E. Winfree, 1986, Aeolian to marine transition in Cambro–Ordovician cratonic sheet sandstones of the northern Mississippi River Valley, U.S.A.: Sedimentology, v. 33, p. 345–367.

Drzewiecki, P, A., 1992, Sedimentology, diagenesis, and geochemistry of the Middle Ordovician St. Peter Sandstone of the Michigan basin: Unpublished M.S. Thesis, University of Wisconsin–Madison, 215 p.

Drzewiecki, P. A., A. Simo, G. Moline, J. M. Bahr, G. C. Nadon, L. D. Shepherd, and M. R. Vandrey, 1991, The significance of stylolitization and intergranular pressure solution in the formation of pressure compartment seals in the St. Peter Sandstone, Middle Ordovician, Michigan basin (abs.): American Association of Petroleum Geologists Bulletin, v. 75, p. 565.

Ells, G. D., 1967, Correlation of Cambro–Ordovician rocks in Michigan: Michigan Basin Geological Society Annual Field Trip Excursion Guidebook, p. 42–57.

Fisher, J. H., and M. W. Barratt, 1985, Exploration in Ordovician of Central Michigan basin: American Association of Petroleum Geologists Bulletin, v. 69, p. 2065–2076.

Fisher, J. H., M. W. Barratt, J. B. Droste, and R. H. Shaver, 1988, Michigan basin, *in* L. L. Sloss, ed., Sedimentary Cover—North American Craton: U. S.: Geological Society of America DNAG, v. D-2, p. 361–382.

Harrison, W. B, III, 1987, Michigan's "deep" St. Peter play continues to expand: World Oil, v. 204, p. 56–61.

Howell, P. D., and B. A. van der Pluijm, 1990, Early history of the Michigan basin: subsidence and Appalachian tectonics: Geology, v. 18, p. 1195–1198.

Hunt, J. M., 1990, Generation and migration of petroleum from abnormally pressured fluid compartments: American Association of Petroleum Geologists Bulletin, v. 74, p. 1–12.

Long, J. D., 1988, Sedimentology of the Glenwood Member of the Middle Ordovician St. Peter Sandstone of southern Wisconsin: Unpublished M.S. Thesis, University of Wisconsin–Madison, 133 p.

Lundgren, C. E., Jr., and D. A. Barnes, 1989, Influence of depositional environments on diagenesis in St. Peter Sandstone, Michigan basin (abs.): American Association of Petroleum Geologists Bulletin, v. 73, p. 384.

Mai, H., and R. H. Dott, Jr., 1985, A subsurface study of the St. Peter Sandstone in southern and eastern Wisconsin: Wisconsin Geological and Natural History Survey, Information Circular 47, 26 p.

Maliva, R. G., and R. Siever, 1988, Diagenetic replacement controlled by force of crystallization: Geology, v. 16, p. 688–691.

Matthews, A., and A. Katz, 1977, Oxygen isotope fractionation during dolomitization of calcium carbonate: Geochimica et Cosmochimica Acta, v. 41, p. 1431–1438.

McCrea, J. M., 1950, On the isotopic chemistry of carbonates and a paleotemperature scale: Journal of Chemical Physics, v. 18, p. 849–857.

Merino, E., P. Ortoleva, and P. Strickholm, 1983, Generation of evenly-spaced pressure solution seams during late diagenesis: a kinetic theory: Contributions to Mineralogy and Petrology, v. 82, p. 360–370.

Moline, G. R., 1992, A multivariate statistical approach to mapping spatial heterogeneity with an application to the investigation of pressure compartments within the St. Peter Sandstone, Michigan basin: Unpublished Ph.D. Thesis, University of Wisconsin–Madison, 159 p.

Moline, G. R., J. M. Bahr, P. A. Drzewiecki, and L. D. Shepherd, 1992, Identification and characterization of pressure seals through the use of wireline logs: a multivariate statistical approach: The Log Analyst, v. 33, p. 362–372.

Nadon, G. C., J. A. Simo, C. W. Byers, and R. H. Dott, Jr., 1992, Sequence stratigraphy: a clue for the distribution of pressure compartmentation within the Middle Ordovician of the Michigan Basin (abs.): Second Symposium on Deep Basin Compartments and Seals (Sept. 29–Oct. 1, 1992), Abstract volume, Gas Research Institute.

Nunn, J. A., N. H. Sleep, and W. E. Moore, 1984, Thermal subsidence and generation of hydrocarbons in Michigan Basin: American Association of Petroleum Geologists Bulletin, v. 68, p. 296–315.

Ortoleva, P., E. Merino, C. Moore, and J. Chadam, 1987, Geochemical self-organization I: reaction-transport feedbacks and modeling approach: American Journal of Science, v. 287, p. 979–1007.

Potter, R. W. III, 1977, Pressure corrections for fluid-inclusion homogenization temperatures based on the volumetric properties of the system NaCl–H2O: Journal of Research, U.S. Geological Survey, v. 5, p. 603–607.

Powley, D. E., 1990, Pressures and hydrogeology in petroleum basins: Earth-Science Reviews, v. 29, p. 215–229.

Radke, B. M., and R. L. Mathis, 1980, On the formation and occurrence of saddle dolomite: Journal of Sedimentary Petrography, v. 50, p. 1149–1168.

Sharma, T., and R. N. Clayton, 1965, Measurements of $^{18}O/^{16}O$ ratios of total oxygen of carbonates: Geochimica et Cosmochimica Acta, v. 36, p. 129–140.

Smith, G. L., C. W. Byers, and R. H. Dott, Jr., 1993, Sequence stratigraphy of the Lower Ordovician Prairie Du Chien Group on the Wisconsin Arch and in the Michigan Basin: American Association of Petroleum Geologists, v. 77, p. 49–67.

Tada, R., and R. Siever, 1989, Pressure solution during diagenesis: Annual Review of Earth and Planetary Sciences, v. 17, p. 89–118.

Taylor, T. R., and D. F. Sibley, 1986, Petrographic and geochemical characteristics of dolomite types and the origin of ferroan dolomite in the Trenton Formation, Ordovician, Michigan Basin, U.S.A: Sedimentology, v. 33, p. 61–86.

Tigert, V., and Z. Al-Shaieb, 1990, Pressure seals: their diagenetic banding patterns: Earth–Science Reviews, v. 29, p. 227–240.

Vandrey, M. R., 1991, Stratigraphy, diagenesis, and geochemistry of the Middle Ordovician Glenwood Formation: Unpublished M.S. Thesis, University of Wisconsin–Madison, 239 pp.

Wanless, H. R., 1977, Limestone response to stress: pressure solution and dolomitization: Journal of Sedimentary Petrology, v. 49, p. 437–462.

Weyl, P. K., 1959, Pressure solution and the force of crystallization—a phenomenological theory: Journal of Geophysical Research, v. 64, p. 2001–2025.

Chapter 14

Permeability and Porosity Estimation by Electrofacies Determination

Gerilynn R. Moline
Jean M. Bahr
Peter A. Drzewiecki
University of Wisconsin—Madison
Madison, Wisconsin, U.S.A.

ABSTRACT

Low-permeability seals associated with abnormal pressures are most commonly identified by examining vertical pressure profiles and noting the depths at which a major change in the pressure gradient occurs. Alternatively, zones of very low permeability that may act as fluid seals may be identified on the basis of core analyses and well tests. Often, however, there are an insufficient number of direct pressure or permeability measurements to adequately identify the depth and lateral extent of these seals. A method has been developed for estimating porosity and permeability through the use of wireline logs. Multivariate statistical techniques are used to segment the logs and group the segments into electrofacies types. Application of this technique to 18 wells within the St. Peter Sandstone of the Michigan basin shows that the electrofacies characterization reflects both the hydraulic and diagenetic characteristics of the formation. Six electrofacies types have been identified, one of which has characteristics similar to those found within seals in other basins.

INTRODUCTION

Low-permeability seals associated with abnormal pressures are of particular interest if they are of diagenetic origin and, as such, provide a type of hydrocarbon trap not normally incorporated into exploration strategies (Powley, 1990; Hunt, 1990; Tigert and Al-Shaieb, 1990). The ability to map the vertical and lateral extent of very low-permeability zones associated with abnormal pressures is critical to understanding the genesis of pressure seals.

In the Michigan basin, anomalously high pressures have been identified within the Middle Ordovician St. Peter Sandstone and overlying Glenwood Formation from vertical pressure profiles and hydraulic head distributions (Bahr et al., this volume). While the regional pressure distributions give general indications of the locations of low-permeability seals, they do not indicate exact locations and lateral extents of the seals. Precise identification of zones of very low permeability in the Michigan basin is complicated by the lack of available core samples and direct pressure and permeability measurements. However, borehole geophysical logs are available for all of the deep Michigan wells, providing both continuous vertical measurements and a wide distribution of data. While the wireline logs do not directly measure permeability, the log responses are a function of the lithology and porosity that also control permeability. This correlation can be used to estimate permeability in wells with no direct permeability measurements.

Conventional log analysis techniques, while useful for discriminating major lithology variations, are not sensitive to the subtle changes that reflect diagenesis in a lithologically uniform formation such as the St. Peter Sandstone. Alternatively, an electrofacies approach has been developed and applied to the St. Peter Sandstone. In this method the multivariate log responses are characterized and grouped into subpopulations, providing a means for estimating the lithologic and hydrologic characteristics of this formation. An electrofacies (Serra and Abbott, 1982; Serra, 1986) is a composite of geophysical log responses that characterizes a part of the formation, permitting it to be distinguished from its surroundings.

Core data and borehole geophysical logs from 18 wells distributed across the basin have been analyzed using multivariate statistical techniques. These wells are located at a variety of depths, both within and outside of the anomalous pressure areas (Figure 1). The data set consists of 1594 core plug and whole core measurements. Permeability and porosity measurements were performed in commercial and industry core labs and are assumed accurate as reported. In this paper, permeabilities are reported as horizontal permeability to air at surface conditions unless otherwise noted.

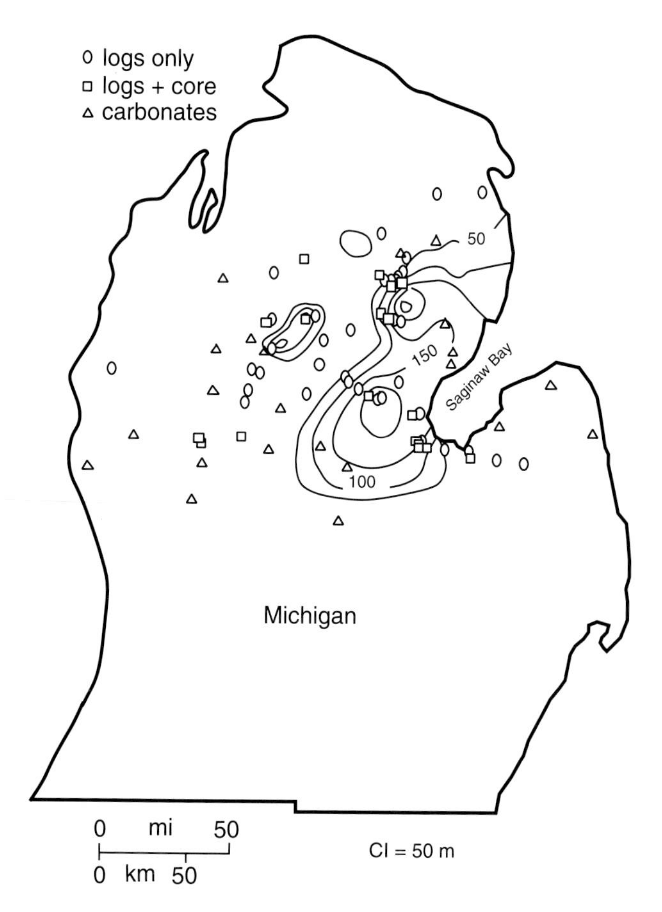

Figure 1. Map of Michigan showing the location of the wells included in this study. Contours indicate hydraulic heads above land surfaces. After Bahr et al. (this volume).

METHOD OF ELECTROFACIES DETERMINATION

Three steps are involved in the electrofacies classification process: reduction of the multivariate problem down to a single variable, log segmentation, and clustering of the log segments into electrofacies types. The statistical techniques utilized in these steps are not new to log analysis; only the method of application is new.

Numerous algorithms for log segmentation can be found in the literature (Testerman, 1962; Kulinkovich et al., 1966; Gill, 1970; Hawkins and Merriam, 1974; Webster, 1973). The majority of these techniques are applicable only to a single trace. Hawkins (1976) presents an algorithm for multivariate segmentation, but the size of the data set is severely restricted. In the method presented here, the multivariate problem is collapsed into a single variable through the use of principal components analysis. This multivariate statistical technique has been successfully applied to problems of zonation and well-to-well correlation (Elek, 1988) and is particularly useful in cases where the input logs are highly correlated, resulting in a concentration of the variance along the first principal component axis.

In Figure 2A, an example of principal component (PC) log construction is shown for the Wolverine Patrick & St. Norwich 2-28 well. Four wireline logs were used in this analysis based on their strong correlation to core permeability: neutron porosity (NPHI), density (RHOB), sonic (DT), and a transform of the microspherically focused resistivity (MSFL). For this well, 86% of the variance occurs along the first principal component.

In the second step of the electrofacies classification process, the univariate PC log is segmented using a maximum likelihood estimation algorithm (Mehta et al., 1990; Radhakrishnan et al., 1991). The algorithm solves for the locations of segment boundaries by maximizing the likelihood function for the log data, assuming a constant signal within segments and superposition of random noise. The program is interactive and the user supplies the length of the window for filtering, the length of the window for computing local variance, and the segmentation threshold. The last parameter controls the coarseness of the segmentation.

The results of segmentation of the PC log are shown in Figure 2B, superimposed on the original log trace. The output of the segmentation algorithm has been edited to remove segments that are less than 3 ft (1 m) thick for consistency with the resolution of the logging tools. Once the segmentation is completed, the segment boundaries are applied to the original log traces and log values are then averaged within the segments. In this manner, a multivariate mean is assigned to each segment.

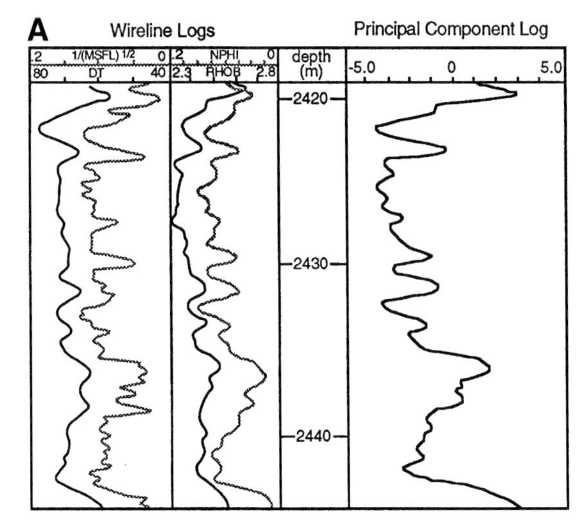

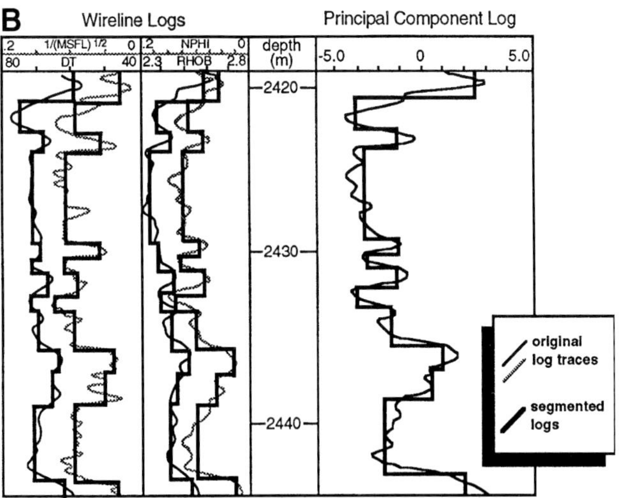

Figure 2. (A) Principal component analysis is applied to the four logs in the left two tracks, resulting in the construction of a first principal component (PC) log, shown on the right. The PC log represents the sum of the variance of the four logs along the first principal component axis. Logs included in the analysis are the sonic (DT), microresistivity (MSFL), density (RHOB), and neutron porosity (NPHI). (B) The PC log is segmented by maximum likelihood estimation, and the segment boundaries are then transferred to the original logs. Segmentation is shown here superimposed on the original log trace.

The final step in the classification process is to group similar log segments into electrofacies types through the use of cluster analysis. The clustering algorithm establishes a hierarchy of distances in multivariate space on the basis of the log averages assigned to each segment. This hierarchy is shown graphically by the dendrogram in Figure 3. Each data point represents a single log segment. Groups are created by cutting the dendrogram along any point with a horizontal line. The desired grouping maximizes the separation of the distributions of core porosities and permeabilities associated with the log segments, as these are the parameters we are interested in estimating. In order to determine optimal grouping, core data are segmented using the same segment boundaries that were applied to the logs.

The dendrogram is cut into successively larger numbers of groups until the best separation of permeabilities and porosities is achieved. For the Michigan wells, optimal grouping resulted in six electrofacies as indicated in Figure 3. The method effectively divides the overall distributions of permeability and porosity measurements into subpopulations, shown graphically by box plots in Figure 4.

As a result of clustering, the 13 log segments in the Patrick & St. Norwich 2-28 well have been assigned to four electrofacies types. In Figure 5, electrofacies classifications are shown superimposed on the wireline log traces and the core measurements. The similarity of log values and of core values within segments demonstrates how effectively the electrofacies distribution reflects the heterogeneity at this scale.

ELECTROFACIES DESCRIPTIONS

Hydraulic Characteristics

The box plots in Figure 4 provide a graphical picture of the distributions of core porosity and permeability associated with the six electrofacies types. Box plots show the median (center line), interquartile range (box), and four times the interquartile range (whiskers) of the distribution. The circles represent statistical outliers. While some overlap occurs, there is a statistically significant separation of the distributions. The wide range within the distributions reflects small-scale variability that is below the resolution of the logging tools. This variability is particularly apparent in the tighter permeability electrofacies types, as evidenced by numerous outliers. Orders-of-magnitude variations in permeability over several scales are a result of diagenetic processes (Drzewiecki et al., this volume; Shepherd et al., this volume).

Summary statistics for the hydraulic properties within each electrofacies type are listed in Table 1. Permeabilities are listed as the natural log of the measurement in millidarcies (md). Horizontal permeabilities (k_h) reflect individual core plug measurements taken at 30 cm (1 ft) intervals. Vertical permeabilities were not measured, but average vertical permeability (k_v) for each segment was calculated by making the assumption that the core plug value represents the entire foot of core. The effective vertical permeability, then, is the segment length divided by the sum of the reciprocals of the horizontal permeability (Freeze and Cherry, 1979, p. 34):

$$\bar{k}_v = \frac{L}{\sum_i \frac{d_i}{k_{h_i}}} = \frac{L}{\sum_i \frac{1}{k_{h_i}}} \quad , \; i = 1, 2, \ldots, L$$

K_v values in Table 1 are segment averages, while k_h values are individual measurements.

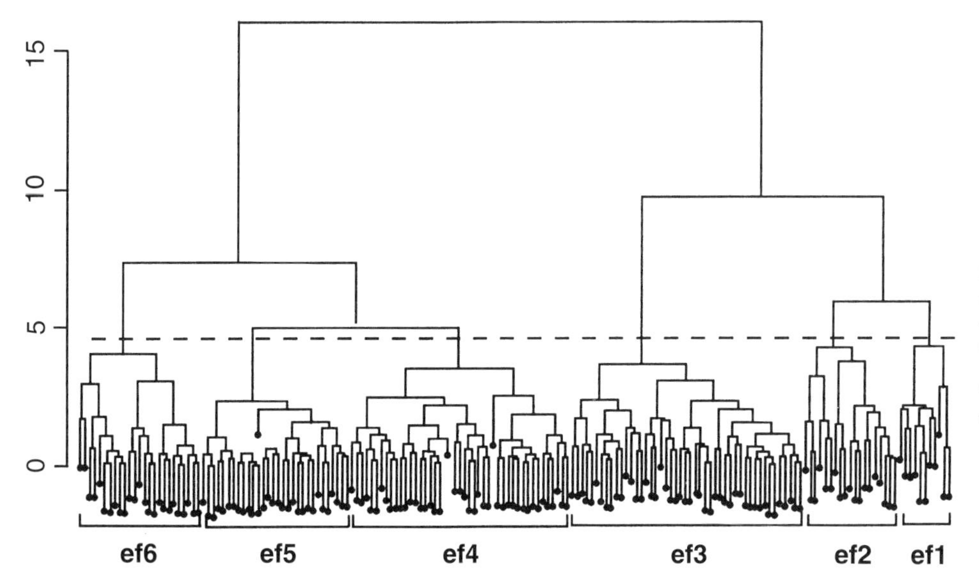

Figure 3. A dendrogram is constructed by cluster analysis of all log segments from the 18 Michigan wells. The hierarchy is a graphical representation of the multivariate distance between groups of log segments. The scale indicates the multivariate distance. Six clusters or electrofacies types (ef1–ef6) have been created by cutting the dendrogram at the location of the dashed line.

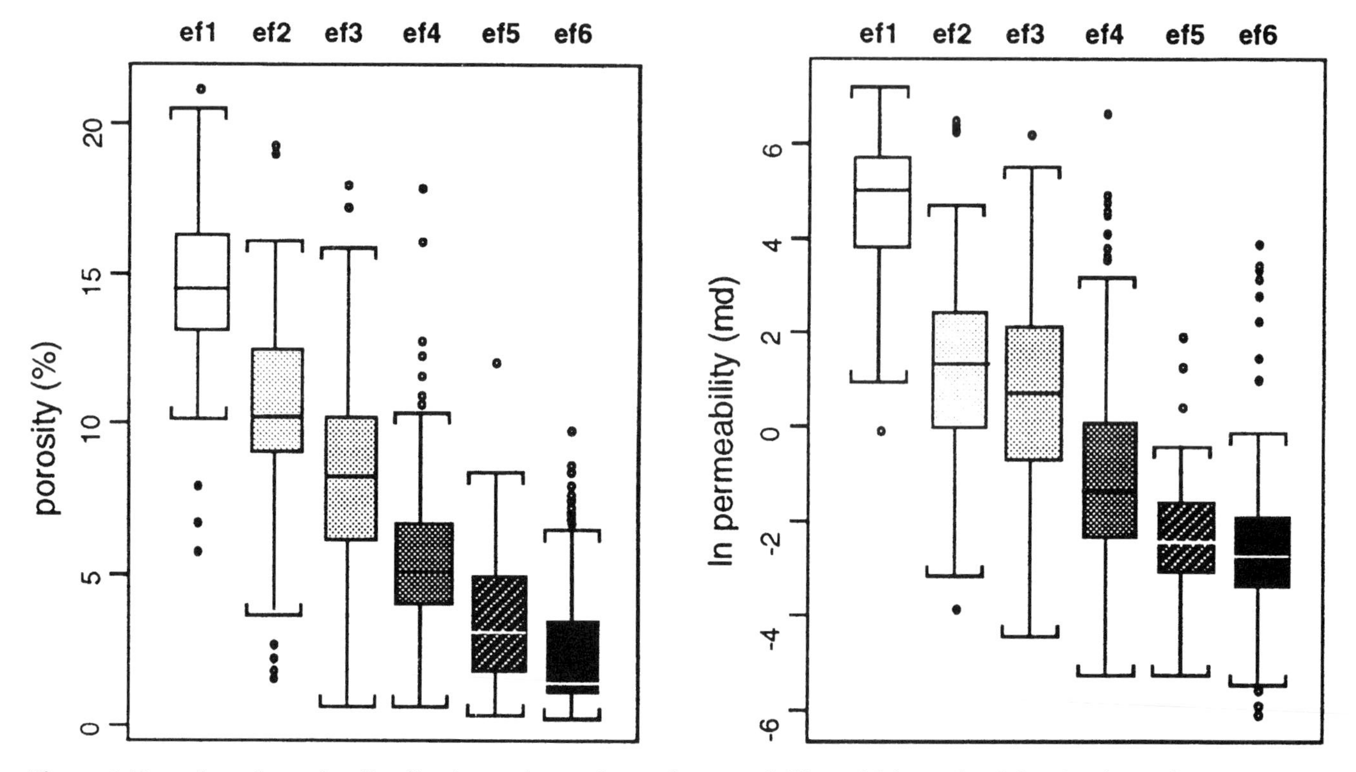

Figure 4. Box plots show the distributions of porosity and permeability within each of the six electrofacies groups determined by clustering the log characteristics (Figure 3). The box indicates the inner two quartiles of the distribution, the line represents the median, the whiskers show the range, and the circles represent statistical outliers.

Average porosity values show a consistent decrease with increasing electrofacies (ef) number.

There is greater overlap in permeability among the electrofacies types, however, particularly ef5 and ef6. These two electrofacies represent different lithologies with similar low permeabilities. Note that in many cases, the effective vertical permeability is even lower than the measured permeability because of variability within segments.

While the lateral distribution of electrofacies types cannot be fully determined with the sparse distribution of core data provided here, some observations can be made. The highest permeability type, ef1, predominates in the wells located in the western portion of the basin (Figure 1). These locations coincide with an area in which pressures within the St. Peter Sandstone appear to be in equilibrium with the regional flow system (Bahr et al., this volume). This

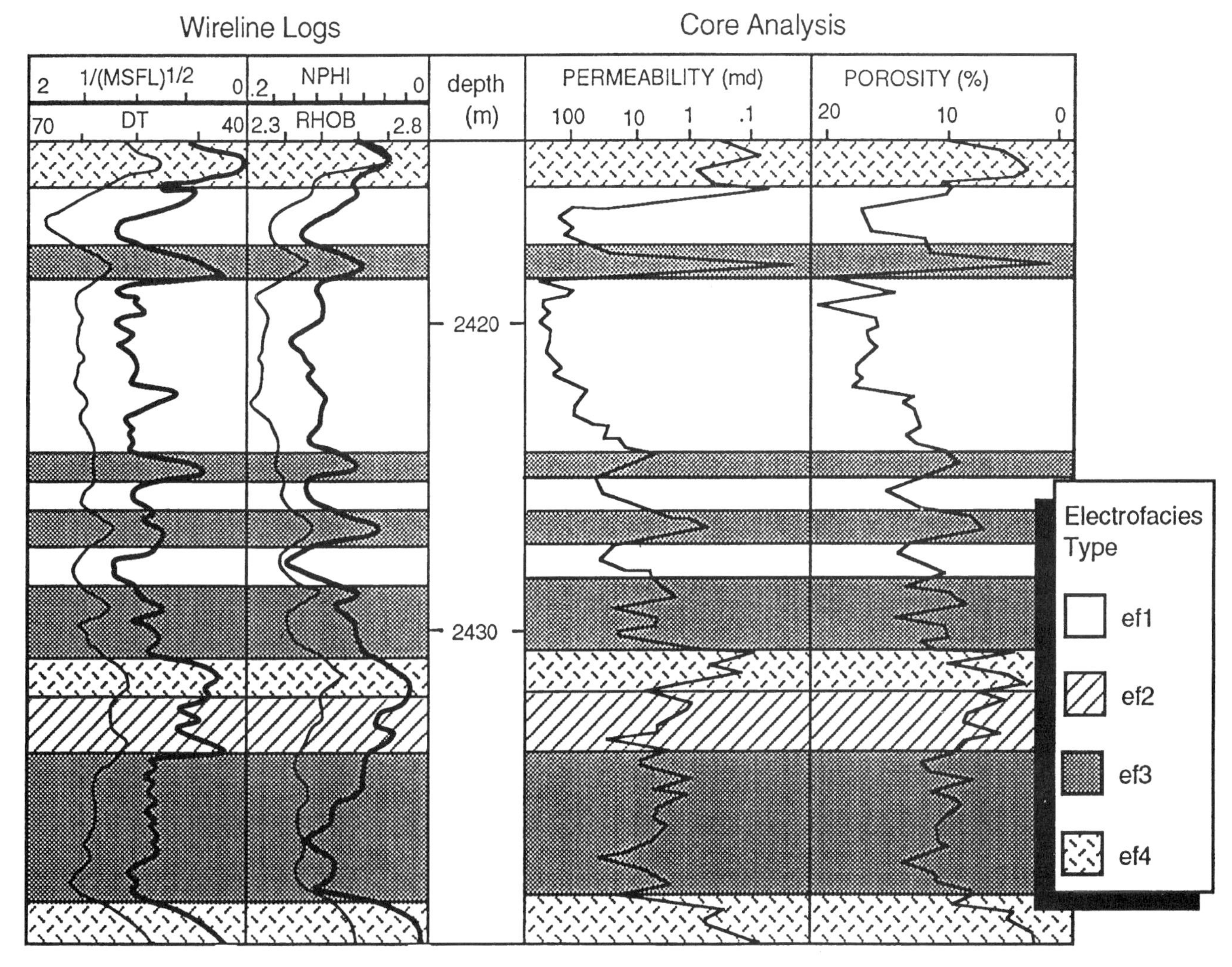

Figure 5. Electrofacies distribution in the Wolverine Patrick & St. Norwich 2-28 well. The wireline logs are shown in the left two tracks and the results from core plug analysis are shown on the right. Log segments are indicated by horizontal lines through the log traces and electrofacies types are indicated by shading.

electrofacies is absent within the deeper portions of the St. Peter except for isolated bands. Electrofacies types ef2 and ef3 are found in all of the wells across the basin. The low-permeability types ef4 and ef5 are abundant within the cores from the northeast and southeast (Saginaw Bay) regions of the basin. Ef6, the lowest permeability electrofacies, occurs predominantly within the Saginaw Bay area. The overall trend is an increase in the low-permeability electrofacies within the cored intervals from the Saginaw Bay and northeast regions.

Geophysical Log Characteristics

Statistics for the wireline log values within each electrofacies type are summarized in Table 2. Generally, as porosity and permeability increase the following trends occur: increased sonic travel time; increased conductivity; increased neutron porosity; and decreased density. Resistivity has been converted to root conductivity (inverse root resistivity) to linearize the relationship between porosity and resistivity as described by Archie's Law (Asquith, 1982).

Variations in log responses between electrofacies types ef1–5 are due to varying degrees of diagenetic alteration of the sandstone facies. Ef6 is a carbonate facies, and the change in log signatures is dominantly lithologic. This electrofacies is easily recognized within the log signatures by an increase in the photoelectric factor (PEF) coupled with a crossover of the density and neutron porosity logs. From a subjective examination of the logs it is apparent that this electrofacies type is abundant within the lower St. Peter in the southeast and disappears to the north and west.

Lithologic Characteristics

Core examination and thin section analyses indicate that the variations in porosity and permeability within the St. Peter Sandstone result predominantly from variations in the type and degree of cementation within an otherwise homogeneous quartz sandstone

Table 1. Summary statistics of the hydraulic characteristics for the six electrofacies types.

	ef1	ef2	ef3	ef4	ef5	ef6
Porosity						
mean	14.567	10.717	7.567	5.156	3.162	1.371
median	14.400	10.800	7.500	4.900	2.600	0.900
std. dev.	2.808	2.643	2.694	2.482	2.181	1.298
In k_h						
mean	4.598	1.573	0.585	−1.375	−2.436	−3.250
median	4.927	1.773	0.710	−1.561	−2.501	−3.244
std. dev.	1.356	1.790	1.886	1.802	1.373	1.117
In k_v						
mean	3.890	0.872	−0.492	−1.966	−2.896	−3.331
median	4.080	1.180	−0.504	−1.921	−2.698	−3.154
std. dev.	1.081	1.629	1.356	1.191	0.781	0.817

Note: All permeabilities are permeability to air under ambient conditions, reported as the natural log in millidarcies. K_h values are horizontal permeabilities and are averaged from individual measurements. K_v values are estimated within segments using the sum of the reciprocals of the horizontal values (k_h) within each segment.

Table 2. Summary statistics of the geophysical characteristics for the six electrofacies types.

	ef1	ef2	ef3	ef4	ef5	ef6
Sonic						
mean	69.734	67.106	62.702	59.454	52.923	48.706
std. dev.	3.318	3.029	3.73	3.19	4.181	4.626
MSFL						
mean	1.041	0.871	0.554	0.344	0.153	0.055
std. dev.	0.212	0.185	0.142	0.16	0.029	0.045
NPHI						
mean	0.126	0.095	0.046	0.022	0.008	0.023
std. dev.	0.017	0.021	0.02	0.016	0.008	0.018
RHOB						
mean	2.399	2.479	2.522	2.566	2.635	2.746
std. dev.	0.04	0.033	0.034	0.036	0.039	0.054

Note: Logs in this study include sonic, microresistivity (MSFL), density (RHOB), and neutron porosity (NPHI). The microresistivity values have been converted to the inverse square root of the resistivity (root conductivity).

(Drzewiecki et al., this volume; Shepherd et al., this volume). Gross lithologic variations are secondary and are restricted to carbonate zones that are on the order of 1–10 m thick. These are the zones that comprise electrofacies type ef6. Drzewiecki et al. (this volume) and Shepherd et al. (this volume) describe a redistribution of cements by diagenetic processes. Resulting diagenetic features are alternating quartz overgrowth- and dolomite-cemented bands, alternating high- and low-porosity bands, pressure solution-dominated bands, and stylolites. Differences in electrofacies types appear to be correlated with the relative abundance or absence of these features.

Electrofacies ef1 appears primarily in the wells in the western portion of the basin within the shallowest of the wells studied. This electrofacies is characterized by medium- to coarse-grained sandstones that have been heavily bioturbated, resulting in the obliteration of most sedimentary structures. Ef1 corresponds most closely to the massive sandstone lithofacies described

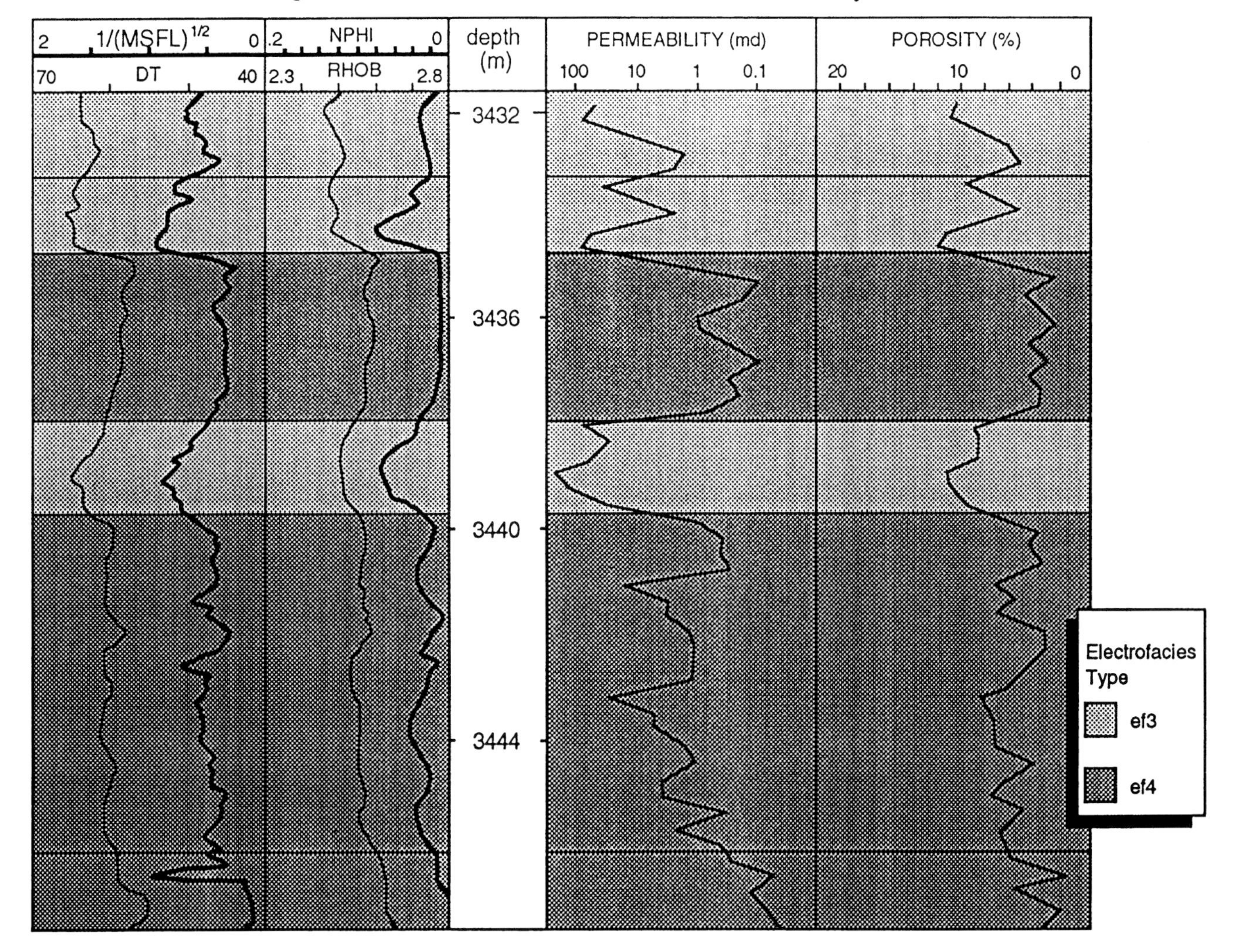

Figure 6. Two electrofacies types appear in the cored interval of the Mobil St. Foster 1-12 well. Diagenetic characteristics of the uppermost ef4 segment are discussed in depth by Shepherd et al. (this volume).

by Drzewiecki et al. (this volume). Cements are sparse to absent in some intervals, resulting in porosities of 20% or greater.

The second most permeable type, ef2, has characteristics similar to type ef1 in the cores in which it has been observed. Ef2 differs from ef1 by a slightly greater degree of quartz cementation. In both of these high-porosity electrofacies types, diagenetic banding and stylolites are rare to absent.

Types ef3 and ef4 are observed within the Mobil St. Foster 1-12 core, which is described in detail by Shepherd et al. (this volume). The electrofacies distribution within this well is shown in Figure 6. Ef3 is widespread throughout the basin and, where observed in cores, has been described as medium- to fine-grained with occasional burrows and sometimes abundant cross- and horizontal bedding. Occasional thin dolomite and quartz-cemented bands occur, but this electrofacies type is differentiated from ef4 in the St. Foster core by the relative absence of banding and a somewhat lesser frequency of stylolites. This electrofacies most likely correlates with the horizontally bedded or cross-bedded sandstone lithofacies (Drzewiecki et al., this volume).

Ef4 has been studied in detail within the St. Foster core by Shepherd et al. (this volume). This electrofacies is characterized by abundant stylolitization and diagenetic banding. This banding consists of millimeter- to centimeter-scale bands of dolomite cement, quartz overgrowth cement, and high porosity. Similar banding is observed in other wells within core intervals corresponding to this electrofacies. In some intervals belonging to this electrofacies the banding is absent and is replaced instead by massive quartz overgrowth cementation. The base of the St. Foster core has this characteristic. This can also be viewed as a large-scale band in itself. The patterns of electrofacies distribution in Figures 5 and 6 reflect heterogeneity on a large scale similar to that observed within individual log segments.

Only limited core samples have been obtained from ef5 type segments. Hydraulically, it is similar to ef4, although somewhat less permeable. Wireline log signatures indicate that this electrofacies is also quartz dominated. While some core segments exhibit massive quartz cementation, still others have PEF signatures that suggest interbedded carbonates or carbonate cement bands at a scale smaller than the resolution of the logging tools. This is supported by the observation that this electrofacies occurs abundantly in association with ef6.

Type ef6 is the only electrofacies that is distinguishable by lithology alone. While it has been only minimally observed in core, the log signatures associated with those observations are easily recognized. As described earlier, these characteristics consist of a high PEF with an associated gamma peak and a distinctive crossover of the density and porosity curves. This electrofacies is associated with the carbonate lithofacies described by Drzewiecki et al. (this volume) and is laterally persistent, with some of the thicker zones being correlative basinwide.

PRESSURE SEALS

The lithologic characteristics associated with some diagenetic seals have been documented. Hunt (1990) observed that the top seal in many sandstones is layered with multiple horizontal bands of carbonate cement. In the sandstones of the Simpson Group in the Anadarko basin, Tigert and Al-Shaieb (1990) observed alternating carbonate-cemented, silica-cemented, and high-porosity intervals within a seal zone. A model that incorporates mechanical and chemical processes has been developed by Dewers and Ortoleva (1990) to explain the formation of banded quartz cementation in sandstones. Thus we might expect to see diagenetic banding in the form of alternating cements and alternating high and low porosity within sandstone intervals that act as pressure seals.

The lithologic characteristics of ef4 within cores that have been studied include diagenetic banding of both cement and porosity types, moderate to abundant stylolitization, and a predominance of quartz-overgrowth cement, often to the total occlusion of porosity. Type ef5 may also possess these characteristics, as the permeabilities are also low and many segments within this electrofacies have PEF values which indicate that they are quartz dominated. This suggests that these electrofacies types are good candidates for pressure seals which are diagenetic in origin. Type ef6 could also act as a seal, although its genesis is depositional rather than diagenetic.

While the diagenetic characteristics of these electrofacies are consistent with those expected, the question arises as to whether or not the permeabilities are sufficiently low enough to maintain a pressure transient over a geologically significant time frame. Here it must be remembered that the permeability values reported in this paper are measured using air (gas) at surface conditions. Core plug tests using water or brine under increased confining pressures show a drop in permeability of 1.5 to 3.5 orders of magnitude relative to air under surface conditions (Hart and Wang, 1992) (Figure 7). This puts the in situ permeabilities in the nanodarcy range at the lower end. Alternating high- and low-permeability electrofacies over hundreds of feet, as indicated by data within the Saginaw Bay and northeast regions of the basin, could provide considerable vertical sealing potential. Gas production is possible from permeable bands within the seals themselves.

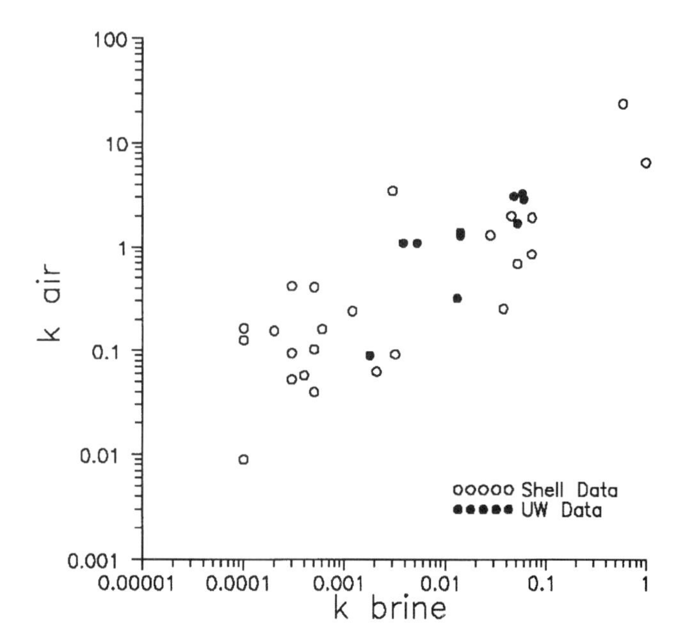

Figure 7. Crossplot of permeability to air under surface conditions versus permeability to brine under in situ conditions shows decreases of 1.5 to 3.5 orders of magnitude. Data were obtained from Shell Western E. and P. and from core tests conducted at the University of Wisconsin–Madison (Hart and Wang, 1992).

CONCLUSIONS

A multivariate analysis of geophysical log data from 18 wells in the Michigan basin has resulted in the identification of six electrofacies types. While the electrofacies classification was based solely on the log responses, it was done in such a way as to optimize the separation of distributions of porosity and permeability associated with each electrofacies type.

Direct examination of cores and lithologic inference from log signatures have shown that the electrofacies types reflect not only the hydraulic characteristics, but the lithologic and diagenetic characteristics as well. In particular, one or more electrofacies types have been identified that have associated formation characteristics consistent with those observed in pressure seal zones in other basins. These characteristics include low permeability, diagenetic banding, abundant quartz-overgrowth cementation, and moderate to abundant stylolitization. While the low-permeability electrofacies are predominantly located within the deeper cores, this probably reflects a bias due to sampling within producing regions. The presence of a type ef4 zone in one of the shallower wells exhibiting diagenetic banding and intense stylolitization within that interval indicates that diagenetic processes responsible for these features are not restricted to the deepest wells.

The electrofacies analysis provides a tool for estimating hydraulic properties in the absence of core data. Characterization of the log responses, porosity, and permeability for each electrofacies type can be used to assign class membership and properties to log segments in wells for which no core data are available. This will provide a more complete picture of the distribution and lateral extent of the low-permeability zones observed in the St. Peter cores. The technique is generally applicable to problems where determining the spatial distribution of hydraulic characteristics is critical but where direct measurements of those properties are insufficient.

ACKNOWLEDGMENTS

The authors would like to thank Mobil Exploration and Producing, Amoco Production, Union Oil Company, Marathon Oil Company, and Shell Research and Development for providing core, core plug analyses, and digital logs. We also thank Western Michigan University for providing core samples. This research was funded by the Gas Research Institute.

REFERENCES CITED

Asquith, G.A., 1982, Basic Well Log Analysis for Geologists: American Association of Petroleum Geologists, Tulsa, Oklahoma, 216 p.

Dewers, T., and P. Ortoleva, 1990, A coupled reaction/transport/mechanical model for intergranular pressure solution, stylolites, and differential compaction and cementation in clean sandstones: Geochemica et Cosmochemica Acta, v. 54, p. 1609–1626.

Elek, I., 1988, Some applications of Principal Component Analysis: well-to-well correlation, zonation: Geobyte, v. 3, no. 2, p. 46–55.

Freeze, R.A., and J.A. Cherry, 1979, Groundwater: Englewood Cliffs, NJ, Prentice-Hall, Inc., 604 p.

Gill, D., 1970, Application of a statistical zonation method to reservoir evaluation and digitized-log analysis: AAPG Bulletin, v. 54, no. 5, p. 719–729.

Hart, D. J., and H. F. Wang, 1992, Core permeabilities of diagenetically-banded St. Peter sandstone from the Michigan basin: EOS, v. 73, no. 43 Supplement, p. 514.

Hawkins, D.M., 1976, FORTRAN IV program to segment multivariate sequences of data: Computers and Geosciences, v. 1, p. 339–351.

Hawkins, D.M., and D.F. Merriam, 1974, Zonation of multivariate sequences of digitized geologic data: Mathematical Geology, v. 6, no. 3, p. 263–269.

Hunt, J.M., 1990, Generation and migration of petroleum from abnormally pressured fluid compartments: AAPG Bulletin, v. 74, no. 1, p. 1–12.

Kulinkovich, A.Y., N.N. Sokhranov, and I.M. Churinova, 1966, Utilization of digital computers to distinguish boundaries of beds and identify sandstones from electric log data: International Geology Reviews, v. 8, p. 416–420.

Mehta, C.H., S. Radhakrishnan, and G. Srikanth, 1990, Segmentation of well logs by maximum-likelihood estimation: Mathematical Geology, v. 22, no. 7, p. 853–869.

Powley, D.E., 1990, Pressures and hydrogeology in petroleum basins: Earth-Science Reviews, v. 29, p. 215–226.

Radhakrishnan, S., G. Srikanth, and C.H. Mehta, 1991, Segmentation of well logs by maximum likelihood estimation: the algorithm and FORTRAN-77 implementation: Computers and Geosciences, v. 17, no. 9, p. 1173–1196.

Serra, O., 1986, Fundamentals of Well-Log Interpretation. 2. The Interpretation of Logging Data: Developments in Petroleum Science, 15B, Elsevier, Amsterdam.

Serra, O., and H.T. Abbott, 1982, The contribution of logging data to sedimentology and stratigraphy: Society of Petroleum Engineers Journal, v. 22, no. 1, p. 117–131.

Testerman, J.D., 1962, A statistical reservoir-zonation technique: Journal of Petroleum Technology, Aug. 1962, p. 889–893.

Tigert, V., and Z. Al-Shaieb, 1990, Pressure seals: their diagenetic banding patterns: Earth-Science Reviews, v. 29, p. 227–240.

Webster, R., 1973, Automatic soil boundary location from transect data: Journal of the International Association of Mathematical Geology, v. 5, p. 27–37.

V. The Powder River Basin

Chapter 15

The Regional Pressure Regime in Cretaceous Sandstones and Shales in the Powder River Basin

Ronald C. Surdam
Zun Sheng Jiao
Randi S. Martinsen
University of Wyoming
Laramie, Wyoming, U.S.A.

ABSTRACT

The Cretaceous shale section in the Powder River basin below a present-day depth of approximately 8000 ± 2000 ft (2400 ± 600 m) typically is overpressured. The top of the transition zone, 500–1000 ft (150–300 m) thick, in the upper portion of the overpressured section occurs within the Steele Formation; and the "hard" overpressured zone, ~2000 ft (600 m) thick, typically begins in the Niobrara Formation, with the base of the zone parallel to the Fuson Shale, the lowermost organic-rich shale in the Cretaceous stratigraphic section. The upper and lower boundaries of the pressure compartment are subparallel to stratigraphic boundaries. Toward the basin margin where the Cretaceous section is at shallow depth (~6000 ft [1800 m]) the overpressured shale section is wedge shaped.

The overpressured Cretaceous shale section in the Powder River basin is a basinwide dynamic pressure compartment. The driving mechanism is the generation of liquid hydrocarbons that subsequently partially react to gas, converting the fluid-flow system to a multiphase regime where capillarity dominates the relative permeability, creating elevated displacement pressures within the shales.

In contrast, many of the Cretaceous sandstones are subdivided into relatively small, isolated pressure or fluid-flow compartments 1 to 10 mi (1.6–16 km) in greatest dimension. The compartmentation is the result of internal stratigraphic elements, such as paleosols along unconformities. These internal stratigraphic elements are low-permeability rocks with finite leak rates in a single-phase fluid-flow system but evolve into relatively impermeable capillary seals with discrete displacement pressures as the flow regime evolves into a multiphase fluid-flow system. This evolution of the fluid-flow system is caused by the addition of hydrocarbons to the fluid phase as a result of continuous burial and increasing thermal exposure. The three-dimensional closure of the capillary seals above, below, and within a sandstone results in isolated fluid-flow or pressure compartments within the sandstone.

Not all the sandstones within the overpressured shale section are at the same pressure as the shales; some are overpressured, some are normally pressured, and some even appear to be underpressured. Those sandstones characterized by compartmentation (three-dimensional closure of capillary seals) are above, at, or slightly below the pressure of the adjacent shales. The sandstones characterized by normal pressure within the overpressured shale section probably represent fluid conduits connecting with the overlying (at 8000 to 9000 ft [2400–2700 m]) or underlying (below Fuson shale) normally pressured fluid-flow regimes.

The major difference between pressure compartmentation in these Cretaceous sandstones and shales is one of scale. In both cases the appearance of hydrocarbons drives the transition from single-phase (water) to multiphase fluid flow (water plus one or more hydrocarbon phases); when the hydrocarbons activate capillary seals, the result is grossly increased displacement pressure. When hydrocarbons saturate the compartment, the integrity of the three-dimensional boundary capillary seals is ensured, and free water is expelled from the system.

In summary, understanding the concept of multiphase fluid flow as it relates to three-dimensional pressure compartmentation will greatly expedite the search for, the discovery of, and the exploitation of new unconventional gas resources.

INTRODUCTION

The main purpose of this chapter is to document the pressure regimes characterizing Cretaceous sandstones and shales in the Powder River basin of Wyoming and Montana (Figure 1). Once the documentation is complete, a secondary objective is to use the Powder River basin as a test case for the pressure compartmentation hypothesis formulated by Bradley (1975) and Powley (1982), and described by Hunt (1990). The Powder River basin was chosen as a test case for three reasons: (1) it is cited by Hunt (1990) as a typical basin with abnormal (i.e., overpressured) pressure compartments; (2) it is a mature hydrocarbon province characterized by a well-documented tectonic and stratigraphic framework, and (3) there are available well histories (scout tickets) for over 15,400 wells, 13,000 drill-stem tests, and cored intervals from 5400 wells. For these reasons, the Powder River basin is an ideal place to test the pressure compartmentation hypothesis.

This chapter consists of four parts: (1) The regional pressure regime in the Cretaceous shales will be evaluated, with emphasis on the causes and timing of overpressuring; (2) the pressure regime within the Cretaceous sandstones will be evaluated, with emphasis on the mechanisms causing abnormal pressures within reservoir sandstones; (3) the relationship between shale and sandstone pressure regimes will be discussed; and (4) the pressure regimes characterizing the Cretaceous stratigraphic interval within the Powder River basin will be used to test the pressure compartmentation hypothesis (Bradley, 1975; Powley, 1982; Hunt, 1990).

PRESSURE REGIMES WITHIN CRETACEOUS SHALES

Trends of transit travel times (μs/ft) on sonic logs are conventional keys to detecting abnormal pressures in shales (Magara, 1976; Powley, 1982). Typically, a marked increase in transit time indicates undercompacted shales (i.e., overpressuring). Many of the pressure compartments cited by Powley (1982) were delineated on the basis of analyzing trends in transit travel times from sonic logs.

The Powder River basin offers some spectacular examples of overpressured rocks that can be detected and delineated by downhole trends in sonic logs. In the example illustrated by Figure 2, the reversal in sonic travel time at approximately 8000 ft (2400 m) is judged to indicate the top of the overpressured section, whereas the bottom of overpressuring is thought to be at approximately 11,500 ft (3500 m), where there is a significant change in the travel-time/depth gradient. During the drilling of the overpressured shale section, the mud weights are adjusted from 8.4 ppg to approximately 9.0 ppg as the transition zone is entered at approximately 8000–9000 ft (2400–2700 m), and from 9.0 up to 15 ppg in the lower portion of the overpressured section (Figure 3). In the case illustrat-

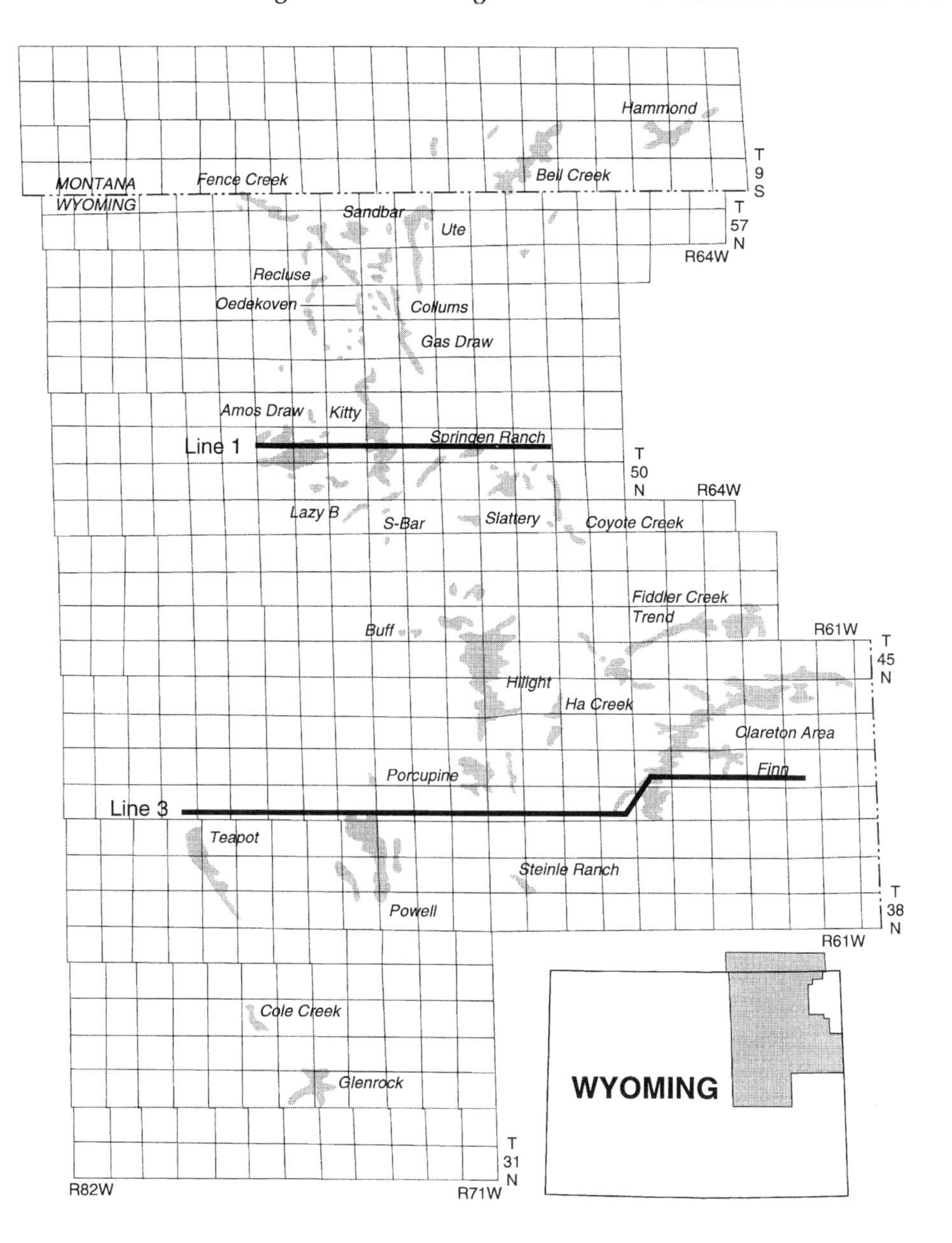

Figure 1. Index map of the Powder River basin. All major fields that produce from the Muddy Sandstone are shown. Lines 1 and 3 indicate the locations of cross sections cited in text. Velocity and pressure profiles along Line 3 are shown in Figures 9–11.

ed in Figure 3, a mud weight of 8.4 ppg is equivalent to a pressure gradient of 0.436 psi/ft, whereas a mud weight of 14.8 ppg is equivalent to a pressure gradient of 0.77 psi/ft. The digitized sonic log shown in Figure 2 is interpreted as indicating a continuously overpressured shale column from depths of approximately 8000 to 11,000 ft (2400–3300 m) in the vicinity of the specific example (the central portion of the Powder River basin). This example is typical of what is found throughout the Powder River basin, except at the basin margin where the Cretaceous section approaches the surface. At the edge of the basin the boundaries of the overpressured shale section are wedge shaped (Figure 4); the base of the overpressured section is parallel with the Fuson Shale, but the top of the overpressured shale cuts across stratigraphy. Throughout the basin, the top of overpressuring in the Cretaceous shales occurs at 8000 ± 2000 ft (2400 ± 600 m), and the base of overpressuring is at the depth of the Fuson Shale.

It is important to note that the Powder River basin was uplifted during and subsequently to the Laramide orogeny. The exact amount of uplift is problematic, but our preliminary work, based on sonic logs and decompaction constants, suggests that the amount of unroofing varied, with the least erosion in the west and with erosion increasing to the east, and that the amount of erosion probably varied from 1000 to 3000 ft (300–900 m). Thus, the variation assigned to the top and bottom of overpressuring within the central portion of the basin may reflect the uncertainty in erosion amounts.

Hunt (1990) suggests that many of the sedimentary basins in the world are characterized by two or more superimposed hydrogeological systems. He suggests that the shallowest fluid-flow systems are at normal hydrostatic pressure and extend down to 10,000 ft (~3000 m). Variations in this depth are judged to be the result of differing thermal gradients and burial histories (Hunt, 1990). Below the normally pressured

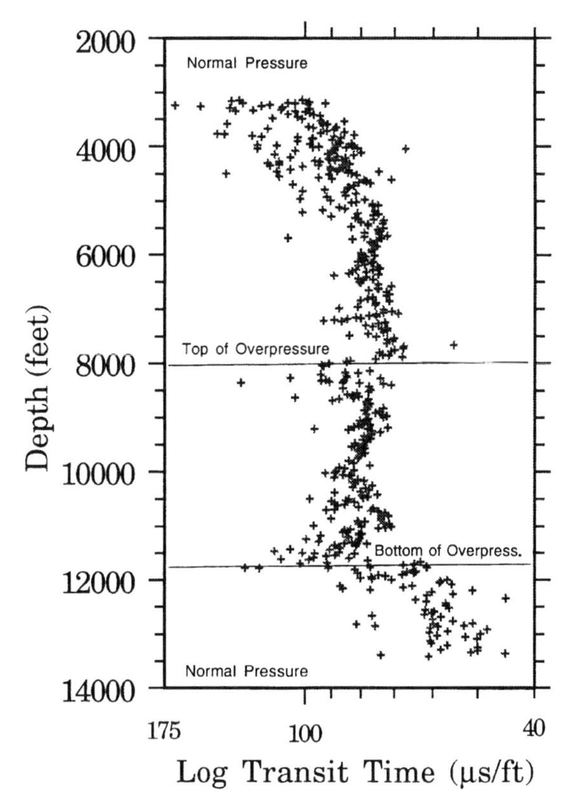

Figure 2. Plot of transit time in shales vs. depth for Federal Means #21-13, Section 13, T46N, R75W. The abnormally pressured zone is marked by the distinctly decreasing transit time between 7800 and 11,500 ft (2400 and 3500 m).

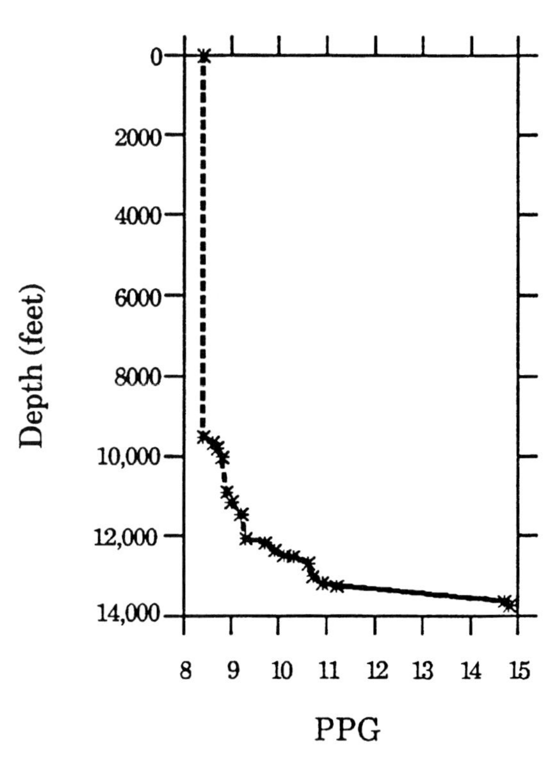

Figure 3. Plot of drilling mud weight vs. depth for Devex Federal #32-11, Section 11, T38N, R74W. The mud weights are adjusted from 8.4 ppg to approximately 9.0 ppg as the transition zone of overpressure is entered at approximately 9000–10,000 ft (2700–3000 m), and from 9.0 up to 15 ppg in the lower portion of the overpressured section.

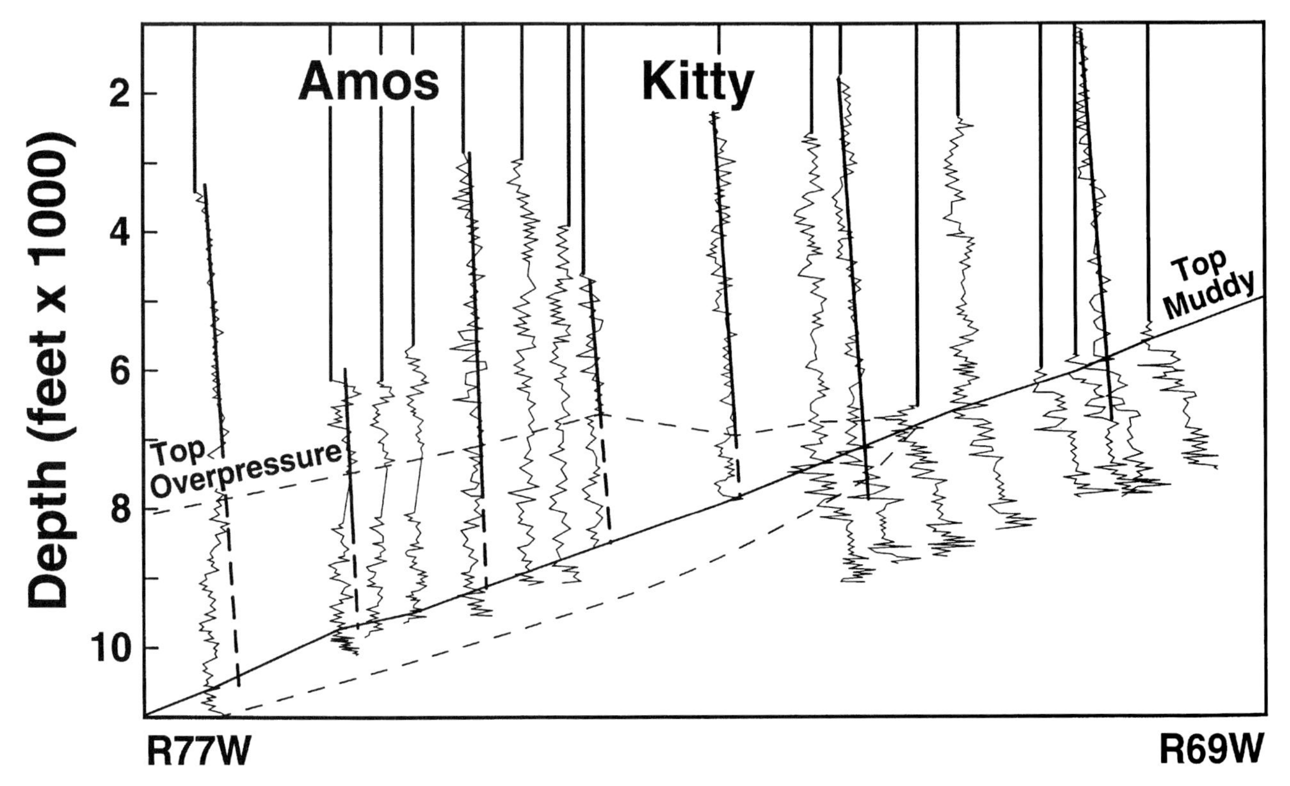

Figure 4. Cross section of velocity logs, showing the overpressured shale section wedging out at the edge of the Powder River basin.

section (the section characterized by hydrostatic gradients) there is a compartmentalized fluid-flow system, mostly overpressured. The tops of many of the overpressured fluid compartments commonly range from 90 to 100°C present-day temperature (Hunt, 1990). How well does this generic description match the overpressured Cretaceous shale section in the Powder River basin?

To answer this question, it is necessary to characterize in more detail the regional pressure regime in the Cretaceous shales. Before pursuing a detailed documentation, it is important to note the following observations: (1) There is a significant change in the vitrinite reflectance/depth gradient at 8000 to 9000 ft (2400–2700 m) present-day depth in the central Powder River basin (Figure 5) (MacGowan et al., this volume; Hunt, 1979); (2) over the same depth interval, the mixed-layer clays in the shale go through a transition from 20% illite to 85% illite, and they become ordered (Figure 6) (Jiao and Surdam, this volume); (3) there is a significant increase in the production index (PI, from anhydrous pyrolysis) at approximately 9000 ft (2700 m) present-day depth in the central Powder River basin (Figure 7) (MacGowan et al., 1994); and (4) most important, at a present-day depth of 8000 ft (2400 m) there is a marked increase in the sealing capacity of the Cretaceous shale for an oil/water and gas/water system (Figure 8) (Schowalter, 1979; Jiao and Surdam, this volume). According to Sneider et al. (1991), the increase in sealing capacity is equivalent to a change from a type C seal (a seal capable of supporting an oil column of ≥100 but <500 ft [≥30 but <150 m]) to a type A seal (a seal capable of supporting an oil column of >1000 ft [>300 m]). Thus, at a present-day depth of approximately 8000 to 9000 ft (2400–2700 m) there is a significant change in the fundamental properties of the Cretaceous shales in the central portion of the Powder River basin. This marked change in fundamental properties coincides with a transition from normally compacted shales resulting from increasing burial to undercompacted shales resulting from hydrocarbon generation, or from a normally pressured section to an overpressured section.

The description of the pressure regime in the Cretaceous shales in the Powder River basin is exemplified in the profiles in Figures 9–11 (along Line 3, Figure 1) (see Maucione et al., this volume, their figures 7, 10–17). These figures show (1) a velocity profile derived from a panel of sonic logs along a specific cross section through the Cretaceous stratigraphic section, (2) the same panel of sonic logs with the velocities corrected for compaction (decompacted sonic logs) per the methods outlined in Maucione et al. (this volume), and (3) the same panel showing relative pressure in the Cretaceous shales derived from pressure values based on the compression curve method discussed in Maucione et al. (this volume) and Dobrynin and Serebryakov (1989).

Several observations can be made from these sets of panels. The top of the overpressured zone occurs within the shales of the Steele Formation. In all areas

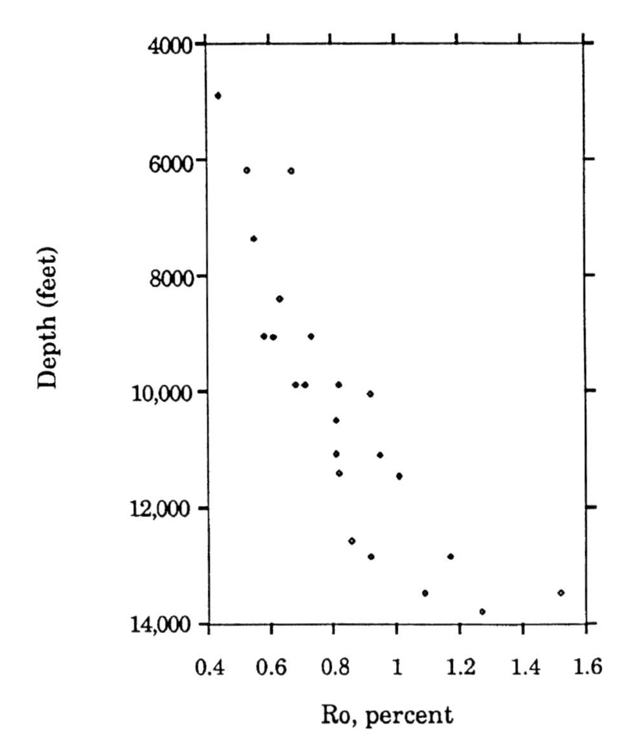

Figure 5. Plot of vitrinite reflectance vs. depth for the Mowry Shale, Powder River basin, Wyoming.

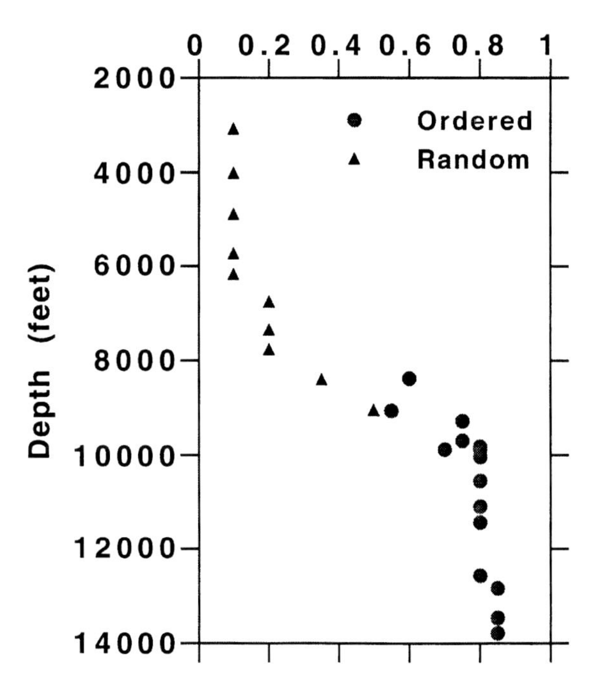

Figure 6. Plot of percentage of illite in I/S (illite/smectite) mixed-layer clays vs. depth for the Mowry Shale, Powder River basin, Wyoming.

studied away from the margin of the basin, the onset of overpressuring is marked by a transition zone (slightly overpressured rocks; mud weights greater than 8.4 ppg) 500–1000 ft (150–300 m) thick. The top of the transition zone is approximately parallel to the

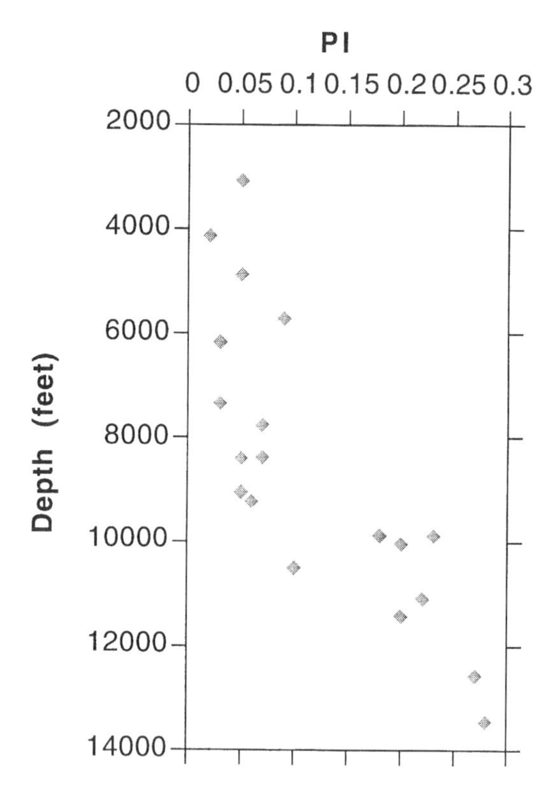

Figure 7. Plot of production index vs. depth for the Mowry Shale, Powder River basin, Wyoming.

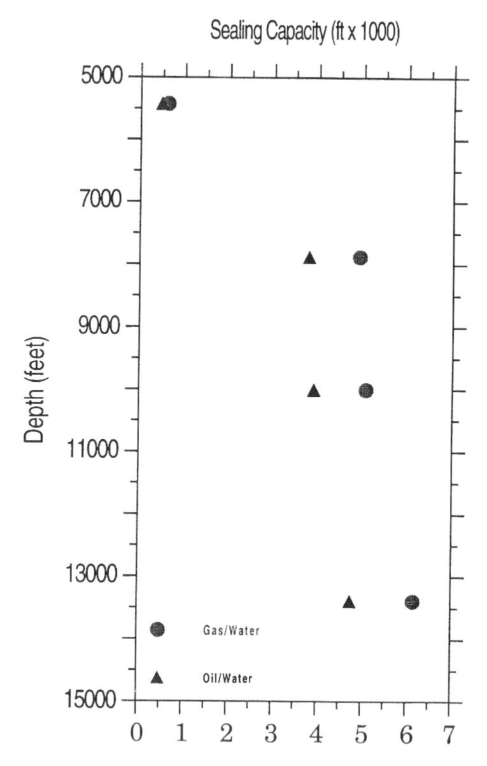

Figure 8. Sealing capacity, in thousands of ft, vs. depth for the Cretaceous Mowry Shale for oil/water (solid triangles) and gas/water (solid circles) systems, showing a marked increase at a present-day depth of approximately 8000 ft (2400 m). The sealing capacity of a hydrocarbon seal is based on petrophysical and flow properties of the rock that have determined the height of hydrocarbon column that can be supported by the seal.

top of the Steele Formation but typically is within the Steele Formation. The boundary between the transition zone and the overpressured zone (mud weights <15 ppg) typically begins in the Niobrara Formation. The bottom of the overpressured shale section is regionally coincident with the Fuson Shale, and if a transition zone exists at the bottom of the overpressured zone, it is much thinner than the transition zone typically found at the top. The total overpressured shale section is approximately 3000 ft (900 m) thick at the basin center. The highest abnormal pressures within the Cretaceous shales typically are found in the lower half of the overpressured column (mud weight >15 ppg; see Figures 9–11). Overpressuring within the Cretaceous shales developed regionally as one large and continuous volume of rock. The only exceptions to the above observations occur along the margin of the basin where the Cretaceous section approaches the surface and where erosion, or unroofing, of the section has been greatest. Along the eastern margin of the basin, the edge of the overpressured shale volume is wedge shaped (Figure 4).

Causes of Overpressuring

As noted by MacGowan et al. (this volume), the produced waters from the Cretaceous section exhibit a significant change in composition over the present-day depth interval of 8000 to 9000 ft (2400–2700 m). Above this depth the total dissolved solids in produced waters average about 10,000 ppm. This value indicates that the marine rocks in the section have been strongly influenced by meteoric water influx from the basin margin. The relatively constant and dilute total dissolved solids content of the water produced from a predominantly marine section suggests that the upper 8000 ft (2400 m) of section in the Powder River basin is in good hydrologic communication. The rapid and significant increase in total dissolved solids (to TDS averages of 35,000 ppm) in produced waters from deeper than 8000 ft (2400 m) present-day depth strongly supports the ideas that there is some type of fluid compartmentation within the basin, and that there is poor hydrologic communication between rocks above and below a present-day depth of 8000 to 9000 ft (2400–2700 m) (see MacGowan et al., this volume, for details).

In hydrocarbon production from sandstones within the overpressured shale section there are no water legs or oil/water contacts (for the Muddy Sandstone, for example, see Table 1). The fields producing hydrocarbons from the overpressured shale section are depletion-drive systems. This fact, plus the large-scale regional aspect of the overpressured section, the fact that there are no detected overpressured pockets of shale above a depth of 8000 ft (2400 m), and the

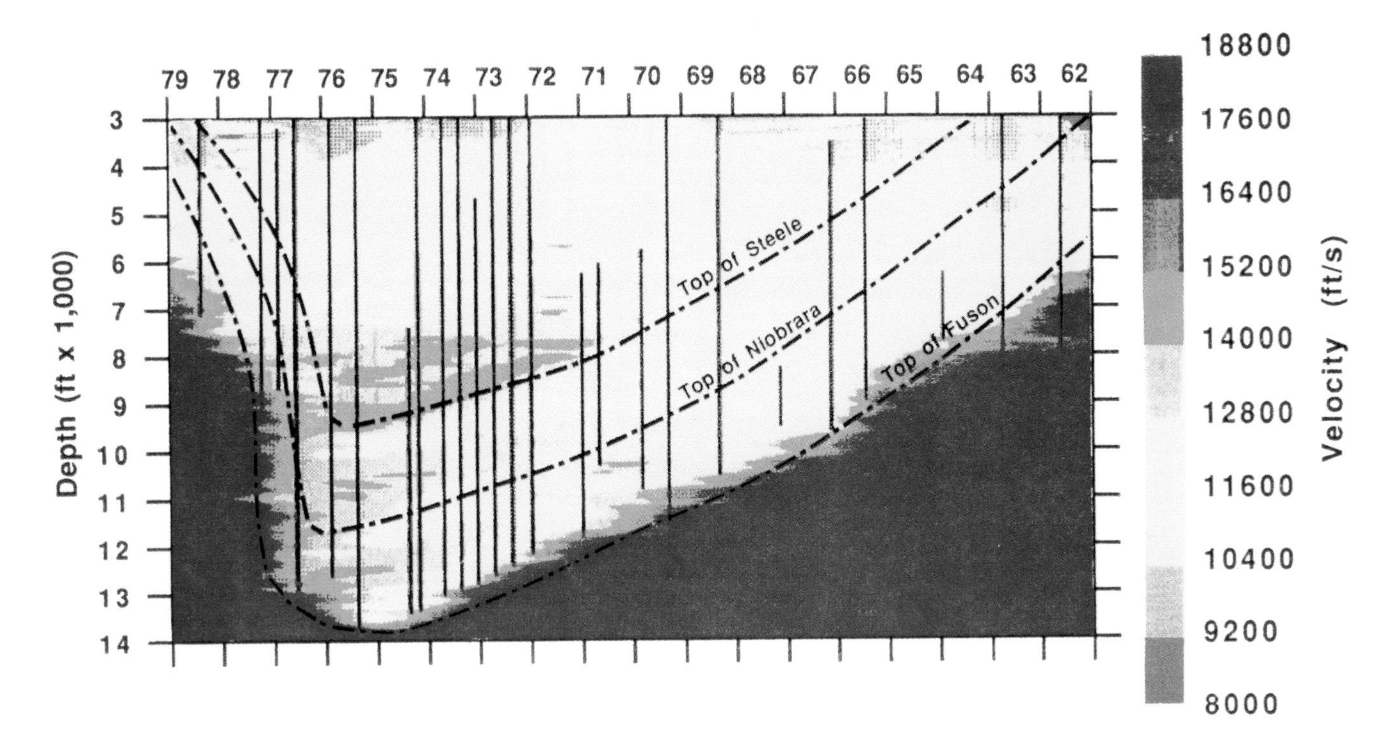

Figure 9. Cross section of sonic logs located in the southern Powder River basin, Wyoming (Figure 1, Line 3), T41–42N, R62–79W. Velocity profile derived from a panel of sonic logs through the Cretaceous stratigraphic section.

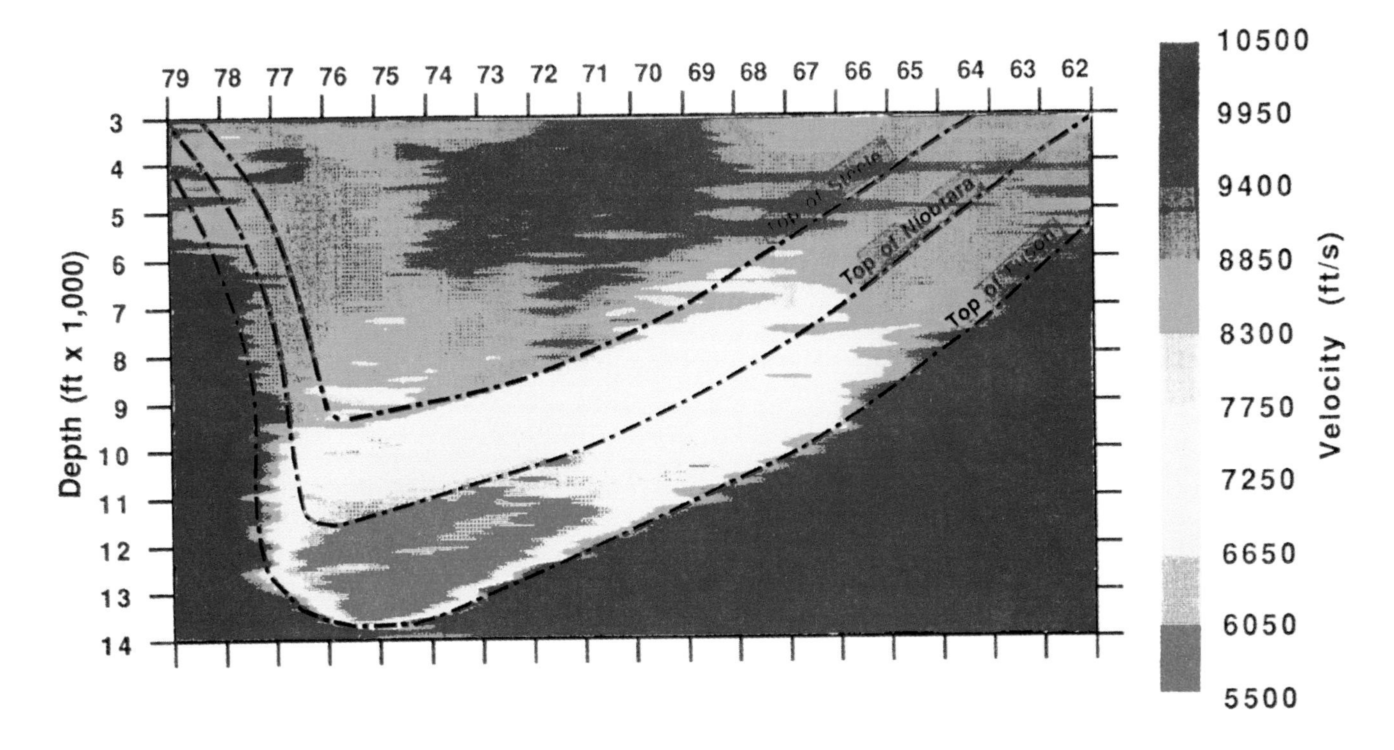

Figure 10. Cross section of sonic logs located in the southern Powder River basin, Wyoming (Figure 1, Line 3), T41–42N, R62–79W. Velocity profile derived from the same panel of sonic logs as in Figure 9, with the velocities corrected for compaction.

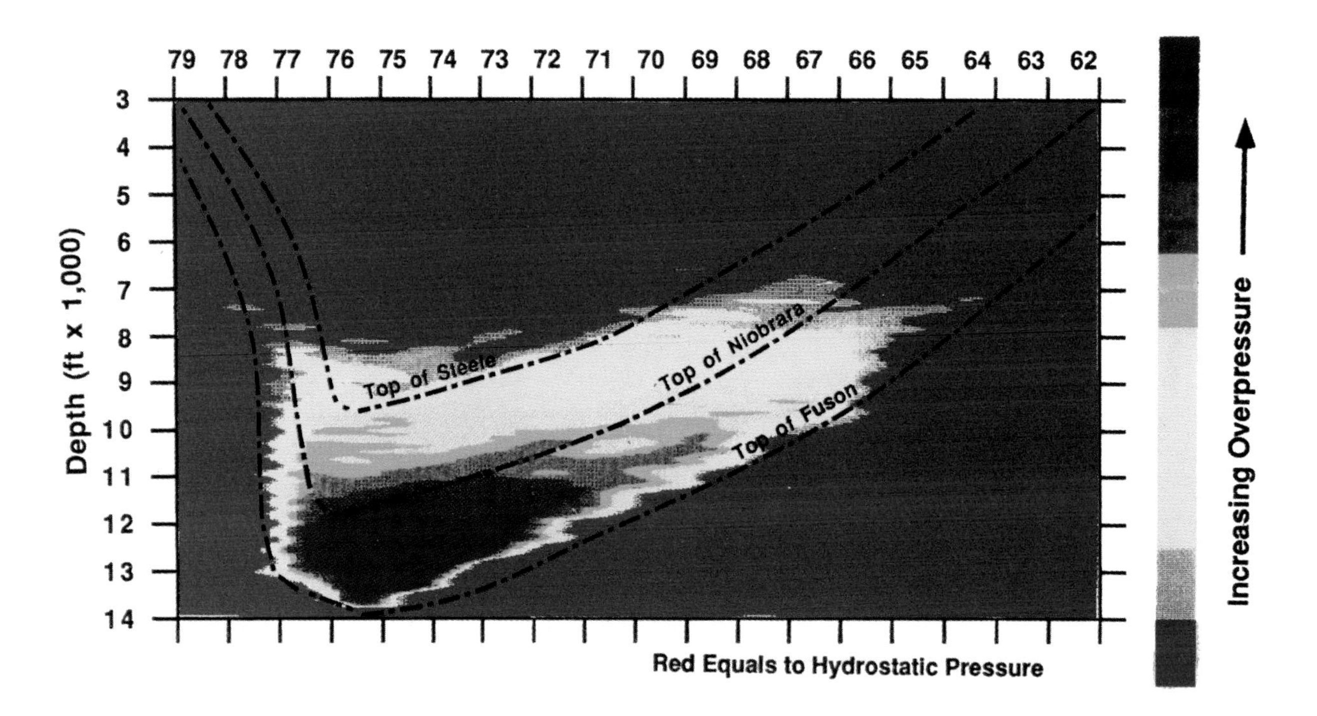

Figure 11. Cross section of sonic logs located in the southern Powder River basin, Wyoming (Figure 1, Line 3), T41–42N, R62–79W. Pressure profile derived from the same panel of sonic logs as in Figure 10, where the derived pressure values are based on the compression curve method discussed in Maucione et al. (this volume). Tops for the Steele, Niobrara, and Fuson shales are shown for stratigraphic reference.

regional trends in hydrologic communication based on the composition of produced waters, leads to the conclusion that disequilibrium compaction is not the primary cause of overpressuring in the Powder River basin.

The vertical trend in the production index [PI = $S_1/(S_1 + S_2)$ where S_1 is the hydrocarbon generated and S_2 is the kerogen remaining in the source rock] from shales in the Powder River basin indicates a marked increase in hydrocarbons within the shales below a present-day depth of 9000 ft (2700 m) (Figure 7). Above a depth of 9000 ft (2700 m) the PI is typically <0.1, but below that depth the PI is >0.1 and rises to values as high as 0.3 at 14,000 ft (4200 m) (Figure 7) (MacGowan et al., this volume). This vertical trend suggests either that above 9000 ft (2700 m) in the shales there is a paucity of hydrocarbon or that below that depth hydrocarbon has been generated in significant quantities. In other words, in the Powder River basin, the upper 8000 ft (2400 m) of section is normally pressured and is dominantly a water-drive system (a single-phase fluid-flow system), whereas below 9000 ft (2700 m) the shales are typically overpressured and characterized by a fluid-flow system dominated by hydrocarbons (a multiphase fluid-flow system). The difference between the production index (PI), or observed remnant hydrocarbon, and the transformation ratio calculated for the reaction of Type II kerogen to liquid hydrocarbon using kinetic parameters from Tissot and Welte (1984) suggests that approximately half of the generated liquid hydrocarbon has been expelled and has migrated out of the source rock (Figure 12). Thus, a significant portion of the liquid hydrocarbon generated and expelled from the source rock has remained in the organic-rich shales below 9000 ft (2700 m).

More insight into the mechanisms overpressuring in the shale section is gained by examining the nuclear magnetic resonance (NMR) patterns for a sequence of the Lower Cretaceous Mowry Shale samples from the present-day depth interval 3000 to 11,000+ ft (900–3300+ m) (Figure 13) (MacGowan et al., this volume). Note that the aliphatic functional groups have been cleaved off the kerogen in the Mowry Shale source rocks by a depth of 9000+ ft (2700+ m). This suggests that a significant percentage of the liquid hydrocarbon in these shales has been generated at burial depths of 9000 ft (2700 m) (equivalent to maximum burial depths of 12,000 ft [3600 m] or less; Figure 14A). As noted, the hydrocarbons are residual oil that has not been expelled (S_1 peak of anhydrous pyrolysis). Yet the shales in the overpressured zone at present have little remaining capacity to generate *liquid* hydrocarbons (Figure 13). It should be noted that using the kinetic parameters developed by MacKenzie and Quigley (1988) for the reactions kerogen-to-gas and oil-to-gas allows an evaluation of the reaction (a calculation of the transformation ratios) for

Table 1. Formation fluid data within Muddy Sandstone, Powder River basin, Wyoming.

Field Name	Location	Reserv. Elev.	Reserv. Geom.	Init. Pressure (psi)	Fluids Produced	HC/W Contact	Reserv. Drive	API Gravity
Big Hand	47N, 71W	–3700	Limited	3490	HC + Water	No	Depletion	42
Boggy Creek	40N, 64W		Discontinuous	2281	Oil + Gas	No	Depletion	42
Chan	56N, 73W	–2950	Discontinuous	2017	HC + Water *	No	Depletion	39
Clareton	43N, 65W	250	Discontinuous		Oil + Gas	No	Depletion	42
Collums	55N, 73W	–3100	Discontinuous	2090	HC + Water *	No	Depletion	
Donkey Creek	49N, 68W		Limited	2300	Oil	No		40
Dry Fork	38N, 73W		Discontinuous	3668	HC + Water **	No	Depletion	46
Duck Creek	55N, 69W	–1300	Discontinuous	1300	Oil + Gas	No	Depletion	40
Dull	40N, 68W	–5046			Oil	No	Depletion	
Fence Creek	58N, 76W		Discontinuous	2300	Oil	No		37
Fenton	41N, 70W	–5400	Continuous	4590	Oil + Gas	No	Depletion	47
Fiddler Creek	46N, 64W	–380	Discontinuous	1970	Oil + Gas	No	Depletion	40
Fly Draw	37N, 75W	–7400			Oil + Gas	No	Depletion	
Frog Creek	41N, 67W	–4017	Discontinuous	3613	Oil + Gas	No	Depletion	42
George Range	45N, 66W	–1900	Discontinuous	2889	Oil + Gas	No	Depletion	39
Hat Creek/ Thunder Creek	43N, 69W	–3900	Limited	3885	Oil + Gas	No	Depletion	40–57
Hilight Field	45N, 71W		Continuous	4010	Oil + Gas	No	Depletion	41
Hines	50N, 73W	–4500	Discontinuous		Oil + Gas	No	Depletion	38–45
Iberlin	46N, 76W	–3000	Limited		Oil + Gas	No	Depletion	40
Interstate	50N, 73W	–4650	Discontinuous		Oil + Gas	No	Depletion	42
Janet	36N, 68W	–5500		5386	Oil + Gas	No	Depletion	40
Kitty	51N, 73W	–4498	Discontinuous	4397	Oil + Gas	No	Depletion	39
Payne	43N, 70W		Discontinuous	2276	Oil + Gas	No	Depletion	42
Prep	57N, 75W	–3675	Limited		HC + Water **	No	Depletion	35
Quest	45N, 67W	–2720	Limited	3298	Oil + Gas	No	Depletion	41
Remington	58N, 77W				Oil + Gas			31
Springen	51N 71W	–3200	Discontinuous	2725	HC + Water *	No	Depletion	40
Tomcat Creek	49N 65W				Oil			
Two V Creek	55N 70W	–1870	Discontinuous	2389	Gas	No	Depletion	38

* Water from secondary water flood production.
** Very little water and possibly completion water.

Source: Wyoming Geological Association Symposium, Powder River Basin Oil and Gas Fields, vols. 1 and 2.

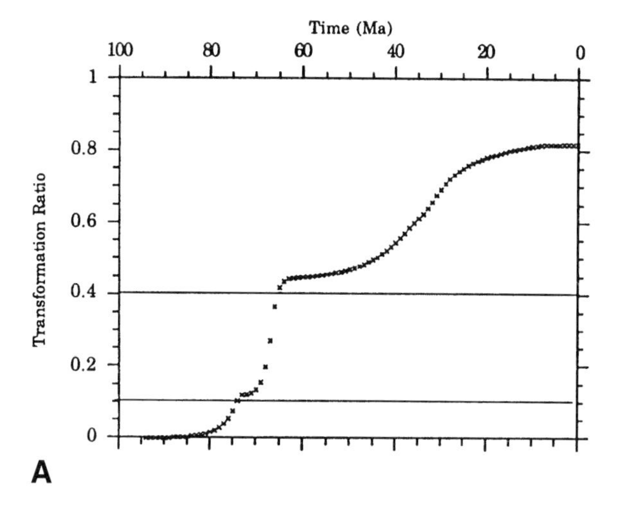

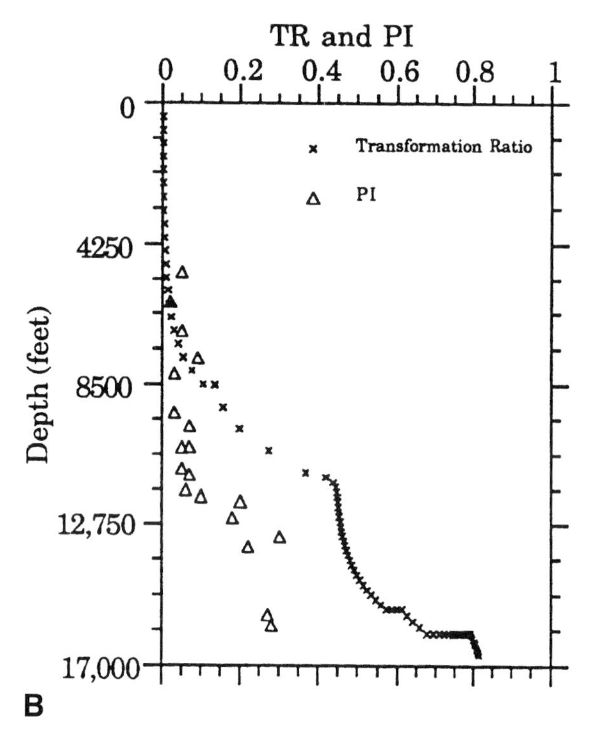

Figure 12. (A) Computed transformation ratios for the Mowry Shale in the central portion of the Powder River basin vs. time using kinetic parameters proposed by Tissot and Welte (1984). (B) Plot of production index (PI) and transformation ratio (TR) for the Mowry Shale. The difference between PI and TR below 9000 ft (2700 m) suggests that a significant portion of the generated hydrocarbon has remained in the organic-rich shale.

the source rocks from typical burial histories from the central Powder River basin (Figure 14B). As can be seen in Figure 14B, the reaction of kerogen to gas will not be significant in the Powder River basin until maximum burial depths of >17,000 ft (>5200 m) are reached. All the overpressured Cretaceous shale sections examined in this study occur at present-day depths of <14,000 ft (<4200 m). So, even if the eroded section is restored, the overpressured sections are well above the depth where the kerogen-to-gas reaction would be significant (Figure 14B). Therefore, the reaction of kerogen to gas is not considered a significant factor in overpressuring the Cretaceous shale system.

The reaction of oil to gas in the Cretaceous shales was evaluated in a similar manner (Figure 14B), using kinetic parameters from MacKenzie and Quigley (1988). The transformation ratio calculated for the oil-to-gas reaction suggests that this reaction will begin to proceed (TR > 0.1) at a maximum depth of burial of approximately 12,000 ft (3600 m). If the amount of unroofing in the basin is 0 to 3000 ft, then this reaction will proceed at 8000 to 11,000 ft (2700–3300 m) present-day depths. Thus, the lower half of the overpressured shale section, where maximum overpressure typically exists, would overlap with the oil-to-gas reaction. We suggest that the generation of liquid hydrocarbons in the organic-rich Cretaceous shales and the subsequent reaction of this oil to gas is the dominant force driving the overpressuring in the shales.

This suggestion is supported by measured displacement pressures in Cretaceous shale samples from the Powder River basin. Figure 15 shows the vertical trend in displacement pressure (breakthrough pressure) for shale samples from a present-day depth interval of 5000 to 13,000 ft (1500–4000 m). Again, there is a pronounced increase (from nearly 0 to 500 psi for an oil/water system, and from ~240 to 2000 psi for a gas/water system) in displacement pressure for the Mowry Shales at approximately 8000 ft (2400 m) present-day depth. Jiao and Surdam (this volume) have shown the relationship between the progress of clay diagenesis and the threshold or displacement pressure in shales. Thus, it is not surprising that there is a significant increase in displacement pressures in the shales at the same depth interval where the percentage of illite in mixed-layer clays changes from 20 to 85% illite and the clays become ordered. Therefore, it is concluded that it is not the absolute permeability of the rock that matters nearly as much as the relative permeability in the multiphase fluid-flow system. In a multiphase fluid-flow system (oil/water or gas/water), the Cretaceous shales will be orders of magnitude more effective in retaining abnormal fluid pressure than in a single-phase (water-dominated) fluid-flow system. In fact, only in a gas/oil/water system are the measured displacement pressures the same order of magnitude as the abnormal pressures measured, or observed, in the Cretaceous system in the Powder River basin. Therefore, it is concluded that the dominant mechanism driving the overpressuring of the Cretaceous shale system in the Powder River basin was the generation of liquid hydrocarbons that converted the fluid-flow system from a water-dominated single-phase regime to a hydrocarbon-dominated multiphase regime. As generated but retained liquid hydrocarbons began to react to gas and the fluid system became gas/oil dominated, the displacement pressures characterizing the shales increased by an additional order of magnitude, firmly establishing a

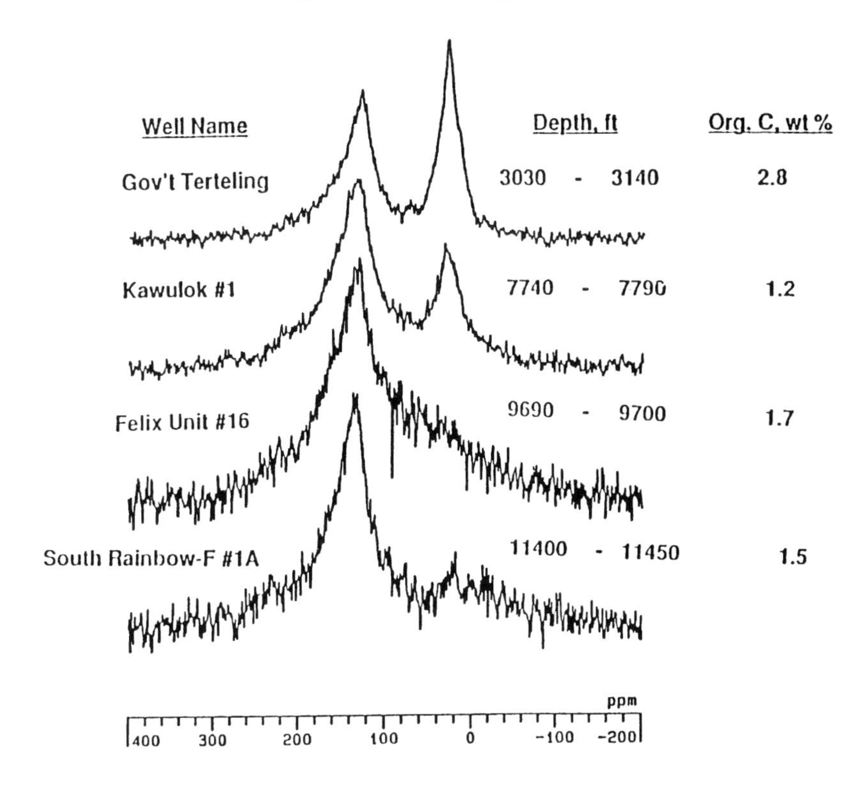

Figure 13. Nuclear magnetic resonance (NMR) patterns for Mowry Shale samples over the present-day depth interval 3030 to 11,450 ft (920–3490) m) in the Powder River basin. The aliphatic functional groups (right-hand major peak) have been cleaved off the kerogen in the Mowry Shale source rocks at a depth of approximately 9000 ft (2700 m).

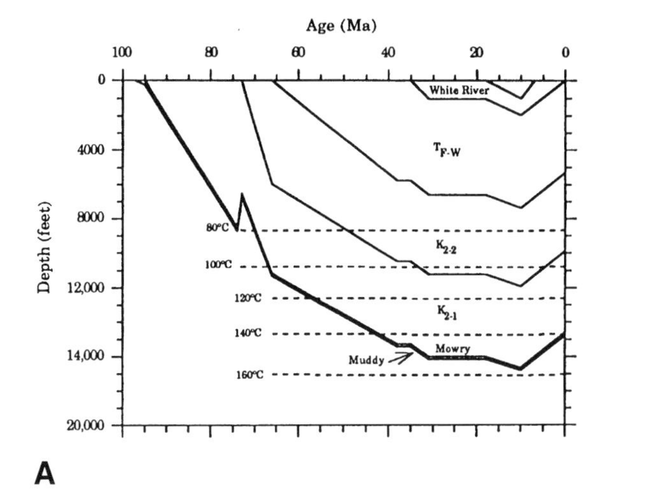

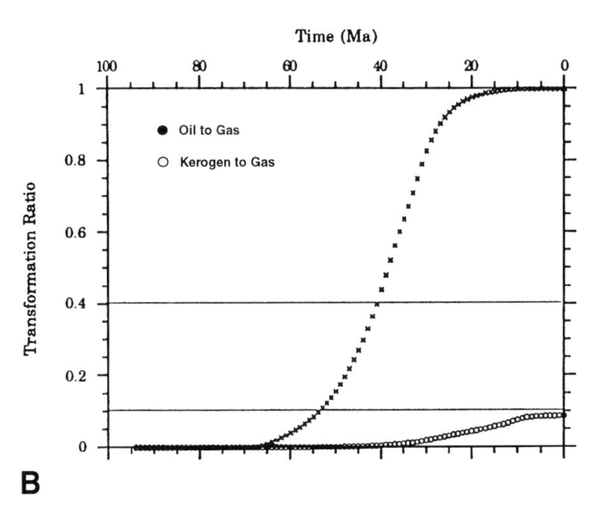

Figure 14. (A) Time-temperature model for the central portion of the Powder River basin. T_{F-W} is the Tertiary Fort Union and Wasatch formations; K_{2-2} is the Upper Cretaceous Mesaverde, Lewis, and Lance formations; K_{2-1} is the Upper Cretaceous Frontier and Niobrara formations and Steele Shale. (B) Computed transformation ratios for oil to gas and kerogen to gas in the central Powder River basin, using kinetic parameters proposed by Mackenzie and Quigley (1988).

multiphase fluid-flow system capable of supporting a basinwide overpressured compartment—the Cretaceous shale section in the Powder River basin.

The overpressured Cretaceous shale section in the Powder River basin bears a remarkable geometric resemblance to the ideal pressure compartment described by Powley (1982) and Hunt (1990). Consider that there is a normally pressured section (i.e., characterized by a hydrostatic gradient) that extends down to 8000 ft (2400 m) (probably close to 10,000 ft [3000 m] maximum burial depth). Below this normally pressured section there is an overpressured section that extends down another 3000 ft (900 m), eventually passing down into normally pressured rock at even greater depth. The overpressured volume of rock within the shale-dominated section wedges out at the margin of the basin where it is at shallower depth (within a depth range of 5000 to 6000 ft [1500–1800 m]). Also, the top of overpressuring in the Powder River basin typically begins at a present-day temperature of 90 to 100°C, very similar to the threshold temperature suggested by Hunt (1990).

It is concluded that the overpressured Cretaceous shale section in the Powder River basin is a very

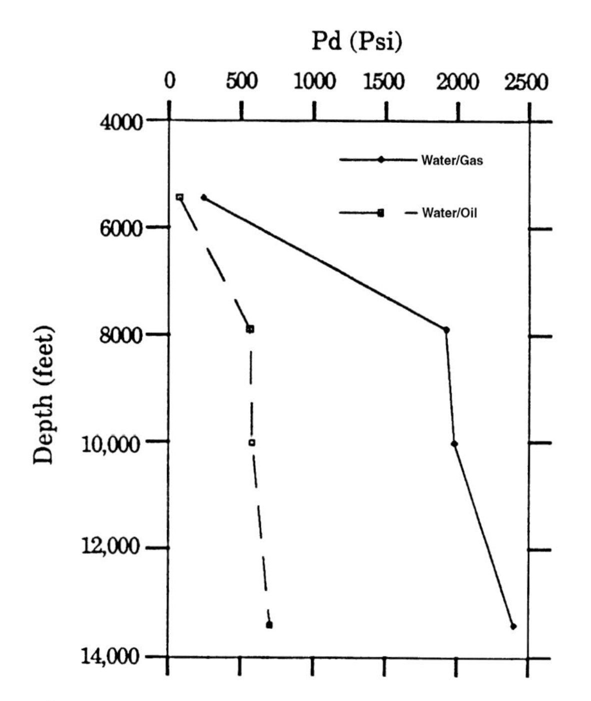

Figure 15. Trend in displacement pressure (Pd) with respect to two fluids vs. depth for Mowry Shale samples taken at a present-day depth interval of 5000 to 13,000 ft (1500–4000 m).

large dynamic pressure compartment. The driving mechanism is the generation of liquid hydrocarbons that convert the fluid-flow system to a multiphase regime in which capillarity controls the relative permeability. This effect is further emphasized as the residual oil in the shales reacts to gas, again decreasing the relative permeability, and increasing the displacement pressures in the shales by an additional order of magnitude. The top and bottom of the large basinwide overpressured volumes of rock within the Powder River basin are approximately parallel to stratigraphic boundaries, except at the basin margin.

Timing of Overpressuring

It is possible to approximate the timing of overpressuring in the Cretaceous shale section in the Powder River basin. The timing of the overpressuring begins with the generation of liquid hydrocarbons and is accentuated by the reaction of liquid hydrocarbon to gas. By comparing depths from Figures 12B and 14B, showing the depth at which these two reactions become important (when TR > 0.1), with a typical burial history diagram for the Powder River basin (Figure 14A), it is possible to place an approximate time on the beginning of overpressuring in the basin. In the Cretaceous shale section, overpressuring began at approximately 70 million years before present (TR of 0.1–0.4 for the kerogen-to-oil reaction) and was firmly developed at approximately 50 million years before present (TR > 0.4 for the oil-to-gas reaction). The reaction of liquid hydrocarbon to gas at 40 million years before present was a significant factor in the maturation history of Cretaceous shales in the Powder River basin (Figure 14). Thus, compartmentalized overpressuring in the Powder River basin is not an ephemeral geological phenomenon, but rather is a long-lived geological process intimately related to hydrocarbon generation and reaction within a rock sequence dominated by organic-rich shales.

PRESSURE REGIMES WITHIN CRETACEOUS SANDSTONES

The pressure regime of Cretaceous sandstones in the Powder River basin is significantly different from the pressure system characterizing the shales. The sandstones are not part of a large basinwide pressure compartment; rather, the fluid-flow and pressure systems of individual sandstones are separated spatially, both vertically and horizontally (Figures 16, 17, and 18) (Heasler et al., this volume). Even within specific sandstones (the Dakota, Muddy, or Frontier sandstones) the rocks are subdivided into relatively small, isolated compartments (Figures 16, 17, and 18) (Martinsen, this volume). The compartmentation is the result of the presence of stratigraphic elements within the individual sandstones, such as lowstand unconformities and paleosols, that are subsequently modified diagenetically (Martinsen, this volume; Jiao and Surdam, this volume). These internal stratigraphic elements are low-permeability rocks in a single-phase fluid-flow system with finite leak rates, which alter diagenetically into relatively impermeable capillary seals as the flow regime evolves into a multiphase fluid-flow system (see especially Berg, 1975; Iverson et al., this volume) by the addition of hydrocarbons as a result of continuous burial and increasing thermal exposure.

Figure 19 illustrates the detailed compartmentation within an individual reservoir facies (in this case the Muddy Sandstone along an east-west cross section from the Amos Draw to the Kitty to the Springen Ranch field). Along this section the Muddy Sandstone is part of a shoreface/valley-fill depositional system in which the younger, valley-fill elements are separated from the older, shore-face sandstones by a paleosol developed along a regionally prominent lowstand unconformity (Martinsen, this volume). As described by Jiao and Surdam (this volume), the paleosol horizons represent low-permeability horizons that in a multiphase fluid-flow system are relatively impermeable to fluid flow until a finite displacement pressure is exceeded. As a consequence, the Muddy Sandstone along the east-west cross section shown in Figure 19 in an oil/water or gas/oil dominated fluid-flow system is externally sealed above and below by the Mowry and Skull Creek shales and is internally separated into flow compartments by impermeable horizons within the sandstone (Figure 20). These flow and pressure relations are shown schematically in Figure 21. The network of sealing boundaries above, below, and within the sandstone determines the three-dimensional closure of seals in a multiphase fluid-flow system; in a single-phase (water) system, the low-permeability horizons have finite leak rates, and fluid compartmentation would be short-lived

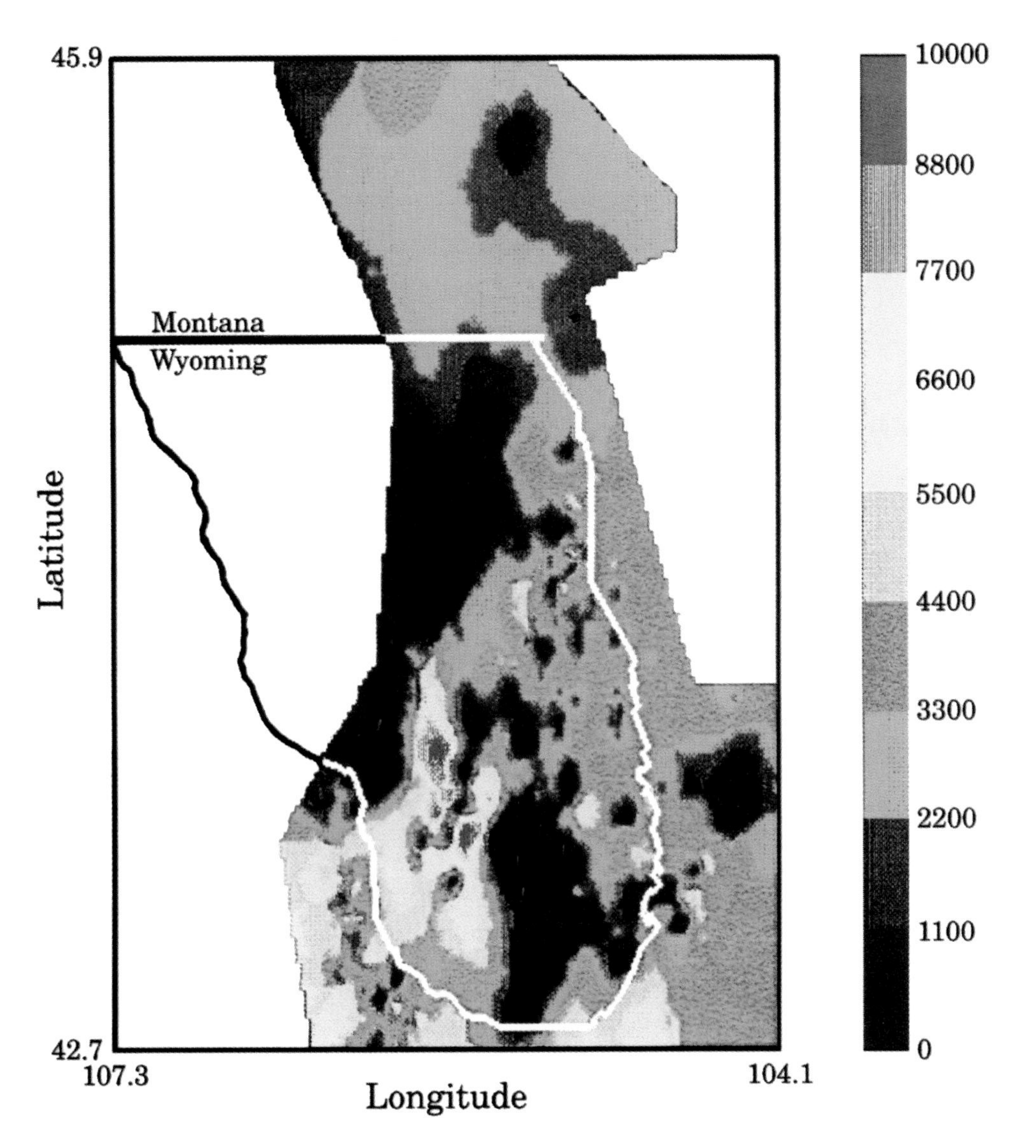

Figure 16. Representation of Dakota Sandstone pseudopotentiometric surface (calculated from DST data, 1157 data points) in the Powder River basin.

(Iverson et al., this volume). The result of three-dimensional closure of the seals is the formation of pressure/fluid compartments completely isolated from the shales above and below, and from the sandstone updip and downdip (Figure 19). It should be noted that along the cross section shown in Figure 19, the inclination of the rocks is 2° regionally, and there is no structural closure along the section.

The pressure anomalies shown in Figure 20 are similar to a multitude of anomalies found in the Cretaceous sandstones in the Powder River basin and described in Heasler et al. (this volume). In each case, the flow anomalies on a potentiometric surface map are characterized by steep sides and relatively flat tops. Heasler et al. (this volume) interpret these anomalies as flow, or pressure, compartments isolated from surrounding rocks by the three-dimensional closure of seals and characterized by insignificant fluid flow within the compartments. Typically in a gas/oil system, the displacement pressures calculated from high-pressure mercury injection test results match nicely with the pressure anomalies observed in the field in drill-stem tests. For example, the internal seals (here, paleosols) and the shales associated with the Muddy Sandstone in the Amos Draw field gave 2000 psi measured and calculated displacement pressures in a gas/oil system, and the gas condensate field is overpressured approximately 2000 psi (Iverson et al., this volume).

RELATIONSHIP BETWEEN THE SANDSTONE PRESSURE COMPARTMENTS AND THE REGIONAL OVERPRESSURING OF SHALES

As described, the overpressured compartments within the Cretaceous sandstone or reservoir facies are typically relatively small (longest dimension 1 to 10 mi [1.6–16 km]), whereas the surrounding shale is

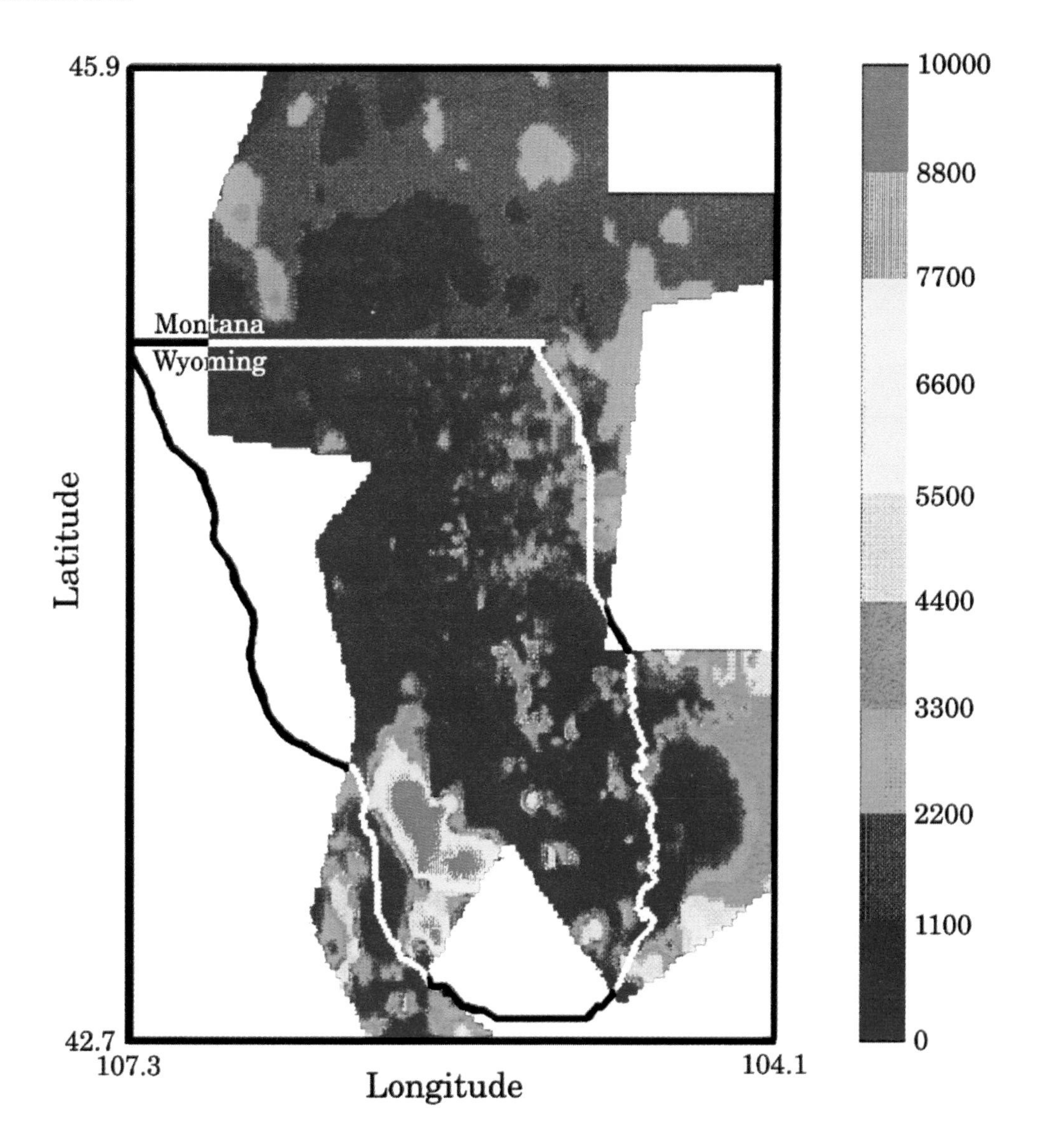

Figure 17. Representation of Muddy Sandstone pseudopotentiometric surface (calculated from DST data, 3895 data points) in the Powder River basin.

overpressured on a basinwide scale. What is the relationship between the sandstone and shale fluid-flow/pressure regimes?

The pressure compartments within the sandstones formed as a result of their fluid-flow system evolving from a single-phase (water) to a multiphase system (oil/water or gas/oil). The source of the hydrocarbons in the sandstones is the surrounding organic-rich shales. Thus, during the initial stages of overpressuring in the shales (primary migration of hydrocarbons into microfractures within the shale) the two flow systems must have been closely related. This close relationship must have persisted until pressure buildup in the shales resulted in hydrocarbon expulsion and migration into sandstone conduits and traps. The expulsion and migration of hydrocarbons from the shales may have occurred once or many times, depending on the nature of the source rocks and how often the displacement pressure in the shales was exceeded by internal pressure buildup.

Once the hydrocarbons migrated updip into stratigraphic/diagenetic traps within sandstones, the fluid-flow system began to evolve into a multiphase system, and low-permeability zones became seals (Figures 21A and 21B). As long as water was a significant component of the fluid phase, the traps remained under water drive, even though the seals in contact with liquid hydrocarbon accumulations were dominated by capillarity (Figure 21C). As the hydrocarbons filled the trap, it is speculated that free water was expelled from the bottom of the trap where the low-permeability horizons were still in contact with a single-phase flow system with finite leak rates (Figure 21D). With additional burial and increased thermal exposure, liquid hydrocarbons within the traps began to react to "pyrobitumen" and gas. At this time in the burial history, the trap was completely surrounded by capillary seals (three-dimen-

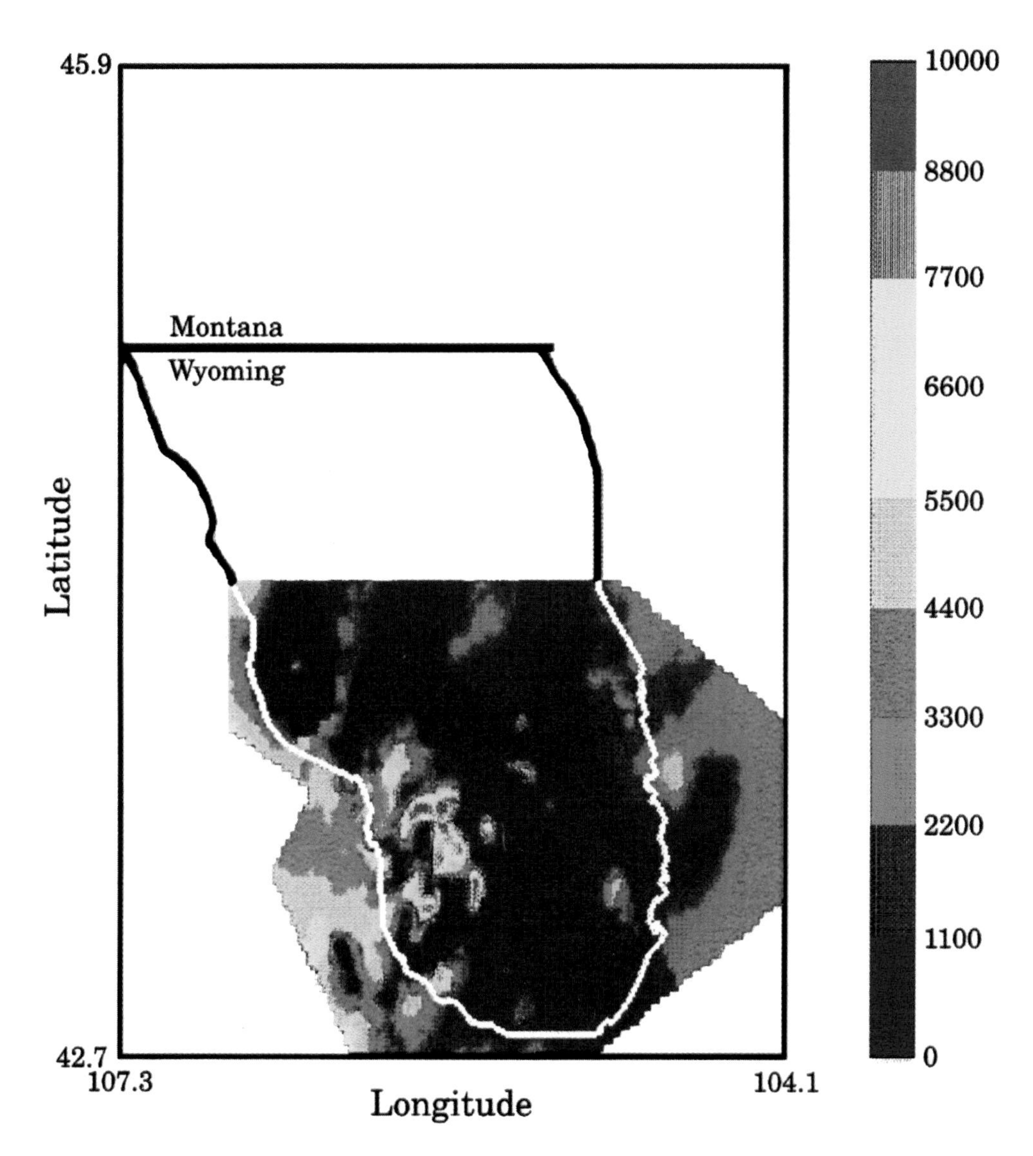

Figure 18. Representation of Frontier Sandstone pseudopotentiometric surface (calculated from DST data, 822 data points) in the Powder River basin.

sional closure), and a pressure compartment was formed (Figure 21D). The pressure could then build up within the compartment until the displacement pressure of the seals was exceeded. Figure 21E is a schematic reconstruction of the fluid-flow system within the Muddy Sandstone (or other sandstone characterized by internal low-permeability stratigraphic/diagenetic horizons) along an east-west cross section on the eastern margin of the Powder River basin (compare Figure 21E with Figure 19).

Once a compartment is formed as a result of three-dimensional closure of the capillary seals, the internal fluid and pressure field is isolated from all adjacent rocks. The pressure/fluid compartment can be destroyed only if one of the following events occurs: (1) pressure within the compartment builds until it exceeds the displacement pressure of the seals (perhaps with additional burial leading to further oil-to-gas reactions); (2) the integrity of the compartment is breached by a fault (the fracture strength of the seal is exceeded, perhaps by pressure buildup); (3) the compartment is uplifted, so that the pressure difference between the inside and the outside of the compartment exceeds the displacement pressure of the seal (the internal pressure remains constant, whereas the external pressure decreases with uplift and erosion); or (4) the compartment undergoes further and significant burial after compartment formation, so that the fixed internal pressure approaches the external pressure (the pressure compartment would be indistinguishable from the surrounding rock because there would be no pressure difference). Unless one of these four events happens, pressure compartments like those in the Powder River basin should last indefinitely.

Three points are emphasized: (1) The hydrocarbon accumulations associated with the type of pressure compartments characterizing the Cretaceous sandstones of the Powder River basin cannot be under water drive;

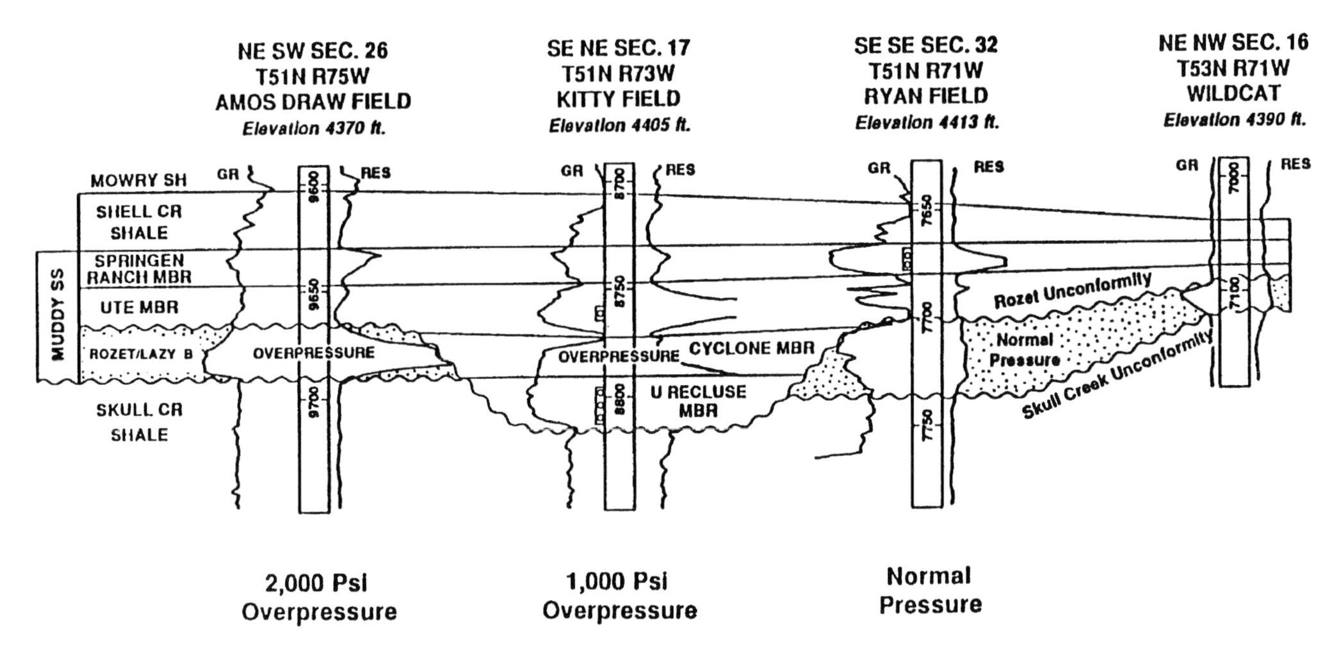

Figure 19. A west-east cross section of the Lower Cretaceous stratigraphic sequence from the Amos Draw to the Kitty to the Springen Ranch (Ryan) field (along Line 1 in Figure 1), then north two townships to a wildcat. The Muddy Sandstone reservoirs along this cross section are in oil/water- or gas/oil-dominated fluid-flow systems and are sealed above and below by the Mowry and Skull Creek shales. The Muddy Sandstone is internally separated into flow compartments by an impermeable barrier, the Rozet unconformity, within the sandstone. The Rozet and Skull Creek unconformities are shown by wavy lines on the cross section.

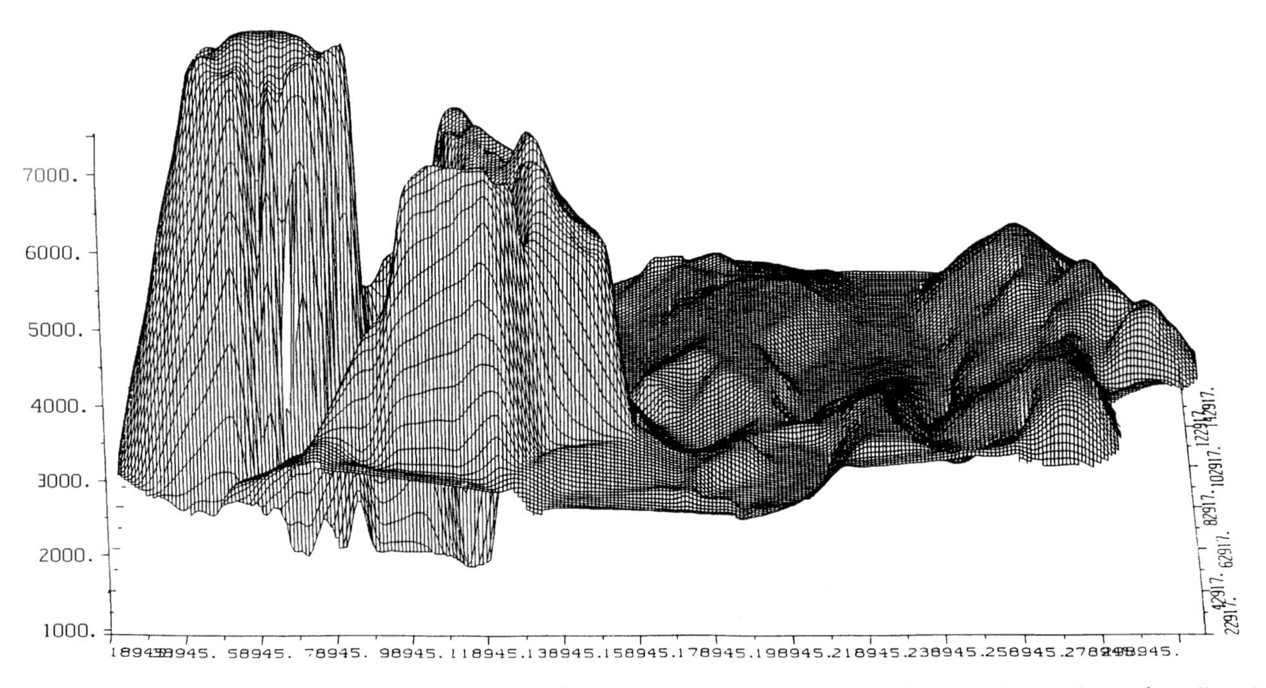

Figure 20. Three-dimensional representation of the Muddy Sandstone pseudopotentiometric surface (head vs. depth) in the vicinity of the Amos Draw and Kitty fields. These two adjacent pressure compartments are separated by the Rozet unconformity flow barrier within the Muddy Sandstone. From Moncur (1993), her figure 34.

they must be under gas-depletion drive. (2) Hydrocarbon generation, expulsion, and migration are absolutely essential to the formation of the type of pressure compartments found in the Powder River basin; this type of pressure compartmentation will always be intimately related to hydrocarbon accumulations. (3) Most important, the pressure/fluid compartmentation found in these sandstones begins with an injection of hydrocarbons from associated shales and evolves as the result of the fluid-flow system undergoing a transition

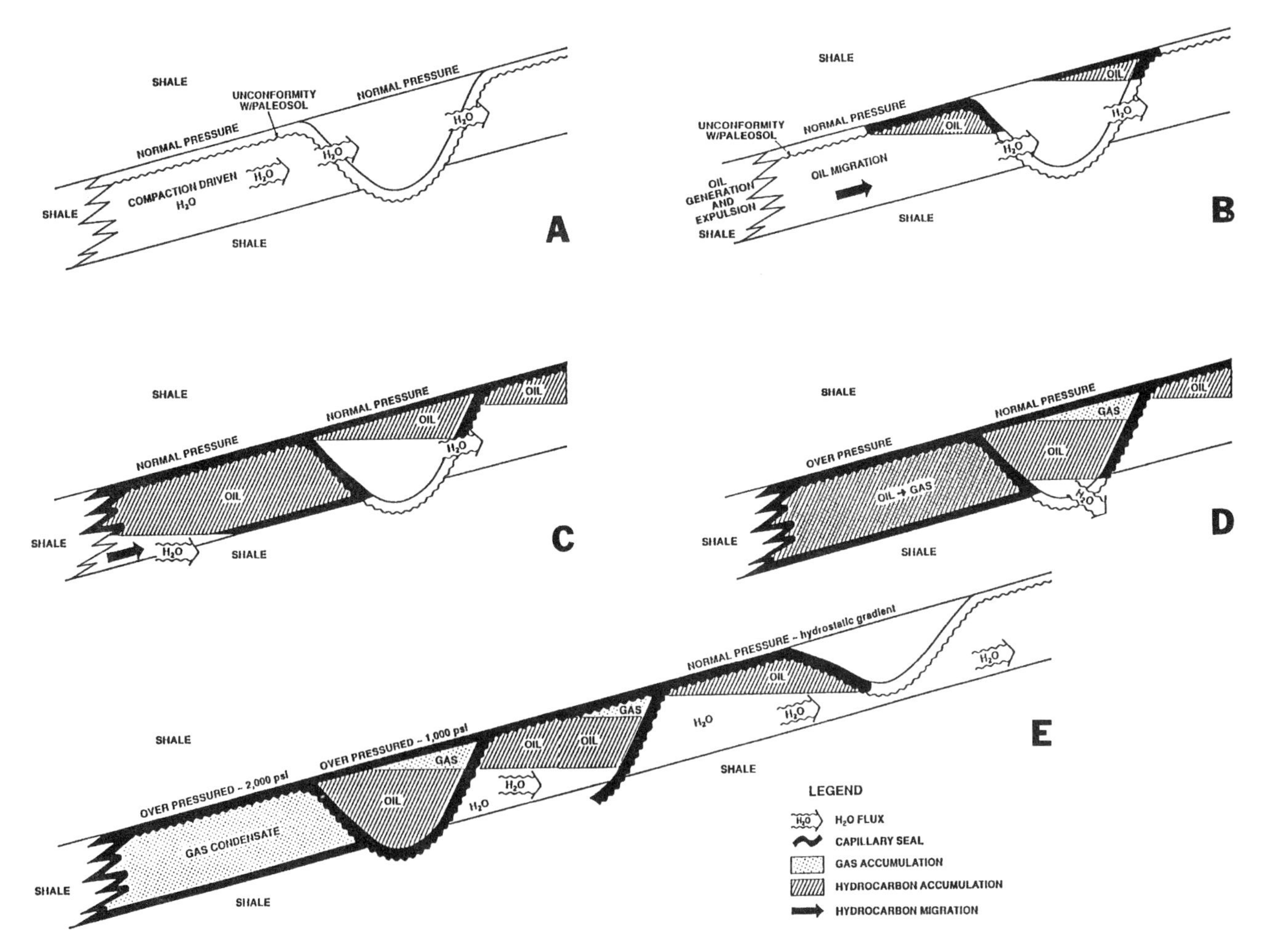

Figure 21. Schematic diagrams showing the evolution of the pressure compartmentation in the Lower Cretaceous sandstones characterized by an internal unconformity in the Powder River basin. (A) The pressure barrier (an unconformity in this case) is low-permeability rock with finite leak rates in a single-phase fluid-flow system. (B) Once the hydrocarbons migrate updip into stratigraphic/diagenetic traps within sandstones, the flow system begins to evolve into a multiphase flow system, and low-permeability zones become pressure seals. (C) As long as water is a dominant fluid component, the traps will remain under water drive. (D) With additional hydrocarbon migration and trapping with "pyrobitumen" or gas generation, the trap is no longer under water drive and is completely surrounded by capillary seals. (E) Updip reconstruction for the eastern flank of the Powder River basin. Compare Figure 21E with Figure 19.

from single-phase to multiphase flow; but once formed, a sandstone compartment is isolated from the rest of the rock system. A significant implication, particularly with respect to drilling procedure, is that the pressure compartments within a sandstone need not be at the same pressure as the surrounding overpressured shales. Supporting this last point are the depth-versus-pressure relationships noted for the Dakota, Muddy, and Frontier sandstones of the Powder River basin (Figure 22). These data show that the Cretaceous sandstones at present-day depths greater than 8000 to 9000 ft (2400–2700 m) can be overpressured, normally pressured, and underpressured, even though the shales at equivalent depths are overpressured.

Figure 23 is a highly schematic east-to-west representation of the pressure compartment system within the Cretaceous section of the Powder River basin. The normally pressured system is dominated by single-phase fluid flow, the overpressured section by multiphase fluid flow. Although overpressuring in the shales varies with depth, overpressuring in the sandstones is a function both of depth (relative to the overpressured shale section) and reservoir geometry.

Presumably, once formed, the isolated compartments within the reservoir facies are less dynamic than the more basinwide overpressured compartment characterizing the shales. At least in the Powder River basin, the shale pressure system is highly dynamic and continuously evolving; for with additional burial, the kerogen-to-gas reaction will significantly affect the pressure regime within the shales but will have no effect on the sandstone compartments, providing that the displacement pressures of the seals are not exceeded.

Another possibility that should be considered for some of the normally pressured sandstones within the

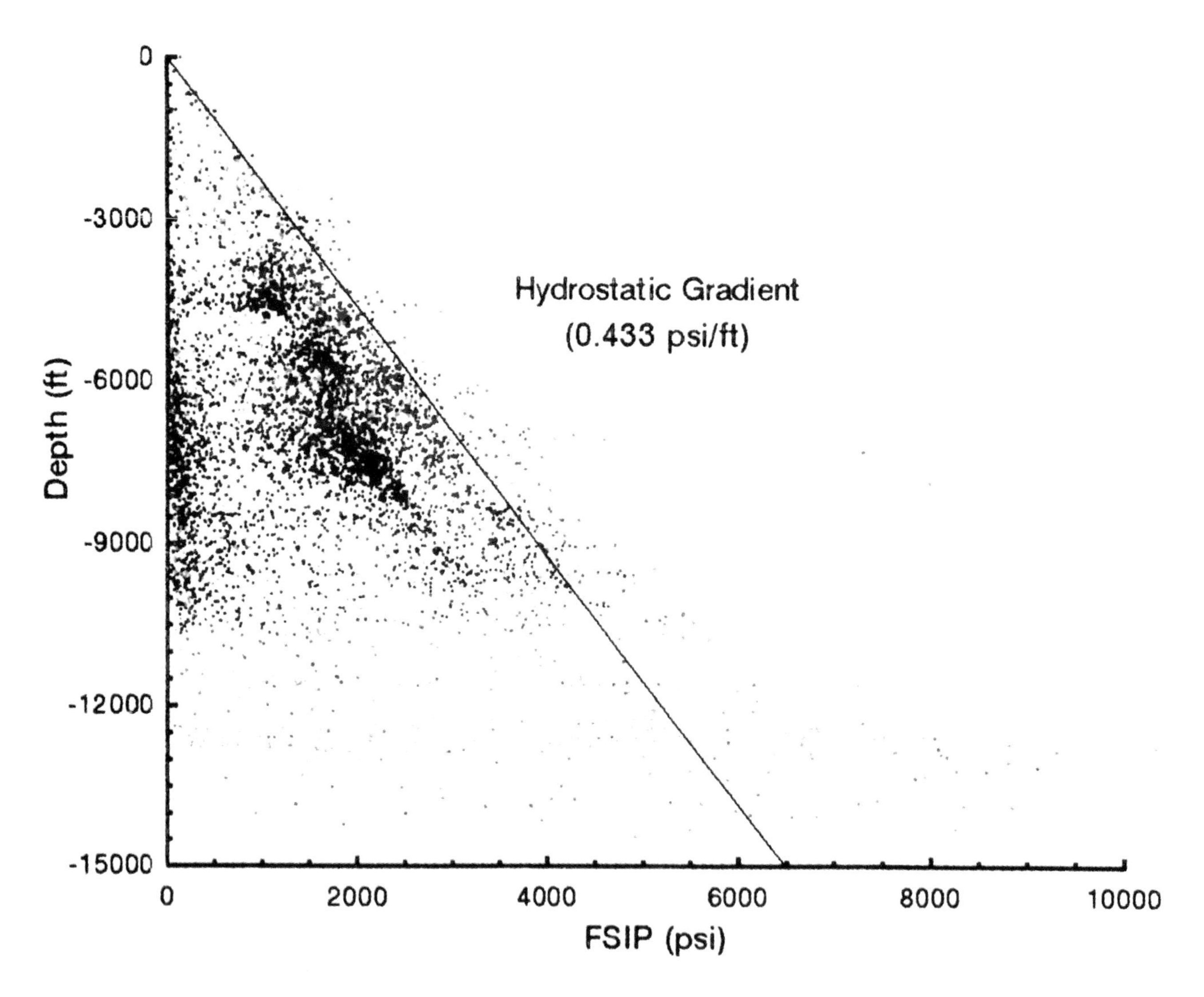

Figure 22. Plot of the DST pressures (final shut-in pressures, FSIP) of the Dakota, Muddy, and Frontier sandstones vs. depth in the Powder River basin. A Cretaceous sandstone at a present-day depth greater than 9000 ft (2700 m) can be overpressured, normally pressured, or underpressured. Blue = 3888 Muddy Formation data values; green = 821 Frontier Formation data values; red = 1157 Dakota Formation data values.

overpressured shale section is that they represent fluid conduits out of the overpressured shale section. If the sandstones were not characterized by three-dimensional closure of the seals, the sandstones would be fluid conduits connecting the shale section to more normally pressured regimes: The sandstones would be normally pressured, or slightly overpressured, whereas the surrounding shales could be significantly overpressured.

THE PRESSURE COMPARTMENT HYPOTHESIS

The pressure compartment that characterizes the Cretaceous shales of the Powder River basin neatly fits the criteria presented by both Powley (1982) and Hunt (1990) for pressure compartmentation. However, the formation of the bounding seals in the Powder River basin is primarily caused by the transition from a single-phase to a multiphase fluid-flow system (with hydrocarbon generation and primary migration). These low-permeability shales become impermeable unless displacement pressures are exceeded. Prior to and concurrent with this transition, clay diagenesis has resulted in reducing absolute permeabilities and thereby significantly enhancing the holding capacity of the seals (Sneider et al., 1991). As MacGowan et al. (this volume) have shown, the top boundary of the overpressured compartment in the shales is at a temperature (~90 to 100° C) where a drop in P_{CO_2}, as would happen if the integrity of the top seal were breached, would result in the precipitation of carbonate. This mechanism might heal any fracture caused by pressure buildup that exceeds rock strength, or by a seal broken by faulting.

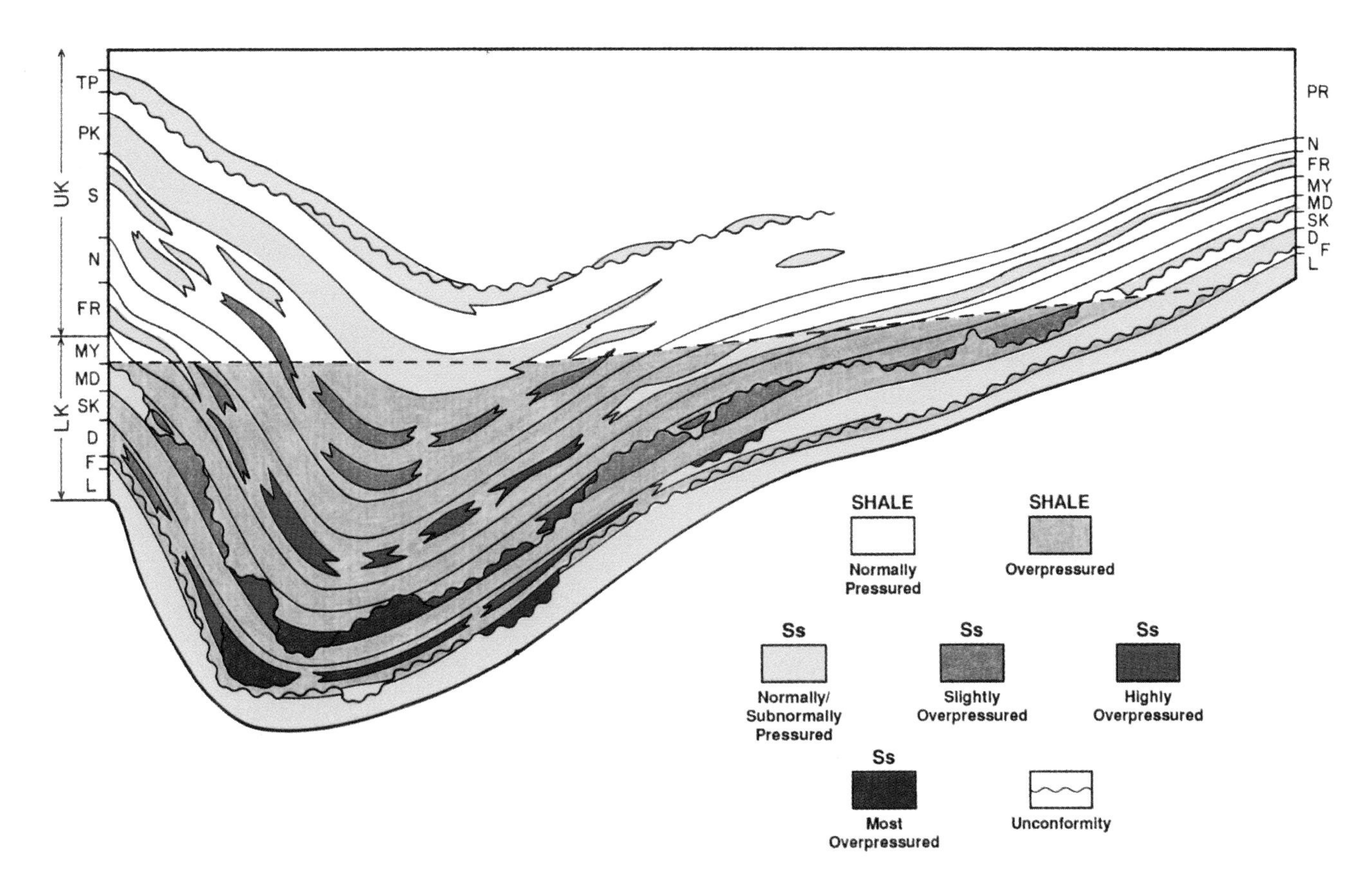

Figure 23. Diagrammatic east-west cross section of the Cretaceous system in the Powder River basin, Wyoming, indicating formations present, gross lithologies, and variations in relative pressure. Not to scale. TP = Teapot Sandstone; PK = Parkman Sandstone; S = Steele Shale; N = Niobrara Formation; FR = Frontier Formation; MY = Mowry Shale; MD = Muddy Sandstone; SK = Skull Creek Shale; D = Dakota Sandstone; F = Fuson Shale; L = Lakota Conglomerate. Note how the top of the overpressuring in the shales is nearly horizontal in the basin center, dips slightly up on the east, and laterally terminates where it intersects the Fuson Shale. Various pressure regimes are present within sandstone reservoirs; note that (1) normally pressured sandstones (which are probably in connection with the surface hydrologic system) extend down into the overpressured shale section, (2) isolated sandstones located entirely outside of the overpressured shale section have normal pressures, (3) isolated sandstones can extend out of the overpressured shale section, yet have slightly elevated pressures, (4) isolated sandstone reservoirs in the basal portions of the overpressured shales display the highest overpressures, (5) because of different geometries, sandstones at approximately the same depth can have different pressure regimes, and (6) because of unconformities, variously pressured reservoir compartments can have nearly vertical boundaries.

The pressure compartments within the Cretaceous sandstones typically are a significantly smaller-scale pressure phenomenon than originally described by Powley (1982) and Hunt (1990), but still fit their definition of a pressure compartment. The major difference between the pressure compartments described herein and those described by Powley (1982) and Hunt (1990) is that the seals surrounding pressure compartments in the Powder River basin are not impermeable seals in a single-phase system but are relatively impermeable seals that result from capillarity in a multiphase fluid-flow system. The pressure seals in the Powder River basin form as a result of the addition of significant quantities of hydrocarbons into the fluid phase, not as a result of a diagenetic mechanism that forms an "impermeable" horizon.

CONCLUSIONS

The differences between the conventional hydrodynamic theory of overpressuring (disequilibrium compaction) and the ideas presented herein concerning three-dimensional compartmentation of overpressured rock are basic and have profound gas exploration/production implications. Conventional hydrologic theory suggests that overpressuring is a function of disequilibrium compaction driven by the imbalance between the rates of sedimentation and fluid flux, or escape. Hydrologic overpressuring is based on single-phase fluid flow; and because all rocks have measurable permeabilities, this type of demonstrated overpressuring is geologically short-lived, tends to occur relatively early in the burial history, and is independent of hydrocarbon generation. It is noteworthy that Darcy fluid-flow theory (single-phase flow) unequivocally demonstrates that significant pressure differences across lithologic barriers, even at extremely low permeabilities like 1 microdarcy, will equilibrate in very short time periods (typically less than 1 million years; Iverson et al., this volume).

In marked contrast, the pressure compartments documented and characterized in the Powder River basin are the result of multiphase fluid flow, coupled with three-dimensional closure of capillary seals. Again, pressure compartmentation in the Powder River basin is genetically related to hydrocarbon generation, migration, and reaction (kerogen to liquid hydrocarbon, and liquid hydrocarbon to "pyrobitumen" + gas). The hydrocarbons are of particular importance, as they are essential to pressure compartment evolution in the Powder River basin and probably elsewhere. First, the appearance of hydrocarbons in the system drives the transition from single-phase (water) to multiphase fluid flow (water plus one or more hydrocarbon phases): The hydrocarbons activate the capillary seals. Second, when hydrocarbons saturate the compartment they ensure the three-dimensional integrity of the bounding capillary seal. In addition, hydrocarbon saturation of the compartment provides a mechanism for expelling free water from the compartment system.

Clearly, hydrologic overpressuring and compartmentation of anomalous pressure regimes are real but fundamentally different processes. The two processes have very different implications for the exploitation of gas resources. Hydrologic single-phase overpressuring is geologically ephemeral; it develops relatively early in the burial history; it can be independent of hydrocarbon generation; and it typically is restricted to specific depth intervals (shallower than compartments related to hydrocarbon generation). In contrast, multiphase pressure compartmentation is long lived; it develops midway to late in the burial history; it is intimately related to hydrocarbon generation; and it can be found over a very wide depth interval. Most important, gas resources associated with multiphase fluid-flow compartments can occur in rocks of any age; they will not spill during structural reorientation (except with rupture by fracturing); and their integrity can only be destroyed by significant uplift when the pressure differential exceeds the displacement pressure, or by deep burial when the pressure differential is eliminated.

At present in the Powder River basin, there is an overpressured compartment characterized by a hydraulic head of 14,000 ft (4200 m) (the ground elevation is approximately 6000 ft [1800 m] and the depth of production is 12,000 ft [3600 m]). The displacement pressure for this reservoir is approximately 4000 psi, and this multiphase fluid-flow pressure compartment has the potential to remain intact as a recognizable pressure anomaly down to a depth of approximately 20,000 ft (6000 m), where no differential pressure will exist. As a consequence, the potential gas production fairway in the Powder River basin, and in similar basins, should be significantly expanded.

Drilling strategies in overpressured regimes have to be carefully considered, for the real possibility exists that some sandstone targets within overpressured shale sequences may be at pressures lower than the pressures in the surrounding shale. It is possible that sandstones within some overpressured shale sequences, such as in the Powder River basin, have been irreversibly damaged by drilling with overcompensated muds.

In summary, the concept and understanding of multiphase fluid flow as it relates to three-dimensional pressure compartmentation will greatly expedite the search for, the discovery of, and the exploitation of new unconventional gas resources. One of the most exciting aspects of this work is that the type of density contrast represented by the pressure compartments and shown in Figures 10, 11, and 12 should be detectable with modern seismic techniques and new pattern recognition technology. Because these pressure anomalies are the direct result of gas accumulation, their detection will result in the discovery of new hydrocarbon resources.

ACKNOWLEDGMENTS

This material covers research performed through November 1991. The original manuscript was reviewed by David Copeland (University of Wyoming), who made numerous helpful suggestions. This study was funded by the Gas Research Institute under contract number 5089-260-1894. RCS wishes to acknowledge very useful discussions with Bob Siebert of Conoco and Alex Marshall of VICO. We also acknowledge the significant contributions made to this chapter by Henry Heasler, William Iverson, Donald MacGowan, Rebecca Moncur, Vladimir Serebryakov, Scott Smithson, and Yue Wang.

REFERENCES CITED

Berg, R. R., 1975, Capillary pressure in stratigraphic traps: American Association of Petroleum Geologists Bulletin, v. 59, p. 939–956.

Bradley, J. S., 1975, Abnormal formation pressure: American Association of Petroleum Geologists Bulletin, v. 59, p. 957–973.

Dobrynin, V., and V. Serebryakov, 1989, Geological-Geophysical Methods: Prediction of Pressure Anomalies: Moscow, 288 p.

Hunt, J. M., 1979, Petroleum Geochemistry and Geology: San Francisco, W.H. Freeman, 617 p.

Hunt, J. M., 1990, Generation and migration of petroleum from abnormally pressured fluid compartments: American Association of Petroleum Geologists Bulletin, v. 74, p. 1–12.

Mackenzie, A. S., and T. M. Quigley, 1988, Principles of geochemical prospect appraisal: American Association of Petroleum Geologists Bulletin, v. 72, p. 399–415.

Magara, K., 1976, Thickness of removed sedimentary rocks, paleopressure, and paleotemperature, southwestern part of Western Canada Basin: American

Association of Petroleum Geologists Bulletin, v. 59, p. 292–302.

Moncur, R. S., 1993, Diagenesis and hydrodynamics of the Lower Cretaceous Muddy Sandstone in the vicinity of Amos Draw, Elk Draw, and Kitty fields, Powder River basin: Ph.D. dissertation, University of Wyoming, 242 p.

Powley, D. E., 1982, Pressures, normal and abnormal: American Association of Petroleum Geologists Advanced Exploration Schools, unpublished lecture notes, 38 p.

Schowalter, T. T., 1979, Mechanics of secondary hydrocarbon migration and entrapment: American Association of Petroleum Geologists Bulletin, v. 63, p. 723–776.

Sneider, R. M., K. Stolper, and J. S. Sneider, 1991, Petrophysical properties of seals: American Association of Petroleum Geologists Bulletin, v. 75, p. 673–674.

Tissot, B. P., and D. H. Welte, 1984, Petroleum Formation and Occurrence: 2nd edition, Berlin, Springer-Verlag, 699 p.

Chapter 16

Pressure Compartments in the Powder River Basin, Wyoming and Montana, as Determined from Drill-Stem Test Data

H. P. Heasler
Ronald C. Surdam
University of Wyoming
Laramie, Wyoming, U.S.A.

J. H. George
Embry-Riddle Aeronautical University
Daytona, Florida, U.S.A.

ABSTRACT

Drill-stem test (DST) pressures from oil and gas wells were analyzed in an attempt to determine the existence of pressure compartments in the Powder River basin. DST data for the entire basin were first sorted by geologic unit for the Mesaverde Formation (984 data values), Sussex Formation (1041 data values), Frontier Formation (821 data values), Muddy Formation (3888 data values), Dakota Formation (1157 data values), and Minnelusa Formation (4470 data values). Initial and final shut-in pressures (ISIP and FSIP) were graphed versus each other and versus depth and elevation to display functional relationships. Potentiometric surfaces were then constructed using the maximum of the ISIP and FSIP.

The pressure-elevation plots and potentiometric surfaces clearly show the existence of anomalously pressured zones in the Frontier, Muddy, and Dakota formations. The anomalously pressured zones as determined from the potentiometric surfaces are discrete areas on the scale of individual oil fields. The boundaries of the anomalously pressured areas as shown on the potentiometric surfaces are characterized by steep hydraulic head gradients of up to 12,000 ft (3600 m) of head difference across small horizontal distances of less than 1 mile. These gradients are interpreted as discontinuities in the fluid-flow regime of the Powder River basin. The internal shape of the anomalies is difficult to determine because data are sparse. However, piecewise continuous least-squares analyses indicate that many of the anomalies contain a nearly horizontal internal potentiometric surface.

Given the discontinuous nature of the constructed potentiometric surfaces and the shape of the pressure anomalies, we conclude that oil-field-size pressure compartments exist in the Powder River basin in the Frontier, Muddy, and Dakota formations.

INTRODUCTION

Conventionally, sequences of sedimentary rocks characterized by abnormal pressure gradients have been explained using hydrodynamics and the presence of an aquitard in the stratigraphic section (Hubbert and Ruby, 1959). (In this chapter, we use the pressure nomenclature of Hunt, 1990. The pressure-depth gradient of a free-standing column of water is assumed to be 0.433 psi/ft [9.79 kPa/m]; and of a saturated salt solution, 0.513 psi/ft [11.9 kPa/m]. The gradient is assumed to extend to the topographic surface. Pressures outside these limits are considered abnormal. The terms *overpressured* and *underpressured* refer to values above and below these limits, respectively.) Abnormally high pressures may originate down flow from the aquitard as a consequence of fluid arriving at the flow barrier at a rate higher than the rate of fluid exiting the barrier. Thus, the development of abnormally high pressures in the stratigraphic section may result from the presence of a low-permeability rock unit in the fluid-flow pathway. Overpressuring behind an aquitard can be significantly enhanced by a variety of geological processes, such as dehydration mineral reactions (smectite conversion to illite); hydrocarbon generation reactions, including the generation of CH_4 or CO_2; thermal expansion of water; and rapid subsidence resulting in undercompaction (low-permeability sediments that expel fluids at a rate lower than necessary for the sediment to normally compact). Overpressuring resulting from an aquitard is a highly dynamic process and typically a short-lived geological phenomenon. Unless the flow barrier is impermeable (a seal, not an aquitard), fluid will flow from regions of high hydraulic head to low hydraulic head at some finite rate, thereby dissipating the initial abnormal pressure. For example, Bethke et al. (1988) have suggested that the deep Gulf Coast stratigraphic section, long considered an outstanding example of overpressured rocks resulting from compaction disequilibrium, became abnormally high pressured only in the last 2 million years.

An alternate hypothesis for explaining abnormally pressured sedimentary rocks has been proposed by David E. Powley and John Bradley (1975) of Amoco. Their work has been summarized and the new hypothesis outlined by Hunt (1990). This hypothesis maintains that sequences of sedimentary rocks characterized by abnormal pressures below depths equivalent to 90° to 100°C commonly are pressure compartments bounded on all sides by seals that hydraulically isolate the abnormally pressured rock prisms from adjacent rocks. Bradley argues that significant pressure differences across the compartmental boundaries cannot be maintained for geologically significant time periods unless the abnormally pressured system is completely encased by seals (hydraulically isolated). Hunt suggests that as many as half of the deep basins in the world are characterized by pressure compartmentation. These sedimentary basins show a layered arrangement of two or more superimposed fluid-flow systems. The deeper portions of the basins are characterized by pressure compartmentation, whereas the upper portions of the basins are characterized by normal pressure and normal hydrologic regimes. In Hunt's pressure compartmentation model, the two hydrologic regimes are separated by an impermeable seal. In contrast to the hydrodynamic development of abnormal pressure resulting from an aquitard, once a pressure compartment is in place it is relatively permanent (static) and can be destroyed only when one of the bounding seals is breached.

Thus, there are two distinct hypotheses to explain pressure compartments in sedimentary rock systems. The first hypothesis uses hydrodynamics to explain abnormal pressure, requiring a fluid source that allows fluid to arrive at a low-permeability barrier at a rate faster than the barrier transfers the fluid. The fluid source may be an active groundwater recharge zone, fluid expelled from compacting sediments (Bethke et al., 1988), fluids created by hydrocarbon generation or other chemical reactions, or perhaps water liberated in the conversion of clay minerals. The alternate hypothesis, as proposed by Hunt (1990), is a larger-scale, more static model for pressure compartments. In Hunt's pressure compartmentation model, the relatively shallow hydrodynamic regime is separated from a deeper, isolated regime by impermeable seals that exist for geologically significant time periods, many millions of years.

It is apparent that the two hypotheses are not mutually exclusive, for the development of both types of abnormally pressured rocks could occur in the same basin. However, the two hypotheses vary in terms of the dynamic and temporal characteristics of the abnormally pressured rocks, particularly when viewed within the context of basin evolution. These variations are manifested in pronounced differences in the reconstruction of fluid-flow histories in sedimentary basins. One model is highly dynamic, requiring a fluid source that continues for the life of the pressure compartment, and also requiring the presence of relatively impermeable zones. The other model is static, allowing existence for a long time but requiring the presence of seals relative to the fluid contained within a compartment.

ESSENTIAL PROBLEM

An important question is: How can the Powley-Bradley-Hunt pressure compartment hypothesis be critically evaluated? At present, this question has not been resolved, for proponents of both hypotheses view the same pressure-depth data as supporting evidence. We suggest that it is impossible to distinguish with certainty between the two hypotheses using existing two-dimensional data sets (pressure-depth plots). By definition, pressure compartments are three-dimensional geological phenomena. Thus, the essential problem is to design a three-dimensional test to differentiate between the two hypotheses. This can be accomplished by mapping the hydraulic head of lithologic units and creating potentiometric surfaces. A potentiometric surface is an imaginary surface, the topography of which reflects the fluid potential of the formation water from place to place within a subsurface reservoir in terms of the elevation to which a column of water originating in the formation would rise above a reference datum within a vertical tube (Dahlberg, 1982). The creation of such a surface allows the three-dimensional comparison of areas of low potential energy with those of high potential energy. Hydraulic head h is calculated as

$$h = \frac{p}{\rho g} + z, \quad (1)$$

where $p/\rho g$ is the pressure potential (p is pressure, ρ is the fluid density, and g is the gravitational acceleration constant), and z is the elevation potential (the elevation of the pressure measurement above some datum).

Thus, hydraulic head represents the total potential energy of fluids contained within a reservoir. For this paper, the pressure potential is standardized to the density of pure water. By using hydraulic head, zones of unusually high potential energy may be defined. Unusually high potential energy is defined as head values greater than the highest groundwater recharge point for the Powder River basin. Thus, any head value greater than 7000 to 8000 ft (2100–2400 m) is considered to be anomalously high. When displayed in map view, such zones of high potential energy indicate possible areas of pressure compartmentation: such high-energy zones must be caused by excess fluid production, the presence of seals relative to the enclosed fluid, or both.

Definitions and Data Sets

Before implementing a test strategy, it is important to clearly define a pressure compartment. In this study, the working definition is: *A pressure compartment is a volume of permeable rock completely surrounded by seals that are impermeable to the contained fluid. Pressure compartments are characterized by a constant hydraulic head, so that the fluids within the compartment exhibit an interior static fluid pressure gradient.* From this definition it is apparent that the analysis of pressure data will be critical in testing the pressure compartment hypothesis.

In most basins, the most readily available source of pressure data is drill-stem tests (DSTs). A DST is a transient pressure test used to estimate reservoir properties prior to well completion (Bair et al., 1985). Most tests consist of an initial flow period and an initial shut-in period followed by a final flow period and a final shut-in period. According to Bair et al. (1985), a complete test should consist of the following:

1. An initial flow period, usually of 5 to 10 minutes, to restore the static reservoir pressure of formation fluids that have been altered by the invasion of drilling muds into the formation.
2. An initial shut-in period, of 30 to 60 minutes, to attain stabilized formation pressure.
3. A second, final flow period, of from 30 minutes to 2 hours, to stabilize flow.
4. A final shut-in period, usually longer than or equal to the second flow period, to retain stabilized formation pressure.

Complete DST records that include time-versus-pressure data can be used to calculate an extrapolated formation pressure and permeability with Horner (1951) diagrams and an analysis similar to the Theis (1935) recovery method (Bredehoeft, 1965; Miller, 1976). However, the database purchased for the present study from Petroleum Information in Denver contained, at most, the interval tested, initial shut-in pressure, initial shut-in time, final shut-in pressure, final shut-in time, and fluid recovery data. This precluded the use of a Horner-type analysis.

The DST pressure database must be converted to hydraulic head and viewed spatially if the Hunt (1990) pressure compartment hypothesis is to be critically evaluated. For example, pressure-depth plots have been used in the past to identify parts of the stratigraphic column characterized by abnormal pressures (for examples, see Hunt, 1990). As discussed above, potentiometric surfaces are essential in determining the three-dimensional aspects of pressure anomalies. Cross sections through those parts of the potentiometric domains containing abnormally pressured rocks are important in documenting the geometry of pressure anomalies and in determining whether or not the anomalies are characterized by constant hydraulic head. A large amount of areally distributed pressure data is essential to document the existence of pressure compartments in a sedimentary basin.

POWDER RIVER BASIN

The Powder River basin of Wyoming and Montana was chosen as our study area for four reasons: (1) it is characterized by a wide diversity of geological settings; (2) it is a highly productive hydrocarbon province; (3) there is a large amount of subsurface data available for the basin (approximately 35,000 wells and 15,000 DSTs); and (4) the Powder River basin experienced uplift and erosion over the last 20 to 10 million years (Love, 1970; McKenna and Love, 1972; Flanagan, 1990). Item 4 is important because the Powder River basin should not be experiencing compaction disequilibrium, since the erosional event has lasted 10 to 20 million years. Also, the current generation of hydrocarbons should not be significant, since it would be precluded by the erosional event and subsequent cooling. Thus, if pressure compartments are found in the Powder River basin, they will have been caused by seals, not by currently ongoing hydrodynamic mechanisms such as hydrocarbon generation or compaction disequilibrium.

Structural Setting

The 25,000 mi^2 (64,700 km^2) Powder River basin of Wyoming and southern Montana (Figure 1) is a foreland structural basin characterized internally by a mildly deformed sedimentary section. The Paleozoic and Mesozoic section exceeds 20,000 ft (6100 m) in

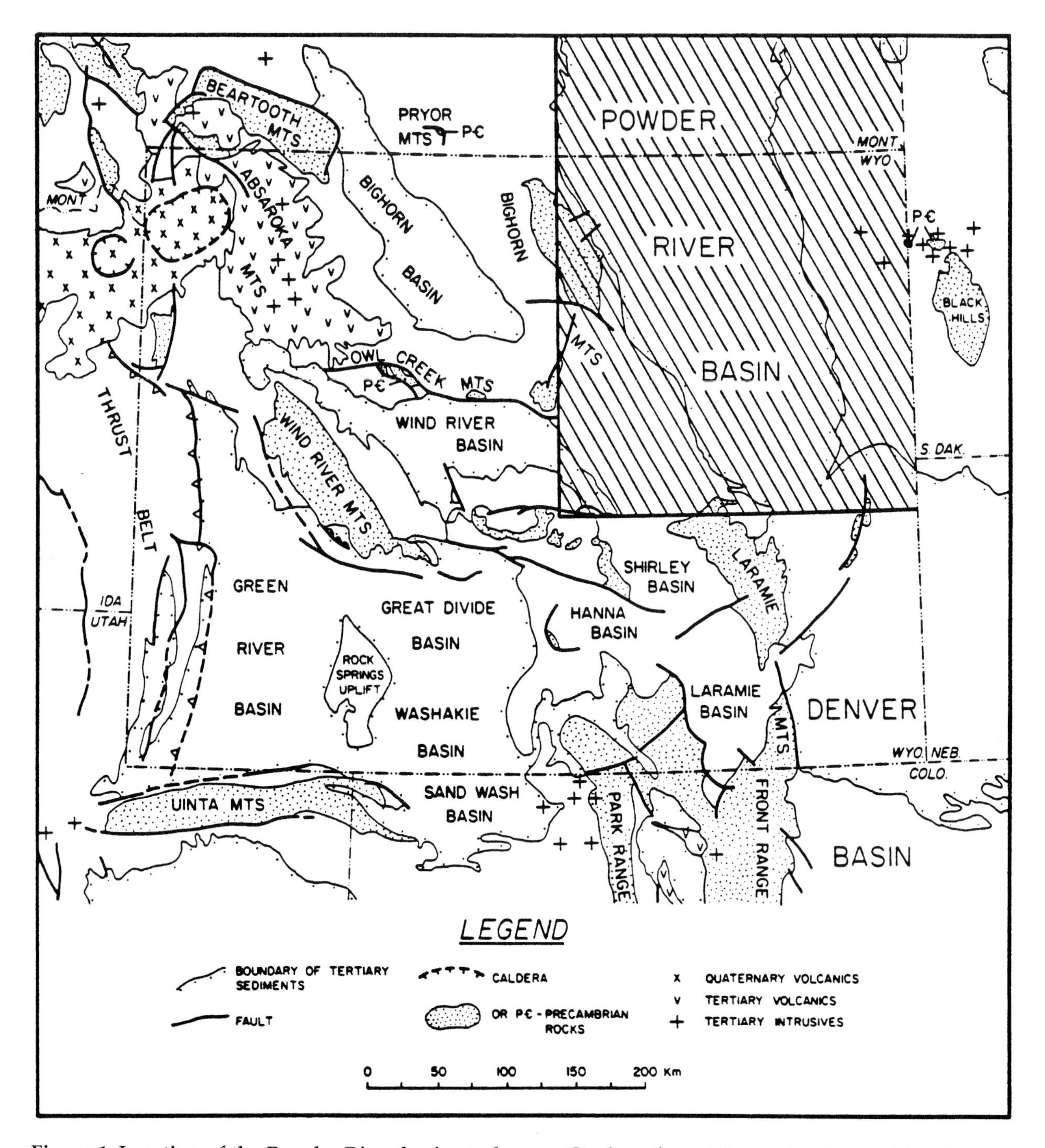

Figure 1. Location of the Powder River basin study area. Geology from King and Beikman (1974).

thickness along the western basin margin adjacent to the southern Bighorn Mountains (Blackstone, 1990). The basin margin is of two types: large-displacement thrust faults that disrupt the lateral continuity of the sedimentary section, and broad homoclines that provide lateral continuity between uplifted basin sediments along the basin margin and the deeply buried sediments in the basin. The basin margin is fault severed along the northern Bighorn Range, across the Shoshone Arch between the southern Bighorn and Laramie ranges, along the Laramie Range, and along the Hartville Uplift between the Laramie Range and the Black Hills (Blackstone, 1981; Huntoon, 1985a). Margin faulting is of the thrust or high-angle reverse type with dip-slip displacements ranging up to 6 to 10 mi (10–16 km) and stratigraphic offsets measuring 1 to 3 mi (1.6–5 km) (Berg, 1962; Sales, 1971; Gries, 1983). The basin margin is homoclinal elsewhere, such as along the southeastern Bighorn Range and along the perimeter of the Black Hills.

Within this larger context, the basin is divided into much smaller blocks bounded by regional lineaments (Marrs and Raines, 1984; Martinsen and Marrs, 1985). Zones of weakness in the basement have been propagated up through the overlying stratigraphic section as the boundaries of a multitude of rectilinear blocks. Stresses within the basin through the Paleozoic, Mesozoic, and Cenozoic eras have been accommodated by movement along these block boundaries. As discussed by Weimer (1984), recurrent movement between these blocks has played an important role in determining the distribution of Cretaceous clastic hydrocarbon reservoirs in the Powder River basin. In addition, these blocks have been determinative in the distribution of coal-rich Tertiary sequences and of modern-day drainage patterns.

Stratigraphy

Over 20,000 ft (6100 m) of dominantly clastic rocks, ranging in age from Middle Cambrian to Oligocene, are preserved in the Powder River basin of Wyoming (Figure 2 and Table 1). An 8000 ft (2400 m) portion of this section, including the Permian–Pennsylvanian Tensleep Sandstone through the Upper Cretaceous Mesaverde Formation, is highly productive of hydrocarbons and has been extensively studied.

Paleozoic formations within the basin are marine shelf deposits with an aggregate thickness ranging from 1300 to 1500 ft (400–460 m) in the east and 1100 to 2000 ft (330–600 m) in the west. Significant hydrocarbon accumulations occur in the Pennsylvanian and Lower Permian section. Cambrian sediments are composed of sandstones, shales, and conglomerates. In the eastern part of the basin, the Cambrian Deadwood Formation is overlain by the Ordovician Winnipeg Formation consisting of siltstone, shale, and sandstone units. A Devonian and Mississippian massive limestone/dolomite/sandy dolomite sequence overlies these units, extending across the entire basin and pinching out in the extreme southeast. Overlying these rocks are the Tensleep Sandstone (western basin) and the Minnelusa Formation (eastern basin). The Tensleep Sandstone consists dominantly of fine-grained, well-sorted sandstone with minor amounts of carbonate, while the Minnelusa is lithologically more heterogeneous and contains more carbonates and shale in addition to sandstone. Evidence of several transgressive-regressive cycles is present within these formations. In general, the Tensleep represents alternating eolian and shallow, open marine deposition, whereas the Minnelusa represents alternating eolian, sabkha, and restricted marine deposition (Desmond et al., 1984; Fryberger, 1984; George, 1984). The upper Paleozoic consists of interbedded shale, siltstone, sandstone, and claystone.

Mesozoic sediments consist of shale, siltstone, claystone, and sandstone, interbedded with small amounts of limestone. The Mesozoic–Paleozoic boundary is located in the interbedded red shale–siltstone beds of the Spearfish and Goose Egg formations in the east and west sides of the basin, respectively. The Goose Egg Formation is overlain by red siltstone, claystone, and fine-grained sandstone of the Chugwater Formation. Overlying this is the Gypsum Spring Formation, composed of white gypsum interbedded with shale and limestone; it is laterally continuous from east to west across the basin and absent in the southern part of the basin. The overlying Jurassic rocks are composed of shales, sandstones, and claystones.

The Lower Cretaceous section consists mostly of shale, sandstone, and some conglomerate. It is 40 to 600 ft (12–180 m) thick and is the major hydrocarbon-producing interval in the basin. The Lower Cretaceous Mowry Shale, an organic-rich, black, siliceous shale present throughout the basin, is believed to be the major hydrocarbon source rock for the Cretaceous reservoirs. The two most important Lower Cretaceous reservoirs are the Muddy Sandstone and the Fall River Sandstone of the Inyan Kara group. Martinsen (this violume) discusses the detailed stratigraphy of the Muddy Sandstone in relation to pressure compartmentation.

The Upper Cretaceous section consists of approximately 5000 ft (1500 m) of shale and subsidiary amounts of sandstone. For the most part these rocks were deposited in a series of very broad, clastic shelf systems that intermittently prograded across the Interior Seaway. The clinoform shelf morphology and stratigraphic relationships of the depositional sequences within this interval have been well documented by Asquith (1970, 1974) and expanded and updated by Martinsen et al. (1984). The Teapot, Parkman (both of the Mesaverde), and parts of the Frontier are mostly shoreline-associated deposits, whereas the Sussex, Shannon (both of the Cody Shale), and other parts of the Frontier have been interpreted as mid- to outer-shelf sand ridges (Gill and Cobban, 1973; Hobson et al., 1982; Coughlan and Steidtmann, 1984; Tillman and Martinsen, 1984, 1987).

The Tertiary section is composed of sandstone, shale, siltstone, claystone, and coal.

Hydrologic Setting

The hydrologic ramification of the faulting along the basin margin is that regional aquifers along the fault-severed perimeter lack hydraulic continuity with potential recharge zones in the adjacent uplifts. In contrast, the homoclinal segments of the margin provide stratigraphic, and thus possibly hydraulic, continuity. Groundwater circulation within the basin interior proximal to a fault-severed boundary is essentially stagnant, whereas a hydrodynamic system is more probable in the vicinity of a homoclinal boundary. Evidence for regional hydrologic partitioning along fault-severed boundaries includes (1) head differences across the boundaries; (2) water quality contrasts across faults, with very poor quality waters occurring in the footwalls; and (3) geothermally heated waters in the footwalls (Huntoon, 1985a).

The presence of a homoclinal margin does not ensure active hydrodynamic circulation between recharge areas and adjacent interior parts of the basin. Huntoon (1985b) demonstrated that permeability decreases markedly basinward in the sedimentary rocks in the downdip parts of most homoclines. This results in the rejection of recharge waters (Mancini, 1976) and their discharge from springs at the toes of the homoclines (Rahn and Gries, 1973). Hydrodynamic

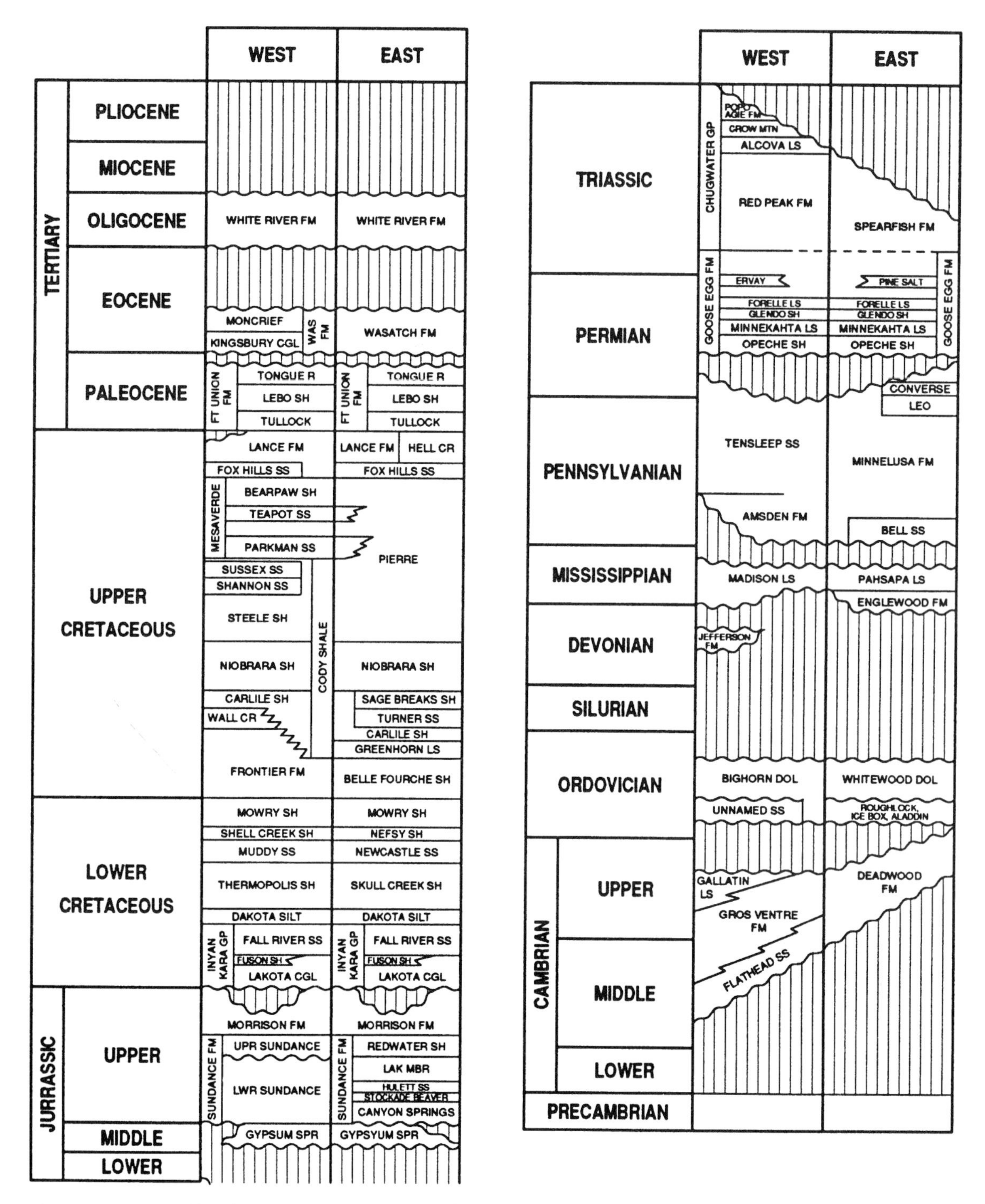

Figure 2. Generalized stratigraphic column for the Powder River basin. Modified from Wyoming Geological Association, 1957, 1958, 1978. Split columns indicate western (left) and eastern (right) facies.

circulation basinward is thus variable, with the tendency for circulation in the Paleozoic carbonates (such as the Madison Formation) to be more active than in overlying, locally and regionally isolated Mesozoic and Paleozoic clastic units. Documented discharge points for aquifers occur primarily from springs along the perimeter of the basin, demonstrating that either (1) circulation of water within the basin interior is negligible, (2) discharge from the basin interior is upward and regionally distributed through regional confining layers, or (3) actual, localized discharge points have not been recognized. The point here is that widespread hydrodynamic circulation systems are rare in the Powder River basin.

Published potentiometric data for various aquifers indicate a regional flow pattern of water moving from the topographic highs into the central parts of the basin (Dana, 1962; Feathers et al., 1981; Hodson, 1971,

Table 1. Generalized lithologic descriptions for the Powder River basin, including hydrologic properties.

	West				East			
	Geologic Formation	Thickness (ft)	Lithologic Description	Hydrologic Properties	Geologic Formation	Thickness (ft)	Lithologic Description	Hydrologic Properties
	Quaternary deposits	0–100	Silt, sand, and gravel deposits	Yields range 2–1000 gpm	Quaternary deposits			
Tertiary	White River Fm.	0–1000	Claystone and siltstone, tuffaceous with coarse-grained sandstone and conglomerate channel deposits	Low yields, 5–20 gpm	White River Fm.			
Tertiary	Wasatch Fm.	1100+	Sandstone, lenticular, fine- to coarse-grained, interbedded with shale and coal, coarser in south and southwest	Yields 10–500 gpm; TDS 500–1500 mg/l; transmissivity 1–5000 gpd/ft	Wasatch Fm.			
Tertiary	Fort Union Fm.	1100–2500+	Sandstone, lenticular, fine- to medium-grained, interbedded with siltstone, coal, and shale	Yields range 1–250 gpm; transmissivity 1–5000 gpd/ft	Fort Union Fm.			
Upper Cretaceous	Lance Fm.	1600–3000	Sandstone, lenticular, fine- to medium-grained, interbedded with sandy shale and claystone	Fox Hills/Lance aquifer system yields up to 350 gpm; transmissivity 1–5000 gpd/ft	Lance Fm.			
Upper Cretaceous	Fox Hills Sandstone	400–700	Sandstone, fine- to medium-grained, containing thin beds of sandy shale		Fox Hills Sandstone			
Upper Cretaceous	Lewis Shale	470	Shale, with sandy shale and thin lenses of sandstone	Major regional aquitard	Pierre Shale	2500–3100	Shale, some sandy shale, and bentonite beds	Major regional aquitard
Upper Cretaceous	Mesaverde Fm.	900+	Sandstone, fine- to medium-grained, massive to thin-bedded with shale, sandy shale, and coal beds	Minor aquifer, yields up to 10 gpm				

Continued on next page

Table 1. Generalized lithologic descriptions for the Powder River basin, including hydrologic properties (continued).

	West				East			
	Geologic Formation	Thickness (ft)	Lithologic Description	Hydrologic Properties	Geologic Formation	Thickness (ft)	Lithologic Description	Hydrologic Properties
	Cody Shale	3000–5000	Shale, calcareous lower part, containing siltstone and sandstone beds	Major regional aquitard				
					Niobrara Shale	100–250	Calcareous shale, shale, and marl, with thin bentonite beds	Major regional aquitard
					Carlile Shale	460–540	Shale, locally sandy	Major regional aquitard
					Greenhorn Ls.	30–70	Shale, limestone, and marl	Major regional aquitard
	Frontier Fm.	900	Sandstone and interbedded shale, conglomeratic sandstone at top	Minor aquifer, yields up to 10 gpm	Belle Fourche Shale	400–850	Shale, containing iron and limestone concretions and bentonite layers	Major regional aquitard
Lower Cretaceous	Mowry Shale	200–300	Siliceous shale with bentonitic beds	Minor aquitard	Mowry Shale			
	Muddy Ss.	20	Sandstone, lenticular, fine-grained	Dakota aquifer system; yields range to 250 gpm; transmissivity 1–900 gpd/ft	Newcastle Ss.	0–100	Sandstone, lenticular, fine-grained	Dakota aquifer system; yields range to 250 gpm; transmissivity 1–900 gpd/ft
	Thermopolis Shale	200	Shale, marine		Skull Creek Shale	180		
	Cloverly Fm.	140	Shale and siltstone with a basal sandstone		Inyan Kara Group	205		
Jurassic	Morrison Fm.	130–220	Shale and claystone with thin beds of limestone and sandstone	Yields to 10 gpm; transmissivity 0–200 gpd/ft	Morrison Fm.			
	Sundance Fm.	300–365	Shale with thin beds of limestone and sandstone	Yields to 50 gpm; transmissivity < 1250 gpd/ft	Sundance Fm.			

Continued on next page

Table 1. (Continued)

	West				East			
	Geologic Formation	Thickness (ft)	Lithologic Description	Hydrologic Properties	Geologic Formation	Thickness (ft)	Lithologic Description	Hydrologic Properties
Jurassic	Gypsum Spring Fm.	0–50	Gypsum, limestone, and shale		Gypsum Spring Fm.			
Triassic	Chugwater Fm.	700–800	Siltstone, claystone, and fine- to medium-grained sandstone	Regional minor aquitard	Spearfish Fm.	550–600	Shale, siltstone, sandstone, and gypsum	
Permian	Goose Egg Fm.	380	Shale and siltstone, interbedded, with limestone and gypsum beds		Minnekahta Ls.	30–50	Limestone and dolomitic limestone	
					Opeche Fm.	50–90	Sandstone, silty and shaley	
Pennsylvanian	Tensleep Ss.	120–500	Sandstone, fine- to medium-grained, massive	Madison aquifer system. Major regional aquifer, yields highly variable. Pumping yields to 1000 gpm; flowing yields > 4000 gpm; transmissivity 0–90,000 gpd/ft and highly dependent on secondary permeability	Minnelusa Fm.	1000	Sandstone, fine- to medium-grained, interbedded with limestone, dolomite, and shale	Madison aquifer system. Major regional aquifer, yields highly variable. Pumping yields to 1000 gpm; flowing yields > 4000 gpm; transmissivity 0–90,000 gpd/ft and highly dependent on secondary permeability
	Amsden Fm.	0–200	Shale with limestone					
Miss.	Madison Ls.	200–400	Limestone and dolomitic limestone, massive		Pahsapa Ls.	250	Limestone and dolomitic limestone, massive	
Camb.	Gallatin Ls., Gros Ventre Fm., Flathead Ss.	90–600	Limestone at top, limestone conglomerate, interbedded with shale, basal sandstone		Deadwood Fm	0–100	Sandstone, interbedded with shale, limestone, dolomite, and siltstone	

Formation thicknesses from Feathers et al. (1981); Wyoming Geological Association (1958). Lithologic descriptions and hydrologic properties from Feathers et al. (1981), Hodson et al. (1973), and Wyoming Geological Association (1958, 1978). Yields are given in gallons per minute (gpm) and transmissivities in gallons per day per foot (gpd/ft).

1974; Lobmeyer, 1980; Swenson et al., 1976). Movement of groundwater into the basin from the Wind River basin via the Casper Arch is possible. Groundwater movement out of the basin is to the north and possibly the southeast (Richter, 1978).

DATA ANALYSIS

The largest source of DST pressure data for the Powder River basin is Petroleum Information, Inc. (PI) in Denver. Other sources of pressure data include published and unpublished reports in the Wyoming Oil and Gas Commission files in Casper, Wyoming. The PI data set was used for this study.

Figure 3 shows the location of 12,361 DST data values extracted from the PI database for the Mesaverde Formation (984 data values), Sussex Formation (1041 data values), Frontier Formation (821 data values), Muddy Formation (3888 values), Dakota Formation (1157 data values), and Minnelusa Formation (4470 data values).

Strategy

The test area in the Powder River basin ranges from latitude 42.7° to 45.9°N and longitude 104.1° to 107.3°W (Figure 1). Within this area, the hydraulic head configuration was evaluated on a regional scale for the Mesaverde, Sussex, Frontier, Muddy, Dakota, and Minnelusa formations. In addition, several areas were evaluated in greater detail. The strategy used to document the existence and geometry, or the absence, of pressure compartments consists of the following steps:

Step 1: Extract the measured pressures, kelly bushing elevations, test depths, and locations from the PI database.

Step 2: Evaluate the pressure-depth data for the six specified formations to determine the consistency of the pressure data as well as any functional relationships (pressure versus elevation, head versus depth).

Step 3: Create contour maps of the well kelly bushing (KB) elevations for the six specified formations. Compare the KB elevation contour maps with actual topographic maps for the Powder River basin to determine if the density of data is adequate for contouring.

Step 4: Create structure contour maps of the six specified formations. Compare these maps with published maps to distinguish spurious data values in the PI database.

Step 5: Convert the pressure values to hydraulic head (h) according to equation 1 where p is the maximum, nonzero shut-in pressure (in pounds per square inch, psi) as reported by PI for formation tests; 0.433 (in psi per ft) is the hydrostatic gradient, ρg in equation 1; and the elevation potential, z in equation 1, is the elevation (in feet) of the midpoint of the DST interval.

Step 6: Contour the h values to generate a potentiometric surface for targeted stratigraphic intervals.

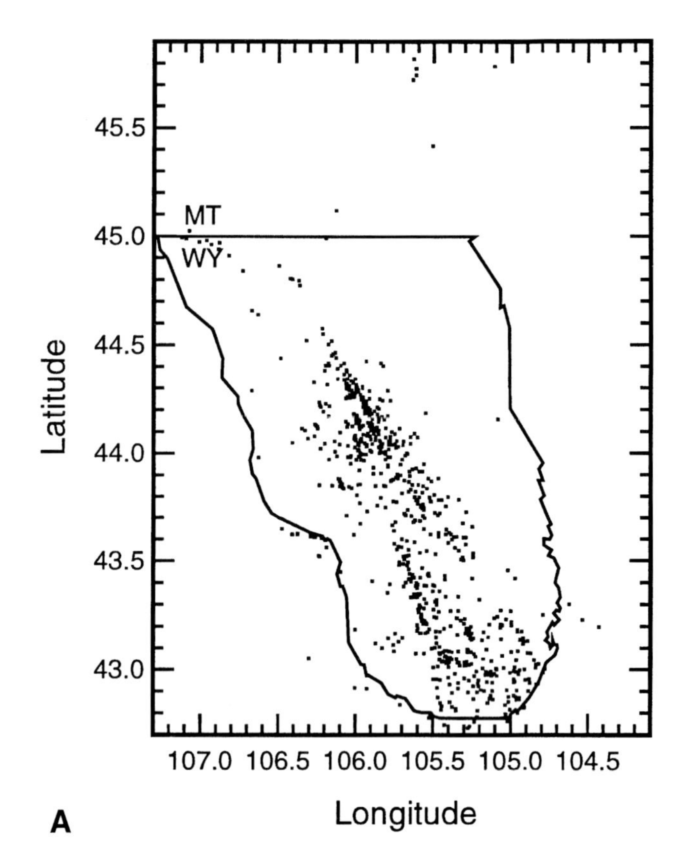

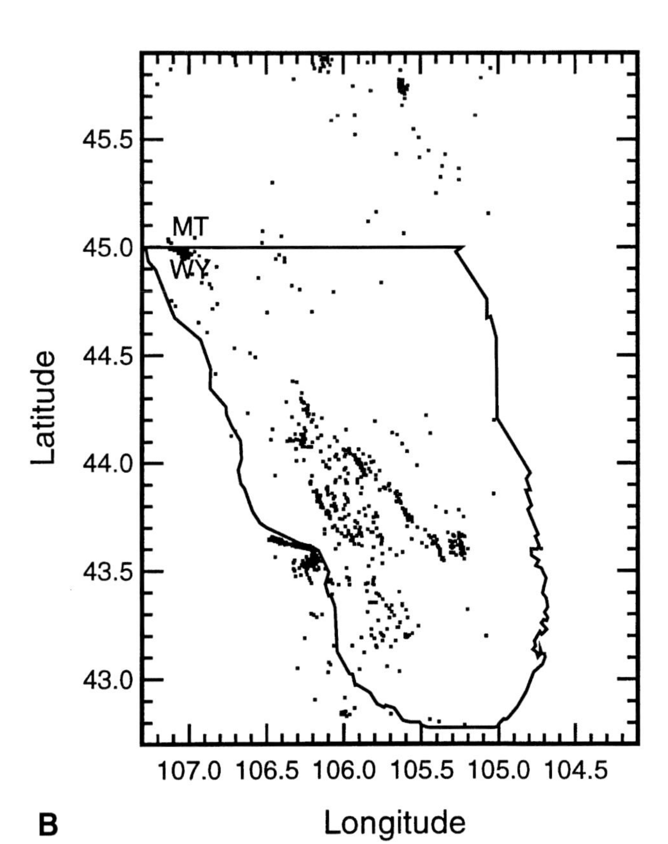

Figure 3. Location of drill-stem test data from the (A) Mesaverde Formation (984 data values), (B) Sussex Formation (1041 data values).

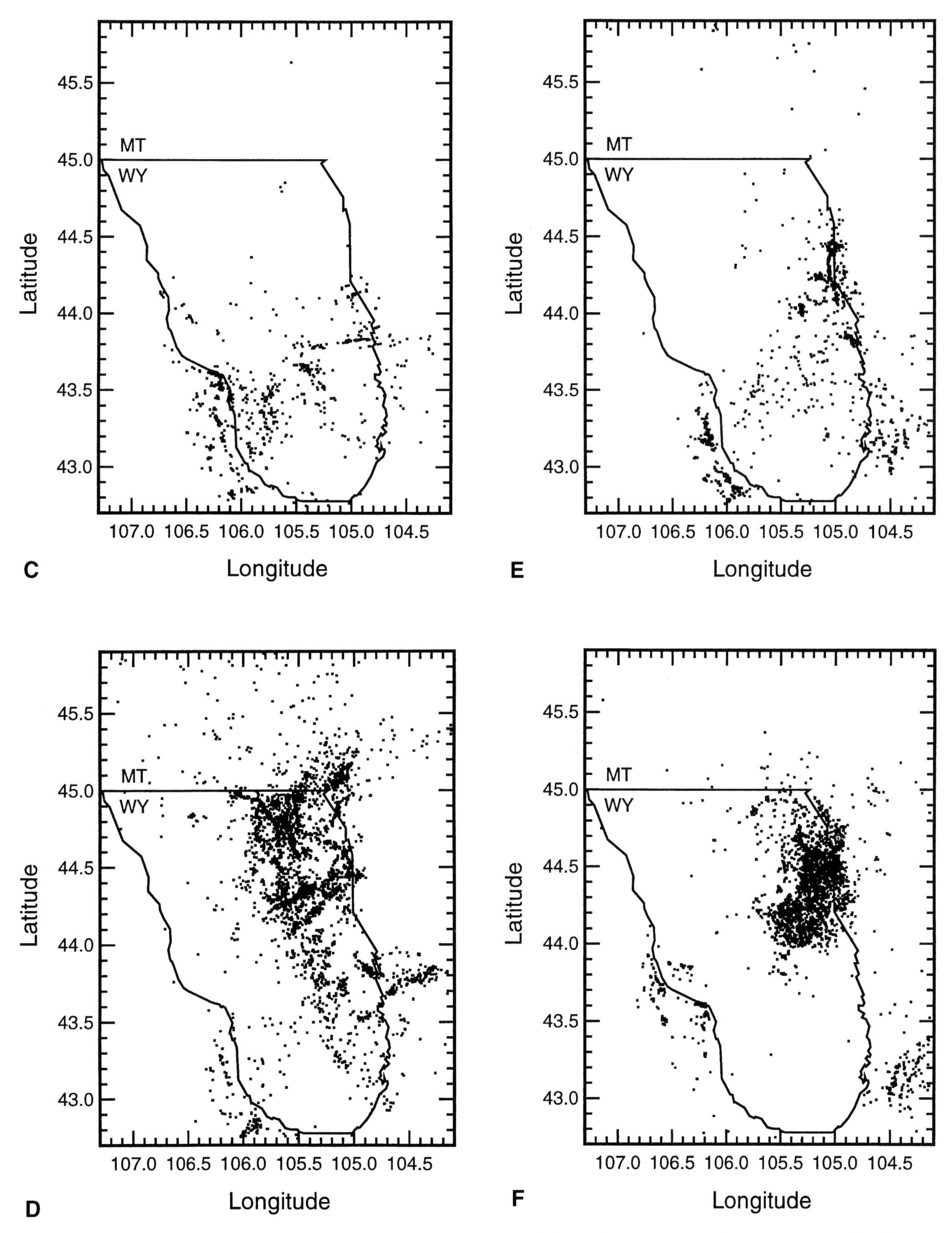

Figure 3 (continued). (C) Frontier Formation (821 data values), (D) Muddy Formation (3888 data values), (E) Dakota Formation (1157 data values), and (F) Minnelusa Formation (4470 data values). Tertiary sediments in Wyoming are outlined.

Step 7: Analyze the contoured potentiometric surface in order to evaluate the presence of pressure anomalies, paying particular attention to any excessive pressure gradient that would require the presence of seals. Additionally, this step allows a comparison of the magnitude and slope of anomalies on the potentiometric surface.

RESULTS

Pressure-Depth Relations

For the six stratigraphic units of interest, the following graphs were plotted: ISIP versus FSIP, FSIP versus depth, and FSIP versus elevation.

Figure 4 shows the relationship between ISIP and FSIP for the Mesaverde, Sussex, Frontier, Muddy, Dakota, and Minnelusa formations. Ideally, the ISIP should equal the FSIP, as shown by the straight line at 45° in Figure 4. This would imply that the DST pressure has reached equilibrium with the formation pressure. Departures from ISIP equaling FSIP could be caused by factors such as mechanical failure of the DST, insufficient pressure buildup time for the initial flow period, or insufficient pressure buildup time for the final flow period. Bair et al. (1985) required ISIP and FSIP to agree within 5% in order to be used for calculation of static hydraulic head values, whereas Doremus (1986), Jarvis (1986), and Spencer (1986) used a value of 25%. In an effort to maximize the number of data values for contouring, the maximum nonzero pressure value of the FSIP and ISIP is used in this study.

Figures 5 and 6 display the FSIP-depth and FSIP-elevation relations for the Mesaverde, Sussex, Frontier, Muddy, Dakota, and Minnelusa formations. Similar graphs were made for ISIP versus depth and elevation; since these graphs demonstrate the same features as the FSIP graphs, only the FSIP graphs are shown.

A line of constant hydrostatic gradient corresponding to fresh water (1.0 gm/cm^3 or 0.433 psi/ft) is shown on Figures 5 and 6. The hydrostatic gradient of 0.433 psi/ft is assumed because it is most appropriate for fresh water (water containing less than 10,000 mg/l dissolved solids; Bair et al., 1985). The Powder River basin, particularly the Cretaceous section, contains abundant fresh water, but many drillers in the region prefer to assume a gradient of 0.465 psi/ft. The scatter of data in Figures 5 and 6 does not justify resolution between a 0.433 psi/ft and a 0.465 psi/ft regional gradient. Only on Figure 5 (FSIP versus depth) can the hydrostatic gradient line be strictly used to define zones of overpressure and underpressure as defined by Hunt (1990). On Figure 6 (FSIP versus elevation) the hydrostatic gradient line is included so that its slope may be compared with trends in the data. The intercept of the hydrostatic gradient line was chosen as 4500 ft (1400 m) for the FSIP versus elevation graphs, although any other elevation between 3000 and 6000 ft (90 and 180 m) also could have been chosen.

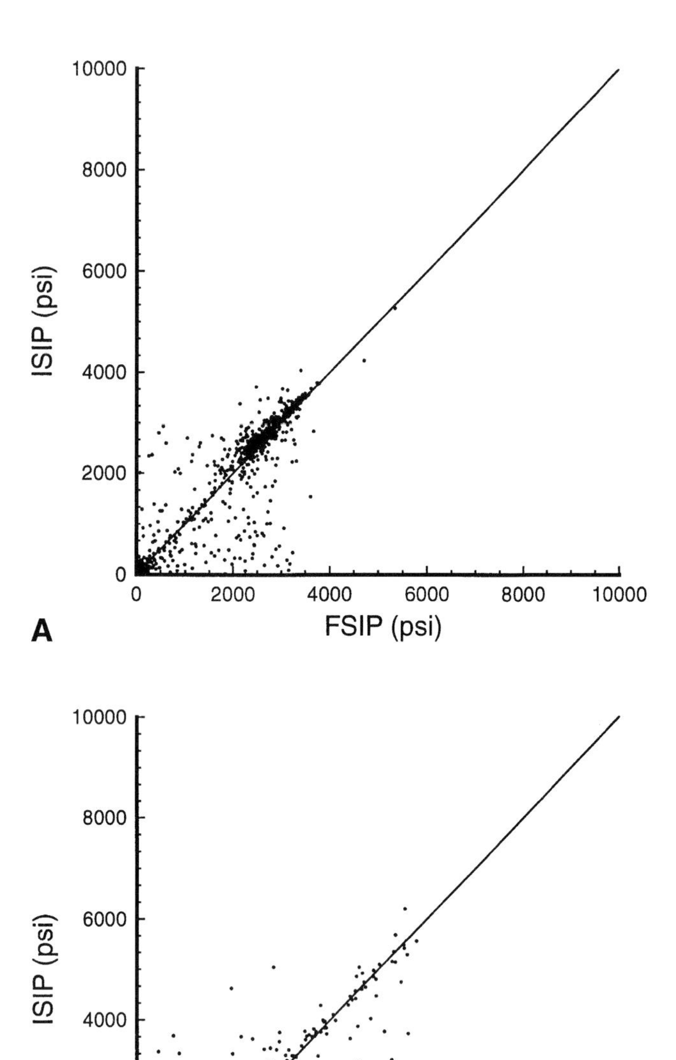

Figure 4. Initial shut-in pressure (ISIP) in pounds per square inch (psi) versus final shut-in pressure (FSIP) of drill-stem test data from the (A) Mesaverde Formation (984 data values), (B) Sussex Formation (1041 data values).

The greatest density of points in Figures 5 and 6 falls along a line parallel to the hydrostatic gradient line, but displaced a constant pressure lower. This pressure offset may be explained in two ways. First, for the active hydrodynamic system, the top of the potentiometric surface may not correspond with the ground surface. This seems to be a possible explanation due to the thousands of feet of aquitards and aquifers between the six formations and the ground surface. The other explanation would be that the DST pressures defining the linear trend are all less than

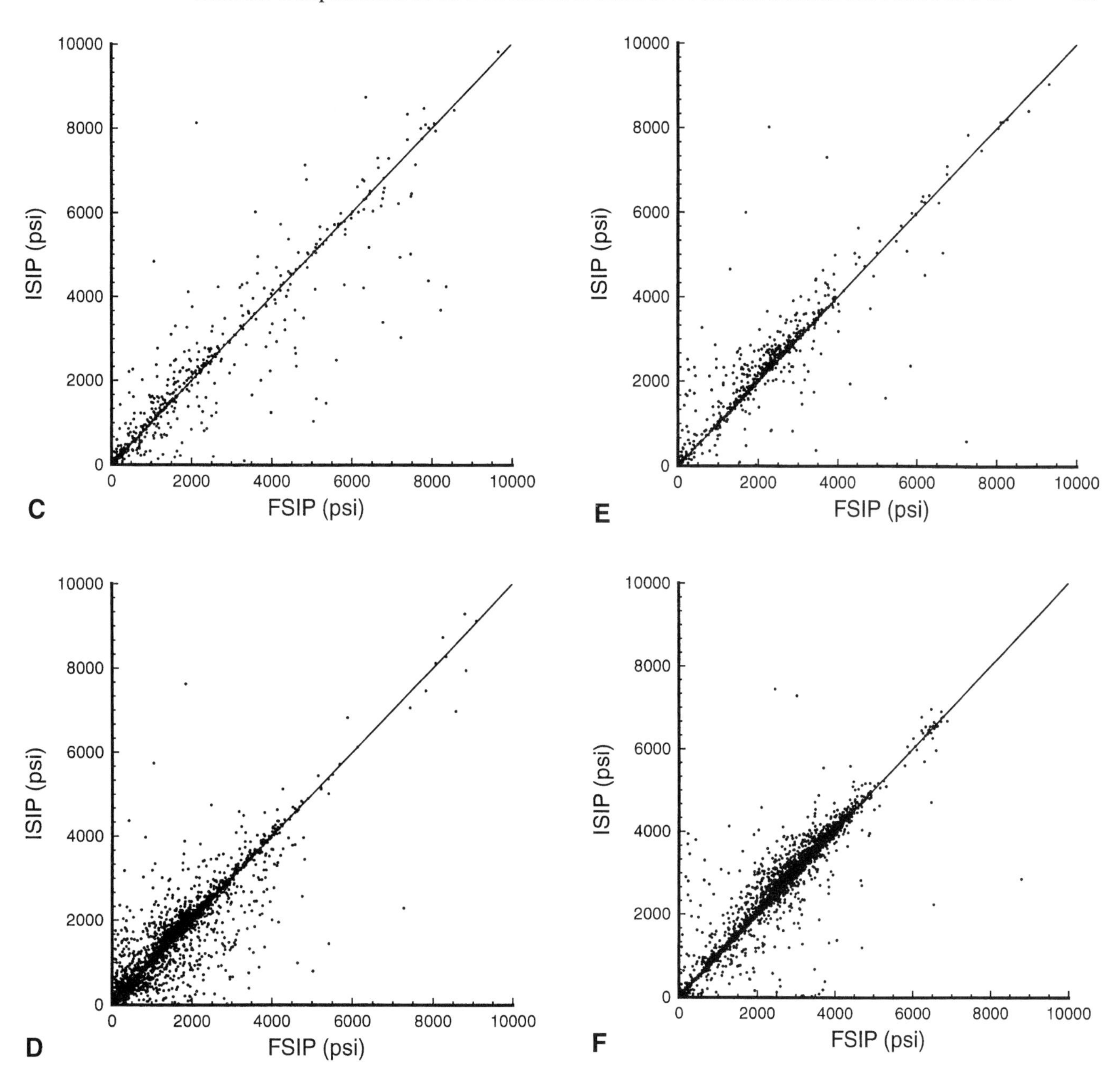

Figure 4 (continued). (C) Frontier Formation (821 data values), (D) Muddy Formation (3888 data values), (E) Dakota Formation (1157 data values), and (F) Minnelusa Formation (4470 data values). The 45° lines indicate where ISIP = FSIP.

equilibrium formation pressures. In Texas, Bair et al. (1985) found that DSTs were typically 43 psi less than Horner-extrapolated pressures, whereas in the Bighorn basin of Wyoming, Doremus (1986), Jarvis (1986), and Spencer (1986) estimated DST pressures to be up to 54 psi different from Horner-extrapolated pressures. In our study area we have found that most Horner plots yield an extrapolated reservoir pressure 0 to 400 psi above the recorded FSIP. Since the majority of the pressure data are within 5% of FSIP equaling ISIP, we believe this explanation to be less significant than the dissimilarity between the potentiometric and ground surfaces.

For FSIP versus depth (Figure 5), notice that most points fall below the hydrostatic gradient line, indicating pressures less than hydrostatic. Data points at moderate to great depths and with low pressures probably correspond to DSTs recorded in very low permeability formations, yielding an apparently low pressure due to an insufficiently long pressure buildup period to reach reservoir pressure. Thus, the data in this region may not represent actual formation pressures, but may indicate zones of low permeability.

Figures 5 and 6 also show zones that are clearly overpressured. Up to 3000 psi overpressure above the hydrostatic gradient exists in these data. The zones of

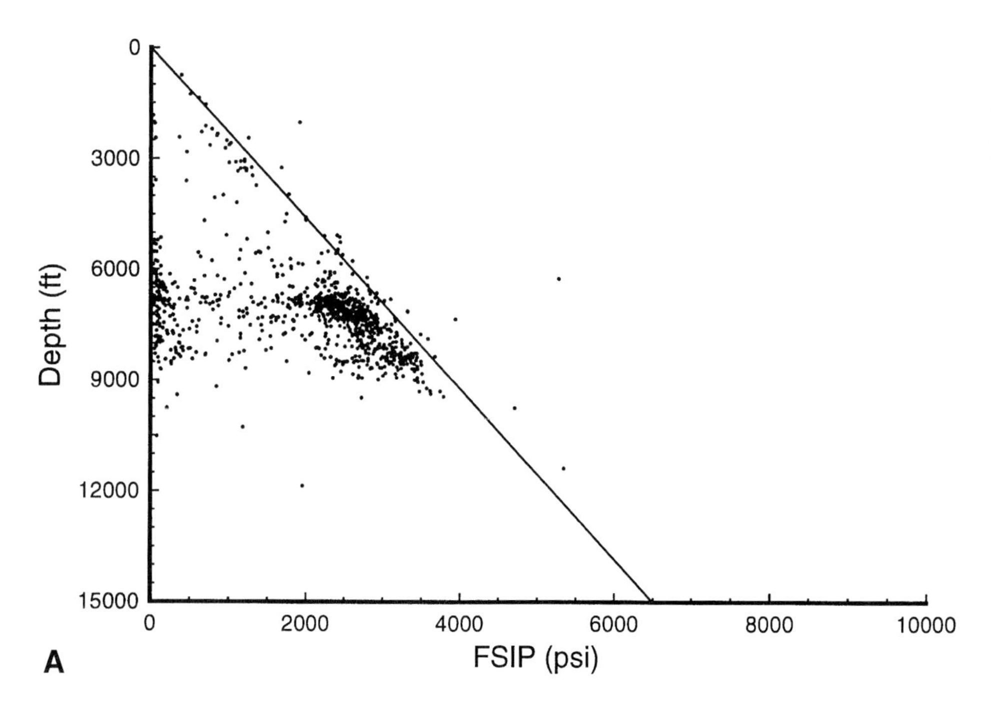

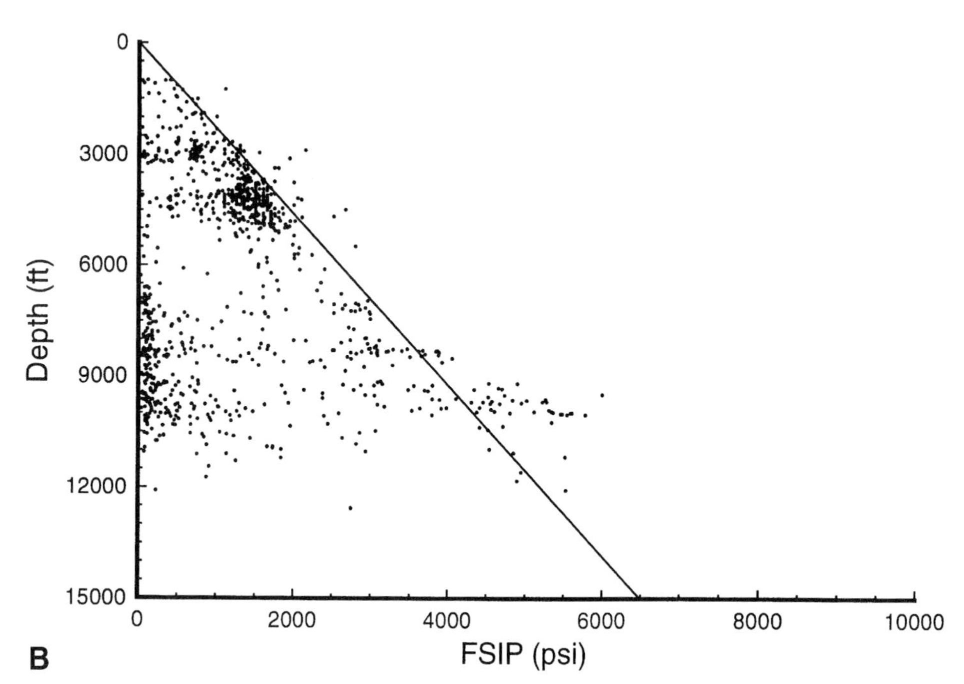

Figure 5. Final shut-in pressure (FSIP) in pounds per square inch (psi) vs. depth for drill-stem test data from the (A) Mesaverde Formation (984 data values). (B) Sussex Formation (1041 data values).

excess pressure for the Frontier, Muddy, and Dakota formations correspond to hydraulic head values of 12,000 to 14,000 ft (3700–4300 m). The overpressured zones begin at approximately 7000 to 9000 ft (2100–2700 m) and attain maximum pressures at about 12,000 ft.

Potentiometric Maps

Potentiometric surfaces were constructed for the six formations using equation 1 to calculate the hydraulic head and contouring these values. Maximum nonzero pressure data from the DSTs were used to construct the potentiometric surfaces. All available pressure data were used in order to maximize the density of data points used to evaluate the pressure anomalies.

Because all nonzero DST pressure data were used, the resulting potentiometric surface does not represent the actual fluid potential. In zones of high permeability, a DST measurement will approach the equilibrium pressure of the formation. Consequently, the calculated hydraulic head value will be close to the potential energy of the fluid. However, in impermeable zones, the DST measurement will be stopped before equilibrium is reached due to the great time

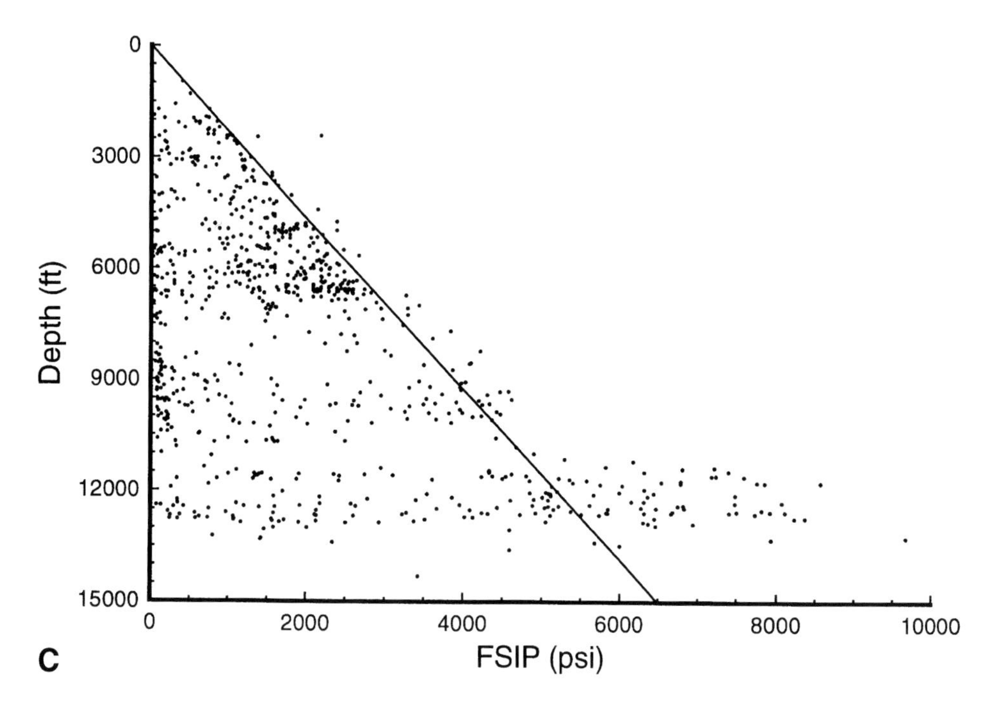

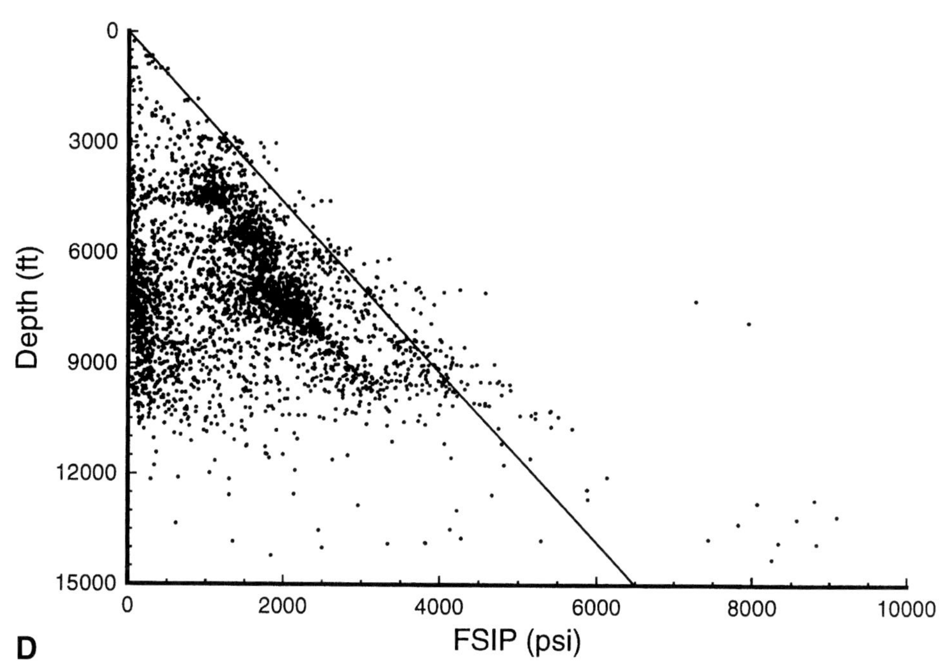

Figure 5 (continued). (C) Frontier Formation (821 data values), (D) Muddy Formation (3888 data values).

period required to reach equilibrium, and thus will underestimate the equilibrium formation pressure. Consequently, because all nonzero pressure data were used, the resulting potentiometric surfaces should be viewed as delineating fluid potential in zones of high permeability and as delineating zones of low permeability in areas of apparently low hydraulic head.

Two primary factors, the density of data values (total number of data points and their spatial distribution) and the accuracy of the data values, affect the accuracy of the calculation of those contour maps. The density of data was roughly checked by contouring the kelly bushing elevation as previously discussed. Also, contours are shown only where there is a well density of at least one well every 6 mi^2 (16 km^2).

The question of the accuracy of individual calculated hydraulic head values is difficult to assess. All nonzero DST data, including very low pressure data, were used to calculate hydraulic head values. In areas of high pressure (interpreted also as zones of high permeability), the calculated hydraulic head (approximately 4000 ft [1200 m] above sea level) is probably accurate to ± 300 ft (90 m). This estimate was determined by comparing Horner-plot DST analyses to hydraulic heads calculated directly from shut-in pressures. In areas of low pressure, the hydraulic head

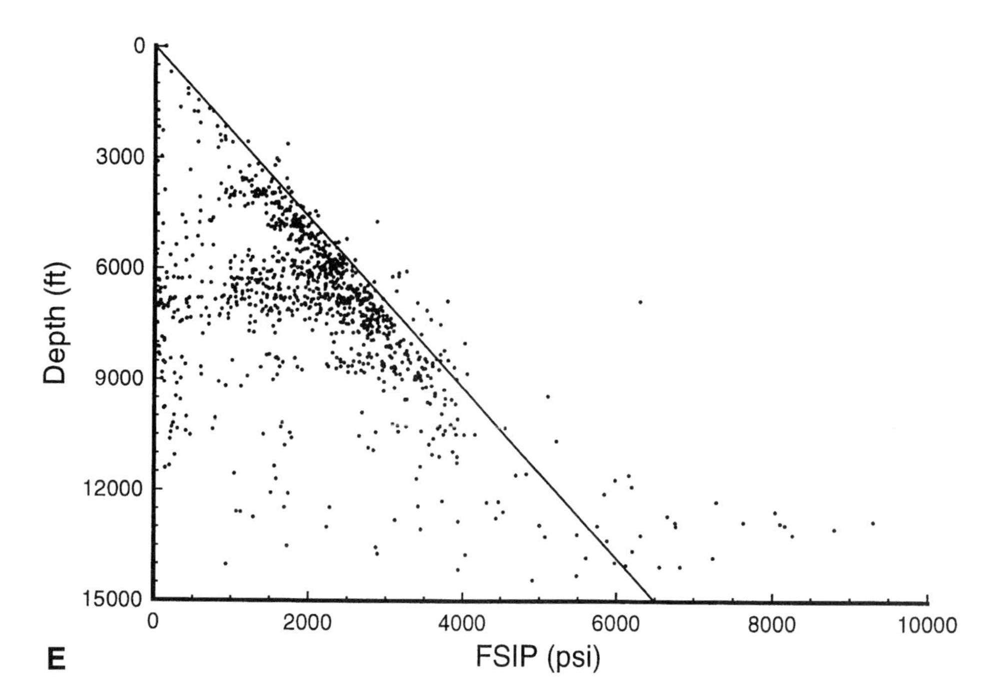

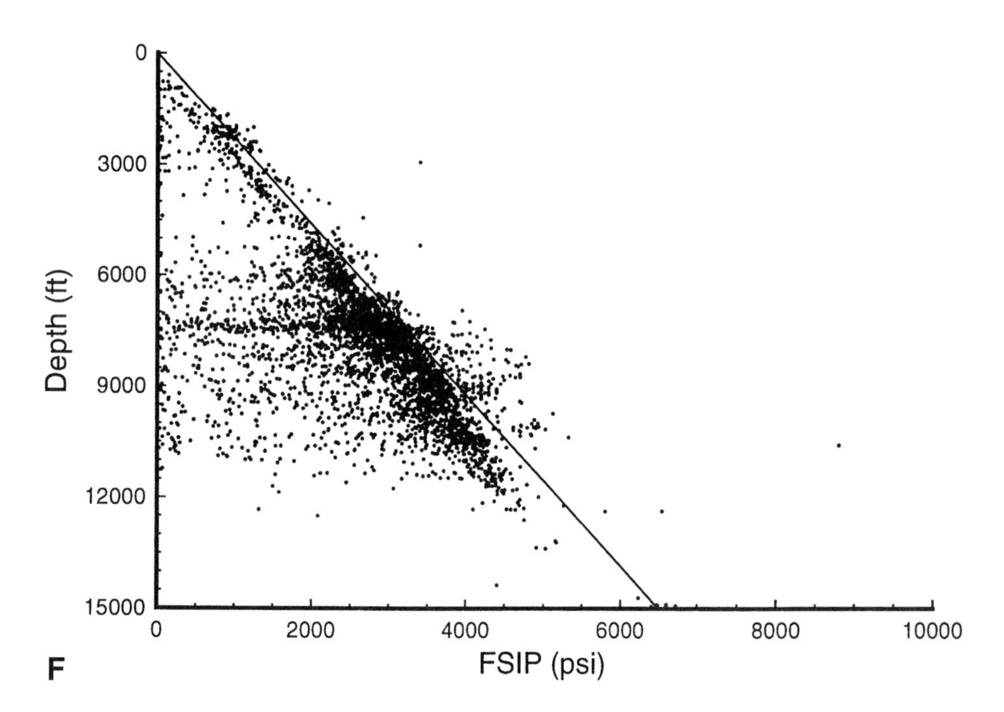

Figure 5 (continued). (E) Dakota Formation (1157 data values), and (F) Minnelusa Formation (4470 data values). The lines shown represent a hydrostatic gradient of 0.433 psi/ft extending from the ground surface to a depth of 15,000 ft (4600 m).

directly calculated from the PI database is unrealistically low (down to 8000 ft [2400 m] below sea level). These extremely low head values are not accurate in their depiction of the potentiometric surface, but they do define potential zones of low fluid transfer (i.e., boundaries or discontinuities in the potentiometric surface).

Production through time at a particular oil field may also affect pressures. Spencer (1986) estimated 370 to 1758 ft (110–540 m) of drawdown for three producing oil fields in the Bighorn basin of Wyoming. Chen et al. (this volume) show that there may be a historical decline in pressure in the Hilight oil field in the Powder River basin, but the trend is difficult to determine. Since the DST data were not corrected for historical drawdown, the resulting potentiometric maps should be considered to represent minimum values.

Regional potentiometric surfaces for the Mesaverde, Sussex, Frontier, Muddy, Dakota, and Minnelusa formations are shown in Figure 7. It is apparent that there are multiple hydraulic head anomalies consisting of both overpressured and underpressured regions. From this set of maps it is possible to evaluate the spatial distribution of the anomalies. Most important, anomalies in the potentiometric surface of the various formations typically are not vertically stacked, but instead are

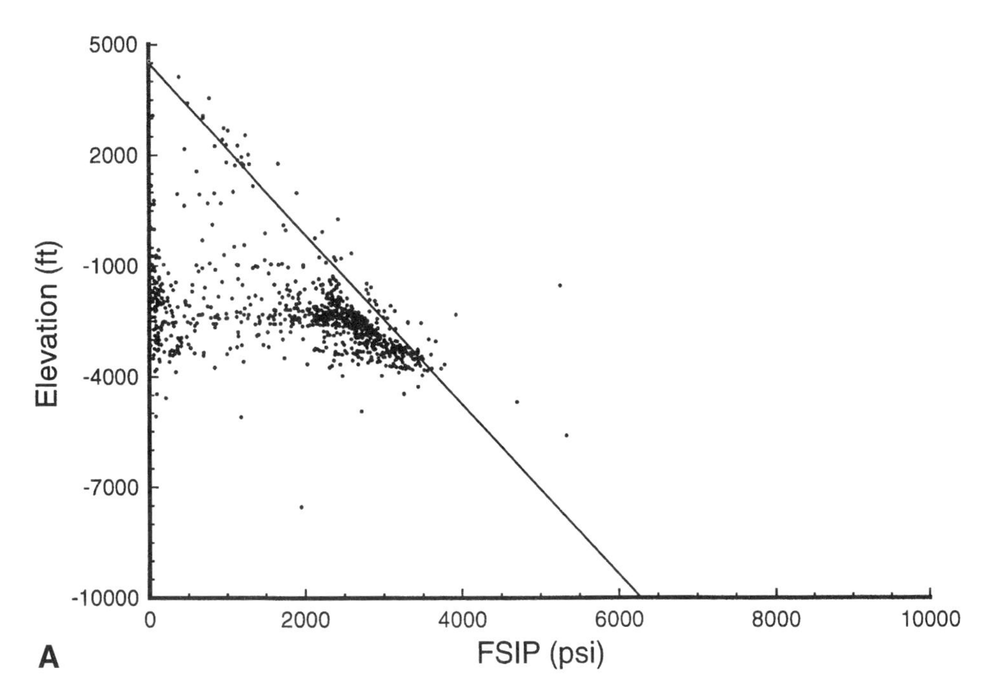

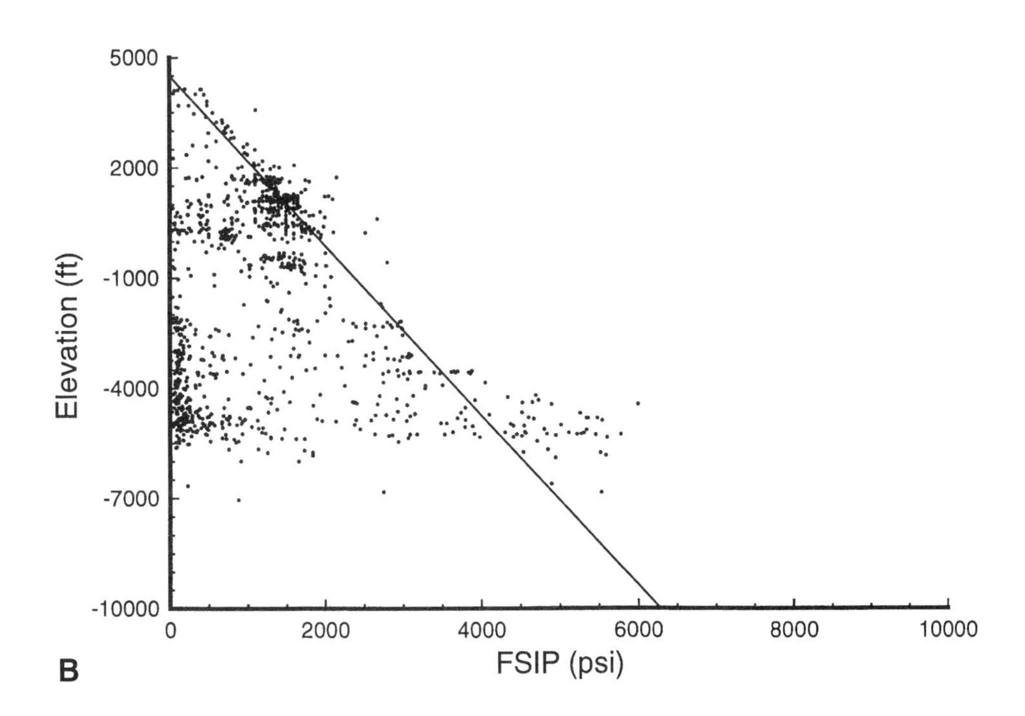

Figure 6. Final shut-in pressure (FSIP) in pounds per square inch (psi) versus elevation for drill-stem test data from the (A) Mesaverde Formation (984 data values), (B) Sussex Formation (1041 data values).

separated from one another by normally pressured rock. Red areas in Figure 7 represent hydraulic heads of approximately 7000 ft (2100 m) or greater elevation, whereas blue areas represent heads of less than 0 ft of elevation.

From this set of diagrams it is possible to evaluate the shape of the pressure anomalies. Several aspects of the diagrams are noteworthy. First, for four of the formations there is no regional slope to the potentiometric surface; rather, the surface appears to consist of discontinuous plateaus dissected by valleys. Only the Mesaverde and Sussex formations appear to have a regional gradient, which is higher near the eastern recharge areas and diminishes basinward. Second, the difference in elevation between the top of the plateaus (anomalously "high pressure") and the bottom of the valleys (anomalously "low pressure") is up to 10,000 ft (3000 m). Last, the differences between the high and low hydraulic head values take place over very short horizontal distances, 1 to 3 mi (1.6–5

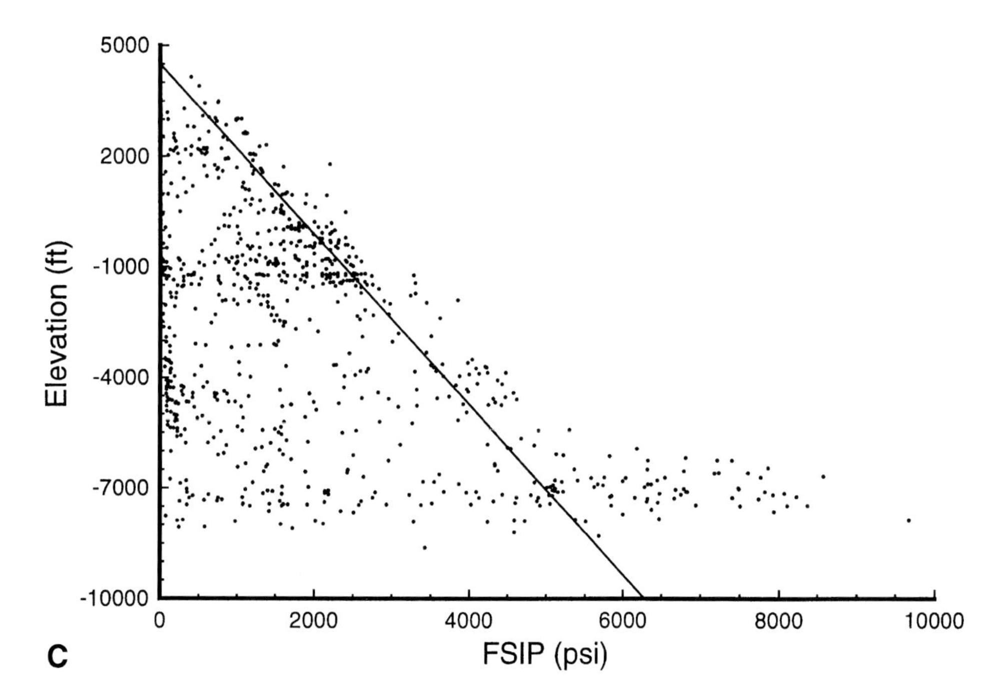

Figure 6 (continued). (C) Frontier Formation (821 data values), (D) Muddy Formation (3888 data values).

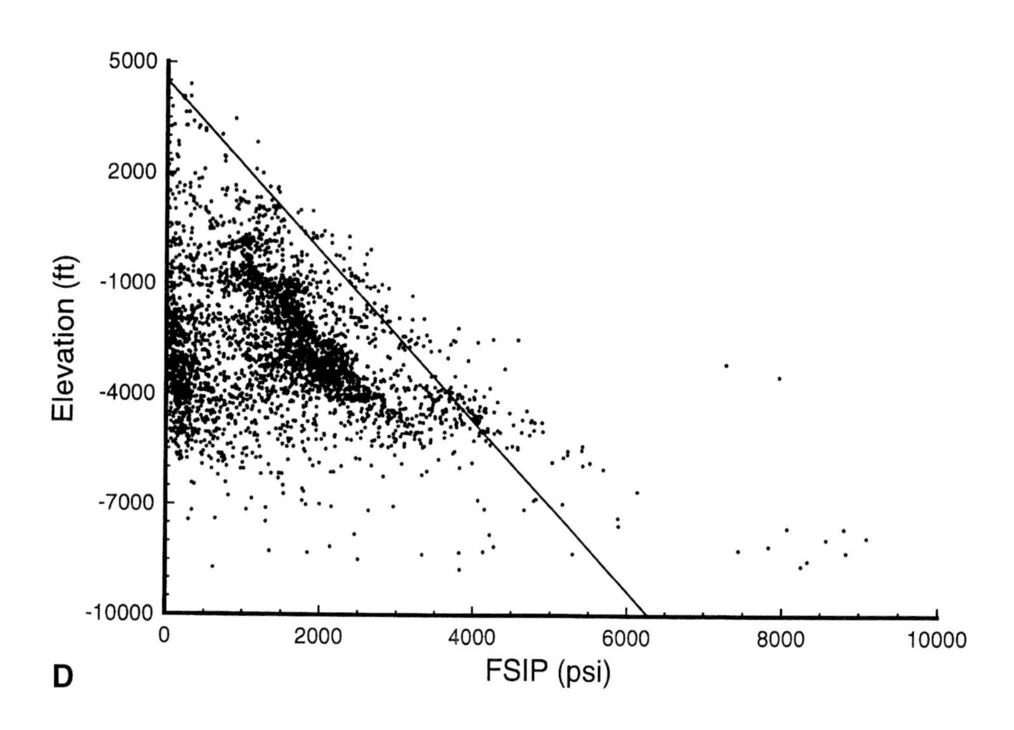

km). These exceedingly steep hydraulic head gradients are interpreted as discontinuities in the potentiometric surface. Since there are no fluid-producing sources in these areas (no current compaction disequilibrium or hydrocarbon generation), the discontinuities are interpreted as indicating seals to the contained fluid. Chen et al. (this volume) have geostatistically analyzed the shape of the pressure anomalies in the area of the Hilight oil field and have found that the hydraulic head anomalies generally are 2.7 mi (4.3 km) wide by 5.3 mi (8.5 km) long and trend in a north-northwest direction.

To examine the positive anomalies in more detail, potentiometric surfaces were constructed for the Muddy, Frontier, and Dakota formations in the vicinity of the Hilight and Powell hydrocarbon fields (Figures 8 through 10). Although the resolution of the anomalies is better than that in Figure 7, the shape, scale, and magnitude of the anomalies remain the same. It is apparent from Figures 8 through 10 that

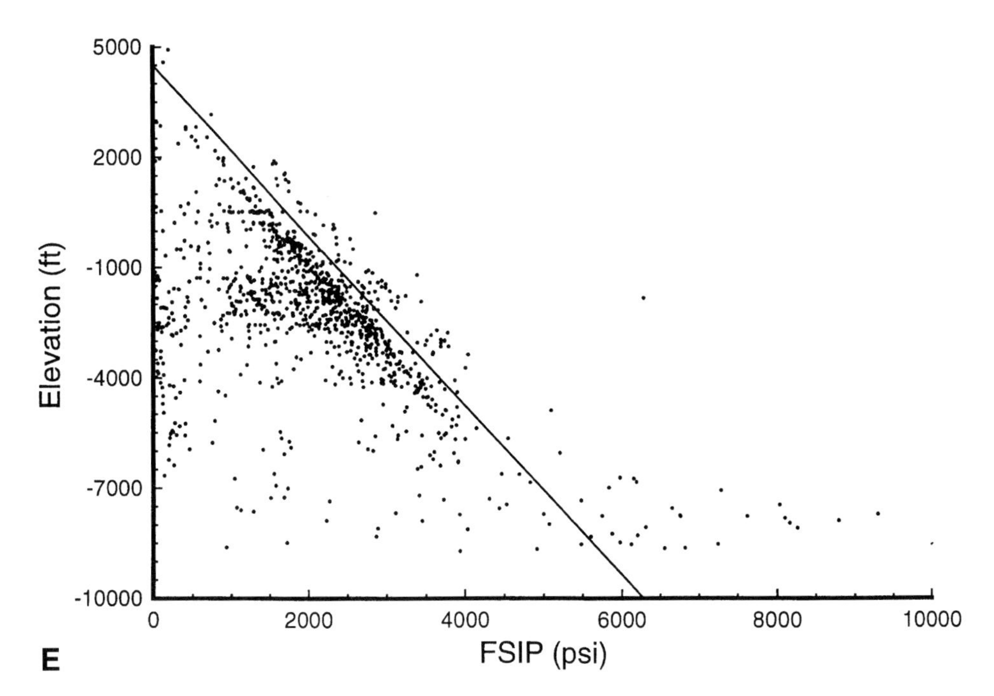

Figure 6 (continued). (E) Dakota Formation (1157 data values), and (F) Minnelusa Formation (4470 data values). The lines shown represent a hydrostatic gradient of 0.433 psi/ft extending from 4500 ft (1400 m) elevation to an elevation of –10,000 ft (–3000 m).

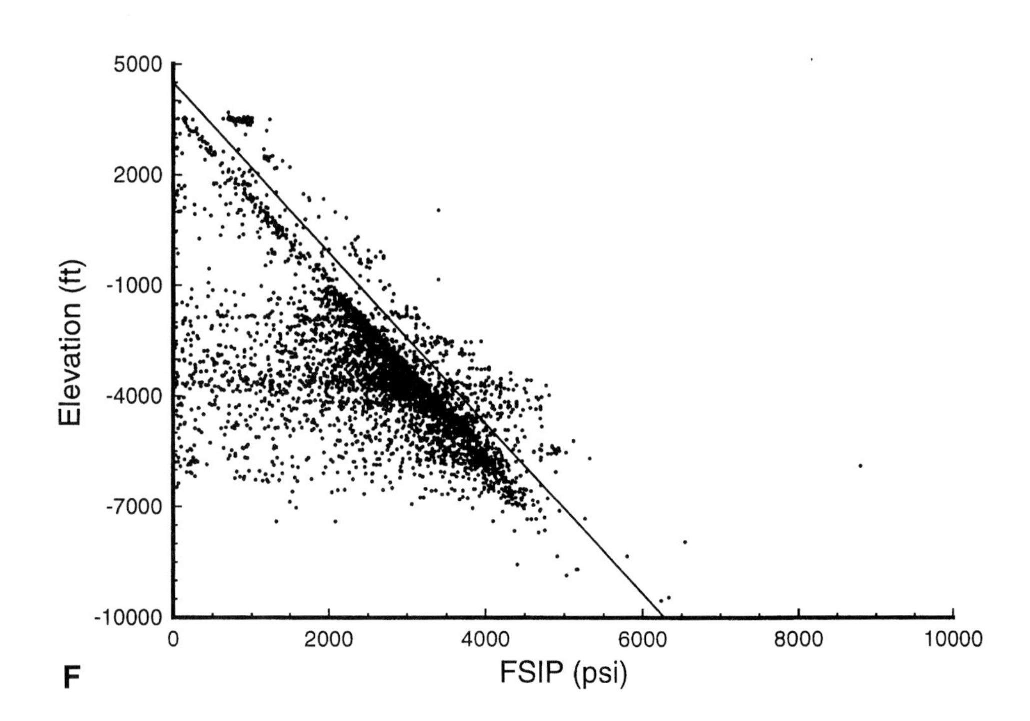

the pressure anomalies occur in different locations and have different magnitudes in the Muddy, Frontier, and Dakota formations.

To further resolve the shape of the anomalies, a series of cross sections showing hydraulic head was drawn through the three formations, as shown in Figures 11 through 13. Cross sections A–A′ and C–C′ (Figures 11 and 13) are drawn through the Powell hydrocarbon field, and B–B′ (Figure 12) is in the vicinity of the Hilight field. Also shown in Figures 11 through 13 are the ground surface, top of the formation, and a piecewise least-squares fit to the potentiometric surface. The piecewise least-squares procedure was designed to minimize the curve-fitting error. Thus, the fitted line segments shown on Figures 11 through 13 represent a linear, segmented "best fit" calculated through the potentiometric surfaces. From these reconstructions it is even more apparent that the boundaries of the anomalies are extremely sharp (characterized by steep gradients), perhaps discontinuous, and characterized by head differences of up to 16,000 ft (4900 m). Most important, note that the tops of most of the anomalies are nearly flat, indicating little or no hydrodynamic flow

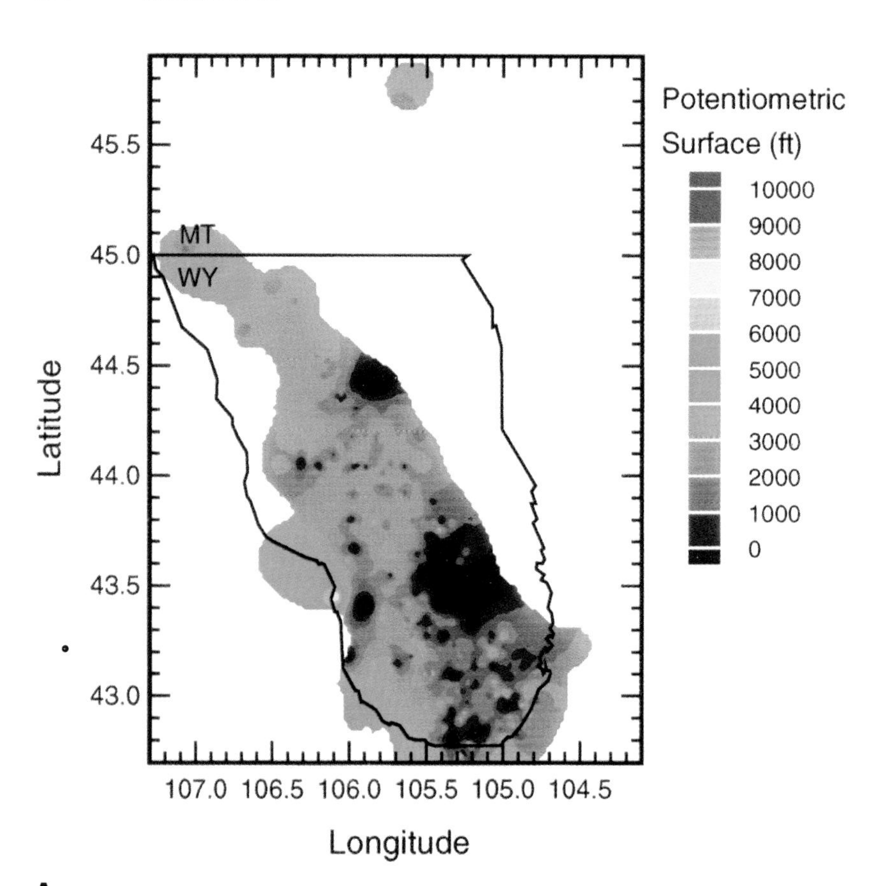

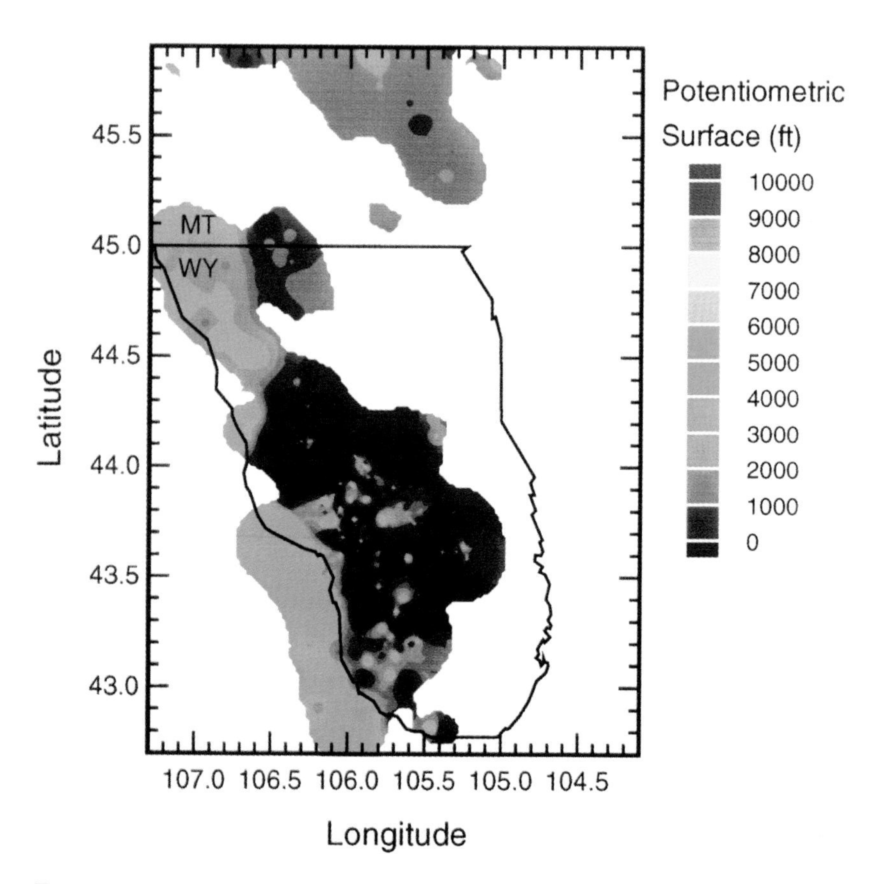

Figure 7. Contoured potentiometric surface of drill-stem test data from the (A) Mesaverde Formation (984 data values), (B) Sussex Formation (1041 data values).

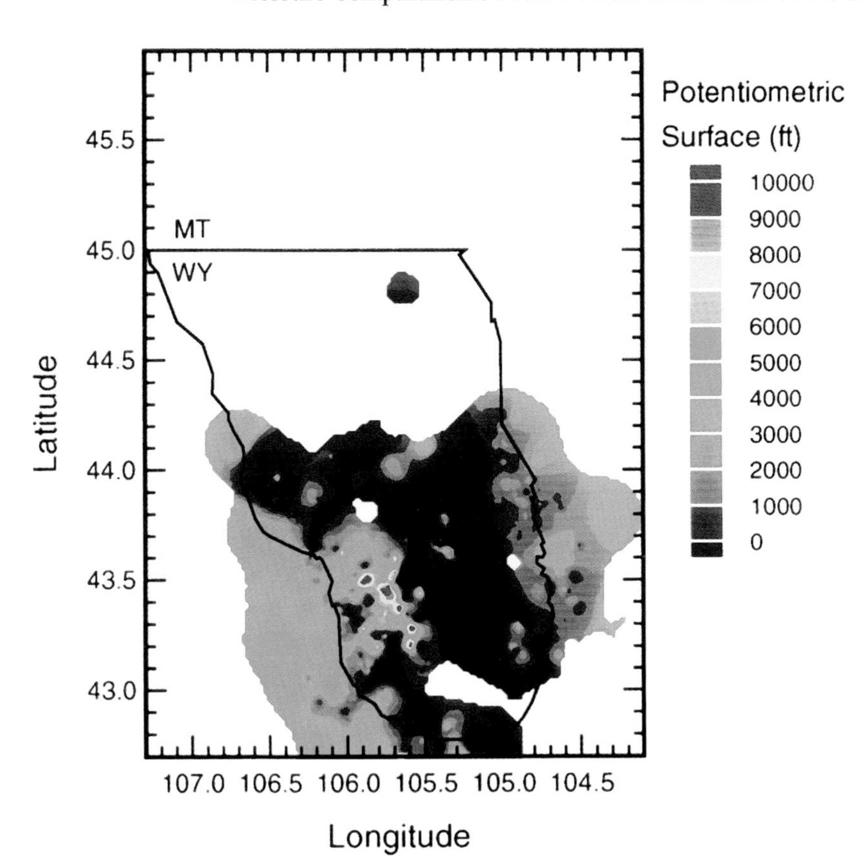

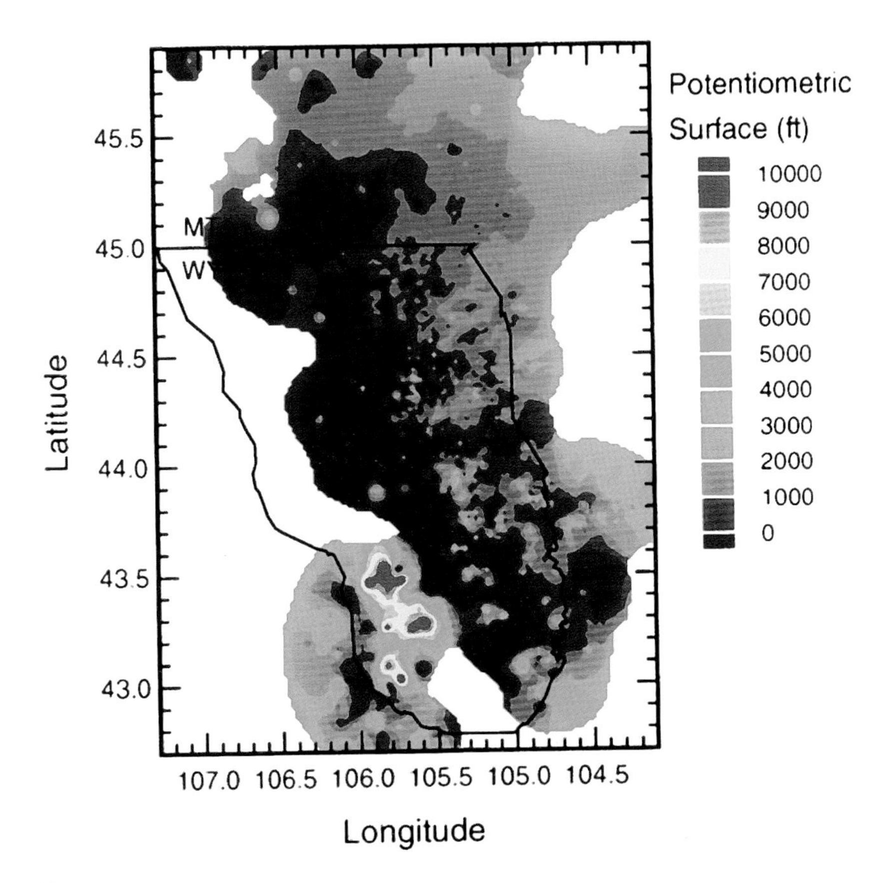

Figure 7 (continued). (C) Frontier Formation (821 data values), (D) Muddy Formation (3888 data values).

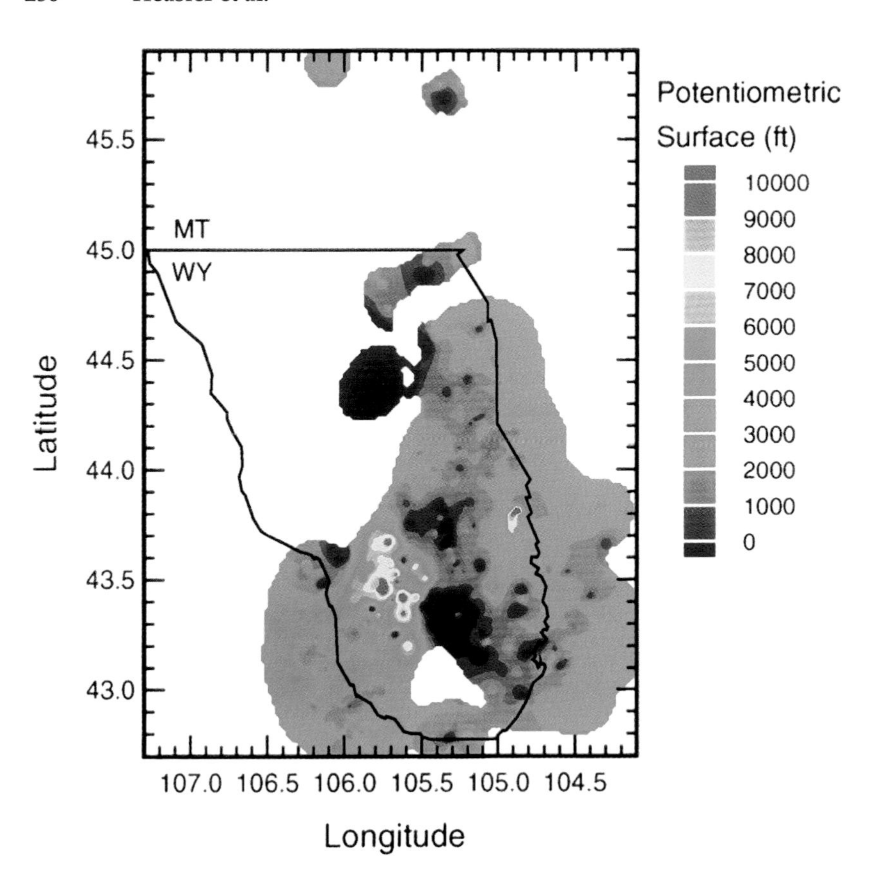

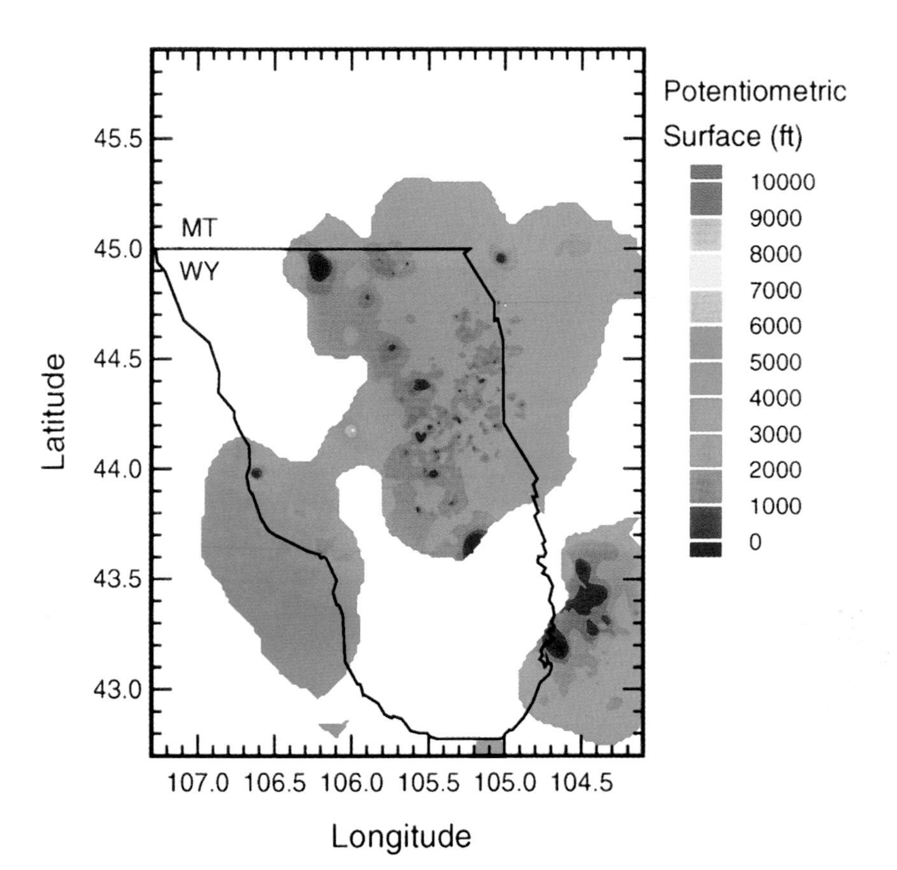

Figure 7 (continued). (E) Dakota Formation (1157 data values), and (F) Minnelusa Formation (4470 data values).

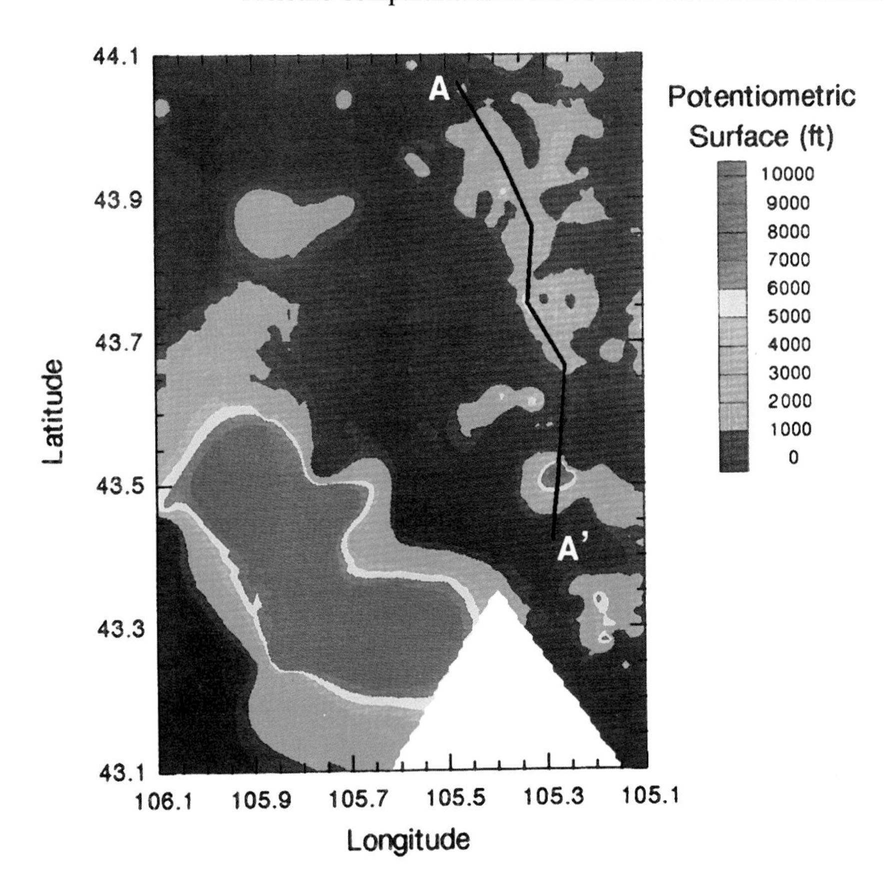

Figure 8. Enlarged potentiometric surface of drill-stem test data from the Muddy Formation covering the Hilight oil field (northeast quadrant of map) and the Powell oil field (southwest quadrant of map).

within the anomalies. This characteristic is one essential aspect of the definition of a pressure compartment.

Figure 12 shows that in the region of the Hilight oil field, the potentiometric surface of the Muddy Sandstone is mostly below the ground surface; consequently, these are underpressured anomalies. In contrast, the hydraulic head anomalies in the Dakota and Frontier formations in the vicinity of the Powell hydrocarbon field are highly overpressured (see Figures 11 and 13); in this example, the tops of some anomalies are overpressured to 6000 ft (1800 m) above the ground surface.

CONCLUSIONS

From the three-dimensional analysis of potentiometric surfaces, two important aspects of the abnormally pressured regions identified in the Powder River basin are apparent. First, the pressure anomalies are laterally characterized by steep hydraulic head gradients (up to 12,000 ft [3700 m] difference). It is common for the potentiometric surface to vary 5000 ft (1500 m) or more across narrow horizontal distances. In order for these hydraulic head gradients to be maintained in the formations, there must be an impermeable barrier (seal) between them. In fact, one possible interpretation of the pressure data is that the particularly low pressure measurements that result in the steep head gradients are the result of making the drill-stem tests in impermeable rock. Thus, the steep hydraulic head gradients are indicating discontinuities either in the hydraulic head or in permeability. Second, from the cross sections through the anomalies in the potentiometric surfaces, it is apparent that they generally are characterized by relatively flat tops. This shape strongly suggests that within these potentiometric surface anomalies there is little or no hydrodynamic flow. The most plausible explanation of the pressure anomalies within those test formations characterized by steep hydraulic head gradients and a lack of hydrodynamic flow is that they are pressure compartments: The pressure anomalies are prisms of rock hydrodynamically isolated from adjacent rocks and characterized by a lack of hydrodynamic flow. If these pressure anomalies are to be explained by aquitards, it would require that they be completely surrounded by aquitards with some continuing source of fluid generation, but both the lack of hydrodynamic flow and the geologic history of the Powder River basin make this hypothesis untenable. It is concluded that those parts of the various reservoir sandstones tested in the Powder River basin that are characterized by abnormally high pressure are examples of pressure compartmentalization as described by Hunt (1990).

The pressure anomalies described in this study are from a wide variety of sandstones: eolian, fluvial, and deltaic (Martinsen, this volume), over a wide stratigraphic interval, and conform to the definition of a

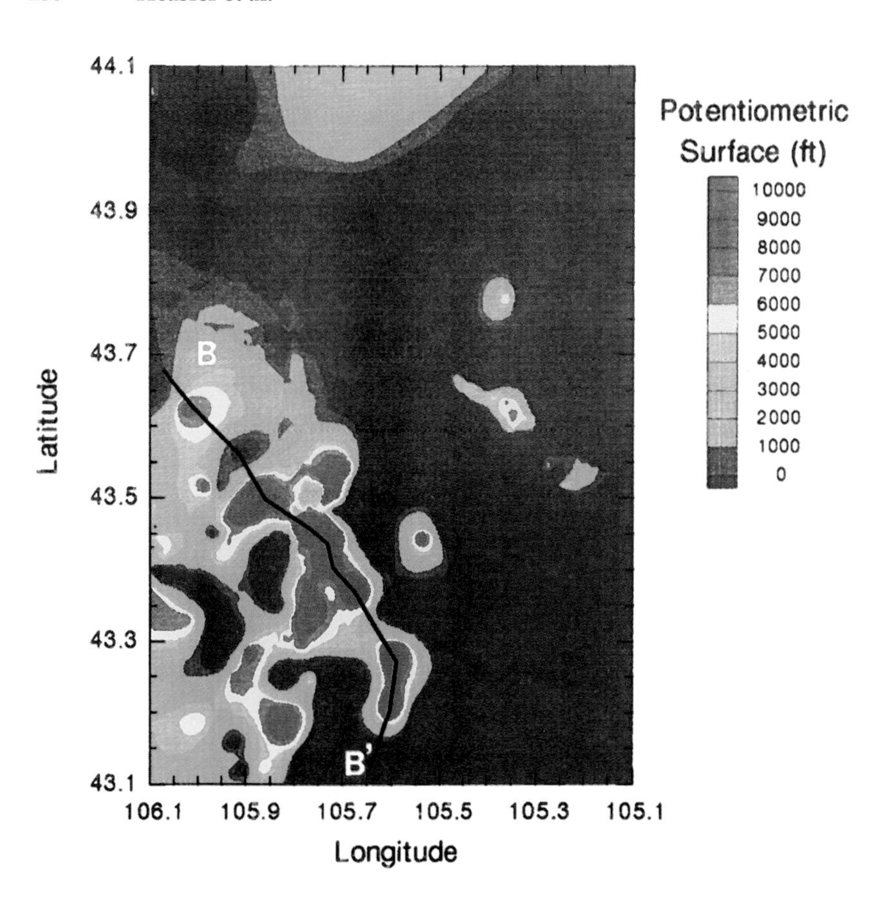

Figure 9. Enlarged potentiometric surface of drill-stem test data from the Frontier Formation covering the Hilight oil field (northeast quadrant of map) and the Powell oil field (southwest quadrant of map).

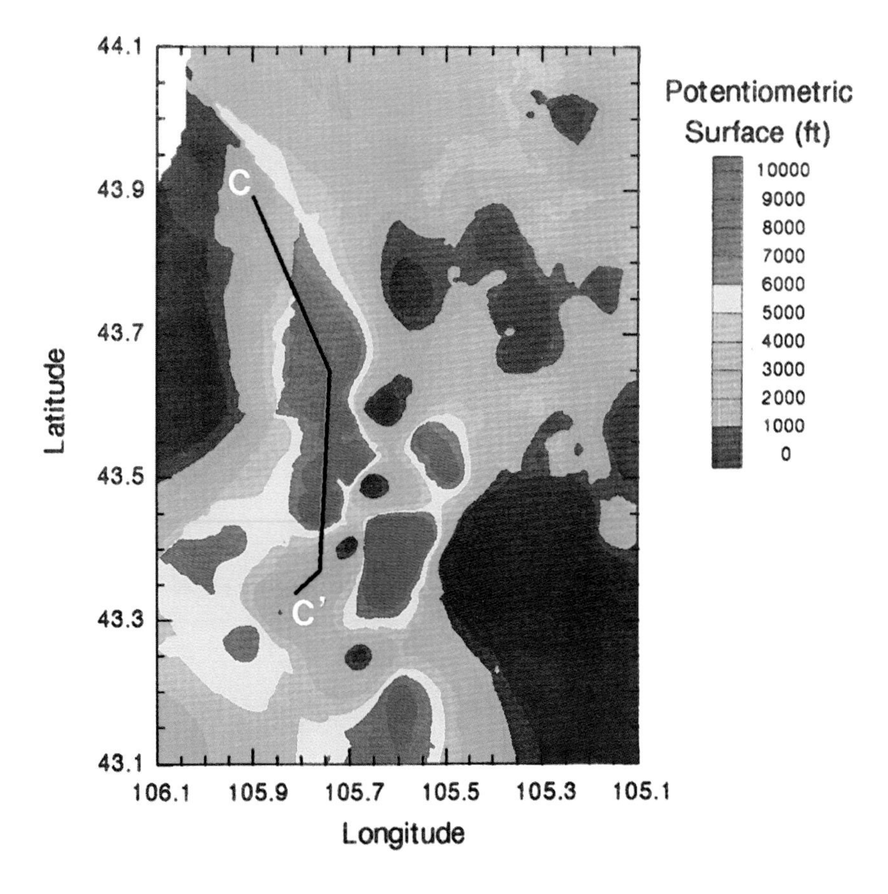

Figure 10. Enlarged potentiometric surface of drill-stem test data from the Dakota Formation covering the Hilight oil field (northeast quadrant of map) and the Powell oil field (southwest quadrant of map).

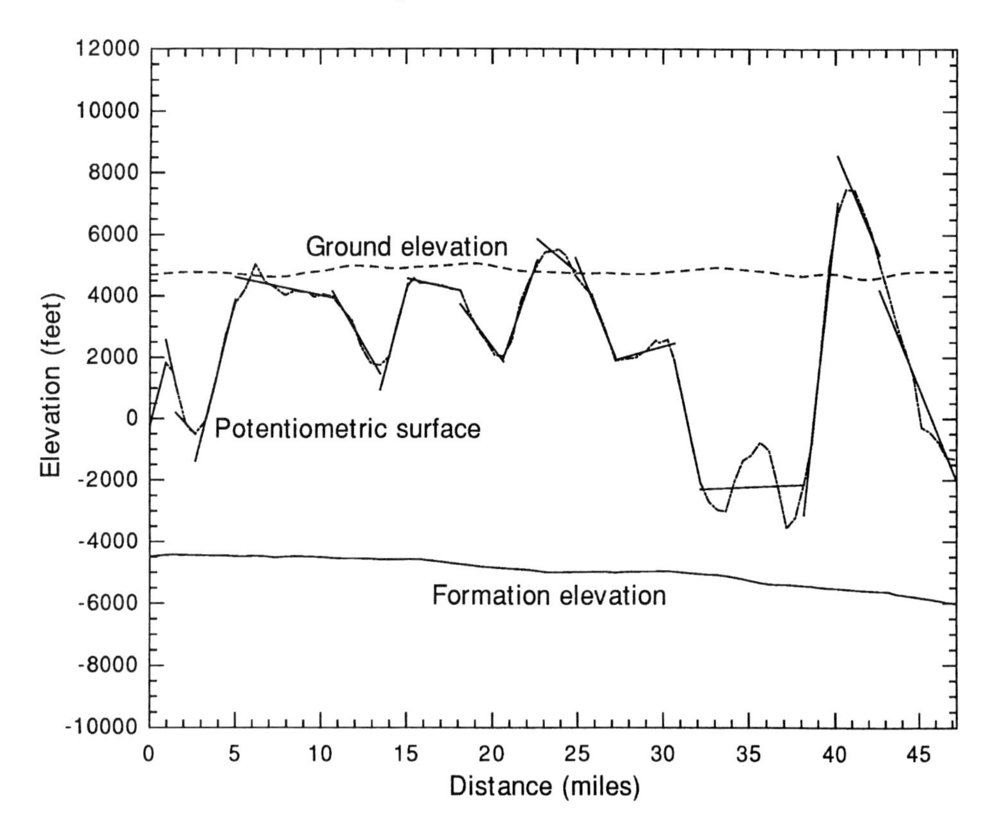

Figure 11. Cross section A–A′ (see Figure 8 for location) through the Hilight oil field showing ground elevation, elevation of the test interval, elevation of the potentiometric surface, and piecewise linear least-squares fit of the potentiometric surface for the Muddy Formation.

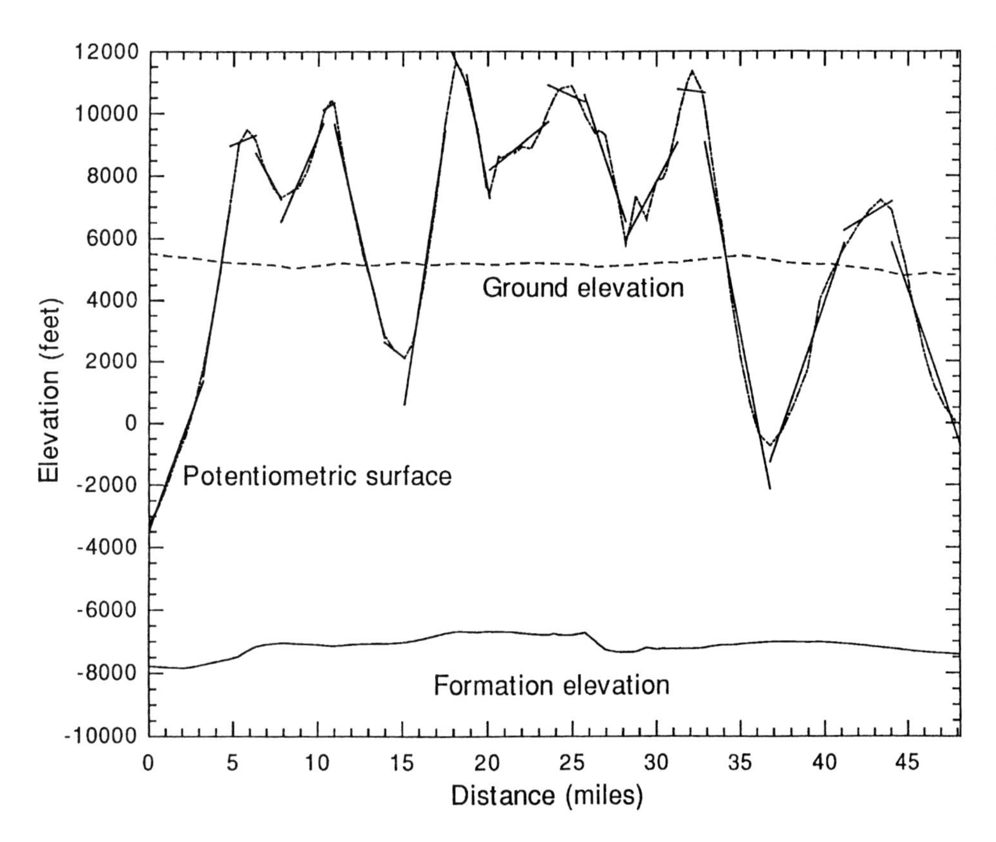

Figure 12. Cross section B–B′ (see Figure 9 for location) through the Powell oil field showing ground elevation, elevation of the test interval, elevation of the potentiometric surface, and piecewise linear least-squares fit of the potentiometric surface for the Frontier Formation.

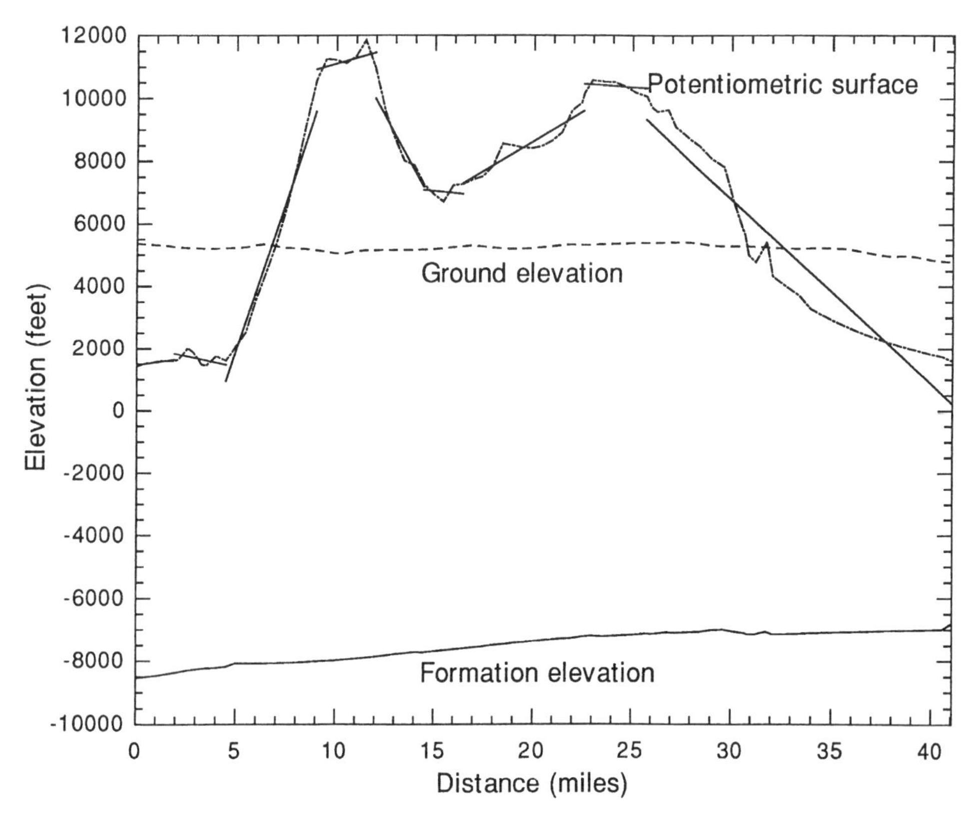

Figure 13. Cross section C–C′ (see Figure 10 for location) through the Powell oil field showing ground elevation, elevation of the test interval, elevation of the potentiometric surface, and piecewise linear least-squares fit of the potentiometric surface for the Dakota Formation.

pressure compartment given earlier. Dimensionally, the pressure compartments described herein tend to be of similar magnitude in anomalous pressure to those described by Hunt (1990), but typically are of smaller scale in volume. None of the pressure compartments delineated in this study can be considered to be part of a basinwide continuous compartment. However, it should be noted that commonly, the individual compartments in a specific stratigraphic unit occur at similar depths on pressure-depth or pressure-elevation plots (Figures 5 and 6). In conclusion, it can be stated that on the basis of an analysis of pressure data, numerous pressure compartments exist in the Frontier, Muddy, and Dakota formations in the Powder River basin.

ACKNOWLEDGMENTS

This material covers research performed through November 1991. The original manuscript was reviewed by David Copeland (University of Wyoming), who made numerous helpful suggestions, and by Peter Huntoon and an anonymous reviewer, whose technical comments greatly improved the quality of the manuscript. Robert Simon was responsible for completion of the graphical depiction of the computer database. This study was funded by the Gas Research Institute under contract number 5089-260-1894.

REFERENCES CITED

Asquith, D. O., 1970, Depositional topography and major marine environments, Late Cretaceous, Wyoming: American Association of Petroleum Geologists Bulletin, v. 54, p. 1184–1224.

Asquith, D. O., 1974, Sedimentary models, cycles, and deltas, Upper Cretaceous, Wyoming: American Association of Petroleum Geologists Bulletin, v. 58, p. 2274–2283.

Bair, E. S., T. P. O'Donnel, and L. W. Picking, 1985, Potentiometric mapping from incomplete drill-stem test data, Palo Duro Basin area, Texas and New Mexico: Groundwater, v. 25, p. 198–211.

Berg, R. R., 1962, Mountain flank thrusting in Rocky Mountain foreland, Wyoming and Colorado: American Association of Petroleum Geologists Bulletin, v. 46, p. 2019–2032.

Bethke, C. M., W. J. Harrison, C. Upson, and S. P. Altaner, 1988, Supercomputer analysis of sedimentary basins: Science, v. 239, p. 261–267.

Blackstone, D. L., Jr., 1981, Compression as an agent in deformation of the east central flank of the Bighorn Mountains, Sheridan and Johnson Counties, Wyoming: Contributions to Geology, University of Wyoming, v. 19, p. 105–122.

Blackstone, D. L., Jr., 1990, Precambrian basement map of Wyoming: outcrop and structural configuration: Geological Survey of Wyoming, Map

Series 27, Scale 1:1,000,000.

Bradley, J. S., 1975, Abnormal formation pressure: American Association of Petroleum Geologists Bulletin, v. 59, p. 957–973.

Bredehoeft, J. D., 1965, The drill-stem test: the petroleum industry's deep-well pumping test: Groundwater, v. 3, p. 31–36

Coughlan, J. P., and J. R. Steidtmann, 1984, Depositional environment and diagenesis of the Teapot Sandstone, southern Powder River Basin, Wyoming: Mountain Geologist, v. 21, p. 91–103.

Dahlberg, E. C., 1982, Applied Hydrodynamics in Petroleum Exploration: New York, Springer-Verlag, 155 p.

Dana, G. F., 1962, Groundwater reconnaissance of the State of Wyoming: Wyoming Natural Resource Board, Cheyenne.

Desmond, R. J., J. R. Steidtmann, and D. F. Cardinal, 1984, Stratigraphy and depositional environments of the Middle Member of the Minnelusa Formation, central Powder River Basin, Wyoming: Wyoming Geological Association, 35th Annual Field Conference Guidebook, p. 213–240.

Doremus, D. M., 1986, Groundwater circulation and water quality associated with the Madison Aquifer in the northeastern Bighorn Basin, Wyoming: M.S. thesis, University of Wyoming, 81 p.

Feathers, K. R., R. Libra, and T. R. Stephenson, 1981, Occurrence and characteristics of groundwater in the Powder River Basin, Wyoming: Water Resources Research Institute, University of Wyoming, report to the U.S. Environmental Protection Agency, 171 p.

Flanagan, K. M., 1990, Late Cenozoic geology of the Pathfinder region, central Wyoming, with tectonic implications for adjacent areas: Ph.D. dissertation, University of Wyoming, 186 p.

Fryberger, S. G., 1984, The Permian Upper Minnelusa Formation, Wyoming: ancient example of an offshore-prograding eolian sand sea with geomorphic facies, and system-boundary traps for petroleum: Wyoming Geological Association, 35th Annual Field Conference Guidebook, p. 241–271.

George, G. R., 1984, Cyclic sedimentation and depositional environments of the upper Minnelusa Formation, central Campbell County, Wyoming: Wyoming Geological Association, 35th Annual Field Conference Guidebook, p. 75–95.

Gill, J. R., and W. A. Cobban, 1973, Stratigraphy and geologic history of the Montana Group and equivalent rocks, Montana, Wyoming, and North Dakota: U.S. Geological Survey Professional Paper 776, 37 p.

Gries, R., 1983, Oil and gas prospecting beneath Precambrian of foreland thrust plates in Rocky Mountains: American Association of Petroleum Geologists Bulletin, v. 67, p. 1–28.

Hobson, J. P., M. L. Fowler, and E. A. Beaumont, 1982, Depositional and statistical exploration models, Upper Cretaceous offshore sandstone complex, Sussex Member, House Creek field, Wyoming: American Association of Petroleum Geologists Bulletin, v. 66, p. 689–707.

Hodson, W. G., 1971, Chemical analyses of ground water in the Powder River Basin and adjacent areas, northeastern Wyoming: Wyoming Department of Economic Planning and Development, Cheyenne, 18 p.

Hodson, W. G., 1974, Records of water wells, springs, oil- and gas-test holes and chemical analyses of water for the Madison Limestone and equivalent rocks in the Powder River Basin and adjacent areas, northeastern Wyoming: U.S. Geological Survey Open File Report, 24 p.

Hodson, W. G., R. H. Pearl, and S. A. Druse, 1973, Water resources of the Powder River Basin and adjacent areas, northeastern Wyoming: U.S. Geological Survey, Hydrologic Investigations Atlas, HA-465, 4 sheets.

Horner, D. R., 1951, Pressure build-up in wells: Proceedings of the Third World Petroleum Congress, Section II, Leiden, Holland, p. 503–521.

Hubbert, M. K., and W. W. Ruby, 1959, Role of fluid pressure in mechanics of overthrust faulting: Geological Society of America Bulletin, v. 70, p. 115–166.

Hunt, J. M., 1990, Generation and migration of petroleum from abnormally pressured fluid compartments: American Association of Petroleum Geologists Bulletin, v. 74, p. 1–12.

Huntoon, P. W., 1985a, Fault severed aquifers along the perimeters of Wyoming artesian basins: Ground Water, v. 23, p. 176–181.

Huntoon, P. W., 1985b, Rejection of water from Madison aquifer along eastern perimeter of Bighorn artesian basin, Wyoming: Ground Water, v. 23, p. 345–353.

Jarvis, T. W., 1986, Regional hydrogeology of the Paleozoic Aquifer System, Southeastern Bighorn Basin Wyoming, With a User Impact Analysis on Hot Spring State Park: M.S. thesis, University of Wyoming, 227 p.

King, P. B., and H. M. Beikman, 1974, Geologic map of the United States, scale 1:2,500,000, United States Geological Survey.

Lobmeyer, D. H., 1980, Preliminary potentiometric-surface map showing fresh water heads for the Lower Cretaceous rocks in the northern Great Plains of Montana, North Dakota, South Dakota, and Wyoming: U.S. Geological Survey Open File Report 80-757, scale 1:500,00.

Love, J.D., 1970, Cenozoic sedimentation and crustal movement in Wyoming: American Journal of Science, v. 258A, p. 204–214.

Mancini, A. J., 1976, Investigation of recharge to ground water reservoirs of northeastern Wyoming (Powder River Basin): For the Old West Regional Commission, Wyoming State Engineer's Office, 111 p.

Marrs, R. W., and G. L. Raines, 1984, Tectonic framework of Powder River Basin, interpreted from Landsat Imagery: American Association of Petroleum Geologists Bulletin, v. 68, p. 1718–1731.

Martinsen, R. S., and R. W. Marrs, 1985, Comparison of major lineament trends to sedimentary rock thicknesses and facies distribution, Powder River Basin, Wyoming: Fourth Thematic Conference in "Remote Sensing for Exploration Geology," San Francisco.

Martinsen, R. S., D. J. P. Swift, and G. C. Gaynor, 1984, Local and regional cross-sections through the Mowry-Teapot sandstone interval, Upper Cretaceous of the Powder River Basin, Wyoming: ARCO Oil and Gas Company Research Report No. 34:24.

McKenna, M.C., and J. D. Love, 1972, High-level strata containing early Miocene mammals on the Bighorn Mountains, Wyoming: American Museum Novitates, No. 2490, 31 p.

Miller, W. R., 1976, Water in carbonate rocks of the Madison Group in southeastern Montana—a preliminary evaluation: United States Geological Survey Water Supply Paper 2043, 51 p.

Rahn, P. H., and J. P. Gries, 1973, Large springs in the Black Hills, South Dakota and Wyoming: South Dakota Geologic Survey Report of Investigations No. 107, 46 p.

Richter, H. R., 1981, Occurrence and characteristics of groundwater in the Wind River Basin, Wyoming: Water Resources Research Institute, University of Wyoming, report to the U.S. Environmental Protection Agency, 149 p.

Sales, J. K., 1971, Structure of the northern margin of the Green River Basin, Wyoming: Wyoming Geological Association, 23rd Annual Field Conference Guidebook, p. 85–102.

Spencer, S. A., 1986, Groundwater movement in the Paleozoic rocks and impact of petroleum production on water levels in the southeastern Bighorn Basin, Wyoming: M.S. thesis, University of Wyoming, 165 p.

Swenson, F. A., W. R. Miller, W. G. Hodson, and F. M. Visher, 1976, Maps showing configuration and thickness and potentiometric surface and water quality in the Madison Group, Powder River Basin, Wyoming and Montana: U.S. Geological Survey Map I-847-C.

Theis, C. V., 1935, The relation between the lowering of the piezometric and the rate and duration of discharge of a well using ground-water storage: American Geophysical Union Transactions, v. 16, p. 519–524.

Tillman, R. W., and R. S. Martinsen, 1984, The Shannon shelf-ridge sandstone complex, Salt Creek Anticline, Powder River Basin, Wyoming, *in* R. Tillman and C. Siemers, eds., Siliciclastic Shelf Sedimentation: Society of Economic Paleontologists and Mineralogists Special Publication 34, p. 85–142.

Tillman, R. W. and R. S. Martinsen, 1987, Sedimentologic model and production characteristics of Hartzog Draw Field, Wyoming, a Shannon shelf-ridge sandstone, *in* R. Tillman and K. Weber, eds., Reservoir Sedimentology: Society of Economic Paleontologists and Mineralogists, Special Publication 40, p. 15–112.

Weimer, R. J., 1984, Relation of unconformities, tectonics, and sea-level changes, Cretaceous of Western Interior, U.S.A., *in* J. Schlee, ed., Interregional Unconformities and Hydrocarbon Accumulation: American Association of Petroleum Geologists Memoir 36, p. 7–35.

Wyoming Geological Association, 1957 (Supplemented 1961), Wyoming Oil and Gas Fields Symposium: 579 p.

Wyoming Geological Association, 1958, Powder River Basin: Wyoming Geological Association 10th Annual Field Conference Guidebook, 341 p.

Wyoming Geological Association, 1978, Resources of the Wind River Basin: Wyoming Geological Association, 30th Annual Field Conference Guidebook, 414 p.

Chapter 17

Geostatistical Methods for the Study of Pressure Compartments: A Case Study in the Hilight Oil Field, Powder River Basin, Wyoming

X. Chen*
H. P. Heasler
L. Borgman
University of Wyoming
Laramie, Wyoming, U.S.A.

ABSTRACT

This chapter introduces the variogram method for analyzing the geometry of pressure compartments. Pressure and hydraulic head data from 192 wells in the Hilight oil field, Powder River basin, Wyoming, were used in this study. Variograms were calculated for various directions. Results indicate that hydraulic head and pressure are spatially correlated, especially in directions parallel to the axes of elongation of compartment cells. The average pressure compartment was estimated via the variogram method to be an ellipse with major and minor axes of 5.3 and 2.7 mi (8.5 and 4.3 km). The major axis trends 15° azimuthally, the minor axis 285°.

INTRODUCTION

Most variables in the earth sciences are functions of spatial position. In hydrogeology, for example, hydraulic head, hydraulic conductivity, and storage coefficient have been shown to be spatially correlated (Hoeksema and Kitanidis, 1985; Marsily, 1986). In general, measurements on two points close to each other in space are more likely to be similar in value than measurements on two points farther apart. In hydrogeology, the spatial structure of aquifer properties has been under investigation for more than a decade (Baker et al., 1978; Mizell et al., 1982; Hoeksema and Kitanidis, 1985).

The estimation of the geometry and orientation of pressure compartments would be an important contribution to the study of oil reservoirs. The individual compartments in an oil stratum are like huge bottles: each one has a thin, impermeable outer seal and an internal volume with effective hydraulic communication (Powley, 1990). An abnormally pressured compartment has an internal pressure higher or lower than that of the surrounding areas. In other words, a compartment is characterized by (1) a hydraulic system within a compartment different from the hydraulic system outside the compartment and (2) fluid communication in the compartment having resulted in the spatial delineation of abnormal pressure or hydraulic head.

These characteristics of pressure compartment cells make it possible to use geostatistical methods to quantify the spatial variation of an oil reservoir. In this study, we use the variogram method to analyze pressure and hydraulic head data and to estimate the heterogeneity, anisotropy, and average size and shape of a set of pressure compartments.

* Present address: University of Nebraska, Lincoln, Nebraska, U.S.A.

REVIEW OF SEMIVARIOGRAM METHOD

The semivariogram and covariance functions are commonly used in geostatistics to determine the spatial structure of a random function. The semivariogram $\gamma(\boldsymbol{h})$ is defined as the half-mean quadratic increment for two points a distance h from one another (Marsily, 1982):

$$\gamma(\boldsymbol{h}) = 0.5E\left\{\left[V(x) - V(x+\boldsymbol{h})\right]^2\right\}$$

and the definition of the covariance function $C(\boldsymbol{h})$ is

$$C(\boldsymbol{h}) = E\left\{\left[V(x) - \mu(x)\right]\left[V(x+\boldsymbol{h}) - \mu(x+\boldsymbol{h})\right]\right\}$$

where $V(x)$ is a random function, $\mu(x)$ is the local mean, x is the vector of space coordinates representing location, $\boldsymbol{h}$ is a vector representing the distance between two locations (defined as lag), and E is the expectation. A semivariogram is often called a variogram for simplicity. Both the covariance function $C(\boldsymbol{h})$ and the variogram $\gamma(\boldsymbol{h})$ are only functions of $\boldsymbol{h}$, but not functions of the location x if a medium in space is held stationary.

In geostatistics, some workers prefer to use the covariance function, while others use the variogram to study the spatial structure of a random function. The relationship between the covariance function and the variogram for a stationary random function can be expressed as:

$$\gamma(\boldsymbol{h}) = C(0) - C(\boldsymbol{h}),$$

where $C(0)$ is the variance of the random function.

Three types of variograms are commonly used: Gaussian, exponential, and spherical. The mathematical expressions are:

$$\text{Gaussian: } \gamma(\boldsymbol{h}) = c\left\{1 - \exp\left[-(\boldsymbol{h}/d)^2\right]\right\}$$

$$\text{Exponential: } \gamma(\boldsymbol{h}) = c\left[1 - \exp(-\boldsymbol{h}/d)\right]$$

$$\text{Spherical: } \gamma(\boldsymbol{h}) = \begin{Bmatrix} c\left[1.5(\boldsymbol{h}/h_0) - 0.5(\boldsymbol{h}/h_0)^3\right] \text{ if } h < h_0 \\ c \quad \text{otherwise} \end{Bmatrix},$$

where c is the height of sill (the limiting value of the variogram), h_0 is the range of influence, and d is a parameter of the variogram associated with h_0. The range of influence is a distance beyond which the variogram or covariance value remains essentially constant (Isaaks and Srivastava, 1989). Two measured values in space at a distance from each other less than a distance equal to the range of influence are considered to be correlated. The ranges of influence for the Gaussian and exponential models are approximately $\sqrt{3}$d and 3d, respectively (Marsily, 1986).

The Gaussian model represents a realization of extreme smoothness at very short separation. The exponential and spherical models represent average smoothness (Matheron, 1963; Isaaks and Srivastava, 1989; Easley et al., 1991). For two-dimensional space, an ellipse is usually used to represent the anisotropic structure of a medium.

Matheron (1963) summarized three essential characteristics of a realization (defined as a set of values for a random function) expressed by the variogram:

1. The behavior of g(h) near the origin reflects the regularity of the realization. A smooth origin represents a variable with continuity, while a discontinuity at the origin, termed a nugget effect, corresponds to a variable with small-scale variability, or to heterogeneity of the variable.
2. The variogram is not only a function of lag length, but also a function of the direction of vector $\boldsymbol{h}$. A difference among directional variograms indicates an anisotropy that may be instructive in geological interpretation.
3. Structural characteristics are also reflected in the variogram through the range of influence, which may be equal to the mean diameter of a geological body. The fact that this range varies with direction makes it possible to determine the geometry of a geological body.

In addition, variograms can also show such other behaviors as periodic structure (Marsily, 1982) and lenticular shape of a realization.

DATA AND METHODOLOGY

The Muddy Sandstone in the Hilight oil field, Powder River basin, Wyoming, was selected for this study. Data from 215 drill-stem tests (DSTs) were retrieved from a database purchased from Petroleum Information, Inc. The study area covers approximately 1152 mi^2 (2984 km^2) (latitude 43.74–44.10°N, longitude 105.05–106.20°W). Major development of the Hilight field started in 1969 and continued until the middle of the 1980s. Data points are unequally distributed, and most are concentrated in the east area, with only 14 in the west area (Figure 1). The surface elevation of the Hilight oil field ranges from 4400 to 4800 ft (1300–1500 m) above sea level. Heasler et al. (this volume) discuss the existence of pressure compartments in this area.

The data consist of the pressure information for a well. The shut-in pressure p of a well from DST data is converted to hydraulic head H using the Bernoulli equation,

$$H = z + \frac{p}{\rho g},$$

where H is hydraulic head, z is elevation head, p is pressure (DST data), ρ is the density of the fluid, and g is the acceleration of gravity. The term $p/\rho g$ can be called pressure head.

Several factors affect the calculation of hydraulic head. They are the fluid density, the measured shut-in pressure values, the production of fluid, and the estimated formation elevation. Among these, the measured pressure values and production of fluid have the most significant effect on the calculated values of hydraulic head. Each drill-stem test has two measurements of pressures in a well, the initial and final shut-in pressures. The initial and final shut-in pressures

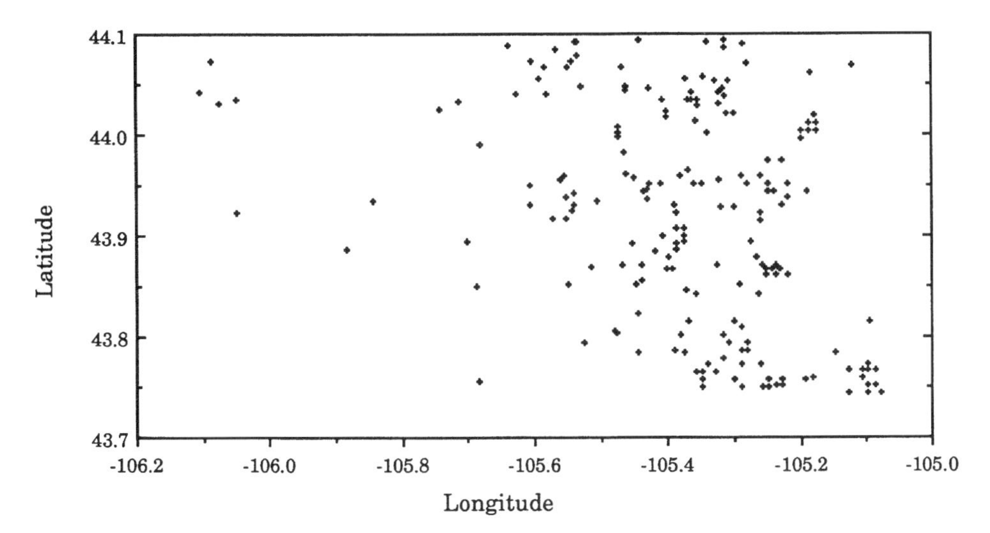

Figure 1. Study area and data distribution.

are not always identical. The difference between them may reach tens to hundreds of pounds per square inch (psi).

The initial and final shut-in pressures in more than 90% of the wells in this study area are actually different. A difference of 100 psi pressure is equal to about 231 ft (70 m) of hydrostatic head. Bair et al. (1985) used the initial shut-in pressure for the calculation of hydraulic head, while Javis (1986) and Spencer (1986) used the higher of the initial and final shut-in pressures for hydrostatic head calculation. Table 1 shows the decrease of pressure in the Muddy Sandstone with time in the study area. The decrease of pressure is mainly due to the production of fluid. Although bias can be brought into the estimation of hydraulic head from DSTs, DST data are still the most important source of information for the construction of potentiometric surfaces and for the study of the fluid dynamics and spatial structure of aquifer properties. The decrease of pressure in later measurements has been caused by production of fluid, but those later data are still usable in providing information about the oil reservoir.

In this study, the maximum shut-in pressures were used for the calculation of hydraulic head. Those data in which both initial and final shut-in pressures are equal to zero were not used; and 192 out of the 215 totally retrieved data points were used to quantify spatial variation in the Muddy Sandstone. The density of water of 1 g/cc was used for the conversion of pressure head. The middle point of the drill-stem test interval was taken as the elevation head. A potentiometric surface map of the Muddy Sandstone in the east study area is shown in Figure 2.

Table 1. Average pressures in Hilight oil field, Powder River basin, Wyoming, at different stages of field development.

		Pressure	
	Number of Data	ISIP (psi)	FSIP (psi)
Before 12/31/74	102	1832	2106
1/1/75–8/31/85	60	1578	1749
After 9/1/85	53	1416	1521
Total	215	—	—
Weighted average	—	1659	1862

STATISTICAL DISTRIBUTION OF PRESSURE AND HYDRAULIC HEAD

Both shut-in pressure and hydraulic head are highly variable. The variation of their values is several orders of magnitude. The average value of hydraulic head is 949 ft (289 m) above sea level, and the average pressure is 2176 psi. Neither average represents the most concentrated part of the data; they lie in the part where the frequency of data is low.

Frequency histograms of pressure and head are shown in Figures 3 and 4. Both show bimodal distribution. This property indicates that two hydraulic regimes exist in the Muddy Sandstone and implies that the aquifer is heterogenous. The right-hand population, with higher values, is probably associated with a pressure compartment cell, while the left-hand population, with lower values, probably represents the area outside the cell.

Although fluid within a compartment cell is in hydraulic continuity, it is unlikely that DST pressures are constant within a pressure compartment. The variation of pressure values in a compartment cell may be caused by (1) historical production of fluid, (2) unequal shut-in DST measurement periods, (3) spatial change of permeability, and (4) differing depths in the DST interval.

Although both pressure and head histograms are bimodally distributed, there are differences between them. The left-hand pressure population is skewed to

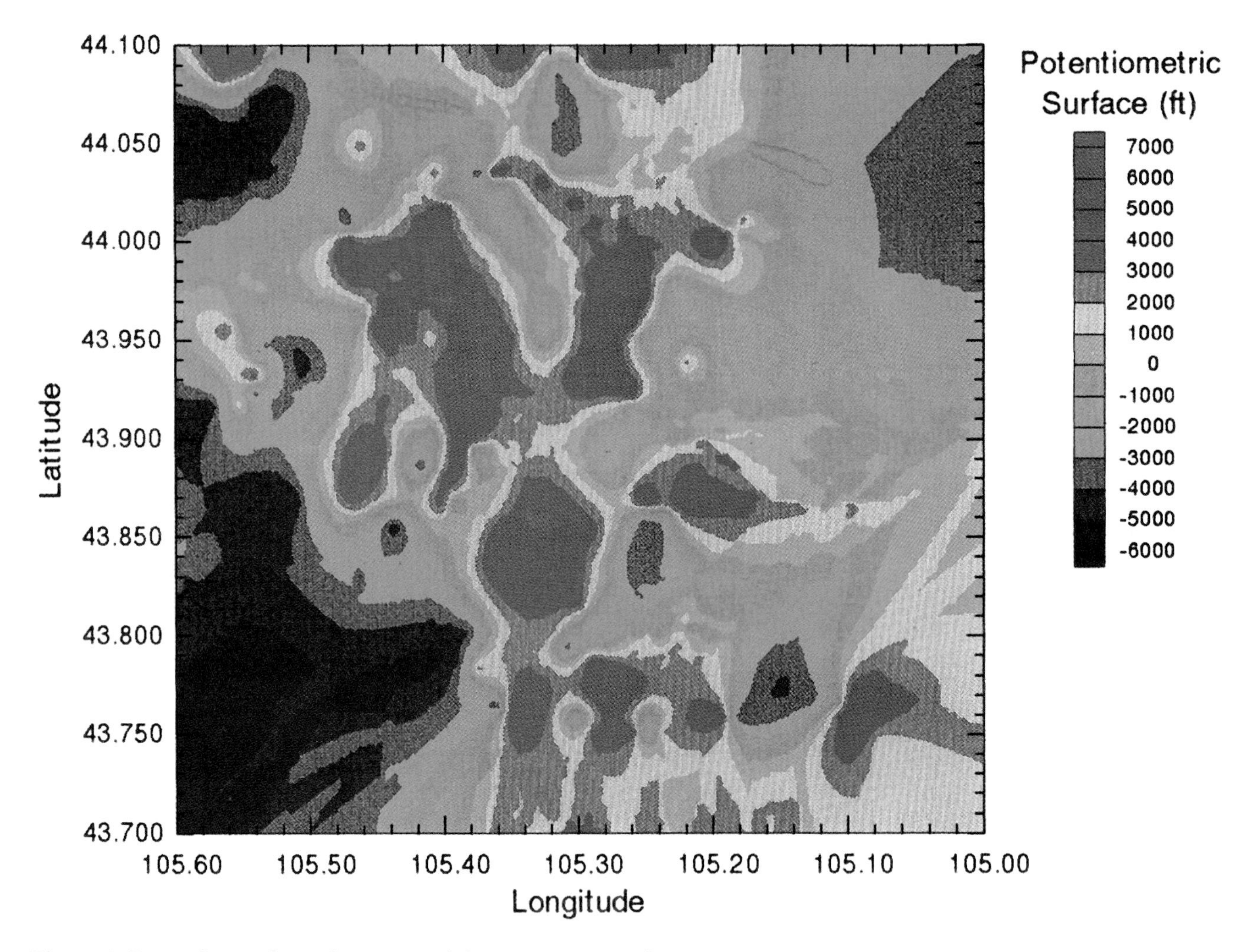

Figure 2. Potentiometric surface map of the eastern part of the study area.

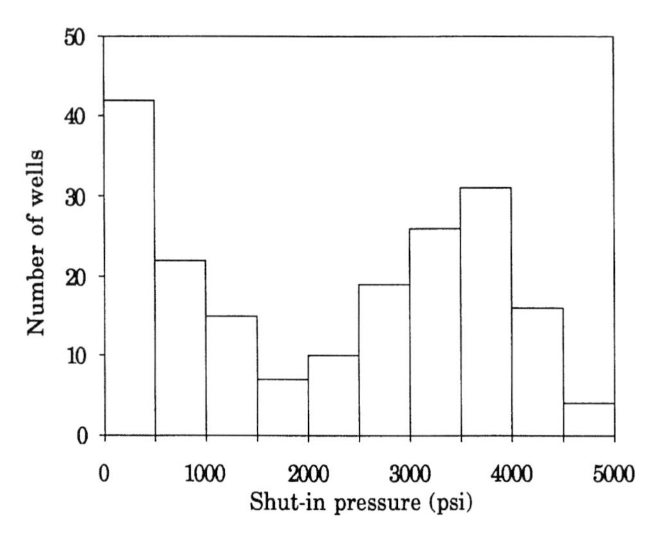

Figure 3. Histogram of maximum shut-in pressure in pounds per square inch (psi) for the study area.

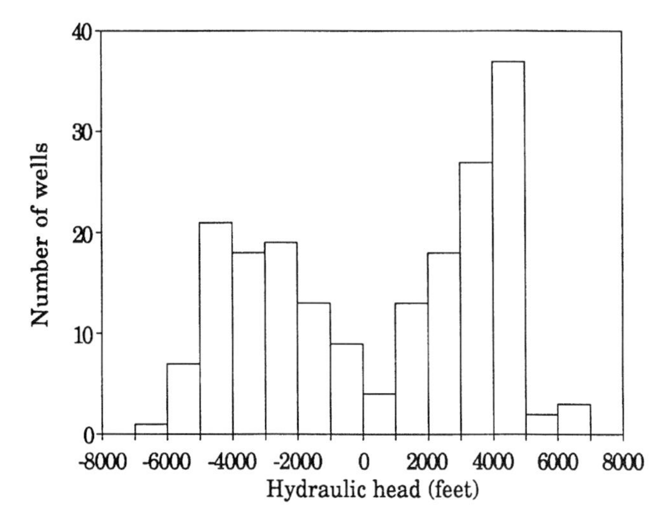

Figure 4. Histogram of maximum hydraulic head in the study area.

the left, while the right-hand head population is skewed to the right. The higher peak of pressure is in the left-hand population, while the higher peak of hydraulic head is in the right-hand population. These differences are probably caused by hydraulic head being the sum of pressure and elevation heads. The Muddy Sandstone in the Hilight field dips southwest about 4°.

The variances of pressure and head are extremely large due to the wide range of data values and the bimodal distribution. However, the variance of each population is about one order of magnitude lower than the variance of the whole data set. Table 2 summarizes the mean, variance, and minimum and maximum values of pressure and hydraulic head of the Muddy Sandstone in the Powder River basin.

EXPERIMENTAL VARIOGRAMS

The experimental variograms of pressure and head are calculated in various directions using the Geo-eas (Geostatistical Environment Assessment Software) program (Englund and Sparks, 1988). They are scatter plots of $\gamma(h)$ versus lag. The parameters of each variogram are (1) direction, (2) variogram model: Gaussian, exponential, or spherical, (3) range of influence, (4) nugget effect, and (5) height of sill. A continuous curve based on the estimated variogram parameters is drawn. Variograms are calculated only for the east area because the west area contains too few data points to obtain parameters with reasonable certainty (Hoeksema and Kitanidis, 1985). In addition, head and pressure variograms for the right-hand populations (Figures 3 and 4) are also calculated. In the calculation, north is defined as 0° and east as 90°. Azimuthal directions are given, in degrees clockwise from true north. One lag unit is equal to 6 mi (10 km) in all variograms.

Variograms of Pressure

The greater of each pair of shut-in pressure values, initial and final, is used for the calculation of the pressure variogram. The pressure variogram appears highly anisotropic. It is observed that not only are the values of range of influence, height of sill, and nugget effect functions of direction, but also that the variogram model changes with direction.

The range of influence in the north direction is generally greater than that in the east direction. The average range of influence in the sector 50 to 70° is 1 lag unit (6 mi [10 km]), and 1.4 lag units (8 mi [13 km]) in the sector 340 to 15°. The nugget effect represents the variance of small-scale fluctuations. It is lowest in the sector 25 to 60° and highest in the sector 270 to 300°. Height of sill expresses the variance of correlated fluctuations. It becomes smaller from 345 to 25°. The nugget effect, height of sill, and other variogram parameters are summarized in Table 3.

The variogram model changes from east to north. In the sector 35 to 80°, the exponential model seems to fit the data, while the spherical model fits the data in the sector 20 to 30° and the Gaussian model fits the sectors 10 to 15° and 340 to 345°. Variograms in the sector 275 to 335° are wavy or periodic.

The highly anisotropic structure of the variogram is probably caused by the property of heterogeneity and by the irregularity of data point spacing. Figures 5A, 5B, and 5C are three variograms along 45°, 15°, and 300° with tolerance 22.5°. Because of high anisotropy, the spatial structure cannot be described by a single variogram model.

Variograms of Head

In general, variograms of hydraulic head are similar to those of pressure. For example, Figures 6A and 6B show the similarity between variograms of head and pressure at 55° with tolerance 22.5°. Both are fitted very well by an exponential model with the same range of influence of 1.0 lag unit (6 mi [10 km]). The only difference between them is that the values of $\gamma(h)$ in the head variogram at different lags are much closer to the fitted curve.

ESTIMATION OF THE AVERAGE SIZE OF THE PRESSURE COMPARTMENT

Experiment for the Range of Influence

An experiment was carried out in order to understand how to use the range of influence for the estimation of

Table 2. Summary of the basic statistical properties of pressure and hydraulic head data for the variogram analyses.

	Minimum			Maximum			Mean			Variance		
Population	both	left	right	both	left	right	both	left	right	both	left	right
Number of data	192	85	107	192	85	107	192	85	107	192	85	107
Pressure	63	63	2021	4895	1998	4895	2176	667	3400	2.2E+6	2.5E+5	4.3E+5
Head	−6744	−6744	250	6590	−207	6590	470	−3081	3475	1.3E+7	2.1E+6	1.7E+6

Table 3. Summary of the parameters of maximum pressure variogram in various directions.

Direction/Tolerance	Nugget Effect	Height of Sill	Range of Influence	Variogram Model
45/22.5	0.7E+6	1.8E+6	1.0	Exponential
70/22.5	1.3E+6	0.9E+6	0.8	Exponential
24/21	1.1E+6	1.4E+6	1.3	Spherical
15/22.5	1.5E+6	1.0E+6	1.6	Gaussian
345/17.5	1.7E+6	0.55E+6	1.4	Gaussian
300/22.5				Periodic structure

the size of pressure compartment cells. The experiment was performed in three steps. First, an area was divided into a 3 × 3 square grid. The length of each cell in the checkerboard was five units. A hydraulic head value was generated for each cell using an independent normal random number (Figure 7). The hydraulic head value was calculated by the formula

$$H(i,j) = avg + sdev \times u$$

where $H(i, j)$ is the hydraulic head for cell (i, j), avg and $sdev$ are mean and standard deviation of hydraulic head of the right-hand population (Table 2), and u is the independent normal random number. Hydraulic head was assigned a constant value within each cell. Second, 180 wells for the whole checkerboard were located using an independent uniform random number. Well locations were randomly distributed on the checkerboard. The number of wells within cells varies. One cell may have more wells than another. Third, variograms were calculated in directions 0°, 45°, and 90° with tolerance 22.5°. The ranges of influence along these three directions are 4.5, 5.5, and 4.5 units. In other words, the range of influence is about 90% of the diameter of a cell in the square grid.

Variograms of Head for Right-Hand Populations

The right-hand population in the hydraulic head histogram is believed to closely approximate the pressure of compartment cells because the pressures of production wells is about 1900 psi higher than those of drilled and abandoned wells in the Hilight oil field. Table 2 shows statistical properties of the right-hand head population.

The variogram of the right-hand head population was used to estimate the average size and orientation of pressure compartment cells. The longest range of influence is found to be 15° azimuthal with a value of 0.8 lag unit (4.8 mi [7.7 km]). The ranges of influence estimated at 285° and 345° are 0.4 and 0.65 lag unit, respectively (2.4 and 3.9 mi [3.9 and 6.3 km]). The variograms in these directions are fitted by the exponential model. We assume that the average compartment cell can be represented by an ellipse constructed on the basis of the ranges of influence in various directions. The ellipse can be viewed as the average size and shape of a compartment cell: The major axis of the ellipse corresponds to the long axis of the compartment cell, and the minor axis corresponds to the short axis of the cell. A single ellipse represents the average orientation of pressure compartments in the study area. According to the information about the ranges of influence derived from the experiment and head variogram, the major and minor diameters of the compartment cell were estimated to be, on average, 5.3 and 2.7 mi (8.5 and 4.3 km), respectively.

The exponential variogram with nugget effect is expressed as

$$\gamma(\mathbf{h}) = c_0 + c\left[1 - \exp\left(-\mathbf{h}/d\right)\right],$$

or approximately

$$\gamma(\mathbf{h}) = c_0 + c\left[1 - \exp\left(-3\,\mathbf{h}/h_0\right)\right],$$

where c_0 is nugget effect; and $\mathbf{h}/h_0$ can be transformed into

$$\left[[x\ y]B[x\ y]^T\right]^{\frac{1}{2}},$$

where B is a matrix that can be expressed by eigenvectors and eigenvalues such that

$$\begin{bmatrix} \cos\theta & -\sin\theta \\ \sin\theta & \cos\theta \end{bmatrix} \begin{bmatrix} \frac{1}{a^2} & 0 \\ 0 & \frac{1}{b^2} \end{bmatrix} \begin{bmatrix} \cos\theta & \sin\theta \\ -\sin\theta & \cos\theta \end{bmatrix},$$

where θ is the angle between the major axis of the ellipse and the positive direction of the x axis in coordinate space, a is the half-length of the long diameter of the ellipse, and b is the half-length of the short diameter of the ellipse. In this case, $\theta = 75°$, $a = 2.65$ mi (4.26 km), $b = 1.35$ mi (2.17 km),

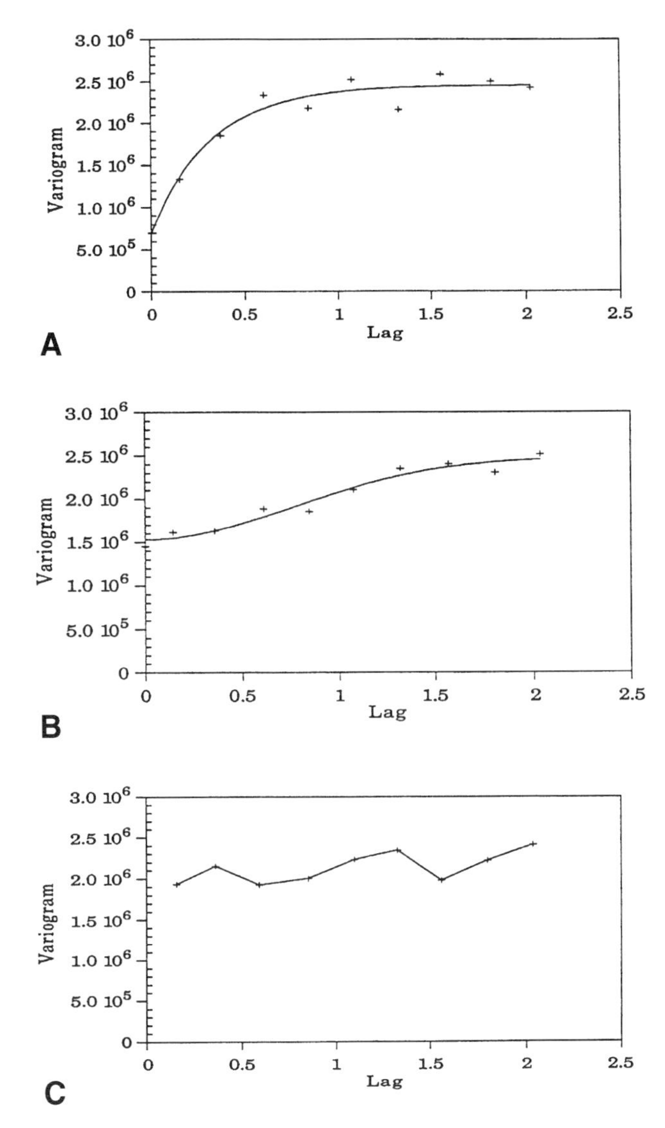

Figure 5. Variograms in different directions: (A) exponential variogram along 45°, (B) Gaussian variogram along 15°, and (C) periodic structure along 300°. See text for discussion.

$$B = \begin{bmatrix} 0.5212 & -0.1016 \\ -0.1016 & 0.1696 \end{bmatrix},$$

$c_0 = 500{,}000$, and $c = 1{,}200{,}000$.

The variogram for hydraulic head in the pressure compartment is

$$\gamma(h) = 1{,}700{,}000 - 1{,}200{,}000$$

$$\exp\left\{ -\left([x\ y] \begin{bmatrix} 4.6933 & -0.9142 \\ -0.9142 & 1.5265 \end{bmatrix} [x\ y]^T \right)^{\frac{1}{2}} \right\},$$

which may be applied to Kriging and simulation.

Variograms of the right-hand pressure population are similar to those of the right-hand head population.

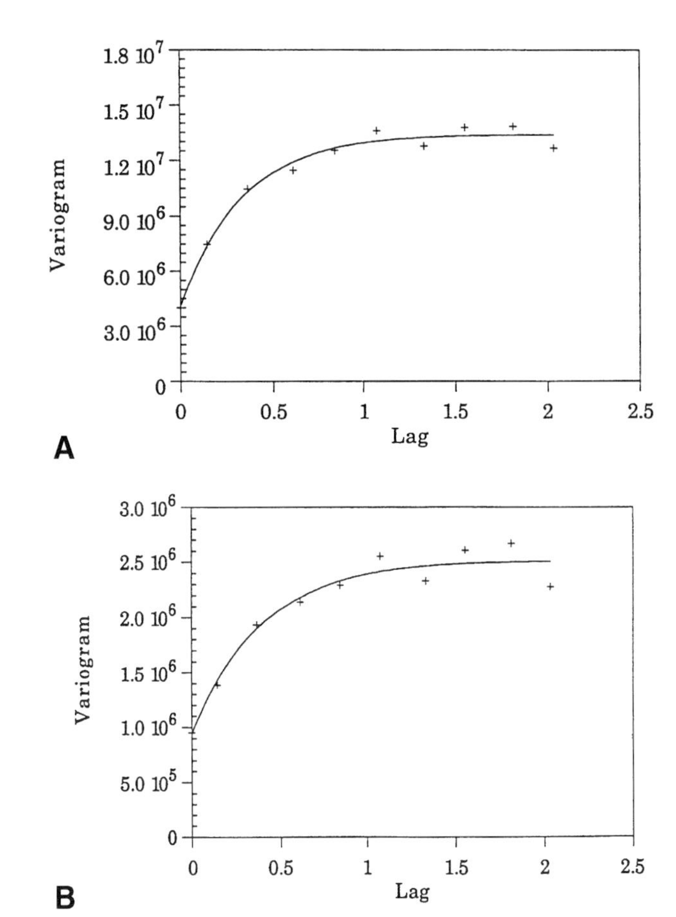

Figure 6. (A) Exponential variogram of hydraulic head, and (B) exponential variogram of pressure, along 55°.

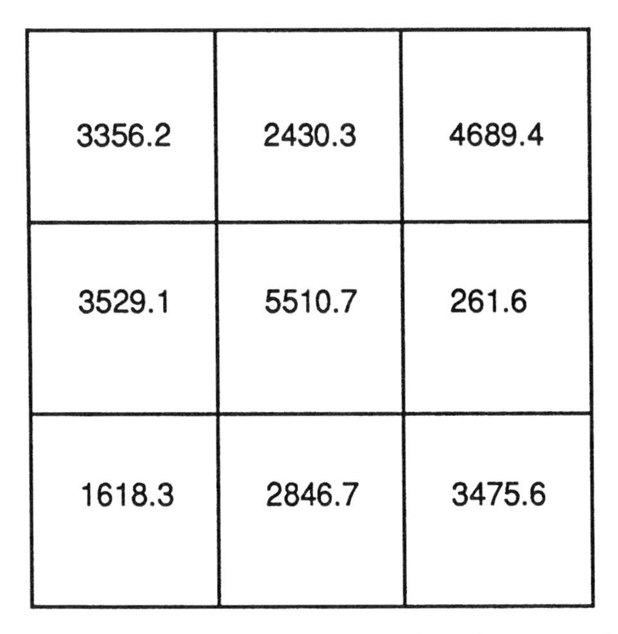

Figure 7. Simulated hydraulic head used to test variogram method. The size of each of the nine cells is 5 × 5 units, and the value of hydraulic head is shown within each cell.

DISCUSSION

It has been shown that variograms describe the spatial structure of hydraulic head and pressure even though fluid production has affected the later data values. The Gaussian variograms of the complete data sets only fit at about 15° and 345° azimuthal. (Ranges of influence are consistently longer and the height of sill is lower along these directions. Variogram models in most other directions include exponential, spherical, and periodic structure.) Since a Gaussian variogram represents less fluctuation of a realization in space, we infer that pressure and hydraulic head are much more uniform in these two directions. It can be seen from the potentiometric surface map (Figure 2) that there exist elongated contours with the highest values along these two directions. According to Powley's (1990) "huge bottle" model of pressure compartments, fluid communicates in the internal volume of pressure compartments. The smoothness of pressure and hydraulic head along these two directions results from fluid communication inside pressure compartments.

The variogram at 300° ± 22.5° (Figure 5C) demonstrates a periodic structure. The periodic structure of the variogram reflects the periodically spatial structure of the realization. Each wave in the variogram probably means that another geologic body of similar properties exists in space and a geologic body of different properties lies between. This phenomenon is consistent with the characteristics of two hydraulic regimes indicated by the bimodally distributed pressure and hydraulic head data.

The range of influence of hydraulic head variograms is greater in the direction normal to the mean fluid flow (Gelhar, 1982; Desbarats and Srivastava, 1991). If this is the case, the average flow direction in the study area is east-west.

The nugget effect is relatively large for both the head and pressure variograms but is significantly lower for the right-hand population. This is because a high nugget effect can be caused by a rapid change between two measured values at a short distance. In the study of pressure compartments, the boundary between the internal volume of a pressure compartment and its seal reflects a sudden change in hydraulic properties. A large nugget effect may thus imply the boundary of a pressure compartment cell.

As pointed out by Philip and Kitanidis (1989), variograms calculated from measurements of hydraulic head in an aquifer often indicate a discontinuity at the origin of the variogram; i.e., the nugget effect. The discontinuity may be caused by measurement error or small-scale variability. The largest error associated with the calculation of hydraulic head in this study is from the measurements of shut-in pressures. Theoretically, initial and final shut-in pressures ought to be nearly the same. In this study of 192 data points, only 38% show a difference between initial and final shut-in pressures within ± 5%. Another source of the nugget effect is probably the nonuniform distribution of data in space. Measurements from two distant wells cannot distinguish small-scale variations between them.

CONCLUSIONS

This study demonstrates the existence of a spatially structured distribution of pressure in the Muddy Sandstone in the Hilight oil field. Variograms appear to be a good method of analyzing such a distribution. The primary conclusions from this study are:

1. Hydraulic head is bimodally distributed, and that distribution implies that the Muddy Sandstone is heterogeneous in the study area. The right-hand population of pressures is interpreted as representing a pressure compartment. This property of the Muddy Sandstone is reflected in the periodic structure of the associated variograms.
2. The oil reservoir is highly anisotropic, with greatest ranges of influence at 15° and 345°. On average, hydraulic head or pressure is more continuous in the north-south direction and fluctuates in the east-west direction. This conclusion is supported by (a) a longer average range of influence and (b) a more continuous variogram model, the Gaussian model, in the north-south direction.
3. The range of influence of the whole data set is almost double the range of influence of the right-hand population. The longest range of influence in the right-hand head population is at about 15°. The average size and shape of a pressure compartment modeled as an ellipse is about 5.3 mi (8.5 km) along 15°, and 2.7 mi (4.3 km) along 285°.
4. The information obtained through the variogram method about anisotropy and periodic structure in the Powder River basin will be important for further prospecting.

ACKNOWLEDGMENTS

This material covers research performed through November 1991. The original manuscript was reviewed by David Copeland (University of Wyoming), who made numerous helpful suggestions. This study was funded by the Gas Research Institute under contract number 5089-260-1894 and U.S. Department of Energy under grant number DE-FG02-91ER75665.

REFERENCES CITED

Bair, E. S., T. P. O'Donnell, and L. W. Picking, 1985, Potentiometric mapping from incomplete drill-stem tests data: Palo Duro Basin Area, Texas and New Mexico: Ground Water, v. 23, p. 198–211.

Baker, A. A., L. W. Gelhar, A. L. Gutjahr, and J. R. MacMillan, 1978, Stochastic analysis of spatial variability in subsurface flows, 1. Comparison of one- and three-dimensional flows: Water Resources Research, v. 14, p. 263–271.

Desbarats, A. J., and R. M. Srivastava, 1991, Geostatistical characterization of groundwater flow parameters in a simulated aquifer: Water Resources Research, v. 27, p. 687–698.

Easley, D. H., L. E. Borgman, and D. Weber, 1991, Monitoring well placement using conditional simulation of hydraulic head: Mathematical Geology, v. 23, p. 1059–1080.

Englund, E., and A. Sparks, 1988, User's guide to Geostatistical Environmental Assessment software: Environmental Monitoring Systems Laboratory, Office of Research and Development, U.S. Environmental Protection Agency, Las Vegas, Nevada 89193-3478.

Gelhar, L. W., 1982, Stochastic analysis of flow in heterogeneous porous media, *in* J. Bear and M. Corapcioglu, eds., Fundamentals of Transport Phenomena in Porous Media: NATO ASI Series, Series E, Applied Sciences, no. 82, p. 673–717.

Hoeksema, R. J., and P. K. Kitanidis, 1985, Analysis of the spatial structure of properties of selected aquifers: Water Resources Research, v. 21, p. 563–572.

Isaaks, E. H., and R. M. Srivastava, 1989, Applied Geostatistics: New York, Oxford University Press, 561 p.

Javis, W. T., 1986, Regional hydrogeology of the Paleozoic aquifer system, southeastern Bighorn Basin, Wyoming, with an impact analysis on Hot Spring State Park: M.S. thesis, University of Wyoming, 227 p.

Marsily, G. de, 1982, Spatial variability of properties, *in* J. Bear and M. Corapcioglu, eds., Fundamentals of Transport Phenomena in Porous Media: NATO ASI Series, Series E, Applied Sciences, no. 82, p. 7193–769.

Marsily, G. de, 1986, Quantitative Hydrogeology, Groundwater Hydrology for Engineers: Orlando, Florida, Academic Press, 440 p.

Matheron, G., 1963, Principles of geostatistics: Economic Geology, v. 58, p. 1246–1266.

Mizell, S. A., A. L. Gutjahr, and L. W. Gekhar, 1982, Stochastic analysis of spatial variability in two-dimensional steady groundwater flow assuming stationary and nonstationary heads: Water Resources Research, v. 18, p. 1053–1067.

Philip, R. D., and P. K. Kitanidis, 1989, Geostatistical estimation of hydraulic head gradients: Ground Water, v. 27, p. 855–865.

Powley, D. E., 1990, Pressures and hydrogeology in petroleum basins: Earth-Science Reviews, v. 29, p. 215–226.

Spencer, S. A., 1986, Groundwater movement in the Paleozoic rocks and impact of petroleum production on water levels in the southwestern Bighorn Basin, Wyoming: M.S. thesis, University of Wyoming, 196 p.

Chapter 18

Stratigraphic Compartmentation of Reservoir Sandstones: Examples from the Muddy Sandstone, Powder River Basin, Wyoming

Randi S. Martinsen
University of Wyoming
Laramie, Wyoming, U.S.A.

ABSTRACT

The Lower Cretaceous Muddy Sandstone (Viking Formation equivalent) is a thin but complex stratigraphic unit that contains a variety of anomalously pressured compartments. One or more lowstand surfaces of subaerial exposure and erosion (LSEs), numerous transgressive surfaces of submarine erosion (TSEs), and varying lithofacies compartmentalize the Muddy Sandstone stratigraphically on at least three levels. The first level of compartmentation is defined by the relief along the LSE surface(s), which is highly variable and physically divides the Muddy, both vertically and laterally, into older and younger sequences. The second level is defined by the intersection of shales above the TSEs with the LSE (either by onlap or truncation). The third level results from variations in lithofacies. Whereas many of the compartments comprise classic stratigraphic traps consisting of shale (seal) encompassing sandstone (reservoir/compartment), compartments exist wherein sand is juxtaposed against sand without benefit of intervening shales to serve as a seal. In these situations, the seal appears to consist of a paleosol developed beneath the LSE. The distribution and geometries of pressure compartments in the Muddy have a high degree of correspondence to the various scales of stratigraphic compartmentation observed. In all probability, similar levels of stratigraphic complexity characterize many basins. Any analysis of the controls on pressure compartment formation and distribution therefore should incorporate these stratigraphic complexities and not assume that stratigraphic systems are characteristically simple.

INTRODUCTION

In order to accurately characterize and model the distribution of pressure compartments in a basin, it is necessary to precisely reconstruct and understand the stratigraphic and structural framework of the basin. Such a framework is needed to determine initially which stratigraphic and structural elements provide seals and under what conditions they do so. Such a framework is also needed in order to determine if seals exist that appear to be unrelated to known structure or stratigraphy.

The existence of an extensive subsurface database, combined with the presence of numerous excellent outcrops, has facilitated construction of such a framework for the Lower Cretaceous Muddy Sandstone of the Powder River basin. The focus of this paper is stratigraphic relationships within the Muddy Sandstone and how these relationships have resulted in a multitiered hierarchy of stratigraphic and fluid compartmentation.

GENERAL SETTING OF THE POWDER RIVER BASIN

The Powder River basin, located in northeastern Wyoming and southeastern Montana, is one of the numerous Rocky Mountain intermontane basins that developed during the Laramide period of deformation (45–75 Ma). The structural basin geometry is highly asymmetric, with the deep axis proximal to the western margin (Figure 1). For the past 10 million years, the basin has been undergoing uplift that has removed more than 2300 ft (700 m) of section. However, over 20,000 ft (6000 m) of dominantly clastic rocks are preserved in the basin and crop out in excellent exposures around the basin margin. The names and ages of the formations present, as well as the major unconformities, are indicated on Figure 2. Although the preserved basin fill ranges in age from Middle Cambrian to Oligocene, more than half consists of Cretaceous-age rocks. The Powder River basin is considered to be in a mature stage in its exploration history, and hydrocarbon production occurs throughout the basin (Figure 1) from many stratigraphic intervals (Figure 2). Within the basin, overpressured, normally pressured, and underpressured fluid compartments have been mapped (Heasler and Surdam, 1992). These compartments are developed in several stratigraphic intervals, almost all of which are within the Cretaceous System. Timing of overpressure generation roughly coincides with the period of Laramide deformation (Jiao and Surdam, this volume).

MUDDY SANDSTONE

The Lower Cretaceous (Albian) Muddy Sandstone crops out around the margin of the Powder River basin and reaches a depth of greater than 13,500 ft (4100 m) in the deep basin. The Muddy is a thin but widespread formation that is recognized throughout most of Wyoming and parts of Colorado, Nebraska, Montana, and South Dakota and is correlative with the Viking Formation of the Alberta basin (Reeside and Cobban, 1960). Because the Muddy and equivalent strata have produced more than 1.5 billion bbl of oil-equivalent hydrocarbons (Dolson et al., 1991), they hold strong economic and scientific interest and have been the focus of numerous studies. These studies have conclusively shown that even though the Muddy is relatively thin, it is stratigraphically complex and contains a wide variety of both marine and nonmarine lithofacies and numerous intraformational unconformities.

Regional Stratigraphy

The Muddy Sandstone is both underlain and overlain by shale, and so may be considered to be stratigraphically bounded by low-permeability rocks. The contact with the underlying shale is at least locally unconformable, and the contact with the overlying shale is disconformable. In the Powder River basin the Muddy is commonly between 30 and 100 ft (9 and 30 m) in thickness, but varies between less than 10 ft (3 m) and as much as 130 ft (40 m). A prominent lowstand surface of subaerial erosion (LSE) divides the Muddy into distinctly separate lower Muddy and upper Muddy sequences. These sequences are further subdivided by several additional unconformities that form the basis for the member-level stratigraphy. Whereas seven members have been informally recognized within the Muddy (Gustason et al., 1988a), no more than five members, and commonly only three or four members, occur at any one locality. Stratigraphic relationships between the members are shown in Figure 3.

Lithofacies and Depositional Environments

Regionally, the Muddy Sandstone can be divided into three genetic packages on the basis of relative stratigraphic position and lithofacies (Dolson et al., 1991): older marine, valley-fill, and transgressive marine.

Older Marine (Rozet and Lazy B Members)

The older marine deposits typically consist of one or more upward-coarsening shale-to-sandstone sequences containing normal open-marine trace fossil assemblages. Sandstones in the lower portion of each sequence display planar laminations and hummocky and (wave-generated) ripple cross-stratification, and in their upper portions are hummocky to high-angle cross-stratified or massive to mottled. These deposits are interpreted as representing deposition in prograding offshore marine to upper shoreface and deltaic environments. Because the Lazy B and Rozet members have similar characteristics, discriminating them is often difficult. In areas where both members are present, however (e.g., Lazy B field), the Lazy B sandstones appear to consist dominantly of lithic arkoses, whereas the Rozet sandstones appear to consist dominantly of quartzarenites, sublitharenites, and litharenites (Figure 4).

Valley-Fill (Upper and Lower Recluse Members)

The valley-fill deposits consist of variable proportions of shale, siltstone, and sandstone and commonly contain thin coal and coaly beds and evidence of roots. The sandstones are composed of quartzarenites and sublitharenites (Figure 4), and commonly contain abundant mudstone rip-up clasts and wood and char-

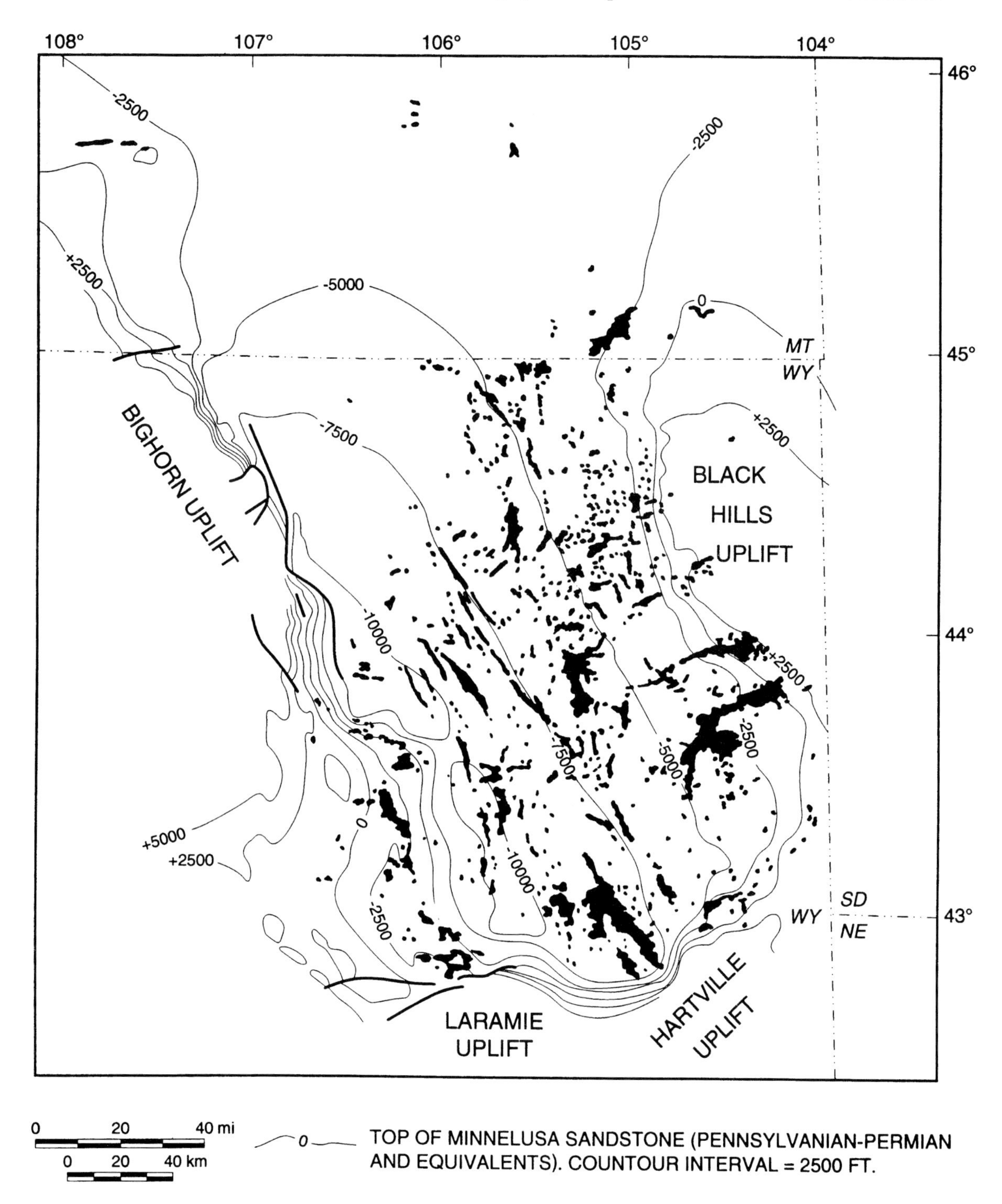

Figure 1. Map of the oil and gas fields of the Powder River basin (solid areas). Generalized structural contours on top of Minnelusa Formation (Pennsylvanian–Permian). Modified from Dolton et al. (1988).

coal fragments (Gustason et al., 1988a). Bedding within the sandstones is massive to irregular, commonly with evidence of soft sediment deformation, and is sometimes rippled or cross-bedded. These deposits are interpreted to represent alluvial to upper estuarine deposition and are geographically confined within valleys developed during the period of relative sea level lowstand that followed deposition of the older marine deposits.

Transgressive Marine (Cyclone, Ute, and Springen Ranch Members)

These deposits generally are quartz rich (Figure 4) and consist of upward-coarsening shale-to-siltstone or

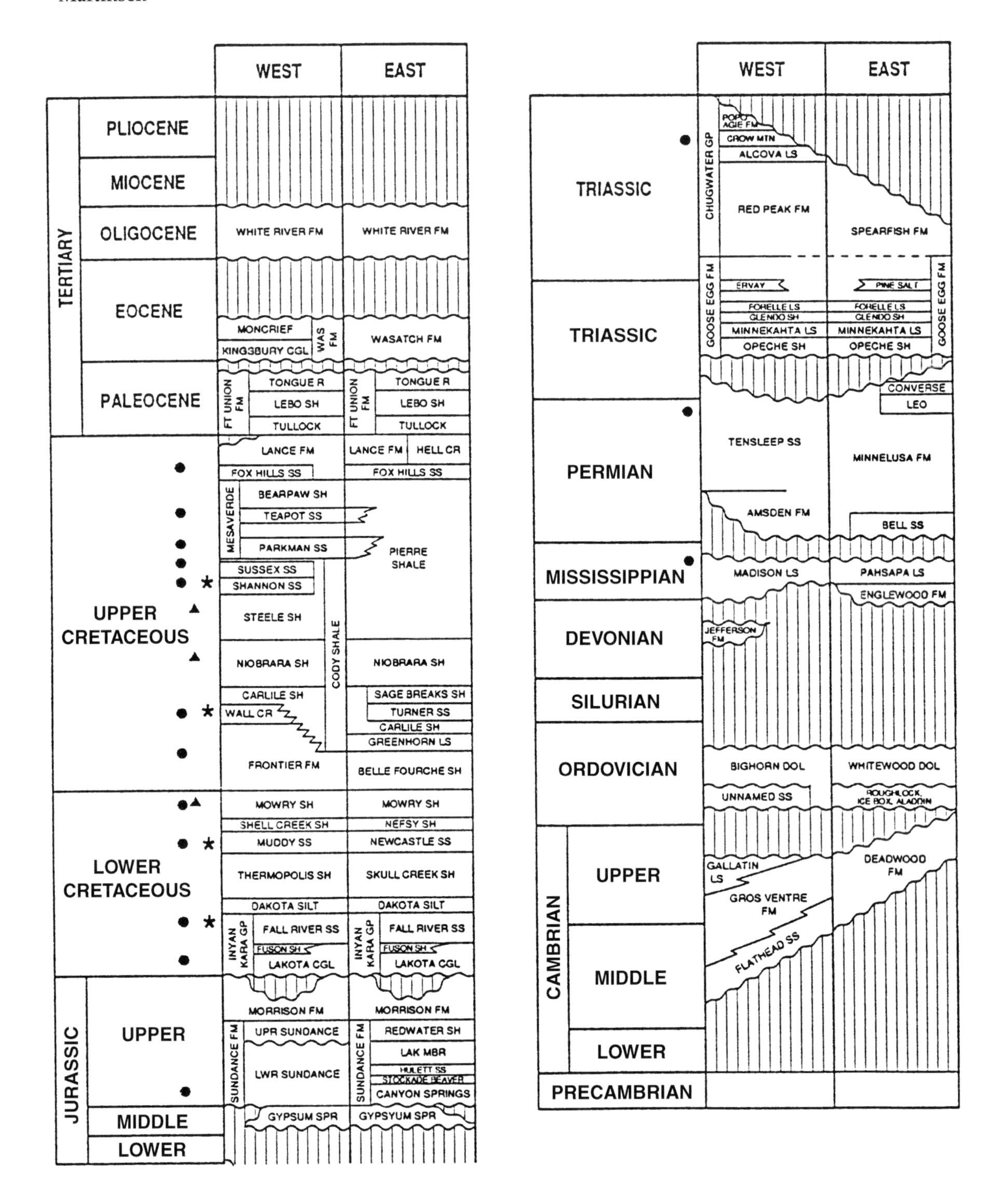

Figure 2. Chronostratigraphic nomenclature chart of the Powder River basin, showing names and relative ages of rocks preserved as well as major unconformities. Modified from Wyoming Geological Association (1991).

shale-to-sandstone sequences that display a wide variety of well-preserved sedimentary and biogenic structures. These deposits represent deposition in a spectrum of open marine to restricted, brackish-water environments, including estuaries, bays, and barrier islands. In general, both the lowest and easternmost deposits are more brackish, indicating transgression from west to east.

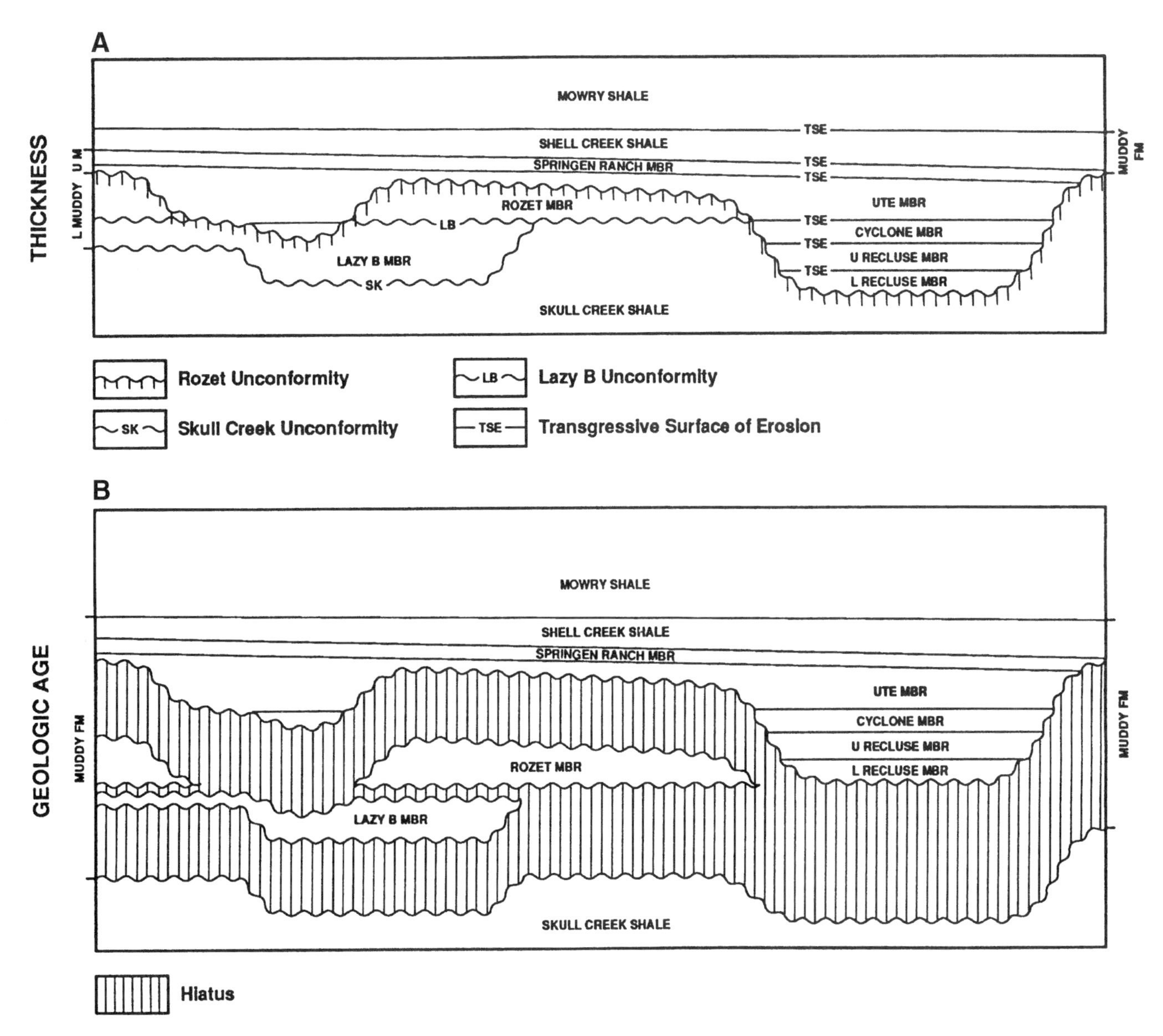

Figure 3. (A) Generalized stratigraphic cross section of the Muddy Sandstone, showing member-level stratigraphy. (B) Generalized chronostratigraphic chart. Only hiatuses associated with LSEs are indicated. Not to scale.

Although the various individual members of the Muddy recognized within the Powder River basin are not generally delineated outside the basin, the various lithofacies represented within each of the members are widely recognized.

Unconformities

Two types of major erosional surfaces are observed within, and at the formation contacts of, the Muddy Sandstone (Weimer, 1991b). The first type, called a *transgressive surface of erosion* (TSE) or *transgressive disconformity,* is the product of shallow marine erosion that occurs during transgressions. TSEs typically are overlain by laterally continuous shales deposited under deeper-water conditions resulting from the rise in relative sea level. The second type, called a *lowstand surface of erosion* (LSE), is related to a lowering of regional base level that results in subaerial exposure and erosion of underlying rock units. LSEs commonly, but not always, are characterized by (1) locally variable relief due to incisement of drainages and (2) paleosol development in the rocks directly below them.

Transgressive Surfaces of Erosion

The transgressive surfaces of erosion in the Powder River basin typically are sharp, may exhibit local relief but generally appear conformable, and are overlain by shale. TSEs separate each of the members of the upper Muddy, and also occur within members. The contact between the Muddy and the overlying

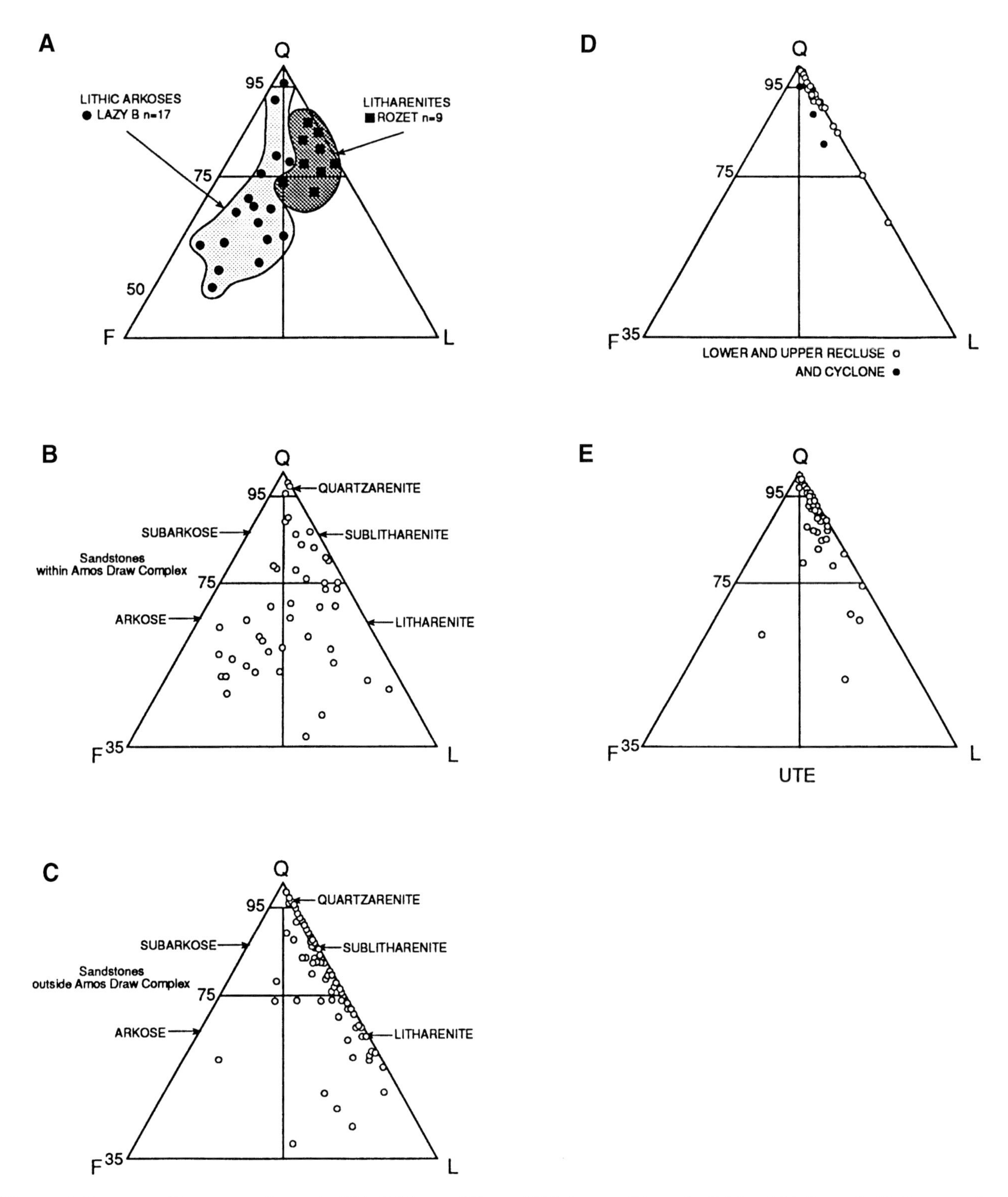

Figure 4. Ternary diagrams of quartz (Q), feldspar (F), and lithic constituents (L) for (A) the sub-K-2 unconformity marine sandstones of the Lazy B and Rozet members in the Lazy B field; (B) the sub-K-2 unconformity marine sandstones in the Amos Draw complex; (C) the sub-K-2 unconformity marine sandstones elsewhere in the Powder River basin; (D) fluvial sandstones of the Recluse and Cyclone members; and (E) barrier island sandstones of the Ute member. Figure A is from Gustason (1988); B and C are compilations from Odland et al. (1988) and Ryer et al. (1990); and D and E are modified from Ryer et al. (1990).

Shell Creek Shale is interpreted as a TSE that has locally caused erosional truncation of the Springen Ranch member beneath it (Wheeler et al., 1988). Drill-stem test pressure data from within several fields indicate that the various members of the upper Muddy are not in pressure communication (Berg et al., 1985; Smith, 1988) and that the transgressive shales are seals.

Lowstand Surfaces of Erosion

At least one LSE, and possibly as many as three, are associated with the Muddy in the Powder River basin. The youngest of these three possible LSEs, known as the Rozet unconformity because it occurred sometime after deposition of the Rozet member, is the best defined and therefore the most commonly recognized and widely accepted unconformity within the Muddy. Studies describing this unconformity include Berg (1976), Dolson et al. (1991), Donovan (1991), Farmer (1981), Gustason et al. (1986, 1988a, b), Honarpour et al. (1988), Larberg (1980), McGookey et al. (1972), Odland et al. (1988), Prescott (1970), Ryer et al. (1990), Stone (1972), Szpakiewicz et al. (1988), Waring (1976), Weimer (1983, 1984, 1985, 1991a), Weimer et al. (1982, 1988), and Wheeler et al. (1988). This unconformity has been correlated with a regionally recognized LSE known as the K-2 unconformity (Dolson et al., 1991); it corresponds to the 98 Ma drop in sea level and is thus considered a sequence-bounding unconformity (after Vail et al., 1977). The Rozet unconformity is associated with (1) regionally extensive erosional truncation and removal of shallow marine sandstones of the Rozet and Lazy B members, (2) west-to-east onlapping above the unconformity of younger and younger members of the upper Muddy, (3) differing sandstone lithologies beneath and above the unconformity (Figure 4), and (4) evidence of subaerial exposure along the unconformity, including roots and a well-developed paleosol.

Unconformities interpreted as LSEs have also been identified in the older marine deposits beneath the Rozet unconformity (Gustason, 1988; Gustason et al., 1986, 1988a; Honarpour et al., 1988; Szpakiewicz et al., 1988). Gustason (1988) and Gustason et al. (1986, 1988a) describe these LSEs as forming after deposition of the Skull Creek Shale (Skull Creek unconformity) and between Lazy B and Rozet deposition. These two unconformities are not as well documented as the Rozet unconformity, however, and the interpretation that they are LSEs is not widely accepted. Reasons and other interpretations include (1) one or both are not LSEs, but are transgressive surfaces of submarine erosion (TSEs); (2) one or both were not regionally extensive, but only of local origin; (3) one or both are not characterized by an incised drainage topography, or are not associated with well-developed soil profiles, and so are more difficult to detect; (4) evidence of their existence was largely removed by truncation associated with the younger Rozet unconformity, such that two or more separate periods of erosion have been merged into a single unconformity surface; and (5) the Rozet unconformity is not regionally extensive, and these other unconformities may have been mistakenly identified as the Rozet unconformity, thus giving the Rozet the appearance of being regionally extensive. If further research confirms the post-Rozet period of erosion to be the only one accompanied by significant and laterally extensive valley incision and soil formation, then perhaps the Rozet is the only sequence-bounding unconformity, and the Skull Creek and Lazy B unconformities are the localized products of either submarine or subaerial erosion along the tops of uplifted basement blocks. If so, then the Rozet unconformity has the potential to be a regional seal to hydrocarbon migration, not just a local phenomenon.

Rozet Unconformity

Geometry

Cross sections and isopachous maps of various intervals within the Muddy all indicate that relief along the Rozet unconformity is highly irregular. Although in places the surface appears flat, in other places it is deeply incised into underlying sediments. Locally, relief along the unconformity exceeds the total preserved thickness of lower Muddy deposits. Because of relief, the Rozet unconformity (in different localities) may occur at the top of the Muddy, within the Muddy, or at the base of the Muddy, and physically divides the formation both vertically and laterally.

The erosional topography produced during the period(s) of subaerial exposure influenced the distribution of subunconformity deposits and the areal distribution of overlying deposits. Upper and lower Recluse, Cyclone, and lower Ute deposits are mostly restricted to valleys. Paleotopographic features such as the Belle Fourche Arch and Hartville Uplift created major drainage divides (Dolson et al., 1991) that physically separated depositional systems. Consequently, while the deposits that later filled the valleys north and south of the Belle Fourche Arch are mapped as upper and lower Recluse, Cyclone, and Ute on the basis of their relative time-stratigraphic positions and lithofacies characteristics, these deposits are not in physical continuity across the arch, but are separated by a wide interfluve area composed mostly of Skull Creek Shale. Furthermore, drainages north and south of the arch may have had different sources and different sand/mud ratios. Although the valleys were eventually filled such that continued transgression and onlap deposited sediments over a more widespread area, some topographic and structural elements continued to restrict deposition and influence lithofacies until the end of Muddy time. For instance, the Springen Ranch member is generally thin and composed entirely of shale over the area of the Belle Fourche Arch (Figure 5), which suggests that the arch was still high at that time. Ultimately, transgression resulted in the deposition of the marine Shell Creek Shale and burial of the Muddy deposits.

Paleosol Characteristics

Sandstones directly beneath the Rozet unconformity have one or more of the following characteristics:

1. A highly distinguishable green to white color that commonly contains small reddish-brown siderite blebs.
2. A massive to highly mottled texture that grades down into well-preserved sedimentary structures.
3. A very high clay content (up to 50% by volume) that decreases with depth below the unconformity (Figure 6).

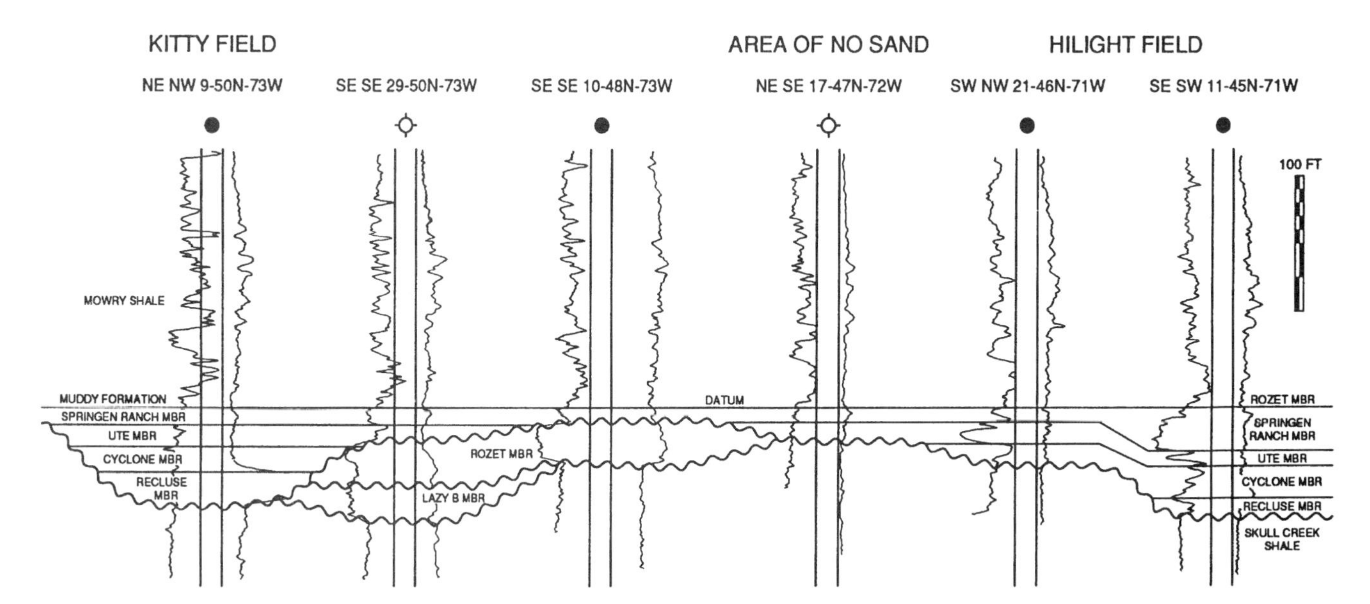

Figure 5. North-to-south gamma-ray and resistivity well-log cross section from Kitty field (T50N, R73W) across the area of the Belle Fourche Arch to Hilight field (T45N, R71W). The top unconformity (wavy line) is the Rozet. Note (1) the paleohill, composed of Skull Creek Shale on the south and Rozet and Lazy B sand on the north, which served as a drainage divide that separated the Kitty and Hilight valley systems, and (2) the lack of sand development within the Springen Ranch member over the arch.

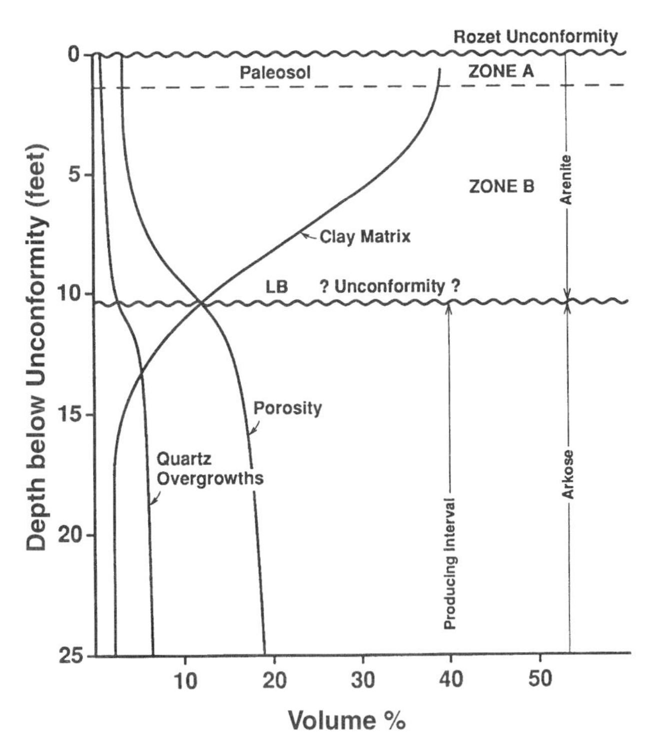

Figure 6. Plot of clay matrix, quartz overgrowths, and porosity in Rozet sandstones vs. depth beneath the unconformity surface. Paleosol zonation is commonly recognized. In many places, the base of the weathering is coincident with the contact between lithic and arkosic quartz sandstones. Modified from Odland et al. (1988).

4. An increase in porosity with depth (inversely proportional to clay volume).
5. Roots, sometimes extending down into layers containing normal marine trace fossils.

The clays consist mostly of kaolinite, illite-smectite, and chlorite, and are the result of various pedogenic processes decomposing unstable grains, and of clays from various ash falls infiltrating into underlying sediments (Figure 7). This texturally and compositionally distinct zone at the top of the lower Muddy varies from a few inches to more than 25 ft (8 m) in thickness and is interpreted as a paleosol. The presence of a paleosol directly over lower and middle shoreface sandstones, with associated deep penetration of roots, indicates that the open marine deposits of the lower Muddy were subaerially exposed and subject to various pedogenic processes for a significant period of time. Exposure must have been at least long enough to replace connate marine waters with meteoric fresh waters in order for the surface to become vegetated with land plants (marine plants do not have well-developed roots, and most land plants cannot tolerate saline waters). Because of relief, the unconformity surface was subaerially exposed for different periods of time in different places. Variations in the thickness of the paleosol beneath the unconformity may be related in part to length of exposure, such that thick soil zones preferentially developed in areas that were exposed for the greatest amount of time or had the least amount of erosion (in other words, the highest paleohills). Depending upon the degree of erosional

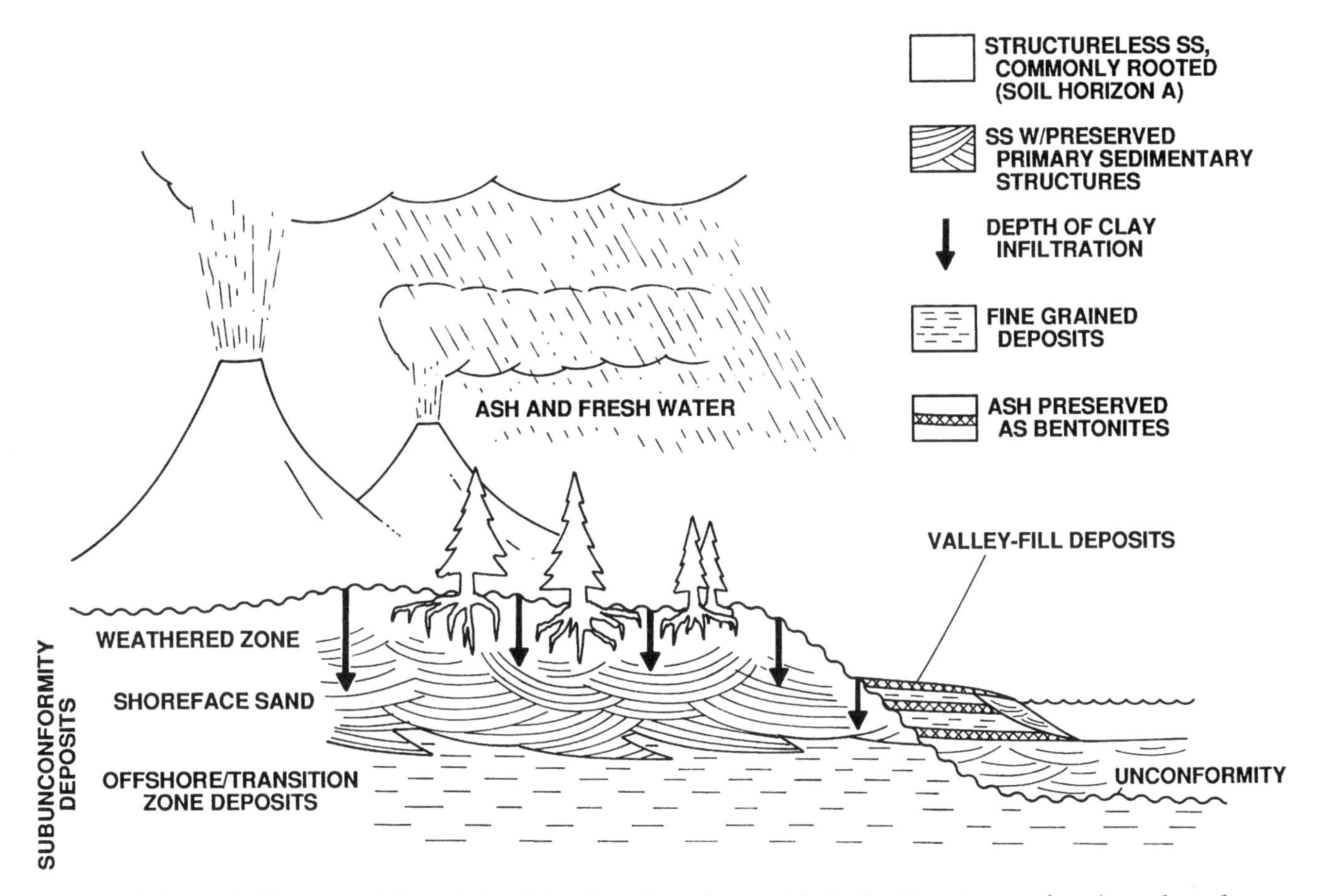

Figure 7. Schematic diagram of the origin of the abundant clay matrix in the Rozet unconformity paleosol. Numerous volcanic ash falls, preserved as bentonites within valley-fill deposits, periodically blanketed the unconformity surface and contributed significantly to the clay matrix. Note that while valleys were being filled, paleohills of subunconformity deposits were still exposed. Modified from Odland et al. (1988).

incision in any area, soil development representing the post-Rozet period of erosion occurred on Rozet, Lazy B, or Skull Creek surfaces (Figure 3).

Core analyses from both the Lazy B and Amos Draw fields indicate that porosity in the weathering zone typically ranges from 5 to 15%, but permeability generally is less than 0.1 md and commonly less than 0.01 md. These data suggest that the paleosol is a potential seal (Jiao and Surdam, this volume). In contrast, core analyses from the producing interval (below the paleosol) in the Amos Draw field indicate 12–18% average porosity and 1–2 md permeability, with a maximum observed permeability of 60 md (Von Drehle, 1985); and analyses from the Lazy B producing interval indicate an average of 14% porosity and 14 md permeability, with a maximum observed permeability of 115 md (Marinovich and Dietrich, 1973). Threshold displacement pressures obtained from samples of the paleosol (see Jiao and Surdam, this volume; Iverson et al., this volume) indicate that it is a highly capable seal to fluid migration, rather than a poor reservoir rock ("waste zone" as described by Schowalter, 1979).

Stratigraphic/Diagenetic Sealing of Pressure Compartments

Four examples of stratigraphic/diagenetic compartmentation in the Muddy are presented here. Two examples are from compartments within upper Muddy valley-fill and transgressive marine deposits (Hilight field area and Kitty field area), and two are from lower Muddy "older" marine sandstones (Amos Draw field area and Lazy B field area) (Figure 8). The Amos Draw and Kitty fields comprise separate, significantly overpressured stratigraphic compartments; Hilight and Lazy B comprise normally pressured to subnormally pressured stratigraphic compartments (see Heasler and Surdam, 1992).

Sub-Rozet Unconformity Compartments (Amos Draw and Lazy B Fields)

Amos Draw and Lazy B both produce from subunconformity sandstones in the lower Muddy, the upper Muddy being of poor reservoir quality in this area (Figure 9). Although these fields produce from similar depths (about 10,000 ft [3000 m]) and are only about

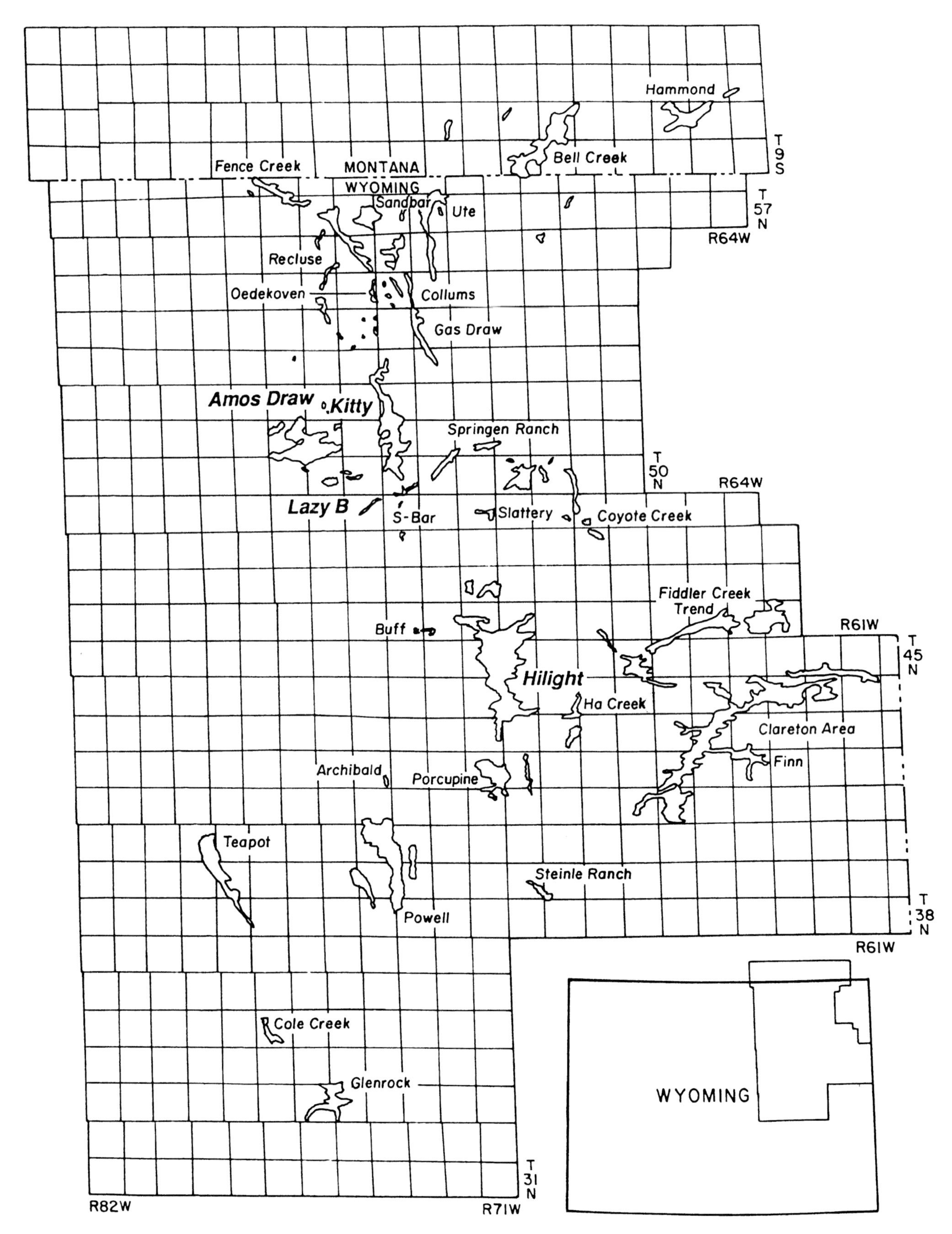

Figure 8. Map of Muddy Sandstone fields in the Powder River basin, Wyoming and Montana, showing the locations of Hilight, Kitty, Amos Draw, and Lazy B fields.

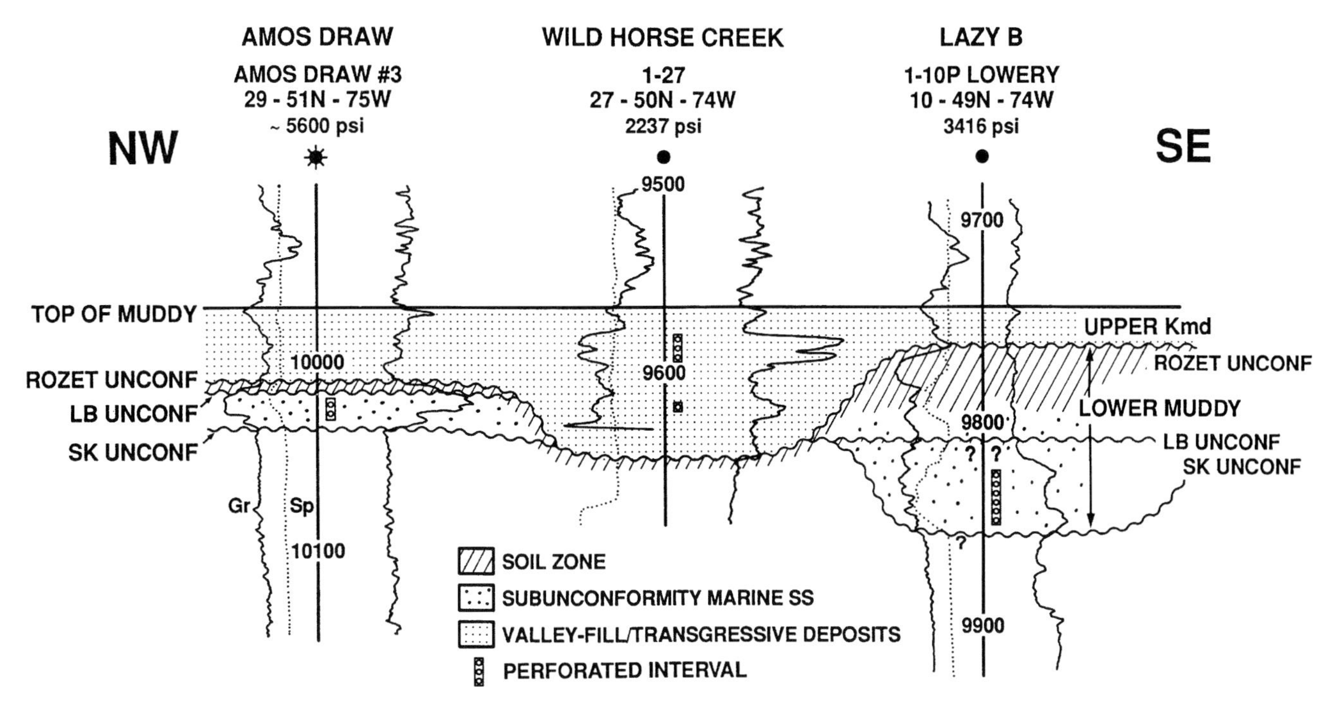

Figure 9. Well-log cross section from Amos Draw (overpressured) through Wildhorse Creek to Lazy B field (normal to subnormally pressured). Note the relief along the unconformity, and how the soil zone is developed beneath the Rozet, Lazy B, and Skull Creek unconformities. Production and reservoir pressures are different in each field.

10 mi (16 km) apart, they are characterized by significantly different pressure regimes. Whereas the original pressure in Amos Draw was about 5600 psi (1300 psi above the expected 4300 psi), the original pressure in Lazy B was only about 3400 psi (900 psi underpressured). This approximately 2000 psi pressure difference is evidence that the two fields are not in pressure continuity.

At *Lazy B field,* the lower Muddy is anomalously thick (up to 100 ft [30 m]), because in addition to the Rozet, a thick section of Lazy B deposits is present (Figure 9). The Lazy B deposits display a highly elongate geometry (roughly 3 mi [5 km] wide and 20 mi [32 km] long) that parallels the northeast–southwest-trending Springen Ranch lineament. This Lazy B thickening also coincides with a thinning in the underlying Skull Creek Shale (Gustason, 1988) that may represent an incised valley developed along the Skull Creek unconformity. Within the Lazy B member, multiple sequences of porous and permeable arkosic sandstones separated by thin marine shales progressively lap onto the underlying Skull Creek from southwest to northeast. Overlying the Lazy B is 20 to 35 ft (6–11 m) of Rozet. In Lazy B field, the Rozet sandstones consist of litharenites and sublitharenites that are clay filled and have low porosity and very low permeability. Apparently, weathering associated with the post-Rozet period of erosion extended down throughout the Rozet member but did not extend down through the Lazy B member. The first level of compartmentation in the Lazy B field, therefore, is the geometry defined by the intersection of the Rozet member (upper seal) with the Skull Creek Shale (bottom and lateral seals). Further compartmentation of the reservoir is defined by the shales that onlap the Skull Creek and separate individual progradational sandstones (parasequences, after Van Wagoner, 1988).

In the *Amos Draw area,* the lower Muddy generally varies between 20 and 35 ft (6–11 m) in thickness. Isopachous maps display a somewhat amoebae-like geometry with a slight northeast-southwest elongation, and with thick sections interpreted to be paleohills that are roughly up to 15 mi (24 km) wide and 40 mi (64 km) long (Odland et al., 1988). Depth of weathering beneath the Rozet unconformity is irregular and varies from only a few feet to more than 20 ft (6 m). As in Lazy B field, the lower Muddy sandstones in Amos Draw include both arkoses (Lazy B member) and arenites (Rozet member) (Odland et al., 1988; Ryer et al., 1990; Von Drehle, 1985; Jiao and Surdam, this volume). Where both compositional groups have been observed in cores, the arenites overlie and are in sharp contact with the arkoses. Usually, only the arenites are tight as a result of paleosol formation, whereas the arkoses display diagenetically enhanced porosity and permeability, and comprise the best reservoirs in both Amos Draw and Lazy B. If porosity and permeability reduced by diagenesis beneath the Rozet unconformity is more prevalent in the Rozet member throughout the basin—perhaps because of its lithology, or simply because it overlies the Lazy B and could have protected it from early diagenesis during

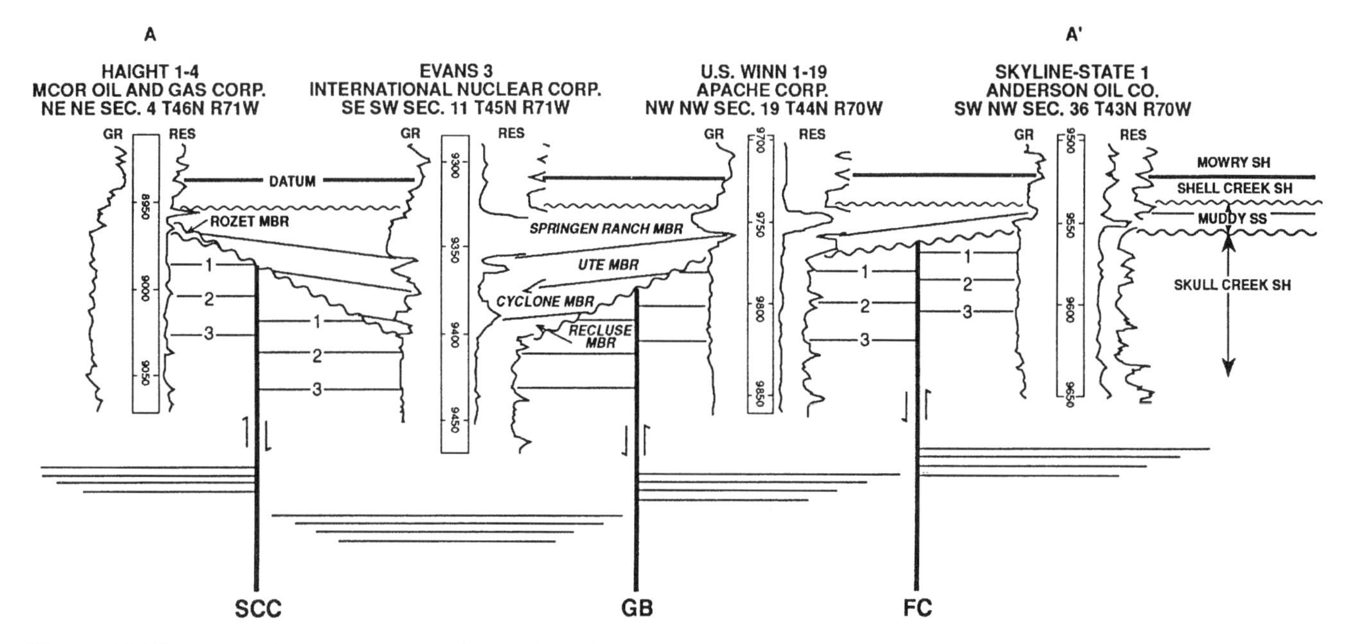

Figure 10. North-south cross section through Hilight Valley, with paleostructural interpretation. Line of section in Figure 12. Northeast-southwest-trending lineaments are the South Coyote Creek (SCC), Gose Butte (GB), and Fiddler Creek (FC); see Figure 11. Layers 1, 2, and 3 are markers. Note northeast–southwest-trending valley in left half of cross section. Faulting prior to, or during, the development of the Rozet unconformity resulted in a highly variable topography. Drainages developed in the lows, and better preservation of both valley-fill and transgressive marine deposits are associated with the lows. Strata deposited over topographic highs were probably depositionally thin and were also partially eroded during ensuing transgressions. Modified from Wheeler et al. (1988).

subareal exposure—then the Rozet member may mostly consist of seal or nonreservoir rock types and the Lazy B member may comprise the best reservoirs. In any event, diagenesis has resulted in the considerably less widespread distribution of porous and permeable sandstone than of the lower Muddy deposits. In Amos Draw, therefore, the first level of compartmentation is defined by the geometry of the intersection of the Rozet unconformity (top and lateral seals) with the Skull Creek Shale (bottom seal and possibly lateral seal). A second level of compartmentation is defined by the intersections of the base of the paleosol with the Skull Creek Shale. In some places, it appears that even though relief along the unconformity did not extend down to the Skull Creek and erosionally isolate paleohills of "older" marine sandstones, diagenesis associated with the paleosol zone did, and created additional lateral seals. For example, the updip limit of Amos Draw field is defined by the updip limit of porous and permeable sand, rather than by the change from sandstone to shale.

Compartments Overlying the Rozet Unconformity (Hilight and Kitty Field Areas)

The *Hilight field* (Figure 8) is located in the area of a large valley (the Hilight Valley system) that extends southward from the Belle Fourche Arch. Figure 5 illustrates the relationship of the Muddy Sandstone north of the Belle Fourche Arch to the Muddy in the Hilight field. Note that the valley floor and most of the valley walls are composed of Skull Creek Shale (Figure 10). The thickness, distribution, and lithofacies of each of the members of the Muddy Sandstone present in the Hilight area were all influenced by the dendritic topography that resulted from valley incision into the Skull Creek Shale.

In the north Hilight area, the Recluse, Cyclone, Ute, and Springen Ranch members all progressively onlap the Skull Creek Shale from west to east within a northeast-trending paleovalley. These members also show a progressive increase upwards in marine influence (from fluvial to barrier island deposition), in sand-to-shale ratio, and in size of producing reservoirs. Furthermore, whereas the members lower than the Springen Ranch are geographically restricted to within the valley, the Springen Ranch member was deposited after valley filling and thus overlies both valley and interfluve areas. As a result, the character of production from the lower members is very different from that of the Springen Ranch (the main producing reservoir). Production from the lower members is from small, thin, stratigraphically isolated sandstone compartments whose overall distribution is defined by the valley geometry (Figure 11). In contrast, production from the Springen Ranch member, which is sand rich, with some individual sands exceeding 40 ft (12 m) in thickness, shows a high degree of lateral continuity (Heasler et al., 1992), the

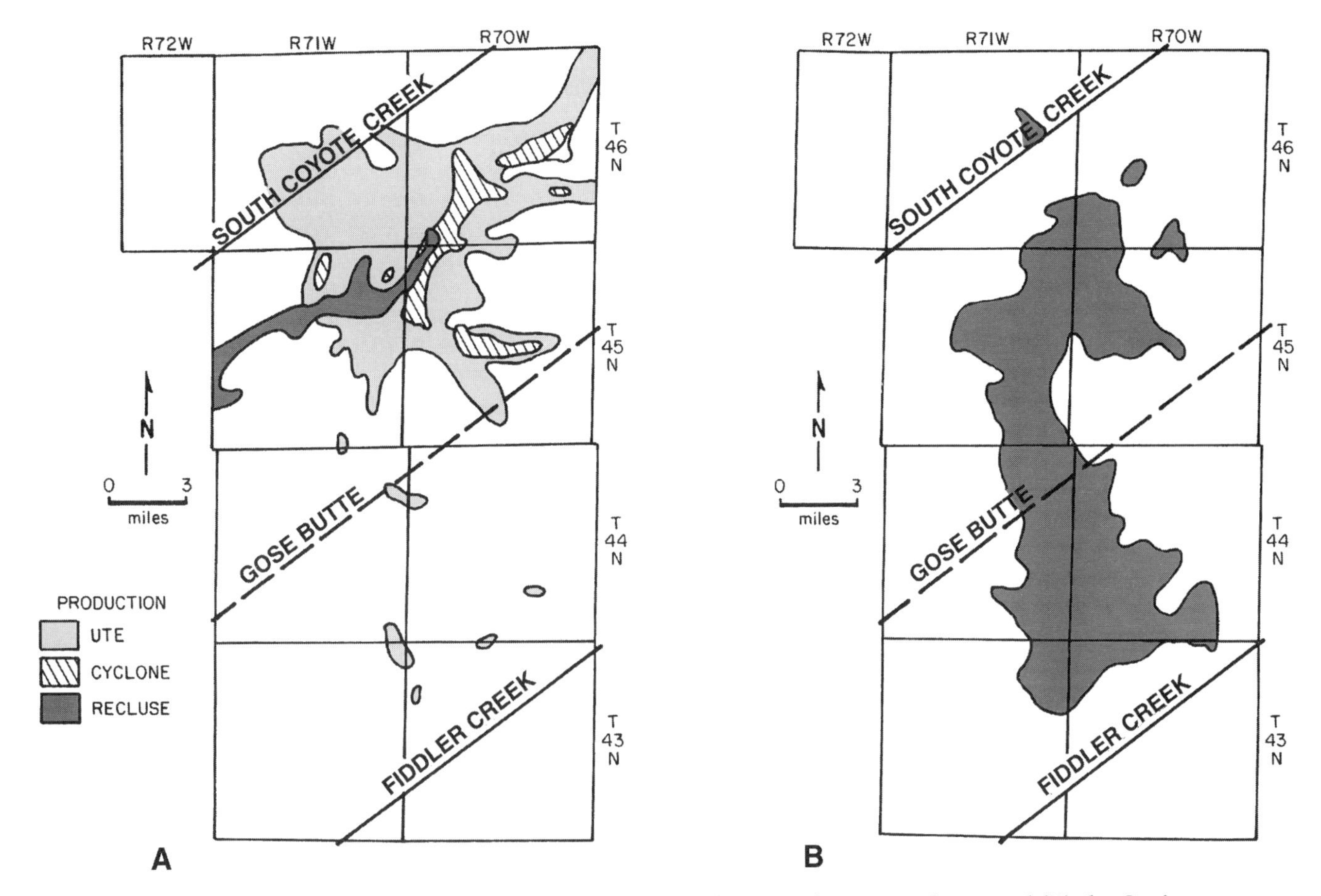

Figure 11. Outlines of production from (A) the Recluse, Cyclone, and Ute members, and (B) the Springen Ranch member. Note how production from the two lower valley-fill members is highly restricted, while production from the marine Springen Ranch member is more widespread and at a different orientation. Note also the location of mapped lineaments. Modified from Wheeler et al. (1988).

trends of which align with depositional facies. This continuity suggests that the Springen Ranch member reservoir comprises a single compartment whose limits are defined by the limits of the barrier island geometry. Although the reservoirs in the Hilight field are stratigraphic, recurrent basement block movement appears to have controlled valley development and sandstone distribution (Wheeler et al., 1988).

Figure 12 is an isopachous map of gross sandstone thickness (as determined from gamma-ray logs) within the Muddy Sandstone, regardless of member-level stratigraphy. Figure 13A is an east-west transect across the south part of Hilight field, where sand development is entirely within the Springen Ranch barrier island deposit, and shows general correspondence between the troughs and peaks of head and sand thickness trends. Figure 13B is a transect from within the northern valley system and shows a lesser degree of correspondence between the troughs and peaks of these trends. This is probably due to the sand thicknesses reported not being from a single member, but from various small, and mostly physically isolated, sandstone lithosomes within the lower alluvial deposits.

In summary, stratigraphic compartmentation occurs on several levels in the Hilight field. For the Recluse, Cyclone, and Ute members, maximum compartment size (even where the system is sand rich) is defined by the valley geometry, with bottom and lateral seals composed of Skull Creek Shale and an upper seal composed of the transgressive shale that separates the Ute from the Springen Ranch. Transgressive shales that separate each of the members provide additional vertical barriers within this valley-defined compartment and divide it into member-level (parasequence) subcompartments. The Springen Ranch compartment geometry—although this compartment is larger than any of the individual compartments in the lower members—is defined by the geometry of the net sandstone distribution within the parasequence.

The *Kitty field* (Figure 8) is located near the head of one of the larger valley systems that developed north of the Belle Fourche Arch. Drainages within Kitty Valley flowed to the north, until they connected with a major east–west-trending trunk drainage system that flowed east and ultimately emptied into the northern foreland basin (Dolson et al., 1991). As in the

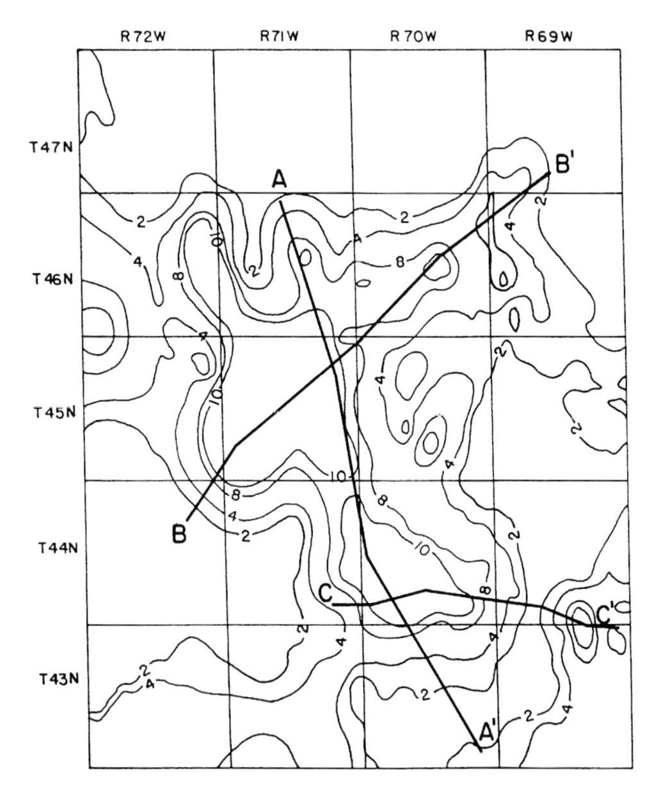

Figure 12. Isopachous map of net sandstone, Hilight field. Contour interval is in feet. Note the two trends of thick sandstone. The northeast-southwest trend is mostly composed of valley-fill sandstones, and the northwest-southeast trend is mostly composed of Springen Ranch strandline sandstones. Also shown are lines of section for Figures 10 (A–A′), 13A (C–C′), and 13B (B–B′).

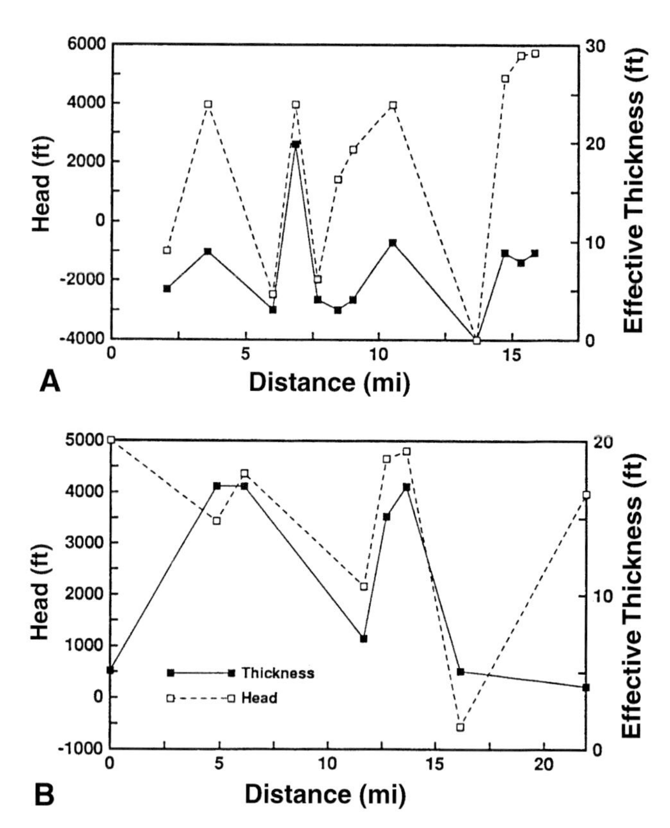

Figure 13. Plot of sand thickness vs. head for two areas of Hilight field: (A) an east-west transect across the southern part of Hilight (Springen Ranch member sandstones only) and (B) a northeast-southwest transect up the northern valley. Lines of section shown in Figure 12.

Hilight Valley, the trends of the valley systems north of the Belle Fourche Arch coincide with mapped lineaments believed to represent the boundaries of basement blocks (Ryer et al., 1990). Unlike the Hilight Valley, the Kitty Valley did not everywhere erode down into the Skull Creek Shale, so that both valley walls and floors are composed in places of subunconformity marine sandstones (Figure 14). Core analyses and petrographic work show that the subunconformity sandstones that border the east and west sides of the Kitty Valley are tight due to abundant clay matrix (Smith, 1988), however, and thus provide a seal between the subunconformity sandstones and the overlying valley-fill sandstones.

The Kitty Valley trends north-south, is up to 6 mi (10 km) wide, and has more than 60 ft (18 m) of relief. First the upper Recluse, then the Cyclone, Ute, and Springen Ranch members lap from north to south onto the Rozet unconformity surface. Upper Recluse deposits are dominantly alluvial, are confined to within the valley, contain up to 30 ft (9 m) of sand, and are highly productive. Because the upper Recluse at Kitty is fairly sand rich, the total sandstone accumulation is greater than at Hilight; and there is a higher probability of sand-on-sand contacts at Kitty than in the Recluse reservoirs at Hilight, and less likelihood of reservoirs in the upper Recluse being segmented into areally restricted stratigraphic compartments. The overlying Cyclone member has some sandstone near the base, but consists mostly of siltstones and shales, and records a transition from dominantly alluvial deposition to estuarine deposition within the valley. The Ute contains two highly productive sandstone zones, separated by shale. The lower zone is alluvial to estuarine and is restricted to within the valley. The upper sand probably represents barrier island deposition and in the Kitty area is oriented north-south, as is the trend of the valley, suggesting that the same fault zones that influenced valley development may have caused stabilization of the strandline during Ute deposition. Tidal channels, oriented roughly east-west, or perpendicular to the barrier island trend, are often tight due to calcite cementation (Rebecca Moncur, 1991, personal communication), and may also provide lateral seals within the Ute reservoir. The Springen Ranch member, although laterally fairly continuous, consists mostly of shales and shaly siltstones and does not appear to contain any reservoir-quality sandstones. Although the Muddy exceeds 100 ft (30 m) in thickness in this area, total net sandstone thickness is less than 60 ft (18 m) (Larberg, 1980). A reservoir pressure-versus-elevation

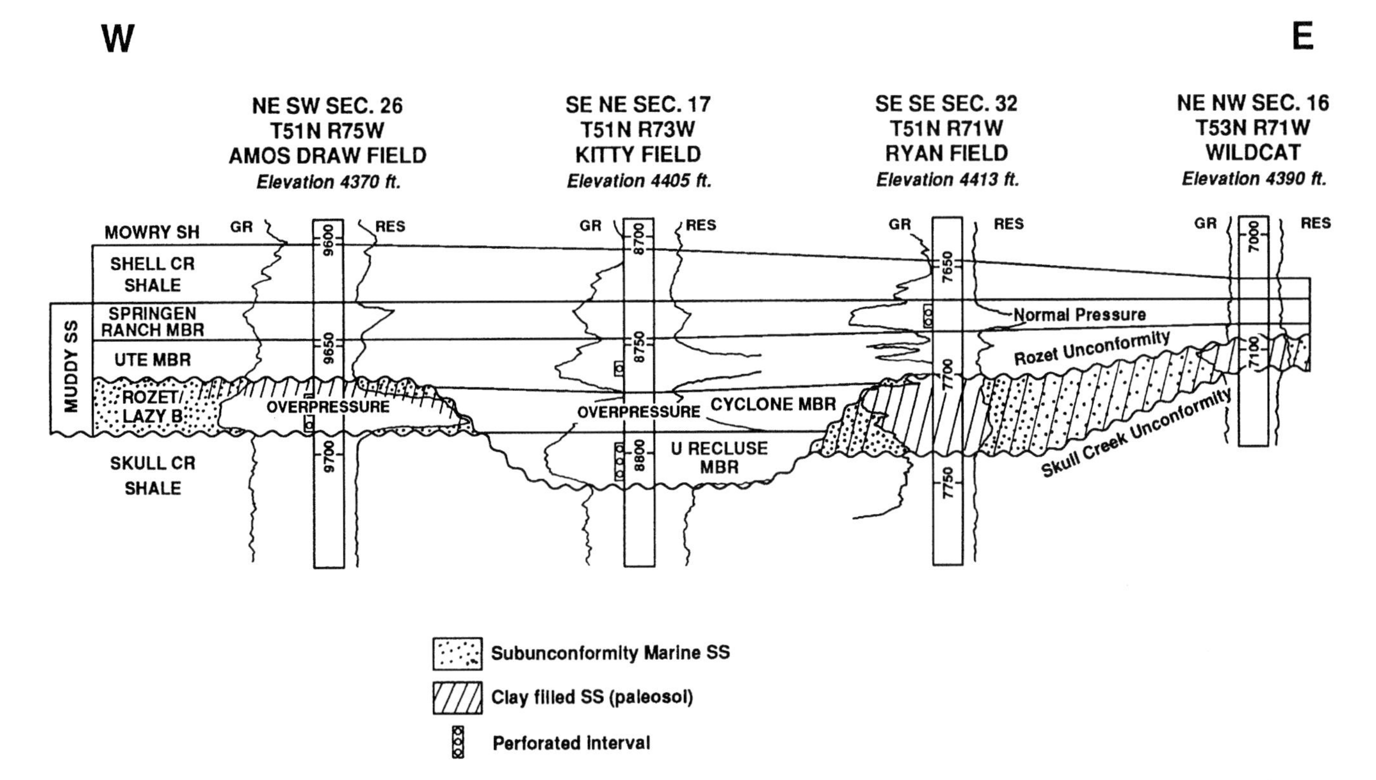

Figure 14. Well-log cross section through Kitty field. Note that the valley walls in Kitty are composed mostly of sandstone; how the depth of weathering varies beneath the unconformity; and that the paleosol apparently forms the lateral seal between overpressured sandstones in Kitty and more highly overpressured sandstones in Amos Draw.

plot suggests that at least some of the transgressive shales separating reservoir sandstones are seals that vertically compartmentalize the Muddy (Figure 15).

As within the Hilight field, several levels of stratigraphic compartmentation exist within the Kitty field, the maximum compartment size being defined by the valley geometry. Even though some of the walls of the Kitty Valley are composed of sandstone, the sandstones are clay filled and tight and are capable of being seals similar to the Skull Creek Shale. Laterally persistent, transgressive shales provide vertical seals between producing sandstones within the valley, and both primary and diagenetically enhanced lithofacies variations provide lateral seals.

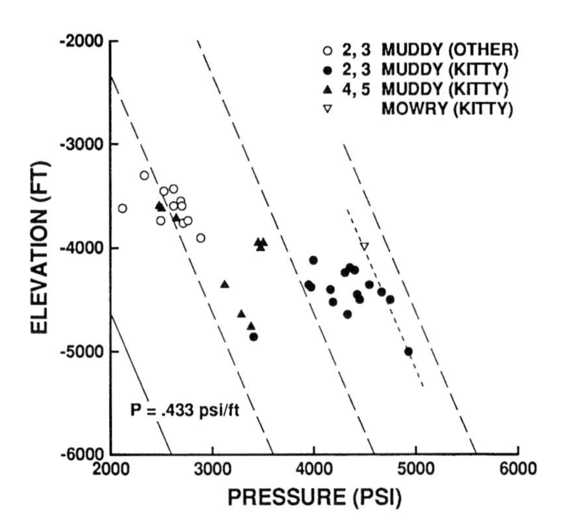

Figure 15. Plot of measured pressures at Kitty field as a function of elevation. Pressure in the Mowry Shale (open triangle) shows a higher gradient than in the Muddy, and hydrostatic heads decrease downward from the Mowry through the upper Muddy (4, 5; Springen Ranch and Ute members ?) to the lower Muddy (2, 3; Cyclone and Recluse members?). Inclined lines represent freshwater pressure gradients of 0.433 psi/ft (hydrostatic head) extending from elevations of zero pressure (from Berg et al., 1985). Two possible interpretations for the varying pressures include (1) pressure compartmentation of the Muddy Sandstone on a member scale and (2) hydrodynamic flow from a high head in the Mowry to a lower head in the lower Muddy.

Hierarchy of Observed Compartmentation

In summary, the interplay between regional changes in base level and local tectonics during Muddy time resulted in a highly complex and highly stratigraphically compartmented rock unit (Figure 16). Although the rock unit is still formally referred to as the Muddy Sandstone member of the Thermopolis Shale, the Muddy is a unit with highly variable lithologies and is characterized by sandstone lithosomes that are generally thin, areally restricted, and both vertically and laterally bounded by shale or mudstone.

At least three levels of stratigraphic compartmentation occur within the Muddy (Figure 17). Each of these three levels appears to coincide with a different scale of pressure compartmentation observed within

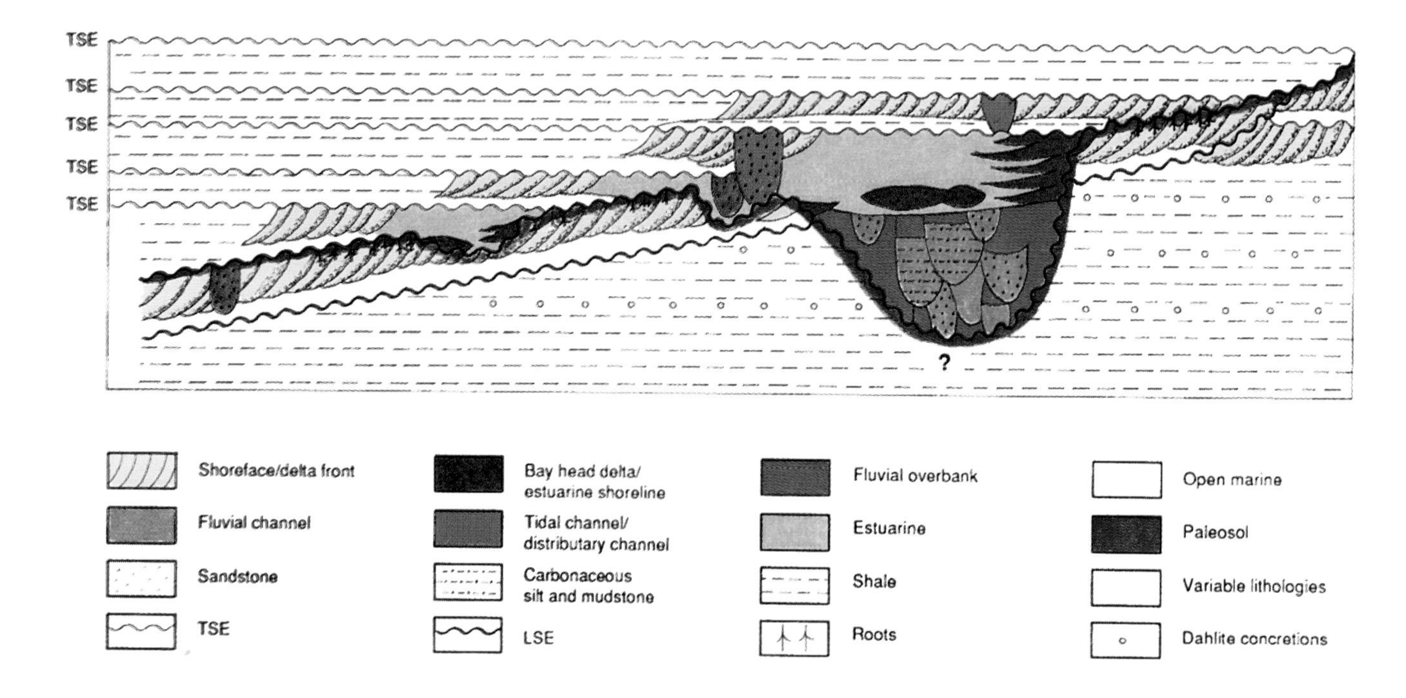

Figure 16. Diagrammatic cross section showing basic facies and stratigraphic relationships of the Muddy Sandstone. Cross section is based on a two-unconformity model of the Muddy, modified from Ryer et al. (1990). Well displayed is the stratigraphic isolation of reservoirs both by primary lithostratigraphic facies changes and by truncation and diagenetic alteration of lithofacies due to the unconformities. Note the highly variable relief along the K-2 unconformity (post-Rozet member); the local merging of the two unconformity surfaces; the extensive development of the paleosol associated with the K-2 unconformity; how the paleosol crosses both lithofacies boundaries and older unconformity surfaces and forms a potential seal along sand-on-sand contacts; and the sharp lithologic boundaries commonly found across the transgressive surfaces of erosion (TSEs).

the formation. The first level of compartmentation is provided by the Rozet unconformity, which stratigraphically divides the Muddy into two separate stratigraphic sequences. Relief along the unconformity surface causes this regionally extensive boundary to be more vertical in one place and more horizontal in another, and thus the seal crosscuts stratigraphy. Depending on the area, this boundary may provide either the top or bottom seal as well as a lateral seal or seals to a compartment. Diagenesis beneath the unconformity created sandstones with excellent sealing capacity, thus making the unconformity a pressure compartment boundary as well as a stratigraphic boundary, even in areas where sand is in contact with sand across the unconformity. Locally within subunconformity deposits, diagenesis altered the entire section of preserved rocks and resulted in subcompartmentation. The overall geometries of the compartments formed by the Rozet unconformity are defined by the intersection of the unconformity or the paleosol with either the underlying Skull Creek Shale or the overlying Shell Creek Shale.

The second level of compartmentation is provided by the shales that overlie the transgressive surfaces of erosion and further divide the stratigraphic sequences into individual parasequences or parasequence sets. The geometries of these compartments are defined by the intersection of the shales (either by onlap or truncation) with the Rozet unconformity or paleosol.

The third level of coincident stratigraphic and pressure compartmentation is provided by the distribution, geometry, and stacking of individual lithosomes within each of the parasequence sets. Various thin zones displaying roots occur within each of the members of the upper Muddy, but the extent of soil development and erosion that may have occurred in association with these zones is not known. Some evidence does suggest, however, that these localized soil horizons at least locally separate individual producing lithosomes and interfere with reservoir performance (Goolsby, 1991; Lin, 1981).

IMPLICATIONS FOR THE DELINEATION AND CHARACTERIZATION OF FLUID AND PRESSURE COMPARTMENTS

Significance of the Valley-Fill Model

The stratigraphic model currently applied to the upper Muddy deposits is known as the "valley-fill"

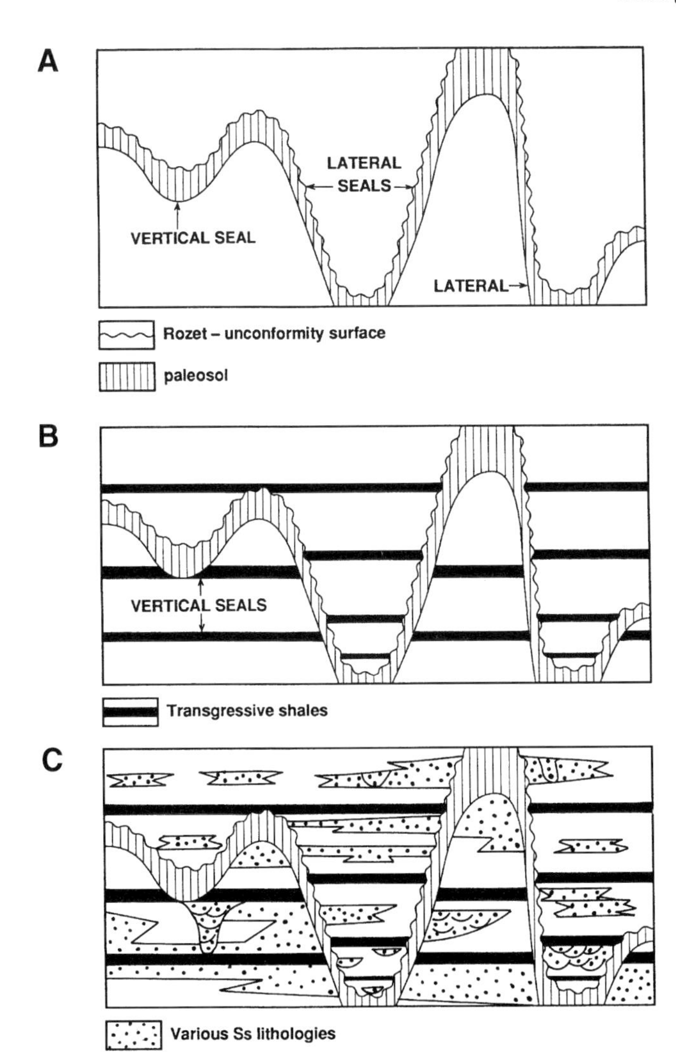

Figure 17. Hierarchy of stratigraphic compartmentation observed in the Muddy. (A) First-level compartment formed by intersection of LSE with impermeable shales that bound Muddy. Note that both vertical and lateral seals to fluid flow exist. Geometry is locally modified by diagenesis within paleosol zone. (B) Second level of compartmentation is the product of transgressive shales. (C) Third level of compartmentation is defined by the distribution of lithofacies.

model and was first described by Harms (1966). Prior to recognition of the widespread existence of a rooted soil zone that physically separates older, exclusively marine deposits of the lower Muddy from younger deposits of varying nonmarine to marine origin, the Muddy was interpreted as a deltaic distributary or barrier island/tidal channel deposit characterized by channel systems that prograded over and cut into slightly older but basically contemporaneous marine deposits (Berg and Davies, 1968; Clark, 1970; McGregor and Biggs, 1968). Because the vertical stratigraphic sequence observed in the Muddy is similar to the vertical sequence within a typical fluvial-deltaic sequence (or upper shoreface barrier island/tidal channel), the presence of the Rozet intraformational unconformity was not recognized for many years. Furthermore, the importance of distinguishing between the valley-fill-deltaic (or valley-fill-shoreface) model and the fluvial-deltaic model in transgressive-regressive transitional marine sequences is just beginning to be recognized. Subsurface correlation interpretations are significantly different for fluvial-deltaic systems and valley-fill-deltaic systems. In the valley-fill-deltaic model, strata of different ages can appear to be laterally equivalent, and older deposits can stratigraphically occur higher than younger valley-fill deposits due to relief along the unconformity separating them. Because of the emphasis placed on facies models and facies relationships in stratigraphic studies, such apparent lateral equivalence commonly leads workers to make erroneous correlations. For example, in the northern Powder River basin, Ute member shoreface sandstones and Rozet shoreface sandstones occupy a similar vertical position within the Muddy (first sandstone above the Skull Creek Shale), have similar electric-log responses, and are commonly miscorrelated (Ryer, 1990, personal communication). In cores, however, the Rozet shoreface sands can readily be distinguished from the Ute by the presence of siderite blebs, roots, and abundant clay matrix in the Rozet.

Whereas the fluvial-deltaic model incorporates the presence of locally formed diastems and soil zones within a sequence, it does not predict the presence of a regionally extensive subaerial erosion surface along which soil formation and extensive diagenesis are likely to occur. The importance of paleosol recognition and characterization has been emphasized by Weimer (1991c, p. 22), who cites "identification by downhole logging of diagenetic traps associated with paleosols beneath sequence boundaries" as one of several areas of future technological breakthrough capable of increasing our domestic exploration and production success. Paleosols can exhibit either decreased or increased porosity and permeability, but early and later diagenesis in the Muddy combined to severely reduce porosity and permeability and produce a regional seal.

Furthermore, because the sandstones above and beneath the unconformity can be lithologically different, they can follow different diagenetic pathways during burial. For example, sandstones overlying the post-Rozet unconformity are compositionally more mature than those underlying the unconformity and have undergone different diagenetic processes (Jiao and Surdam this volume; Higley and Schmoker, 1989; Suchecki et al., 1988).

Lastly, the valley-fill model indicates the presence of relief along the unconformity. Locally, this relief, as discussed above with respect to the Muddy, oftentimes provides lateral as well as vertical seals to fluid migration.

Applicability of the Model to Other Systems

Although numerous unconformities are indicated in Figure 2, none are intraformational, and none are shown as occurring within the Cretaceous System. Numerous studies, however, many of them recent (Dolson et al., 1991; Gill and Cobban, 1966; Gustason, 1988; Gustason et al., 1986, 1988a; Martinsen, 1985; Martinsen et al., 1984; Merewether and Cobban, 1985; Merewether et al., 1979; Odland et al., 1988; Weimer, 1983, 1984, 1985, 1991a, b; Weimer and Flexer, 1985; Weimer et al., 1982, 1988; Wheeler et al., 1988), describe the occurrence of many unconformities within the Cretaceous of the Powder River basin. Figure 18, a modified version of Figure 2, indicates the position of unconformities that have been recognized to date within the Cretaceous. Table 1 lists studies that identify or corroborate the unconformities identified on Figure 18. These unconformities, for the most part, have shorter hiatal intervals than those indicated on Figure 2, have only slight angularity variations across them, are intraformational and do not always separate contrasting lithologies, probably are not all subaerial in origin, and may or may not have associated weathering profiles. For these reasons, the unconformities within the Cretaceous have been difficult to recognize and map, and for the most part they are still not delineated on standard stratigraphic nomenclature charts of the basin. It is quite possible that if the Cretaceous were not one of the major hydrocarbon-producing intervals in Wyoming and had not therefore been intensively studied and explored, many if not all of these unconformities might still be unrecognized today. Intraformational unconformities have also been identified in the Permo-Pennsylvanian section (Maughn, 1990), another rich producer of hydrocarbons. Therefore, it may be that a level of stratigraphic complexity similar to that observed in the Muddy Sandstone may be present throughout the Powder River basin section.

The question remains, however: How applicable are the stratigraphic/diagenetic complexities observed in the Muddy, or even the entire rock record of the Powder River basin, to the rest of the world? Within the context of sequence stratigraphy, valley-fill systems are beginning to be widely recognized. For example, valley-fill systems have been recognized in several formations in the Alberta basin, including the Viking Formation (Bhattacharya, 1991; Cox, 1991; Hayes, 1988; Leckie, 1991a, b; Leckie and Hayes, 1991; Pattison, 1988, 1991a, b; Posamentier and Chamberlain, 1991; Strobl, 1988; Walker, 1988; Walker and Boreen, 1991; Walker and Davies, 1991), and in the Morrow of the Anadarko basin (Sonnenberg et al., 1990; Blakeney-DeJarnett and Krystinik, 1991; Krystinik and Blakeney-DeJarnett, 1991). It is interesting that, like the Muddy Sandstone, many of these formations are characterized by anomalous pressures and are used as examples to document the existence of pressure compartments (Powley, 1982; 1987, personal communication).

CONCLUSIONS

Although it is relatively thin, the Muddy Sandstone of the Powder River basin is a very complex stratigraphic unit. The high degree of complexity is a product of the interaction between one or more regional changes in base level, differential basement block movement, and varying depositional environments. Construction of an accurate stratigraphic framework for the Muddy requires both detailed lithofacies analyses and the recognition and delineation of various stratigraphic surfaces, especially lowstand surfaces of subaerial erosion (LSEs) and transgressive surfaces of erosion (TSEs).

The stratigraphic framework presented here indicates at least three levels of reservoir compartmentation. The first level of compartmentation is defined by the relief along the LSEs, which in places exceeds the preserved thickness of underlying deposits and physically divides the Muddy, both vertically *and* laterally, into older and younger sequences. The second level is defined by the intersection of shales that overlie the TSEs with the LSE (either by onlap or truncation). The third level results from variations in lithofacies. That (1) the bounding lithologies of the three levels of stratigraphic compartments are either shale or diagenetically tight sandstone and that (2) the geometries of reservoir compartments are of the same magnitude as observed fluid/pressure compartments suggest that fluid compartmentation is strongly stratigraphically controlled. The delineation of fluid compartments in the Muddy should also be complex and show a similar hierarchical distribution.

The complex stratigraphic model beginning to be recognized in the Muddy is the outcome of numerous studies and probably would not have been developed were it not for the extensive subsurface database available in the Powder River basin and the existence of numerous excellent Muddy Sandstone outcrops. In all probability, the complexities being unraveled in the Powder River basin are not atypical of the rock record, but are characteristic of many basins. This leads to the conclusion that simple stratigraphic models may be more a result of insufficient data than of the simplicity of the systems.

ACKNOWLEDGMENTS

This material covers research performed through November 1991. The author is especially grateful to John Dolson, Tom Ryer, Gus Gustason, and Bob Weimer for the numerous discussions and field trips during which they shared their knowledge and concepts of the Muddy Sandstone. Thanks are also given to Rod Tillman, Frank Ethridge, Jim Steidtmann, Uwe Strecker, Z. S. Jiao, Paul Valasek, Hank Heasler, Amy Ryan, Rebecca Moncur, Karen Porter, Michael Szpakiewicz, and Don Boyd for their help. The assistance of the United States Geological Survey Core Research Center, the Wyoming Geological Survey, and Amoco Production Company in obtaining data is also acknowledged. Al Deiss is thanked for his drafting assistance. David Copeland provided much-needed editorial assistance. This study was funded by Gas Research Institute under contract number 5089-260-1894.

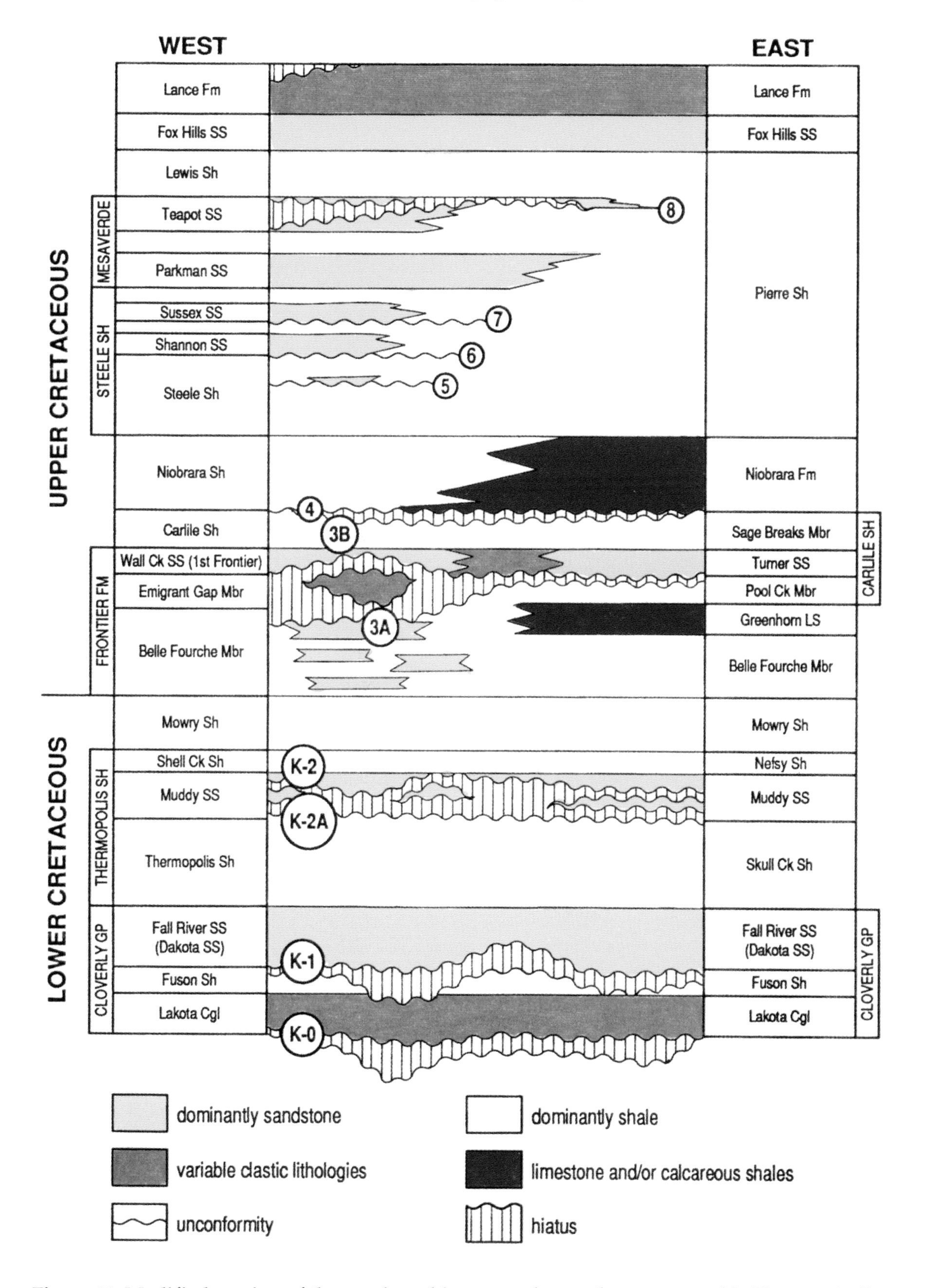

Figure 18. Modified version of the stratigraphic nomenclature chart presented in Figure 2, indicating the locations of additional unconformities, mostly intraformational, that have been identified within the Cretaceous System of the Powder River basin. Studies that identify these unconformities are listed in Table 1.

Table 1. List of studies that identify or corroborate the unconformities indicated on Figure 18.

K-0	**CRETACEOUS–JURASSIC BOUNDARY**
	Fox, 1986
	McGookey et al., 1972
K-1	**BASAL DAKOTA**
	Dolson et al., 1991
	Haerter, 1990
	Weimer, 1983, 1984
K-2	**INTRAFORMATIONAL (POST-ROZET) UNCONFORMITY**
	Clark, 1970
	Dolson et al., 1991
	Dolton et al., 1988
	Donovan, 1991
	Emme, 1981
	Farmer, 1981
	Fox, 1986
	Gustason, 1988
	Gustason et al., 1986, 1988
	Honarpour et al., 1988
	McGookey et al., 1972
	Odland et al., 1988
	Prescott, 1970
	Ryer et al., 1990
	Stone, 1972
	Szpakiewicz et al., 1988
	Waring, 1976
	Weimer, 1983, 1984, 1985, 1991a, 1991b
	Weimer and Flexer, 1991
	Weimer et al., 1982, 1988
	Wheeler et al., 1988
K-2A	**UNCONFORMITY(IES) EITHER WITHIN (LAZY B) OR AT BASE OF MUDDY SANDSTONE (SKULL CREEK)**
	Gustason, 1988
	Gustason et al., 1986, 1988
	Honarpour et al., 1988
	Odland et al., 1988
	Ryer et al., 1990
	Szpakiewicz et al., 1988
	Wheeler et al., 1988
3	**FRONTIER INTRAFORMATIONAL BASE OF EMIGRANT GAP**
	Martinsen et al., 1984
	Merewether and Cobban, 1985
	Merewether et al., 1977, 1979
	Weimer, 1983, 1984
	Weimer and Flexer, 1991
4	**FRONTIER INTRAFORMATIONAL BASE OF WALL CREEK**
	Fox, 1986
	Merewether and Cobban, 1985
	Merewether et al., 1979
	Merewether et al., 1977
	Weimer, 1984
	Weimer and Flexer, 1985

Continued on next page

Table 1 (Continued).

5	**BASE OF NIOBRARA**
	Chamberlain, 1991
	Sonnenberg et al., 1991
	Fox, 1986
	Martinsen et al., 1984
	Weimer, 1984
	Weimer and Flexer, 1985
6	**STEELE SHALE INTRAFORMATIONAL—BASE OF FISHTOOTH SANDSTONE**
	Martinsen et al., 1984
7	**STEELE SHALE INTRAFORMATIONAL—BASE OF SHANNON SANDSTONE**
	Crews et al., 1976
	Martinsen et al., 1984
	Weimer, 1984
8	**STEELE SHALE INTRAFORMATIONAL—BASE OF SUSSEX SANDSTONE**
	Crews et al., 1976
	Martinsen et al., 1984
9	**TEAPOT SANDSTONE INTRAFORMATIONAL**
	Gill and Cobban, 1966
	Martinsen et al., 1984
	Weimer, 1984

REFERENCES CITED

Berg, R. R., 1976, Hilight Muddy Field—Lower Cretaceous transgressive deposits in the Powder River basin, Wyoming: The Mountain Geologist, Rocky Mountain Association of Geologists, Denver, v. 13, p. 33–45.

Berg, R. R., and D. K. Davies, 1968, Origin of Lower Cretaceous Muddy Sandstone at Bell Creek field, Montana: American Association of Petroleum Geologists Bulletin, v. 52, p. 1888–1898.

Berg, R. R., R. M. Larberg, and L. D. Recker, 1985, Hydrodynamic flow in the Lower Cretaceous Muddy Formation, Northeast Powder River basin, Wyoming, *in* G. Nelson, ed., The Cretaceous Geology of Wyoming: Wyoming Geological Association Guidebook, p. 149–156.

Bhattacharya, J., 1991, The recognition and significance of marine flooding surfaces and erosional surfaces in core; examples from Upper Cretaceous clastic units in the Alberta foreland basin, *in* 1991 NUNA Conference on High-Resolution Sequence Stratigraphy Guidebook: Geological Association of Canada, p. 1–2.

Blakeney-DeJarnett, B. A., and L. F. Krystinik, 1991, Core-derived high-resolution sequence stratigraphy of the Pennsylvanian Morrow Formation, E. Colorado and W. Kansas, *in* 1991 NUNA Conference on High-Resolution Sequence Stratigraphy Guidebook: Geological Association of Canada, p. 136a–136g.

Chamberlain, C. K., 1991, Stratigraphy, petrology, sedimentology, and source rock of the Niobrara Formation, *in* G. Lawyer and C. Chamberlain, eds., Niobrara Reservoir Core Workshop Short Course Notes: Denver, Colorado, Exploration Methods, Inc.

Clark, C. R., 1970, Kitty Field, Campbell County, Wyoming, *in* Wyoming Sandstones Symposium: Wyoming Geological Association 22nd Annual Field Conference Guidebook, p. 79–84.

Cox, J., 1991, The Blood Gas Field (Bow Island Formation), a Lower Cretaceous estuarine valley-fill complex in southwest Alberta, *in* 1991 NUNA Conference on High-Resolution Sequence Stratigraphy Guidebook: Geological Association of Canada, p. 105–114.

Crews, G. C., J. A. Barlow, and J. D. Haun, 1976, Upper Cretaceous Gammon, Shannon and Sussex Sandstones, central Powder River basin, Wyoming, *in* R. Laudon, ed., Geology and Energy Resources of the Powder River: Wyoming Geological Association 28th Annual Field Conference Guidebook, p. 9–20.

Dolson, J. C., D. S. Muller, M. J. Evetts, and J. A. Stein, 1991, Regional paleotopographic trends and production, Muddy Sandstone (Lower Cretaceous), central and northern Rocky Mountains: American Association of Petroleum Geologists Bulletin, v. 75, p. 409–435.

Dolton, G. L., J. E. Fox, and J. L. Clayton, 1988, Petroleum geology of the Powder River basin, Wyoming and Montana: U.S. Geological Survey Open-File Report 88-450 P, 64 p.

Donovan, A. D., 1991, Sequence stratigraphy of the Muddy Sandstone: Powder River basin, Wyoming, *in* 1991 NUNA Conference on High-Resolution Sequence Stratigraphy Guidebook: Geological Association of Canada, p. 11–12.

Emme, J. J., 1981, Tectonic influence on sedimentation, Lower Cretaceous strata, Osage–Newcastle area, Powder River basin, Wyoming: M.S. thesis, Colorado School of Mines, Golden, 171 p.

Farmer, C., 1981, Tectonics and sedimentation, Newcastle Formation (Lower Cretaceous), southwestern flank Black Hills uplift, Wyoming and South Dakota: M.S. thesis, Colorado School of Mines, Golden, 195 p.

Fox, J. E., 1986, Stratigraphic cross-sections showing electric logs of Upper Cretaceous and older rocks, Powder River basin, Wyoming: U.S. Geological Survey Open-File Reports 86-0465A through 86-0465V.

Gill J. R., and W. A. Cobban, 1966, Regional unconformity in Late Cretaceous, Wyoming, *in* Geological Survey Research 1966: U.S. Geological Survey Professional Paper 550-B, p. B20–B27.

Goolsby, J. E., 1991, Enhanced recovery from lower Muddy fluvial sandstone reservoirs of the northern Powder River basin, Wyoming: The Contact, Newsletter of the Wyoming Geological Association, v. 38, no. 11, p. 2–4.

Gustason, E. R., 1988, Depositional and tectonic history of the Lower Cretaceous Muddy Sandstone, Lazy B Field, Powder River basin, Wyoming, *in* R. Diedrich, M. Dyka, and W. Miller, eds., Eastern Powder River basin—Black Hills: Wyoming Geological Association 39th Field Conference Guidebook, p. 129–146.

Gustason, E. R., T. A. Ryer, and S. K. Odland, 1986, Unconformities and facies relationships of Muddy Sandstone, northern Powder River basin, Wyoming and Montana: American Association of Petroleum Geologists Bulletin, v. 70, p. 1042.

Gustason, E. R., T. A. Ryer, and S. K. Odland, 1988a, Stratigraphy and depositional environments of the Muddy Sandstone, northwestern Black Hills, Wyoming: Wyoming Geological Association Earth Science Bulletin, v. 20, p. 49–60.

Gustason, E. R., D. A. Wheeler, and T. A. Ryer, 1988b, Structural control on paleovalley development, Muddy Sandstone, Powder River basin, Wyoming: American Association of Petroleum Geologists Bulletin, v. 72, p. 871.

Haerter, J., 1990, Sequence stratigraphy of the Fall River Sandstone, northern Powder River basin, Wyoming and Montana: M.S. thesis, Colorado State University, Fort Collins.

Harms, J. C., 1966, Stratigraphic traps in valley fill, western Nebraska: American Association of Petroleum Geologists Bulletin, v. 50, p. 2119–2149.

Hayes, B. J. R., 1988, Incision of a Cadotte member paleovalley-system at Noel, British Colombia—Evidence of a late Albian sea-level fall, *in* D. James and D. Leckie, eds., Sequences, Stratigraphy, Sedimentology: Surface and Subsurface: Canadian Society of Petroleum Geologists Memoir 15, p. 97–106.

Heasler, H.P., and R.C. Surdam, 1992, Pressure compartments in the Mesaverde Formation of the Green River and Washakie basins, as determined from drill stem test data, *in* C. Mullen, ed., Rediscover the Rockies: Wyoming Geological Association 43rd Field Conference Guidebook, Casper, Wyoming, p. 207–220.

Heasler, H., X. Chen, R. Simon, and R. C. Surdam, 1992, Geometry of pressure compartments in sandstones in the Powder River basin, Wyoming and Montana, as determined from drill stem test data: Final Report (Draft), Gas Research Institute contract no. 5089-260-1894, Multidisciplinary Analysis of Pressure Chambers in the Powder River Basin, Wyoming and Montana, p. 35–64.

Higley, D. K., and J. W. Schmoker, 1989, Influence of depositional environment and diagenesis on regional porosity trends in the Lower Cretaceous "J" Sandstone, Denver basin, Colorado, *in* E. Coalson, ed., Petrogenesis and Petrophysics of Selected Sandstone Reservoirs of the Rocky Mountain Region: Rocky Mountain Association of Geologists, Denver, p. 183–196.

Honarpour, M. M., et. al., 1988, Integrated geologic/engineering model for Barrier Island deposits in Bell Field, Montana: Paper Number 17366: Society of Petroleum Engineers/Department of Energy Enhanced Oil Recovery Symposium, Tulsa, Oklahoma, p. 491–511.

Krystinic, L. F., and B. A. Blakeney-DeJarnett, 1991, Sequence stratigraphy and sedimentologic character of valley fills, Lower Pennsylvanian Morrow Formation, Eastern Colorado and Western Kansas, *in* 1991 NUNA Conference on High-Resolution Sequence Stratigraphy Guidebook: Geological Association of Canada, p. 24–26.

Larberg, G. M., 1980, Depositional environments and sand body morphologies of the Muddy Sandstone at Kitty field, Powder River basin, Wyoming, *in* R. Enyert and W. Curry, eds., Symposium on Early Cretaceous rocks of Wyoming: Wyoming Geological Association 31st Annual Field Conference Guidebook, p. 117–135.

Leckie, D., 1991a, Middle Albian paleosols in the Boulder Creek Formation and Peace River Formation (Paddy member): What are the sequence stratigraphic implications?, *in* 1991 NUNA Conference on High-Resolution Sequence Stratigraphy Guidebook: Geological Association of Canada, p. 93–101.

Leckie, D., 1991b, Sequence stratigraphic significance of the Boulder Creek Formation paleosols, northeastern Alberta, *in* 1991 NUNA Conference on High-Resolution Sequence Stratigraphy Guidebook: Geological Association of Canada, p. 26–27.

Leckie, D., and B. J. Hayes, 1991, Regional overview of the Albian Peace River Formation: evidence for a major erosion surface, *in* 1991 NUNA Conference on High-Resolution Sequence Stratigraphy Guidebook: Geological Association of Canada, p. 27–29.

Lin, J. T. C., 1981, Hydrodynamic flow in Lower Cretaceous Muddy Sandstone, Gas Draw Field, Powder River basin, Wyoming: The Mountain Geologist, v. 18, no. 4, p. 78–87.

Marinovich, M. A., and E. S. Dietrich, 1973, Waterflood feasibility study, Lazy "B" Field, Campbell County, Wyoming: unpublished report, Wyoming Oil and Gas Commission files, Casper, Wyoming.

Martinsen, R. S., 1985, Tectonic control of Upper Cretaceous clastic shelves and associated shelf sand ridges, Western Interior Seaway, Wyoming, *in* Proceedings, IAS-SEPM Conference on Foreland basins: Friborg, Switzerland.

Martinsen, R. S., D. J. P. Swift, and G. C. Gaynor, 1984, Local and regional cross-sections through the Mowry—Teapot sandstone interval, Upper Cretaceous of the Powder River basin, Wyoming: ARCO Oil and Gas Company Research Report #84-24, 21 p.

Maughn, E. K., 1990, Summary of the ancestral Rocky Mountains epeirogeny in Wyoming and adjacent areas: U. S. Geological Survey Open-File Report OF 90-447, 8 p.

McGookey, D. P., et al., 1972, Cretaceous System, *in* W. Mallory, ed., Geologic Atlas of the Rocky Mountain Region: Denver, Rocky Mountain Association of Geologists, p. 190–228.

McGregor, A. A., and C. A. Biggs, 1968, Bell Creek Field, Montana: a rich stratigraphic trap: American Association of Petroleum Geologists Bulletin, v. 52, p. 1869–1887.

Merewether, E. A., and W. A. Cobban, 1985, Tectonism in the mid-Cretaceous Foreland, southeastern Wyoming and adjoining areas, *in* G. E. Nelson, ed., The Cretaceous Geology of Wyoming: Wyoming Geological Association Guidebook, p. 67–74.

Merewether, E. A., W. A. Cobban, R. M. Matson, and W. J. Magathan, 1977, Stratigraphic diagrams with electric logs of Upper Cretaceous rocks, Powder River basin, Wyoming: U.S. Geological Survey Oil and Gas Investigation Charts, OC-73, OC-74, OC-75, OC-76.

Merewether, E. A., W. A. Cobban, and E. T. Cavanaugh, 1979, Frontier Formation and equivalent rocks in eastern Wyoming: The Mountain Geologist, v. 16, p. 67–102.

Odland, S. K., P. E. Patterson, and E. R. Gustason, 1988, Amos Draw field: a diagenetic trap related to an intraformational unconformity in the Muddy Sandstone, Powder River basin, Wyoming, *in* R. Diedrich, M. Dyka, and W. Miller, eds., Eastern Powder River basin—Black Hills: Wyoming Geological Association 39th Field Conference Guidebook, p. 147–160.

Pattison, S. A. J., 1988, Transgressive, incised shoreface deposits of the Burnstick member (Cardium "B" sandstone) at Caroline, Crossfield, Garrington and Lochend; Cretaceous western interior seaway, Alberta, Canada, *in* D. James and D. Leckie, eds., Sequences, Stratigraphy, Sedimentology: Surface and Subsurface: Canadian Society of Petroleum Geologists Memoir 15, p. 155–166.

Pattison, S. A. J., 1991a, Crystal, Sundance and Edson valley-fill deposits, *in* 1991 NUNA Conference on High-Resolution Sequence Stratigraphy Guidebook: Geological Association of Canada, p. 44–47.

Pattison, S. A. J., 1991b, Viking Formation overview, *in* 1991 NUNA Conference on High-Resolution Sequence Stratigraphy Guidebook: Geological Association of Canada, p. 40–43.

Posamentier, H. W., and C. J. Chamberlain, 1991, Sequence stratigraphic analysis of Viking Formation lowstand beach deposits at Jorcam Field, Alberta, Canada, *in* 1991 NUNA Conference on High-Resolution Sequence Stratigraphy Guidebook: Geological Association of Canada, p. 75–92.

Powley, D., 1982, Pressures, normal and abnormal: American Association of Petroleum Geologists Advanced Exploration Schools unpublished lecture notes, 48 p.

Prescott, M. W., 1970, Hilight field, Campbell County, Wyoming, *in* R. Enyert, ed., Symposium on Wyoming sandstones: their economic importance—past, present and future: Wyoming Geological Association 22nd Annual Field Conference Guidebook, p. 89–103.

Reeside, J. B. and A. C. Cobban, 1960, Studies of the Mowry Shale (Cretaceous) and contemporary formations in the United States and Canada: U.S. Geological Survey Professional Paper 355, 5 p.

Ryer, T. A., E. R. Gustason, and S. K. Odland, 1990, Depositional history and sea-level fluctuations, Lower Cretaceous Muddy Sandstone, Powder River basin, Wyoming, *in* T. Ryer, ed., Stratigraphic Concepts and Methods for Exploration in Clastic Facies: Rocky Mountain Association Geologists Short Course Notes.

Schowalter, T. T., 1979, Mechanics of secondary hydrocarbon migration and entrapment: American Association of Petroleum Geologists Bulletin, v. 63, p. 723–760.

Smith, D. A., 1988, The integration of hydrodynamics and stratigraphy, Muddy Sandstone, northern Powder River basin, Wyoming and Montana, *in* R. Diedrich, M. Dyka, and W. Miller, eds., Eastern Powder River basin—Black Hills: Wyoming Geological Association 39th Field Conference Guidebook, p. 179–189.

Sonnenberg, S., L. Shannon, K. Rader, W. Von Drehle, and L. Martin, (eds.), 1990, Morrow Sandstones of Southeastern Colorado and Adjacent States, Rocky Mountain Association of Geologists Guidebook, 236 p.

Sonnenberg, S., J. Dolson, and K. Porter, 1991, Codell and basal Niobrara surfaces—submarine, subaerial or both—what to think?, *in* J. Dolson et al., eds., Unconformity Related Hydrocarbon Exploitation and Accumulation in Clastic and Carbonate Settings: Rocky Mountain Association of Geologists Short Course Notes, p. 159–186.

Stone, W. D., 1972, Stratigraphy and exploration of the Lower Cretaceous Muddy Formation, northern

Powder River basin, Wyoming and Montana: The Mountain Geologist, v. 9, p. 355–378.

Strobl, R. S., 1988, The effects of sea-level fluctuations on prograding shorelines and estuarian valley-fill sequences in the glauconitic member, Medicine River Field and adjacent areas, *in* D. James and D. Leckie, eds., Sequences, Stratigraphy, Sedimentology: Surface and Subsurface: Canadian Society of Petroleum Geologists Memoir 15, p. 221–236.

Suchecki, R. K., G. R. Baum, S. Phillips, and J. S. Hewlett, 1988, Role of stratigraphic discontinuities in development of reservoir quality, Muddy and Skull Creek Sandstones, Denver basin, *in* D. James and D. Leckie, eds., Sequences, Stratigraphy, and Sedimentology: Surface and Subsurface: Canadian Society of Petroleum Geologists Memoir 15, p. 584.

Szpakiewicz, M., et. al., 1988, Geological and engineering evaluation of barrier island and valley-fill lithotypes in Muddy Formation, Bell Creek field, Montana, *in* E. Coalson et al., eds., Petrogenesis and Petrophysics of Selected Sandstone Reservoirs of the Rocky Mountain Region: Rocky Mountain Association of Geologists, Denver Colorado, p. 159–182.

Vail, P. R., R. M. Mitchum, Jr., and S. Thompson III, 1977, Seismic stratigraphy and global changes of sea level from coastal onlap, *in* C. Payton, ed., Seismic Stratigraphy—Applications to Hydrocarbon Exploration: American Association of Petroleum Geologists Memoir 26, p. 63–81.

Van Wagoner, J. C., 1988, Sequences and parasequences in siliciclastic rocks, *in* D. James and D. Leckie, eds., Sequences, Stratigraphy, Sedimentology: Surface and Subsurface: Canadian Society of Petroleum Geologists Memoir 15, p. 572.

Von Drehle, W. F., 1985, Amos Draw field, Campbell County, Wyoming: Wyoming Geological Association 36th Field Conference Guidebook, p. 11–31.

Walker, R. G., 1988, The origin and scale of sequences and erosional bounding surfaces in the Cardium Formation, *in* D. James and D. Leckie, eds., Sequences, Stratigraphy, Sedimentology: Surface and Subsurface: Canadian Society of Petroleum Geologists Memoir 15, p. 573.

Walker R. G., and T. Boreen, 1991, High-resolution allostratigraphy in the Viking Formation at Willesdon Green: description of all members and their bounding discontinuities, *in* 1991 NUNA Conference on High-Resolution Sequence Stratigraphy Guidebook: Geological Association of Canada, p. 57–58.

Walker R. G., and S. D. Davies, 1991, Control of prograding and incised-transgressive shoreface deposits by rapid fluctuations of relative sea level; Viking Formation in the Caroline-Garrington Area, *in* 1991 NUNA Conference on High-Resolution Sequence Stratigraphy Guidebook, Geological Association of Canada, p. 58–59.

Waring, J., 1976, Regional distribution of environments of the Lower Cretaceous Muddy Sandstone, southeastern Montana, *in* R. Laudon, ed., Geology and Energy Resources of the Powder River basin: Wyoming Geological Association 28th Annual Field Conference Guidebook, p. 83–96.

Weimer, R. J., 1983, Relation of unconformities, tectonics and sea level changes, Cretaceous of the Denver basin and adjacent areas, *in* R. Laudon, E. Reynolds, and E. Dolly, eds., Mesozoic Paleogeography of West–Central United States: Society of Economic Paleontologists and Mineralogists, Rocky Mountain Section, Rocky Mountain Paleogeography Symposium 2, p. 359–376.

Weimer, R. J., 1984, Relation of unconformities, tectonics, and sea-level changes, Cretaceous of Western Interior, U.S.A., *in* J. Schlee, ed., Interregional Unconformities and Hydrocarbon Accumulation: American Association of Petroleum Geologists Memoir 36, p. 7–35.

Weimer, R. J., 1985, New age interpretation of Bell Creek Sandstone, Powder River basin, Montana and Wyoming: American Association of Petroleum Geologists Bulletin, v. 69, p. 870.

Weimer, R. J., 1991a, Sequence stratigraphy of the Lower Cretaceous Muddy Formation, U.S.A., *in* 1991 NUNA Conference on High-Resolution Sequence Stratigraphy Guidebook: Geological Association of Canada, p. 59–60.

Weimer, R. J., 1991b, Sequence stratigraphy of the Muddy (Viking) Sandstone (Lower Cretaceous), Rocky Mountain region, United States of America: Illustrated by core data from the Denver basin, Colorado, *in* 1991 NUNA Conference on High-Resolution Sequence Stratigraphy Guidebook: Geological Association of Canada, p. 116–135.

Weimer, R. J., 1991c, President's Column: Fading Exploration: American Association of Petroleum Geologists Explorer, v. 12, p. 122–123.

Weimer, R. J., and A. Flexer, 1985, Depositional patterns and unconformities, Upper Cretaceous, Eastern Powder River basin, Wyoming, *in* G. Nelson, ed., The Cretaceous Geology of Wyoming: Wyoming Geological Association Guidebook, p. 131–148.

Weimer, R. J., J. J. Emme, C. L. Farmer, L. O. Anna, T. L. Davis, and R. L. Kidney, 1982, Tectonic influence on sedimentation, Early Cretaceous east flank, Powder River basin, Wyoming and South Dakota: Colorado School of Mines Quarterly, v. 73, 62 p.

Weimer, R. J., C. A. Rebne, and T. L. Davis, 1988, Geologic and seismic models, Muddy Sandstone, Lower Cretaceous, Bell Creek—Rocky Point area, Powder River basin, Montana and Wyoming, *in* R. Diedrich, M. Dyka, and W. Miller, eds., Eastern Powder River basin—Black Hills: Wyoming Geological Association 39th Field Conference Guidebook, p. 147–160.

Wheeler, D. M., E. R. Gustason, and M. J. Furst, 1988, The distribution of reservoir sandstone in the Lower Cretaceous Muddy Sandstone, Hilight Field, Powder River basin, Wyoming, *in* A. Lomando and P. Harris, eds., Giant Oil and Gas Fields: Society of Economic Paleontologists and Mineralogists Core Workshop No. 12, p. 179–228.

Chapter 19

Stratigraphic/Diagenetic Pressure Seals in the Muddy Sandstone, Powder River Basin, Wyoming

Zun Sheng Jiao
Ronald C. Surdam
University of Wyoming
Laramie, Wyoming, U.S.A.

ABSTRACT

Sandstones in the Rozet unconformity zone in the Muddy Sandstone are characterized by abundant clay matrix (up to 55% of the rock volume), absence of intergranular pores, and very low permeability. The diagenesis of clay minerals in the Rozet unconformity sandstone and overlying Mowry Shale includes smectite altering to illite in mixed-layer smectite/illite clays (I/S) and kaolinite reacting to chlorite. The I/S composition changes with progressive burial from approximately 20% illite in the mixed-layer smectite/illite clays at 900 m (3000 ft) to 85% illite at 4200 m (13,500 ft).

High-pressure mercury injection tests were performed on the sandstone samples from the Rozet unconformity zone. Pore throats for those samples are primarily in the subnano and nano categories (<0.01 to 0.05 μm), and permeabilities are from 0.02 to 0.08 md. Such sandstones can hold a differential pressure of 1800 psi, which is the same as the differential pressure in the Amos Draw overpressured compartment from which the samples were taken.

There is a direct correlation between the diagenesis of clay minerals in the Rozet unconformity zone and the maturation of the Mowry Shale, and the sealing capacity or displacement pressure of the pedogenic units. The sealing capacity of the sandstone associated with the unconformity is derived from primary pedogenic processes and from diagenetic enhancement during progressive burial. The diagenetic processes can increase the sealing capacity of sandstone along the unconformity by an order of magnitude, or from a type C seal to a type A seal. The transition of the fluid-flow system from single phase to multiphase results in converting the low-permeability rocks along the unconformity to fluid/pressure seals capable of withstanding >1800 psi pressure differentials.

The recognition of the presence of widespread subaerial unconformities in the Muddy Sandstone is important in understanding abnormally-pressured compartments within the reservoir facies.

INTRODUCTION

An indispensable condition for the development of an abnormally pressured compartment in sandstone is the three-dimensional closure of relatively low permeability rock units that serve as barriers in the fluid-flow pathway. Such a low-permeability barrier can result from the original depositional environment, from later diagenetic processes, or from a combination of the two. The Rozet unconformity zone in the Powder River basin has undergone intensive pedogenic and diagenetic development and provides a potential top and lateral seal for pressure compartmentation in the Muddy Sandstone.

Subaerial exposure of rock along an unconformity may cause the breakdown of the original rock components and fabric. The final products of the pedogenic process are new fabrics and mineral assemblages that are stable under surface conditions. Two contrary processes, enhancement and reduction of porosity and permeability, can both occur along an unconformity, depending on the geologic setting. Most workers agree that porosity and permeability may be enhanced below an erosional unconformity because rocks are exposed to meteoric water-undersaturated CO_2. In this case, a net gain in porosity and permeability near the unconformity is caused by the abundant supply of fresh meteoric water and the relatively unrestricted transport of dissolved constituents from the site of weathering. Such porosity and permeability enhancement in hydrocarbon-bearing sandstone reservoirs has been observed, e.g., in the Alaskan North Slope (Shanmugan and Higgins, 1988) and northeastern China (Guangming and Quanheng, 1982). On the other hand, porosity and permeability also can be significantly reduced below an erosional unconformity because of abundant clay infiltration and cementation during subaerial exposure and progressive burial. There are numerous examples of an unconformity causing a significant decrease in the porosity and permeability of sandstone associated with the unconformity and forming a seal for hydrocarbon accumulation (Levorsen, 1934; Al-Gailani, 1981). In a compilation of giant hydrocarbon fields by Halbouty et al. (1970), 42 percent of the gas fields were determined to have an unconformity as an essential factor of the sealing capacity. In most petroliferous grabens, the basic trapping mechanism of individual oil and gas fields is a seal along an unconformity (North, 1985). Most "tar sand" deposits in the world are intimately associated with unconformities. In the Parkman field, Williston basin, Canada, the top and lateral seals are clearly formed by sandstones associated with an unconformity (Miller, 1984).

Production from the Lower Cretaceous Muddy Sandstone in the Rocky Mountain area is controlled principally by unconformities formed during a relative lowstand of sea level (Dolson et al., 1991). As a trapping mechanism, a sandstone associated with an unconformity within the Muddy Sandstone has high potential for forming seals of abnormally pressured compartments at depths of more than 2500 m (8000 ft) present-day burial (Surdam et al., this volume). Our recent research integrating the hydrodynamics, stratigraphy, petrology, and petrophysics of the Muddy Sandstone has revealed that the sandstone associated with the Rozet unconformity forms pressure seals for the abnormally pressured compartments developed in the Muddy Sandstone in the Powder River basin. At the Amos Draw field, an abnormally high pressure compartment has developed in the Rozet and Lazy B members of the Muddy Sandstone. The Rozet unconformity defines the top and east boundaries of the Amos Draw overpressured compartment (Surdam et al., this volume).

The objectives of this chapter are:

1. To summarize the factors that promote porosity and permeability reduction when rocks are subaerially exposed.
2. To delineate the petrographic and petrophysical characteristics of sandstones in the unconformity zone.
3. To characterize the sealing capacity of the Rozet unconformity.
4. To discuss the formation of the Rozet unconformity pressure seal as the result of organic maturation in the Mowry Shale and sandstone diagenesis in the Muddy Sandstone.

PEDOGENESIS AND PERMEABILITY CHANGES AT EROSIONAL UNCONFORMITIES

Porosity and Permeability Changes

Hydroxide and clay-mineral precipitation and clay infiltration are the two major causes of significant reduction of porosity and permeability at erosional unconformities. Soil formation on an unconformity is an exceedingly complex process that affects the redistribution of rock porosity and permeability. In temperate, humid climates, soil usually develops a characteristic set of horizons (Buol, et al., 1980; Retallack, 1990) (Figure 1). The O horizon is the organic matter accumulation zone. Below the O horizon is the A horizon, a coarse-grained layer from which iron and aluminum have been leached. The process by which iron and aluminum are removed from the A horizon is probably dissolution as organic complexes. Such leaching may cause an increase in porosity and permeability. Beneath the A horizon is the B horizon, a layer in which iron and aluminum have accumulated as fine-grained, poorly crystallized hydroxides and clay minerals. The accumulation of these authigenic minerals is subsequently important from the hydrocarbon exploration standpoint, for it can cause a significant decrease in porosity and permeability in the B horizon. In less-humid environments, leaching is less important as a pedogenic process, and nodules and cemented layers of calcium carbonate (caliche) tend to form in the soil profile because of insufficient water flow through the soil to remove the calcium released by weathering.

ERRATUM

BASIN COMPARTMENTS AND SEALS

AAPG Memoir 61

The enclosed page is a replacement for page 299 in the chapter "Stratigraphic/Diagenetic Pressure Seals in the Muddy Sandstone, Powder River Basin, Wyoming," by Zun Sheng Jiao and Ronald C. Surdam. The figures in the book on page 299 were incorrectly repeated from a previous chapter.

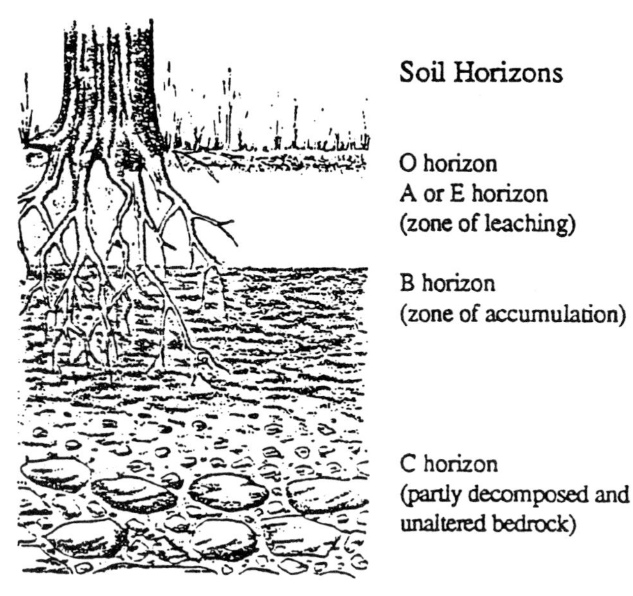

Figure 1. Schematic diagram of soil horizon formed in a temperate, humid climate. From Drever (1988).

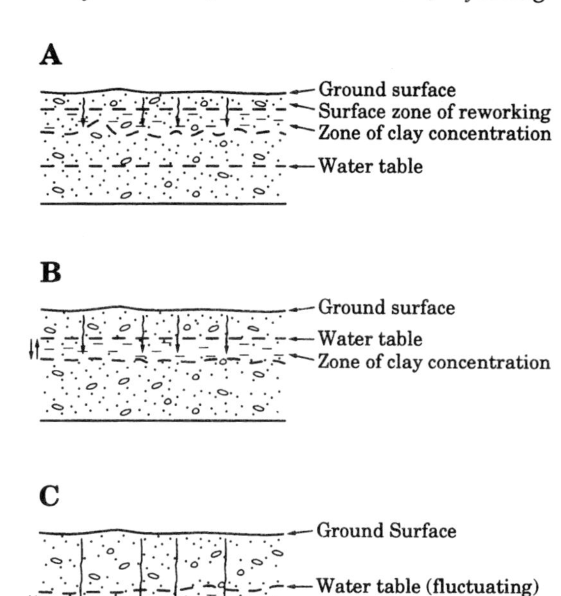

Figure 2. Mechanisms and sites of clay infiltration. Modified from Walker (1976). (A) Clay accumulation due to evaporation of the infiltrating water within the vadose zone. (B) Clay accumulation related to the loss of velocity as the infiltrating water reaches the water table. (C) Clay accumulation due to the development of perched water tables over impermeable barriers.

The other important process affecting porosity and permeability during pedogenesis is clay infiltration. The infiltration of alluvial clay or loess into coarse continental sediments is a well-known process (Bull, 1968). Along the Rozet unconformity zone, the infiltrated clays are mainly kaolinite resulting from the weathering of material in the drainage basin. Besides weathered clay, volcanic ash fall during the Rozet lowstand time also provided an abundance of infiltrated clays. Walker (1976) described three mechanisms that lead to the accumulation of significant amounts of infiltration clays (Figure 2): (1) concentrations in the vadose zone (Figure 2A), (2) concentration near the potentiometric surface (Figure 2B), and (3) concentration above impermeable barriers (Figure 2C). It is believed that the B horizon produced a basic barrier for the concentration of infiltration clays in the Rozet unconformity zone. Even though thin infiltration clay cutans may preserve intergranular porosity, permeability can be severely reduced by the obstruction of pore throats. For instance, in zones of high clay concentration, in which the clay matrix may constitute up to 35% of rock bulk volume, infiltration clays create permeability barriers in many Sergi Formation reservoirs (Moraes, 1991).

Authigenic Minerals

Subaerial exposure of a rock leads to the development of a new mineral assemblage in equilibrium with surface conditions. Fairbridge (1967) proposed the term *epidiagenesis* for the development of new fabrics and minerals during subaerial exposure. Development of authigenic minerals at unconformities is caused by the abrupt changes in pH and Eh that result from subaerial exposure, and the consequent accessibility of the surface and near-surface rocks to percolating water. The precipitation of different minerals depends on the fluid pH, the chemical composition of the rock material, and the concentration of ions in the fluid (Drever, 1988). The most common authigenic minerals that precipitate in unconformity zones are kaolinite, overgrowth quartz, carbonate, gypsum, and smectite.

Humic acids developed during the disintegration of organic matter in the soil cause acidity to increase in the water. Percolation by acidic water may lead to an increase in the solubility and dissolution of silica and Al_2O_3 and to increased porosity and permeability at the top of the unconformity (A horizon). As mentioned above, such leached material will accumulate in the B horizon, where porosity and permeability significantly decrease. Precipitation in the B horizon may be caused by the loss of organic complexing agents through bacterial decomposition and absorption. In the case of alkaline water percolation, such as accompanies marine transgression, the mixing of alkaline waters with acidic indigenous waters decreases the acidity of the fluid within the pore system, resulting in decreased solubility of silica and aluminum. This results in the precipitation of kaolinite and quartz overgrowths (Fairbridge, 1967). Precipitated kaolinite

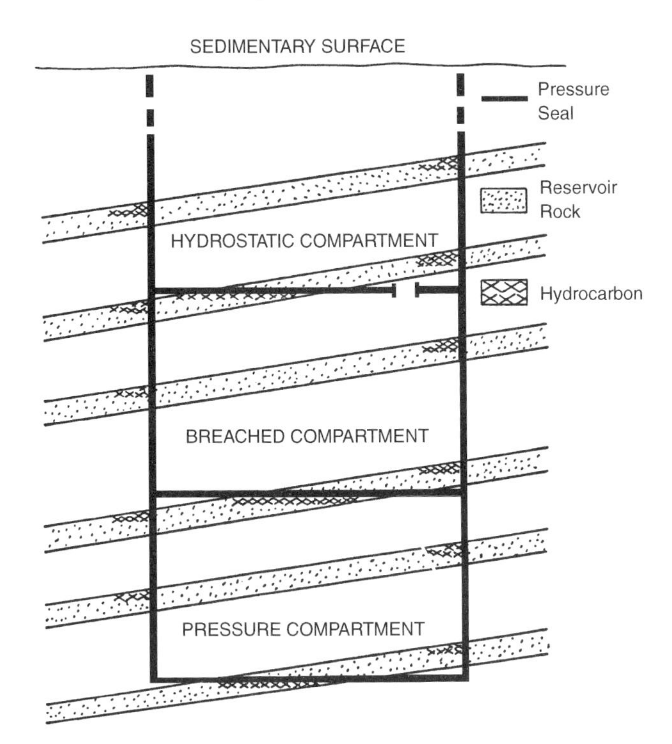

Figure 1. Schematic diagram of soil horizon formed in a temperate, humid climate. From Drever (1988).

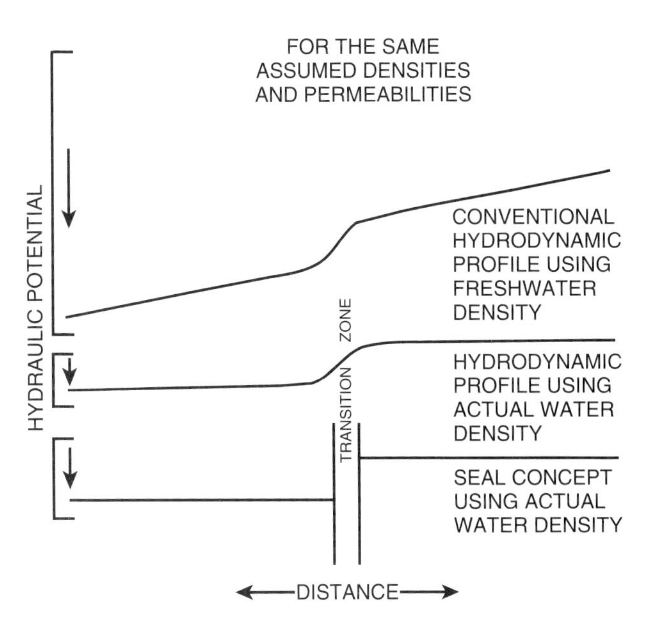

Figure 2. Mechanisms and sites of clay infiltration. Modified from Walker (1976). (A) Clay accumulation due to evaporation of the infiltrating water within the vadose zone. (B) Clay accumulation related to the loss of velocity as the infiltrating water reaches the water table. (C) Clay accumulation due to the development of perched water tables over impermeable barriers.

The other important process affecting porosity and permeability during pedogenesis is clay infiltration. The infiltration of alluvial clay or loess into coarse continental sediments is a well-known process (Bull, 1968). Along the Rozet unconformity zone, the infiltrated clays are mainly kaolinite resulting from the weathering of material in the drainage basin. Besides weathered clay, volcanic ash fall during the Rozet lowstand time also provided an abundance of infiltrated clays. Walker (1976) described three mechanisms that lead to the accumulation of significant amounts of infiltration clays (Figure 2): (1) concentrations in the vadose zone (Figure 2A), (2) concentration near the potentiometric surface (Figure 2B), and (3) concentration above impermeable barriers (Figure 2C). It is believed that the B horizon produced a basic barrier for the concentration of infiltration clays in the Rozet unconformity zone. Even though thin infiltration clay cutans may preserve intergranular porosity, permeability can be severely reduced by the obstruction of pore throats. For instance, in zones of high clay concentration, in which the clay matrix may constitute up to 35% of rock bulk volume, infiltration clays create permeability barriers in many Sergi Formation reservoirs (Moraes, 1991).

Authigenic Minerals

Subaerial exposure of a rock leads to the development of a new mineral assemblage in equilibrium with surface conditions. Fairbridge (1967) proposed the term *epidiagenesis* for the development of new fabrics and minerals during subaerial exposure. Development of authigenic minerals at unconformities is caused by the abrupt changes in pH and Eh that result from subaerial exposure, and the consequent accessibility of the surface and near-surface rocks to percolating water. The precipitation of different minerals depends on the fluid pH, the chemical composition of the rock material, and the concentration of ions in the fluid (Drever, 1988). The most common authigenic minerals that precipitate in unconformity zones are kaolinite, overgrowth quartz, carbonate, gypsum, and smectite.

Humic acids developed during the disintegration of organic matter in the soil cause acidity to increase in the water. Percolation by acidic water may lead to an increase in the solubility and dissolution of silica and Al_2O_3 and to increased porosity and permeability at the top of the unconformity (A horizon). As mentioned above, such leached material will accumulate in the B horizon, where porosity and permeability significantly decrease. Precipitation in the B horizon may be caused by the loss of organic complexing agents through bacterial decomposition and absorption. In the case of alkaline water percolation, such as accompanies marine transgression, the mixing of alkaline waters with acidic indigenous waters decreases the acidity of the fluid within the pore system, resulting in decreased solubility of silica and aluminum. This results in the precipitation of kaolinite and quartz overgrowths (Fairbridge, 1967). Precipitated kaolinite and quartz overgrowths can eventually fill pore spaces to such an extent that a rock becomes practically impermeable (Al-Gailani, 1981).

Near-surface carbonate cements have been documented by many investigators (Curtis, 1978; Berner, 1981; Surdam et al., 1989a, b). Curtis (1978) proposed that in an environment where microbial sulfate reduction is accompanied by Fe or Mn reduction, there is a high probability that a carbonate phase will precipitate. These early carbonate cements may irreversibly destroy porosity and permeability if they completely occlude all effective fluid migration channels.

The initial mineralogy of infiltration clay is most likely kaolinite and smectite. The main weathering product in a warm, humid climate (dominant climate during the Early Cretaceous in the Powder River basin) is kaolinite (Buol et al., 1980). The infiltration clays from the volcanic ash falls in the Powder River basin were probably kaolinite and smectite. Dethier et al. (1981) found that groundwater percolating through the ash and pumice deposits of the 1980 Mount St. Helens eruption had a water chemistry within the stability field of kaolinite.

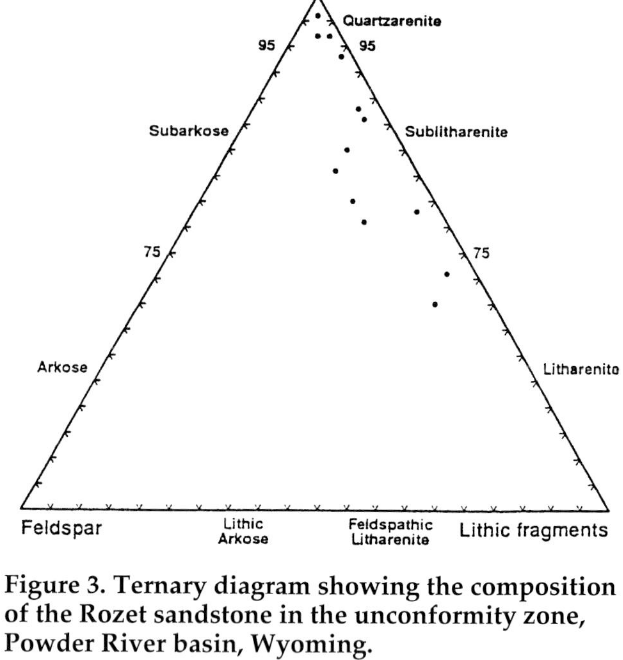

Figure 3. Ternary diagram showing the composition of the Rozet sandstone in the unconformity zone, Powder River basin, Wyoming.

PETROGRAPHY AND SEALING CAPACITY OF THE ROZET UNCONFORMITY SANDSTONE

Petrography

The Rozet unconformity developed after the deposition of the Rozet member of the Muddy Sandstone. The Rozet member is an upward-coarsening shoreface deposit. It consists of a poorly to moderately sorted, clay-rich sandstone that includes quartzarenite, sublitharenite, and litharenite (Figure 3). Framework grains are typically subangular to subrounded and range in size from very fine to medium sand. Quartz is usually the most abundant framework constituent. Lithic fragments consist primarily of chert, and to a lesser extent mica, shale, and volcaniclastics. Plagioclase is the most common feldspar component.

As observed in thin section, scanning electron microscope (SEM) analysis, X-ray diffraction (XRD) analysis, and core examination, the uppermost portion of the Rozet sandstone is mottled and contains a high percentage of clay matrix. The clay matrix constitutes about 20% of the rock volume, on average; and in some samples the clay matrix is up to 55% of the rock volume (Figure 4A). There is a lower content of clay matrix in the lower portion of the Rozet member

A

B

Figure 4. (A) Photomicrograph showing abundant clay matrix and absence from intergranular porosity in the Rozet sandstone in the unconformity zone. (B) Photo shows much less clay matrix and higher porosity in the underlying Lazy B sandstone. Scale bars on both photos equal to 76 μm.

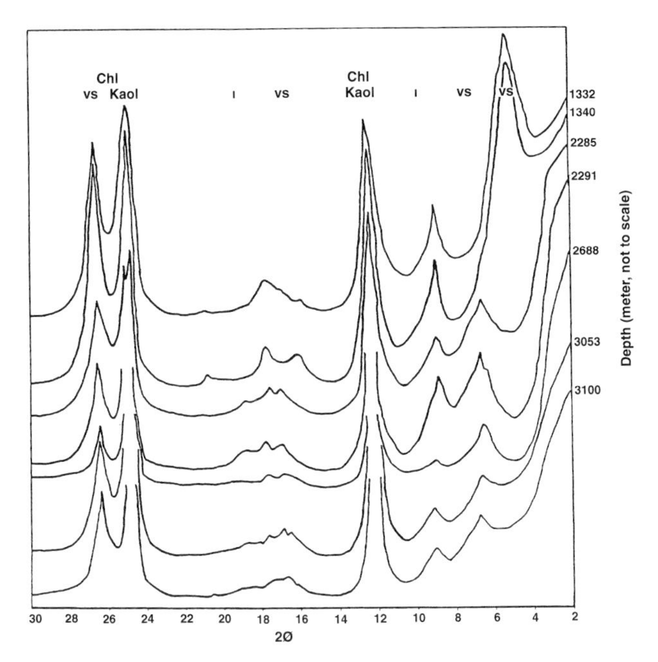

Figure 5. X-ray diffractograms of ethylene glycolate-treated samples (< 0.5 μm fraction) of the Muddy Sandstone from the Rozet unconformity zone, Powder River basin. Smectite progressively alters to illite in the mixed-layer smectite/illite clays with increasing burial depth. The structure of the mixed-layer smectite/illite clays changes from random to ordered at approximately 2500 m. The content of chlorite also increases with burial.

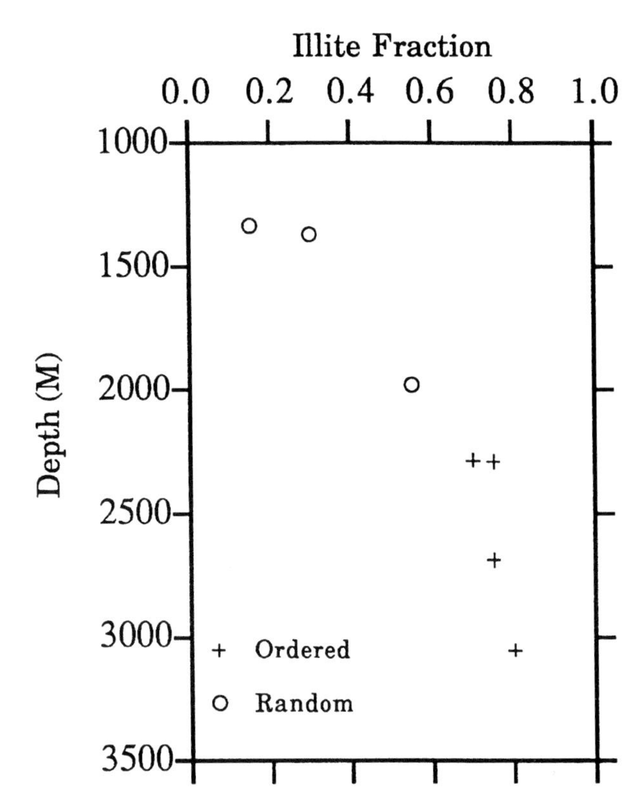

Figure 6. Fraction of illite layers and ordering in mixed-layer smectite/illite clays vs. depth for the < 0.5 μm fraction in sandstones from the Rozet unconformity zone, Powder River basin.

(Figure 4B). Semiquantitative XRD analysis indicates that kaolinite constitutes more than half the clay matrix; next in abundance are interstratified illite-smectite (hereafter I/S), and chlorite (Figure 5). With increasing burial depth, the percentage of illite in the interstratified I/S steadily increases from 20% at about 1200 m present-day depth to 80% at 3000 m (Figure 6). Chlorite content also increases with burial. Thin-section examination reveals that a curved texture (by compaction of clay matrix) is common in the upper portion of the unconformity zone (Figure 7). The typical kaolinite booklet-morphology is absent. Figure 8 shows the mixed morphological nature of the clay matrix. Vermicular kaolinite is embedded in clay matrix rich in I/S and chlorite. Figure 8 also shows the authigenic nature of the kaolinite.

Cementation is limited by the high amount of clay matrix present. In samples where matrix does not completely fill all pore space, quartz overgrowths can be seen. If calcite is present, it commonly replaces clay matrix (Figure 9). Authigenic kaolinite occurs as pore-filling and as replacement of plagioclase. Chlorite occurs typically as pore-filling. In some localities siderite concretions are abundant in the soil zone; they may be examples of glaebules, a common feature of ancient soil horizons (Brewer, 1964).

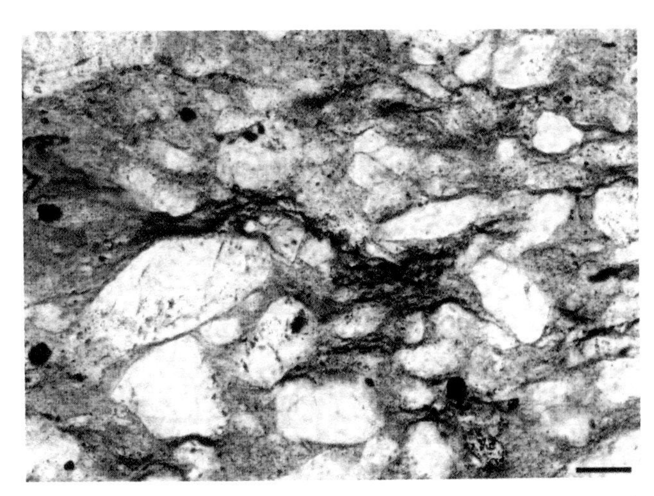

Figure 7. Photomicrograph showing common curved textures of clay matrix in the uppermost portion of the Rozet sandstone, suggesting intense compaction of this sandstone in the unconformity zone. Amos Draw 3. Scale bar equal to 76 μm.

Porosity and Permeability of the Sandstone Associated with the Rozet Unconformity

Figure 10 plots clay matrix, quartz cement, and porosity in the Rozet sandstone versus depth beneath

Figure 8. SEM photomicrograph showing the mixed morphological nature of clays in sandstones from the Rozet unconformity zone. Scale bar equal to 76 µm. Amos Draw 3.

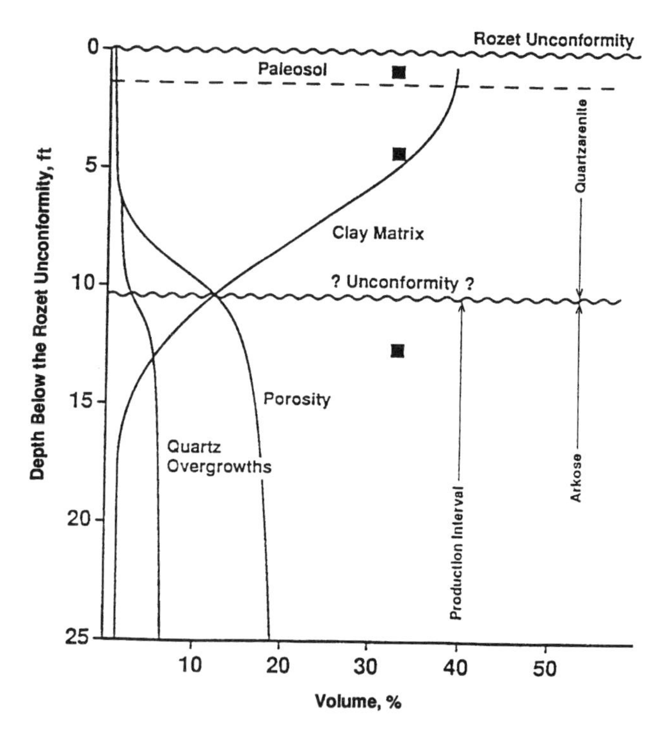

Figure 10. Plot of clay matrix, quartz cement, and porosity of the Rozet sandstone vs. depth beneath the unconformity surface. The volume is determined by point counting thin sections.

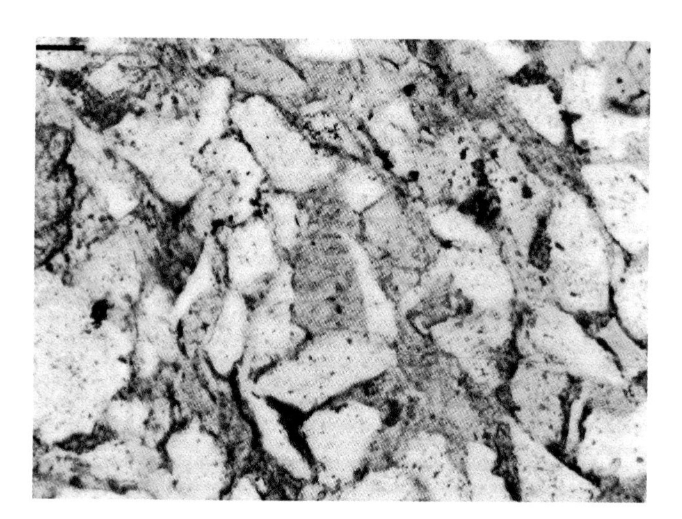

Figure 9. Photomicrograph showing calcite replacing clay matrix in the Rozet sandstone. Scale bar equal to 76 µm. Sagebrush Federal 2-1.

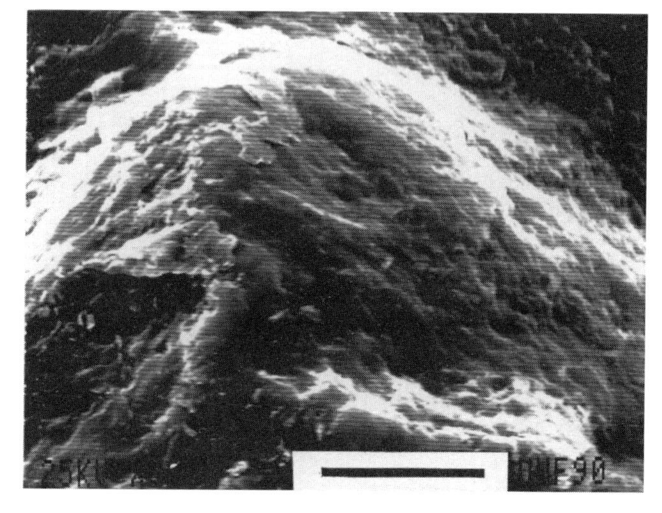

Figure 11. SEM photomicrograph showing the tightly packed clay matrix (mainly kaolinite) and abundant microporosity of the sandstone in the Rozet unconformity zone. Amos Draw 3. Scale bar is 10 µm.

the unconformity surface reference. As clay matrix rapidly increases upward toward the unconformity, the porosity decreases from about 18% in the reservoir sandstone to 5% in the sandstone associated with the unconformity.

SEM photomicrographs show that intergranular pore space is absent in the sandstone in the unconformity zone, but that microporosity is common in the matrix (Figure 11). An example of microporosity associated with abundant clay is found in core samples from the Amos Draw 3 well: two core porosities measured in the sandstone associated with the Rozet unconformity zone are 7 and 7.4%, but permeabilities are only 0.02 to 0.08 md (Table 1). Again, a soil-zone sandstone sample taken from the Federal 1-33 well (Section 16, T50N, R74W) has an average of 10.4% porosity, but permeability is below the detection limit. This low permeability results from an almost complete lack of interconnected pores in the clay matrix.

A decrease of permeability in the unconformity zone is also shown in Figure 12. Samples 1 and 2 are

Table 1. Sealing capacity of the Rozet unconformity sandstones.

Sample No.	Depth (m)	Porosity (%)	k (md)	Pd(psi) gas/water	Sealing capacity (ft, gas column)
1	3051	7.4	0.08	1800	5000
2	3052	7.0	0.02	1000	2600
3	3055	17.1	12	3	(reservoir sandstone)
4	2684	7.3	0.1	130	350
5	1689	9	0.2	60	160

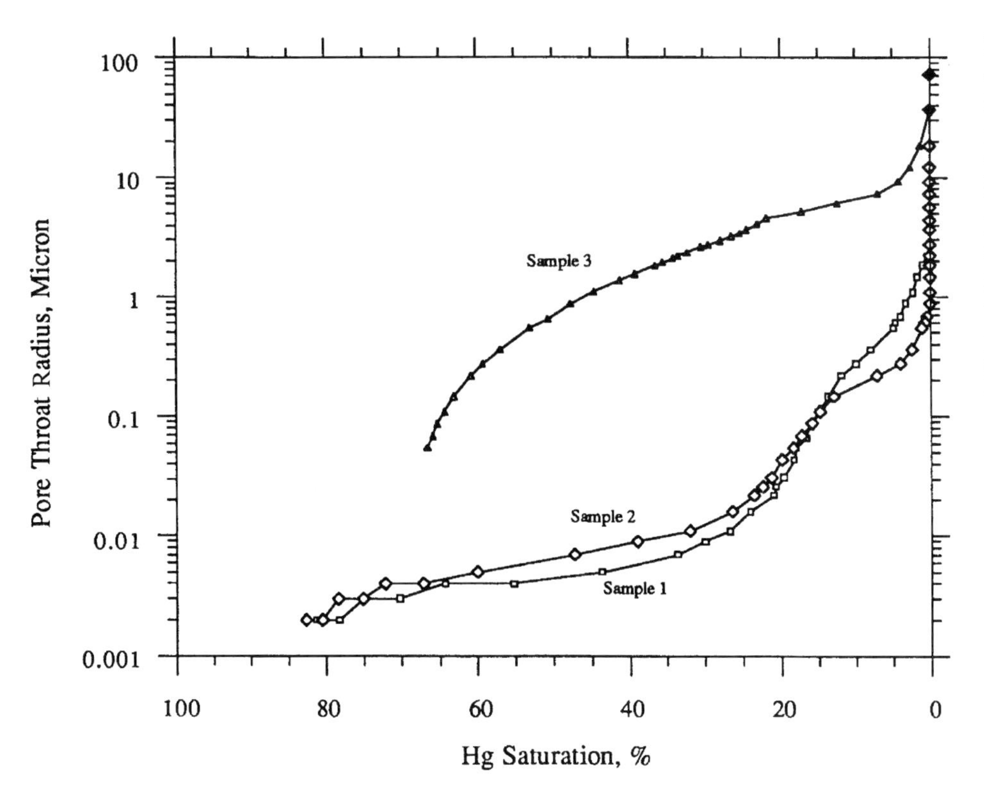

Figure 12. Pore throat radius distribution determined from the results of the mercury injection tests on samples described in Figure 13. Samples 1 and 2 contain pore throats primarily in the subnano and nano categories (<0.01 to 0.05 μm), while Sample 3 contains pore throats primarily in the meso and macro categories (0.25 to 2.5 μm).

from the unconformity zone, and pore throats are primarily in the subnano and nano categories (<0.01 to 0.05 microns), and permeability from 0.02 to 0.08 md. Sample 3, 12 ft (4 m) down from Sample 1, has pore throats primarily in the meso and macro categories (0.25 to 2.5 microns), and permeability is three orders of magnitude higher at 12 md. This last sample is from the hydrocarbon-producing interval in the Amos Draw field.

Sealing Capacity of the Rozet Unconformity

The sealing capacity of a hydrocarbon seal is based on petrophysical and flow properties of the rock that determine the height of the hydrocarbon column that can be supported by the seal. Sealing capacity is a function of pore radius and fluid characteristics. Smith's (1966) equation is used in this study for calculating the sealing capacity H of an unconformity seal:

$$H = \frac{P_{dB} - P_{dR}}{.433(\rho_w - \rho_h)}$$

where H is the maximum hydrocarbon column in feet that a seal can hold; P_{dB} is the subsurface displacement pressure in psi of the seal rock; P_{dR} is the subsurface displacement pressure in psi of the reservoir rock; ρ_w is the subsurface density of water in g/cm^3; ρ_h is the subsurface density of the hydrocarbon in g/cm^3; and 0.433 is a units conversion factor.

In our investigation, three bulk sandstone samples from the Rozet unconformity zone (Figure 13) were run in a high-pressure mercury injection test. Data from 32 mercury pressure tests on the Muddy sandstones also were assembled from the USGS core library. The displacement pressure, the minimum pressure to start mercury moving through a sample, was determined by extrapolating the injection pres-

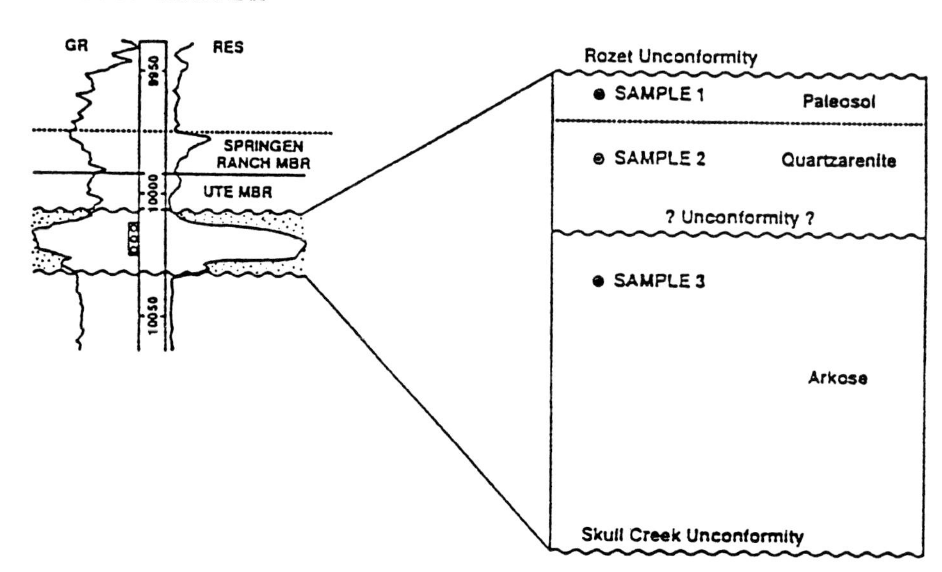

Figure 13. High-pressure mercury injection tests were performed on three sandstone samples from the Amos Draw 3 well. Samples 1 and 2 were taken in the Rozet unconformity zone, and Sample 3 was from reservoir sandstone in the Lazy B member.

sure curve (plateau portion) to 0% mercury saturation (Figure 14, Sample #1). This pressure was then converted to displacement pressure for the subsurface gas/water and oil/water systems using Schowalter's nomograms (1979). Calculated results are listed in Table 1.

Sneider et al. (1991) proposed a quasilogarithmic classification of hydrocarbon seals based on sealing capacity (Table 2). According to the Sneider classification, our Samples 1 and 2 are type A seals, and Samples 4 and 5 are type C seals. Thus, the sealing capacity of the Rozet unconformity zone increases an order of magnitude with increasing depth of burial.

Ibrahim et al. (1970), in a gas storage study, showed that displacement pressure (sealing capacity) is a function of the properties of the porous medium of interest. Ibrahim investigated shale, evaporite, and carbonate seals. Our data extend Ibrahim's investigation to include sandstone. Our results corroborate Ibrahim's correlation trend (Iverson et al., this volume). We conclude that an unconformity developed in a sandstone may form a hydrocarbon seal that is as good as a seal in a shale or evaporite. For example, the results of a mercury injection test on a sandstone sample taken from the paleosol in the Lakota Formation indicate that the sandstone associated with the unconformity at approximately 3300 m can withstand a 2000 psi pressure differential in the water-gas system, a differential pressure higher than the differential pressure characterizing the associated Fuson siltstone (Figure 15).

Burial and Thermal Histories

As mentioned above, the sealing capacity of the sandstone in the Rozet unconformity zone increases exponentially with depth of burial. The increase in the sealing capacity of this sandstone is dependent on the diagenetic processes that occurred during burial. The maturation of organic material in hydrocarbon source rocks and inorganic diagenetic reactions in sandstones are natural consequences of a prism of sedimentary rock containing organic material undergoing burial (Surdam et al., 1989a). In order to evaluate the porosity and permeability characterizing the unconformity zone, it is necessary to reconstruct the burial and thermal histories of the Rozet unconformity zone and overlying Mowry Shale.

A burial history diagram is the first approach to reconstructing the time-temperature profile of a basin. The accuracy of the reconstruction depends on the accuracy of the input data: the present-day thickness of the stratigraphic units, lithologies, ages of horizons, estimated porosities, and paleobathymetry. The ages of unconformities and the lithologies and thicknesses of eroded intervals must also be considered.

In this study, the thicknesses of strata were taken from well reports (Wyoming Geology Survey). The ages of horizons are adapted from McGookey et al. (1972), Kauffman (1977), and Obradovich and Cobban (1975). Most of the stratigraphic units contain several lithologies, and average porosities are used.

Although many minor unconformities in the strata overlying the Muddy Sandstone are documented in the literature, only two unconformities have had significant effect on the final curves of the burial history of the Muddy Sandstone. The earlier unconformity, documented by Gill and Cobban (1966) and Weimer (1983, 1984), involved the erosion of 200 m of an unnamed shale that lies between the Parkman and the Teapot sandstones and has a 73 m.y. age. The second and more significant unconformity is at the present-day surface, where erosion has been dominant since major uplift began in the late Miocene or early Pliocene (Curry, 1971; Trimble, 1980). Although uplift may have occurred as a series

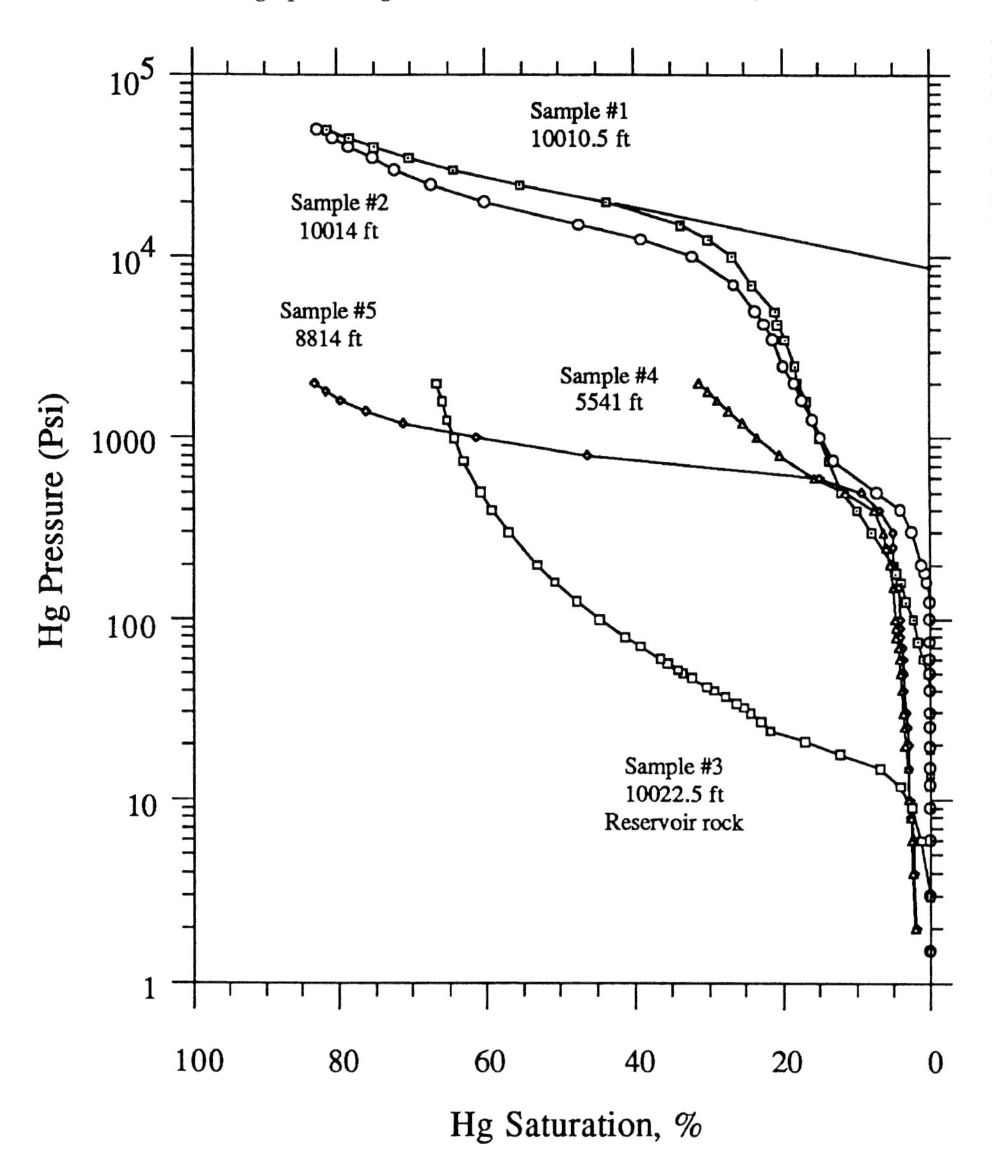

Figure 14. Mercury injection pressure curves. Data on Samples 1, 2, and 3 are from this study. Data on Samples 4 and 5 were assembled from the USGS Core Library.

Table 2. A classification of hydrocarbon seal type by sealing capacity.

Type	Sealing Capacity
A	H ≥ 300 m (≥1,000 ft)
B	H ≥ 150 m but <300 m (≥500 ft but <1000 ft)
C	H ≥ 30 m but <150 m (≥100 ft but <500 ft)
D	H ≥ 15 m but <30 m (≥50 ft but <100 ft)
E	H < 15 m (<50 ft)
F	Waste Zone Rocks

From Sneider et al. (1991).

of minor uplifts over a long period, a single period of steady uplift, beginning 10 million years ago, was assumed in the burial history diagrams. It is estimated that at least 600 m of sediments at the center and 1000 m at the margins of the basin have been eroded from the surface (Curry, 1971; Trimble, 1980; Nuccio, 1990).

A steady-state, one-dimensional thermal model was used in this study. The heat flow is assumed to have been constant through time. The thermal gradient is calculated from bottom-hole temperatures (Figure 16). The average thermal gradient is 27.2°C/km (14.9°F/1000 ft). Using thermal conductivities reported in a study of geothermal resources of the southern Powder River basin (Kenneth et al., 1986) and a thermal gradient of 27.2°C/km, a heat flow of 60.2 mw/m^2 was calculated from Fourier's equation.

The time-temperature profiles for the central portion of the Powder River basin are shown in Figure 17. The modeled bottom-hole temperatures are very close to the least-squares regression line of the observed bottom-hole temperatures (Figure 16). Figure 17 shows that the Muddy Sandstone in the central portion of the Powder River basin passed the 150°C isotherm during maximum burial.

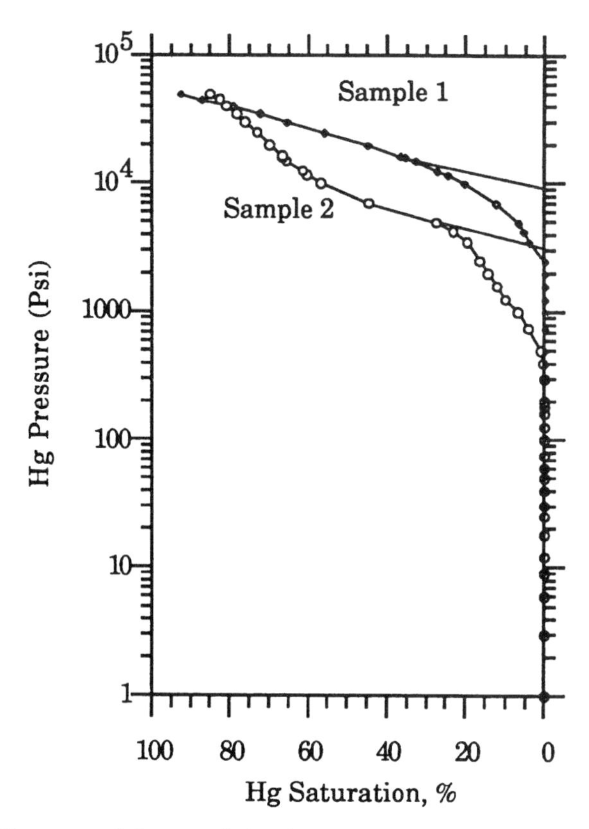

Figure 15. Mercury injection pressure curves. Sample 1 was sandstone from the Lakota unconformity zone. Sample 2 was from the overlying Fuson Shale. Allite State 1.

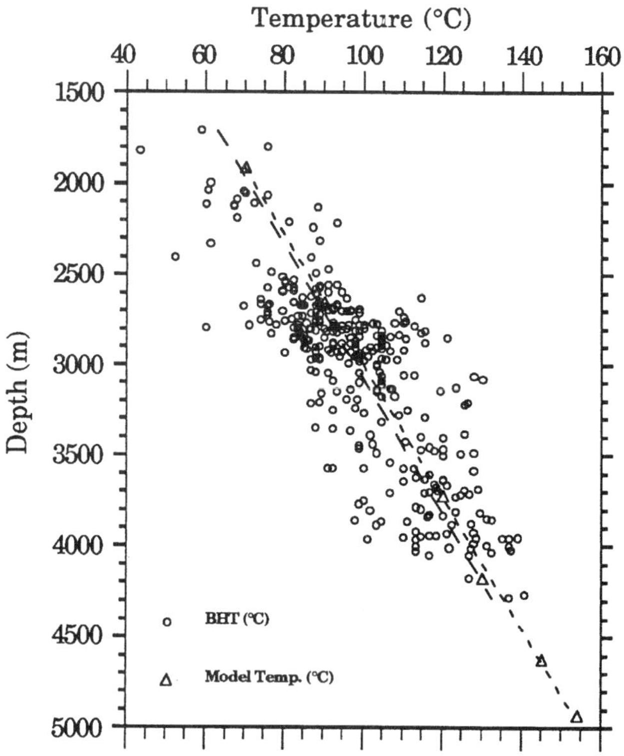

Figure 16. Present-day temperature vs. depth of the Muddy Sandstone in the Powder River basin, based on bottom-hole temperatures. Temperatures calculated from the thermal model presented herein are also shown.

CLAY DIAGENESIS AND THERMAL MATURATION OF THE MOWRY SHALE

The diagenesis of clay minerals in the Mowry Shale and Muddy Sandstone associated with the Rozet unconformity includes smectite altering to illite in interstratified illite-smectite (I/S), and kaolinite reacting to chlorite (Figure 18).

The diagenesis of interstratified I/S clays during the progressive burial of a sedimentary sequence is widely recognized and is considered to be an important empirical diagenetic geothermometer (Burst, 1969; Hower et al., 1976; Boles and Franks, 1979; Hower, 1981; Pytte and Reynolds, 1989; Elliott et al., 1991). It has also been demonstrated that the diagenetic trend in I/S clay is temperature dependent and may be related to regional hydrocarbon generation (Bruce, 1984; Hagen and Surdam, 1984). Some of the workers above have documented that the main compositional and structural changes in I/S burial diagenetic sequence are (1) an increase in illite layers, (2) an increase in interlayer potassium, (3) an increase in aluminum substituted for silicon in the tetrahedral layer, and (4) the release of Mg^{2+}, Fe^{2+}, Ca^{2+}, Si^{4+}, Na^{+}, and water. Such water can make up about 35% of the volume of a smectite crystallite (Perry and Hower, 1972). Two important points can be made concerning these changes: (1) the structural changes and water release have a significant effect on the porosity and permeability of the shale, and (2) the released elements may alter the properties of adjacent sandstone, by depositing quartz, chlorite, kaolinite, and late carbonate, for example.

Figure 18 illustrates the I/S diagenetic profile for the Mowry Shale in the Powder River basin. The I/S composition changes with progressive burial from approximately 20% illite in I/S at a present-day depth of 900 m (3000 ft) to 85% illite in I/S at 4200 m (14,000 ft) (Figure 19). This trend can also be shown for I/S clay in the sandstone samples from the Rozet unconformity zone (Figure 6); however, compared with the Mowry Shale samples, the depth of the ordering of illite/smectite in the sandstone of the unconformity zone appears to occur about 400 m (1300 ft) shallower. It is interesting that an increased sealing capacity of the Rozet unconformity coincides with progressive clay diagenesis as the structure of mixed-layer I/S changes from random to ordered at approximately 2300 m (7500 ft); the sealing capacity of the Rozet unconformity jumps from type C to type A (Table 1).

XRD data also show that diagenesis of the I/S clays is closely correlated with hydrocarbon generation in the Mowry Shale. As shown in Figure 20, a production index of 0.2 to 0.3 corresponds to 70% to 85%

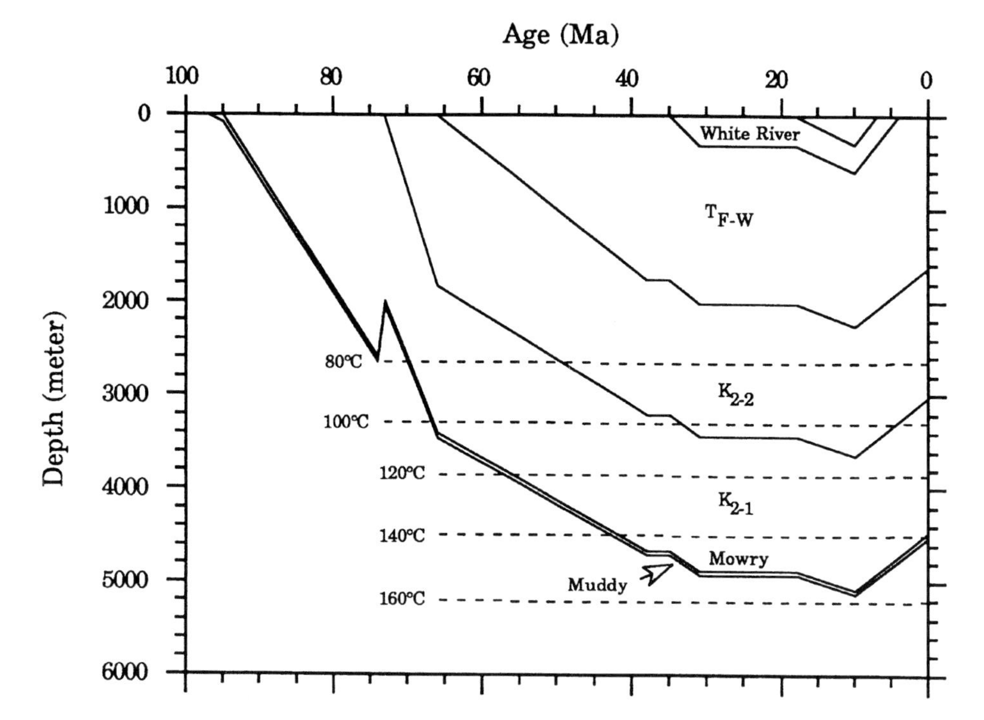

Figure 17. Time-temperature model for the central portion of the Powder River basin. K_{2-1} represents the Frontier, Niobrara, and Steele formations. K_{2-2} represents the Mesaverde, Lewis, and Lance formations. T_{F-W} represents the Fort Union and Wasatch formations.

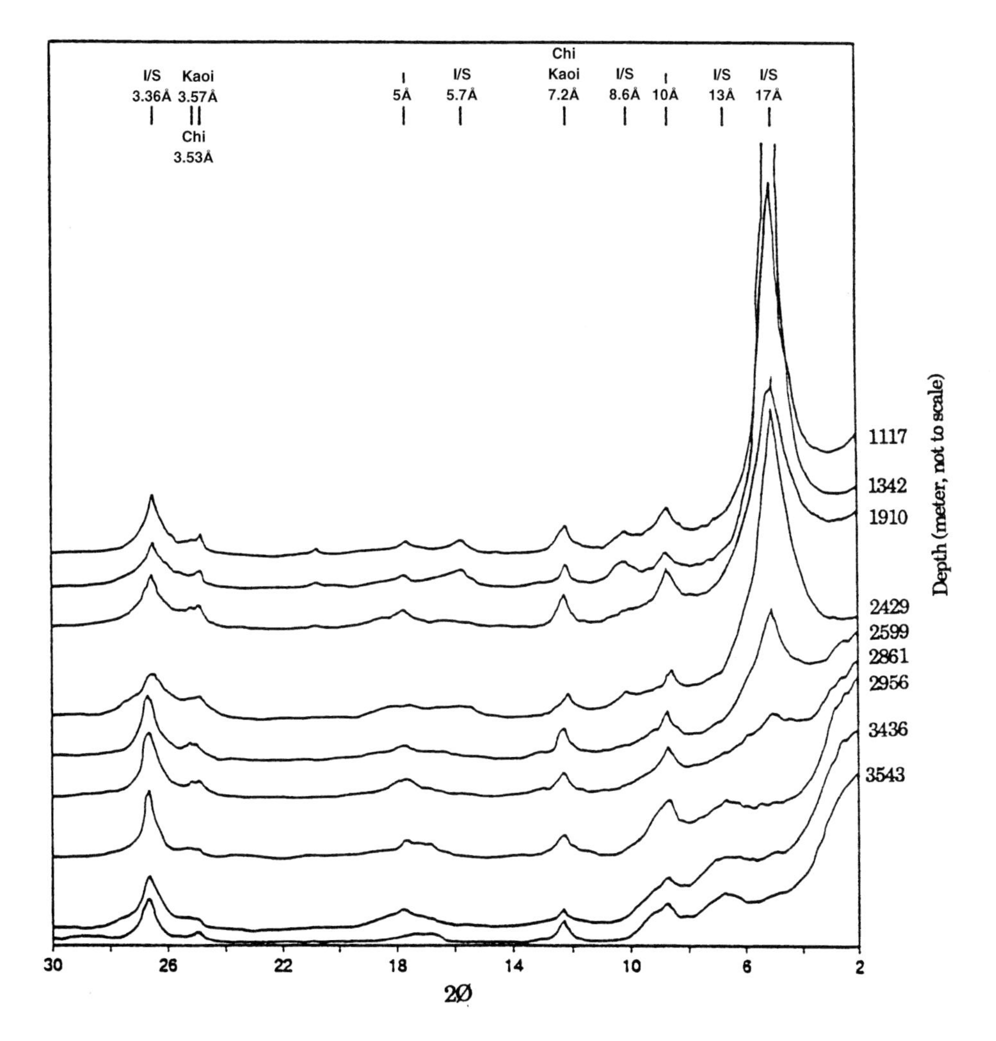

Figure 18. X-ray diffractograms of ethylene glycolate-treated samples (< 0.5 µm fraction) of the Mowry Shale, Powder River basin. With increasing burial depth, smectite alters to illite in the mixed-layer smectite/illite clays (the 8.6Å peak shifts to 10Å, the 5.7Å peak to 5Å). The structure of the mixed-layer smectite/illite clays changes from random to ordered at approximately 3000 m. The content of chlorite also increases with burial.

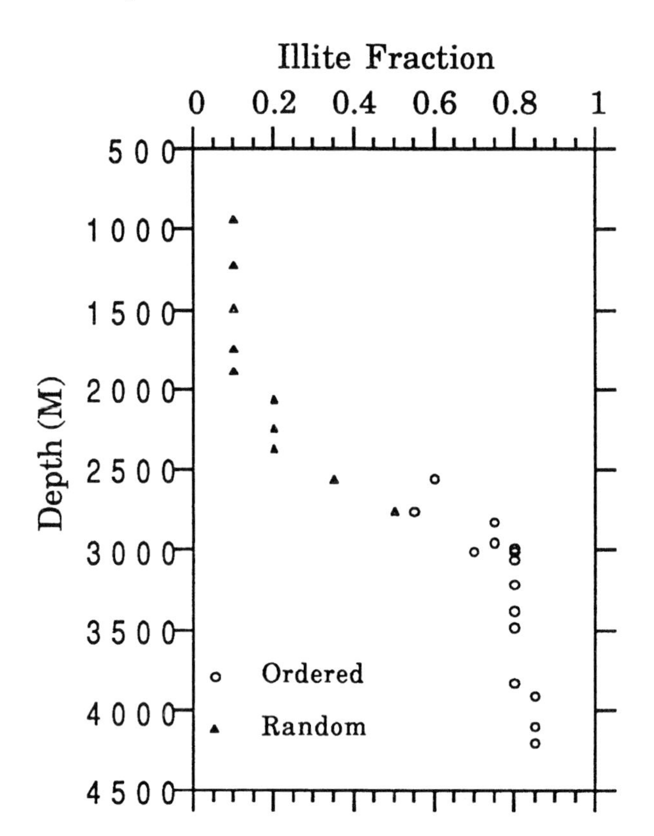

Figure 19. Plot of illite fraction in the mixed-layer smectite/illite clays vs. depth for the Mowry Shale, Powder River basin, Wyoming.

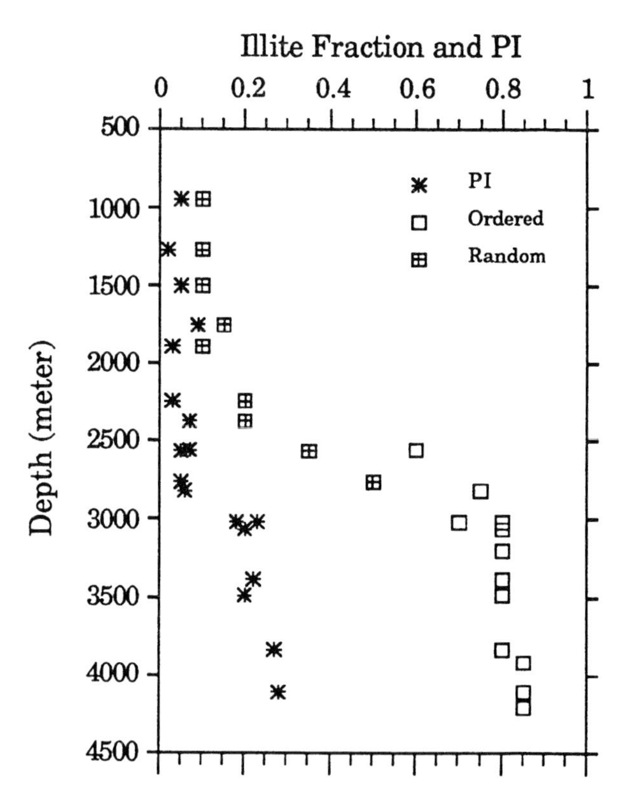

Figure 20. Plot of illite fraction in the mixed-layer smectite/illite clays and production index (PI) vs. depth for the Mowry Shale, Powder River basin, Wyoming.

illite in I/S clays. Furthermore, as interstratified I/S changes from random to ordered, the production index correspondingly jumps one order magnitude, from 0.06 to 0.3. Clearly, the Mowry Shale, a major hydrocarbon source rock in the Powder River basin, has passed the oil window, and wet gas has been generated where the depth of burial of the Mowry Shale is more than approximately 2700 m (9000 ft) in the Powder River basin. Vitrinite reflectance data further support this analysis, for vitrinite reflectance ranges from $R_o = 0.4$ at about 1000 m (3500 ft) to 1.6 at about 4400 m (13,500 ft) (MacGowan et al., this volume, their Figure 6). The classical liquid oil generation window is generally thought to lie between R_o of about 0.5 to 0.7 and 1.0 to 1.3 (Tissot and Welte, 1984). Above 1.3 vitrinite reflectance, hydrocarbon production is mainly wet gas. Such a fluid phase change has a significant effect on the capillary pressure, or sealing capacity, of the Rozet unconformity seal (Iverson et al., this volume).

Increased vitrinite reflectance and transformed clay are irreversible reaction products (Hower, 1981; Pytte and Reynolds, 1989). Data assembled in this study reveal that observed present-day temperatures typically are lower than the temperatures indicated by clay transformation data and vitrinite reflectance. There is evidence that the temperature difference is due to erosion during post-Laramide uplift. In general, the I/S clay reaction is largely controlled by temperature. Numerous studies (Hower et al., 1976; Hoffman and Hower, 1979; Boles and Franks, 1979; Hower, 1981) suggest that the conversion of random to ordered I/S clays takes place at approximately 100°C, and that the conversion of smectite to illite begins at approximately 60°C. By applying these temperatures to the I/S clay composition in the Powder River basin and comparing them with present-day formation temperatures, we find that the present-day temperatures are lower than the maximum diagenetic temperatures deduced from clay composition. This implies that some amount of sedimentary rock at the top of the sequence has been eroded following maximum burial.

Vitrinite reflectance data on the Mowry Shale in the Powder River basin may be related empirically to temperatures by employing I/S clay temperatures (Hower, 1981). Figure 21 shows the suggested correlation of the I/S temperature with vitrinite reflectance in the Powder River basin. The intersection of the regression line with shallowest sample (depth = 0.94 km, $R_o = 0.4$) occurs at a temperature of 60°C. With a mean annual surface temperature of 7°C for the Powder River basin and using a gradient of 27.2°C/km as calculated above, there is a 27°C temperature anomaly at this depth. This suggests that approximately 1000 m (3300 ft) of sediment have been removed from the present surface at the east margin of the Powder River basin.

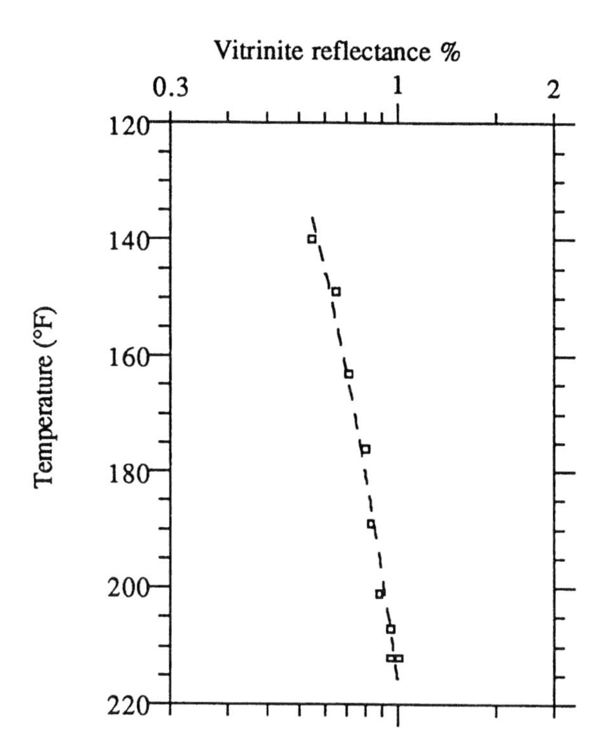

Figure 21. Vitrinite reflectance data for the Mowry Shale in the Powder River basin on a semilog plot; related to maximum diagenetic temperatures derived from the mixed-layer smectite/illite clays in the same samples.

SYNOPSIS OF PROCESSES FORMING THE PRESSURE SEAL ASSOCIATED WITH THE ROZET UNCONFORMITY

Petrographic and petrophysical data discussed above indicate that the sandstone associated with the Rozet unconformity at shallow depth has a fair sealing capacity (type C seal), even though porosity and permeability have been significantly reduced by pedogenic processes. But with progressive burial, the sealing capacity is significantly increased. Therefore, the formation of the Rozet unconformity pressure seal can be divided into two stages: pedogenic processes at the surface and diagenetic processes during progressive burial. Both processes resulted in the decreased porosity and permeability documented in the unconformity zone.

Pedogenic processes in the Rozet unconformity zone must have occurred shortly after the deposition of the Rozet member during a sea-level lowstand that resulted in subaerial exposure of the strandplain deposit and the development of a regionally prominent unconformity. The pedogenic zone was well developed on the Rozet paleohills and the upper slopes of the paleovalleys; most of the Rozet sediments at these locations remained at or near the surface during stream incision and later valley aggradation.

The decrease of sandstone porosity and permeability mainly took place in the paleosol zone developed on the unconformity as a result of the redistribution of material. Authigenic minerals at the unconformity formed as a result of abrupt changes in pH and Eh accompanying surface-water percolation. With continued downward percolation and leaching, the soil water was cation enriched. Below the leaching horizon, silica and alumina precipitated as fine-grained, poorly crystallized hydroxides or clay minerals. As a result, porosity and permeability in this lower zone decreased.

The other important porosity- and permeability-decreasing process occurring during pedogenesis was clay infiltration, similar to loess infiltration in modern soils. The infiltrated clay consisted mainly of kaolinite and smectite. As discussed above, the infiltrated clays were concentrated above the accumulation horizon (B horizon) of the paleosol, further enhancing the permeability barrier. The sources of the infiltrated clays were intensive in situ weathering and the volcanic ash falls that frequently covered the drainage area during the Cretaceous. Pedogenic processes acting in concert with infiltrated clay produced an abundant clay matrix in the sandstones along the unconformity that occluded most of the original porosity.

The porosity and permeability decrease that occurred during burial diagenesis mainly resulted from mechanical compaction during shallow burial and the precipitation of authigenic clay minerals during intermediate burial. The high-clay matrix, up to 55% of rock volume, increased the ductility of the rock; as a result, compaction played an important role in the evolution of porosity and permeability of the sandstones along the unconformity zone.

Thermal modeling indicates that the Rozet unconformity zone and Mowry Shale entered the zone of intense diagenesis (80–100°C) at about 70 Ma. In this zone, the alkalinity of the formation water was probably dominated by carboxylic acids and anions (Surdam et al., 1989a, b). As deduced from petrography, the potential reactions in this zone are feldspar and carbonate dissolution and chlorite precipitation. Computer simulation (SOLMINEQ.88) of the zone of intense diagenesis predicts that both carbonate and feldspar will be destabilized, and chlorite will be stable. XRD data from the Rozet unconformity zone clearly show an increase in chlorite in the clay fraction in this depth interval. An increase of clay minerals at the reservoir/trap contact has been mentioned by several authors. Sullivan and McBride (1991), in a study of the diagenesis of sandstones in contact with shales and diagenetic heterogeneity, suggested that authigenic chlorite is likely to be concentrated near the contact between reservoir and trap rocks. The Fe, Si, and Al ions released by the dissolution of unstable minerals (feldspars and lithic grains) are the source of material needed for chlorite precipitation.

Ordering of mixed-layer I/S, late kaolinite precipitation, and carbonate cementation also contributes to the reduction of porosity and permeability along the unconformity zone. As mentioned above, the increase in sealing capacity of the Rozet unconformity coincided with progressive clay diagenesis during increasing

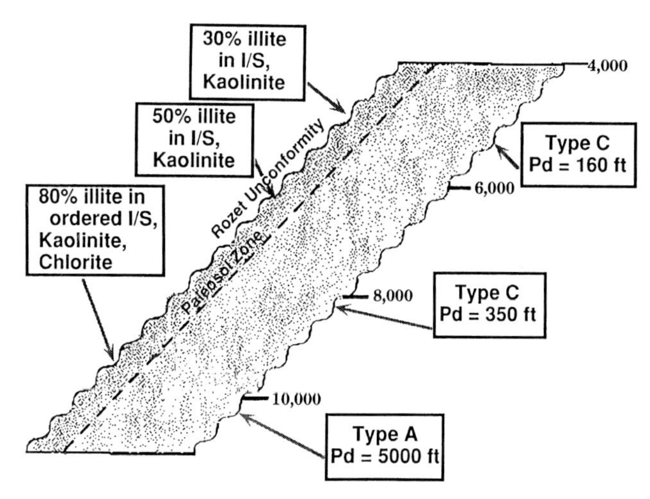

Figure 22. Schematic diagram showing that the sealing capacity of the Rozet unconformity pressure seal progressively increases with increasing burial depth. With increasing burial depth, the proportion of the illite in the mixed-layer smectite/illite clays increases from 30% to 80%, the sealing capacity of the low-permeability rocks associated with the unconformity increases by an order of magnitude from 135 ft to 1536 ft (41 to 468 m). A type A seal was formed when the unconformity zone was buried more than 9000 ft (2700 m) deep.

burial. We believe that the diagenesis of clay minerals in both the shale and sandstone had a significant effect on the enhancement of the sealing capacity of the sandstones and shales associated with the Rozet unconformity in the Powder River basin. With progressive burial, the sandstone pore-throat radii in the Rozet unconformity zone decreased from the submicro category (0.05 to 0.25 microns) at about 1500 m (5000 ft) to the subnano category (<0.01 microns) at approximately 2700 m (9000 ft). Correspondingly, this zone at these two depths can withstand pressure differentials of 20 psi and 2000 psi in a water/gas system, respectively. A 2000 psi differential is of the same magnitude as that observed within many of the overpressured fields in the Powder River basin (Surdam et al., this volume).

During subsequent burial, the pore throats remaining after pedogenesis were further occluded as a result of mineral reactions associated with burial diagenesis. Primarily as a result of clay diagenesis, type C seals were converted to type A seals at a burial depth of approximately 3000 m (10,000 ft), at perhaps 67 Ma. At that time the fluid-flow system was converted from a single-phase (water) to a multiphase fluid-flow system (water/oil/gas) by hydrocarbon generation and alteration. Consequently, capillarity became a significant aspect of the sealing capacity of these low-permeability rocks. Our research shows that presently, with increasing burial depth, the proportion of illite in I/S clays increases from 20% to 80%, and the chlorite/kaolinite ratio increases; the sealing capacity of the low-permeability rocks associated with the unconformity increases by an order of magnitude; and, most important in a gas/water system, the displacement pressures of the seals increases from 600 to 1800 psi (see Figure 22).

CONCLUSIONS

This study suggests the following conclusions:

1. There is a direct correlation between the diagenesis of clay minerals in the Rozet unconformity zone and the maturation of the Mowry Shale and the sealing capacity or displacement pressures of the pedogenic units. The sealing capacity or displacement pressure of rocks along the Rozet unconformity is a function of thermal exposure that occurred at approximately 67 Ma.
2. The sealing capacity of unconformity rocks was derived from primary pedogenic processes and from diagenetic enhancement during progressive burial. The diagenetic processes increased the sealing capacity of sandstone along the unconformity by an order of magnitude, from 45 m (150 ft) to as much as 1500 m (5000 ft) of the gas column. The presence of widespread subaerial unconformities in the Muddy Sandstone is an important factor in the formation of abnormally pressured compartments within the reservoir facies.
3. The transition of the fluid-flow system from single-phase to multiphase resulted in converting the low-permeability rocks along the unconformity to fluid/pressure seals capable of withstanding >2000 psi pressure differentials.

ACKNOWLEDGMENTS

This material covers research performed through November 1991. The authors wish to thank David Copeland (University of Wyoming), who reviewed the original manuscript and made numerous helpful suggestions. Steven Boese, Department of Geology and Geophysics, University of Wyoming, performed the anhydrous pyrolysis (for PI) and TOC analyses. Discussions with Randi Martinsen, Henry Heasler, and Donald MacGowan (all of the University of Wyoming) provided useful input to this study. This study was funded by the Gas Research Institute under contract number 5089-260-1894.

REFERENCES CITED

Al-Gailani, M. B., 1981, Authigenic mineralization at unconformities, implication for reservoir characteristics: Sedimentary Geology, v. 29, p. 89–115.

Berner, R. A., 1981, A new geochemical classification of sedimentary environments: Journal of Sedimentary Petrology, v. 51, p. 359–365.

Boles, J. R., and S. G. Franks, 1979, Clay diagenesis in Wilcox sandstones of southwest Texas, implications of smectite diagenesis on sandstone cementation: Journal of Sedimentary Petrology, v. 49, p. 55–70.

Brewer, R., 1964, Fabric and Mineral Analysis of Soils: New York, Wiley, 470 p.

Bruce, C. H., 1984, Smectite dehydration—its relation to structural development and hydrocarbon accumulation in northern Gulf of Mexico basin: American Association of Petroleum Geologists Bulletin, v. 68, p. 673–683.

Bull, W. B., 1968, The alluvial fan environment: Progress in Physical Geology, v. 1, p. 222–270.

Buol, S. W., F. D. Hole, and R. T. McCracken, 1980, Soil Genesis and Classification: Ames, Iowa University, 406 p.

Burst, J. R., Jr., 1969, Diagenesis of Gulf Coast clay sediments and its possible relationships to petroleum migration: American Association of Petroleum Geologists Bulletin, v. 53, p. 487–502.

Curry, W. H., III, 1971, Laramide structural history of the Powder River basin, Wyoming: Wyoming Geological Association, Twenty-third Annual Field Conference Guidebook, p. 49–60.

Curtis, C. D., and M. L. Coleman, 1986, Controls on the precipitation of early diagenetic calcite, dolomite, and siderite concretions in complex depositional sequence, *in* D. Gauter, ed., Roles of Organic Matter in Sediment Diagenesis: Society of Economic Paleontologists and Mineralogists, Special Publication 38, p. 23–34.

Dethier, D. P., D. R. Pevear, and D. Frank, 1981, Alteration of new volcanic deposits, *in* P. Lipman and D. Mullineaux, eds., The 1980 Eruptions of Mount St. Helens, Washington: U. S. Geological Survey Professional Paper 1250, p. 649–665.

Dolson, J. C., D. S. Muller, M. J. Evetts, and J. A. Stein, 1991, Regional paleotopographic trends and production, Muddy Sandstone (Lower Cretaceous), Central and Northern Rocky Mountains: American Association of Petroleum Geologists Bulletin, v. 75, p. 409–435.

Drever, J. I., 1988, The Geochemistry of Natural Water: Englewood Cliffs, New Jersey, Prentice Hall, 437 p.

Elliott, W. C., J. L. Aronson, G. Matisoff, and D. L. Gautier, 1991, Kinetics of the smectite to illite transformation in the Denver basin, clay mineral, K-Ar data, and mathematical model results: American Association of Petroleum Geologists Bulletin, v. 75, p. 436–462.

Fairbridge, R. W., 1967, Phases of diagenesis and authigenesis, *in* G. Larsen and G. Chilinger, eds., Diagenesis in Sediments: Amsterdam, Elsevier, p. 19–89.

Gill, J. R., and W. A. Cobban, 1966, Regional unconformity in the Late Cretaceous, Wyoming: U.S. Geological Survey Professional Paper 550-13, p. 1320–1327.

Guangming, Z., and Z. Quanheng, 1982, Buried-hill oil and gas pools in the north China basin, *in* M. Halbouty, ed., The Deliberate Search for the Subtle Trap: American Association of Petroleum Geologists Memoir 32, p. 317–335.

Hagen, E. S., and R. C. Surdam, 1984, Maturation history and thermal evolution of Cretaceous source rocks of the Big Horn basin, Wyoming and Montana, *in* J. Woodward, F. Meisser, and J. Clayton, eds., Hydrocarbon Source Rocks of the Greater Rocky Mountain Region: Denver, Rocky Mountain Association of Geologists, p. 321–338.

Halbouty, M. T., R. E. Meyerhoft, R. E. King, et al., 1970, World giant oil and gas fields, geologic factors affecting their formation and basin classifications, *in* M. Halbouty, ed., Geology of Giant Petroleum Fields: American Association of Petroleum Geologists Memoir 14, p. 502–556.

Hoffman, J., and J. Hower, 1979, Clay mineral assemblages as low grade metamorphic geothermometers: application to the thrust faulted disturbed belt of Montana, U.S.A., *in* P. Scholle and P. Schluger, eds., Aspects of Diagenesis: Society of Economic Paleontologists and Mineralogists Special Publication 26, p. 55–80.

Hower, J., 1981, Shale diagenesis, *in* F. Longstaff, ed., Short Course in Clay Diagenesis and the Resource Geologist: Toronto, Mineralogical Association of Canada, p. 60–80.

Hower, J., E. V. Eslinger, M. E. Hower, and E. A. Perry, 1976, Mechanism of burial and metamorphism of argillaceous sediments. 1. Mineralogical and chemical evidence: Geological Society of America Bulletin, v. 87, p. 725–737.

Ibrahim, M. A., M. R. Tek, and D. L. Katz, 1970, Threshold Pressure in Gas Storage: University of Michigan Press, 309 p.

Kauffman, E. G., 1977, Geological and biological overview—western interior Cretaceous basin, *in* E. Kauffman, ed., Cretaceous Facies, Faunas, and Paleoenvironments across the Western Interior Basins: Mountain Geologist, v. 14, p. 75–99.

Kenneth, L. B., H. P. Heasler, and B. S. Hinckley, 1986, Geothermal resources of the southern Powder River Basin, Wyoming: Report of Investigations no. 36, Wyoming Geological Survey, 32 p.

Levorsen, A. I., 1934, Relation of oil and gas pools to unconformities in the Mid Continent region, *in* Problems of Petroleum Geology: American Association of Petroleum Geology, 1073 p.

McGookey, D. P., et al., 1972, Cretaceous system, *in* W. Mallory, ed., Geologic Atlas of the Rocky Mountain Region: Rocky Mountain Association of Geologists Guidebook, p. 190–242.

Miller, E. G., 1984, Parkman field, Williston basin, Saskatchewan, Canada: American Association of Petroleum Geologists Memoir 16, p. 502–510.

Moraes, M. A. S., 1991, Diagenesis and microscopic heterogeneity of lacustrine deltaic and turbiditic sandstone reservoirs (Lower Cretaceous), Potiguar basin, Brazil: American Association of Petroleum Geologists Bulletin, v. 75, p. 1758–1771.

North, F. K., 1985, Petroleum Geology: Boston, Allen and Unwin, 607 p.

Nuccio, V. F., 1990, Burial thermal and petroleum generation history of the Upper Cretaceous Steele member of the Cody Shale (Shannon sandstone bed horizon), Powder River basin, Wyoming: U.S. Geological Survey Bulletin 1917-A, p. A1–A17.

Obradovich, J. D., and W. A. Cobban, 1975, A time scale for the Late Cretaceous at the western interior of North America, *in* W. Caldwell, ed., The Cretaceous System in the Western Interior of North America: Geological Association of Canada Special Paper 13, p. 31–54.

Perry, E. A., Jr., and J. Hower, 1972, Late stage dehydration in deeply buried pelitic sediments: American Association of Petroleum Geologists Bulletin, v. 56, p. 2013–2021.

Pytte, A. M., and R. C. Reynolds, 1989, The thermal transformation of smectite to illite, *in* N. Naesser and T. McColloh, eds., Thermal History of Sedimentary Basins: New York, Springer-Verlag, p. 133–140.

Retallack, G. J., 1990, Soils of the Past, an Introduction to Paleopetrology: Boston, Unwin Hyman, 520 p.

Schowalter, T. T., 1979, Mechanics of secondary hydrocarbon migration and entrapment: American Association of Petroleum Geologists Bulletin, v. 63, p. 723–760.

Shanmugan, G., and J. B. Higgins, 1988, Porosity enhancement from chert dissolution beneath Neocomian unconformity, Ivishak Formation, North Slope, Alaska: American Association of Petroleum Geologists Bulletin, v. 72, p. 523–535.

Smith, D. A., 1966, Theoretical considerations of sealing and non-sealing faults: American Association of Petroleum Geologists Bulletin, v. 50, p. 363–374.

Sneider, R.M., K. Stolper, and J. S. Sneider, 1991, Petrophysical properties of seals: American Association of Petroleum Geologists Bulletin, v. 75, p. 673–674.

Sullivan, K. B., and E. F. McBride, 1991, Diagenesis of sandstone at shale contacts and diagenetic heterogeneity, Frio Formation, Texas: American Association of Petroleum Geologists Bulletin, v. 75, p. 121–138.

Surdam, R. C., L. J. Crossey, E. S. Hagen, and H. P. Heasler, 1989a, Organic-inorganic interactions and sandstone diagenesis: American Association of Petroleum Geologists Bulletin, v. 73, p. 1–23.

Surdam, R. C., T. L. Dunn, D. B. MacGowan, and H. P. Heasler, 1989b, Conceptual models for the prediction of porosity evolution, with an example from the Frontier sandstone, Bighorn basin, Wyoming, *in* E. Coalson, ed., Petrogenesis and Petrophysics of Selected Sandstone Reservoirs of the Rocky Mountain Region: Rocky Mountain Association of Geologists, p. 7–28.

Tissot, B. P., and D. H. Welte, 1984, Petroleum Formation and Occurrence: New York, Springer-Verlag, 539 p.

Trimble, D. E., 1980, Cenozoic tectonic history of the great plains contrasted with that of the southern Rocky Mountains: Mountain Geologist, v. 17, n. 3, p. 59–69.

Walker, T.R., 1976, Diagenetic origin of continental red beds, *in* H. Falke, ed., The Continental Permian in Central West and South Europe: Dordredcht, D. Reidel, p. 240–482.

Weimer, R. L., 1983, Relation of unconformities, tectonics, and sea-level changes, Cretaceous of the Denver basin and adjacent areas, *in* M. Reynolds and E. Doly, eds., Mesozoic Paleogeography of the West-Central United States: Rocky Mountain Section, Society of Economic Paleontologists and Mineralogists, p. 356–376.

Weimer, R. L., 1984, Relation of unconformities, tectonics, and sea-level changes, Cretaceous of western interior, U.S.A., *in* J. Schlee, ed., Interregional Unconformities and Hydrocarbon Accumulation: American Association of Petroleum Geologists Memoir 36, p. 7–35.

Chapter 20

Pressure Seal Permeability and Two-Phase Flow

W. P. Iverson
Randi S. Martinsen
Ronald C. Surdam
University of Wyoming
Laramie, Wyoming, U.S.A.

ABSTRACT

Pressure compartment seals all have permeability to single-phase flow. Complete sealing can occur only in a multiphase fluid environment. For physical properties typical of the Powder River basin, Wyoming, Darcy flow allows single-phase leak rates such that observed pressure compartments would leak off in about 1 million years. Pressure compartments can be held indefinitely, however, under multiphase flow. Muddy sandstones of anomalously high threshold displacement pressure, about 2000 psi, appear to contain gas reservoirs at high pressure. Such high displacement pressures correlate well with those of classic carbonate and shale seals. The Muddy, however, contains sandstones capable of sealing adjacent reservoir sandstones. Sealing sandstones correlate with zones of unconformities between sandstones of good reservoir quality. Capillary sealing, as observed here, is certainly a worldwide phenomenon but is not the only mechanism of holding a pressure compartment. Other pressure compartments might be actively leaking (e.g., Gulf Coast type) and geologically temporary. Conversely, the capillary seal is permanent up to the threshold displacement pressure, which is the observed pressure in Muddy pressure compartments.

INTRODUCTION

Pressure compartments must be sealed by some sort of permeability restriction, or else the pressure would bleed off in a relatively short period of time. How rapidly a pressure compartment can leak, and how low permeability must be to maintain a differential pressure, are the two major topics of this paper.

In general, pressure compartments can be subdivided into two very different categories: disequilibrium compaction compartments and sealed pressure compartments. The pressure compartments created by disequilibrium compaction are characterized by large volumes of rock compacted more rapidly than fluids can completely escape. Such pressure compartments do not necessarily require the existence of low-permeability rocks; they simply need to contain more fluid than can escape at the rate allowed by the existing permeability. When connate water volume is large and compaction rates rapid, as is the case in some parts of the U.S. Gulf Coast, overpressured compartments are common. Conversely, pressure compartments of the Powder River basin of Wyoming and Montana are not experiencing such compaction and apparently have been in existence for geologically significant periods of time (MacGowan et al., this volume), requiring that they be isolated by some type of seal.

Completely impermeable rocks do not exist in nature, yet sealing of oil and gas reservoirs is a commonly recognized phenomenon. Seals are formed by the combination of low permeability and high capillary pressure due to interfacial tension between two fluids (Schowalter, 1979). An excellent example is shale, with sufficient permeability to expel connate water during compaction, and measurable permeability to gas when dried and placed in a laboratory permeameter. Shale, however, is a very efficient seal when two fluid phases are present. Properties of two-phase flow create an *apparently* impermeable medium. Another example is concrete, a very permeable material with respect to water, yet capable of holding significant water pressure behind a dam because of capillarity in a two-phase system (air/water).

Darcy's Law of fluid flow, when applied to subsurface pressure compartments, indicates that abnormal pressures can exist only for finite periods of time. Furthermore, considering the high pressure gradients revealed by subsurface pressure measurements (Heasler et al., this volume), the leaking of pressure compartments due to Darcy flow should be significant. Organic geochemical data, however, indicate that overpressuring has existed for approximately 40 million years in the Muddy Sandstone of the Powder River basin. As an initial step in the analysis and evaluation of pressure seals, the leakage rate of a pressure compartment must be estimated. Appropriate geometries and physical properties for the types of pressure compartments in the Muddy Sandstone of the Powder River basin as described by Heasler et al. (this volume) will be considered.

PERMEABILITY MEASUREMENTS

Figure 1 shows a selection of orders-of-magnitude variations in permeability for the Muddy Sandstone in the Powder River basin. Permeability is measured either by drill-stem test (DST) data or by conventional core analysis in the laboratory. Although thousands of DSTs have been recorded in the Muddy, only a handful have both DST and core permeabilities. DST permeability is determined from the slope of the pressure buildup curve (Horner plot), as described by Earlougher (1977). If the slope of the Horner plot is m psi per logarithmic cycle of Horner buildup time, then

$$(kh/\mu) = 162.6\, QB/m,$$

where Q is average flow rate (bbl/day from fluid recovery divided by flow time), and B is the oil formation volume factor (STB/RB, about 1.0). The term kh/μ represents the flow capacity of a sandstone and is controlled primarily by the average sand permeability k (millidarcys), which varies by many orders of magnitude. Sandstone thickness h (feet) and fluid viscosity μ (cp) change very little compared with variations in permeability k. The solid symbols in Figure 1 represent those DST data for which a clear pressure buildup curve is defined by a Horner plot, and the open circles represent tests which encountered extremely tight sandstones that flow very little fluid. The former data set, solid symbols in Figure 1, would generally be considered good well tests where sufficient fluid is recovered and a clear linear pressure buildup trend develops on the Horner plot. The latter data set, open circles in Figure 1, required more innovative analysis before a comparison with core permeability was possible (Iverson, 1990). Such DSTs record very low apparent formation pressure due to the fact that a low-permeability formation takes a very long time to build up to true reservoir pressure. These low-pressure DSTs are characterized by very low fluid recovery (less than one barrel) and a nonlinear pressure buildup curve at unreasonably low downhole pressure (typically less than 100 psi). If given sufficient time, days to months, the pressure would eventually build up to formation pressure. Regardless of the true formation pressure, the Horner plot would exhibit a very steep slope m. This large value of m can be estimated by projecting the final buildup pressure up to an assumed hydrostatic level, then using that m to calculate kh/μ from the equation above.

Core permeability in Figure 1 is determined from conventional core analysis, using gas (He) injected through the core in a direction consistent with horizontal reservoir flow. Permeability is measured from core plugs at 1 ft increments, and a summation is taken over the DST depth intervals to yield a core-derived kh/μ. Ideally, the core kh/μ should match the DST kh/μ flow capacity. A consistent deviation, however, of about two orders of magnitude exists, with DST flow capacity lower than core-determined flow capacity. This discrepancy is easily explained by two dominant physical conditions—the lack of confining pressure on laboratory cores and the lack of multiphase flow in laboratory core measurements with gas—which cause laboratory measurements of permeability to be much higher than DST measurements.

The relevance of these data to the study at hand is the verification of the permeability range for Muddy sandstones. Although some very low permeability values are calculated for the low-pressure DST cases, no permeability is calculated at zero. This fact is verified by a general correlation with laboratory-derived permeability. All sandstones have some permeability and will permit the passage of fluids when a pressure gradient exists in single-phase flow. The rate at which these "tight" rocks can leak is analyzed below.

LEAKING SIDES

A high-pressure compartment characterized by single-phase fluid flow will slowly leak gas and eventually will bleed off the overpressuring. If only single-phase flow is considered, then the rate of gas leakage from a high-pressure compartment is controlled by Darcy's Law for radial flow (Amyx et al., 1960). A pressure compartment is modeled as a cylinder of

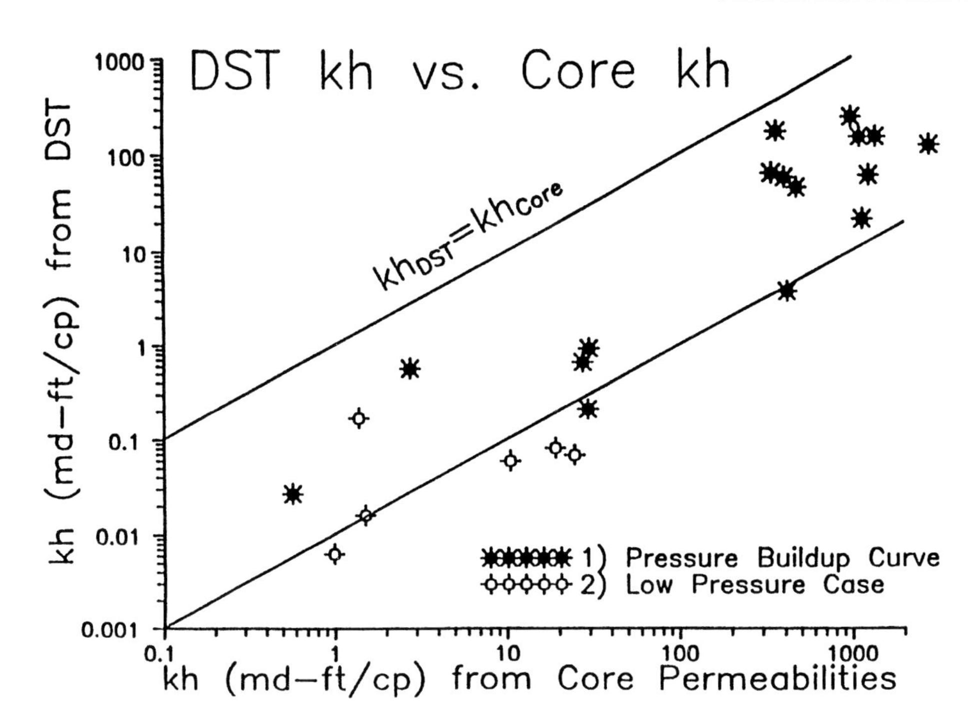

Figure 1. Drill-stem test and core-derived permeabilities for Muddy sandstones.

radius r_1 with internal pressure P_1, as shown in Figure 2A. Flow is radial, at a rate Q, where

$$Q = 1.4 \frac{\left(P_1^2 - P_2^2\right)}{\ln(r_2 / r_1)} \frac{k_s h}{u},$$

where pressure P is in psia, seal permeability k_s in darcys, sandstone thickness h in feet, fluid viscosity μ in centipoise (cp), reservoir radius r_1 and seal thickness $(r_2 - r_1)$ in feet, and flow rate Q in units of standard cubic feet per day (scf/day). The volume of gas in this reservoir is given by

$$V_g = \frac{\pi r_1^2 h}{B_g}, \quad B_g = \frac{P_{sc} T_1 Z}{P_1 T_{sc}},$$

where B_g is the gas volume factor compensating for gas expansion from reservoir conditions T_1 and P_1 to standard conditions T_{sc} and P_{sc}, and Z is the gas deviation factor from the ideal gas law. The time it takes to leak off the high pressure P_1 in the gas reservoir is approximately gas volume divided by flow rate,

$$t \approx \frac{V_g}{Q_g} = 2.3 \frac{r_1^2 \ln(r_2)}{B_g\left(P_1^2 - P_2^2\right)} \frac{\mu}{k_s},$$

with t in days. This is only an approximation, as the entire gas volume V_g does not need to leak out to achieve pressure equalization, so V_g is too large. Flow rate Q_g is also too large because the pressure difference $P_1 - P_2$ slowly decreases as gas leaks out. Therefore, the ratio V_g/Q_g can be considered a good order-of-magnitude approximation for the time to reach pressure equalization.

In order to proceed with this analysis, it is necessary to make some reasonable approximations of the constants in this last equation. Because we are attempting to estimate the length of time a pressure compartment can be contained, the conservative approach would be to estimate all parameters to yield the highest reasonable leak time t. For example, the largest time t will be obtained by considering the lowest permeability data presented in Figure 1. The minimum observed $k_s h/\mu$ is 0.006 md-ft/cp. Decrease this permeability value slightly to compensate for relative permeability, then divide by h, and a minimum estimate of 10^{-6} darcy/cp is obtained. The pressure gradient terms are approximated by $(P_1^2 - P_2^2) \approx 10^7$ psi, which corresponds to about 5000 psi inside and 3000 psi outside of the pressure compartment; these values are consistent with the data presented for the Muddy by Heasler et al. (this volume). Then, for an average pressure of 4000 psi and $B_g \approx 5 \times 10^{-3}$ ft^3/scf, the above equation for radial flow reduces to

$$t \approx 50\, r_1^2 \ln(r_2 / r_1) \text{ days}.$$

If r_1 is too close to r_2, then $\ln(r_2/r_1) \approx 0$ and the pressure compartment cannot exist for any length of time. A reasonable assumption would be to assume a 200 ft transition from high to low pressure for the lateral seal, such that $r_2 \approx r_1 + 200$ ft. Then, for a high-pressure compartmentalized gas reservoir of approximately 50,000 acres, $r_1 \approx 5$ miles, and we obtain

$$t \approx 0.7 \text{ million years}.$$

As an order-of-magnitude approximation, a gas reservoir leaking out the sides will lose its overpressuring in 1 million years or less, a geologically short period of time.

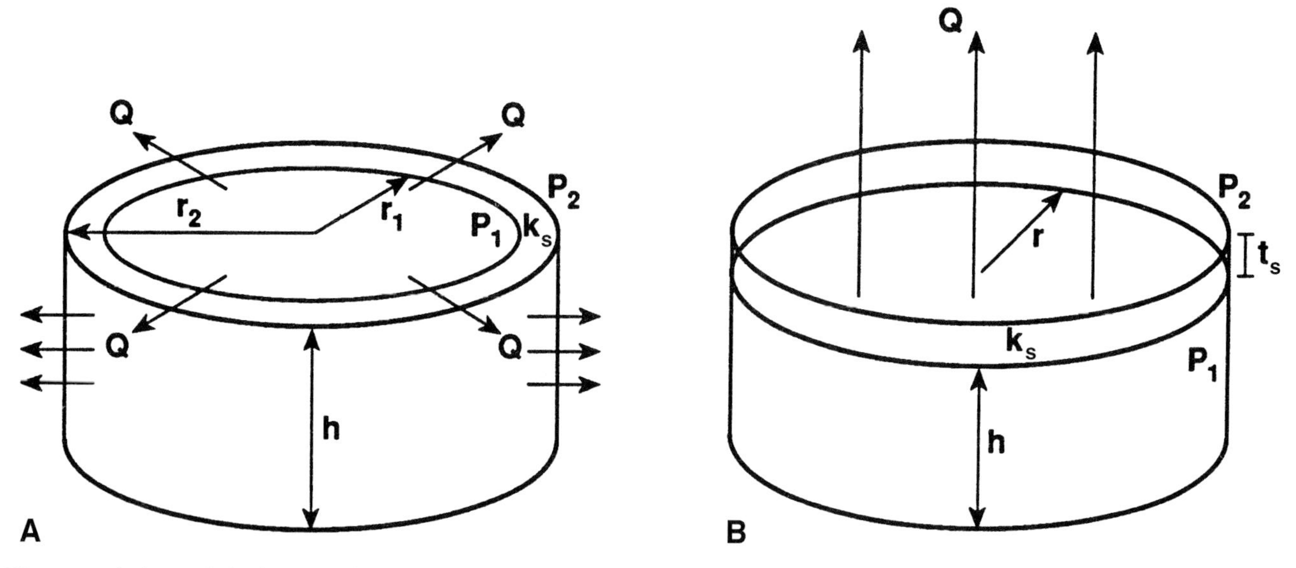

Figure 2. (A) Model of radial flow system, simulating a pressure compartment leaking out the sides through adjacent sandstones. (B) Model of areal flow, simulating a pressure compartment leaking out the top through an overlying shale.

LEAKING TOP

The escape of gas from an abnormally high-pressured compartment is accelerated even more when a leaking top is considered. If the top (or bottom) of a gas reservoir leaks, then the gas flow rate is controlled by a slightly different set of equations. Figure 2B shows the geometry and parameter definition for areal flow. Darcy's Law for flow of gas across an areal surface of area A is

$$Q_g = 0.22 \frac{\left(P_1^2 - P_2^2\right)}{t_s} \frac{k_s A}{\mu} \text{scf/day}$$

(Amyx et al., 1960), and the volume of gas in the reservoir is

$$V_g = \frac{Ah}{B_g}, \text{ where } B_g = \frac{P_{sc} T_1 Z}{P_1 T_{sc}}.$$

The approximate time to lose overpressuring is

$$t \approx \frac{V_g}{Q_g} = 4.6 \frac{h t_s}{B_g \left(P_1^2 - P_2^2\right)} \frac{\mu}{k_s} \text{days.}$$

Now, we make the same approximations for the physical constants for the top seal that we made for the lateral seal; $(P_1{}^2 - P_2{}^2) \approx 10^7$ psi, $B_g \approx 5 \times 10^{-3}$ ft^3/scf. The permeability term k_s/μ should be much smaller, because top seals are commonly shale with very low permeability, say $k_s/\mu \approx 10^{-8}$ darcy/cp. Finally,

$$t \approx 9266\, h t_s \text{ days,}$$

where reservoir height h and seal thickness t_s are both in feet. As before, assuming a 200 ft transition from pressure P_2 to P_1 such that the overlying seal is 200 ft thick (t_s = 200 ft) and for a thick Muddy Sandstone reservoir thickness h of 100 ft,

$$t \approx 0.5 \text{ million years.}$$

When both the sides and tops of an overpressured compartment leak, the time required to bleed out pressure is even shorter. In any case, even the smallest permeability we have found from the Powder River basin will leak out the observed overpressuring in less than 1 million years. Evidence (i.e., pressure anomalies) that overpressuring in the Muddy Sandstone has been in existence for 40 to 70 million years (MacGowan et al., this volume) requires a more efficient sealing mechanism.

TWO-PHASE FLOW

Abnormally pressured compartments in the Powder River basin are associated with hydrocarbon accumulations. It is only logical, therefore, to analyze the sealing mechanism of pressure compartments with respect to two-phase flow. Darcy's Law predicts that flow rate is proportional to pressure gradient. In the case of two-phase flow, this is not necessarily true. If the interfacial tension is sufficiently high between the two phases in a porous medium, then no Darcy flow will occur. A certain threshold pressure must be reached in order to initiate flow through a porous medium containing multiple phases. This threshold pressure, at which displacement will begin, is here called the displacement pressure.

Most of the petroleum industry still uses mercury injection techniques, as originally described by Purcell (1949), to determine capillary pressure and

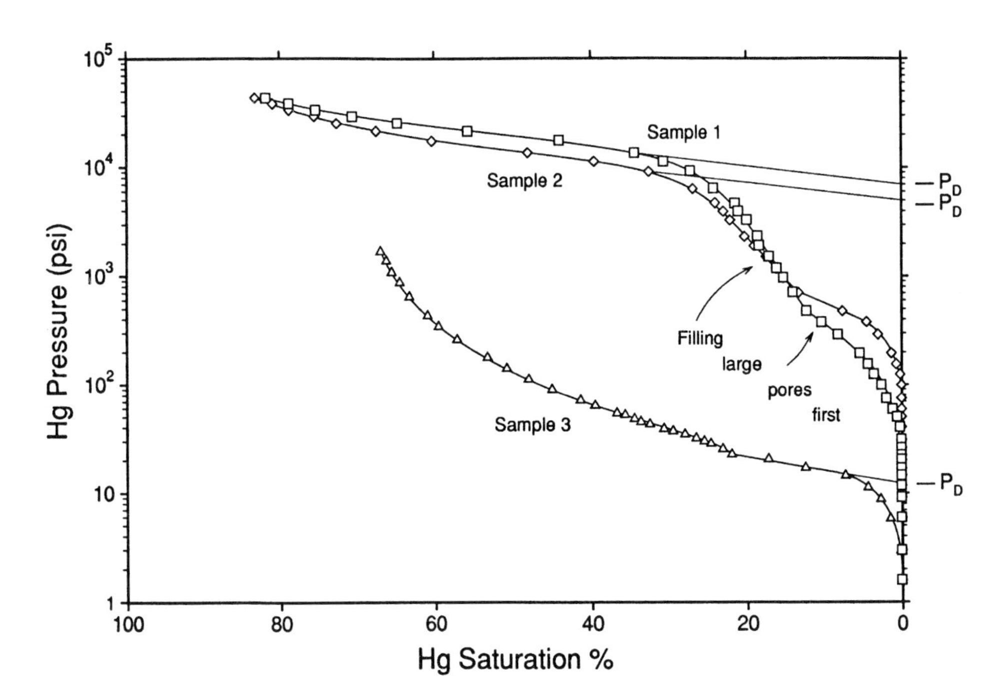

Figure 3. Mercury injection capillary pressure curves for three Muddy Sandstone samples.

displacement pressure. The pressure required to inject mercury is plotted against the mercury saturation as shown in Figure 3. The experiment begins with a dry core; mercury pressure is increased, and initially nothing happens: No mercury enters the core, so the mercury saturation remains zero. Eventually, some mercury finds large pores on the entrance side of the core, and partial mercury saturation occurs. As the mercury pressure is increased further, many more pores begin to fill. At this point, mercury injection is actually proceeding, and the core rapidly increases in mercury saturation with very little increase in mercury pressure. The displacement pressure is taken as an extrapolation of this flat portion of the curve onto the pressure axis (Amyx et al., 1960; Thomas, 1967). The rationale for this procedure is to avoid the effects of isolated filling of large pores on the injection end and to project a more realistic threshold for initiating Darcy flow through the core. There are experimental data to verify that this is a proper extrapolation point (Thomas, 1967).

Mercury injection data from the Muddy Sandstone have been collected from selected stratigraphic intervals, as described by Surdam et al. (this volume). These data are presented in Figure 3. Samples 1 and 2 were taken from relatively tight Muddy Sandstone intervals (believed to be compartment seals), whereas Sample 3 is from a typical reservoir rock. The tight sands have gas permeabilities of only 10 microdarcys (μd), whereas the reservoir rock has 10 millidarcys (md). These differences in permeability are reflected in the capillary pressure curves. Over 10,000 psi is needed to inject mercury into the 10 μd rocks, whereas only 20 psi is needed for the reservoir rock. These displacement pressures can be converted to equivalent gas-water values using procedures developed by workers at the University of Michigan (Katz, 1970).

Katz and co-workers, in studies applied to natural gas storage, have collected relevant data that verify the extrapolation of capillary pressure curves to displacement pressure values. As part of that work, Thomas (1967) constructed a core holder that measured displacement pressures directly. Thomas saturated cores with water and then attempted to inject gas at incremental pressures until displacement of water began. Displacement was judged by a micropipette connected to the outlet of the core, and stabilization was allowed over a period of days to weeks. All rocks exhibited some finite displacement pressure; no perfectly impermeable rock was found; and all rocks could support some pressure without flow initiating.

Mercury injection displacement pressures (into air-filled core) can be readily compared with gas injection displacement pressures (into water-filled core) by correcting for the differences in interfacial tension. The mercury-air system has an interfacial tension about five times that of an air-water system, so a mercury-derived displacement pressure is converted to a gas displacement pressure by dividing by five. Purcell (1949), Thomas (1967), and many other laboratory studies have verified this factor-of-five difference between mercury-air and gas-water capillary pressures. Our Muddy Sandstone data, therefore, showing a displacement pressure of about 10,000 psi mercury-air, is equivalent to about 2000 psi in a gas-water system. This indicates that the low-permeability Muddy Sandstone is capable of supporting a 2000 psi pressure differential before flow is initiated through the core. This is approximately equivalent to the amount of overpressuring observed in the field (Amos Draw) where the Muddy cores were extracted (Heasler et al., this volume).

Figure 4 shows displacement pressure plotted against permeability for numerous samples. Notice

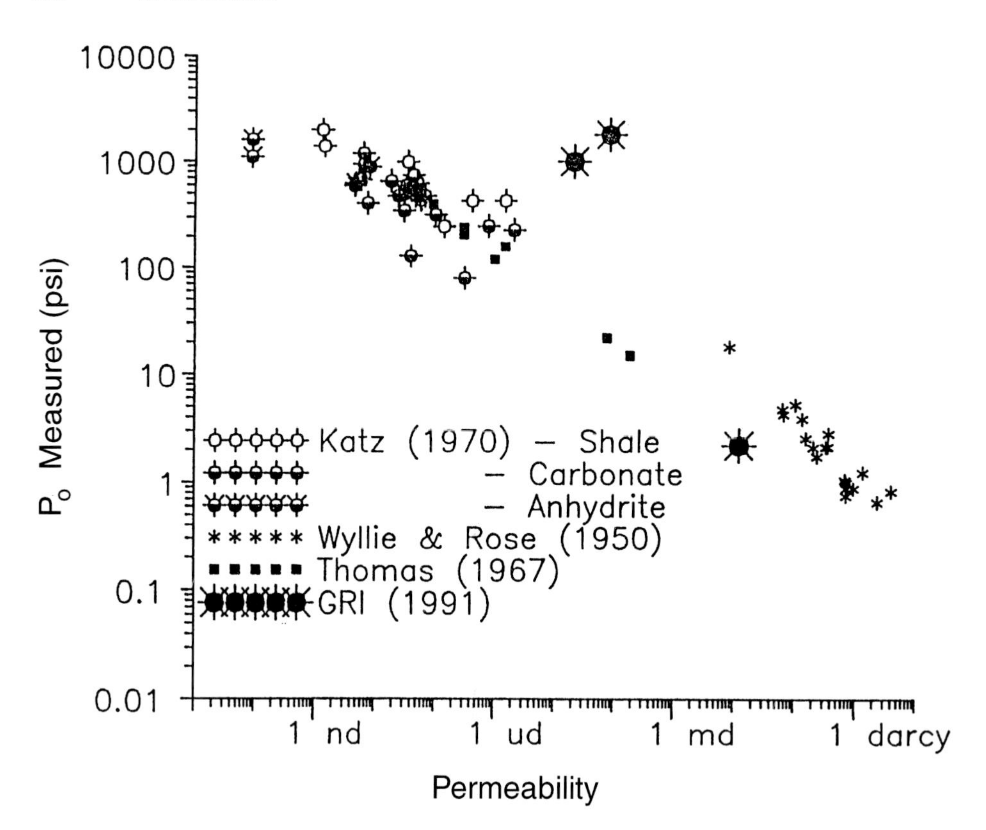

Figure 4. Displacement pressure vs. permeability for a wide variety of rocks. GRI points are from the Muddy Sandstone for this study.

the general separation into two groups of data points, the high displacement pressures characteristic of seals and the low displacement pressures characteristic of reservoir sandstones. Our Muddy Sandstone data are compared with previously published data from the University of Michigan Underground Gas Storage Project as summarized by Ibrahim et al. (1970), Wyllie and Rose (1950) using some high-permeability oil reservoir sands, and Thomas (1967) for assorted sands and carbonates from Michigan. The three data points corresponding to our data from Figure 3 are labeled GRI (1991). All three of these data points are from sandstones in the Muddy Sandstone, yet the two low-permeability samples have displacement pressures equivalent to those of rocks widely accepted as seals. For a detailed discussion of the stratigraphic, mineralogic, and diagenetic aspects of these rocks, see Martinsen (this volume).

CONCLUSIONS

Pressure compartments in the Muddy Sandstone of the Powder River basin are surrounded by lithologic units of very low permeability and very high threshold displacement pressures. These tight lithologic units (identified as paleosols by Martinsen, this volume) are barriers to flow under any conditions, but they are not seals that can maintain overpressuring within a compartment under single-phase flow conditions. The seal is created by the presence of two fluid phases. A single-phase system would behave according to Darcy's Law, and the pressure difference would dissipate within a geologically short period of time. Therefore, the rocks themselves are not the seal; it is the interaction of rock and fluid properties that creates the pressure seal. This observation becomes particularly clear when one considers that overpressuring is associated with the occurrence of hydrocarbons in the Powder River basin. If there are no hydrocarbons, then there is no capillarity to create a seal, and pressure compartments will not form. It is capillarity and a resultant anomalously high displacement pressure that creates a seal surrounding these hydrocarbon reservoirs.

Such a capillary pressure seal is mechanically equivalent to the classic hydrocarbon trapping mechanisms as described by Schowalter (1979) and in other publications. The trapping of hydrocarbons is not caused by impermeable rocks; there are essentially no impermeable rocks in nature. Hydrocarbons are trapped whenever the displacement pressure for the two-phase system (hydrocarbon and water) is higher than the pressure in the hydrocarbon phase. At this point, migration ceases and a hydrocarbon trap is formed. Normally, such hydrocarbon traps are created by shales, faults, or other geologic structures. Within the Muddy Formation in the Powder River basin, the sandstone acts as both the reservoir rock and the hydrocarbon seal. Martinsen (this volume) has shown that the Muddy is composed of many distinct depositional units. Some units have good permeability and act as reservoir rock, whereas other units have low permeability and act as seals in the presence of two-phase fluid flow. Most important, the displacement pressure of these low-permeability sandstones is very high, as high as that of shales and anhydrites.

Capillary sealing is a universal phenomenon that occurs in every petroleum reservoir of the world. Capillary sealing of overpressured sandstone compartments has been demonstrated only for the Muddy Sandstone of the Powder River basin. Although capillary sealing of pressure compartments is undoubtedly important for many other regions, it must be approached cautiously. Regions of rapid compaction and overpressured aquifers (e.g., the Gulf Coast) are probably leaking pressure due to Darcy flow. Such sands have very high permeability and probably very low displacement pressures. For the Muddy Sandstone, there is evidence that these pressure compartments have been in existence for millions of years (MacGowan et al., this volume). Such longevity of a high-pressure compartment can only be explained by capillary sealing in low-permeability rocks.

ACKNOWLEDGMENTS

This material covers research performed through November 1991. The original manuscript was reviewed by David Copeland (University of Wyoming), who made numerous helpful suggestions. This study was funded by Gas Research Institute under contract number 5089-260-1894.

REFERENCES CITED

Amyx, J. W., D. M. Bass, and R. L. Whiting, 1960, Petroleum Reservoir Engineering: McGraw-Hill Book Company, 610 p.

Earlougher, R. C., 1977, Advances in well test analysis: Society of Petroleum Engineers Monograph, v. 5, 264 p.

Ibrahim, M. A., M. R. Tek, and D. L. Katz, 1970, Threshold Pressure in Gas Storage: University of Michigan Press, 309 p.

Iverson, W.P., 1990, Horner analysis of drill-stem tests: Annual Report, Gas Research Institute 92/0483.

Katz, D. L., 1970, Underground storage of natural gas: American Gas Association Research Report, University of Michigan, 309 p.

Purcell, W. R., 1949, Capillary pressures—their measurement using mercury and the calculation of permeability therefrom: Transactions of the American Institute of Mining Engineers, v. 186, p. 39.

Schowalter, T. T., 1979, Mechanics of secondary hydrocarbon migration and entrapment: American Association of Petroleum Geologists Bulletin, v. 63, p. 723–776.

Thomas, L. K., 1967, Threshold pressure phenomena in porous media: Ph.D. dissertation, University of Michigan, 148 p.

Wyllie, M. R. J., and W. E. Rose, 1950, Application of the Kozenay equation to consolidated porous media: Transactions of the American Institute of Mining Engineers, v. 187, p. 127.

Chapter 21

Formation Water Chemistry of the Muddy Sandstone and Organic Geochemistry of the Mowry Shale, Powder River Basin, Wyoming: Evidence for Mechanism of Pressure Compartment Formation

D. B. MacGowan
State University of New York
College of Fredonia
Fredonia, New York, U.S.A.

Zun Sheng Jiao
Ronald C. Surdam
University of Wyoming
Laramie, Wyoming, U.S.A.

F. P. Miknis
Western Research Institute
Laramie, Wyoming, U.S.A.

ABSTRACT

In the Powder River basin, pressure compartmentation has been linked to the establishment of multiphase fluid-flow systems. The transition from a single-phase to a multiphase fluid-flow system is driven by liquid hydrocarbon generation and its subsequent reaction to gas. As a consequence, pressure compartments in this basin should be related to changes in formation water chemistry, thermal maturation of organics, clay diagenesis, and other geochemical reactions associated with progressive burial. To test this, measured and calculated pressure anomalies were studied in relation to changes in formation water chemistry, clay mineralogy, kerogen structure, carbon aromaticity, vitrinite reflectance, and organic-matter production indices. The results indicate that fundamental changes in formation water chemistry, rock inorganic geochemistry, and organic geochemistry occur between about 8000 and 10,000 ft (2400 and 3000 m) present-day burial depth, coincident with a major change in the formation pressure regime, the onset of abnormal pressure, in the Muddy Sandstone. The results also indicate that the onset of abnormal pressure is coincident with the generation, migration, and reaction to gas of liquid hydrocarbons. Thermal modeling, organic geochemistry, and pressure measurements suggest that abnormal pressures have existed in the Muddy and Mowry formations for a geologically significant time (>40 m.y.). Further, geochemical modeling suggests that the rupture of boundary seals accompanied by fluid migration—formation water mixing, temperature drop, and pressure drop—or degassing can cause calcite precipitation and, consequently, seal restoration. These results differentiate the type of pressure anomalies seen in the Muddy Sandstone from those resulting from either compaction or hydrodynamic disequilibria.

INTRODUCTION

Pressure compartments in sedimentary basins may be related to variations in formation water chemistry in several ways. The formation pressure differential associated with compartmentation implies variations in fluid flux through stratigraphic units containing pressure compartments. These variations should be reflected in the formation water chemistry. Additionally, some models for the formation of some pressure compartment seals call upon precipitation of diagenetic cements; diagenetic pathways are critically controlled by formation water chemistry (see MacGowan and Surdam, 1990; Surdam et al., 1989b). Also, various cementation reactions that may be involved in seal enhancement or restoration after rupture frequently occur in zones of formation water mixing or at the interface between waters of differing chemical composition. Therefore, knowledge of the distribution of heterogeneities in formation water chemistry within a stratigraphic interval containing pressure compartments is of use in studying pressure compartmentation.

Additionally, temperature anomalies are frequently associated with zones of overpressure (Hunt, 1979, p. 539). These accelerated geothermal gradients can have a great effect upon temperature-dependent organic and inorganic diagenetic reactions and, if overpressure is present for a geologically significant period of time, on kinetically controlled organic and inorganic diagenetic reactions (e.g., MacGowan et al., 1990).

This chapter reports on the analysis of 230 wells from which formation water chemical analyses were available from the Muddy Sandstone, and on the X-ray diffraction analysis of clay-sized (<2μ) minerals, total organic carbon (TOC) analysis, nuclear magnetic resonance (NMR) determinations of kerogen functionality and carbon aromaticity, vitrinite reflectance (R_o) analysis, and anhydrous pyrolysis (PI) analysis of cuttings from the overlying Mowry Shale. The locations of the samples for organic geochemistry are given in Jiao and Surdam (this volume; their Figure 1). The data presented below indicate that a fundamental change in the rock-fluid system occurs between about 8000 ft (2400 m) and 10,000 ft (3000 m) present burial depth, coincident with a major change in the pressure regime (onset of abnormal pressure) in the Muddy Sandstone of the Powder River basin (Heasler et al., this volume). The organic geochemical data indicate that the onset of abnormal pressure is coincident with the conversion of liquid hydrocarbons to gas; and X-ray diffraction of clays in the Mowry suggests that illitization of smectite becomes important over this depth interval. Thermal modeling, organic geochemistry, and pressure measurements suggest that abnormal pressures have existed in the Muddy Sandstone and Mowry Shale for geologically significant periods of time (at least 40 million years). The facts that (1) overpressuring resulting from either hydrodynamic or compaction disequilibria may be geologically ephemeral (e.g., Bethke et al., 1988); (2) some rocks thought to be seals in the Muddy Sandstone are, in fact, moderately permeable to single-phase flow (Iverson et al., this volume); and (3) the onset of abnormal pressure is associated with the conversion of oil to gas in the Muddy and Mowry suggest that the sealing mechanism for pressure compartmentation in this instance is capillarity resulting from the differential in interfacial tension between water, gas, and oil associated with multiphase flow (see Iverson et al., this volume).

RESULTS AND DISCUSSION

Muddy Formation Water Chemistry

Amoco Production in Tulsa, Oklahoma, donated to the Pressure Compartmentalization Research Group at the University of Wyoming a database consisting of approximately 1800 formation water analyses from all producing intervals in the Powder River basin. Statistical screening of the Muddy Sandstone water samples was performed at Amoco, and several screening steps were taken at the University of Wyoming. For this study, analyses were used only from wells that perforated exclusively within the Muddy Sandstone; that gave a charge balance within ±10% of total dissolved solids; and that had data for total dissolved solids, Na^+, Ca^{2+}, Cl^-, HCO_3^-, CO_3^{2-}, SO_4^{2-}, pH, latitude, longitude, and depth of perforations. Passing these screenings were 230 Muddy

Sandstone samples from across the basin. It should be noted that the Amoco database contains all available data collected from many sources, obtained by the use of a variety of analytical techniques, gathered over a period of several decades. Analysis of the Muddy Sandstone samples indicates that significant variations exist in Muddy Sandstone water chemistries, basin-wide. For instance, the total dissolved solids (TDS) varies from about 2600 ppm to greater than 55,000 ppm, the alkalinity (as bicarbonate) varies from 230 ppm to >5000 ppm, and the pH varies from 7.2 to 8.5.

Figures 1A and 1B show the distribution of TDS in Muddy Sandstone waters with depth and with location in the Powder River basin, respectively. The depositional environments in which the Muddy Sandstone was deposited should give connate waters exhibiting a wide range of compositions, from fresh (fluvial) to brackish to moderately saline (marine) (Martinsen, 1992). Figure 1A shows that TDS increases slowly with depth to about 8000 ft (2400 m) present burial, and then begins to increase rapidly. Since the modern basin margin roughly conforms to the Muddy depositional boundary (Martinsen, 1992), more shallowly buried portions of the Muddy should exhibit fresher water, with more saline waters in the basin center. Formation water salinities in excess of about 35,000 ppm TDS indicate that the water is either connate and originates from a very saline environment (silled basin, saline lake, etc.), or is in hydrologic communication with a deeper, evaporite-bearing unit such as the Minnelusa (see MacGowan et al., 1993). The depositional environments that characterized Muddy and Mowry deposition (Martinsen, 1992) would lead us to expect neither hypersaline connate waters nor evaporate minerals. Therefore, the few samples that exhibit TDS higher than normal marine water TDS should be attributed either to formation water mixing with waters from deeper, evaporite-bearing units or to errors in reported perforation intervals or TDS measurements in the Amoco database. As will be shown, the high-TDS samples are geographically bunched and roughly coincide with other anomalies (pH, alkalinity, pressure); it is therefore likely that the values are valid and represent formation water mixing with deeper, saline formation waters or multiple production intervals (but reported as exclusively Muddy). Shallow samples with low TDS and high bicarbonate alkalinity along the basin margin indicate that the Muddy may be in hydrologic communication with carbonate groundwater aquifers (such as the Madison Limestone) at shallow depths along the basin margin. Most trends in formation water chemistry are regular and mappable, at least in the Muddy Sandstone. Depth vs. pH and depth vs. bicarbonate alkalinity plots show similar trends to TDS vs. depth plots (Figure 2 shows depth vs. alkalinity). These quantities also change slowly with depth from about 3000 ft (900 m) to about 8000 ft (2400 m) present burial; below about 8000 ft, the change with depth becomes very rapid.

Comparing contour plots in latitude-longitude space of TDS, alkalinity, and pH with similar plots of

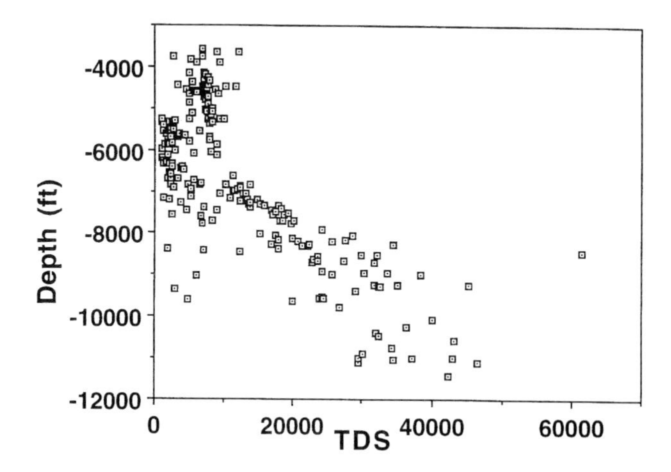

Figure 1A. Depth vs. total dissolved solids (TDS) for Muddy Sandstone water samples, Powder River basin.

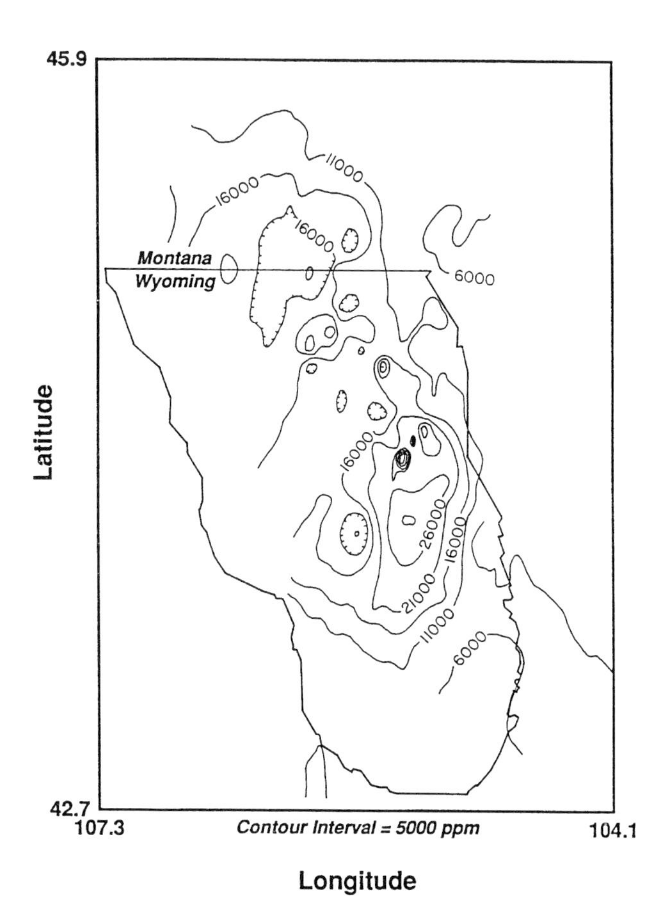

Figure 1B. Contour map of Muddy Sandstone water total dissolved solids (TDS). Contour interval = 5000 ppm. Powder River basin, 230 data points.

hydrodynamic head in the Muddy Sandstone (Heasler et al., this volume) indicates that anomalies in formation water chemistry generally coincide with anomalies in head wherever the pressure and formation water chemistry databases overlap geographical-

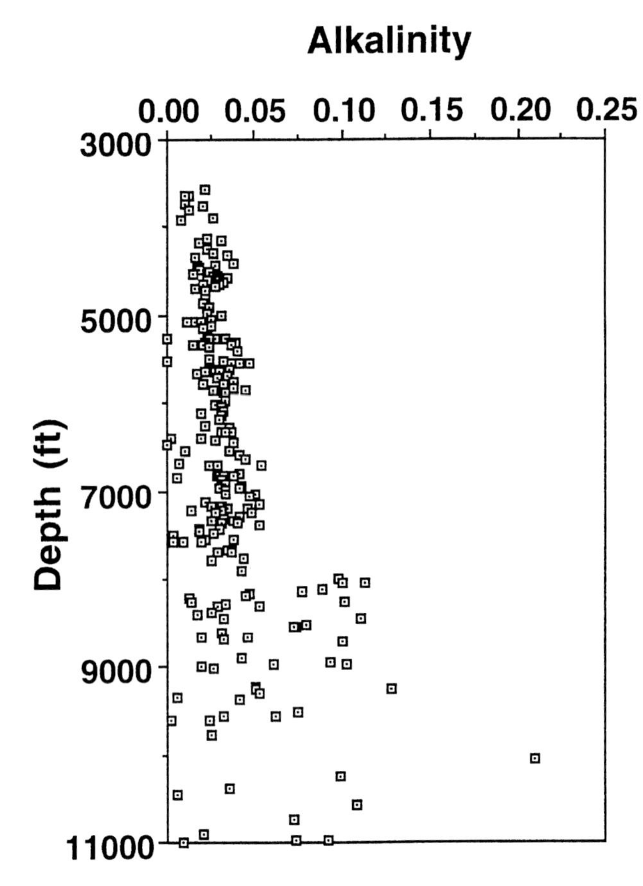

Figure 2. Depth vs. alkalinity, expressed as moles HCO_3^- per liter, for Muddy Sandstone water samples, Powder River basin.

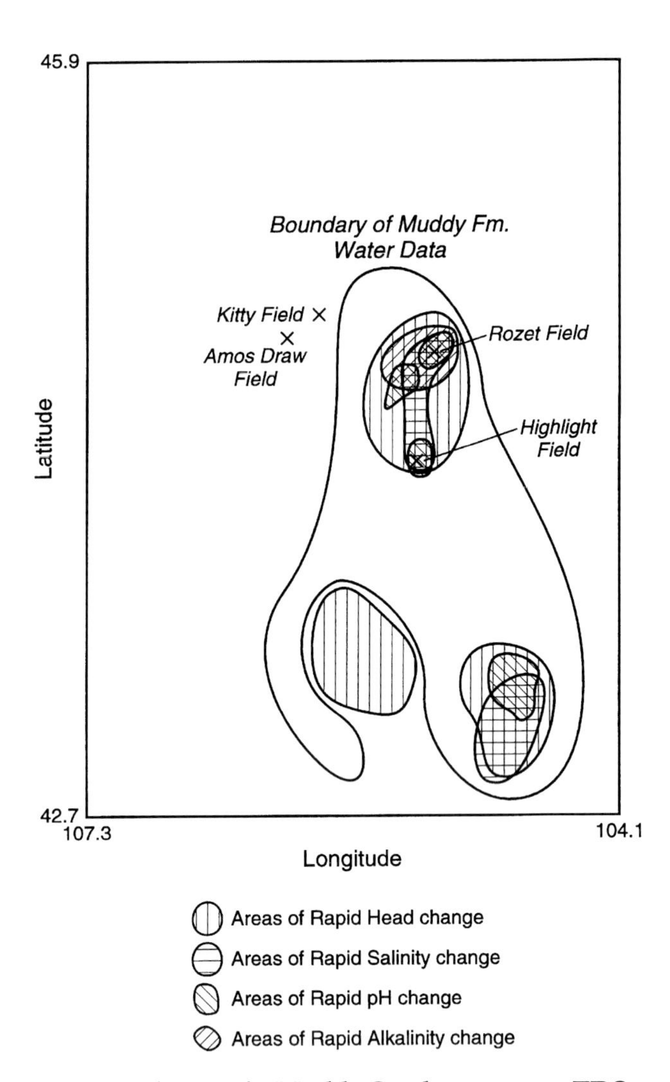

Figure 3. Changes in Muddy Sandstone water TDS, alkalinity, pH, and head in the Powder River basin. In areas where there is data from both head calculations and chemical analyses of formation waters, areas of rapid change in head generally coincide with areas of rapid change in TDS, pH, and alkalinity, especially the Rozet and Hilight fields.

ly. Figure 3 shows the areas of high rate of change (delineated by closely spaced contour lines) in TDS, pH, alkalinity, and calculated head (from Heasler et al., this volume) in latitude-longitude space. Areas of rapid head change (which may represent pressure compartment boundaries) generally coincide with areas of rapid change in formation water chemistry (especially at Hilight and Rozet fields) where the databases overlap. This coincidence suggests that the boundaries of pressure cells are related to changes in formation water chemistry.

In summary, formation water samples from above about 8000 to 9000 ft (2400 to 2700 m) present burial depth have generally between 5000 and 10,000 ppm TDS. Marine rocks in this depth interval would initially have had about 35,000 ppm TDS. Therefore, these formation waters must have undergone significant mixing with meteoric water. This suggests that the marine units of the Muddy, above about 8000 ft (2400 m) present burial, are in hydrologic communication with the surface. Below about 8000 to 10,000 ft (2400 to 3000 m) present burial the TDS increases rapidly and averages about 35,000 ppm TDS, approximately the composition of sea water (i.e., relatively unmodified connate water). The rapid rise in TDS at about the same depth as the onset of abnormal formation pressure suggests that the Muddy below about 8000 to 10,000 ft (2400 to 3000 m) present burial is not in hydrologic communication with more shallowly buried Muddy, or with the surface. The few very high TDS formation water samples likely represent communication with a deeper, evaporite-bearing unit (such as the Minnelusa Formation; see MacGowan et al., 1993). As these samples are geographically bunched, this may be due to faulting, to mixed production reported as exclusively Muddy, or to the rupture of a deeper pressure cell.

X-Ray Diffraction of Mowry Samples

X-ray diffraction analyses of the clay-sized fraction of cuttings from the Mowry Sandstone in the Powder River basin show kaolinite, mixed-layer illite/smec-

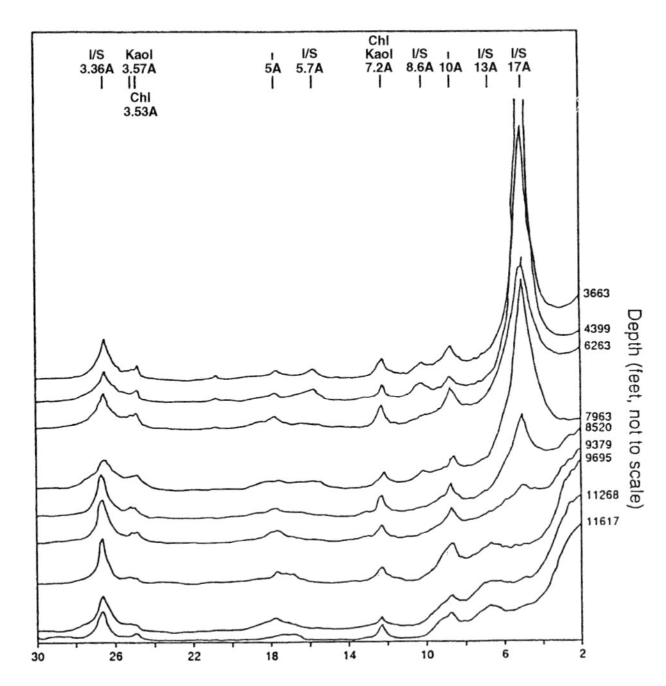

Figure 4A. X-ray diffraction (XRD) traces of the <2μ fraction of Mowry Shale samples from the Powder River basin vs. depth. Illitization of smectite begins at about 8000 ft (2400 m) and is largely complete by about 10,000 ft (3000 m).

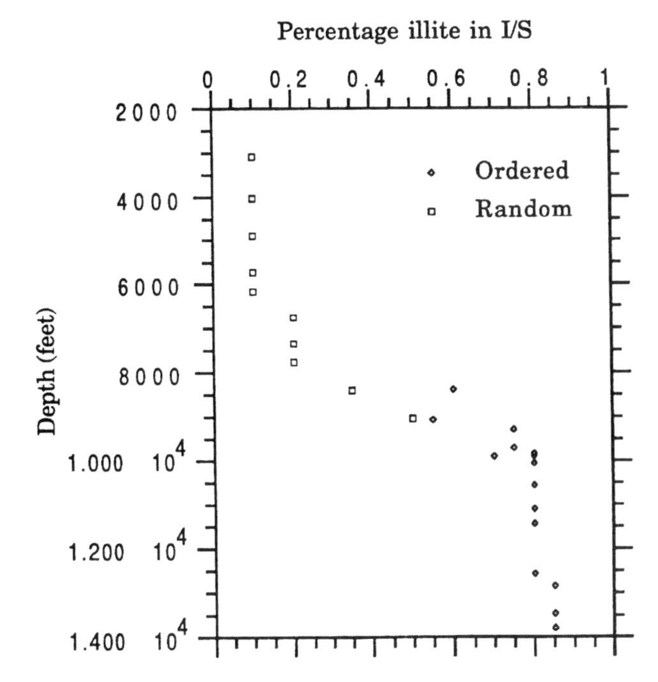

Figure 4B. Percent illite (as a fraction; 0.8 = 80%) in illite-smectite mixed-layer minerals vs. depth in the Mowry Shale, from XRD data.

tite, chlorite, and quartz, with minor feldspar. Figure 4A shows the X-ray diffraction patterns of the clay-sized fraction from the Mowry Shale between about 3600 and 11,600 ft (1100 and 3500 m) present burial; Figure 4B shows percentage illite in illite-smectite minerals with depth. Burial diagenesis, as demonstrated by a sample suite from about 3000 to 13,700 ft (900 to 4200 m), is expressed in the smectite-to-illite transformation and the ordering of illite-rich mixed-layer clays at depth. The illitization of smectite begins to become important at about 8000 ft (2400 m) and is largely complete by 10,000 ft (3000 m) present burial, where the mixed-layer illite/smectite clay contains 85% illitic layers (Figures 4A and 4B) (for details, see Jiao and Surdam, this volume). The conversion of smectite to illite is a kinetically controlled process that requires either very high temperature or significant geologic time to proceed (see discussions in Surdam et al., 1989a; Pytte and Reynolds, 1989; Elliot et al., 1991).

Organic Geochemistry of Mowry Samples

Organic geochemical analyses performed on cuttings from the Mowry Shale included total organic carbon (TOC); vitrinite reflectance (R_0); anhydrous pyrolysis, reported as production index, PI (Tissot and Welte, 1984); and solid-state ^{13}C nuclear magnetic resonance (NMR), reported as NMR spectra and carbon aromaticity analysis.

In the samples studied, TOC, a measure of organic richness, ranges from 1.2 to 4.2 wt% (moderately lean to moderately rich source rock), but shows no trend with depth (Table 1).

Vitrinite reflectance (R_0), a measure of organic-matter thermal maturity, ranges from 0.47 at about 3500 ft (1100 m) to 1.6 at about 13,700 ft (4200 m) (Figure 5). The liquid oil generation window is generally thought to occur between about 0.5–0.7% and 1.0–1.3% reflectance (Tissot and Welte, 1984). Between R_0 = 1.3 and 2.0%, oil is thermally cracked to wet gas; the aliphatic side chains all have been cleaved from the

Table 1. Total organic carbon in Mowry Formation samples.*

Well Name	Depth, ft	Org. C, wt%
Gov't Terteling	3030–3140	2.8
Walker #1	4000–4020	4.2
Kummerfield #1	5700–5740	1.6
Fartin #7	6740–6760	2.1
Kawulok #1	7740–7790	1.2
State #1	8370–8385	1.4
Annie #1	9220–9240	1.4
Felix Unit #16	9690–9700	1.7
Christnick State #2	9770–9870	3.1
Little Burgher Draw	11,080–11,000	1.6
South Rainbow-F #1A	11,400–11,450	1.5

* Sample locations are shown in Figure 1 of Jiao and Surdam (this volume).

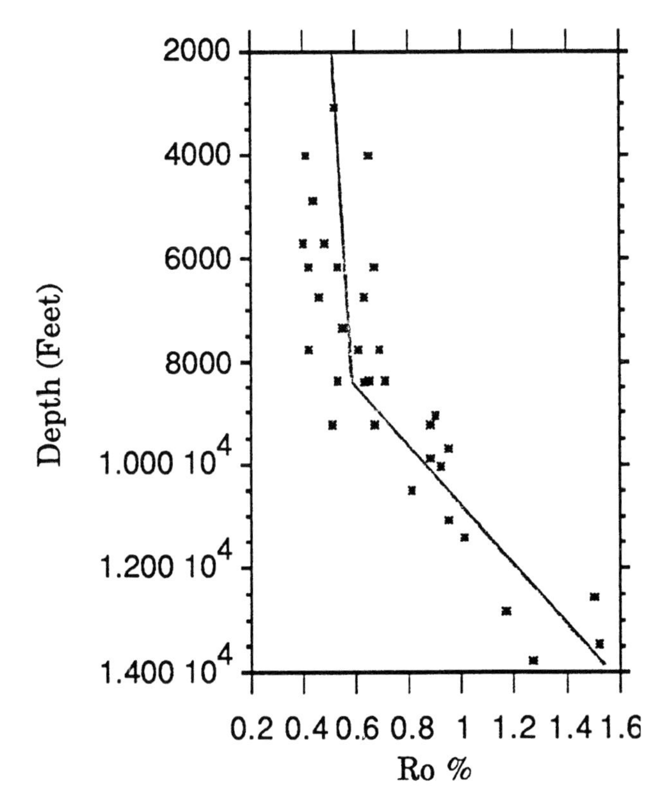

Figure 5. Vitrinite reflectance (R_o) of Mowry Shale source rocks from the Powder River basin vs. depth. There is a marked steepening of the gradient below about 8000 ft (2400 m). Trench line is placed by eye.

kerogen, and the basic ordering of kerogenic units into anthracitic structures has begun (Tissot and Welte, 1984). Vitrinite reflectance values >2.0 generally indicate thermal exposure to temperatures in excess of 150°C (Tissot and Welte, 1984).

Thermal modeling by Jiao and Surdam (this volume) of the temperature history of the Powder River basin gives Muddy temperatures in excess of 140°C at maximum burial. A preliminary survey of bottom-hole temperatures in the Muddy Sandstone (below the Mowry) indicates that present-day, in situ formation temperatures range from 75 to 130°C (Jiao and Surdam, this volume). Apparently, the vitrinite reflectance data are recording exposure to higher paleotemperatures. This is to be expected, given the post-Laramide-uplift history of the Powder River basin. As shown in Figure 5, the vitrinite reflectance increases slowly with depth from 0.4 to 1.0 between 3500 and about 9000 ft (1100 and 2700 m). A rapid increase in thermal maturity of the Mowry, as recorded by R_o, occurs from 9000 to 11,500 ft (2700 to 3500 m) present burial: R_o increases from about 1.0 to 1.6. This implies that the organic material above about 9000 ft (2700 m) present burial matured in a different thermal regime (different geothermal gradient) than did the organic material below it (see discussions in Middleton, 1983; Hunt, 1979; MacGowan et al., 1990).

Hunt (1979, p. 339) shows a very similar trend in R_o from the Powder River basin. He attributes the acceleration in the R_o-vs.-depth curve to an increase in geothermal gradient associated with abnormally high formation pressures. The maturation of vitrinite has been shown to be a kinetically controlled process (see Price, 1983), so geologic time as well as temperature is important in the rate of maturation.

In order for the observed trend in R_o to have developed, the accelerated geothermal gradient must have existed for a geologically significant period of time. The kinetics of vitrinite maturation and the association of accelerated geothermal gradient with abnormal pressure suggest that the abnormal formation pressures have also existed for geologically significant periods of time. Since the vitrinite reflectance records exposure to higher, pre-Laramide-uplift temperatures, and the change in the current pressure regime is still associated with the change in thermal regimes recorded in the vitrinite reflectance values, it is evident that these pressure anomalies must have originated in pre-Laramide-uplift times.

Rock Eval-type anhydrous pyrolysis of Mowry Shale samples was performed; the results are expressed as production index, PI (Figure 6). Production indices are calculated as the amount of hydrocarbon generated but not expelled (S_1 peak) divided by the sum of that amount (S_1) and the amount of additional hydrocarbon that potentially can be generated (S_2 peak) (Tissot and Welte, 1984). The PI in the Mowry Shale (Figure 6), as measured by anhydrous pyrolysis, ranges from 0.03 to 0.3 between 3000 and 11,500 ft (900 and 3500 m) present burial. Like R_o, the production index increases slowly between 3000 and 9000 ft (900 and 2700 m) from 0.03 to 0.1, then increases rapidly to 0.3 between 9000 and 11,500 ft (2700 and 3500 m). The results of the anhydrous pyrolysis analyses also appear to indicate both an accelerated geothermal gradient below about 9000 ft (2700 m) present burial and exposure to higher formation temperatures in the past. Also, the PI data suggest that below about 9000 ft (2700 m) present burial, the organic matter present in the shales is hydrocarbon that has been generated but not expelled.

Organic material in samples from the Mowry Shale was also analyzed by solid-state ^{13}C nuclear magnetic resonance (NMR) utilizing cross-polarization with magic-angle spinning (MAS) and high-power decoupling (see *Experimental Conditions*). Samples having low TOC were analyzed without performing kerogen isolation procedures; these procedures are time-consuming and can alter the composition and structure of the kerogen (as discussed in Vandergrift et al., 1980). Instead, the samples were pre-washed in an HCl acid solution before analysis. The acid washing removes oxides, hydroxides, and carbonates, which can contain paramagnetic nuclei that, where present, reduce the signal-to-noise ratio. Additionally, use of a large-volume sample spinner facilitated the analysis of low-TOC samples without performing kerogen isolation.

Figure 7 shows the ^{13}C NMR spectra from the Mowry Shale samples from depths of about 3000 to

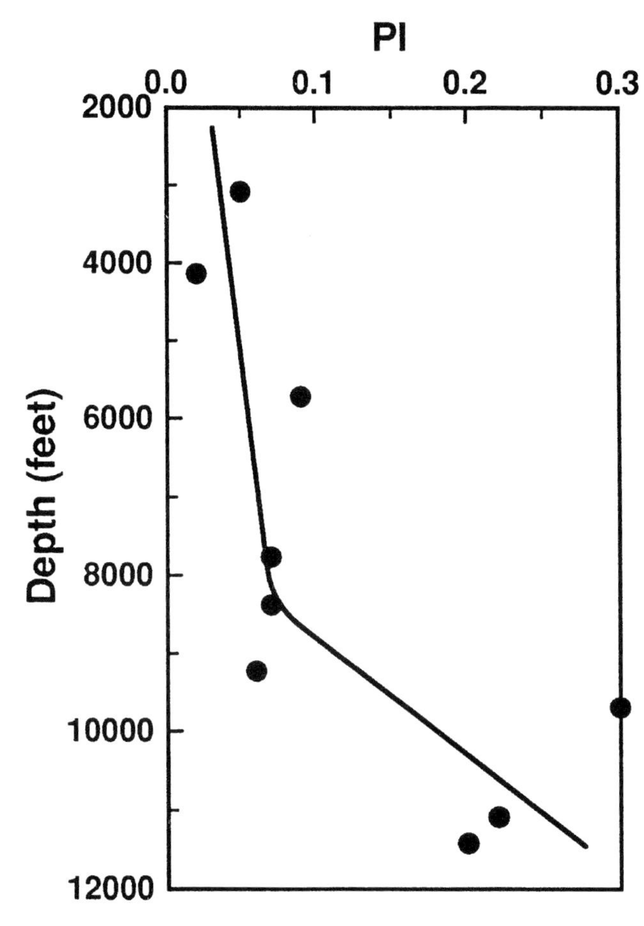

Figure 6. Production index (PI) from anhydrous pyrolysis of the Mowry Shale source rocks from the Powder River basin vs. depth. There is a marked steepening of the gradient with depth below about 8000 ft (2400 m). Trench line is placed by eye.

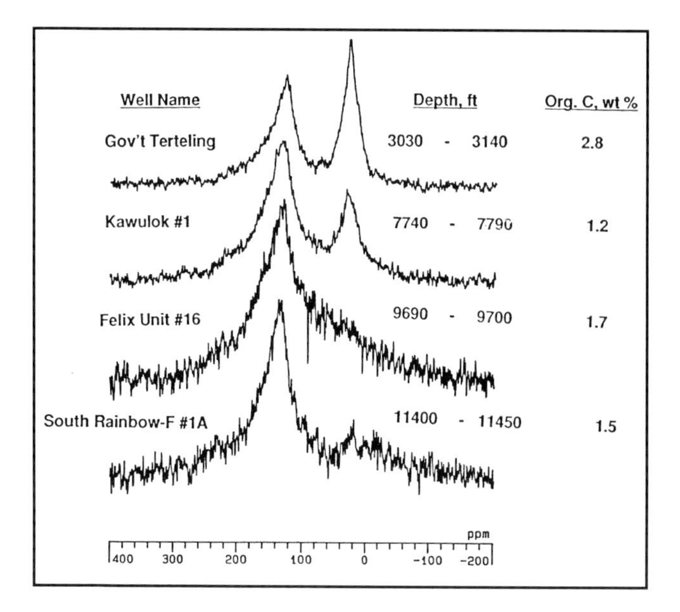

Figure 7A. ^{13}C NMR spectra vs. depth for Mowry Formation source rocks from the Powder River basin. Note the disappearance of the aliphatic carbon peak (large upfield peak) between 8400 ft and 9300 ft and the continued sharpening of the aromatic peak (large downfield peak) with depth. This indicates that these source rocks have lost their capacity to generate liquid hydrocarbons by about 9300 ft present burial depth.

11,500 ft (900 to 3500 m). The x-axis scale (ppm) is the chemical shift, a measure of the magnetic field strength required to cause the paramagnetic nucleus—^{13}C in this case—to precess and give a resonance signal.

In typical kerogen NMR spectra (see review in Miknis et al., 1993), the broad band between 0 and 60 ppm is associated with aliphatic carbons in branched and straight chains and with naphthenic structures. The broad band between 100 and 200 ppm is attributable to aromatic and carbonyl carbon structures. Bands centered around 128 ppm are due to protonated aromatic carbons; shoulders at ~140 ppm are due to substituted aromatic carbons; bands at ~150 ppm are due to phenolic carbons and aromatic ethers; and the bands at ~180 ppm and 210 ppm are due to carbonyl carbons in various types of functional groups (acids, esters, ketones, aldehydes). Thus, NMR can be used to study the abundance of ^{13}C nuclei in various structural and functional groups in kerogen; and when applied to a suite of samples from various depths, it can be used to study the changes in kerogen structure associated with thermal maturation (see discussion in Surdam and Crossey, 1985). This assumes that all the samples from the suite were the same type of organic matter to start with, as is apparently the case with these samples.

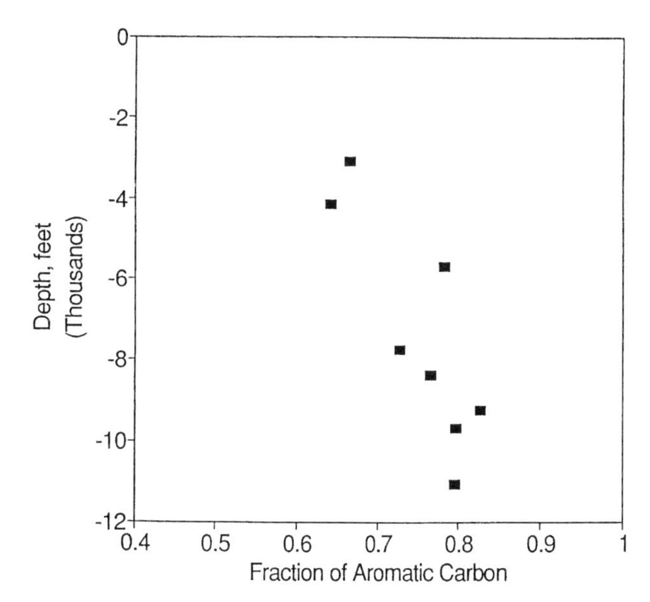

Figure 7B. Carbon aromaticity vs. depth, as determined by NMR of Mowry Formation kerogen.

In Figure 7A, the first major upfield (righthand) peak is the resonance signal from the carbon in aliphatic functional groups. The major downfield (lefthand) peak represents carbon in aromatic functional groups; on the downfield shoulder of this peak are signals from carbon in oxygenated functional groups: phenolic and carbonyl, respectively, downfield. The ^{13}C NMR spectra in Figure 7A show dominant broad bands due to aliphatic and aromatic carbon structures. Signals from the oxygenated functional groups are small and not readily discernible, inasmuch as most of these functional groups have already been released by the kerogen during thermal maturation. The trend with depth in the NMR spectra is one of decreasing peak area of the aliphatic carbon peak and of increasing peak area and sharpness of the aromatic carbon peak. This trend reflects changes in the kerogen structure due to thermal maturation (loss of aliphatic functional groups as liquid and gaseous hydrocarbons) and increasing aromaticity of the kerogen. By 9250 ft (2820 m) present depth, the aliphatic carbon peak is largely gone, and the kerogen has virtually no capacity to generate liquid hydrocarbons and little capacity to generate gaseous hydrocarbons. This also indicates exposure to significantly higher temperatures in the geologic past. The carbon aromaticity (Figure 7B), as measured by NMR, ranges from 63% to about 80% between about 3000 and 8000 ft (900 and 2400 m) present depth and remains fairly constant at 80–83% from 8000 to 11,500 ft (2400 to 3500 m), indicating that the kerogen has released about as much liquid hydrocarbon as it can.

It is interesting that the changes in inorganic formation water geochemistry of the Muddy Sandstone and the changes in organic geochemistry of the Mowry Formation are both characterized by fairly low gradients with depth to about 8000 to 10,000 ft (2400 to 3000 m) present burial, where a rapid steepening of all gradients occurs. This is also the depth at which there is a rapid steepening of the pressure/depth gradient in the Muddy Sandstone (Heasler et al., 1992; Jiao and Surdam, 1992). Thus, the changes in geochemistry and the onset of abnormal pressure appear to be genetically linked.

Further, inasmuch as (1) PI data suggest that the organic material present in the shales below about 9000 ft (2700 m) present burial is dominantly hydrocarbons that have been generated by the kerogen but not expelled, and (2) NMR data suggest that the kerogen below about 9000 ft (2700 m) present burial is no longer capable of generating liquid or gaseous hydrocarbons, it is evident that the gas currently being generated in the Mowry is from the oil-to-gas reaction. This is supported by geochemical modeling of the oil-to-gas and kerogen-to-gas reactions (see Surdam et al., this volume). That this reaction is taking place over the same depth interval as the onset of abnormal pressure (Heasler et al., this volume) suggests that overpressure may be maintained by capillarity resulting from a multiphase flow system (Surdam and Jiao, this volume; Iverson et al., this volume). The establishment of the multiphase flow system (and thus the genesis of pressure compartmentation) likely began in the geological past with the generation and expulsion of liquid hydrocarbons (Jiao and Surdam, this volume). The establishment and maintenance of the multiphase flow system by the generation and expulsion of liquid hydrocarbons, and their subsequent reaction to gas, explain how a pressure compartment can exist for a geologically significant period of time (see Iverson et al., this volume), as is suggested by the geochemical data (XRD, NMR, PI, and R_o).

Geochemical Modeling

Geochemical modeling, using the computer code SOLMINEQ.88 (Kharaka et al., 1988), was performed to study the effect of the mixing of Muddy Sandstone waters on calcite mineral stability. In Muddy Sandstone rocks, calcite is the only important diagenetic phase that may act as a seal (Jiao and Surdam, this volume) and for which the database is sufficient for this type of modeling. Two water samples, representative of fairly fresh and fairly saline Muddy Formation water chemistries, were chosen to illustrate schematically the effect of formation water mixing: one water sample is fairly fresh (2640 ppm TDS) and the other, more saline, is similar to seawater composition (35,400 ppm TDS). Figure 8 shows the results of this modeling. In calcite saturation index [SI = log (ion activity product / equilibrium constant)] vs. temperature space, calcite stability is shown as dashed lines for the dilute and moderately saline waters. A positive SI indicates that the water is oversaturated with respect to the mineral; a negative SI indicates undersaturation. Additionally, the calcite-stability mixing points for a 50/50 mixture of 60°C dilute water and 125°C saline water mixture, as well as for a 30% 60°C dilute water and 70% 125°C saline water mixture, are shown. The modeling shows that mixing the two Muddy Sandstone waters (at least in the ratios shown) will result in calcite (or other carbonate mineral) precipitation. Therefore, if a pressure cell ruptures and the pore fluid in it migrates out and mixes with water of different composition and temperature, calcite may precipitate and restore the seal. Such a situation is illustrated in Figure 9.

In addition to formation water mixing, other processes may cause carbonate mineral precipitation to heal pressure-cell ruptures. As a fluid migrates along a fracture from an abnormally pressured compartment to the normally pressured section, it will encounter a modest drop in temperature and a continuous drop in pressure. A sufficient pressure drop will cause the fluid to degas (exsolve CO_2, CH_4, etc.), which will cause the pH to rise and, consequently, calcite to become thermodynamically stable and precipitate; this may lead to healing of the fracture and restoration of the seal. This process was modeled using SOLMINEQ.88 and a moderately saline (similar to marine water), high-bicarbonate-alkalinity water from the Muddy Sandstone database. Figure 10, a plot of pressure vs. calcite saturation index, shows the path of this process: under initial conditions (water

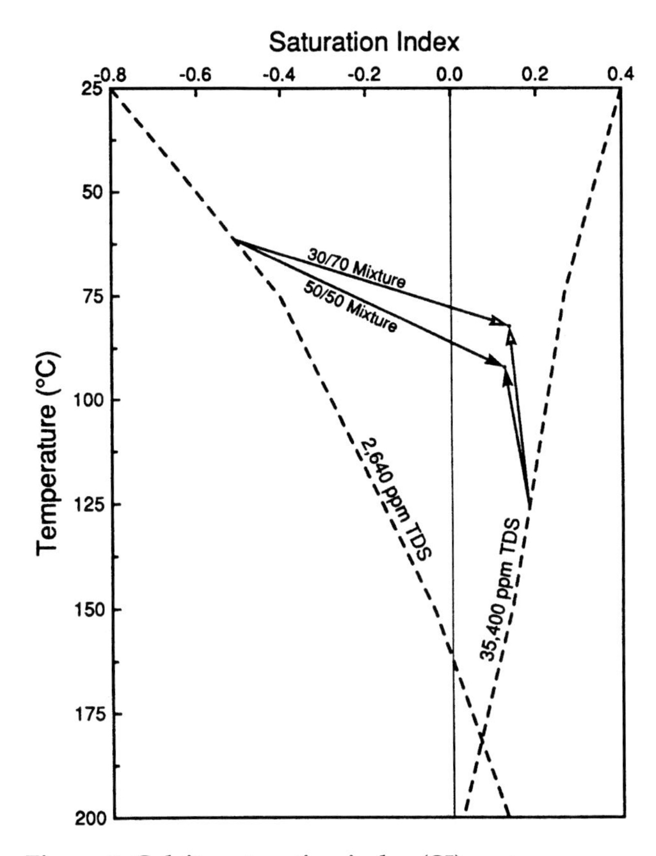

Figure 8. Calcite saturation index (SI) vs. temperature for two samples of Muddy Sandstone water: a relatively dilute sample (2640 ppm TDS) and a more saline sample (35,400 ppm TDS). Also shown are the saturation indices for a mixture of 50% dilute water at 60°C with 50% saline water at 125°C as well as a mixture 30% 60°C dilute water with 70% 125°C saline water.

undersaturated with respect to calcite; 100°C, 400 bars, point A), as the pressure drops by 140 bars (about 2000 psi) to saturation with calcite (point B), and as the solution is cooled to 85°C (point C). As shown by the path from point A to C on Figure 10, release of pressure and subsequent degassing followed by cooling leads to a solution oversaturated with respect to calcite. Figure 11 is a photomicrograph of the Muddy Sandstone near a seal, showing calcite-filled fractures. It is hypothesized that the calcite fracture-filling was precipitated by a process similar to the modeled process described above.

SUMMARY AND CONCLUSIONS

In summary, we have observed inorganic and organic geochemical evidence of a fundamental change in the rock/fluid system of the Muddy and Mowry formations occurring between current burial depths of 8000 and 10,000 ft (2400 and 3000 m). This change is reflected in variations in formation water chemistry (TDS, pH, and alkalinity); changes in organic geochemistry of the source rock related to thermal maturation and oil generation, expulsion, and conversion to gas (R_o, PI, NMR spectra, carbon aromaticity); changes in the level of clay diagenesis in the <2µ minerals associated with the source rock (X-ray diffraction); and changes in the formation pressure (DST measurements, head calculations; Heasler et al., this volume). These observations and data lead us to the following conclusions:

1. The fundamental changes in the rock/fluid system that occur between 8000 and 10,000 ft (2400 and 3000 m) present burial are related to pressure compartmentation and to a rise in geothermal gradient within the compartment (see Hunt, 1979; Mac-Gowan et al., 1990). This conclusion is consistent, at least in the Muddy Sandstone of the Powder River basin, with capillary sealing of the compartment arising from the emplacement of a multiphase fluid-flow system (Iverson et al., this volume). The change from a single-phase fluid-flow system (water) to a multiphase fluid-flow system (water-gas-oil) is associated with the thermal maturation of the source rock, expulsion of oil, and reaction of liquid petroleum to gas.

2. Organic geochemical and geological data and geohistory modeling (Jiao and Surdam, this volume) suggest that this compartmentation occurred prior to

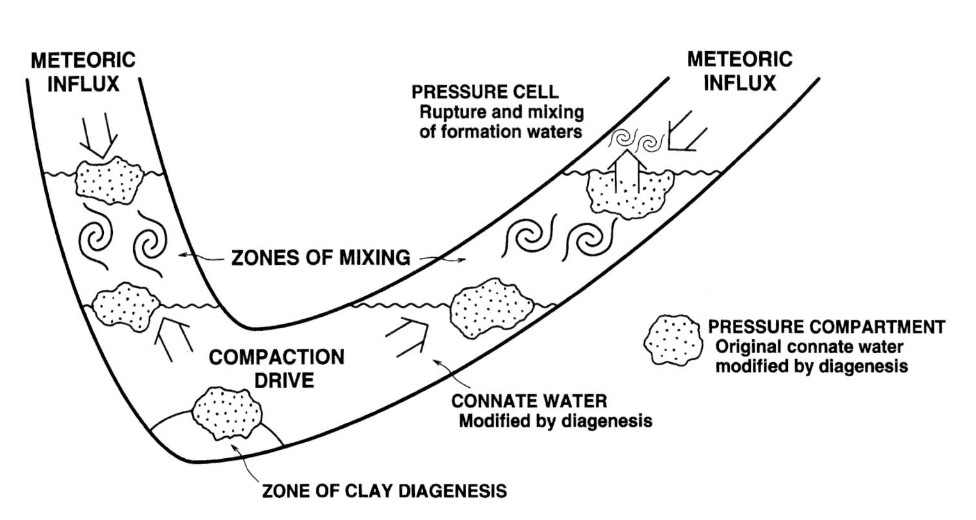

Figure 9. Cartoon showing formation water regimes in a formation containing pressure cells in a foreland basin. This rendering assumes no evaporite minerals.

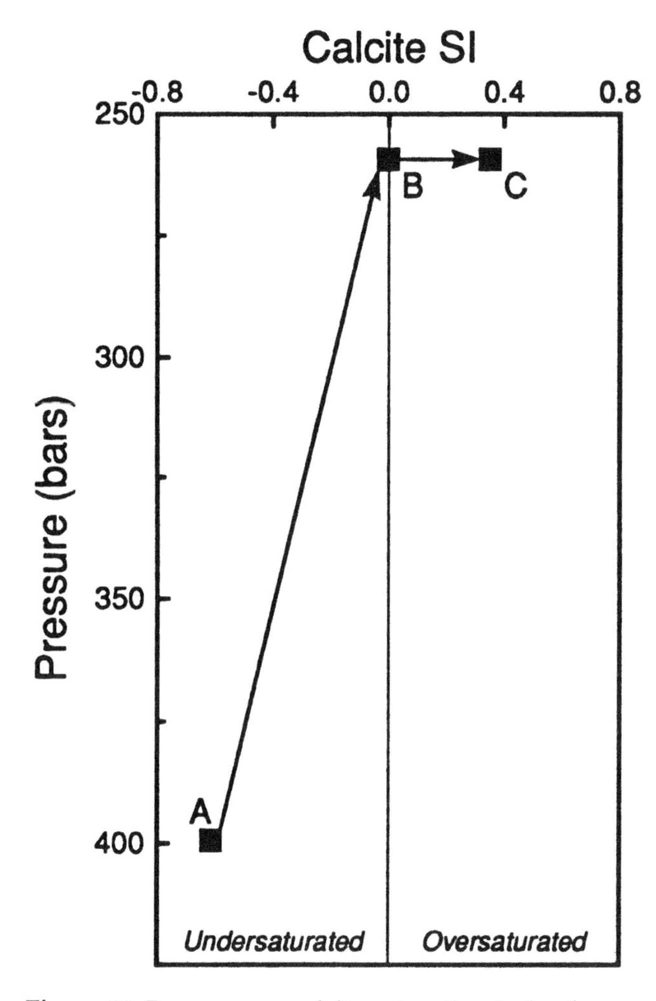

Figure 10. Pressure vs. calcite saturation index for a formation water initially undersaturated with respect to calcite at 100°C, 400 bars (point A), after dropping the pressure and degassing to equilibrium with calcite (point B), and after cooling to 85°C (point C).

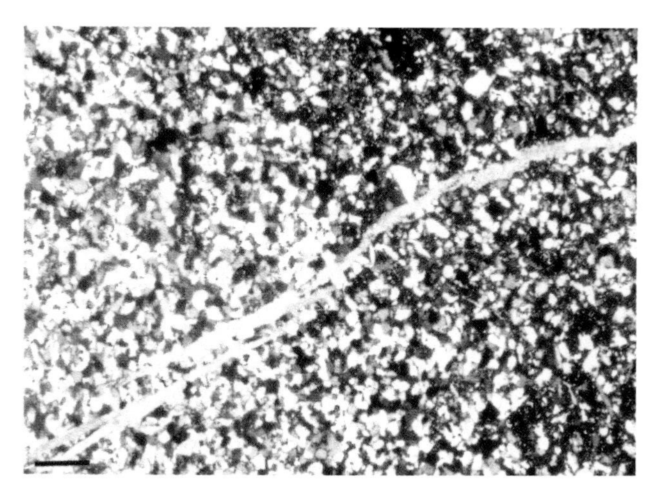

Figure 11. Photomicrograph of Muddy Sandstone from near a seal, showing a calcite-filled fracture. Scale bar = 76 micrometers.

or during maximum burial, perhaps 40–50 Ma (Surdam et al., this volume). Temperatures at maximum burial must have been significantly higher than present-day temperatures, according to the trends observed in the NMR spectra, production index, and vitrinite reflectance.

3. Compartmentation appears to have resulted from organic maturation, the generation of liquid hydrocarbons, and their conversion to gas, phenomena that occurred at or before maximum burial. For the kinetically controlled diagenetic reactions (clay transformation, kerogen maturation, vitrinite maturation) that we have documented to have occurred, the currently overpressured regions in the Muddy Sandstone of the Powder River basin must have been compartmentized for a geologically significant period of time. This distinguishes a type of compartment that is different from one whose overpressuring has resulted from hydrodynamic or compaction disequilibria, phenomena that have been shown to be short-lived (Bethke et al., 1988; Iverson et al., this volume).

EXPERIMENTAL CONDITIONS

Solid-state NMR measurements were made using a Chemagnetics CMX 100/200 solids NMR spectrometer. Carbon aromaticity measurements were made at a frequency of 25 MHz using the technique of cross polarization (CP), with magic-angle spinning (MAS) and high-power decoupling. Because of the low levels of organic carbon in the shales (1–4%), a large-volume sample spinner (2.5 ml) was used (Zhang and Maciel, 1989; Miknis et al., 1993). Even so, 64,800 transients were recorded to obtain an NMR spectrum with a reasonable signal-to-noise ratio. Sample spinning rates were between 3.4 and 3.8 kHz. Other NMR parameters were a 90° pulse width of 6.4 μsec, a contact time of 1 msec, and a pulse delay of 1 sec.

ACKNOWLEDGMENTS

This material covers research performed through November 1991. The authors wish to thank J. Berton Fisher of Amoco Production, Tulsa, Oklahoma, for providing the formation water chemistry data set for the Powder River basin. Dr. Steven W. Boese, Department of Geology and Geophysics, University of Wyoming, performed the anhydrous pyrolysis and TOC analyses. Discussions with Dr. Thomas L. Dunn, Randi S. Martinsen, and Dr. William P. Iverson (all of the University of Wyoming), and Dr. Jean Whelan (Woods Hole Oceanographic Institute) provided useful input to this study. The original manuscript was reviewed by David Copeland (University of Wyoming), who made numerous helpful suggestions. The solid-state NMR analyses were provided, in part, by DOE University Research Instrumentation grant No. DE-FG05-89ER75506. Such support does not, however, constitute endorsement by the DOE of the views expressed in this article. This study was funded by the Gas Research Institute under contract number 5089-260-1894.

REFERENCES CITED

Bethke, C. M., W. J. Harrison, C. Upson, and S. P. Altaner, 1988, Super-computer analysis of sedimentary basins: Science, v. 239, p. 233–237.

Elliot, C. E., J. L. Aronson, G. Matisoft, and D. L. Gautier, 1991, Kinetics of the smectite–illite transformation in the Denver basin: clay mineral, K-Ar data, and mathematical model results: American Association of Petroleum Geologists Bulletin, v. 25, p. 436–462.

Hunt, J. M., 1979, Petroleum Geochemistry and Geology: San Francisco, W. H. Freeman, 617 p.

Kharaka, Y. K., W. D. Gunter, P. K. Aggarwal, E. H. Perkins, and J. D. DeBraal, 1988, SOLMINEQ.88: A computer program for geochemical modeling of water-rock interactions: U.S. Geological Survey, Water Resources Investigations Report 88-4227, 420 p.

MacGowan, D. B., and R. C. Surdam, 1990, Carboxylic acid anions in formation waters, San Joaquin Basin and Louisiana Gulf Coast, U.S.A.: implications for clastic diagenesis: Applied Geochemistry, v. 5, p. 687–701.

MacGowan, D. B., R. C. Surdam, and R. E. Ewing, 1990, The effect of carboxylic acid anions on the stability of framework mineral grains in petroleum reservoirs: SPE #17802, Society of Petroleum Engineers Formation Evaluation, June 1990, p. 161–166.

MacGowan, D. B., Z. S. Jiao, R. C. Surdam, and F. P. Miknis, 1993, Normally vs. abnormally pressured sandstones in the Powder River basin, Wyoming: a comparative study of the Muddy Sandstone and the Minnelusa Formation, *in* S. Andrews and B. Strook, eds., Wyoming Geological Association 50th Anniversary Guidebook, p. 281–295.

Martinsen, R. S., 1992, Stratigraphic controls on the development and distribution of pressure compartments in the Powder River basin, Wyoming: Final Report (Draft) on Contract No. 5089-260-1894, Multidisciplinary Analysis of the Pressure Chambers in the Powder River Basin of Wyoming and Montana, Gas Research Institute, Chicago, p. 64–91.

Middleton, M. F., 1983, Tectonic history from vitrinite reflectance: Geophysical Journal of the Royal Astronomical Society, v. 68, p. 121–132.

Miknis, F. P., Z. S. Jiao, D. B. MacGowan, and R. C. Surdam, 1993, Solid state NMR characterization of Mowry Formation shales: Organic Geochemistry, v. 20, p. 339–347.

Price, L. C., 1983, Geologic time as a parameter in organic metamorphism and vitrinite reflectance as an absolute paleogeothermometer: Journal of Petroleum Geology, v. 6, p. 5–38.

Pytte, A. M., and R. C. Reynolds, Jr., 1989, The thermal transformation of smectite to illite, *in* N. Naeser and T. McCulloch, eds., Thermal History of Sedimentary Basins: New York, Springer-Verlag, p. 133–140.

Surdam, R. C., and L. J. Crossey, 1985, Organic-inorganic reactions during progressive burial: key to porosity/permeability enhancement and/or preservation: Philosophical Transactions of the Royal Society of London, v. 315, ser. A, p. 172–232.

Surdam, R. C., T. L. Dunn, D. B. MacGowan, and H. P. Heasler, 1989a, Conceptual models for the prediction of porosity evolution, with an example from the Frontier Sandstone, Bighorn Basin, Wyoming, *in* E. Coalson, S. Kaplan, C. Keighin, C. Oglesby, and J. Robinson, eds., Petrogenesis and Petrophysics of Selected Sandstone Reservoirs of the Rocky Mountain Region: Rocky Mountain Association of Geologists Guidebook, p. 7–28.

Surdam, R. C., D. B. MacGowan, and T. L. Dunn, 1989b, Diagenetic pathways of sandstone and shale sequences: Contributions to Geology, University of Wyoming, v. 27, p. 21–32.

Tissot, B. P., and D. H. Welte, 1984, Petroleum Formation and Occurrence, 2nd edition: Berlin, Springer-Verlag, 699 p.

Vandergrift, G. F., R. E. Winans, R. G. Scott, and E. P. Horwitz, 1980, Quantitative study of the carboxylic acids in the Green River oil shale bitumen: Fuel, v. 59, p. 627–633.

Zhang, M., and G. E. Maciel, 1989, Large-volume MAS system for improved signal-to-noise ratio: Journal of Magnetic Resonance, v. 85, p. 156–161.

Chapter 22

A Sonic Log Study of Abnormally Pressured Zones in the Powder River Basin of Wyoming

Debra Maucione
Vladimir Serebryakov
Paul Valasek
Yue Wang
Scott Smithson
University of Wyoming
Laramie, Wyoming, U.S.A.

ABSTRACT

Most hydrocarbon production from the Powder River basin in northeastern Wyoming is from an abnormally pressured Cretaceous section. The preliminary identification and delineation of abnormally pressured zones by surface seismic methods would greatly enhance hydrocarbon recovery. The present velocity study shows the usefulness of a detailed sonic log analysis to find abnormally pressured zones. Because the conclusion of the log response study positively indicated the presence of abnormal pressure, surface seismic data will also show these anomalies. Velocity profiles, pressure profiles, and abnormal pressure-gradient calculations all indicate that the Cretaceous shales are overpressured throughout the basin, whereas the Cretaceous sandstones are present in overpressured, underpressured, and normally pressured regimes. Pressure compartment boundaries appear to be stratigraphically controlled. Three profiles in different parts of the basin—T46N, R69–76W (Hilight field area), T51N, R69–77W (Amos Draw–Kitty field area), and T41N, R69–76W (north of the Powell field area)—delineate the top of overpressure near the Parkman Sandstone–Steele Shale contact and the bottom seal along the Fuson Shale.

INTRODUCTION

Surface mapping of pressure compartments using seismic reflection methods can be important for hydrocarbon exploration. We have studied well logs (Wang, 1992) and drill-stem test (DST) data in order to develop a velocity model that allows the prediction of seismic response in abnormally pressured regions. This study shows that well-log data, especially sonic logs, can be used to identify zones of abnormal pressure from velocity anomalies in the Powder River basin.

Much of the eastern portion of Wyoming is covered by the Powder River basin (see Heasler et al., this volume). The Amos Draw and Kitty fields in the northern Powder River basin and the Powell field in the southern Powder River basin (Figure 1) are two areas with abnormal pressures in the Cretaceous section (Figure 2) (Heasler et al., this volume). Abnormal pressure is defined as pressure significantly above or below that established by the hydrostatic gradient. Potentiometric surface maps (Heasler et al., this volume) show that the pressure in the Early Cretaceous

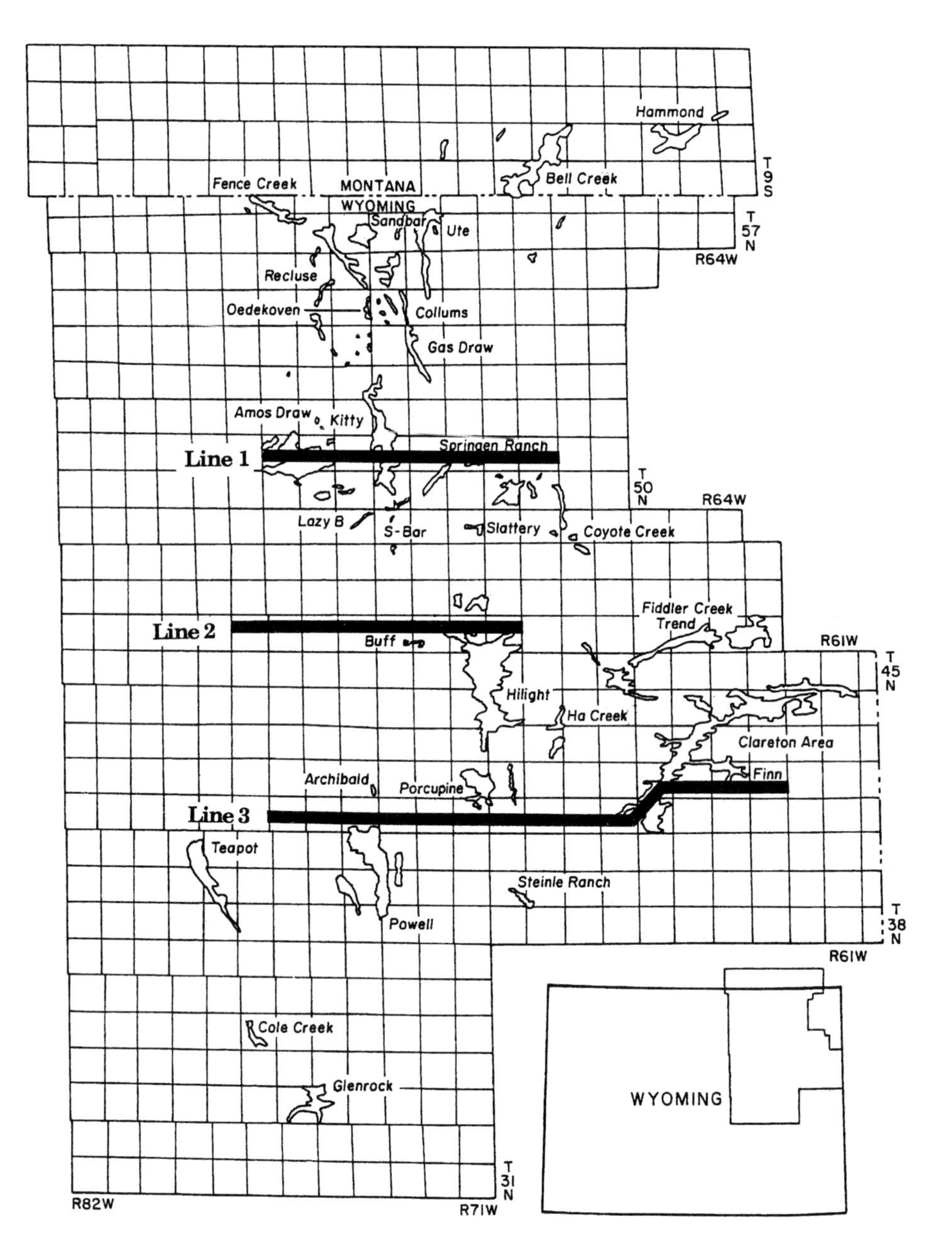

Figure 1. Location map of the study areas, lines of section for the velocity and pressure profiles, and major producing fields in the Powder River basin.

Muddy Sandstone (see stratigraphic column in Martinsen, this volume) has a head 3000 ft (900 m) greater in the Amos Draw and Kitty fields than in surrounding areas (Figure 3; Surdam et al., this volume). In the south, the head in the Muddy Sandstone is 12,000 ft (3700 m), 9000 ft (2700 m) greater than normal. The head at the level of the older Dakota Formation in the Powell field is 6000 ft (1800 m), 3000 ft (900 m) higher than normal. In the mid-Cretaceous Frontier Formation, the head is 8000 ft (2400 m), 5000 ft (1500 m) above normal. Overpressure diminishes and finally disappears in successively higher stratigraphic intervals, as is apparent from potentiometric surface maps of the Late Cretaceous Mesaverde and Sussex formations. At these levels, the area is predominantly normally pressured or underpressured (Surdam et al., this volume). The objective of this study is to determine the velocity characteristics of abnormally pressured zones using sonic (velocity) logs. This will form a basis for subsequent research into the delineation of anomalous pressures using seismic methods. In this chapter, pore pressure will represent the pressure in the shales calculated using estimation techniques (Dobrynin and Serebryakov, 1989), and formation pressure is that measured during DSTs or formation tests (the sandstone pressures).

LOG RESPONSE

In normally pressured zones, the sonic, resistivity, and density log responses form straight lines when plotted semilogarithmically (the vertical axis is linear

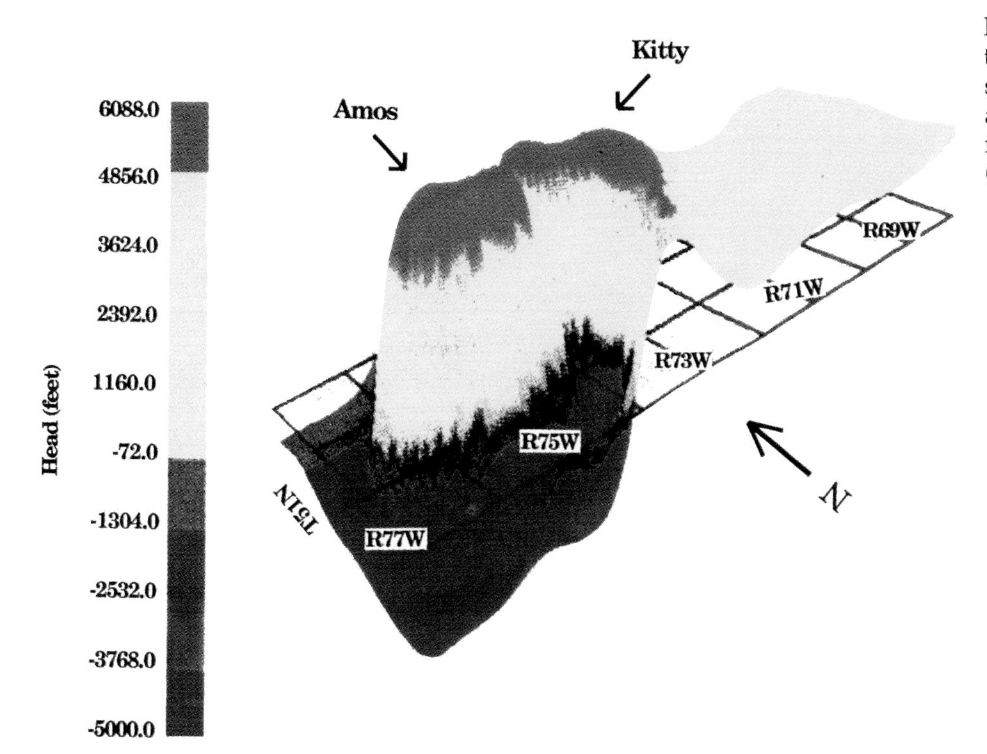

Figure 2. 3-D perspective of the Muddy potentiometric surface in the Amos Draw and Kitty fields. Zero head = normal hydrostatic pressure (overburden).

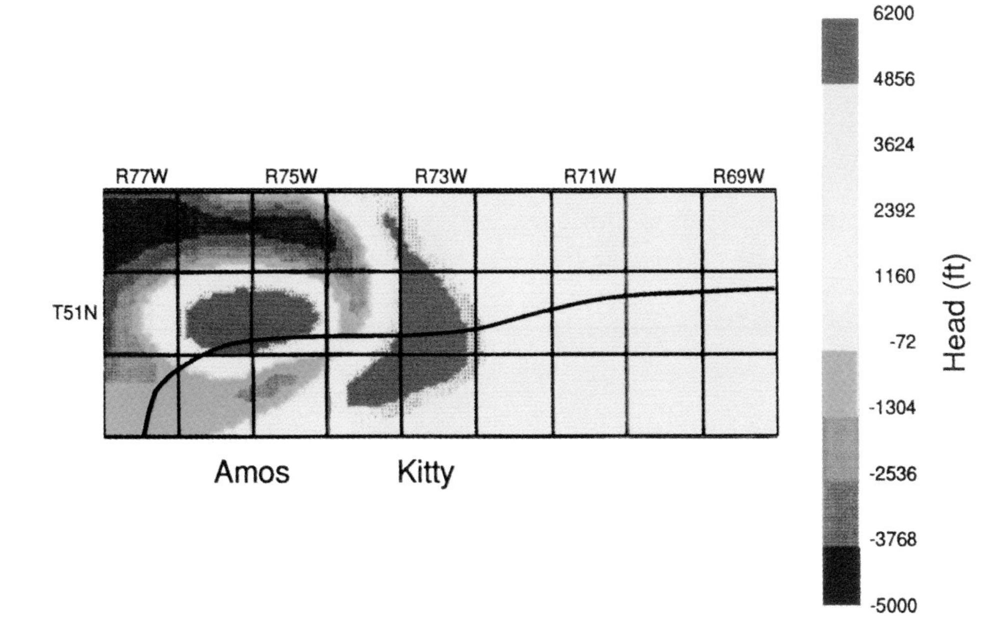

Figure 3. Muddy potentiometric surface in the Amos Draw and Kitty fields. The line indicates the location of the velocity profiles shown in Figures 7A, 7B, 10A, and 10B.

depth and the horizontal axis is log x, where x is resistivity, sonic, density, gamma ray, etc.). However, in zones of abnormally high pressure, the magnitudes of the log responses decrease significantly. Therefore, the abnormal pressures can be detected on wireline logs (MacGregor, 1965). Figure 4 gives an example of well-log response to an abnormally pressured zone in a well located in the Powell field of the southern Powder River basin. In this study we used well logs edited to include shale intervals only. When using the normal compaction trend, the trend should be determined for shales by looking only at shale intervals, or for sandstones by looking only at sandstone intervals. The resistivity, sonic, and density logs show significant changes in trend below 8000 ft (2400 m). This change in trend represents the top of the zone of

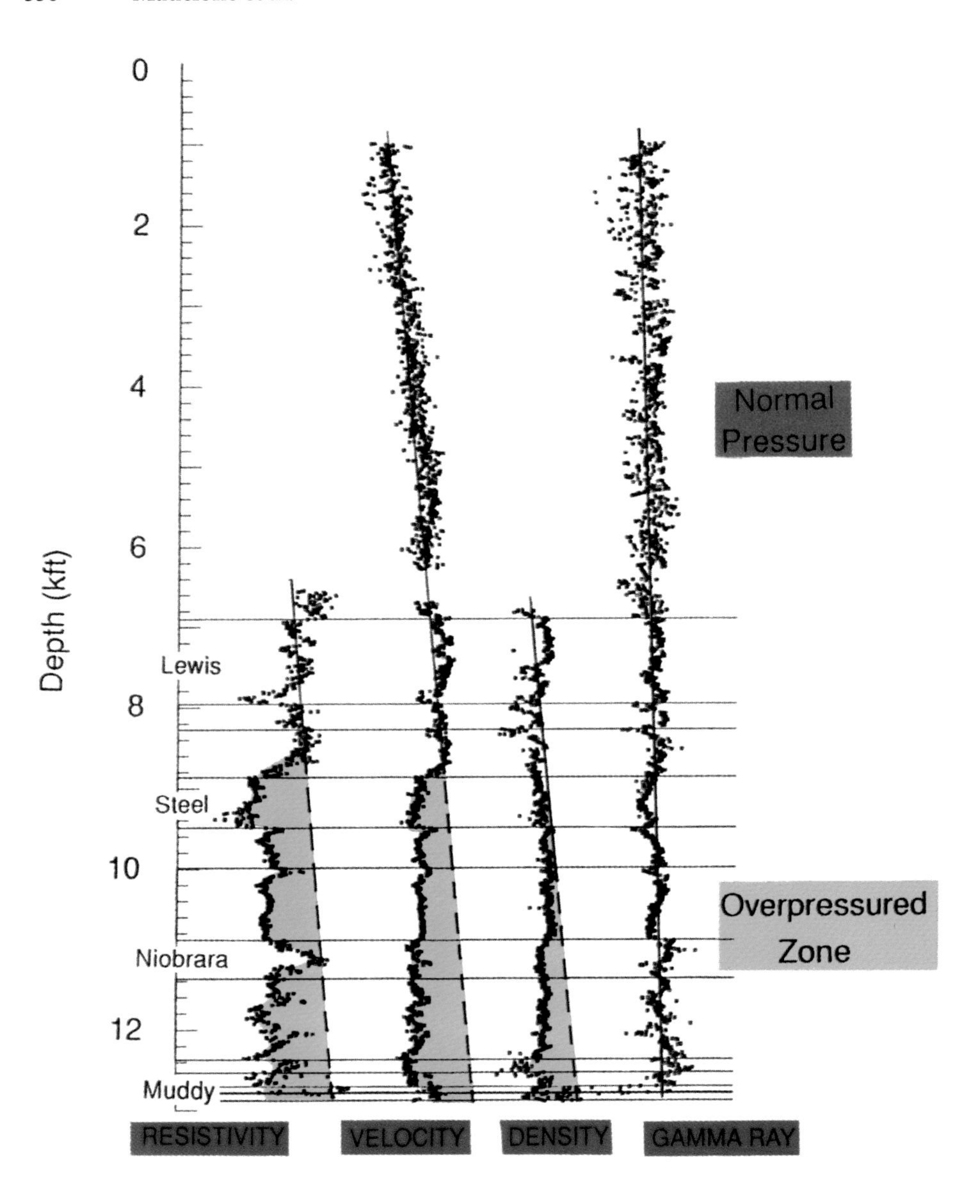

Figure 4. Well-log responses in an overpressured zone, Powell field, sec. 16, T39N, R73W. Note the decreased log response in the overpressured zone at approximately 9000 ft.

undercompaction. From DST and formation tests, it has been shown that sandstones in undercompacted shale sections are typically overpressured. It follows that the decrease in log trend indicates abnormally high pressure existing below 8000 ft (2400 m). This was confirmed by drill-stem test (DST) measurements and by estimating the pressure using the normal compaction trend method (Dobrynin and Serebryakov, 1989). The change in log response is due to a change in physical rock properties within the undercompacted, overpressured zone. It should be noted that log responses are not the only indicators of a significant change in rock properties at the transition. It has been proved repeatedly by drill-stem tests that the areas of undercompaction are the zones of abnormal pressure. There are other changes occurring at the transition between normal and overpressured regimes. A change in the slope of vitrinite reflectance vs. depth occurs at this transition (Jiao and Surdam, this volume; Hunt, 1979). The water above the abnormally pressured zone is typically fresh, whereas within the zone produced water is dilute marine (MacGowan et al., this volume). The production index also increases significantly below the transition zone (MacGowan et al., this volume). At the transition from normal pressure, the clay diagenesis is characterized by significant change. The smectite-to-illite reaction proceeds to 80% illite and the structure to ordered at the transition (Jiao and Surdam, this volume).

The method of normal compaction trend estimation (Dobrynin and Serebryakov, 1989) was used to determine zones of abnormally high pressure in the Powder River basin from well logs. When the system responds to normal hydrostatic pressure, the normal compaction trend is affected by depth of burial and variation in rock properties such as porosity, density, or resistivity. These properties can be determined using well-log data, drilling data, core-sample data, etc. In particular, the physical properties of pure shale depend primarily on the degree of compaction. This relationship is observed in nature as the exponential relationship between the porosity, density, or resistivity of normally compacted rocks and depth of burial. On semilogarithmic plots, these exponential dependencies plot as straight lines. In this study, numerous well logs were acquired and edited for clean shale intervals. Clean shale intervals were determined as those sections with low resistivity, positive SP, and high gamma-ray response. The edited logs were used to determine the normal compaction trend. A deviation from this straight line indicates the top of the zone of abnormal pore pressure. For estimating values of abnormal pore pressure P_a, equation 1 was used:

$$P_a = P_n \pm \frac{g(\rho_r - \rho_w)}{\log \frac{x_2}{x_1} \pm \frac{\alpha(x)}{2.3}\tau} \log \frac{x_n}{x_a} \quad (1)$$

where P_n is normal hydrostatic pressure; g is the acceleration of gravity; ρ_r is the average density of the rocks; ρ_w is the average density of water; $\alpha(x)$ is the temperature coefficient for the geophysical property x of interest (± depends on the property); τ is the geothermal gradient for the depth interval Δh; x_1, x_2 are the values of geophysical properties such as density, resistivity, or travel time at depths h_1 and h_2, respectively; and x_n, x_a are values of the geophysical property within the normal compaction trend (x_n) and within the zone of abnormal pore pressure (x_a).

It is possible to estimate abnormally high and low pressure with equation 1. In the case of abnormally high pressure, the second term in equation 1 is positive; it is negative for abnormally low pressure. Another method, that of compression curves (Dobrynin and Serebryakov, 1989), has also been used to estimate abnormal pressures (equations 2–4):

$$P_a = \sigma - \frac{\log x \pm \frac{\alpha(x)\tau(h - h_1)}{2.3} - bx}{k_x} \quad (2)$$

$$b_x = \frac{\log x_1(\sigma_2 - p_2)}{(\sigma_2 - p_2)(\sigma_1 - p_1)} - \frac{\left[\log x_2 \pm \frac{\alpha(x)\tau(h_2 - h_1)}{2.3}\right](\sigma_1 - p_1)}{(\sigma_2 - p_2) - (\sigma_1 - p_1)} \quad (3)$$

$$k_x = \frac{\log \frac{x_2}{x_1} \pm \frac{\alpha(x)\tau(h_2 - h_1)}{2.3}}{(\sigma_2 - p_2) - (\sigma_1 - p_1)} \quad (4)$$

where σ is the overburden pressure at the depth of interest h; σ_2, σ_1 are the overburden pressures at depths h_2, h_1; x is the value of the geophysical property of interest; x_2, x_1 are as in equation 1; p_2, p_1 are the values of formation (pore) pressure at depths h_2, h_1; $\alpha(x)$ is the temperature coefficient for the geophysical property of interest x (a correction factor to account for the effect of temperature with depth on rock properties such as resistivity or transit time; see Dobrynin and Serebryakov, 1989); τ is the mean geothermal gradient over the depth range; and b_x, k_x are calculated parameters of the compression curve. This method is the more efficacious because it is based on the dependence of physical properties on effective stress. This method is also advantageous because it enables the estimation of the thickness of eroded deposits (see Maucione et al., 1992a). All data will plot as a straight line for both normal and abnormal pressure values when plotted on a semilogarithmic graph with the x axis representing log x, as in the normal compaction trend plots, and the y axis representing linear effective stress. If the data do not plot in a straight line, the presence of unconformities (discordant bedding) or a fault is indicated (see Figure 5).

Calculations of this type were performed for several wells in the Powell area (Tables 1 and 2). The results are similar, whether the calculation is made using sonic logs or resistivity logs (see Figure 6). Resistivity, density, velocity, and gamma-ray well-logging data were used (Figure 4) for estimating zones of abnormally high pressure in the Powder River basin. Zones of abnormally high pressure could be estimated with all of these methods, except the gamma-ray method. The predicted decrease in gamma-ray log response does not occur when the log enters the overpressured zone, as has been observed in the Gulf Coast (Martinez et al., 1991). This could be due to the fact that the Gulf Coast has experienced hydrologic or compaction disequilibria (see Fertl, 1976), whereas different mechanisms of abnormal pressurization may be operating in the Powder River basin due to multiphase flow (Surdam et al., this volume; Iverson et al., this volume). The abnormal pressure coefficient $K = P_a/P_H$, where P_a is the value of abnormal pressure and P_H is the value of normal hydrostatic pressure. All calculated abnormal pressure coefficients (K) plot below the maximum lithostatic pressure when a calculation error is considered (±7% as documented in Dobrynin and Serebryakov, 1989). The maximum lithostatic pressure value corresponding to abnormal pressure coefficient K is 2.29. This is determined using an average pressure gradient of 0.433 psi/ft (Surdam et al., this volume).

Estimates of the abnormally high pressure in the Powder River basin show that the values of abnormal pore pressure are greater than the values of abnormal formation pressure (Figure 6). For example, abnormal pore pressure was estimated in the Powell Coquina ETAL #1 well at a depth of 11,461 ft (3493 m). The value of this pressure is 71.9 MPa (coefficient of abnormal pressure K = 2.06). The formation pressure

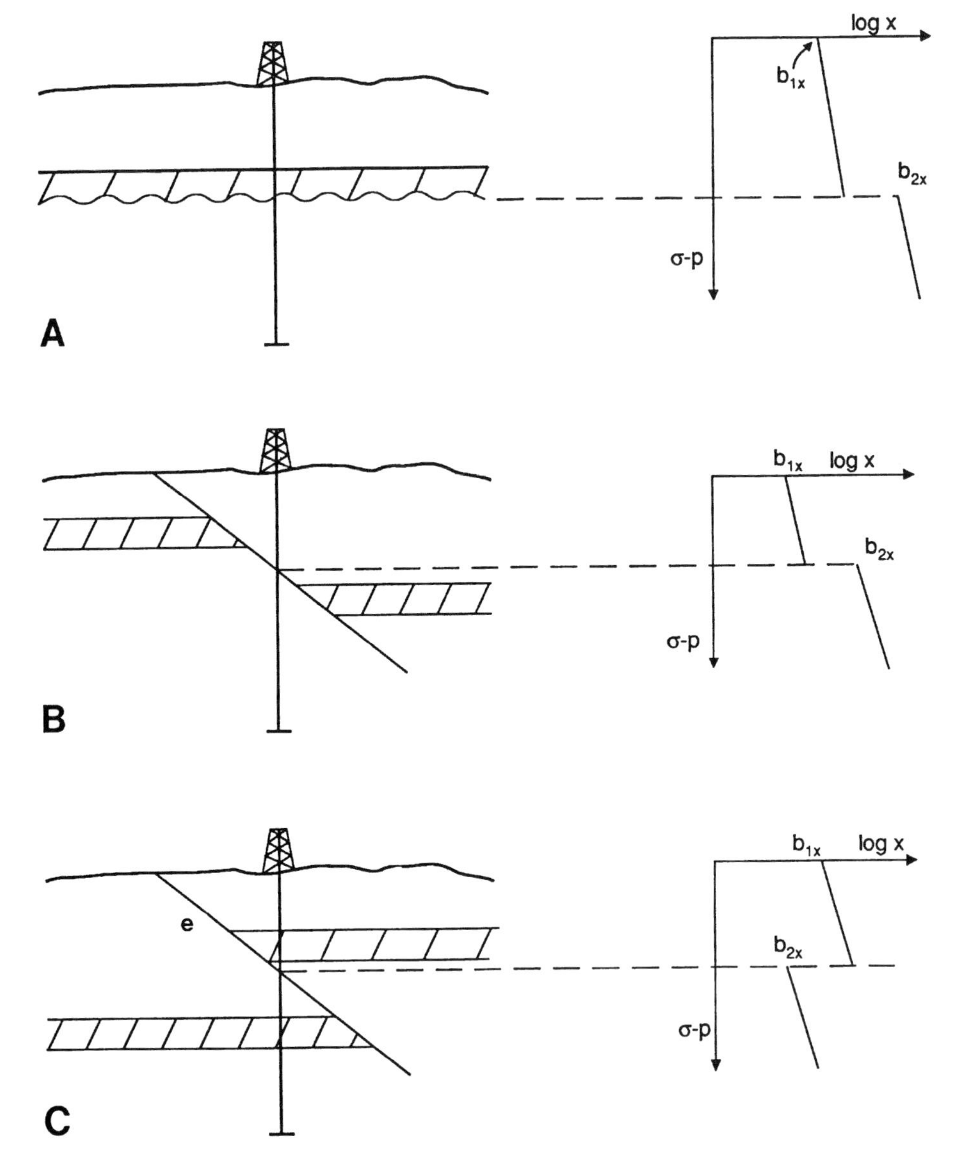

Figure 5. The effects of a discontinuity (stratigraphic or structural) on a compressional curve plot. When using the method of compressional curves to estimate pressures, all log values should plot along a straight line. If they do not, the plot may show (A) log response indicating an unconformity; (B) log response produced by crossing a normal fault; or (C) log response produced by crossing a reverse fault. See Dobrynin and Serebryakov (1989).

test at a depth of 12,205 ft (3716 m) yielded a value of 53.9 MPa (K = 1.46). The same results were obtained in several other wells in this region (Table 1). One should note two items when looking at Figure 6. First, there appear to be multiple trends in the pressure gradients of the Powell area. Second, there is not a constant difference between the pore pressure and formation pressure in the individual wells.

In order to use seismic methods to predict seismic response to the abnormal pressure zones in the study areas, an accurate velocity model must be developed. Figure 3 shows the location of a velocity profile that was obtained from sonic logs. The profile consists of 21 wells, 3 of which contain data from only several hundred feet above and below the Muddy Sandstone. Figure 7A shows a plot of 17 velocity logs along the cross section. On the eastern side of the section, the velocity increases linearly, indicating that the pressure is normal. Toward the west, a large pressure chamber is delineated by gradual deviations or decreases in velocity from the normal trend. The abnormal pressure chamber starts between Ranges 71 and 72 West and extends to the western edge of the section. The upper boundary of the pressure chamber appears to be subhorizontal at a depth of approximately 7500 ft. The lower boundary cannot be delin-

Table 1. Normal compaction trend.

Name/ Number	Depth (ft)	Depth (m)	Pressure (psi)	Pressure (MPa)	K= P_a/P_H	Depth (ft)	Depth (m)	Pressure (psi)	Pressure (MPa)	K
	Estimations Using Resistivity Logs							Tests		
Powell Coquina ETAL #1	11,460	3493	10,163	71.9	2.06	12,205	3716	7664	53.9	1.46
						12,440	3784.9	7749	54.9	1.44
Powell Spearhead Ranch #2	13,038	3974	12,434	87.6	2.20	12,002	3658	6317	44.5	1.21
						12,738	3883	7336	51.7	1.33
Powell Spearhead Ranch #18	12,116	3693	10,447	73.6	1.99	12,375	3768	7474	52.7	1.45
						12,425	3673	7774	54.8	1.45
Powell Hartley Federal #1	12,923	3939	11,696	82.4	2.10	12,337	3760	8343	58.8	1.56
Powell Anderson Federal #1	12,077	3681	8815	67.2	1.82	11,675	3559	4376	30.8	0.86
Powell M&M #33	10,768	3282	8545	60.2	1.84	11,494	3656	6452	45.5	1.25
						12,598	3540	8368	59.0	1.53
Powell MDU #1	12,480	3804	11,724	82.6	2.17	12,698	3870	6829	48.1	1.24
	Estimations Using Sonic Logs							Tests		
Powell MDU #1	12,480	3804	10,956	77.0	2.02	12,698	3870	6829	48.1	1.24
Powell Spearhead Ranch #2	13,022	3969	9309	65.5	1.65	12,738	3883	7336	51.7	1.33
Powell Spearhead Ranch #5	12,500	3810	10,985	77.2	2.03	13,290	4051	9673	68.1	1.68
Powell Spearhead Ranch #18	12,201	3719	10,090	70.9	1.91	12,375	3768	7474	52.7	1.45
Irwin #44-8	9501	2896	6868	48.3	1.67	9790	2979	4852	34.2	1.15
Rudesill-Federal A #1	9984	3043	7053	49.6	1.63	10,134	3084	4907	34.6	1.12
Diamond Back Unit #1	11,001	3353	9777	68.7	2.05	11,376	3462	5934	41.8	1.20
Pine Tree Unit #2-68	12,979	3956	11,893	83.6	2.11	13,294	4052	7688	54.2	1.34
Harris Marion #1	12,979	3956	12,248	86.1	2.18	13,193	4021	8900	62.7	1.56
Moore 12A-13 #1	12,979	3956	12,248	86.1	2.18	13,156	4010	9098	64.1	1.60
Cross Roads II Unit #1	12,979	3956	12,024	84.5	2.14	10,750	3277	5034	35.5	1.08
Taylor Unit #8	11,981	3652	9296	65.4	1.79	12,524	3817	6220	43.8	1.15
Union-Federal #1	13,501	4115	10,743	75.5	1.83	12,583	3835	7232	50.9	1.33

eated because it was not crossed by the well logs. The change in velocity below the Muddy Sandstone may be explained by differential erosion across an unconformity (see Martinsen, this volume).

To further examine the velocity changes, vertical velocity profiles were generated by gridding and contouring the velocity information from 19 logs along the profile. This process makes it easier to relate the sonic log velocities to the interval velocities that can be obtained from analysis of seismic data. Figure 7B shows the gridded velocity profile in the vicinity of the Amos Draw (R75W–R76W) and Kitty (R73W–R74W) fields. The velocities shown in red, in the shales surrounding the Muddy Sandstone, are lower than the velocities in the Muddy Sandstone: This suggests that the pressure in the shales is abnormally high.

In order to remove the effects of normal increase in velocity with depth due to compaction, a decompaction correction was applied to the velocity profiles. Figure 8A shows a typical velocity log in an abnormal pressure regime. In the shallow section, a normal compaction trend corresponds to a linear increase in velocity. The velocity increase is represented by $V = V_0 \exp(cz)$, where V_0 is the velocity at the surface, z is depth, and c is a constant called the compaction fac-

Table 2. Compressional curves.

Name	Depth ft (m)	K_x	B_x	Pressure (psi)	Pressure (MPa)	K	Depth ft (m)	Pressure (psi)	Pressure (MPa)	K
	Estimations Using Sonic Logs						Tests			
S08 T41N R69W	9501 (2896)	0.007	3.48	6868	48.3	1.67	9790 (2984)	4852	34.2	1.15
S11 T41N R70W	9984 (3043)	0.0059	3.5	7053	49.6	1.63	10,134 (3089)	4907	34.6	1.12
S07 T41N R71W	10,984 (3348)	0.00188	3.6	9778	68.7	2.05	11,376 (3467)	5934	41.8	1.20
S16 T41N R71W	9501 (2896)	0.0018	3.6	8316	58.4	2.05				
S06 T41N R72W	11,483 (3500)	0.0062	3.5	8912	62.7	1.79				
S14 T41N R73W	11,982 (3652)	0.0026	3.54	8671	61.0	1.67				
S16 T41N R73W	11,982 (3652)	0.0022	3.59	10,261	72.1	1.97				
S18 T41N R73W	12,500 (3810)	0.00175	3.61	11,467	80.6	2.12	11,494 (3503)	6452	45.5	1.30
							12,598 (3840)	8368	59.0	1.54
S13 T41N R75W	12,979 (3956)	0.0035	3.61	12,248	86.1	2.18	13,193 (4021)	8900	62.7	1.56
S13 T41N R75W	12,979 (3956)	0.0035	3.61	12,248	86.1	2.18	13,156 (4010)	9098	64.1	1.60
S02 T41N R75W	12,979 (3956)	0.00385	3.60	11,893	83.6	2.11	13,294 (4052)	7688	54.2	1.34
S14 T41N R75W	12,979 (3956)	0.0035	3.61	12,021	84.5	2.14	10,750 (3277)	5034	35.5	1.08
S20 T41N R76W	11,982 (3652)	0.0027	3.55	9296	65.4	1.79				
S23 T41N R76W	13,501 (4115)	0.0032	3.54	1074	75.5	1.83				
S18 T39N R74W	13,022 (3969)	0.003	3.6	11,382	80.0	2.02				
S22 T39N R74W	12,201 (3719)	0.003	3.62	10,090	70.9	1.91				
S11 T39N R75W	13,501 (4115)	0.0064	3.4	10,985	77.2	1.88				
S36 T40N R71W	11,001 (3353)	0.0077	3.43	8760	57.4	1.71				
S16 T39N R73W	12,480 (3804)	0.0073	3.53	10,956	77.0	2.02				

tor. From an analysis of each sonic log along the profile (see Figure 9), an average compaction factor of 0.000045 was measured. A decompacted velocity is obtained by multiplying the velocity V by the factor $\exp(-cz)$. In a normally pressured zone, this results in a constant velocity of V_0. Any deviation from this background decompacted velocity may be related to abnormal pressure. For example, the velocity in the shallow part of Figure 8B is decompacted, and the blue area in the deep part of the plot indicates a significant velocity decrease caused by abnormally high pressures.

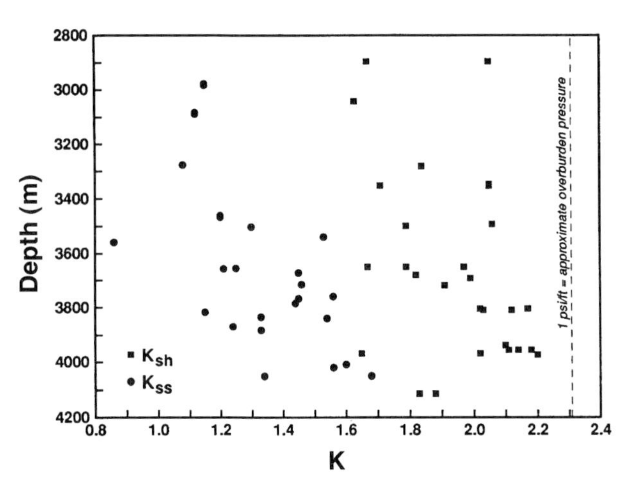

Figure 6. The values of abnormal pressure in Tables 1 and 2 at given depths. $K = P_a/P_H$, where P_a is the abnormal pressure, P_H is the hydrostatic pressure, and K is the calculated abnormal pressure coefficient. K_{sh} are the calculated values of abnormal pore pressure (pressure in the shales) as determined from both resistivity and sonic logs using both the normal compaction trend and the compressional curves methods. K_{ss} are the values of formation pressure obtained in sandstones from DSTs. The line represents approximate overburden pressure. Error in calculations ± 7%. Plotted values have gas correction applied. From Maucione et al. (1992b).

Figure 10A shows the velocity profile in Figure 7B after a decompaction correction was applied. The velocity ranges from 5000 to 9400 ft/sec. The red areas in the vicinity of the Amos Draw and Kitty fields around the Muddy Sandstone show the velocity decrease more clearly than does Figure 7B. The red area delineates an abnormally high pressured region similar to that in Figure 7B. Figure 10B is an abnormal pressure profile of the same line. However, this profile is calculated from the decompacted sonic profile using the normal compaction trend method. It should be taken into account that there has been differential uplift and erosion in the basin. Therefore, as the profile nears the basin flank, where there has been greater uplift and erosion, the abnormal pressures are found nearer the surface (see Figure 7A).

Using the same methods, two more profiles were created. One extends across the Hilight field area, T46N, R69–76W (Figures 11–13); the other is in the southern part of the basin, extending across the northern portion of the Powell field, T41N, R69–76W (Figures 14–16). Velocity profiles were made, a decompaction correction was applied, and pressure sections were calculated using the normal compaction trend method. These three different sections from different parts of the basin illustrate that abnormally pressured zones can be delineated by constructing detailed velocity profiles. Compartmentation is highlighted in the abnormal-pressure profiles. This repre-

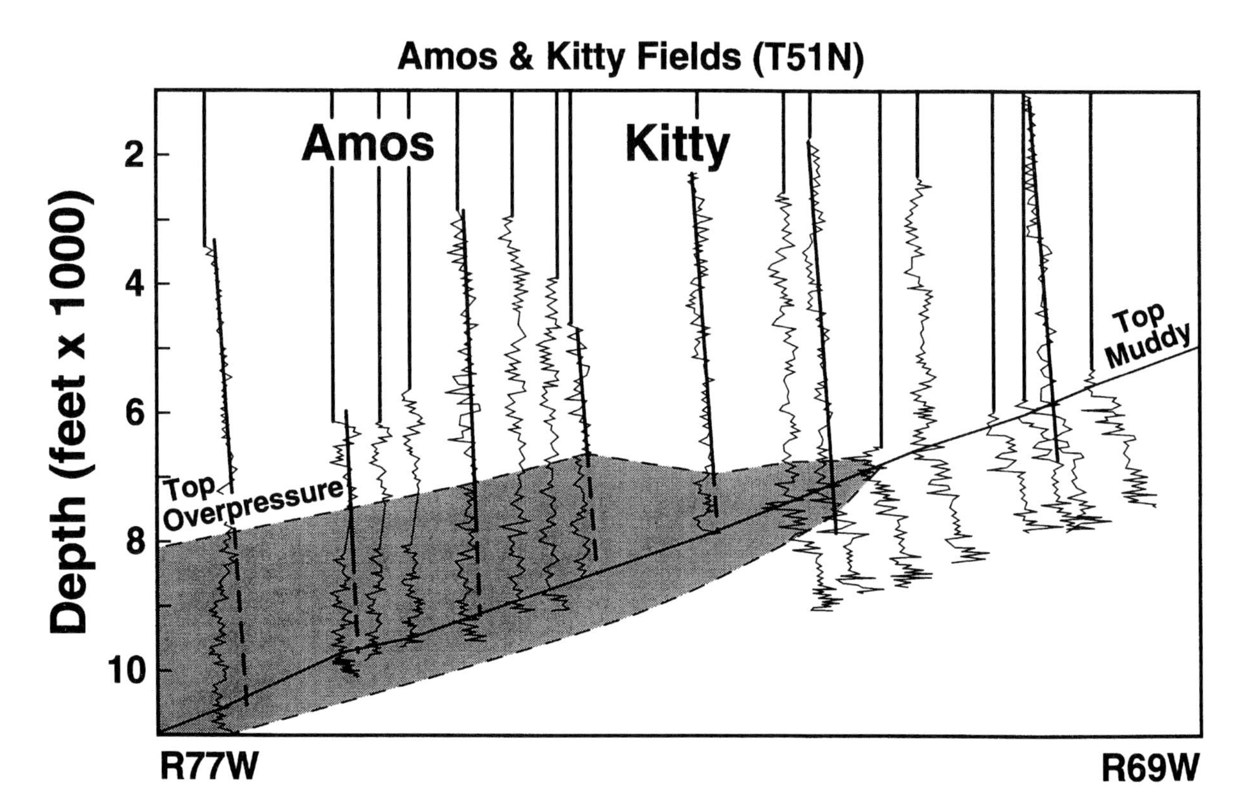

Figure 7A. Velocity profile across the Amos Draw and Kitty fields constructed from digitized velocity (calculated from sonic logs). Note that the overpressured region (stippled) diminishes in thickness to the east.

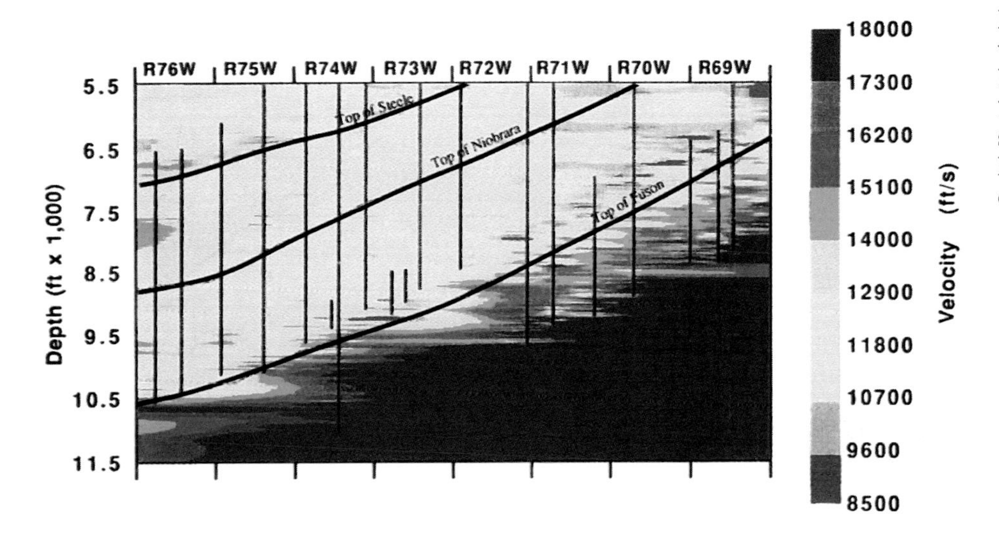

Figure 7B. Color-contoured velocity profile in the Amos Draw–Kitty field area, Line 1 of Figure 1. The vertical straight lines indicate the locations of wells with sonic data.

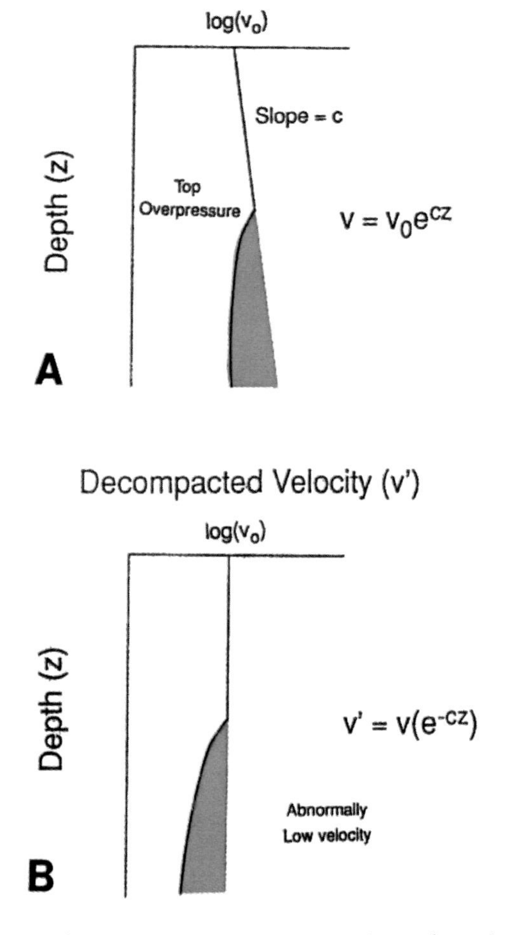

Figure 8. (A). Schematic representation of a velocity log in an overpressured zone. (B) The same after a decompaction correction has been applied.

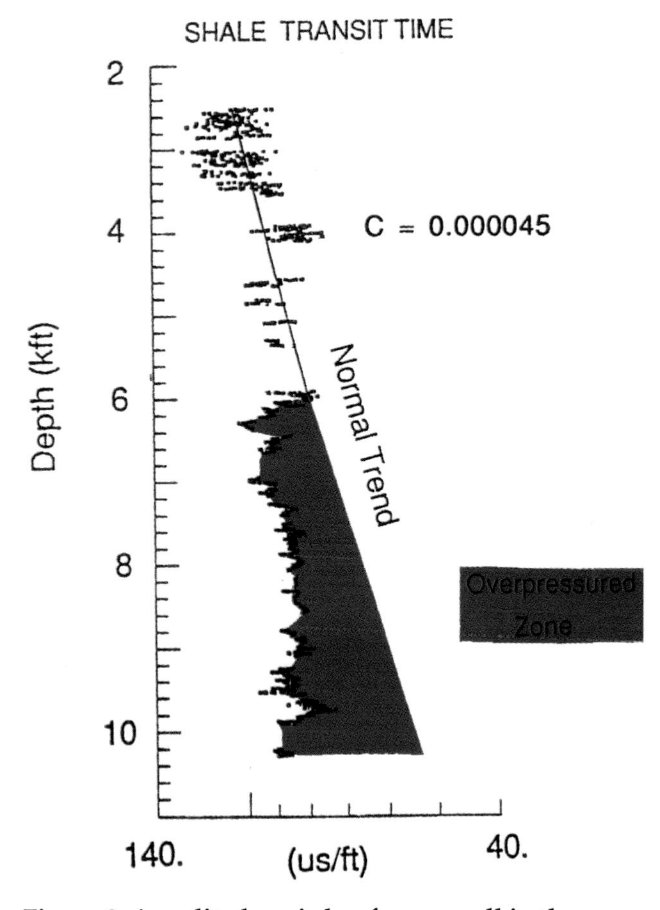

Figure 9. An edited sonic log from a well in the Amos Draw area showing the determination of the compaction factor C. Note that the top of the overpressured zone is clearly seen.

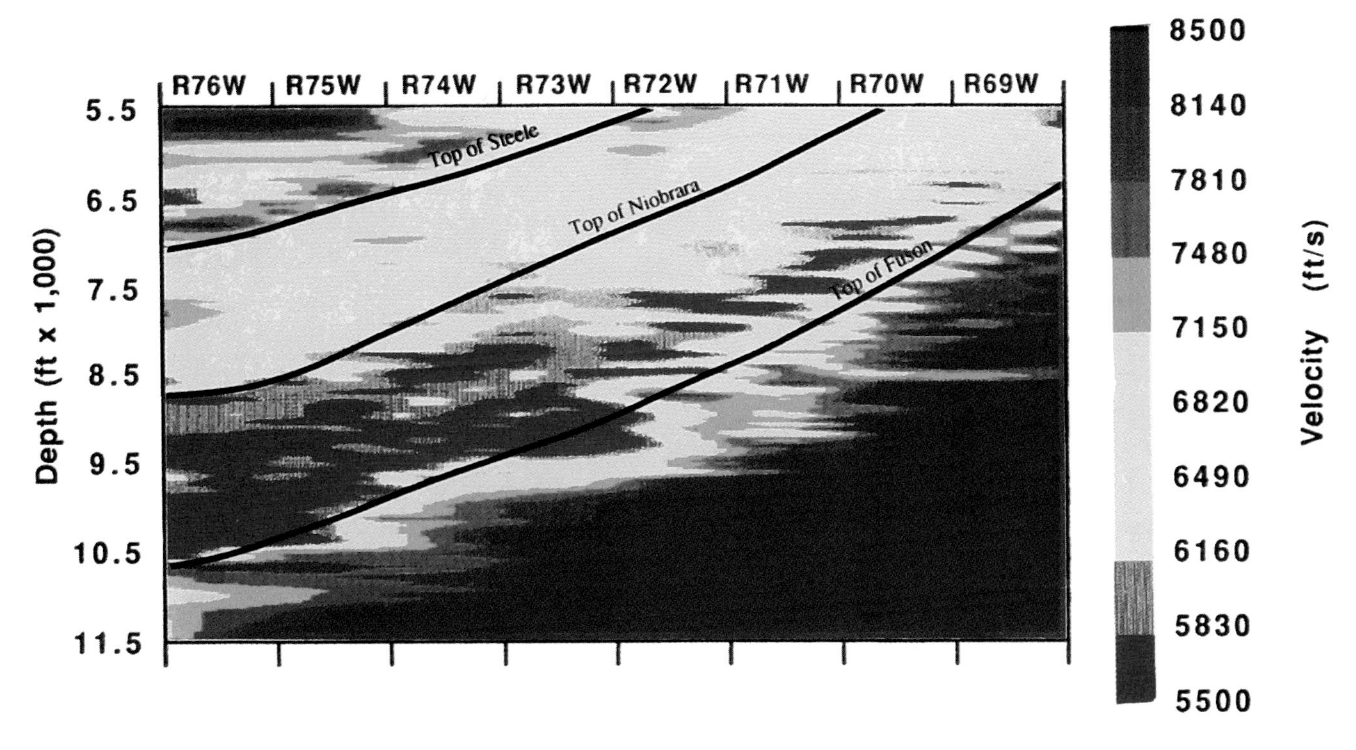

Figure 10A. Color-contoured velocity profile from Figure 7B after a decompaction correction has been applied. Red represents low velocities due to high pressures.

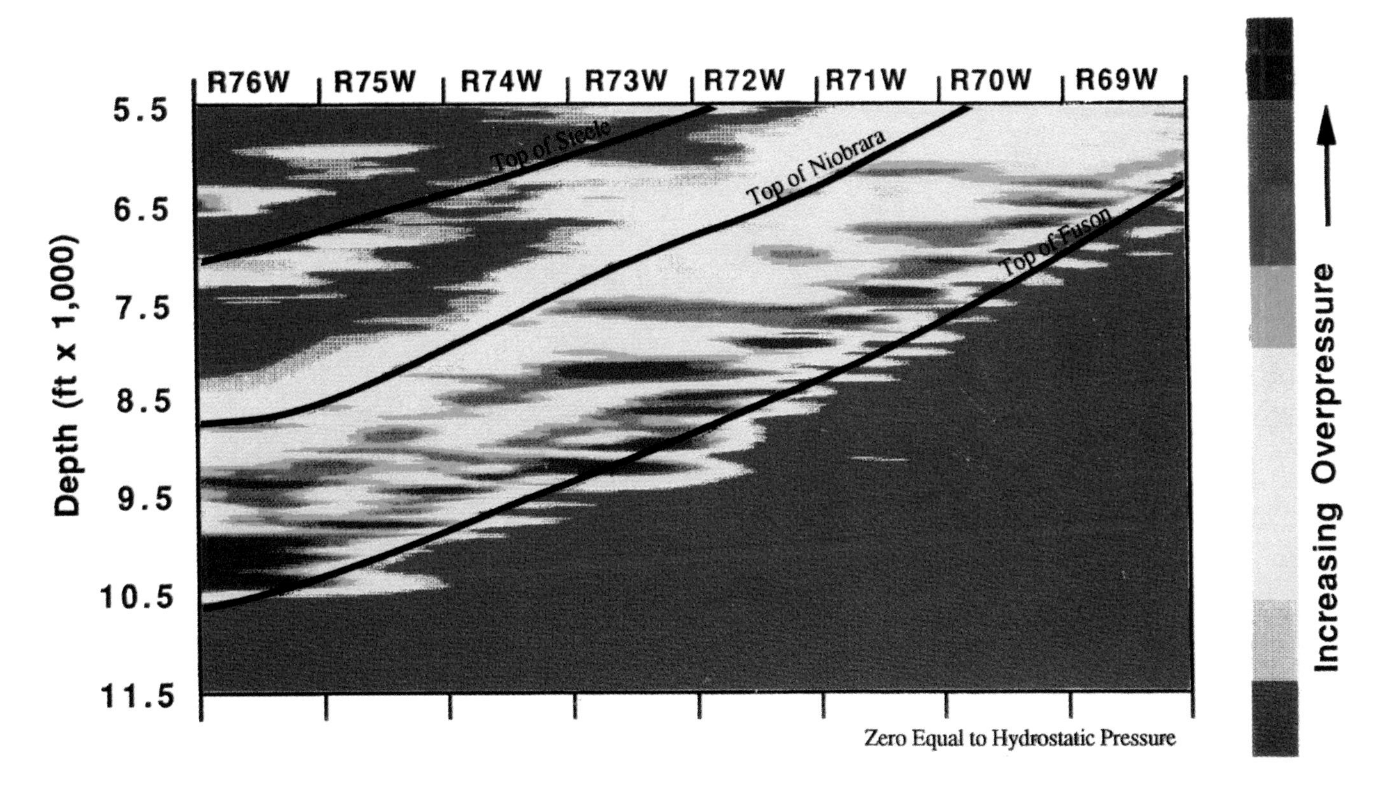

Figure 10B. Color-contoured pressure profile across the Kitty–Amos Draw field area. This section was created using the normal compaction trend method. Red indicates normal hydrostatic gradient, yellow-green indicates abnormal pressure, and blue indicates extreme abnormal pressure.

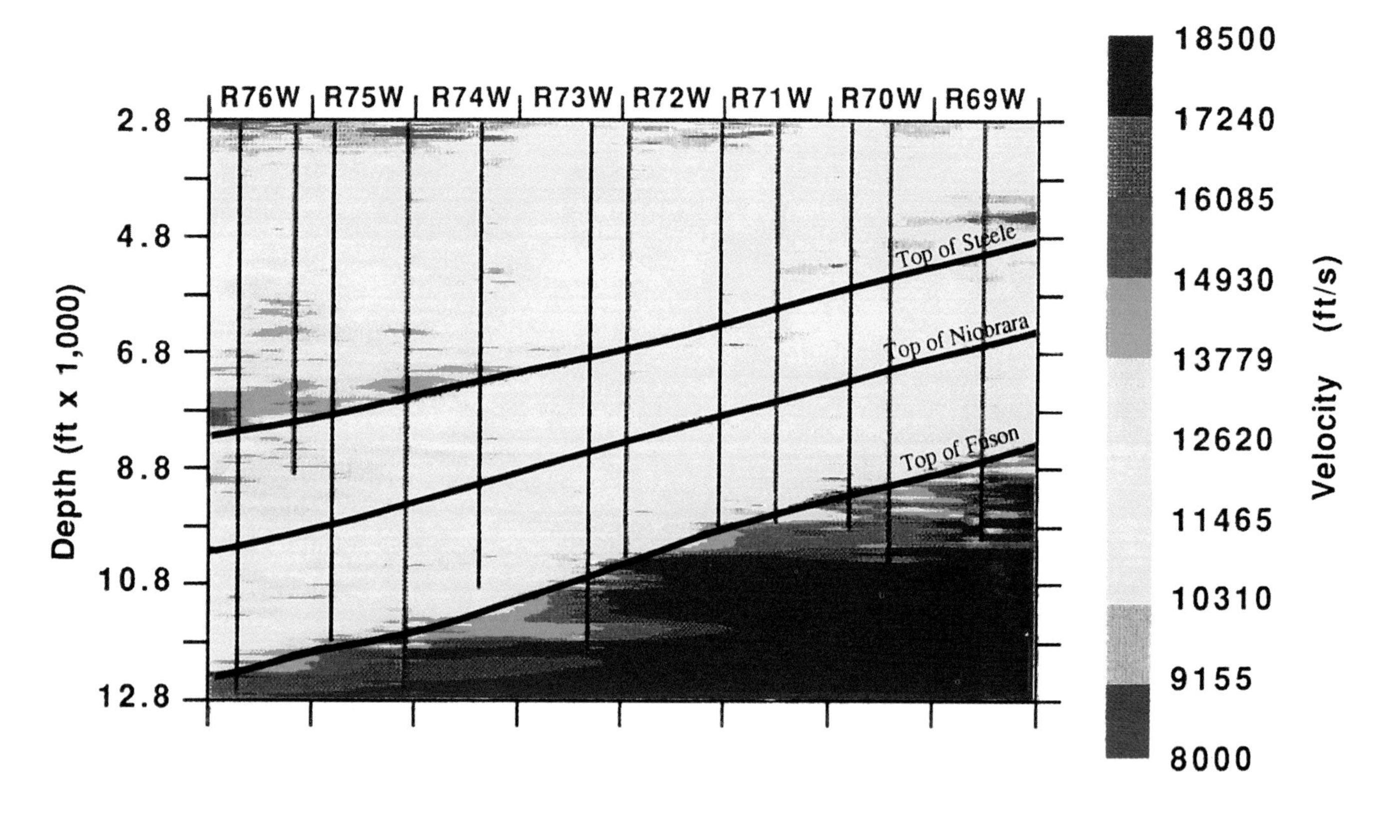

Figure 11. Color-contoured velocity profile across the Hilight field area, Line 2 of Figure 1. A velocity anomaly below the Steele Shale is evident. The velocity increases abruptly below the Fuson Shale.

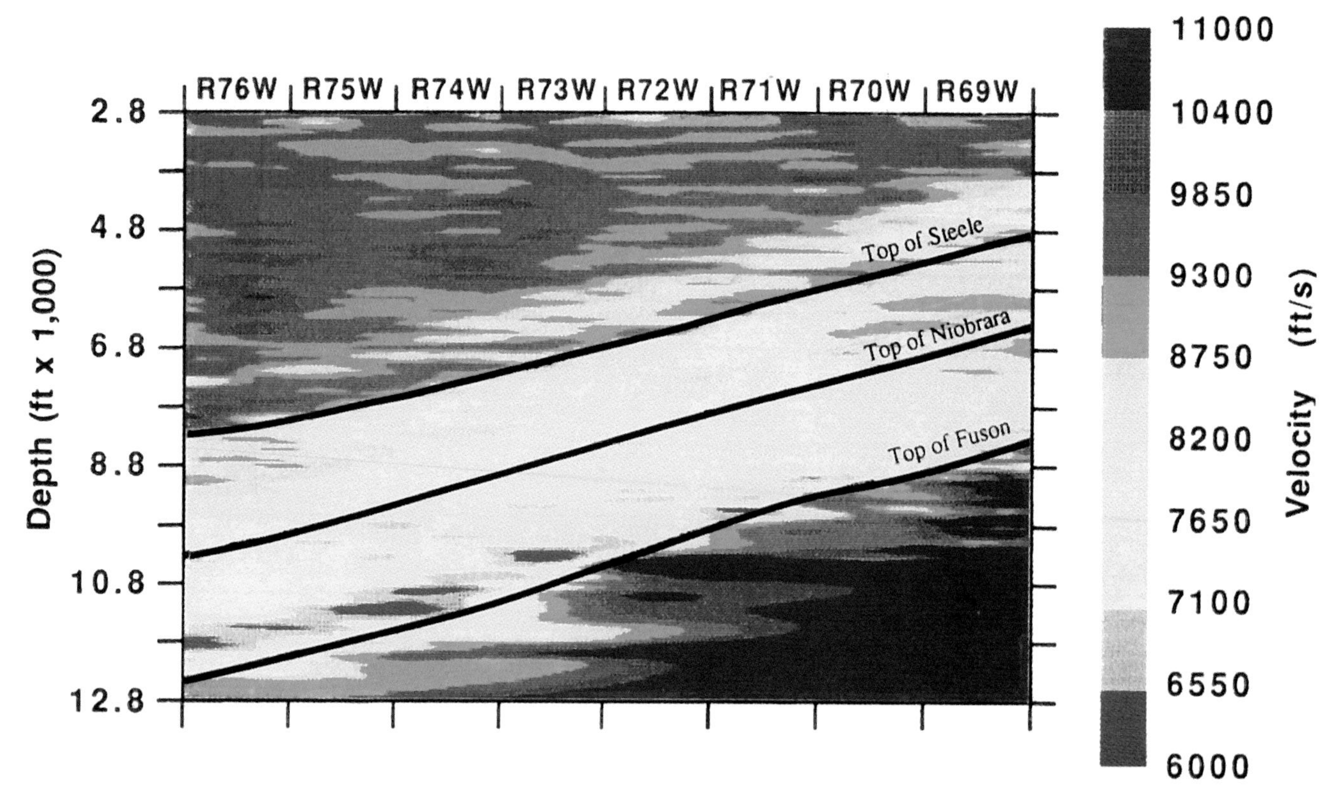

Figure 12. Color-contoured velocity profile in Figure 11 after a decompaction correction has been applied. The velocity anomaly between the Steele and Fuson shales is more evident, clarifying that the anomaly generally parallels the stratigraphy.

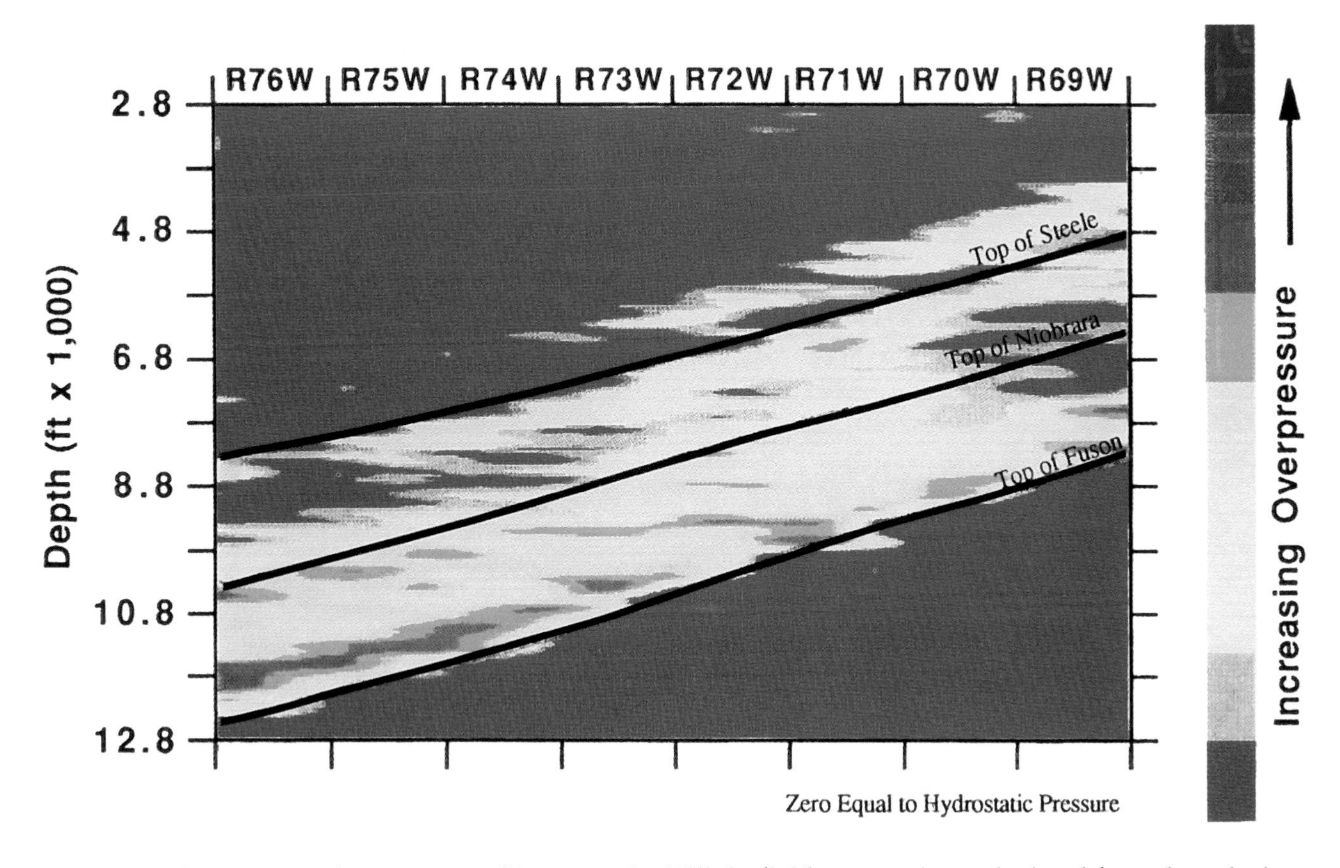

Figure 13. Color-contoured pressure profile across the Hilight field area section, calculated from the velocity profile by the method of normal compaction trend. The figure shows that the Cretaceous shales between Steele and Fuson are overpressured across the basin. The top diagonal line represents the top of the Steele Shale. The lower diagonal line represents the top of the Fuson.

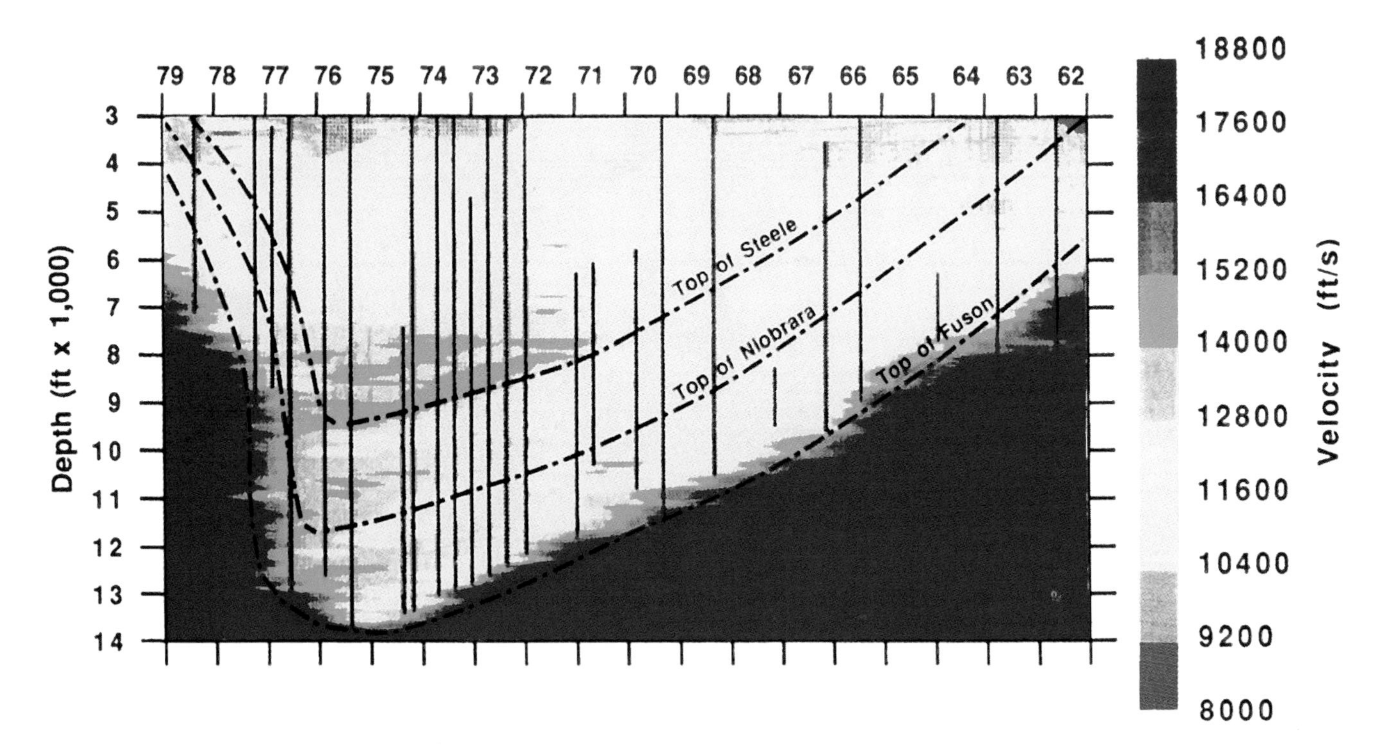

Figure 14. Color-contoured velocity profile across the northern portion of the Powell field, part of Line 3, Figure 1. The velocity anomaly here is also found between the Parkman and Fuson shales.

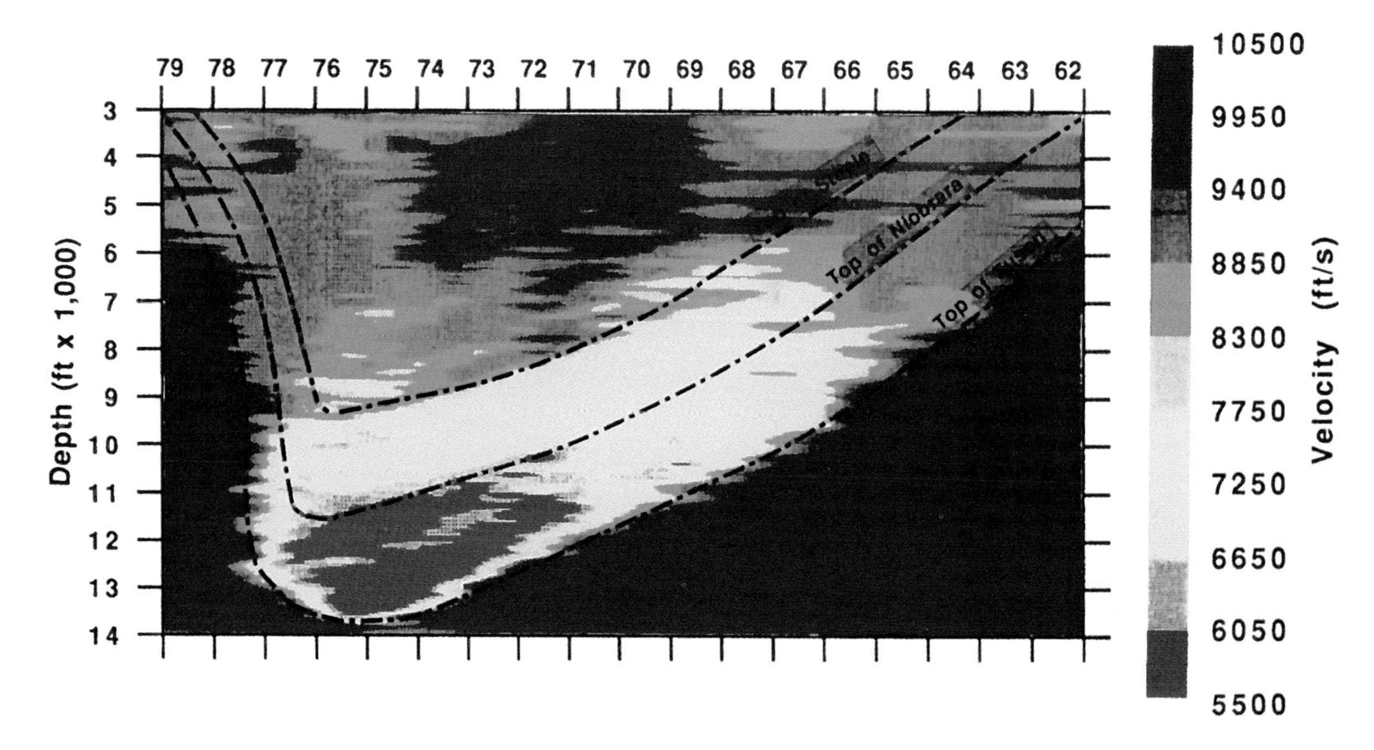

Figure 15. Color-contoured velocity profile in Figure 14 after a decompaction correction has been applied. The velocity anomaly between the Steele and Fuson shales is more evident.

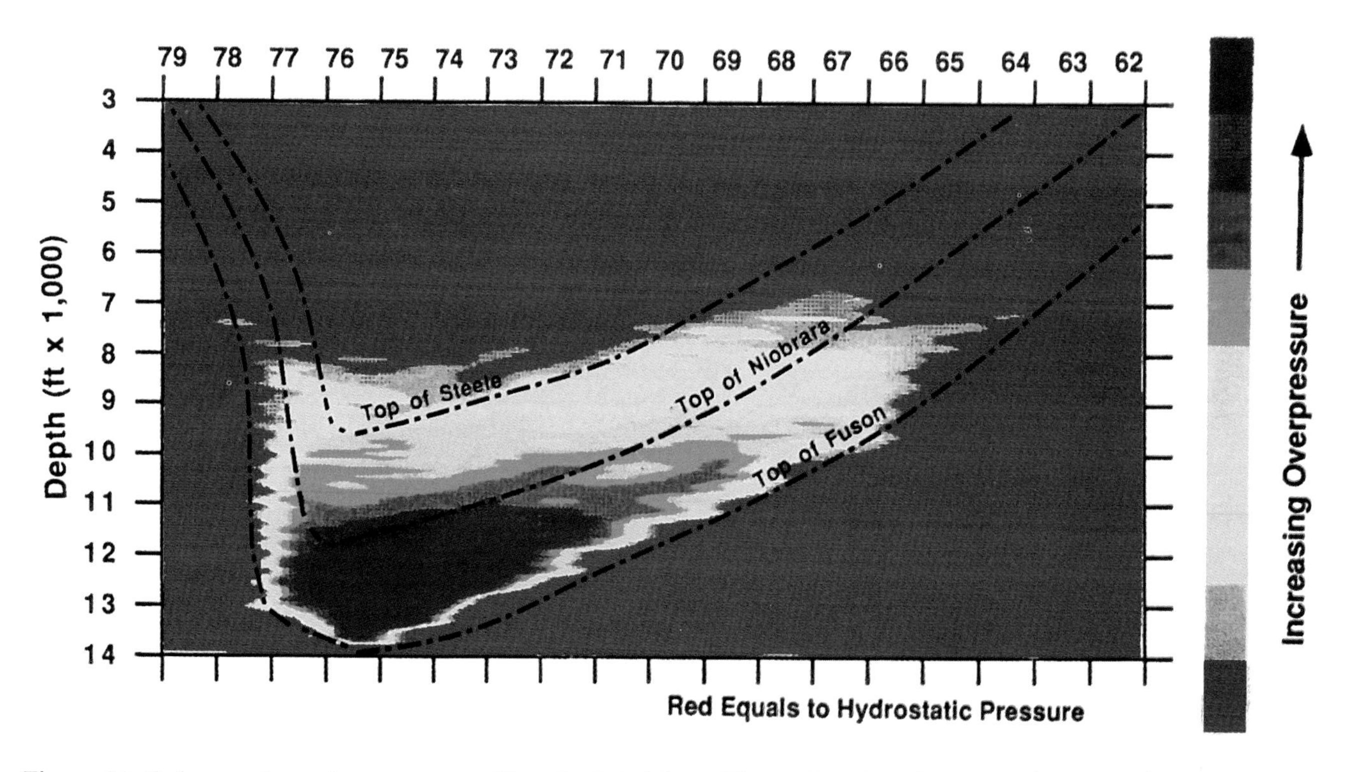

Figure 16. Color-contoured pressure profile calculated from Figure 15 using the normal compaction trend method. The figure shows that the Cretaceous shales are overpressured across the basin. The upper diagonal line represents the top of the Steele Shale. The lower diagonal line represents the top of the Fuson Shale.

sentation also shows that Cretaceous shales are overpressured throughout the basin, but that the sandstones are not always overpressured (see Maucione et al., 1992a). It also shows that the seal boundaries coincide with stratigraphic boundaries (for the stratigraphic control of seal boundaries, see Martinsen, this volume; Jiao and Surdam, this volume).

CONCLUSIONS

Velocity anomalies that may be associated with zones of abnormal pressure have been detected in vertical velocity profiles. These velocity anomalies are more evident after the application of a sediment decompaction correction. This study provides a solid foundation for further work using surface seismic methods to detect velocity anomalies such as those observed in the Powder River basin (Maucione et al., 1992b).

Zones of abnormally high pore pressure can be estimated in this region using resistivity, sonic, and density logs. Pore pressure values represent the shale pressure values are consistently higher (greater abnormal pressure) than the formation pressure values. The margin of error for these calculations is generally difficult to determine, because formation tests are not performed routinely in overpressured shales.

In the Powder River basin, pore pressure and formation pressure values differ significantly; there have been similar cases documented in eastern Siberia (Serebryakov and Dobrynin, 1989). The difference between pore pressure and formation pressure varies from well to well; therefore, this difference is not likely to be caused by a drilling/logging phenomenon. These differences between pore pressure and formation pressure can also be seen in the velocity and pressure profiles. These differences support the idea that the Cretaceous shales are overpressured throughout the basin, whereas the sandstones are present in all pressure regimes. This is likely controlled by the complex stratigraphy of the section (see Martinsen, this volume).

The different trends shown in Figure 6 may indicate migration paths of hydrocarbons. Since it is apparent from other work (MacGowan et al., this volume; Surdam et al., this volume; Iverson et al., this volume) that the presence of abnormally pressured compartments in the Powder River basin is associated with the generation of hydrocarbons, the difference in gradient of the abnormal pressure coefficients may illustrate the flow paths of hydrocarbons within the system. Velocity anomalies in profiles can be the basis for prediction of abnormal pressure using seismic techniques.

Figures 7B and 10–16 show that the velocity anomaly caused by overpressuring is related to the stratigraphy in the Powder River basin. The top seal is located within the Parkman Sandstone–Steele Shale interval. The bottom seal is along the stratigraphic boundary of the Fuson Shale. As stated above, all Cretaceous shales from the Steele-Fuson are overpressured, whereas the sandstones are variably overpressured, underpressured, and normally pressured. The compartmentation of the sandstones is complex and is likely controlled by stratigraphy and hydrocarbon maturation (Surdam et al., this volume). However, the delineation of hydrocarbon accumulations is possible with seismic methods. Once an accurate velocity model is developed, synthetic seismograms can be made. The synthetics can then be used for calibration with seismic data that have been specially analyzed and processed to identify subtle features associated with abnormally pressured zones. They can also be used to determine the resolution in seismic data necessary to make these techniques usable. Once these techniques have been refined, it will be possible to predict the location of abnormally pressured regions with seismic data prior to drilling. The seismic delineation of abnormally pressured compartments in the Powder River basin, and in other similar basins, would then be an effective exploration method in finding subtle hydrocarbon accumulations associated with abnormal pressure compartments. Also, the mapping of abnormal pressure coefficients could be used to determine pressure-determined (rather than structural) migration routes.

In summary, results presented for several fields in the Powder River basin show strong velocity anomalies with decreases of at least 2000 ft/sec associated with overpressured zones. These seismic anomalies correspond to geological and geochemical features determined in other studies and also with hydrocarbon reservoirs. One important conclusion is that the disequilibrium compaction model of the Gulf of Mexico does not explain the pressure anomalies found in the Powder River basin. The Powder River basin has complex stratigraphic, burial, and diagenetic histories that include multiphase flow systems (see Iverson et al., this volume). The abnormal pressure anomalies coincident with hydrocarbon accumulations (detectable velocity anomalies) must be viewed within this complicated framework. These velocity anomalies are large enough to allow detection from seismic reflection measurements on the surface. Such results are highly encouraging for the application of seismic reflection methods to the direct detection of hydrocarbon reservoirs and of geologic and geochemical features associated with pressure compartments. Pattern recognition (artificial intelligence) approaches may be used to detect such features by analysis of more subtle changes in seismic attributes.

ACKNOWLEDGMENTS

This material covers research performed through November 1991. The original manuscript was reviewed by David Copeland (University of Wyoming), who made numerous helpful suggestions. This study was funded by Gas Research Institute under contract number 5089-260-1894.

REFERENCES CITED

Dobrynin, V., and V. Serebryakov, 1989, Geological–Geophysical Methods, Prediction of Pressure Anomalies: Moscow, 288 p.

Fertl, W.H., 1976, Abnormal Formation Pressures: New York, Elsevier, 382 p.

Hunt, J. M., 1979, Petroleum Geochemistry and Geology: San Francisco, W.H. Freeman, 617 p.

MacGregor, R. D., 1965, Quantitative determination of reservoir pressures from conductivity log: American Association of Petroleum Geologists Bulletin, v. 49, p. 1502–1511.

Martinez, R. D., J. D. Schroeder, and G. A. King, 1991, Formation pressure prediction with seismic data from the Gulf of Mexico: Society of Petroleum Engineers, Formation Evaluation, p. 27–32.

Maucione, D., W. P. Iverson, and V. A. Serebryakov, 1992a, Pressure analysis using resistivity and sonic logs: Final Report (Draft) on Contract No. 5089-260-1894, Multidisciplinary Analysis of Pressure Chambers in the Powder River Basin of Wyoming and Montana, Gas Research Institute, Chicago, p. 145–152.

Maucione, D., S. Kubickek, V.A. Serebryakov, M. Speece, P. Valasek, Y. Wang, and S. Smithson, 1992b, Seismic reflection studies of abnormally pressured zones in the Powder River Basin of Wyoming using sonic logs: Final Report (Draft) on Contract No. 5089-260-1894, Multidisciplinary Analysis of Pressure Chambers in the Powder River Basin, Wyoming and Montana, Gas Research Institute, Chicago, p. 153–170.

Wang, Y., 1992, Velocity study of overpressured zones in the Powder River Basin of Wyoming using sonic logs: M.S. thesis, University of Wyoming, 92 p.

VI. Banded Pressure Seals

Chapter 23

The Banded Character of Pressure Seals

Zuhair Al-Shaieb
James O. Puckette
Azhari A. Abdalla
Vanessa Tigert
Oklahoma State University
Stillwater, Oklahoma, U.S.A.
Peter J. Ortoleva
Indiana University
Bloomington, Indiana, U.S.A.

ABSTRACT

Pressure seals in the Anadarko basin contain distinct diagenetic bands that play a significant role in the isolation of high-pressure domains. Both sand-rich and clay-rich clastic rocks exhibit banding patterns. Banding in sandstones results primarily from the processes of pressure solution, pore-fluid interaction, and precipitation. Some banding patterns in clay-rich rocks appear to form independently of sedimentary textures while others result from the enhancement or modification of sedimentary features. All of these bands are noticeably absent in rocks that were never buried deeply enough to be overpressured.

Diagenetic bands in sandstones consist of silica- and carbonate-cemented layers that are separated by clay-coated porous layers. Stylolites and other pressure-solution features such as penetrating grain boundaries suggest a mechanism for the derivation of silica cements.

Diagenetic bands in clay-rich rocks appear to have two distinctly different origins. In the Pink/Red Fork interval, diagenetic chlorite bands exhibit no apparent relationship to sedimentary features. The chlorite bands have less porosity and smaller pore-aperture radii than the surrounding shale (Powers, 1991). In the Woodford Shale, silica- and clay-rich bands develop via the diagenetic modification or enhancement of sedimentary features.

INTRODUCTION

Many seal rocks display unique diagenetic banding structures that formed due to the interplay of stress-induced mineral reactions, pore-fluid interactions, mass transport, and precipitation. This banding is observed in rocks that were buried deep enough to enter the "seal window" in the Anadarko basin (1829–3000 m). Rocks from shallower intervals clearly lack these banding features. These patterns have been simulated using reaction-transport models that further verified their significance in generating effective pressure seals (Dewers and Ortoleva, 1990a, 1988; Qin and Ortoleva, this volume).

Tigert and Al-Shaieb (1990) recognized the importance of diagenetic banding as a mechanism of converting sandstone intervals into effective pressure seals. Al-Shaieb et al. (1991) indicate similar banding patterns exist in overpressured shaly intervals as well, such as the Woodford Shale and the shaly Pink Limestone/Red Fork sandstone interval.

Cementation bands are economically significant geologic phenomena since they may play an important role in trapping hydrocarbons. Intrafacies cementing can compartmentalize relatively compositionally homogeneous sandstones. Permeability contrasts created by diagenetically banded rock fabrics serve as both traps and reservoirs in the Simpson (Tigert and Al-Shaieb, 1990) and Springer (Al-Shaieb et al., 1993) sandstones in the Anadarko basin.

BANDING PATTERNS AND THEIR GENESIS IN SANDSTONES

Silica cement bands (Figures 1 and 2) are the most common bands in sandstones and consist of zones of enhanced quartz overgrowths alternating with bands of preserved porosity. The silica was apparently derived from quartz-grain pressure solution in the adjacent band (Figure 2). Porous regions often contain thicker clay coatings on grains, suggesting clay-inhibition of quartz cementation in these areas. Silica bands may occur relatively early (around 2000 m and 65°C) in the rocks' diagenetic history and reflect the mechano-chemical processes associated with burial compaction, dissolution, and precipitation.

Dewers and Ortoleva (1993, 1990a), Ortoleva et al. (in press, 1993, 1987), and Qin and Ortoleva (this volume) have described several mechano-chemical (chemical compaction) processes that involve stress-mediated dissolution and reprecipitation at free faces in contact with pore fluid. One process is porosity feedback where higher grain stresses increase dissolution in more porous rock (pillar effect) and encourage silica precipitation in neighboring lower-porosity areas. Another is contact-area feedback where dissolution is slower at larger grain-to-grain contacts than smaller ones for grains of similar volume. Perhaps the most easily discerned mechano-chemical process in sandstones is clay-coating–mediated feedback. Clay coatings tend to inhibit precipitation on free faces. Solutes derived from dissolution at grain contacts precipitate faster in areas of thinner clay coatings.

Within the diagenetically banded pressure seal in the Ordovician-age Simpson Group in Oklahoma (Figure 3), the alternation of permeable and cemented sandstones is a manifestation of these processes. The compositionally homogeneous First Bromide Sand-stone member contains silica bands that were apparently generated by local dissolution and precipitation. Sandstones with thin or nonpervasive clay-grain coatings (Figure 4) compacted and became likely silica sources. Adjacent intervals received imported silica and became silica-cemented bands. The cemented bands often contain fewer pressure-solution features (Figure 5), suggesting that the overgrowths increased grain-to-grain contact and slowed dissolution. On the other hand, higher grain stresses accelerated dissolution in the more porous zones. Clay coatings here inhibited silica precipitation on the adjacent free faces (Figure 4), porosity was preserved, and the silica was exported to precipitate in adjacent bands. Clay-coating–mediated feedback is also manifested by very porous oil-stained reservoirs adjacent to silica-cemented sandstones (Figure 6) and detrital matrix-rich rocks where framework grain morphology is preserved (Figure 7).

Stylolites cross-cutting quartz overgrowths (Figure 8) indicate additional pressure solution occurred after initial silica cementation. Stylolites are more prevalent in slightly porous sandstones with thinner clay-grain coatings and minor amounts of clay matrix. It is suggested by Dewers and Ortoleva (1991) that pressure solution progressed in these areas because the clay coatings inhibited overgrowth development on the free faces adjacent to grain-to-grain contacts. This prevented the grain contacts from increasing in size, which would decrease the stress per unit area and slow dissolution. Stylolites that developed after syntaxial-quartz overgrowth formed as a result of the increased stress generated by additional burial. Silica liberated during this phase may have precipitated as the postsyntaxial-overgrowth intergranular quartz cements in the silica bands. Silica band/porous band alternations occur on several scales. They range from a series of 4 to 5 bands that are 1–3 mm thick in a 10 cm interval (see Figure 1) up to 2 or 3 bands that are 10 to 15 cm thick within a 20 to 30 cm interval.

The amount of silica cement contained in the bands is considerably less than that apparently liberated by dissolution. Low-porosity silica-cemented sandstones exhibit approximately equal amounts of dissolution and cement and appear to be silica balanced. On the other hand, more porous sandstones with significant clay coatings have apparently exported considerably more silica than can be accounted for as cement in the Bromide interval.

A second type of band we have observed in sandstones is composed of carbonate that postdates the silica bands. Carbonate-cementation bands often consist of calcite and/or dolomite that alternate with porous bands in silica-cemented sandstones (Figure 9). Carbonate bands occur on scales similar to those observed for silica bands. They may form a collection of millimeter-scale bands within a thin (10 cm) interval or be composed of thicker (5–10 cm) bands within a larger (20–30 cm) interval.

The distribution of silica and carbonate bands across a few meters of a seal interval is depicted by the schematic diagrams of the seal zone that accompany the core photographs.

In the Bromide sandstone interval, the earliest diagenetic calcite cements were apparently derived from compaction dissolution of carbonates found within or adjacent to the sandstones. These potential sources include limestone beds and fossil-rich zones within the sandstones, and the overlying Bromide and Viola carbonates. It has been suggested that calcite grains in a limestone are stress supporting and at a higher free energy than calcite cement precipitated in an adjacent

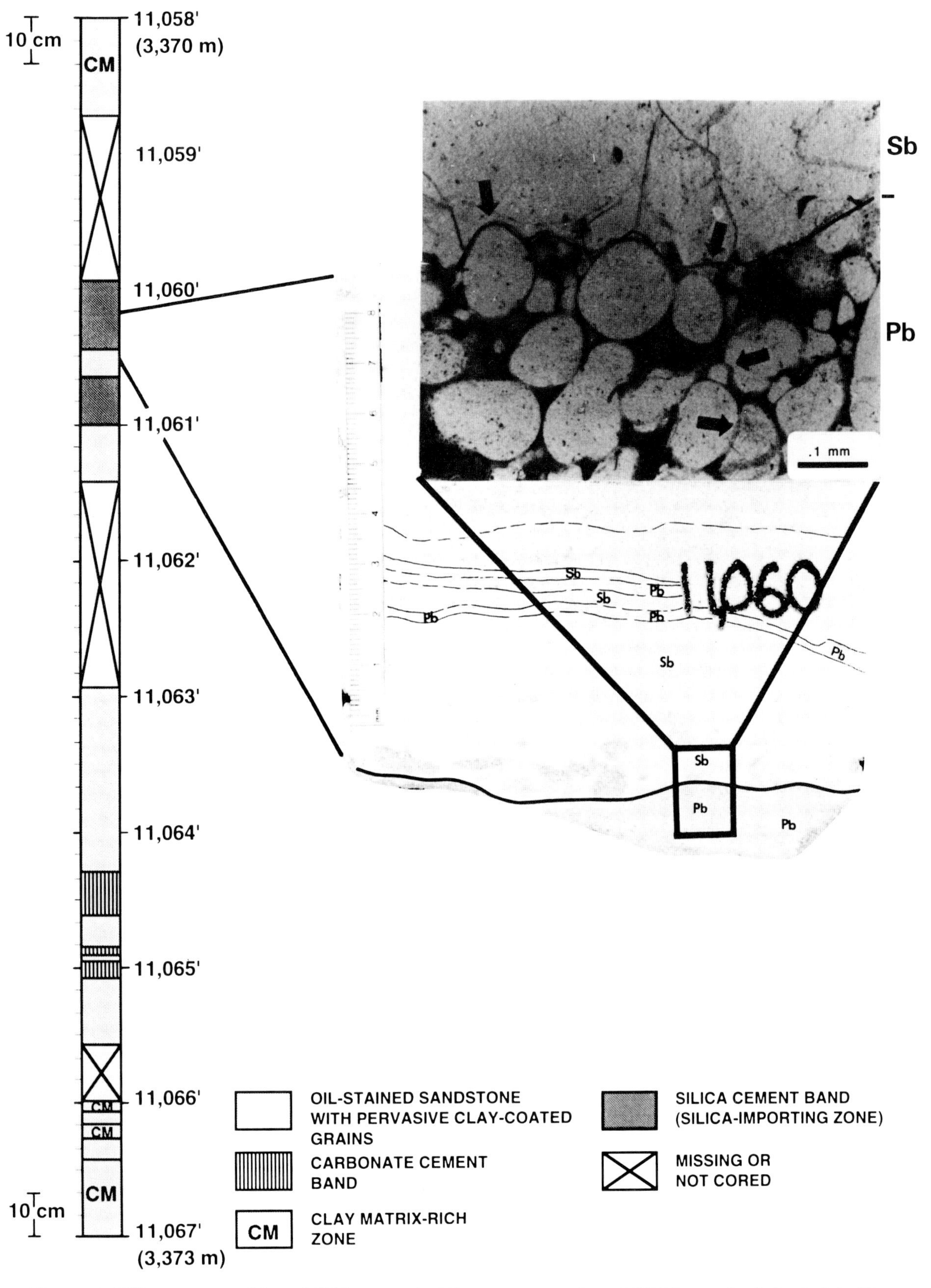

Figure 1. Silica-cemented band (Sb) separated from a porous band (Pb) by an incipient stylolite or pressure-solution boundary. Note the stress-induced pressure solution along grain contacts (arrows). Simpson sandstone, Gulf, Weaver No. 1. McClain County, Oklahoma. Depth 3371 m (11,060 ft).

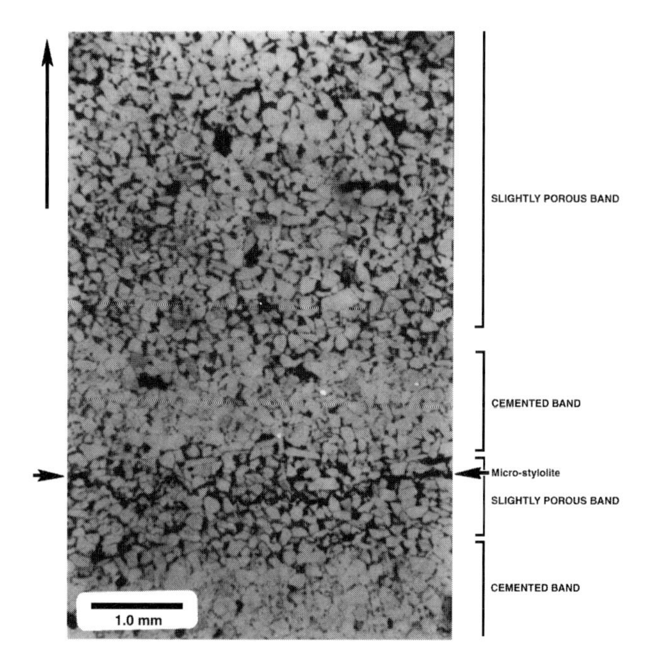

Figure 2. Alternating porous and silica-cemented bands in the Springer Lower Cunningham Sandstone. Microstylolite is manifestation of pressure solution. Silica liberated by dissolution was apparently precipitated as adjacent cemented bands. Gulf, Miller No. 1. Caddo County, Oklahoma. Depth 5022 m (16,475 ft).

sandstone. This presumes the precipitate is subject only to fluid pressure while overgrowth on the free faces of the calcite grains in the carbonate is inhibited by clay coatings or other factors. Even after the sandstone becomes tightly cemented, the calcite is bathed in fluid more or less at equilibrium with stressed calcite contacts in the limestone. As a result, the calcite in the sandstone can continue to grow at the expense of the quartz grains, yielding a poikilotopic texture (Ortoleva et al., in press, 1993; Qin and Ortoleva, this volume).

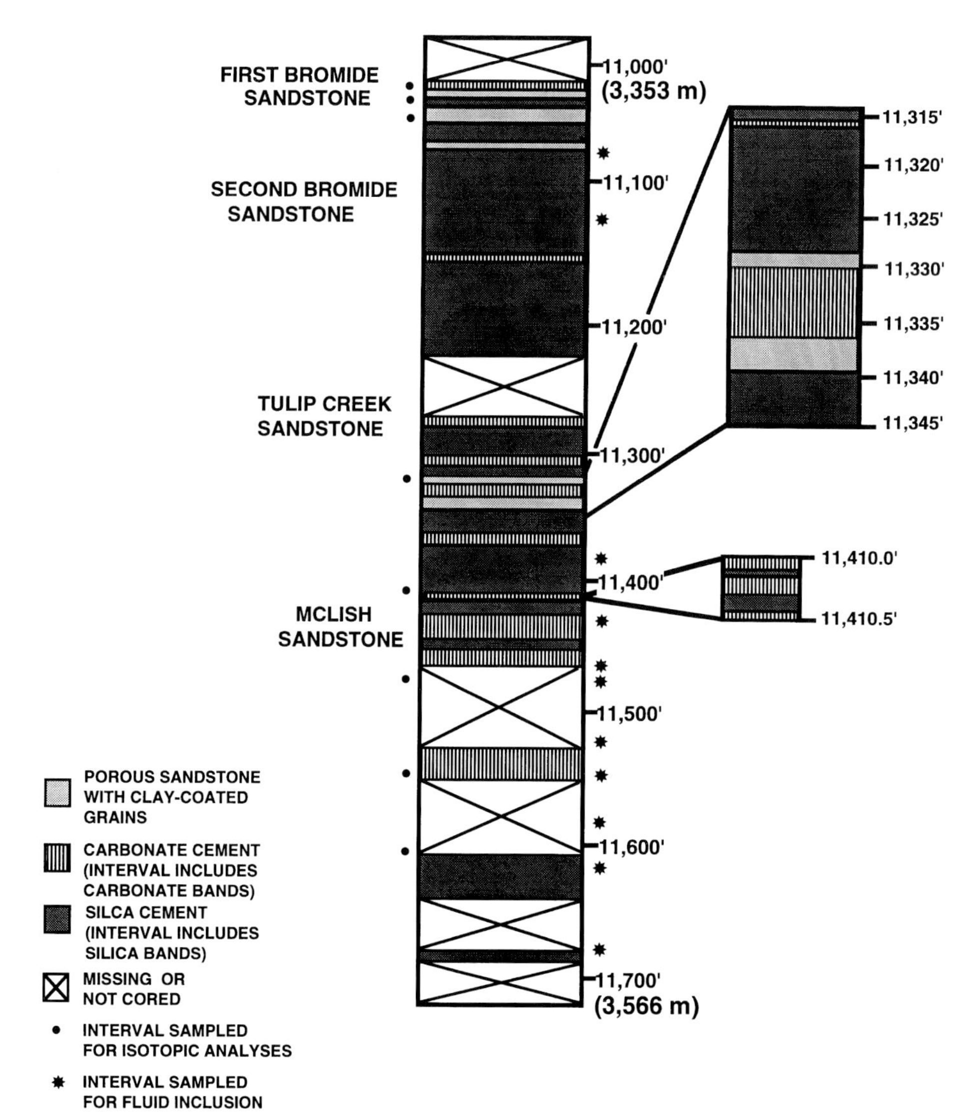

Figure 3. Schematic diagram of a banded pressure seal zone in the Simpson sandstone, Anadarko basin, Oklahoma. After Tigert (1989).

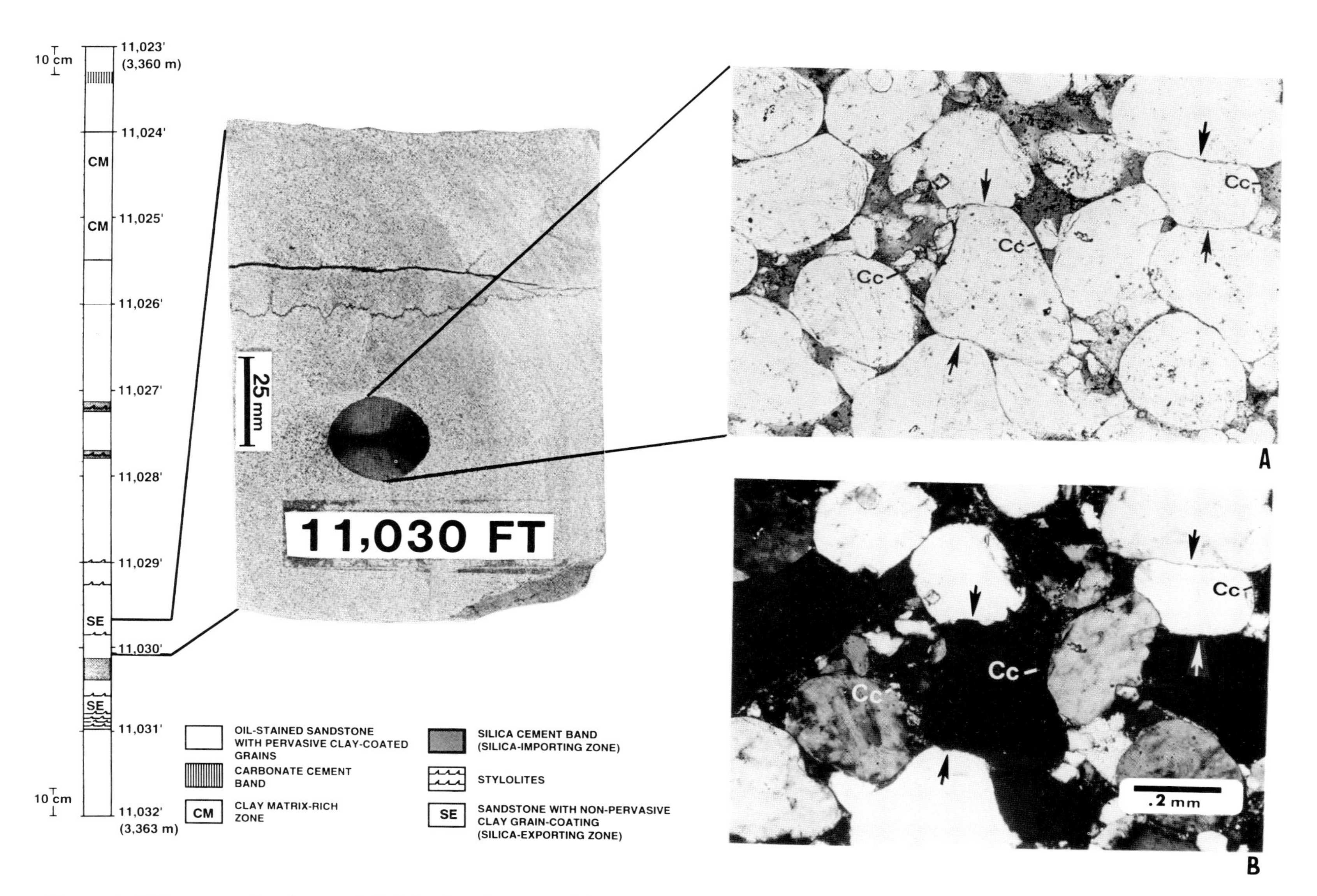

Figure 4. Silica-exporting interval (SE) in the First Bromide Sandstone. Dissolution features include penetrating grain contacts (arrows) and stylolites (core photograph). Clay coatings (Cc) inhibited silica cementation on free faces adjacent to grain contacts. Schematic diagram indicates location of the described feature within the seal zone. Gulf, Weaver No. 1. Depth 3362 m (11,029.8 ft). (A) Plane-polarized light (PPL). (B) Cross-polarized light (CPL).

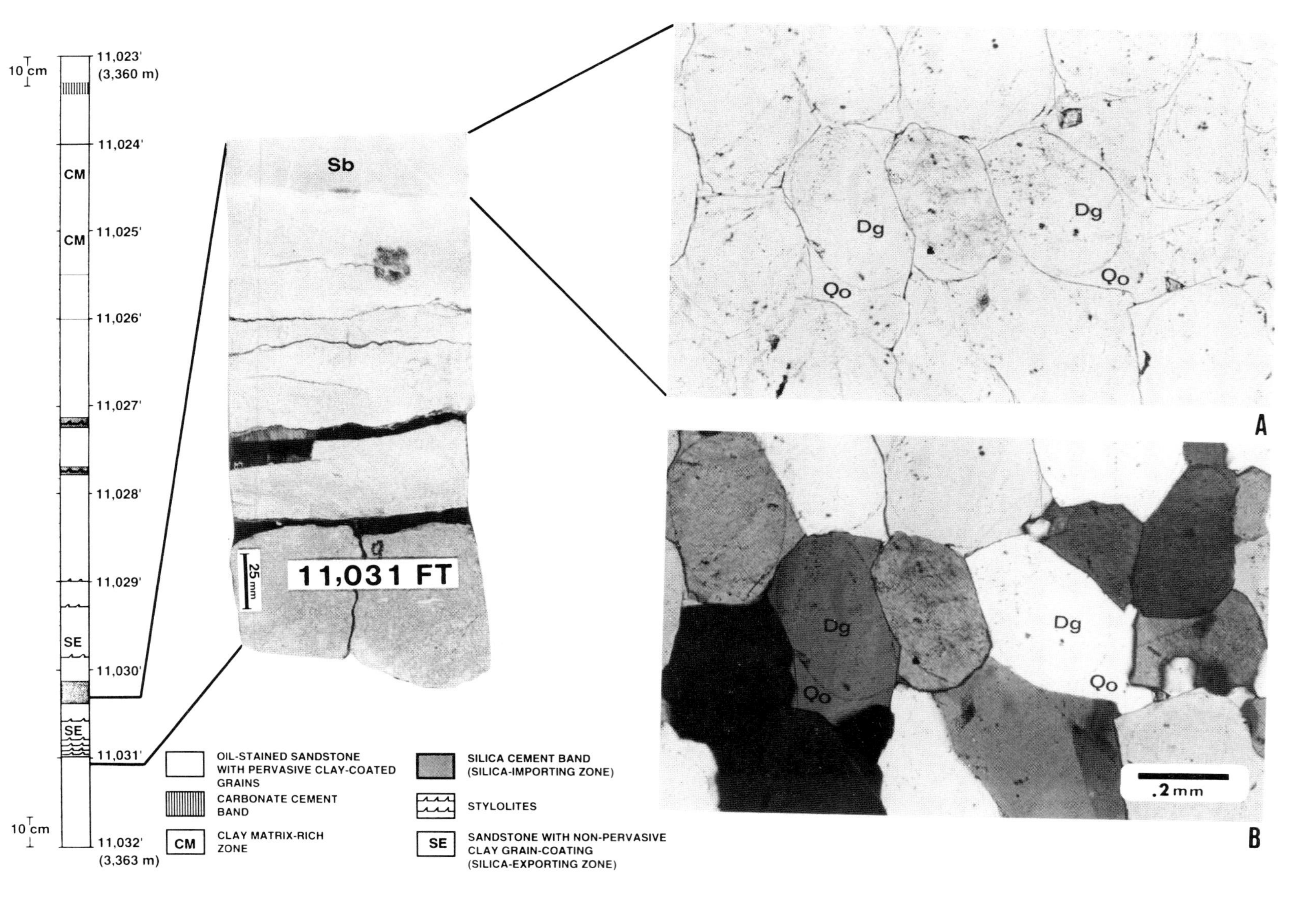

Figure 5. Silica-cemented band (Sb) exhibiting extensive quartz overgrowth (Qo). Detrital grain (Dg) morphologies are preserved suggesting this zone imported silica and did not experience significant stress-induced dissolution. Gulf, Weaver No. 1. Depth 3362 m (11,030.5 ft). (A) PPL. (B) CPL.

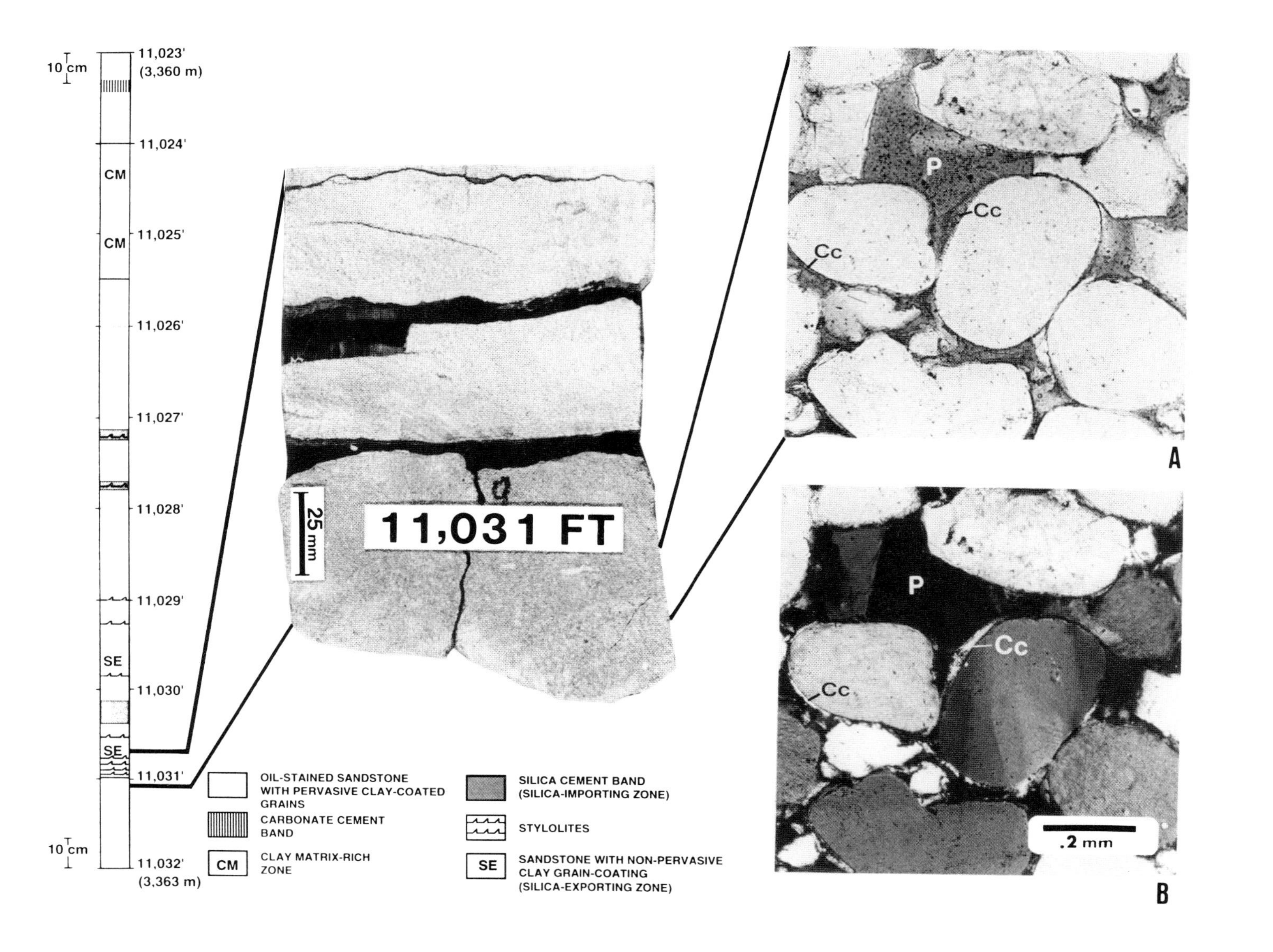

Figure 6. Contact between silica-cemented sandstone (light) and very porous oil-stained sandstone (dark). Pervasive clay-grain coatings (Cc) inhibited silica precipitation and preserved porosity (P) in the reservoir rock. Gulf, Weaver No. 1. Depth 3363 m (11,031 ft). (A) PPL. (B) CPL.

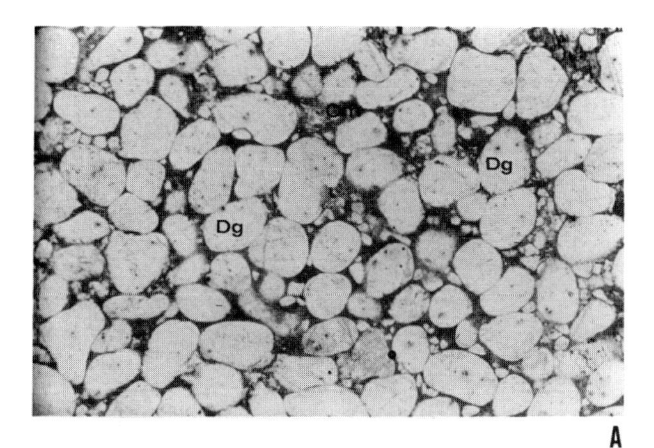

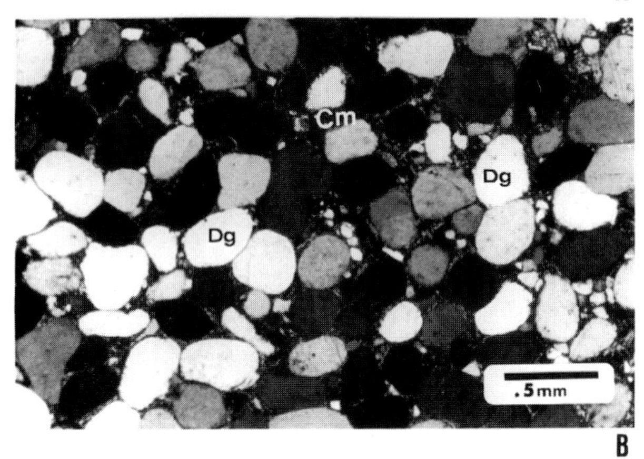

Figure 7. Photomicrograph of allogenic clay matrix (Cm) that preserved detrital grain (Dg) morphology by inhibiting quartz overgrowth nucleation. Gulf, Weaver No. 1. Depth 3369 m (11,052 ft). (A) PPL. (B) CPL.

In the Anadarko basin, poikilotopic calcite cements represent an apparent intermediate stage between initial pore-filling cement and later fracture-filling cements in sandstone seals. A final stage of calcite cement occurs both as fracture filling and bands.

In the Simpson sandstone seal, $\delta^{13}C$ and $\delta^{18}O$ values of several carbonate cements range from –5 to –9 and from –7 to –11, respectively. These isotopic compositions indicate that in these cements, the carbon was partially derived from an organic source while the oxygen isotope ratios indicate relatively higher temperature fluids. On the other hand, the $\delta^{13}C$ values of most carbonate cements (0 to –4) suggest a reworked marine source for the carbon. A cross-plot of the carbon and oxygen isotopic compositions is shown in Figure 10.

Fluid inclusions within the silica and carbonate cements in the Ordovician Simpson Group sandstone seal interval were examined to determine the timing, temperature, and nature of fluids that resulted in generation of diagenetic cements. These analyses were performed under the supervision of David I. Norman at New Mexico Tech. Inclusion homogenization temperatures (T_h) of four cementation episodes (C–1, C–2, Q–1, and Q–2) are shown in Figure 11. The first episode consists of two-phase aqueous fluid inclusions from silica cement (Q–1) with (T_h) values ranging from 70 to 100°C. The second episode consists of two-phase aqueous fluid inclusions from calcite cement (C-1) with (T_h) values concentrated between 80 and 110°C. The third episode consists of petroleum-bearing and aqueous inclusions from a second calcite (C–2) cement. The (T_h) values for the hydrocarbon-bearing inclusions range from 95 to 140°C. The fourth episode is a later silica (Q–2) cement that contains hydrocarbon inclusions. The (T_h) values of this second silica cement range from 110 to 120°C. The genesis of petroleum-free, higher-temperature calcite (C–1) inclusions (T_h > 120°C) may be related to hot, basinal fluids. A burial-history curve, modified after Schmoker (1986), was constructed for the southeastern part of the Anadarko basin. Burial temperatures were calculated using Schmoker's paleogeothermal gradient and the burial-history curve. The homogenization temperatures of the various cement phases were plotted on the curve (Figure 12) to depict the timing relationships of cementation phases during burial history. The relationships of the cement stratigraphy may be important in the interpretation of seal evolution.

Figure 12 indicates a silica-cement phase was precipitated as early quartz overgrowths at approximately 1829 m (6000 ft). Next, a calcite cement with aqueous fluid inclusions was precipitated approximately between 2667 and 3353 meters (8750 and 11,000 ft). The following episodes of cementation occurred after the rock entered the oil window. The inclusions contained in the second calcite cement and a late silica cement contain both hydrocarbon-bearing and aqueous inclusions. The petrographic evidence and geologic history of the basin suggest the following plausible interpretation. The occlusion of porosity was initiated at approximately 1829 m (6000 ft) during the precipitation of early quartz overgrowths. The silica was generated by quartz-grain dissolution due to pressure solution. This early quartz cementation results in what was termed a "protoseal" (Al-Shaieb et al., 1990). This protoseal likely developed in the Simpson seal zone around 305 Ma, shortly after the onset of the Pennsylvanian Wichita-Ouachita-Arbuckle Orogeny (Figure 12). During the rapid burial phase of the basin (300 to 270 Ma), the protoseal was probably strengthened by the precipitation of additional silica generated by stress-induced pressure solution. Porosity occlusion and seal formation progressed further during the initial stage of calcite cementation that accompanied the rapid burial phase. Toward the completion of this phase, the rocks entered the oil window and this seal zone was essentially completed by the mid-Permian. The higher temperatures associated with the later stage carbonate-cementation phases (C-1) and (C-2) (Figure 12) indicate that some inclusions may have been associated with heated basinal fluid migration. These cements

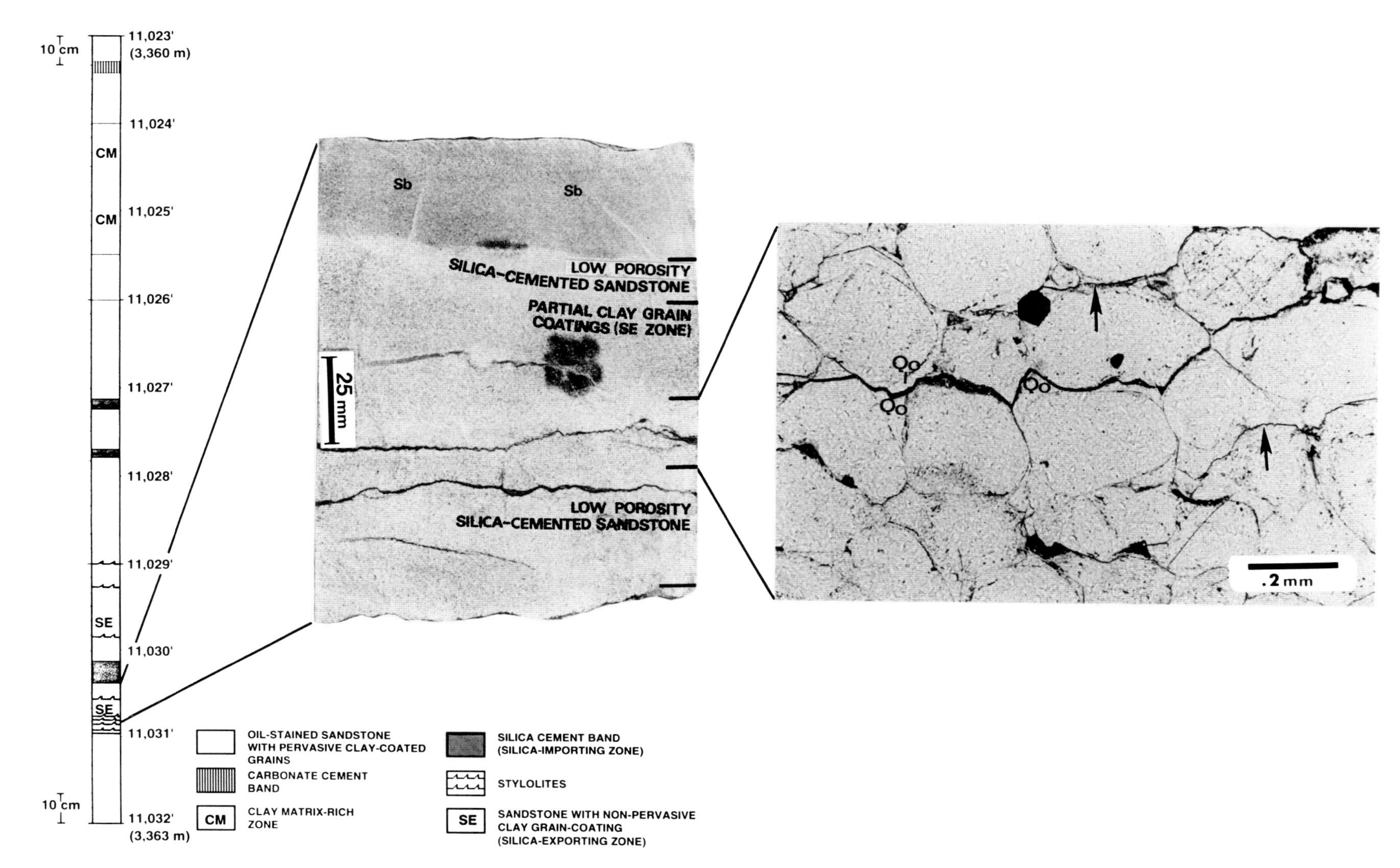

Figure 8. Photomicrograph of stylolites cross-cutting quartz overgrowths (Qo), indicating additional stress-induced dissolution followed overgrowth cementation. Incipient stylolites (arrows) appear parallel to the primary stylolite.

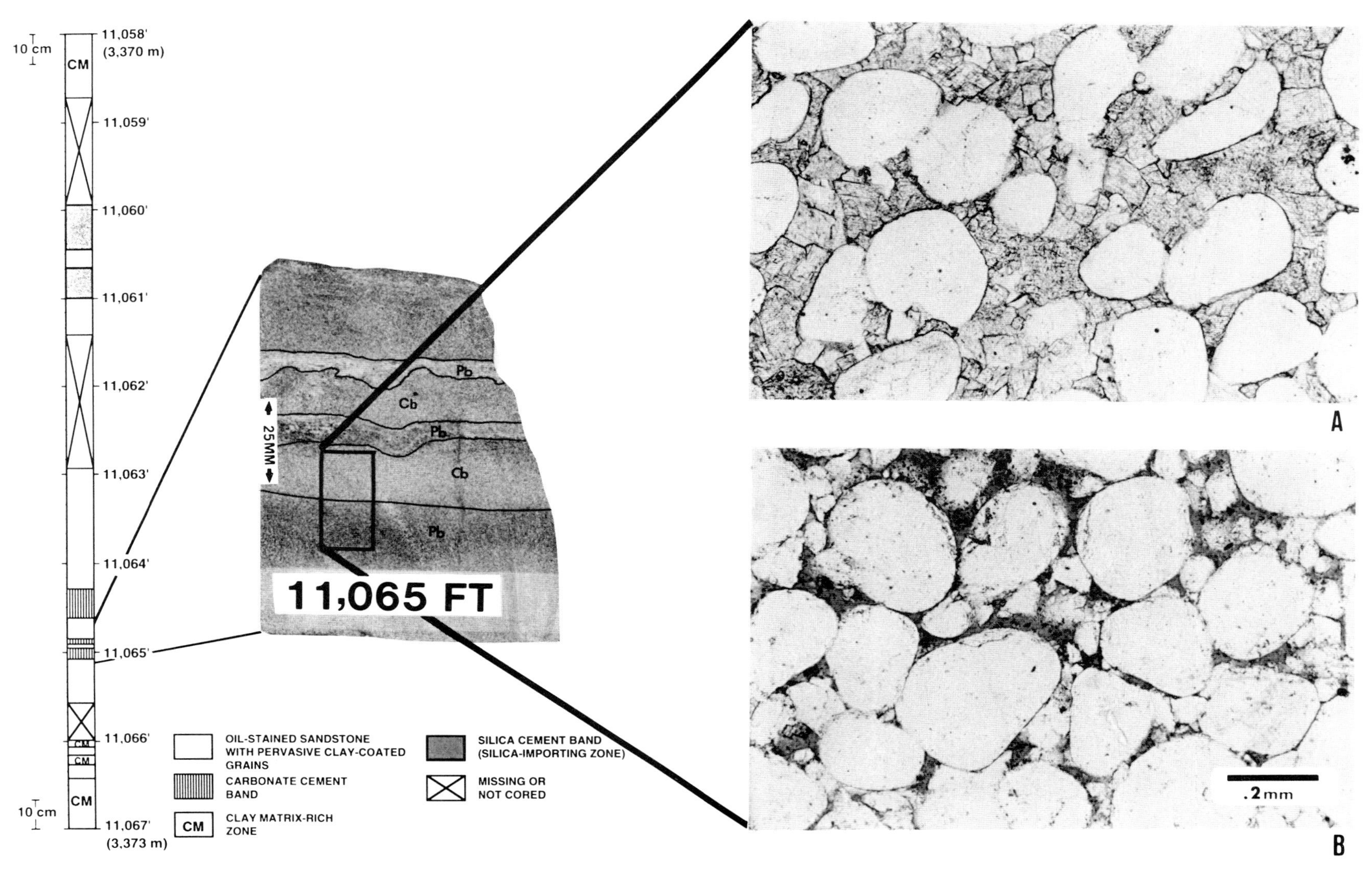

Figure 9. Calcite-cemented bands (Cb) alternating with porous bands (Pb). Photomicrographs: (A) Carbonate band; (B) porous band. Gulf, Weaver No. 1. Depth 3372 m (11,065 ft).

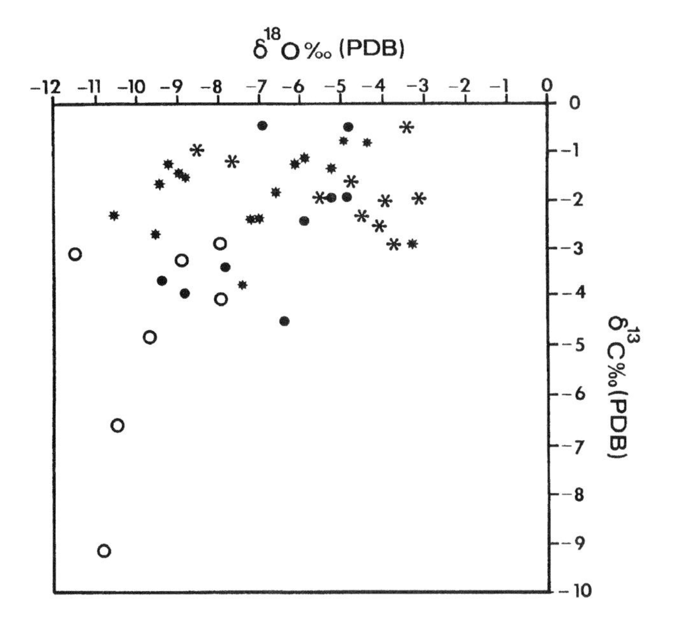

Figure 10. Cross-plot of carbon and oxygen isotope composition for carbonate cements in the Simpson sandstone seal interval. Data are from Al-Shaieb et al. (1992), Mitcheltree (1991), and Pitman and Burrus (1989). ❍ = calcite, Weaver Unit No. 1; ✳ = dolomite, Weaver Unit No. 1; ● = calcite, Costello and Mazur; ⁕ = dolomite/ankerite, Costello and Mazur.

commonly filled fractures but also augmented existing carbonate bands. Higher-temperature quartz cements (Q-2) precipitated during this stage filled intergranular voids between the quartz overgrowths. Evolution of the seal is believed to have taken approximately 30 million years. The hydrocarbon-generation processes and continued burial likely led to overpressuring of the reservoirs below the sealed intervals.

BANDING IN CLAY-RICH ROCKS

Shales and siltstones of seal intervals contain distinct diagenetic bands. Comparative analyses of cores and outcrops of shales and siltstones reveal this banding is noticeably absent in rocks that were never buried deeply enough to become overpressured. In the Pennsylvanian Pink Limestone/Red Fork Sandstone interval, diagenetic bands apparently developed independently of any depositional facies. On the other hand, banding in the Devonian Woodford Shale commonly occurs as an enhancement and/or modification of existing sedimentary features.

The Pink Limestone/Red Fork Sandstone seal interval consists of calcareous black shales and thinly laminated siltstones and shales. The frequency of diagenetic bands and the lateral extent of the banded interval (Al-Shaieb et al., 1991) suggest that this is an

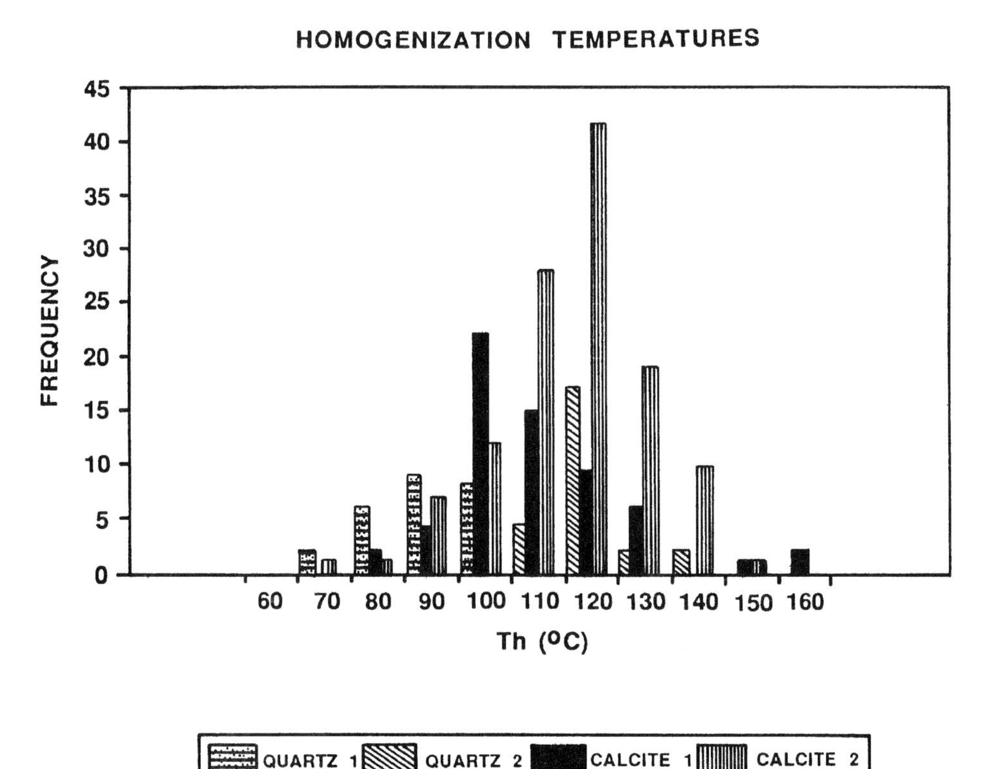

Figure 11. Homogenization temperatures (T_h) of various cement episodes within the Simpson sandstone seal, Gulf, Weaver No. 1.

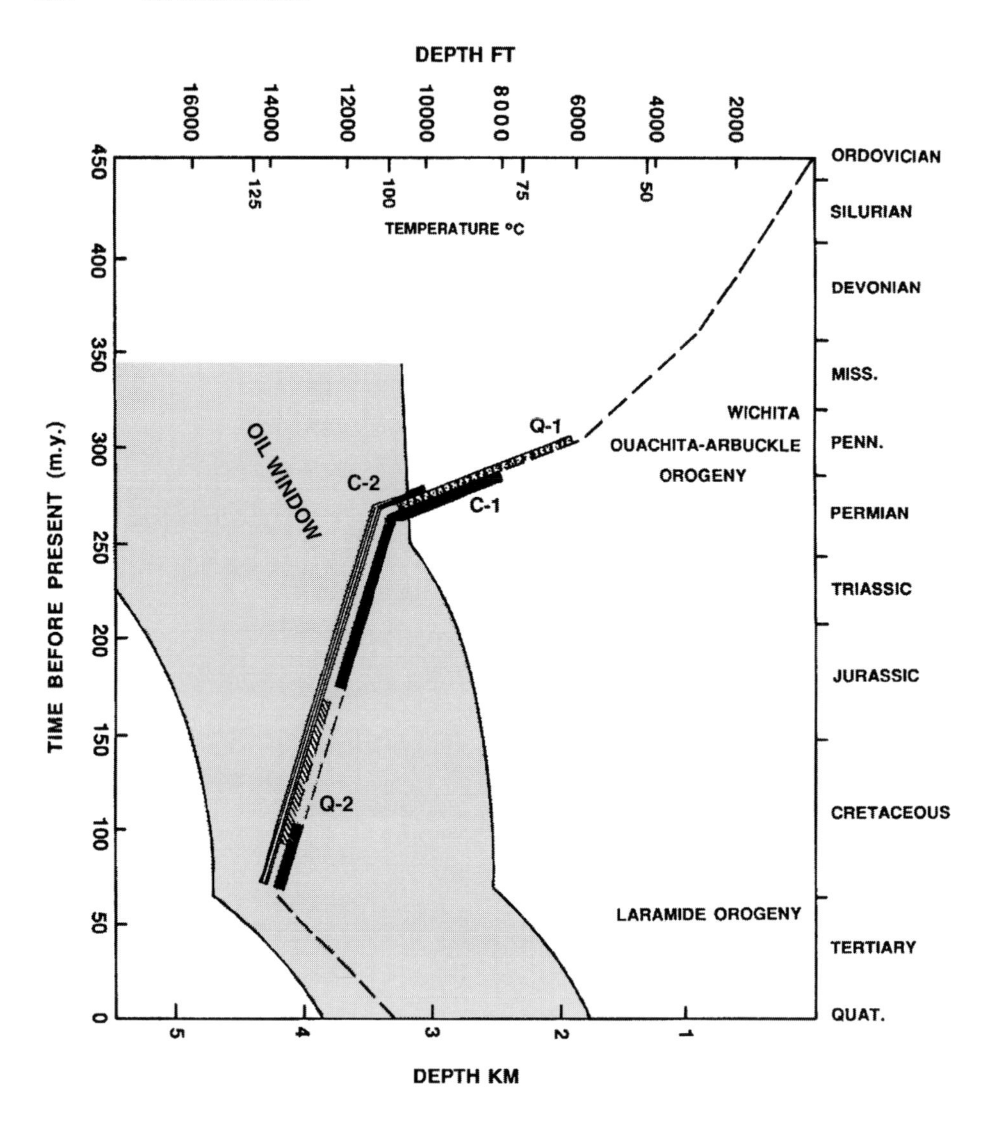

Figure 12. Timing of top seal cementation phases in relation to burial history of the Simpson interval, Gulf, Weaver No. 1 well.

effective seal for the underlying overpressured Red Fork reservoirs. The limited core data indicate the seal interval consists of at least 6 m (20 ft) of black shale that contains repetitive diagenetic bands. These bands have specific morphological characteristics. One group has sharp boundaries between the band and the surrounding shale (Figure 13). These bands are generally (1) of uniform thickness and flat or (2) of irregular thickness and hummocky or concretionary in appearance. The bands are typically chlorite rich. They generally range from 3 to 10 cm (1.25 to 4 in.) in thickness and are separated by finely laminated, illitic black shale. Some of the bands are cross-cut by fractures that are calcite cemented.

A second group of bands is characterized by gradational boundaries with the host rock (Figure 14). In some instances, a series of 0.6 to 1.3 cm (0.25 to 0.5 in.) thick bands within the shale creates a banded zone approximately 15.3 cm (6 in.) thick (Figure 15). These bands are also chlorite rich and apparently formed as a result of the diagenetic alteration of the shale. The origin of the chlorite cement is somewhat problematic, but it may have formed from siderite as suggested by Powers (1991).

Mercury injection tests indicate that the chlorite bands have much lower porosity and smaller pore throats than the surrounding shales (Powers, 1991). Band porosity is approximately 0.6% compared to 6.6% for the host shale. Most pore throat radii in the bands are less than 0.004 microns while those in the shale are between 0.011 and 0.004 microns (Powers, 1991). These measurements suggest that repetitive bands restrict fluid flow much more effectively than unaltered shale and are critical to seal competence.

The Woodford Shale basal seal interval contains diagenetic banding that is primarily facies controlled. The sand- and silt-rich intervals of the lower Woodford Shale facies are highly calcite cemented, and visible porosity appears completely occluded. Diagenetically enhanced sedimentary lithologies are also prominent in the middle Woodford euxinic shale. Silica-rich siltstone bands that are 0.63 to 10 cm (0.25 to 4.0 in.) thick have abrupt contacts with the enclosing black shale (Figure 16). These bands are usually

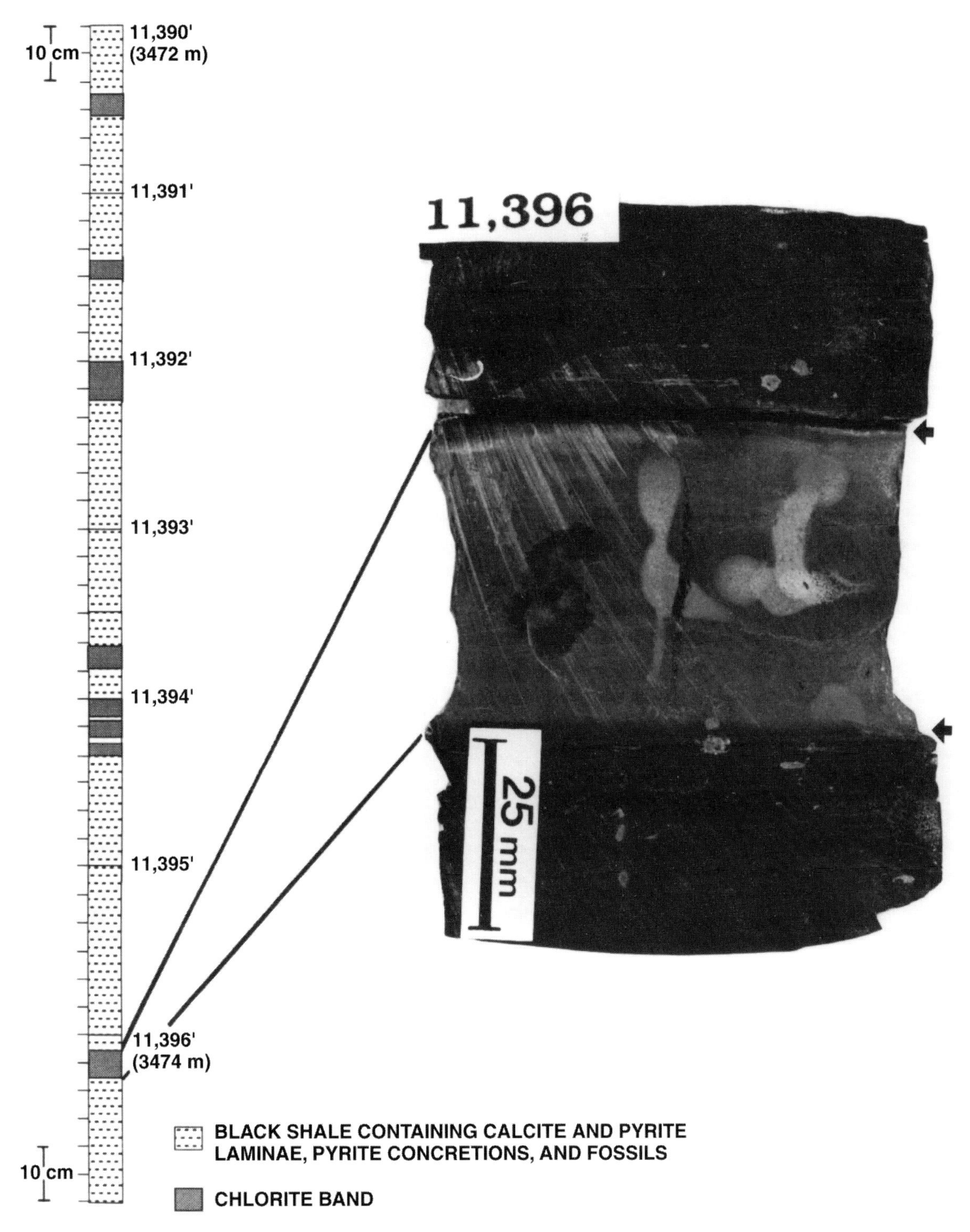

Figure 13. Diagenetic band exhibiting sharp boundaries (arrows) with shale host rock. Fractures are calcite cemented. Schematic diagram indicates location of the described feature within the seal zone. Woods, Switzer No. 1. Roger Mills County, Oklahoma. Depth 3474 m (11,396 ft).

cut by calcite-cemented fractures. The upper Woodford Shale contains laterally continuous, phosphate-rich chert bands 2.5 to 10.0 cm (1.0 to 4.0 in.) thick that exhibit abrupt contacts with dark shale (Figure 17).

Organization of the upper Woodford into alternating chert and shale bands reflects the diagenetic alteration of silica-rich deeper-water sediments. Silica and phosphate in the pre-altered sediment may have been supplied by the degradation of radiolarian tests

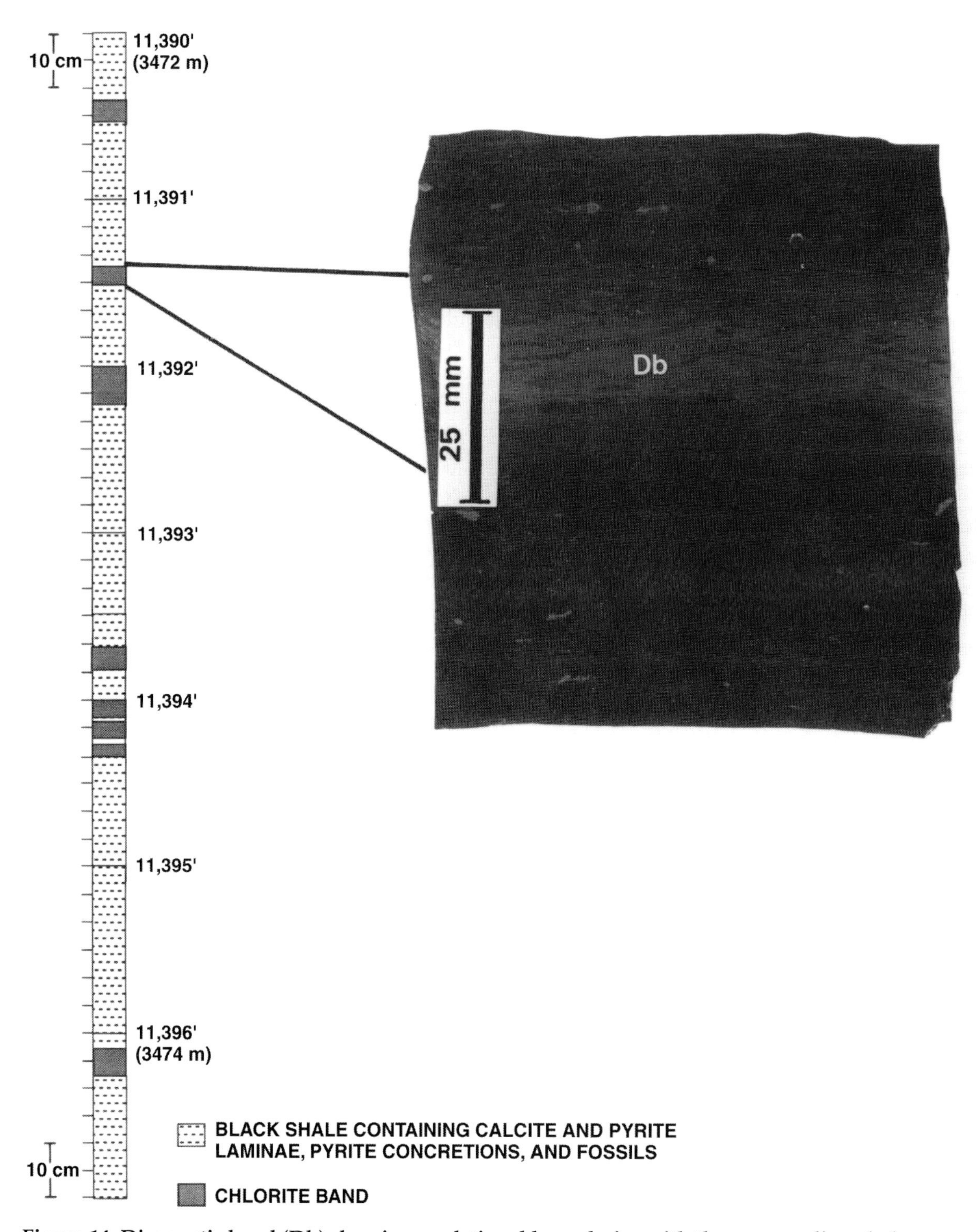

Figure 14. Diagenetic band (Db) showing gradational boundaries with the surrounding shale. This suggests the band formed as a diagenetic alteration of the shale. Woods, Switzer. Roger Mills County, Oklahoma. Depth 3471 m (11,391 ft).

(Spesshardt, 1985). The scarceness of chert in the lower and middle Woodford suggests they were deposited in shallower water where siliceous phytoplankton were less abundant and little chert accumulated (Spesshardt, 1985). With burial and compaction, we believe that silica and phosphate became concentrated during the spontaneous differentiation of the alternating silica/phosphate and shale bands.

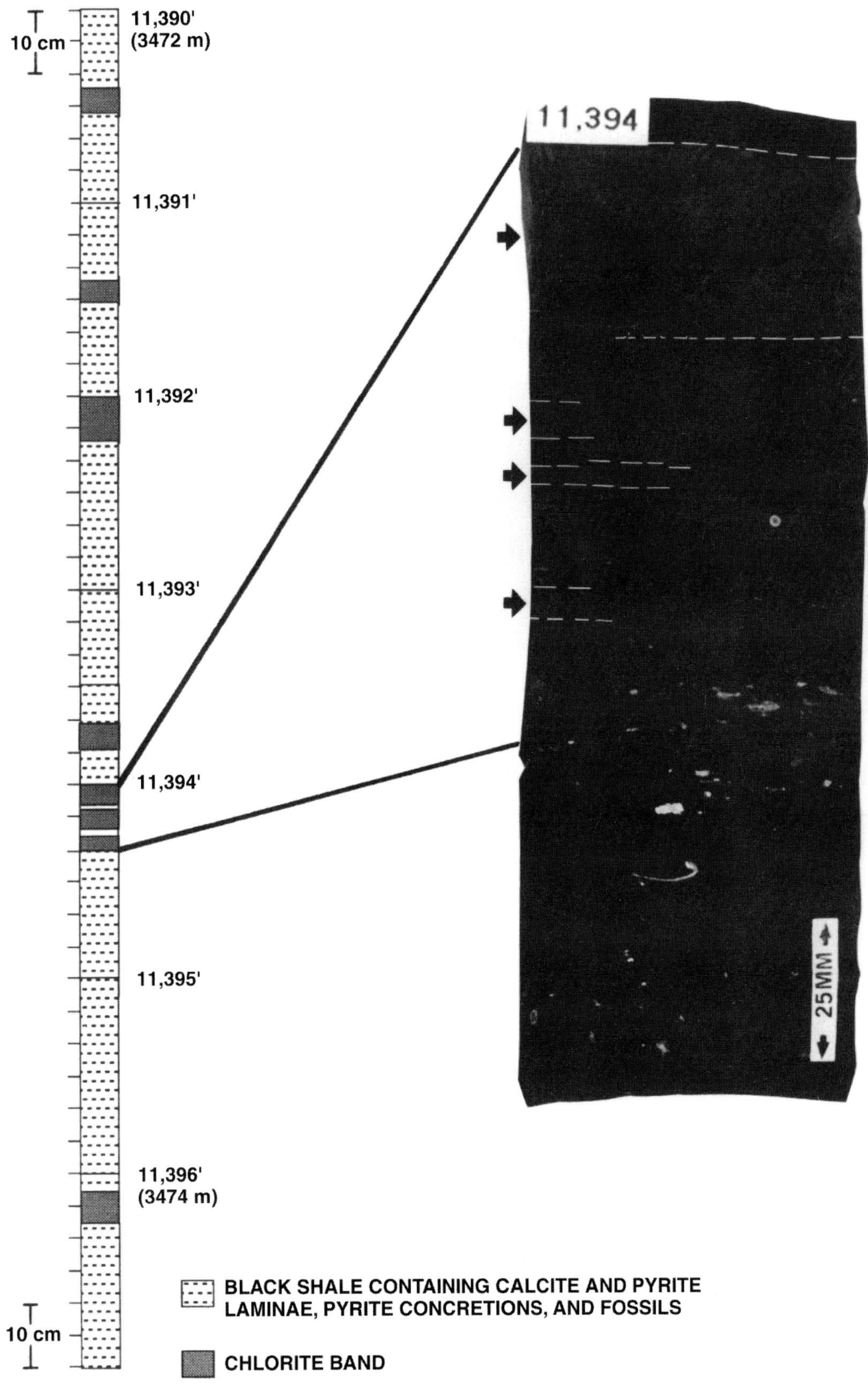

Figure 15. Series of gradational bands (arrows and dashed lines) that form a banded zone. Woods, Switzer. Depth 3470 m (11,394 ft).

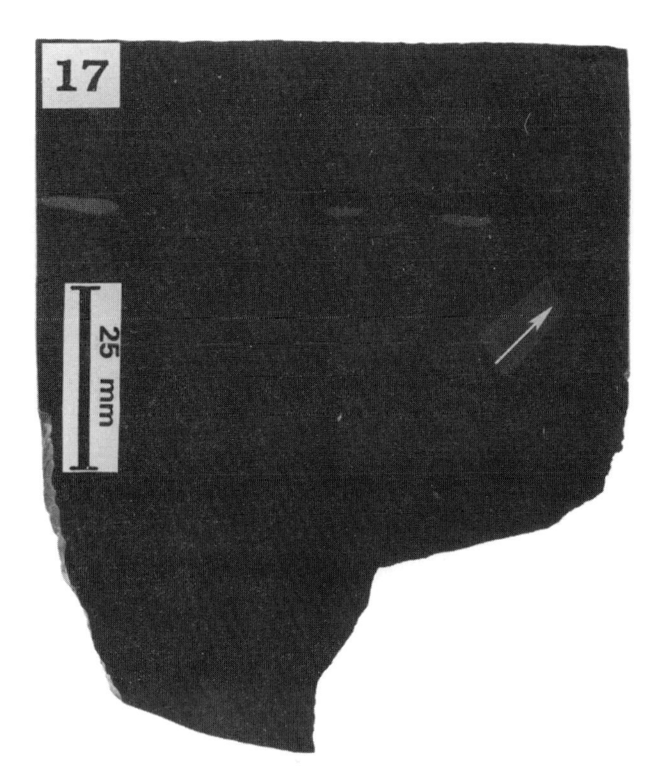

Figure 16. Diagenetic modification of middle Woodford Shale. Silica band with pyrite. Cross-cutting fracture (arrow) is calcite cemented. Universal, Dannehl. Canadian County, Oklahoma. Depth 2626 m (8617 ft).

CONCLUSIONS

The examination of banded pressure seals in the Anadarko basin integrated detailed petrographic analyses with depth-burial history curves to decipher band morphology and genesis. The following conclusions were formulated using the findings of this study.

- Banded pressure seals are significant geological phenomena since they may form a trapping mechanism for petroleum.
- In rocks that entered the seal window, the mechano-chemical processes of pressure-solution–induced dissolution and precipitation generated cementation bands.
- Silica imported from adjacent porous sandstones augmented quartz cementation in silica bands while band-generating carbonate was apparently imported from nearby carbonate units.
- Porous regions in sandstones often contain clay-grain coatings that inhibited cement precipitation.
- Porous clay-rich bands that were subjected to stress-induced dissolution often served as the source of silica for the cemented bands.
- Bands in sandstones are multi-episodic and formed within a variety of temperature and fluid domains occurring during different stages of basin evolution.
- Clay-rich rocks contain bands that developed independently of depositional features and others that resulted from the modification of existing features.
- Chlorite bands appear to have formed in part by the alteration of the surrounding shale and have significantly lower porosity and smaller pore-aperture radii.
- Banding in shales and mudstones was influenced by host rock composition and produced under burial conditions similar to those of sandstone bands.

Finally, our observations indicate that the diagenetic banding arose through processes of differentiation from depositionally uniform or less-differentiated sediment through feedback processes. Such self-organization phenomena have been observed in a variety of systems and are reviewed in Ortoleva et al. (1987) and Ortoleva (in press). Differentiated banding through mechano-chemical processes has been quantitatively characterized by Ricken (1986) in the context

Figure 17. Diagenetic banding in the upper Woodford Shale. Chert bands (Ch) contain phosphate nodules (Ph) and exhibit sharp contacts with adjacent dark shales (Sh). Arbuckle Mountains, Oklahoma. Camera lens cover is 50 mm in diameter.

of marl/limestone sequences and has been modeled via reaction-transport-mechanical equations by P. Ortoleva and co-workers (Ortoleva et al., 1982, 1993; Merino et al., 1983; Dewers and Ortoleva, 1990a, b, 1993; Qin and Ortoleva, this volume).

ACKNOWLEDGMENTS

The authors gratefully acknowledge the Gas Research Institute for funding this research through Contract No. 5089-261805.

REFERENCES CITED

Al-Shaieb, Z., J. Puckette, R. Ely, and A. Abdalla, 1990, Fluid inclusion analysis and genesis of a top seal: Quarterly Report prepared for the Gas Research Institute, Chicago, Illinois.

Al-Shaieb, Z., J. Puckette, P. Ely, A. Abdalla, and A. Rice, 1991, Banding patterns in clay-rich clastic rocks: Quarterly Report prepared for the Gas Research Institute, Chicago, Illinois.

Al-Shaieb, Z., J. Puckette, A. Abdalla, and R. Ely, 1992, Megacompartment complex in the Anadarko basin: a completely sealed overpressured phenomenon: Annual Report prepared for the Gas Research Institute, Chicago, Illinois.

Al-Shaieb, Z., J. Puckette, A. Rice, and A. Abdalla, 1993, Preferential cementing and auto-isolation in the Lower Cunningham Sandstone: Quarterly Report prepared for the Gas Research Institute, Chicago, Illinois.

Dewers, T., and P. Ortoleva, 1988, The role of geochemical self-organization in the migration and trapping of hydrocarbons: Applied Geochemistry, v. 3, p. 287–316.

Dewers, T., and P. Ortoleva, 1990a, A coupled reaction/transport/mechanical model for intergranular pressure solution, stylolites, and differential compaction and cementation in clean sandstones: Geochimica et Cosmochimica Acta, v. 54, pt. 2, p. 1609–1625.

Dewers, T., and P. Ortoleva, 1990b, Geochemical self-organization III: a mechano-chemical model of metamorphic differentiation: American Journal of Science, v. 290, p. 473–521.

Dewers, T., and P. Ortoleva, 1991, Nonlinear dynamics in chemically compacting porous media: Annual Report prepared for the Gas Research Institute, Chicago, Illinois.

Dewers, T., and P. Ortoleva, 1993, Formation of stylolites, marl/limestone alternations, and clay seams through unstable chemical compaction of argillaceous carbonates, *in* K. H. Wolf and G. V. Chilingarian, eds., Diagenesis Volume IV: Amsterdam, Elsevier, p. 155–216.

Merino, E., P. Strickholm, and P. Ortoleva, 1983, Generation of evenly spaced pressure-solution seams during (late) diagenesis: a kinetic theory: Contributions to Mineralogy and Petrology, v. 82, p. 360–370.

Mitcheltree, D. B., 1991, Diagenetic pressure seal analysis using fluid inclusions and stable isotopes, the Simpson Group (Middle Ordovician), Anadarko basin, Oklahoma: Unpublished M.S. thesis, New Mexico Institute of Mining and Technology, 107 p.

Ortoleva, P., in press, Geochemical Self-Organization: New York, Oxford University Press.

Ortoleva, P., E. Merino, and P. Strickholm, 1982, Kinetics of metamorphic layering in anisotropically stressed rocks: American Journal of Science, v. 282, p. 617–643.

Ortoleva, P., E. Merino, C. Moore, and J. Chadam, 1987, Geochemical self-organization I: Reaction-transport feedbacks and modeling approach: American Journal of Science, v. 287, p. 979–1007.

Ortoleva, P., T. Dewers, and B. Sauer, 1993, Modeling diagenetic bedding, stylolites, concretions, and other mesoscopic mechano-chemical structures, *in* R. Rezak and D. Lavoie, eds., Carbonate Microfabrics: New York, Springer-Verlag, p. 291–300.

Ortoleva, P., Z. Al-Shaieb, and J. Puckette, in press, Genesis and dynamics of basin compartments and seals: American Journal of Science.

Pitman, J. K., and R. C. Burrus, 1989, Diagenesis of hydrocarbon-bearing rocks in the Middle Ordovician Simpson Group, southeastern Anadarko basin, Oklahoma, in Anadarko Basin Symposium, 1988: Oklahoma Geological Survey Circular 90, p. 134–142.

Powers, G. R., 1991, A petrographic study of lithologies forming the top seal in the Anadarko basin: Unpublished M.S. Project Report, University of Tulsa, 46 p.

Ricken, W., 1986, Diagenetic Banding: Berlin, Springer-Verlag, 210 p.

Schmoker, J. W., 1986, Oil generation in the Anadarko basin, Oklahoma and Texas: modeling using Lopatin's method: Oklahoma Geological Survey Special Publication 86-3, 40 p.

Spesshardt, S. A., 1985, Late Devonian–early Mississippian phosphorite-bearing shales, Arbuckle Mountain region, south-central Oklahoma: Unpublished M.S. thesis, Texas Tech Univ., 106 p.

Tigert, V., 1989, Identification and characterization of a pressure seal in south-central Oklahoma: Unpublished M.S. thesis, Oklahoma State Univ., 83 p.

Tigert, V., and Z. Al-Shaieb, 1990, Pressure seals: their diagenetic banding patterns: Earth-Science Review, v. 29, p. 227–40.

Chapter 24

Silica Budget for a Diagenetic Seal

Lisa D. Shepherd*
Peter A. Drzewiecki
Jean M. Bahr
J. Antonio Simo
University of Wisconsin — Madison
Madison, Wisconsin, U.S.A.

ABSTRACT

Diagenetic banding commonly occurs in association with zones of abnormal fluid pressures that have been identified as pressure compartments. Although diagenetically banded intervals may contain layers of moderately high porosity, the bands act collectively as a low-permeability unit and are therefore important as potential low-permeability seals for pressure compartments. This study focuses on a diagenetically banded interval in the Middle Ordovician St. Peter Sandstone of the Michigan basin. This interval is located within a large area of anomalous pressures identified by Bahr et al. (this volume) in the deep Michigan basin and is composed of a seal-forming lithology. The banded interval is characterized by millimeter- and centimeter-scale diagenetic banding, with alternating quartz-cemented bands, pressure solution-dominated bands, and porous bands.

Point counting techniques and an image analysis system were used to quantify porosity, textural properties, and quartz cement. A theoretical model was used in conjunction with these data to estimate the amount of silica dissolved by intergranular pressure solution. Porosity variations in the St. Peter Sandstone are controlled by the combined effects of quartz cementation and intergranular pressure solution. A silica budget calculated for the banded interval indicates that more silica was dissolved by intergranular pressure solution than is present as quartz cement, suggesting that pressure solution alone could have produced enough silica to account for the banded quartz cement. On a local scale, the banded interval served as an exporter of silica. However, a larger-scale silica budget analysis computed for another well in the same region of the basin indicates that the St. Peter may actually be balanced on a regional scale.

Results of this study were used to investigate the controls on diagenetic band formation. No significant correlation exists between porosity and grain size or porosity and sorting in the banded interval, suggesting that depositional textural parameters are not important in controlling the distribution of porosity and cement within the banded interval itself. However, original

*Present address: RMT Inc., Madison, Wisconsin.

depositional textures do appear to be important in controlling the localization of diagenetic banding, with banding developing preferentially in intervals that are relatively fine-grained and well-sorted. The occurrence of banding in the St. Peter Sandstone is qualitatively consistent with predictions of models developed by Ortoleva et al. (1987) and Dewers and Ortoleva (1990a, b, c). However, net export of silica over meter-scale intervals implies that advective flux should be included in the models in addition to diffusional transport.

INTRODUCTION

Diagenetic banding has been observed in a number of sedimentary basins (Heald and Anderegg, 1960; Hunt, 1990; Tigert and Al-Shaieb, 1990), commonly occurring in association with zones of abnormal fluid pressures that have been identified as pressure compartments (Hunt, 1990; Tigert and Al-Shaieb, 1990). The banding consists of alternating layers of different cement types and/or of significantly different porosities. The main processes responsible for the formation of banding include intergranular pressure solution, dissolution, and cementation. Bands are typically horizontal and extend laterally across the width of the cores. The vertical thickness of individual bands ranges from millimeters to meters. In the St. Peter Sandstone there appears to be a lithologic control on the location of banded intervals (Drzewiecki et al., this volume), with banding being best developed in sandstones that were originally horizontally bedded. The planar textural variations in this lithofacies appear to promote the development of bands and stylolites. Although high-porosity zones can occur within a banded interval, an abundance of horizontal bands in which porosity is completely eliminated by cement can result in a very low effective vertical permeability for the interval as a whole. Thus, banded intervals can act as seals for pressure compartments and as diagenetic traps for hydrocarbons.

Examination of cores from the Middle Ordovician St. Peter Sandstone in the deep Michigan basin has revealed widespread banding of five major types: quartz-cemented bands, dolomite-cemented bands, clay-cemented bands, pressure solution-dominated bands, and porous bands. In a mature quartzarenite like the St. Peter, clay- and dolomite-cemented bands require an external source for the cementing agent. However, alternating quartz-cemented, pressure solution-dominated, and porous bands could be the result of local redistribution of silica following dissolution and pressure solution of detrital quartz and feldspar. Stylolites and textures diagnostic of intergranular pressure solution are abundant in zones containing quartz-cemented bands. Both intergranular pressure solution and stylolitization produce silica that could be redistributed, leading to the formation of diagenetic bands. Furthermore, the banding provides planar textural heterogeneities that encourage further stylolitization and band development. Thus, intergranular pressure solution, stylolitization, and diagenetic band formation may be integrally linked.

This chapter presents the results of a detailed investigation of a 3 m diagenetically banded interval from the St. Peter Sandstone in the Mobil St. Foster 1-12 core. The core is located in a region of the Michigan basin that is characterized by vertically stacked pressure compartments (Bahr et al., this volume). A silica budget calculation was used to examine the relative importance of quartz cementation and pressure solution in the formation of millimeter-scale diagenetic bands. More limited analyses of the silica budget computed for additional core from this well and from another well in the same region of the basin were used to assess the relative importance of compaction, cementation, and open system fluid flow on porosity reduction in the St. Peter Sandstone. Statistical analysis was used to investigate the relationships among porosity, grain size, sorting, quartz cement, and intergranular pressure solution in the banded interval. Results from this analysis were used to examine the main controls on porosity variations and diagenetic band formation in the St. Peter Sandstone and to evaluate theoretical models for banding formation.

REGIONAL GEOLOGY

The Middle Ordovician St. Peter Sandstone in the Michigan basin is situated stratigraphically between the Lower Ordovician Prairie du Chien Group below and the Middle Ordovician Glenwood Formation above. On a macroscopic scale the St. Peter is a relatively homogeneous quartz sandstone, characterized by fine- to medium-grained, well-rounded, and moderately to well-sorted sand grains. It is composed predominantly of quartz but also contains minor amounts of feldspar, concentrated mainly in the finer-grained sandstones. The dominant cement in the St. Peter Sandstone is authigenic quartz, in the form of syntaxial overgrowths. Minor amounts of dolomite and clay (mainly illite) cements are also present. The St. Peter reaches a maximum thickness of over 320 m in the north-central part of Michigan (Nadon et al., 1991) and thins toward the basin margins. In the

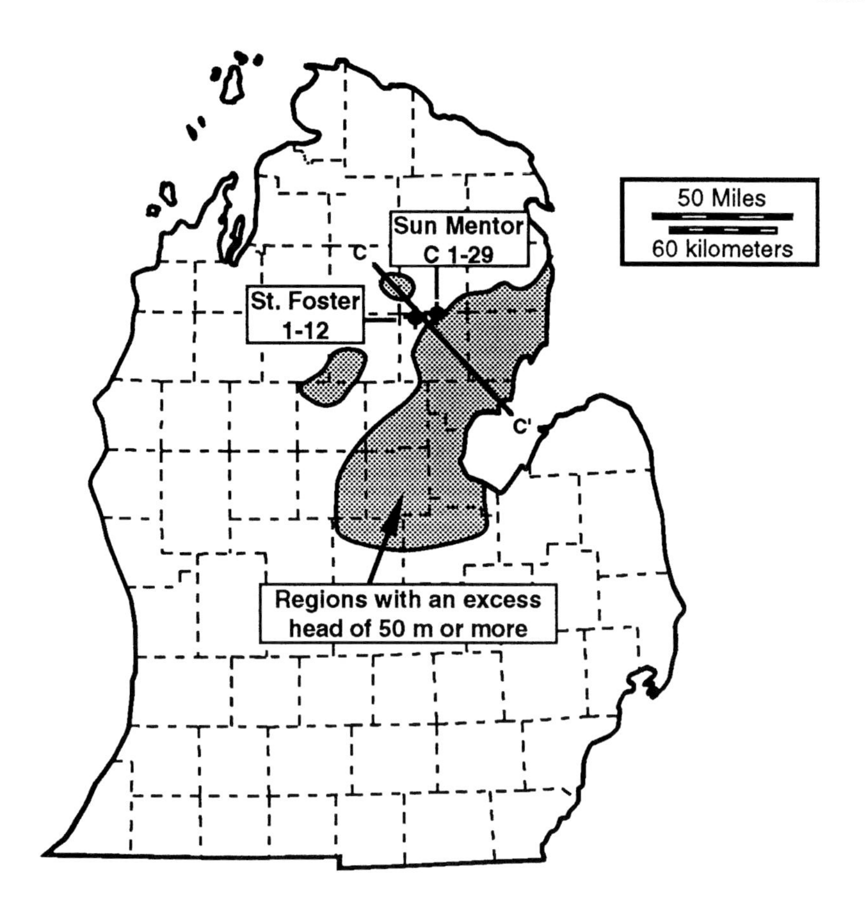

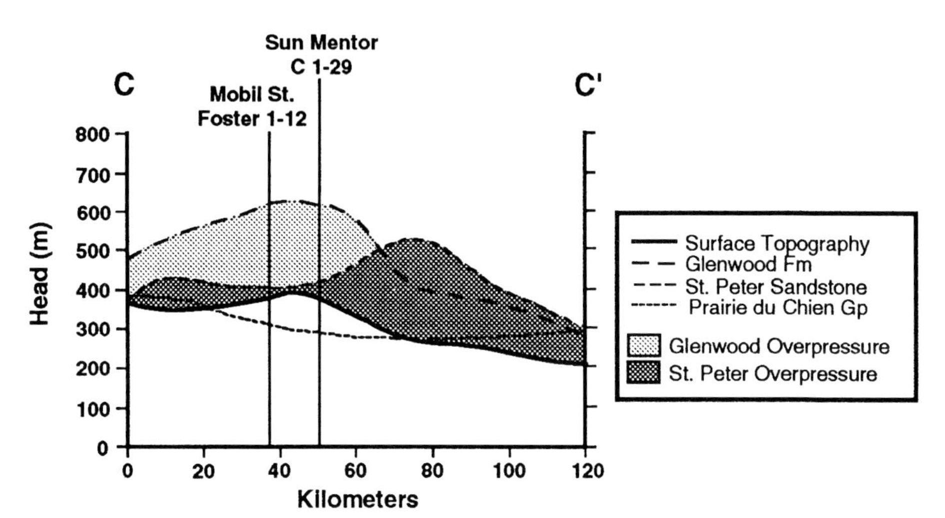

Figure 1. Map of the lower peninsula of Michigan showing regions where computed heads in the St. Peter Sandstone exceed surface elevation by at least 50 m (shaded regions). A cross section through the anomalously pressured regions shows the head distribution in the Glenwood Formation, St. Peter Sandstone, and Prairie du Chien Group relative to surface topography. The shaded areas in the cross section show the distribution of excess head in the St. Peter and Glenwood. Locations of the Mobil St. Foster 1-12 and Sun Mentor C 1-29 wells are shown on the map and cross section.

southern portion of Michigan, it has been removed by erosion (Catacosinos and Daniels, 1991).

Mobil St. Foster 1-12 core

A 15 m core from the Mobil St. Foster 1-12 well was chosen for this study. The cored interval in this well is from the base of the St. Peter, just above the contact with the Prairie du Chien Group. A 3 m interval located in the upper part of the core was chosen for detailed analysis because it is characterized by millimeter-scale diagenetic banding and therefore is ideal for studying the processes that control banded cementation in the St. Peter. The Mobil St. Foster 1-12 well lies within a large area of anomalous pressures in the St. Peter Sandstone (Figure 1) identified on the basis of computed hydraulic heads and gradients in the St. Peter Sandstone and the overlying and underlying formations (Bahr et al., this volume). Bahr et al. (this volume) used RFT measurements from the Mobil St. Foster 1-12 well and DST measurements from neighboring wells to identify an overpressured zone that extends into the lowermost St. Peter Sandstone (Figures 1 and 2). A decrease in overpressure occurs at the base of the St. Peter, between 2860 m and 3020 m below sea level. This indicates that the lower St.

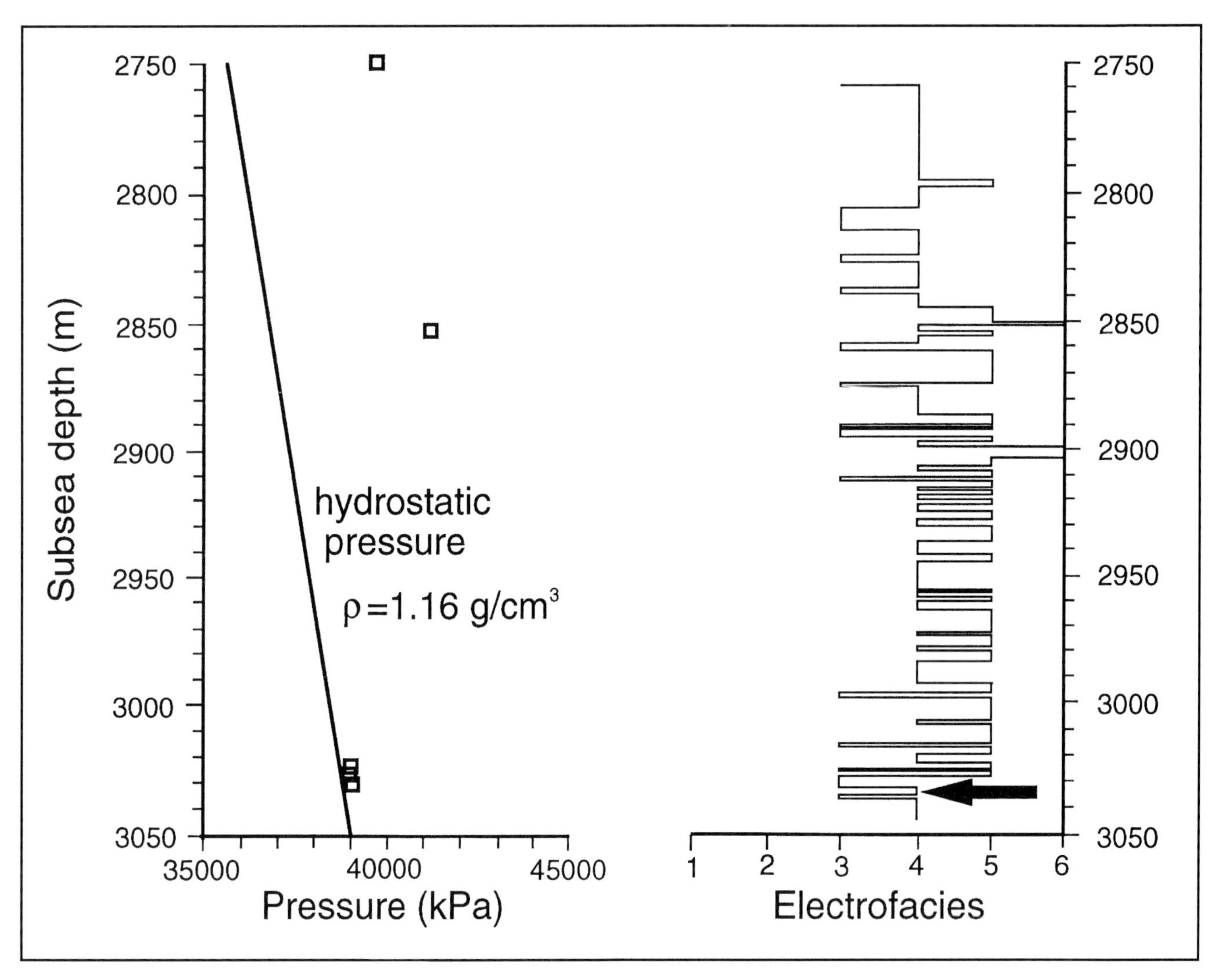

Figure 2. Pressure and electrofacies distribution in the lower St. Peter Sandstone of the Mobil St. Foster 1-12 well. A pressure compartment seal exists between depths of approximately 2860 m (where overpressure occurs) and 3020 m (where hydrostatic pressure exists). This seal is characterized by electrofacies ef4 and ef5. The arrow indicates the location of the 3 m banded interval (electrofacies ef4) that is the focus of this investigation.

Peter Sandstone contains a basal seal to the overlying overpressured zone.

The St. Peter Sandstone in the Mobil St. Foster 1-12 well has been divided into electrofacies based on log responses and porosity and permeability characteristics (Moline et al., this volume). Electrofacies ef4, ef5, and ef6 are characterized by low porosities and permeabilities. The basal seal of the overpressured compartment in this well is composed primarily of electrofacies ef4 and ef5 (Figure 2). A 3 m banded interval in the Mobil St. Foster core, which lies just below the seal, has also been identified as electrofacies ef4. Assuming the cored interval has lithologic characteristics similar to the electrofacies ef4 zones identified in the seal, a detailed investigation of this interval will provide information on the genesis of the seal-forming ef4 electrofacies. Electrofacies ef5 was not present in the cored interval and could not be investigated.

The 3 m banded interval examined in this study is located near the upper end of the Mobil St. Foster core (Figure 3) and has some of the lowest permeabilities measured in core plugs, ranging from 0.08 to 1 md. Thin-section analysis indicates that this interval is characterized by low average porosity and abundant quartz overgrowth cement (Figure 3). Stylolites (described in Drzewiecki et al., this volume) and millimeter- to centimeter-scale bands are abundant throughout the interval and are visible both in hand specimen and in thin section (Figure 4). Furthermore, dolomite and clay cements are negligible in the 3 m interval, making this core ideal for an investigation of the processes resulting in the redistribution of silica.

There are over 200 bands in the 3 m interval of the Mobil St. Foster core. These bands have been divided into two main types, based on porosity differences: (1) nonporous bands and (2) porous bands. Thickness of the nonporous bands ranges from a few millimeters to

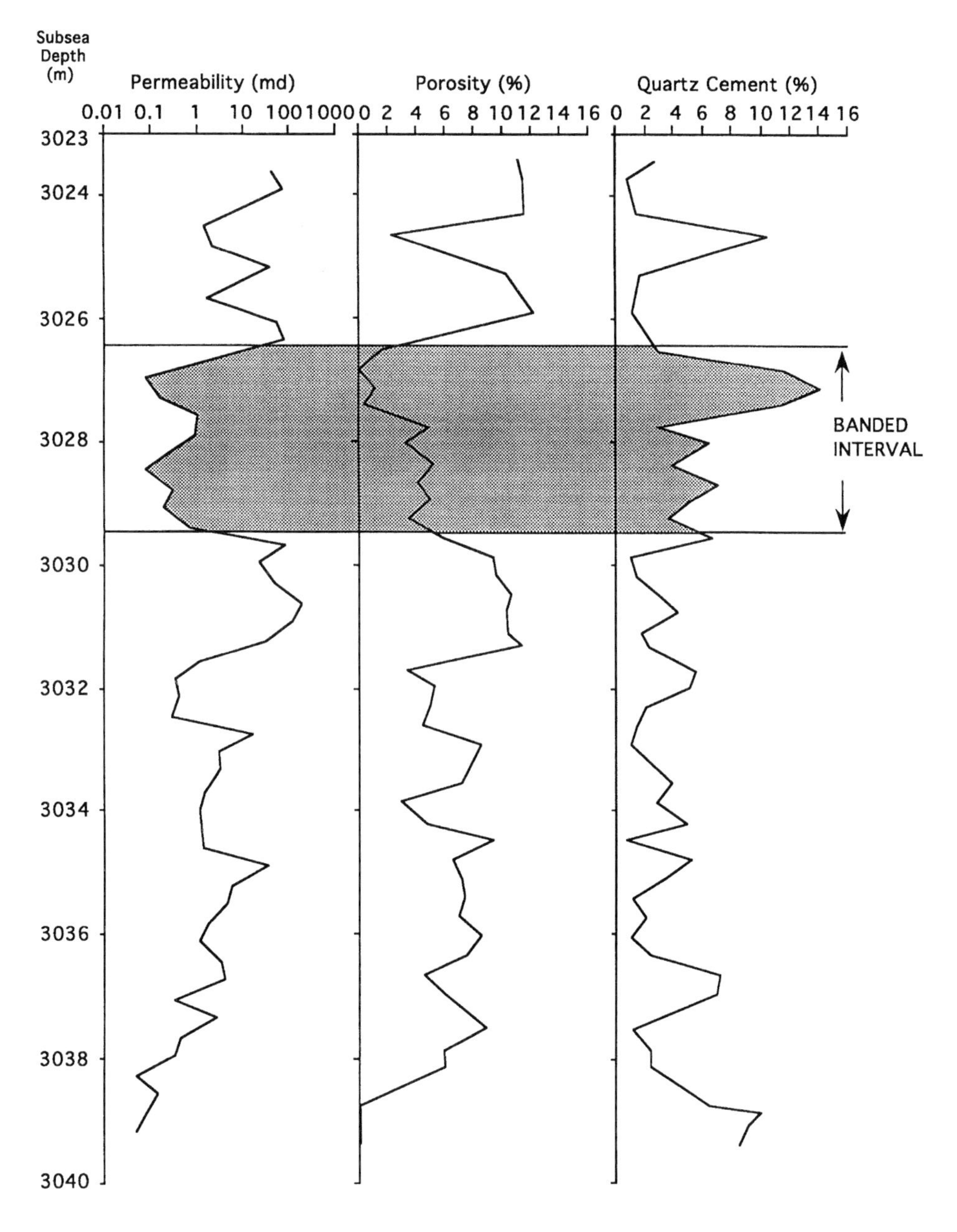

Figure 3. Variation in permeability, porosity, and quartz overgrowth cement with depth for the St. Peter Sandstone in the Mobil St. Foster 1-12 core. The shaded area shows the location of the 3 m banded interval discussed in this study. Moline et al. (this volume) have identified this interval as a low-permeability electrofacies based on wireline log signatures.

7 cm, with an average thickness of 1.2 cm. Nonporous bands have porosities ranging from 0 to 3% and are of two types: quartz-cemented bands and pressure solution-dominated bands. Diagenetic studies (Drzewiecki et al., this volume) indicate that these bands formed contemporaneously during burial diagenesis, suggesting that intergranular pressure solution may have been a source of quartz cement. Quartz overgrowths are typically well developed in quartz-cemented bands, making up about 10% of the rock by volume. In general, quartz-cemented bands appear to have undergone very little dissolution by intergranular pressure solution. Pressure solution–dominated bands contain very little quartz cement and generally have porosities of <1%. Porous bands have porosities ranging from 3 to 15% and range in thickness from a few millimeters to over 13 cm. The porous bands generally have only a minor amount of quartz cement and have undergone a wide range of dissolution by pressure solution, resulting in a wide range of porosities. Preliminary scanning electron microscope studies indicate that quartz grains in the porous bands are often coated with a thin lining of clay, which may have inhibited the development of quartz overgrowths. Porous bands generally form sharp contacts with both the quartz-cemented bands (Figure 4B) and the pressure solution–dominated bands. Some porous bands contain primary porosity inherited from the time of deposition; others were formed or enhanced by secondary dissolution of either quartz or dolomite (Drzewiecki et al., this volume).

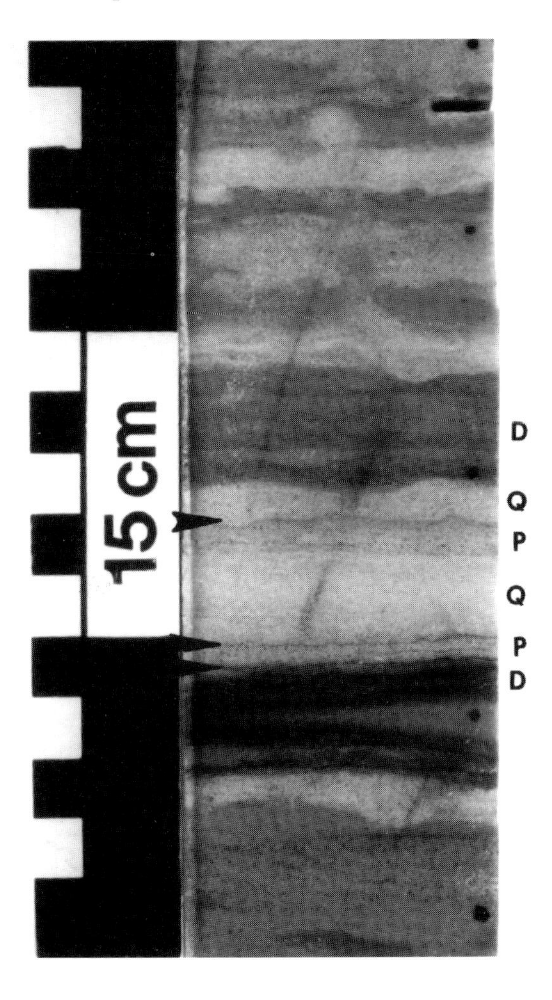

A

Figure 4. Diagenetic banding in the Mobil St. Foster 1-12 core is visible both in hand specimen and in thin section. (A) Piece of the Mobil St. Foster 1-12 core showing millimeter- to centimeter-scale diagenetic banding. The individual band types are labeled as follows: Q = quartz-cemented band, P = porous band, and D = dolomite-cemented band. Note the stylolites (arrows) associated with this banded interval.

B

Figure 4 (continued). (B) Photomicrograph showing millimeter-scale diagenetic banding in the Mobil St. Foster 1-12 core. Photograph shows a nonporous, quartz-cemented band adjacent to a more porous band. Dark areas represent porosity. Note the sharp contact between the two band types. Scale bar is 1 mm.

METHODS

Measurement of Porosity, Textural Parameters, and Quartz Cement

A series of 88 samples representing continuous coverage of the 3 m banded interval were collected. In addition, samples were collected approximately every 0.3 m throughout the remainder of the core. Core samples were vacuum-impregnated with a low-viscosity blue-dyed epoxy and cut perpendicular to bedding before being made into thin sections.

A computerized image analysis system was used to estimate porosity, grain size, and sorting (characterized by the standard deviation of grain size) for each thin section. For each image, the area of the pore-filling epoxy was quantified as a measure of sample porosity. Between 20 and 65 images were analyzed per thin section, depending on the size of the thin section. The fields of view were evenly spaced throughout the thin section, forming a two-dimensional grid of points with a distance of 5 mm between each image. Grain size and sorting were quantified using measurements of the apparent long axis of approximately 200 grains per thin section. A random selection of grains was obtained by measuring the apparent long axes of all grains that crossed a horizontal line drawn across the image.

The image analysis system could not be used to quantify the amount of quartz overgrowth cement because the system was not able to distinguish between the color of the quartz overgrowths and the color of the detrital quartz grains. Other workers (Sibley and Blatt, 1976; Houseknecht, 1984, 1987, 1988, 1991; Murphy, 1984) have used cathodoluminescence microscopy to distinguish nonluminescent quartz overgrowths from red, brown, or blue luminescent quartz grains. However, the majority of quartz grains in samples analyzed from the St. Peter did not luminesce under our luminoscope, and for those that did it was difficult to distinguish between the dark luminescent detrital quartz and the nonluminescent overgrowths. For these reasons, the amount of quartz cement was estimated by a point counting method using a standard petrographic microscope. Over 600 points were counted for most thin sections.

Mean values of porosity, grain size, sorting, and quartz overgrowth cement were computed for each thin section. In addition to these thin-section averages, data were recorded as averages over 5 mm vertical intervals in order to capture the millimeter- to centimeter-scale variations in porosity associated with the banding.

Estimates of Quartz Dissolved by Intergranular Pressure Solution

Unlike volumes of porosity and cement, which can be measured directly by thin-section techniques, the amount of quartz that has been removed from a rock by intergranular pressure solution is relatively difficult to estimate. Both indirect measurements and theoretical models have been used to quantify intergranular pressure solution. Initial studies focused on geochemical controls (Weyl, 1959; Sprunt and Nur, 1976) and packing models of spherical grains (Kahn, 1956a, b). Graphical reconstruction of original grain shape, by projecting grain boundaries across areas of dissolution, has provided the basis for more recent estimates of dissolution employing cathodoluminescence and point counting techniques (Sibley and Blatt, 1976; Houseknecht, 1984, 1987, 1988, 1991; Murphy, 1984).

In this study we used a version of the theoretical model developed by Mitra and Beard (1980). Their model is an extension of those developed by Rittenhouse (1971) and Manus and Coogan (1974), which predict changes in intergranular volume, defined as the sum of porosity and cement, for cubic and orthorhombic packing arrangements subjected to intergranular pressure solution. These models are limited in that they restrict the value of initial porosity that can be used. In order to account for a more realistic, random packing of grains, Mitra and Beard simulated six stable packing arrangements and then combined them in various proportions to account for a large range of initial porosities. Mitra and Beard constructed plots of porosity versus vertical shortening for initial porosities of 30, 32.5, 35, 37.5, and 40%. Given an initial and final porosity of a sandstone, the appropriate curve can be used to determine the amount of vertical shortening and the amount of cement generated by intergranular pressure solution.

Mitra and Beard (1980) simulated two end-member cases for each initial porosity: one based on the assumption that all dissolved quartz is exported from the system and one based on the assumption that all dissolved quartz is precipitated locally. Assuming that real systems are likely to fall somewhere between these two end-members, this model is useful for placing upper and lower limits on the bulk volume reduction that can occur as a result of intergranular pressure solution.

The model of Mitra and Beard (1980) requires an estimate of post-mechanical compaction porosity, since this model is assumed to represent conditions following mechanical compaction during burial. The initial depositional porosity of the St. Peter Sandstone was estimated at 40%. This value agrees with studies of natural and artificially packed, well-sorted sands (Beard and Weyl, 1973; Pryor, 1973) and is consistent with estimates of initial depositional porosity of sandstones used by Houseknecht (1987), Sibley and Blatt (1976), and Ehrenberg (1990). Blatt (1979) reports that a bulk volume reduction of 10 to 15% can occur as a result of rearrangement and chipping of grains, corresponding to a post-mechanical compaction porosity of 25 to 30%. Atwater and Miller (1964), Surdam et al. (1989), and Wood (1989) found a range of about 25 to 30% for post-mechanical compaction porosity for deeply buried sands. In light of these studies, 25 to 30% appears to be a reasonable estimate of the post-mechanical compaction porosity for the St. Peter Sandstone and was assumed for our evaluation of intergranular pressure solution.

The model of Mitra and Beard (1980) was used to generate plots of vertical shortening versus porosity, assuming post-mechanical compaction porosities of 26% (the lower limit of the model) and 30% (Figure 5). From these curves the volume of cement generated for a given amount of vertical shortening was determined. For each initial porosity, three curves were computed as shown in Figure 5. Curve "a" represents the amount of vertical shortening assuming that all of the dissolved quartz is exported from the system. Curve "b" shows the amount of vertical shortening predicted if all the dissolved material is precipitated locally. Curve "c" shows the volume of cement generated for a given amount of vertical shortening.

These curves were used in conjunction with thin-section measurements of porosity and cement in order to estimate minimum and maximum amounts of intergranular pressure solution. Using the intergranular volume as an estimate of the maximum porosity with curve "a," a maximum amount of vertical shortening was determined based on the assumption of complete export of dissolved silica. The actual porosity for the sample was then used to determine the minimum vertical shortening from curve "b." The maximum and minimum values represent the upper and lower limits on the bulk volume reduction resulting from intergranular pressure solution. For each sample, the average of these two extremes was assumed to represent a more realistic value since some of the dissolved quartz was probably exported while some was precipitated as cement. This average value was used with curve "c" to determine the amount of silica generated from pressure solution.

SILICA BUDGET

Studies reported in the literature indicate that quartz cement in clastic rocks can be derived locally from processes such as intergranular pressure solution, stylolitization, clay mineral transformation, dissolution of detrital quartz and feldspar grains, kaolinitization of feldspars, and replacement of quartz grains by carbonates (McBride, 1989; Sibley and Blatt, 1976; Houseknecht, 1988), or can be transported into the formation by seawater, hypersaline brines, or meteoric water (McBride, 1989). The relative importance of local versus external silica sources can be assessed by comparing the estimated volume of dissolved silica with the amount of quartz cement observed in each sample. The difference between these two values provides an estimate of the volume of silica that has entered or exited the system. In this study the silica budget was used to determine (1) if

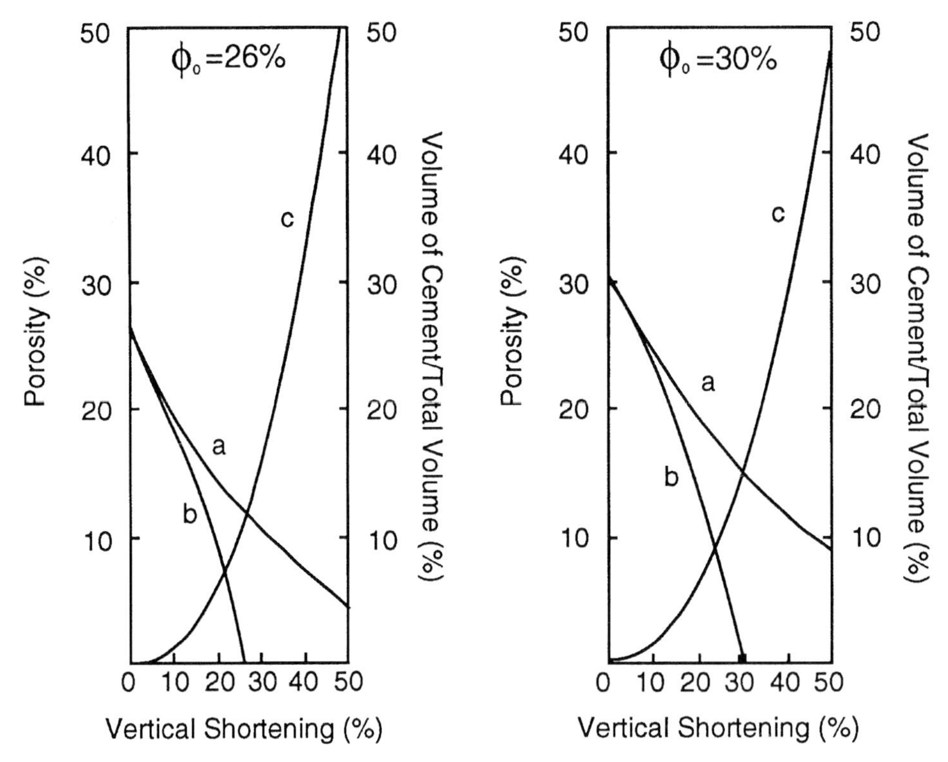

Figure 5. Graphs used to calculate the amount of vertical shortening and the volume of silica dissolved by intergranular pressure solution for samples from the Mobil St. Foster 1-12 and Sun Mentor C 1-29 cores. Curve "a" represents the amount of vertical shortening expected if all dissolved silica is transported out of the system. The intergranular volume is used to calculate this value. Curve "b," which uses the actual porosity in the sample, shows the amount of vertical shortening predicted if all dissolved silica is precipitated as cement. Curve "c" shows the volume of cement generated for a given amount of vertical shortening. Curves "a," "b," and "c" are provided for initial porosities of both 26% and 30%. Graph of 26% initial porosity modified after Mitra and Beard (1980); graph of 30% initial porosity taken from Mitra and Beard (1980).

pressure solution alone could have produced the observed bands in a relatively closed system, (2) if an external source of silica was required to account for the cemented bands, or (3) if the banded interval generated excess silica that was removed by advective or diffusive transport.

A silica budget was calculated for a total of 481 samples throughout the vertical extent of the banded interval, each sample consisting of properties averaged over a 5 mm vertical distance. In addition, thin-section–scale silica budgets were calculated for 46 additional samples from the Mobil St. Foster well and for 28 samples from the Sun Mentor C 1-29 well. These results were used to assess the relative importance of pressure solution and quartz cementation in the formation of diagenetic banding in the St. Peter Sandstone.

The relative importance of local versus external silica sources in the banded interval was assessed by plotting dissolved quartz versus quartz cement using techniques of Houseknecht (1988) (Figure 6). The 1:1 line represents a balance between quartz cement and quartz dissolved by intergranular pressure solution. Samples that plot above the line are importers of silica while samples that plot below the line are exporters. The majority of the samples in Figure 6 plot below the 1:1 line, indicating that more quartz has been dissolved by intergranular pressure solution than is present as quartz cement. Of all the samples analyzed, 382 were exporters, 88 were importers, and 11 were balanced, demonstrating that the amounts of pressure solution and cementation vary considerably throughout the interval and on a relatively small scale. Therefore silica dissolved in zones of intense pressure solution become a potential source of cement for quartz-cemented bands, suggesting that redistribution of silica on a local scale may be the cause of millimeter- and centimeter-scale banding.

The average silica budget for the banded interval is represented by the triangle on Figure 6. The average amount of quartz dissolved by pressure solution was 15% versus an average of 6% quartz cement, indicat-

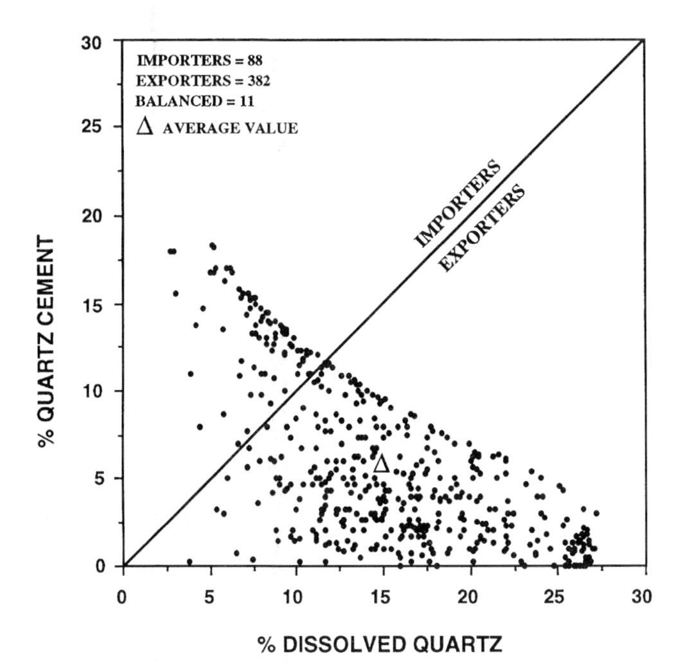

Figure 6. Plot of dissolved quartz vs. quartz cement for the 3 m banded interval in the Mobil St. Foster 1-12 core. Points plotting below the 1:1 line on this diagram are characterized by abundant intergranular pressure solution and little quartz cement. They represent silica exporters. Silica importers plot above the 1:1 line on the diagram. Samples that plot on the line are balanced with respect to silica flux. The open triangle is the average of all the points. The near linear limit of the data is an artifact of using a 26% initial porosity to calculate intergranular pressure solution, using the curves developed by Mitra and Beard (1980). Data plotted using the techniques of Houseknecht (1984, 1988).

ing that the 3 m interval has exported an average of approximately 9% silica by volume as a result of intergranular pressure solution. Thus, local pressure solution represents a viable source for the quartz cement within the banded interval and no external source of silica is needed. In addition, advective fluid flux is required in order to export the observed amount of silica. It should be noted that the above estimate of exported silica represents a minimum amount for the banded interval since it was based on a post-mechanical compaction porosity of 26%. Using a post-mechanical compaction porosity of 30% would indicate an even greater volume (average of 22%) of dissolved quartz produced by pressure solution. In addition, the effects of stylolites, which are another important source of dissolved silica within the banded interval, have not been accounted for in these budget calculations.

Houseknecht (1988) found that the Bromide (Ordovician, Oklahoma), upper Minnelusa (Permian, Wyoming), and Nugget (Triassic–Jurassic, Utah) sandstones are all exporters of silica. James et al.

Table 1. Silica budget calculations for the St. Peter Sandstone.

	Quartz Cement	Dissolved Quartz	Silica Flux
Banded interval	5.7	15.1	–9.4
Mobil St. Foster 1-12	4.2	9.4	–5.2
Sun Mentor C 1-29	5.9	5.9	0

(1986) also concluded that the Nugget is an exporter, but only if the silica released by stylolites is included in the budget. Other studies have found that silica budgets and the role of intergranular pressure solution can vary considerably within a single formation. Houseknecht (1984) found that in the Hartshorne Sandstone in the Arkoma basin silica was exported from some locations and imported into others. However, the unit was balanced on a regional scale, assuming lateral transfer of silica over distances of up to 240 km by formation water (Houseknecht, 1984). Sibley and Blatt (1976) found that intergranular pressure solution can only account for 30 to 35% of pore-filling quartz cement in the Tuscarora Sandstone in Pennsylvania, whereas Houseknecht (1988) concluded that on a regional scale the Tuscarora Sandstone is balanced.

In light of these regional studies, a silica budget was calculated for the entire cored interval of the St. Peter Sandstone in the Mobil St. Foster 1-12 well and in the nearby Sun Mentor C 1-29 well (Figure 1) in order to examine the balance between pressure solution and cementation on a larger scale. Results of these silica budget calculations are shown in Table 1.

The entire Mobil St. Foster 1-12 cored interval was an exporter of silica overall, with a total of 37 exporters, 5 importers, and 4 balanced (Figure 7A). The average volume of quartz cement in these samples is 4% compared to an average dissolved quartz volume of approximately 9%. As in the case of the banded interval, about twice as much silica was generated by pressure solution as can be observed as local cement. However, the magnitude of the silica export is somewhat lower (5% for the full core versus 9% for the banded interval), probably because of the lower average intensity of intergranular pressure solution. In contrast to the Mobil St. Foster well, the Sun Mentor silica budget is approximately balanced (Figure 7B). The dominant cement type in the Sun Mentor is quartz-overgrowths, with minor amounts of clay and carbonate cement. Millimeter-scale banding is visible in the lower part of the core, consisting of interbedded white, quartz-cemented zones and darker, porous zones. Results from the Sun Mentor silica budget indicate a total of 15 exporters, 10 importers, and 3 balanced. Although silica redistribution is apparent on a local scale, processes of dissolution and precipitation appear to balance at the scale of the core.

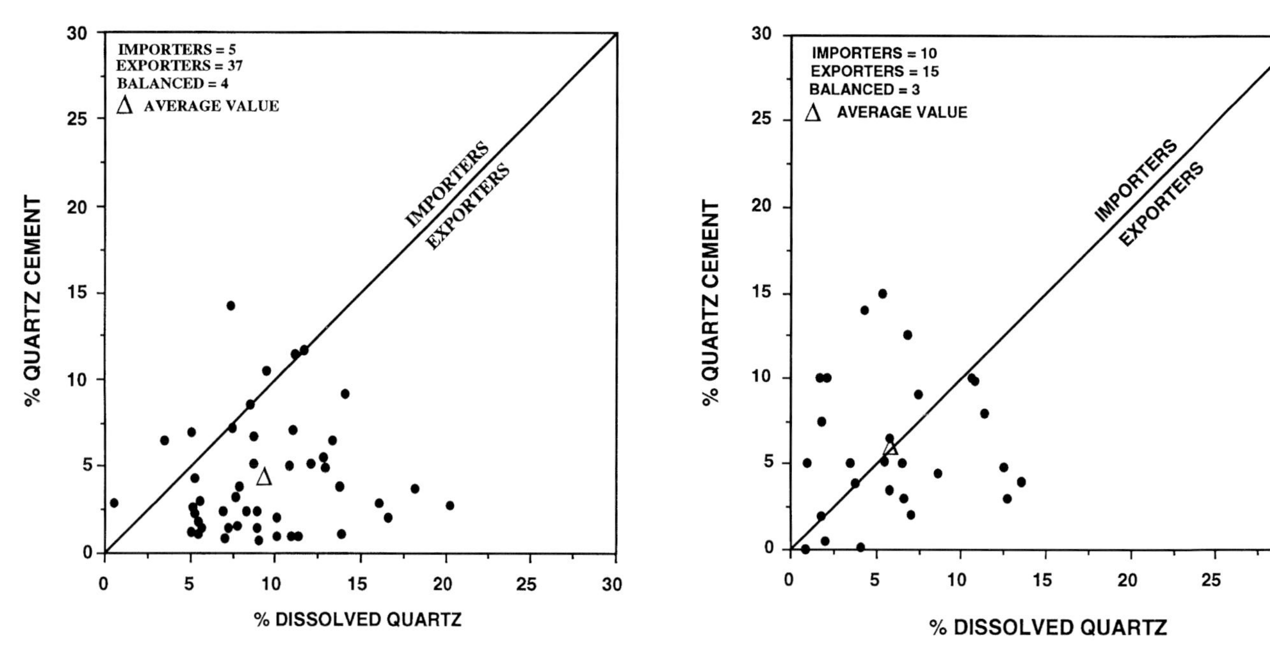

Figure 7. Plots of dissolved quartz vs. quartz cement for the entire Mobil St. Foster 1-12 core and the Sun Mentor C 1-29 core. Points plotting below the 1:1 line on this diagram are exporters of silica and points plotting above the line are importers. Samples that plot on the line are balanced with respect to silica flux. The open triangle is the average of all the points. Data plotted using the techniques of Houseknecht (1984, 1988). (A) Mobil St. Foster 1-12 core. (B) Sun Mentor C 1-29 core.

Silica budget results for the St. Peter Sandstone in the Mobil St. Foster 1-12 core indicate that net export of silica occurred at least over meter-scale intervals, requiring transport by advective fluid flux through the sandstone. In contrast, the Sun Mentor well was balanced with respect to silica migration. This indicates that there are regional variations in the amount of silica flux in the St. Peter. Although advective export of silica may have been important on a local scale, the system may actually have remained closed with respect to silica migration at a larger scale. This is consistent with the concept of a pressure compartment that is internally open but relatively isolated from the surrounding system.

CONTROLS ON DIAGENETIC BAND FORMATION

Average and extreme values of porosity, grain size, sorting, quartz cement, and pressure solution measurements for the 3 m banded interval are listed in Table 2. Although the average porosity is relatively low (3.6%), porosity is highly variable on a millimeter scale, ranging from 0 to 15%. The most significant cement in this interval is authigenic quartz, ranging from 0 to 18%, with a mean of approximately 6%. The amount of pressure solution varies considerably from sample to sample. Most samples have undergone a moderate amount of pressure solution, although there are some zones in which intense pressure solution has completely eliminated porosity. Assuming a post-mechanical compaction porosity of 26%, the average amount of quartz dissolved by intergranular pressure solution is 15%, with a range of 3 to 27%. Since this represents an upper limit on the amount of mechanical compaction, the amount of quartz dissolved by intergranular pressure solution represents a minimum.

Textural Controls on Porosity and Cement Distribution

Table 3 summarizes the relationships among grain size, sorting, quartz cement, porosity, and quartz dissolved by intergranular pressure solution for the banded interval. Results of this study indicate that there is no significant relationship between grain size and porosity (r = 0.09), grain size and intergranular pressure solution (r = 0.09), or grain size and quartz cement (r = 0.15).

A similar lack of correlation between grain size and intergranular pressure solution was found by Sibley and Blatt (1976) for the Tuscarora Sandstone. However, most studies (Heald, 1956; Houseknecht, 1984, 1988; Houseknecht and Hathon, 1987; Murphy, 1984; Tada and Siever, 1989) indicate a significant inverse relationship between grain size and intergranular pressure solution, with finer-grained sandstones undergoing more intense pressure solution than coarser-grained sandstones. Such an inverse relationship could be the result of variations in clay content or grain size, or of diffusion kinetics that increase with decreasing grain size (Weyl, 1959). However, a

Table 2. Summary data for the banded interval.

	Mean	Standard Deviation	Minimum	Maximum
Porosity (%)	3.6	2.7	0	15
Intergranular volume (%)	9.3	3.7	1.5	22
Quartz cement (%)	5.7	4.6	0	18
Dissolved quartz (ϕ_0 = 26%) (%)	15.1	6.0	3	27
Dissolved quartz (ϕ_0 = 30%) (%)	22.4	5.2	6	29
Grain size (mm)	0.22	0.06	0.12	0.45
Grain size (phi)	2.30	0.35	1.19	3.10
Sorting (phi)	0.53	0.16	0.14	1.05

Table 3. Correlation matrix showing relationships between porosity, texture, quartz cement, and dissolved quartz for the banded interval.

	Porosity	Quartz Cement	Dissolved Quartz	Grain Size (phi)	Sorting (phi)
Porosity	—	–0.59*	–0.22*	0.09	0.12
Quartz cement	–0.59*	—	–0.67*	–0.15	–0.15
Dissolved quartz	–0.22*	–0.67*	—	0.09	0.07
Grain size (phi)	0.09	–0.15	0.09	—	–0.14
Sorting (phi)	0.12	–0.15	0.07	–0.14	—

* Correlation coefficient is significant at a probability level of 0.95.

recent study by Gratz (1991) suggests that the diffusion path length may actually be independent of grain size, being controlled instead by island and channel topography that characterizes intergranular contacts. The effect of this is to reduce the dependence of the effective rate law on grain size. In addition, the correlation between grain size and intergranular pressure solution can potentially be masked by certain depositional and diagenetic properties, such as sorting, clay coatings on grains, presence of ductile grains, earlier cements, depth and rate of burial, time, temperature, fluid flow, and overpressures (Houseknecht and Hathon, 1987).

A lack of correlation between grain size and quartz cement has been found in other studies of quartzose sandstones (Houseknecht, 1988; Murphy, 1984). Overall, previous studies do not reveal any consistent influence of grain size on quartz cementation. McBride (1989) reports that the majority of the formations examined in the literature show an increase in quartz cement with decreasing grain size, but Fox et al. (1975), McBride (1984), and Houseknecht (1984) report that quartz cement is more common in coarser-grained sandstones.

No correlation was found between sorting and porosity (r = 0.12), sorting and quartz cement (r = 0.15), or sorting and intergranular pressure solution (r = 0.07). The lack of correlation between sorting and pressure solution is consistent with results of several studies reported in the literature (Sibley and Blatt, 1976; Houseknecht, 1988).

Diagenetic Controls on Porosity and Cement Distribution

The lack of significant correlation between depositional factors and porosity, cement, or intergranular pressure solution suggests that the dominant control on porosity and cement variations in the banded interval is diagenetic. In the banded interval, there is an inverse correlation between the amount of porosity and quartz cement (r = 0.59; Figure 8), supporting this hypothesis. Porosity is also weakly correlated to intergranular pressure solution (r = 0.22), with lower porosity in zones of greater pressure solution. The scatter of data shown on Figure 8, particularly in the low-porosity portion, is a result of the fact that porosity can be reduced by both cementation and pressure solution. Even though there is a significant inverse correlation between quartz cement and porosity, a low-porosity zone may have very little quartz cement if intergranular pressure solution is the dominant porosity-reducing mechanism. The inverse relationship between porosity and quartz cement can also be

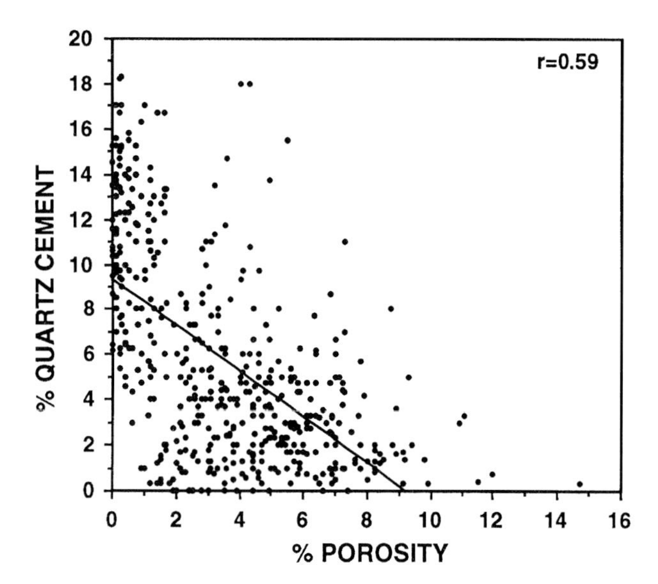

Figure 8. Relationship between porosity and quartz-overgrowth cement.

seen on Figure 9. This figure shows variations in porosity and quartz cement with depth for the banded interval along with a core description, which shows the alternation between quartz-cemented and porous bands. In general, quartz-cemented bands observed in hand specimen have lower porosity and larger percentages of quartz-overgrowth cement than macroscopic porous bands.

Intergranular pressure solution and quartz cementation have a significant inverse correlation (r = 0.67), suggesting that quartz is exported, at least locally, from regions of intense intergranular pressure solution. This type of relationship has been reported by several other researchers (Heald, 1956; Sibley and Blatt, 1976; Houseknecht, 1984; Tada and Siever, 1989) and can be explained in part by a reduction in pore space available for cementation in regions of intense intergranular pressure solution (Tada and Siever, 1989). Furthermore, quartz cementation inhibits the effects of pressure solution by distributing stress over larger grain contacts (Heald, 1956; Sibley and Blatt, 1976) and restricting the flow of interstitial water (Heald, 1956).

Compaction and cementation are important mechanisms in controlling porosity and cement distribution. One way of assessing the relative importance of compaction versus cementation as porosity-reduction mechanisms involves a comparison of intergranular volume and quartz cement. The amount of intergranular volume that remains in a sample is a function of the initial porosity and how much volume has been destroyed by mechanical and chemical compaction. Assuming an initial porosity of 40% for the St. Peter Sandstone, one can determine the amount of original porosity that was destroyed by compaction versus cementation.

Data plotted using techniques of Houseknecht (1987) (Figure 10) show the amount of original porosity that was destroyed by compaction versus the amount occluded by cementation. The dashed diagonal line represents the case in which equal amounts of porosity have been destroyed by compaction and cementation. Samples that plot above the line are dominated by cementation while those below the line have experienced greater compaction. The majority of samples from the banded interval plot in the lower lefthand corner, indicating that the dominant control on porosity in these rocks is compaction. On average, over 75% of the original porosity has been destroyed by compaction (both mechanical and chemical) and 15% has been occluded by cementation.

Controls on Localization of Diagenetic Banding

Modeling studies by Ortoleva et al. (1987) and Dewers and Ortoleva (1990a, b, c) suggest that diagenetic bands can develop in sandstones as a result of feedback between mechanical compaction, quartz precipitation–dissolution reactions, and transport. The data required to constrain these models consist of detailed profiles of textural parameters (size and shapes of grains), porosity, and cement, both for the final banded formation and for the sandstone prior to diagenesis. The patterns and scale of banding observed in the St. Peter Sandstone are similar to those predicted by such models. The models can be further constrained by estimates of the overall silica budget for the interval in question. Although the models predict that a sandstone can be divided internally into a series of small-scale silica exporters and importers, export out of the simulated interval is limited because transport in the fluid phase is controlled by diffusion. The result is that on a scale of a number of bands (i.e., tens of centimeters), mass is approximately balanced and the system is essentially closed. The net export determined by our silica budget calculations implies that advective transport must be included as a process in mechano-chemical models of diagenetic banding.

The mechano-chemical models require initial variations in texture or clay content to initiate the feedback that produces diagenetic banding. Although textural parameters do not appear to be important in controlling the distribution of porosity within the banded interval itself, they do appear to be important in controlling the initial localization of the banding. When the banded interval was compared to the remainder of the Mobil St. Foster 1-12 core, which has a much more uniform distribution of cement types and only minor banding, a significant difference in grain size and sorting was found between the banded and nonbanded lithologies. Banded sandstones are finer grained and better sorted on average than nonbanded sandstones. In addition, the banded sandstones have considerably more quartz cement and have undergone significantly more dissolution by intergranular pressure solution than sandstones from the nonbanded intervals. These relationships were confirmed by a Mann Whitney statistical test. Drzewiecki et al. (this volume) note that diagenetic banding is best devel-

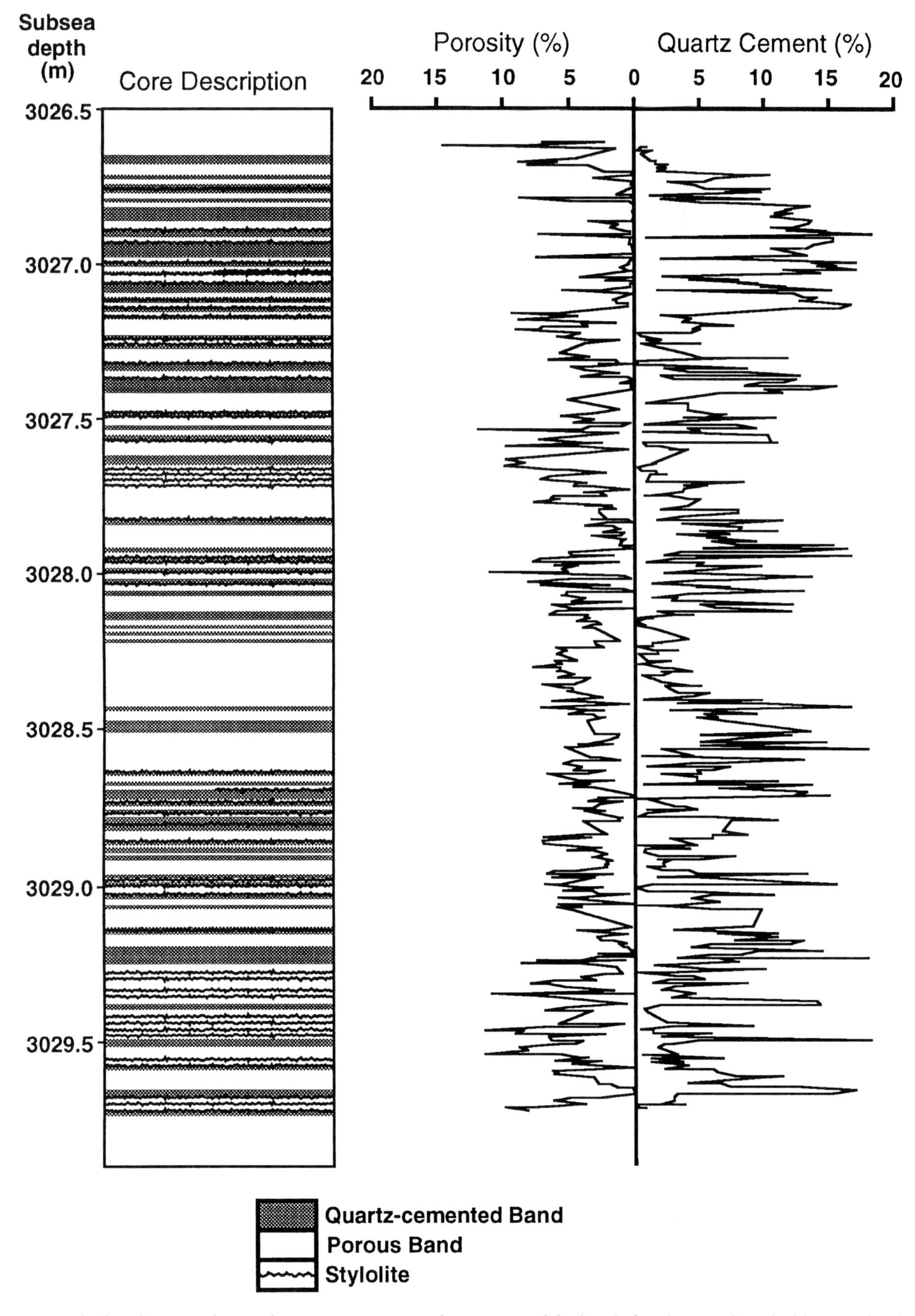

Figure 9. Variation in porosity and quartz-overgrowth cement with depth for the 3 m banded interval. Also shown is a core description based on visual inspection that shows the alternation between quartz-cemented bands and porous bands. Note the abundant stylolites throughout the interval that often form at the contacts between adjacent bands.

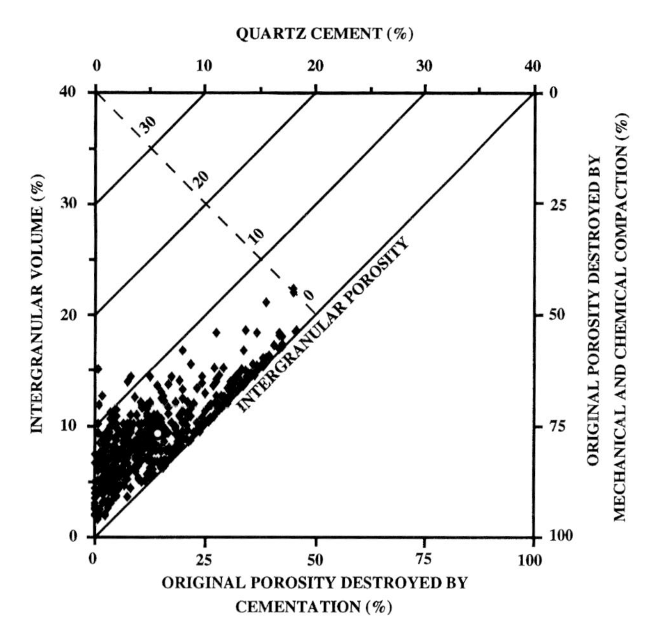

Figure 10. Plot of intergranular volume vs. quartz cement for the 3 m banded interval of the Mobil St. Foster 1-12 core, showing the relative amount of porosity lost by compaction vs. the relative amount lost by cementation. Dashed line separates samples on the basis of whether compaction (lower left) or cementation (upper right) has been more important in reducing porosity. The open circle is the average of all the points. Data plotted using the techniques of Houseknecht (1987).

oped in samples that contain small-scale horizontal laminations which served as a template for the localization and initiation of diagenetic banding. Such grain-size laminations are also present within the 3 m diagenetically banded interval from the Mobil St. Foster core. These results indicate that the location of banded intervals in the St. Peter may be controlled by textural parameters, with banding preferentially developing in finer-grained, better-sorted intervals. This conclusion is confirmed by Houseknecht (1984) and Murphy (1984), who suggest that pressure solution in the fine-grained sandstones creates a source of cement that is precipitated in coarse-grained sandstones.

Another control on the distribution of pressure solution and quartz cementation is the presence of clay coating on grains, which can act both to promote pressure solution and to inhibit quartz cementation (Houseknecht, 1984; Tada and Siever, 1989; Houseknecht and Hathon, 1987). In the 3 m banded interval, clay is not visible in thin section but has been found as a thin coating on grains with a scanning electron microscope. Preliminary analysis indicates that clay linings are thickest in the coarser-grained, porous samples (Drzewiecki, 1992). Other depositional controls may also be important in the localization of diagenetic banding in the St. Peter, such as detrital mineralogies (clay, carbonates) and early cementation (Drzewiecki et al., this volume).

Rearrangement of grains and reduction of pore throat size through mechanical compaction is another possible control on differential intergranular pressure solution and quartz cementation. Pore throat size affects the rate of advective flow and therefore regulates the rate of dissolved silica transport. Differences in pore throat size and grain arrangement between adjacent sandstone laminae may initiate diagenetic band formation.

CONCLUSIONS

Diagenetic processes can result in the formation of banded intervals in sandstones by the local redistribution of silica. A banded interval such as that examined in this study can act collectively to form a low-permeability seal to vertical flow, even though individual bands within the interval may have a relatively high porosity and horizontal permeability.

Silica budget calculations for the Mobil St. Foster 1-12 core indicate that intergranular pressure solution and local redistribution of silica can account for all quartz cement present. Petrographic evidence suggests that intergranular pressure solution and quartz cementation were contemporaneous (Drzewiecki et al., this volume), lending support to this argument. The Mobil St. Foster core was an exporter of silica, whereas the Sun Mentor C 1-29 core was balanced with respect to silica. These results indicate that there are regional-scale variations in silica flux across the Michigan basin, suggesting that the St. Peter Sandstone may actually behave as a closed system on a basinwide scale.

Porosity modification of the St. Peter Sandstone in banded intervals appears to be diagenetically controlled. No significant correlation exists between porosity and grain size or porosity and sorting, suggesting that depositional textures did not directly affect porosity modification. However, there does appear to be an important link between the localization of banding and depositional textural parameters. Drzewiecki et al. (this volume) conclude that diagenetic banding is best developed in sandstones with horizontal depositional textures (horizontal grain-size laminations). Although there is no correlation between porosity and grain size within the banded interval itself, analyses of the entire Mobil St. Foster core indicate that banded intervals are finer grained and better sorted on average than nonbanded intervals.

ACKNOWLEDGMENTS

This paper has benefited immensely from reviews by David Houseknecht and two anonymous reviewers and from valuable discussions with Greg Nadon and Geri Moline about general Michigan basin geology. We would also like to thank Bill Harrison and Dave Barnes from Western Michigan University Core

Lab for providing samples and thin sections from the Sun Mentor core. Mobil Oil Corporation provided the St. Foster 1-12 core used in this investigation. This work was supported by a contract from the Gas Research Institute (5089-260-1810).

REFERENCES CITED

Atwater, G. I., and E. E. Miller, 1964, The effect of decrease in porosity with depth on future development of oil and gas reserves in south Louisiana (abst.): American Association of Petroleum Geologists Bulletin, v. 49, p. 334.

Beard, D. C., and P. K. Weyl, 1973, Influence of texture on porosity and permeability of unconsolidated sand: American Association of Petroleum Geologists Bulletin, v. 57, p. 349–369.

Blatt, H., 1979, Diagenetic processes in sandstones, *in* P. A. Scholle and P. Schluger, eds., Aspects of Diagenesis: Society of Economic Paleontologists and Mineralogists, Sp. Pub. 26, p. 141–157.

Catacosinos, P. A., and P. A. Daniels, 1991, Stratigraphy of Middle Proterozoic to Middle Ordovician formations of the Michigan Basin: Geological Society of America Special Paper 256, p. 53–71.

Dewers, T., and P. Ortoleva, 1990a, A coupled reaction/transport/mechanical model for intergranular pressure solution, stylolites, and differential compaction and cementation in clean sandstones: Geochimica et Cosmochimica Acta, v. 54, p. 1609–1625.

Dewers, T., and P. Ortoleva, 1990b, Differentiated structures arising from mechano-chemical feedback in stressed rocks: Earth–Science Reviews, v. 29, p. 283–298.

Dewers, T., and P. Ortoleva, 1990c, Interaction of reaction, mass transport, and rock deformation during diagenesis: mathematical modeling of intergranular pressure solution, stylolites, and differential compaction/cementation, *in* I. Meshri and P. Ortoleva, eds., Prediction of Reservoir Quality Through Chemical Modeling: American Association of Petroleum Geologists Memoir 49, p. 147–160.

Drzewiecki, P.A., 1992, Sedimentology, diagenesis, and geochemistry of the Middle Ordovician St. Peter Sandstone of the Michigan Basin: Unpublished MS thesis, University of Wisconsin, Madison, Wisconsin, 215 p.

Ehrenberg, S. N., 1990, Relationship between diagenesis and reservoir quality in sandstones of the Garn Formation, Haltenbanken, mid-Norwegian continental shelf: American Association of Petroleum Geologists Bulletin, v. 74, p. 1538–1558.

Fox, J. E., P. W. Lambert, R. F. Mast, N. W. Nuss, and R. D. Rein, 1975, Porosity variation in the Tensleep and its equivalent, the Weber Sandstone, western Wyoming: a log and petrographic analysis, *in* D. W. Bolyard, ed., Symposium on Deep Drilling Frontiers in the Central Rocky Mountains: Rocky Mountain Association of Geologists, Denver, Colorado, p. 185–216.

Gratz, A. J, 1991, Solution-transfer compaction of quartzites: Progress toward a rate law: Geology, v. 19, p. 901–904.

Heald, M. T., 1956, Cementation of Simpson and St. Peter Sandstones in parts of Oklahoma, Arkansas, and Missouri: Journal of Geology, v. 64, p. 16–30.

Heald, M. T., and R. C. Anderegg, 1960, Differential cementation in the Tuscarora sandstone: Journal of Sedimentary Petrology, v. 48, p. 568–577.

Houseknecht, D. W., 1984, Influence of grain size and temperature on intergranular pressure solution, quartz cementation, and porosity in a quartzose sandstone: Journal of Sedimentary Petrology, v. 54, p. 348–361.

Houseknecht, D. W., 1987, Assessing the relative importance of compaction processes and cementation to reduction of porosity in sandstones: American Association of Petroleum Geologists Bulletin, v. 71, p. 633–642.

Houseknecht, D. W., 1988, Intergranular pressure solution in four quartzose sandstones: Journal of Sedimentary Petrology, v. 58, p. 228–246.

Houseknecht, D. W., 1991, Use of cathodoluminescence petrography for understanding compaction, quartz cementation, and porosity in sandstones, *in* C. E. Barker and O. C. Kopp, eds., Luminescence Microscopy and Spectroscopy: Society of Economic Paleontologists and Mineralogists Short Course No. 25, p. 59–66.

Houseknecht, D. W., and L. A. Hathon, 1987, Petrographic constraints on models of intergranular pressure solution in quartzose sandstones: Applied Geochemistry, v. 2, p. 507–521.

Hunt, J. M., 1990, Generation and migration of petroleum from abnormally pressured fluid compartments: American Association of Petro-leum Geologists Bulletin, v. 74, p. 1–12.

James, W. C., G. C. Wilmar, and B. G. Davidson, 1986, Role of quartz type and grain size in silica diagenesis, Nugget Sandstone, south–central Wyoming: Journal of Sedimentary Petrology, v. 56, p. 657–662.

Kahn, J. S., 1956a, The analysis and distribution of the properties of packing in sand-size sediments. I. On the measurement of packing in sandstones: Journal of Geology, v. 64, p. 385–395.

Kahn, J. S., 1956b, Analysis and distribution of packing properties in sand-sized sediments. II. The distribution of the packing measurements and an example of packing analysis: Journal of Geology, v. 64, p. 578–606.

Manus, R. W., and A. H. Coogan, 1974, Bulk volume reduction and pressure solution derived cement: Journal of Sedimentary Petrology, v. 44, p. 466–471.

McBride, E. F., 1984, Rules of sandstone diagenesis related to reservoir quality (expanded abstract): Transactions of the Gulf Coast Association of Geological Societies, v. 34, p. 137–139.

McBride, E. F., 1989, Quartz cement in sandstones: a review: Earth–Science Reviews, v. 26, p. 69–112.

Mitra, S., and W. C. Beard, 1980, Theoretical models

of porosity reduction by pressure solution for well-sorted sandstones: Journal of Sedimentary Petrology, v. 50, p. 1347–1360.

Murphy, T. B., 1984, Diagenesis and porosity reduction of the Tuscarora Sandstone, central Pennsylvania: Unpublished MS thesis, University of Missouri, Columbia, 104 p.

Nadon, G. C., A. Simo, C. W. Byers, and R. J. Dott, Jr., 1991, Controls on deposition of the St. Peter Sandstone (Middle–Late Ordovician), Michigan basin (abst.): American Association of Petroleum Geologists Bulletin, v. 75, p. 1388–1389.

Ortoleva, P., E. Merino, C. Moore, and J. Chadam, 1987, Geochemical self-organization I: reaction–transport feedbacks and modeling approach: American Journal of Science, v. 287, p. 979–1007.

Pryor, W. A, 1973, Permeability–porosity patterns and variations in some Holocene sand bodies: American Association of Petroleum Geologists Bulletin, v. 57, p. 162–189.

Rittenhouse, G., 1971, Pore-space reduction by solution and cementation: American Association of Petroleum Geologists Bulletin, v. 55, p. 80–91.

Sibley, D. F., and H. Blatt, 1976, Intergranular pressure solution and cementation of the Tuscarora orthoquartzite: Journal of Sedimentary Petrology, v. 46, p. 881–896.

Sprunt, E. S., and A. Nur, 1976, Reduction of porosity by pressure solution: experimental verification: Geology, v. 4, p. 463–466.

Surdam, R. C., T. L. Dunn, H. P. Heasler, and D. B. MacGowan, 1989, Porosity evolution in sandstone/shale systems, *in* I. E. Hutcheon, ed., Short Course in Burial Diagenesis: Mineralogical Association of Canada, Short Course v. 15, p. 61–134.

Tada, R., and R. Siever, 1989, Pressure solution during diagenesis: Annual Review in Earth and Planetary Sciences, v. 17, p. 89–118.

Tigert, V., and Z. Al-Shaieb, 1990, Pressure seals: their diagenetic banding patterns: Earth–Science Reviews, v. 29, p. 227–240.

Weyl, P. K., 1959, Pressure solution and the force of crystallization—a phenomenological theory: Journal of Geophysical Research, v. 64, p. 2001–2025.

Wood, J. R., 1989, Modeling the effect of compaction and precipitation/dissolution on porosity, *in* I. E. Hutcheon, ed., Short Course in Burial Diagenesis: Mineralogical Association of Canada, Short Course v. 15, p. 311–362.

Chapter 25

Banded Diagenetic Pressure Seals: Types, Mechanisms, and Homogenized Basin Dynamics

C. Qin
Peter J. Ortoleva
Indiana University
Bloomington, Indiana, U.S.A.

ABSTRACT

The structure, mechanisms of formation, and key role of diagenetically banded pressure seals are reviewed. A difficulty in predicting the genesis time and location of these seals is that they are apparently affected by local stresses and fluid pressures, but, in turn, they affect the basin-scale distribution of these factors. We show that this very formidable multiple-scale basin modeling problem can be solved via a computational homogenization technique.

Banded pressure seals are classified in two complementary ways. First we distinguish those generated through self-organization from those directly of sedimentary origin. They are then classified according to the specific processes (pressure solution, nucleation, diffusion, flow, coupled mineral reactions, and grain comminution) by which they emerge. The argument is made that feedback in the network of reaction, transport, and mechanical processes underlies the development of many diagenetic seals.

The technical problem of the need to simulate submeter-scale banding phenomena in basin-scale models is introduced and addressed. A computational homogenization scheme is shown to simultaneously capture phenomena on these two scales by treating them both in a way that preserves their characteristics. We believe that this approach is a great advance in basin modeling more generally in that it allows one to capture phenomena on multiple spatial scales arising from complex sedimentary features and their reworking through diagenesis.

The following are demonstrated by simulations carried out for pure quartz sandstones undergoing stress-induced reactions, diffusion, and fluid flow. Seals are very likely to form within finer grain beds or at some specific depth within macroscopically uniform grain-size sediments. The depth and the time needed to form a seal depend upon the intensity of overpressure, fluid flow, geothermal gradient, subsidence velocity, grain size, the sediment sequence, and tectonic environment. High overpressure or higher geothermal gradient tends to decrease the depth of the seal. Lower fluid pressure is favorable for the differentiation of the rock texture (and the formation of the seal). Once a seal starts to develop, the process is self-accelerated and self-enhanced and the sediments become more heterogeneous and anisotropic.

BACKGROUND

In layered (or banded) rocks the texture (grain size, shape, packing, mineral identity) is organized into roughly parallel, highly elongate domains within which the texture is relatively constant. Layered rock has been observed to serve as efficient pressure seals (as pointed out by Bradley and Powley, 1986, personal communication; Dewers and Ortoleva, 1988; Tigert and Al-Shaieb, 1990; Al-Shaieb et al., this volume; Shepherd et al., this volume).

Layered rock may make a very efficient barrier to flow normal to the layering plane. If a few layers are of very low permeability, the flow-through normal to layering is repressed. Consider flow normal to the layering in the case of layers of infinite lateral extent. Let k be the local permeability and k_{eff} be the effective permeability for the overall flow averaged out over many layers. Then we have

$$k_{eff} = \frac{1}{\langle 1/k \rangle} \tag{1}$$

a result of an averaging or "homogenization" approach (see below), where $\langle \cdots \rangle$ implies an average over many layers. This shows that the effective (i.e., averaged over many layers) permeability is dominated by the presence of the few low-permeability domains. For example, if 1% of the rock is occupied by 10^{-9} darcy material and the rest is millidarcy material, then $k_{eff} = 10^{-7}$ darcy.

A main point of the present paper is that intense layering can arise through diagenesis even in the absence of appreciable textural contrast from sedimentary bedding or lamination. In particular, the coupling of chemical, mechanical, and transport processes at depth can lead to a feedback that amplifies the intensity of an initial textural pattern (Ortoleva et al., 1987; Dewers and Ortoleva, 1988, 1989a, 1990b, 1994). Specifically the state of uniform compaction is shown to be unstable to the development of lenticular or layered patterns.

This textural instability is very interesting in the context of seals as noted above. Through it layering of weak contrast can be amplified into a very intense pattern wherein mass is transferred between neighboring layers. A subset of layers can become tightly cemented at the expense of grains in neighboring layers via dissolution of the latter and reprecipitation in the neighboring layers of accumulating overgrowth or other (permeability-destroying) precipitation (as observed by Tigert and Al-Shaieb, 1990; Al-Shaieb et al., this volume; Shepherd et al., this volume). When layered rocks have been so differentiated in such a way as to serve as barriers to layer-normal flow they are termed diagenetically banded seals.

It is likely that diagenetically layered seals have fracture development and healing properties that make them resistive to rupture or, if rupture takes place, allows for time delays for fractures to heal before appreciable loss takes place. The former arises because of the heterogeneous (layered) nature of the tensile strength. The time delay for fluid escape in the fractured layered rock likely arises because fractures in layered media have a strong tendency to meander through the medium by following the lower tensile strength layers. Both the hydraulic and fracture mechanic properties of layered rock are a rich source of future research.

The viability of layered rock as a seal depends on a number of factors including:

- The intensity of the layering—i.e., are the low-permeability layers sufficiently tight
- The widths and frequency of the low-permeability layers
- The total thickness of the layered domain
- The capacity of the layered rock to resist fracturing or to heal fractures once formed
- The lateral continuity of the layers

Observed layered seals typically involve hundreds or thousands of individual layers. As the mechanochemical processes we believe to underlie a number of aspects of layer development involve the dynamics of the fluid pressure and as the latter varies over the seal and is in turn affected by the layering, it is necessary to simulate the development of many layers simultaneously in order to predict conditions favorable for their genesis and location within basins.

The layered seal system presents us with a two-scale problem. Layer spacing and thickness occur on the millimeter to meter length scale. In contrast, seals are of 10–100 meter-scale thickness and the compartments of interest are also on a scale of hundreds of meters to kilometers or, in the case of the megacompartment, basin scale (tens to hundreds of kilometers).

One might take the approach that the seal layering problem should be solved separately from the seal and compartment problem. However, we believe that the mechanisms of layering involve the local fluid pressure, matrix stresses, fluid composition and temperature. These are all affected by the presence of a layered seal—i.e., basin-scale fluid pressure reflects the presence of seals while the pressure affects the distribution of stresses and the flow of fluids (and hence of local fluid composition and temperature). Thus in order to develop a predictive theory of layered seals, we must solve the full two-scale problem.

Capturing thousands of features in a two or three spatial dimensional domain is prohibitive computationally. Consider a layered seal of a thickness of 100 m and an area of 1 km^2. If in every vertical transect there are 10^3 layers and we need at least 10 calculational grid points per layer (i.e., 10^4 grid points/10^2 m = 10^2 grid points/m), then to describe the full seal with a regular grid requires 10^{12} grid nodes. If we also wish to describe the underlying compartment and the lateral and basal seals and the involvement of millimeter-scale layering, the problem becomes even more formidable.

The above-cited difficulties with the simulation of layered seals turn out to be the key to its resolution. Homogenization or, more generally, multiple-scale analysis (Nayfeh, 1973; Keller, 1977; Bensoussan et al., 1978; Araki et al., 1987; Hornung, 1991), exploits this separation of scales.

This paper first discusses the types and mechanisms of the banded seals. The mechano-chemical model of a particular type of repetitive porosity band is then described as an example of mechano-chemical banding. To capture the coupled two-scaled aspect of the system for basin dynamics, homogenization theory is applied to the repetitive porosity band problem. The analysis yields two sets of coupled equations. One set of "macroscopic" equations describes basin-scale dynamics, and a set of "microscopic" equations treats local variations. Finally, a numerical scheme and our parallel computational approach and some simulation results are presented.

CLASSIFICATION OF BANDED DIAGENETIC SEALS: MECHANO-CHEMICAL MODELS

Consideration of the physical chemistry of grain growth/dissolution and transport processes in porous media leads one to conclude that there can be a number of distinct pathways whereby rocks may become banded via diagenesis. Here we attempt to classify them and then briefly review examples of diagenetically banded rocks that have become seals. We conclude with conjectures on which mechanisms might play a key role in a basin's overall strategy to compartmentize itself.

Three General Classes of Mechanisms

First let us distinguish banding that is a direct reflection of sedimentary features from that which arises autonomously by a process of differentiation from unbanded rock.

Fully Templated Banding

Here the role of diagenesis is passive. With burial-associated changes in stress, fluid pressure, temperature, and composition, different minerals respond differently and hence the nature and degree of progress of diagenetic reactions simply follows the sedimentary layering precursor.

For example, limestone beds likely compact faster than sandstones. Also, a calcite-supersaturated fluid passing through a system will tend to precipitate calcite cement beds rich in calcite nucleation sites.

Fully Differentiated Banding

The general theme here is that within the network of diagenetic reaction-transport–mechanical (RTM) processes there may be a feedback loop that tends to amplify any heterogeneity into an intense pattern of textural contrast (Ortoleva et al., 1987; Ortoleva, 1994). The initial disturbance that started the feedback loop may have little resemblance to the final pattern in terms of the intensity of the contrast between layers or the nature of the layers. This type of banding can appear to spontaneously emerge from a rock whose sedimentary history left it essentially uniform in composition and texture as shown in the simulation of Figure 1.

Physico-chemical processes that can lead to such differentiated layering include the following (Ortoleva, 1994):

1. Liesegang banding (Liesegang, 1913; Ostwald, 1925; Prager, 1956; Ortoleva et al., 1987)
2. Unstable coarsening (Feinn et al., 1978; Lovett et al., 1978; Feeney et al., 1983)
3. Unstable compaction arising from the coupling of the stress-mediating effect of grain–grain contacts, the water film diffusion (or related) mechanism of pressure solution, and diffusion of solutes on the supra-grain scale (Dewers and Ortoleva, 1990b-d, 1992, 1994; Ortoleva et al., 1993)
4. Differentiation due to a coupling involving the overall effect of porosity or grain strain-induced dissolution at free faces (and possibly free-face–pressure dissolution at contacts) and supra-grain-scale solute diffusion (Dewers and Ortoleva, 1990b, c, 1992, 1994)
5. A clay-mediated compactional instability wherein clay promotes compaction through pressure solution with the resultant accumulation of clay
6. A compositional differentiation in low porosity rocks involving the effect of variations of medium mechanical properties with mineral content coupled to supra-grain-scale diffusion (Dewers and Ortoleva, 1989b, 1992, 1994)

Unlike the other mechanisms, (1) is driven by an influx of reactants from outside the system boundaries (although even in that case the resulting banded precipitation pattern has no direct correspondence to any pattern imposed at the boundary).

Our studies show that in all the above cases, the systems can support "symmetry breaking instability"; thereby even infinitesimal initial nonuniformities from the unpatterned (i.e., uniform) state are amplified into a pattern with strong textural contrast between bands of alternating properties.

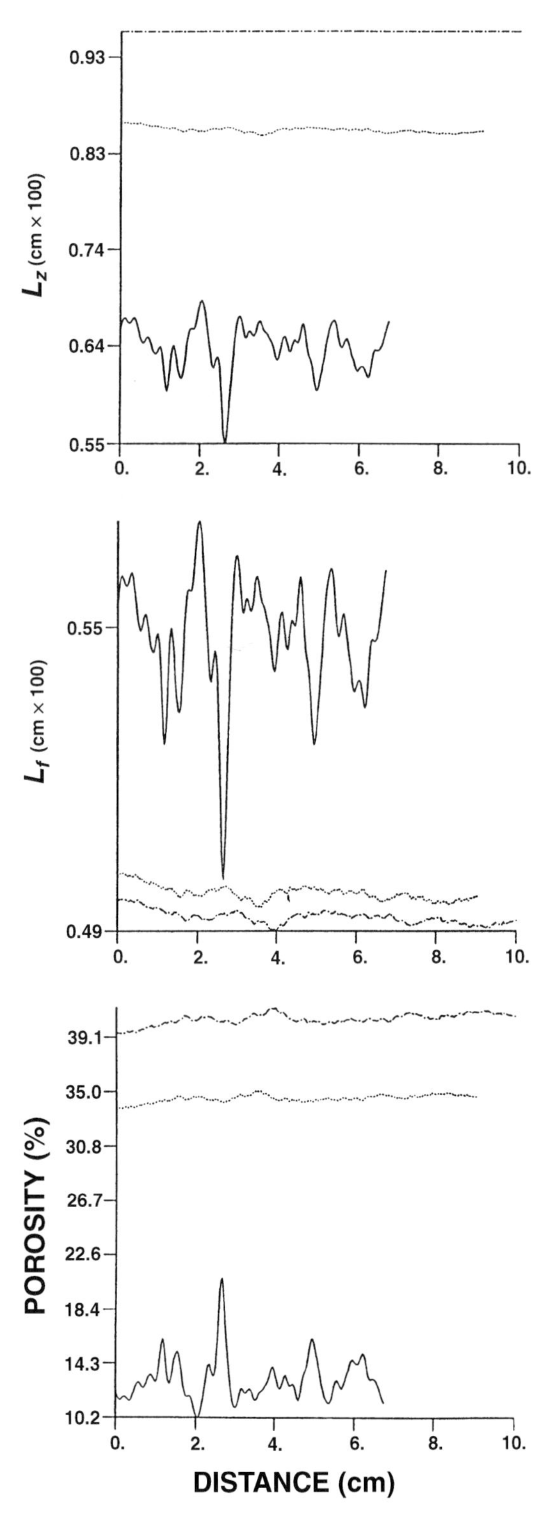

Figure 1. Predictions from the Dewers and Ortoleva (1990b) model of compaction in a sandstone. Note that as compaction proceeds (and hence porosity decreases), small heterogeneities are amplified into a differentiated layering. The layering involves zones of low porosity, high overgrowth, and low compaction (i.e., larger grain height) alternating with zones of opposite character. Such bands are apparently one of the types observed in banded seals.

Diagenetically Enhanced Sedimentary Bedding

While the differentiated layering of the previous subsection is an important idealization, real sediments are seldom ever roughly homogeneous. To understand observed banding in rocks we must understand how the differentiation mechanisms can operate in real—i.e, initially nonuniform—sediments.

The least interesting end-member response is a small modification of the effects of the sedimentary templating. In the most dramatic effect the induced amplification of textural contrast can be very strong, and sedimentary correlations among variables may be completely reversed.

An example of the latter is well documented in the case of differentiated marl/limestone sequences (Ricken, 1986). There, sedimentary processes suggest that porosity should be anticorrelated with clay fraction—i.e., regions of greater fraction of clay have more clay in the pore space and therefore reduce porosity with respect to relatively clay-free regions. However, after diagenesis, regions of maximum porosity are in regions of maximum clay fraction. An equally dramatic correlation that develops in this system is that regions of maximum compaction are those of maximum porosity. These relations have been predicted using a fully coupled RTM pressure solution model (Dewers and Ortoleva, 1994).

From the above, we suggest that while the presence of an early sedimentary prepattern may be documentable, the effect of a diagenetic feedback and pattern-forming instability may be profound in determining the ultimate fate of the banded patterns.

Other Key Banding Processes

There are several other band-producing mechanisms that we have encountered in our analysis of banded seals (see also below).

Displaced Pressure Solution–Derived Precipitation

If grain free faces are kinetically inhibited (through clay coatings or surface-attached monolayers of growth poisons), the pore fluid can become saturated with respect to the minerals in their state of stress at grain–grain contacts. When such a bed communicates with another without such free-face inhibition, overgrowth can occur and porosity can become occluded (Sauer et al., 1991; Ortoleva et al., 1993). This can lead, for example, to the development of carbonate bands in sandstones; the location of these bands would depend on the presence and magnitude of flow from the free-face–inhibited carbonate rock to the sandstone and the possible existence of layers in the sandstone where carbonate nucleation was preferred. In the carbonate-banded sandstone problem, one might even expect that carbonate cementation could proceed to the extent of causing dissolution of the quartz matrix due to a force of crystallization effect imposed by the high calcite or dolomite free energy source—i.e., at equilibrium with stressed carbonate grain–grain contacts.

Comminution Bands

If stresses born by a grain exceed a critical value, grain breakage can occur. Such collapse horizons can reduce permeability normal to the bands and serve as effective seals and have been observed in the Anadarko basin. These comminution bands can follow sedimentary features or may have differentiated via the compaction-augmented porosity/cementation banding noted above.

In principle, collapse can occur under a number of circumstances:

- Burial of relatively uncompacted rock
- Mechano-chemical diagenesis leading to mass transfer to adjacent beds or, through flow, further away
- Dissolution of supporting matrix through the influx of undersaturated fluids
- Reduction of pore fluid pressure leading to increased stress at grain–grain contacts
- Increase in stress due to changes in tectonic movement

Pressure Gradient Autocatalysis

In many cases mineral dissolution reactions take place with a decrease in volume because most ions have negative partial molar volumes. Thus as fluids migrate down a pressure gradient such minerals tend to be precipitated out. As the rate of this reaction is enhanced by a larger pressure gradient, precipitation tends to be within regions where the pressure gradient is highest—i.e., where the permeability is lowest. But this increases the local pressure gradient—a positive feedback situation. Thereby a pressure gradient tends to intensify (autocatalyze) itself.

The effect can be even more pronounced as an overpressuring mechanism drives fluid through such an autocatalytic region. This leads to greater overall pressure across the region containing the autocatalytic region. Thus overpressuring and sealing get caught in a runaway feedback (see Y. Chen et al., this volume).

Observed Constituents of Diagenetically Banded Seals

A rich class of layered diagenetic structures has been identified:

- Evenly spaced banded calcite cements in sandstones (Levandowski et al., 1973; Boles, 1987; Bjorkum and Walderhaug, 1990)
- Arrays of stylolites in limestones and sandstones (Tada and Siever, 1989)
- Small-scale lamination in chalk (Ekdale and Bromley, 1988)
- Diagenetically differentiated bedding in marl/limestone sequences (Bathurst, 1986; Ricken, 1986)
- Layers of enhanced intergranular pressure solution in sandstones (Heald and Anderegg, 1960)
- Mica laminations in very fine grained sandstones from the Pennsylvanian Coffeyville Formation (Al-Shaieb, 1988, unpublished)
- Spaced cleavage (Alvarez et al., 1976; Engelder and Marshak, 1985)
- Stylolites in a variety of rocks including sandstones, limestones, dolomites, and coals
- Fine-scale diagenetically differentiated lamination in shales (Z. Al-Shaieb and J. Puckette, 1990, personal communication)
- Comminution collapse bands in sandstones (Tigert and Al-Shaieb, 1990)

These phenomena are not completely distinct but rather may grade into each other. We suggest that they are end-member responses of a more general process of diagenetic mechano-chemical differentiation leading to roughly planar features.

Observations from sandstones and shales in the Anadarko basin (Tigert and Al-Shaieb, 1990; Ortoleva et al., in press; Al-Shaieb et al., this volume; Shepherd et al., this volume) appear to illustrate many of the phenomena noted above. These include stylolite arrays, banded compaction–porosity/cementation alternations, carbonate bands in sandstones with quartz, presence of crystallization-induced removal of quartz grains by precipitating carbonate grains, and diagenetic comminution bands. While more-detailed petrological and geochemical work needs to be done to make a more convincing case for the mechanisms of band development, the above-cited lists of mechanisms and examples provide a starting point for an analysis of these systems.

MECHANO-CHEMICAL MODEL OF REPETITIVE POROSITY BANDING

To demonstrate multiscale dynamics of the development of banded seals and basin-scale compartmentation, we consider the problem of porosity/cementation band alternations in sandstones. This phenomenon was adopted because of the strongly coupled nature of the problem. Fluid pressure is affected by the distribution in space of banded seals. In turn, the pressure affects the dynamics of the genesis of the seals and also affects the distribution of stresses in the pressurized porous medium (that itself in turn affects the genesis of the bands). Furthermore, the problem involves the evolution of many variables, each of which obeys rather different types of equations.

Qualitative Picture

The model adopted is basically that proposed by Dewers and Ortoleva (1990b) for the diagenesis of monomineralic sediments. In the model, grain dissolution and growth kinetics are coupled to solute transport and texture-dependent effective stress. This model is briefly as follows:

1. The system is saturated with water and is composed of quartz that dissolves into aqueous silica.
2. A representative grain is used as a texture model of the rock. This grain is taken to be a truncated sphere described by its radius L_f and three trunca-

tion lengths L_x, L_y, and L_z, x, y, and z being the three directions of the axes of the coordinate system; z is in the vertical direction. The rock is represented by a three-dimensional lattice of such grains. The rock texture is described by the distribution of L_x, L_y, L_z, L_f in space. The number of grains per unit volume, the porosity and other auxiliary textural variables can be obtained in terms of the basic texture variables L_x, L_y, L_z, and L_f.

3. Pressure solution plays a central role in the model. The dissolution at the horizontal (z) contact faces causes compaction. In this study we neglect all compaction for simplicity. This is likely reasonable when grain–grain contacts are clean and conditions do not support an appreciable water film between contacts. However, the effect of stress on the kinetics of the free face is accounted for through the influence of stress on grain free energy.
4. The free energy associated with the grain faces in contact with the pore fluid depends on the local stress and temperature.
5. The macroscopic (i.e., supragrain-scale) stresses are determined by force balance. These vary with depth and depend on the spatial distribution of texture.
6. Temperature increases linearly with depth and affects the grain dissolution/precipitation kinetics through the rate coefficient and the equilibrium constant.
7. Mass transfer occurs by diffusion and advection.

Basic Equations

A more detailed discussion of the equations for the dynamics of the system can be found in Dewers and Ortoleva (1990b) and Ortoleva (1993). The following briefly reviews the equations for the model described in the previous subsection.

The basic descriptive variables needed are:

L_f	grain radius (L_x, L_y, L_z are fixed)
w_x, w_y, w_z	horizontal and vertical macro-elastic displacements
p	fluid pressure
c	aqueous species concentration in the pore fluid

The velocity of the rock flow $(\vec{u})$ is composed of the velocity due to subsidence $(\vec{u}_s)$, the velocity due to compaction $(\vec{u}_c)$, and the velocity due to elastic strain $(\vec{u}_e)$. The latter is ignored because it is very small. For simplicity, we further assume no growth and dissolution at the grain–grain contacts so that the velocity of compaction is zero. With these assumptions, we have

$$\vec{u} = u_s\vec{z} \tag{2}$$

where subsidence is assumed in the vertical (z) direction and $\vec{z}$ is a unit upward-pointing vector.

We transform from a fixed coordinate system (x,y,z) to a coordinate system (x',y',z'), moving downward with velocity u_s, via

$$z' = z - u_s t,\ y' = y,\ x' = x. \tag{3}$$

The equations for the dynamics of the system are based on the kinetics of grain growth/dissolution, mass conservation, and force balance. Dropping the prime on x, y, and z for simplicity, one obtains

$$\frac{\partial \phi c}{\partial t} = \vec{\nabla} \bullet \left(D\phi\vec{\nabla}c - \phi c\vec{v}\right) - n\rho A_f G_f \tag{4}$$

$$\frac{\partial L_f}{\partial t} = G_f \tag{5}$$

$$\phi\frac{\partial \rho_w}{\partial t} = \Gamma G_f - \vec{\nabla} \bullet \left(\rho_w \phi \vec{v}\right) \tag{6}$$

$$\rho_w = \bar{\rho}_w\left[1 + \beta(p - \bar{p}) - \alpha(T - \bar{T})\right] \tag{7}$$

$$\phi\vec{v} = -\omega\left(\vec{\nabla}p + \Gamma_f\vec{z}\right) \tag{8}$$

$$\sum_{j=1}^{3}\frac{\partial \sigma_{ij}^m}{\partial x_j} - \left[\phi\Gamma_f + (1-\phi)\Gamma_s\right]\delta_{j3} = 0. \tag{9}$$

In these equations, ϕ is porosity; D is the effective diffusion coefficient; n is the number density of grains; ρ is the mole density of the mineral; A_f is the free face area of a grain; β and α are the compressibility and thermal expansivity of the fluid, respectively; T is temperature; Γ is $\partial\phi/\partial L_f$ times ρ_w; ω is permeability divided by fluid viscosity; $\vec{v}$ is the velocity of fluid flow; and Γ_f is the gravitational acceleration times the mass density of the fluid (and similarly for Γ_s for the solid skeleton).

The macroscopic stress $\underline{\underline{\sigma}}^m$ and strain $\underline{\underline{e}}^m$ tensors are related by the tensor of elastic compliances $\underline{\underline{\underline{\underline{C}}}}^m$, fluid pressure p, and an effective stress coefficient $\underline{\underline{\alpha}}^m$ according to

$$\underline{\underline{\sigma}}^m + \alpha^m p\underline{\underline{I}} = \underline{\underline{\underline{\underline{C}}}}^m e^m, \tag{10}$$

and

$$e_{ij}^m = \frac{1}{2}\left(\frac{\partial w_i}{\partial x_j} + \frac{\partial w_j}{\partial x_i}\right) \tag{11}$$

where $\underline{I}$ is the identity matrix. The elements of $\underline{\underline{\underline{\underline{C}}}}^m$ for the (assumed) elastically isotropic rock are in the form

$$C_{ijkl}^m = \mu^m\left(\delta_{ik}\delta_{jl} + \delta_{il}\delta_{jk}\right) + \lambda^m\delta_{ij}\delta_{kl}. \tag{12}$$

The shear modulus (μ^m) and Lamé's constant (λ^m) of the porous medium are taken to depend on texture via Eshelby's (1957) formulae:

$$\mu^m = \mu\left[1 + \frac{15\phi(14\mu - 3k)}{82\mu - 15k}\right]^{-1}, \tag{13}$$

$$\kappa^m = \kappa\left[1+\frac{(4\mu+3k)\phi}{4\mu}\right]^{-1}, \quad (14)$$

$$\lambda^m = \kappa^m + 2\mu^m/3. \quad (15)$$

while α^{μ} is expressed by (Zimmerman et al., 1986)

$$\alpha^m = 1-\kappa^m/\kappa. \quad (16)$$

In the above, κ and μ are the bulk and shear moduli of the grains (assumed elastically isotropic), respectively.

The reaction rate G_f is taken in the form (Dewers and Ortoleva, 1990a, b; Ortoleva, 1994)

$$G_f = k\left(c-c^{eq}\right) \quad (17)$$

$$c^{eq} = Ke^{\gamma/RT} \quad (18)$$

where k is the rate coefficient for free face overgrowth, and c^{eq} is the concentration of aqueous silica in equilibrium with the stressed solid. $K(p,T)$, equilibrium constant, is obtained by a functional least-square fit of experimental data (Bowers et al., 1984), and the dependence of γ on $\underline{\underline{\sigma}}$, p, and T can be found elsewhere (Dewers et al., 1990b). The value of stress on the grain ($\underline{\underline{\sigma}}$) used for the evaluation of c^{eq} is obtained via a simple volume weighted average in terms of $\underline{\underline{\sigma}}^m$ and p:

$$\underline{\underline{\sigma}}^m = -\phi p\underline{\underline{I}} + (1-\phi)\underline{\underline{\sigma}} \quad (19)$$

The free energy needed to obtain γ in c^{eq} is calculated using p as the normal stress and $\underline{\underline{\sigma}}$ to calculate the strain energy using Kamb's (1961) formulae.

Based on the data of Chilingarian (1963) and Qin (1992), the relationship between permeability and porosity used in our simulation is

$$\text{permeability} = 1.51\times10^4\phi^6(\text{darcy}). \quad (20)$$

These equations are the basis of our homogenization approach. Note their strong coupling and nonlinearity.

HOMOGENIZED BASIN DYNAMICS

To simulate the model of the previous section in a basin-scale domain we must capture both the basin-wide trends of stress, pressure, and other factors and their submeter-scale variations due to sedimentary input patterns or diagenesis. Consider a basin of dimensions 100 by 20 km^2 and depth 5 km, constituting a 10,000 km^3 volume. There can be 10^{13} meter-scale features in such a domain. As diagenetic and sedimentary features can even be on the submeter scale, there is no computer now or in the foreseeable future that is large or fast enough to simulate the evolution of such a system using straightforward methods.

The homogenization or multiple-scale method (Nayfeh, 1973; Keller, 1977; Bensoussan et al., 1978; Araki et al., 1987) has been widely used to analyze overall behavior of heterogeneous systems such as composite media (Douglas and Arbogast, 1990; Ene, 1990; Amaziane et al., 1992). To simulate our basin-scale dynamics without losing the small-scale phenomena, we propose a homogenization technique that involves

- Making some assumptions about the regularity or statistics of the short-scale variations
- Deriving equations for the short-scale dynamics for given local average stresses, pressures, etc., yielding the short-scale fluctuations of these quantities
- Equations for the basin-scale variations of the locally averaged variables that involve effective compliances, permeabilities, and other quantities that are given explicitly in terms of averages of the short-scale fluctuations

Let us now present our homogenization approach more explicitly.

An Illustrative Problem

Consider the following model problem as would be suggested by the Navier equations for one-dimensional elastic displacement:

$$\frac{\partial}{\partial z}\left(\lambda\frac{\partial w}{\partial z}\right) = g \quad (21)$$

where λ and g are functions of space that vary on a short scale. The factors λ and g depend on both short and long scales ζ and Z, respectively, defined via:

$$\zeta = z,\ Z = \varepsilon z,\ \varepsilon << 1 \quad (22)$$

where e is the ratio of the characteristic length of small-scale variation to the characteristic length of large-scale phenomena. With these definitions, z will change by a large amount for a distance encompassing many small-scale phenomena while Z changes by only one unit. For the geologic problem we seek solutions on a domain on the scale of e–1 in length (the basin scale). From equation 21 we see that w behaves as z2, i.e.,

$$w \approx z^2 = Z^2/\varepsilon^2 \quad (23)$$

Therefore it is most natural to introduce a scaled displacement $\tilde{w}$ such that

$$\tilde{w} = \varepsilon^2 w. \quad (24)$$

With this we have the scaled problem

$$\left(\frac{\partial}{\partial\xi}+\varepsilon\frac{\partial}{\partial Z}\right)\left[\lambda\left(\frac{\partial}{\partial\xi}+\varepsilon\frac{\partial}{\partial Z}\right)\tilde{w}\right] = \varepsilon^2 g. \quad (25)$$

Our approach is to solve this equation via the scheme

$$\tilde{w} = W(Z) + \sum_{q=1}^{\infty}\varepsilon^q\tilde{w}_q(\xi,Z) \quad (26)$$

using a perturbation development.

The theory unfolds to various orders of ε. Let

$$\Lambda = \frac{\partial}{\partial \xi}\left(\lambda \frac{\partial}{\partial \xi}\right). \tag{27}$$

With this the lowest order equation becomes

$$\Lambda W = 0 \tag{28}$$

which is satisfied trivially since W is independent of ζ. To $O(\varepsilon)$ we get

$$\Lambda \tilde{w}_1 + \frac{\partial \lambda}{\partial \xi}\frac{\partial W}{\partial Z} = 0. \tag{29}$$

Thus if

$$\tilde{w}_1 = Q\frac{\partial W}{\partial Z} \tag{30}$$

we have

$$\Lambda Q + \frac{\partial \lambda}{\partial \xi} = 0. \tag{31}$$

The theory closes to $O(\varepsilon^2)$. Here we get

$$\Lambda \tilde{w}_2 + \frac{\partial}{\partial Z}\left(\lambda \frac{\partial \tilde{w}_1}{\partial \xi}\right) + \frac{\partial}{\partial Z}\left(\lambda \frac{\partial W}{\partial Z}\right) = g. \tag{32}$$

To make further progress on the homogenization equations one usually assumes that the short scale (ζ) variations have some simplifying character. The usual cases treated are that these variations are either periodic or are stochastic with some relatively simple statistical character. We adopt the former, assuming variables are periodic in ζ with some unit cell of repetition.

Averaging equation 32 over a unit cell by integrating it with respect to ζ and dividing the result by the length of the unit cell, we get, denoting this process by <…>,

$$\frac{\partial}{\partial Z}\left(\lambda^* \frac{\partial W}{\partial Z}\right) = \langle g \rangle \tag{33}$$

$$\lambda^* = \left\langle \lambda\left(1 + \frac{\partial Q}{\partial \xi}\right)\right\rangle. \tag{34}$$

The first term in equation 32 vanishes when integrated due to the periodic feature of w_2 and λ on the short scale. Explicit solution of equation 31 for Q can be used to show $\lambda^* = 1/\langle 1/\lambda \rangle$, a basic result of the theory of layered media.

The question arises as to the boundary conditions to which we should subject W and $\tilde{w}_1$. Note that from its equation, $\tilde{w}_1$ is only defined up to a constant (or more precisely a function of Z). A similar comment can be made about $\tilde{w}_q$, $q > 1$. This "freedom" allows us to take the convention that W represents the local average displacement; we get

$$\langle \tilde{w}_q \rangle = 0, q = 0. \tag{35}$$

Thus, for example, Q satisfies periodic boundary conditions and the additional constraint

$$\langle Q \rangle = 0. \tag{36}$$

This completes our theory of this example problem. Equation 31 is solved on space ζ to obtain the small-scale fluctuations and equation 33 is solved on large-scale space Z for the overall (basin-scale) trend of displacement.

The Full Mechano-Chemical Model

Because diffusion of the aqueous species through the rock and the change of L_f are very slow in the processes of interest, we emphasize this explicitly via the formulation

$$t = t'/\varepsilon^3, G_f = \varepsilon^2 G_f', D = \varepsilon^2 D', \text{and } \omega = \varepsilon^2\omega' \tag{37}$$

for the ratio ε of the characteristic short-scale length to the basin-scale length. The system is to be understood on two distinct length scales. If $\vec{r}$ specifies the position within the system, the appropriate two position variables are

$$\vec{r}_0 = \vec{r}, \text{and } \vec{r}_1 = \varepsilon\vec{r}. \tag{38}$$

Here $\vec{r}_0$ is analogous to ζ and $\vec{r}_1$ is analogous to Z of the previous subsection. Furthermore, the pressure, concentration, and elastic displacements can be divided into a macroscopic part P, C, W_i ($i = x,y,z$) and the remainder such that

$$p = \tilde{p} = P(\vec{r}_1, t) + \sum_{q=1}^{\infty} \varepsilon^q \tilde{p}_q(\vec{r}_0, \vec{r}_1, t), \tag{39}$$

$$c = C(\vec{r}_1, t) + \sum_{q=1}^{\infty} \varepsilon^q c_q(\vec{r}_0, \vec{r}_1, t), \tag{40}$$

$$\varepsilon^2 w_i = \tilde{w}_i = W_i(\vec{r}_1, t) + \sum_{q=1}^{\infty} \varepsilon^q \tilde{w}_{iq}(\vec{r}_0, \vec{r}_1, t) \tag{41}$$

$$(i = x, y, z \Leftrightarrow 1,2,3).$$

The macroscopic parts P, C, and W_i capture the overall trends while the remainders $\tilde{w}_{iq}$, c_q, and $\tilde{p}_q$ capture the small-length scale variations (local deviations from the trends). Even to lowest order, L_f varies on both scales, so

$$L_f = \sum_{q=1}^{\infty} \varepsilon^q L_{fq}(\vec{r}_0, \vec{r}_1, t). \tag{42}$$

All other dependent variables, such as D, ϕ, c^{eq}, ω and elastic moduli (μ^m,λ^m) have the same behavior as L_f and hence are expanded in a similar way.

The application of the method described in the above subsection to the full mechano-chemical problem is carried out briefly as follows. The equation for the fluid pressure is treated similarly to the above approach for w. It and the two elastic displacements yield a number of Q-like functions that are all coupled (there are 16 such factors for a two spatial dimensional problem). As a result, a set of microscopic equations for small-scale variations of the variables and

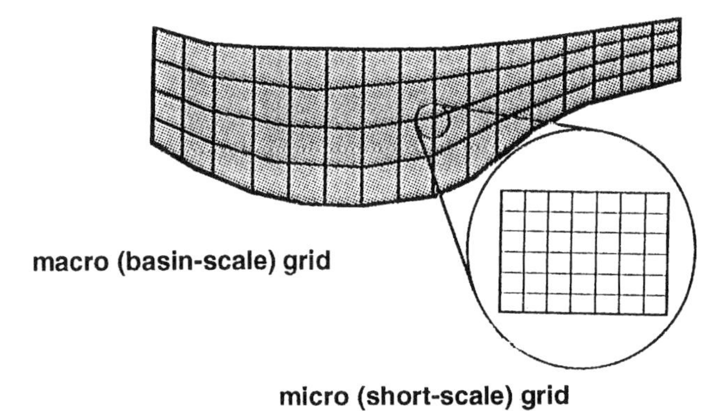

Figure 2. Illustration of macrogrid and one of the microgrids associated with a node of the macrogrid. This two embedded grid approach is the basis of our computational homogenization approach.

homogenized macroscopic equations for overall pressure and elastic displacements are obtained. These homogenization equations are the basis of the simulations of the following section (since the derivations and results of the equations are lengthy, they will not be presented here).

NUMERICAL SIMULATIONS

Numerical Scheme and Parallel Computational Approach

The homogenized problem as set forth above can be simulated numerically via the following scheme. Consider the two-dimensional system of Figure 2. The macroscopic spatial variable, i.e., the $\vec{r}_1 = (x_1, z_1)$ plane is divided into a macrogrid specified by indices I_1, J_1. At each macrogrid node we associate a microgrid with indices I_0, J_0. The homogenized problem is then to be solved by obtaining numerical solutions of the equations for L_{f0}, c_1 and the Qs for $\tilde{p}_1$ and $\tilde{w}_{x1}$ and $\tilde{w}_{z1}$ on each of the microgrids. These quantities are used to obtain the coefficients in the P, W_x, and W_z equations by taking an appropriate average over the microgrids. Then the P, W_x, and W_z equations are solved on the macrogrid. The newly obtained P, W_x, and W_z are then used to obtain a better estimate for the solutions of the microscopic equations on all the microgrids. This loop is iterated until convergence. The system is extrapolated to the next time in a similar fashion via an implicit method. At each new time step the iteration is started using the values at the previous time as the first estimate.

Based on the structure of the equations, the following procedure is used in our simulation:

- take the previous estimate for P, W_x, and W_z, use initial L_{f0}, solve the microscopic equations for all the Qs
- solve the microscopic equation for c_1
- calculate L_{f0}
- solve the macroscopic equations for P, W_x, and W_z by using all the Qs and L_{f0}.

Repeat this process in the same time step until convergence, then proceed to the next time step.

A very important aspect of the homogenization calculation is its highly parallel nature. On a massively parallel platform all microgrid problems can be solved simultaneously consistent with the current level of approximation of the macroproblem. Based on this property, an efficient massively parallel code for this problem was developed on an Intel iPSC/860 machine. Most of our calculations were carried out on a 64 processor iPSC/860 at Intel Supercomputer Division headquarters courtesy of Intel.

Preliminary Simulation Results

Development of a Seal within a Uniform Sandstone

Differentiated layered structures decrease the permeability normal to the layers and therefore can act as seals. The pattern of the spatial distribution of the amplitude of the structures arises from the interplay of a number of factors. The increase of stress and temperature with depth enhances the growth of the cementation/porosity alternations. Overpressuring inhibits the feedback underlying the amplification of textural patterns. With this, at some specific depth the overall rate of the texture differentiation may reach its maximum value, and eventually a seal forms there. This is demonstrated by the simulation result shown in Figure 3. The porosity deviation is much more greatly intensified at about a third of the way from the bottom of the domain after 12.6 Ma of burial while the initial porosity deviation was the same everywhere (Figure 3A). The homogenized permeability normal to the layers decreases more than two orders of magnitude in that zone (Figure 3B). The vertical fluid flux decreases with time due to the formation of the intensively differentiated region (Figure 3D). As a result, overpressure is developed beneath the zone, and the pressure increases dramatically across it (Figure 3C). This indicates that a pressure seal has been formed in that region. Note that the permeability normal to the layers decreases everywhere across the system; this is because the fluid flows through it from the bottom, and aqueous silica precipitated everywhere since lower temperature and pressure decrease the solubility of silica (Bowers et al., 1984).

Development of a Seal across a Finer Grain Bed

The texture changes faster within a finer grain bed than within a coarse-grain one because the total reactive surface area is greater in the fine bed. This is demonstrated in Figures 3–5. The simulations shown in Figures 3 and 4 use the same initial data except that the grain size in the case of Figure 3 is three times that in Figure 4. Therefore only 7.4 m.y. is required for the seal in Figure 4 to form while 12.6 m.y. is required for formation of comparable seal in Figure 3. The grain-size effect is more clearly shown by the basin-scale simulation in Figure 5. The initial grain-size distribution is as in Figure 5A. The intensity of initial porosity fluctuation is uniform in the basin (Figure 5F).

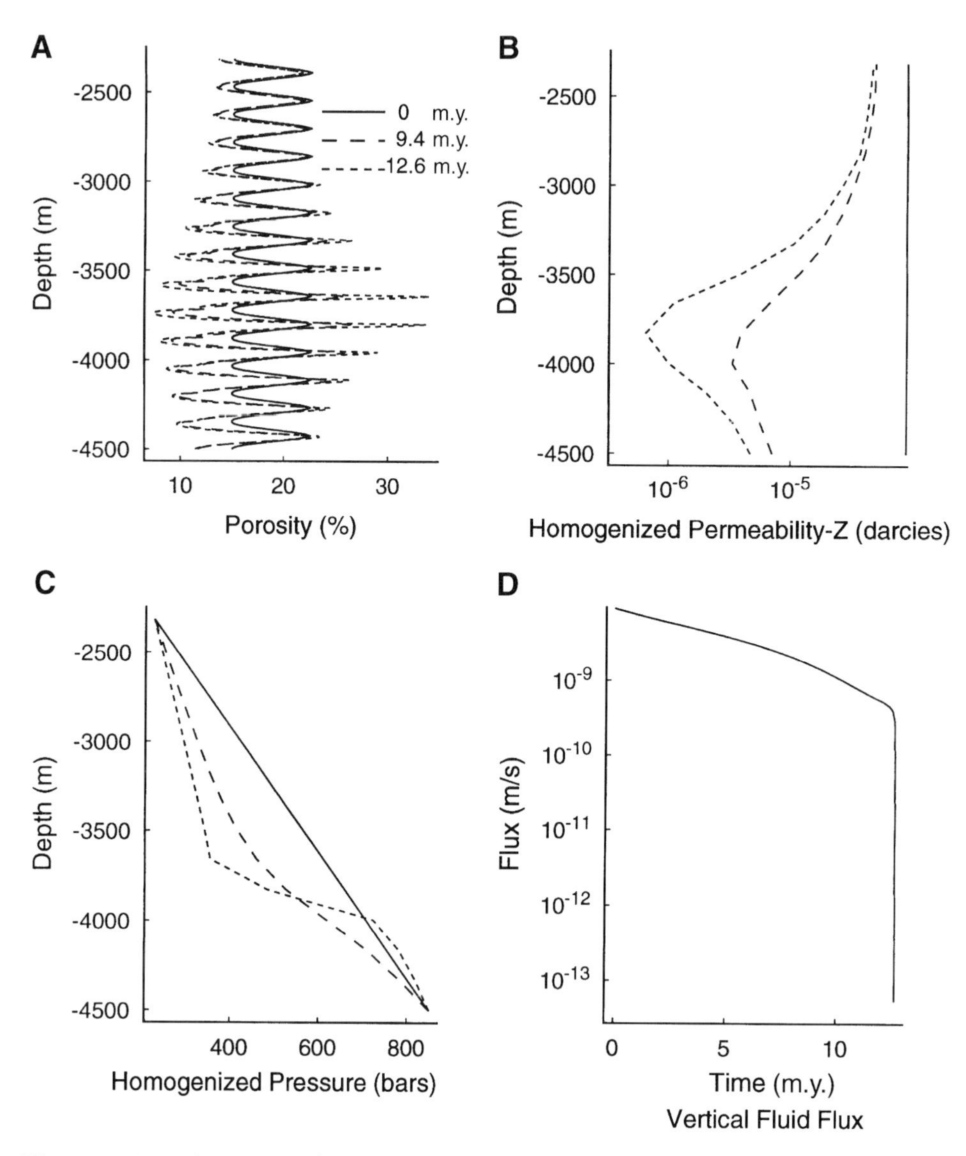

Figure 3. Development of a seal within a uniform grain-size sandstone. The initial grain size, porosity, and porosity deviation is the same macroscopically everywhere in the system. Hydrostatic pressure is set at the top of the domain while an overpressure condition is set at the bottom of the domain. The geothermal gradient is 35°C/km. After 12.6 m.y. of burial, the porosity deviation is much more greatly amplified at about a third of the way from the bottom via mechano-chemical feedback and that zone acts as a pressure seal. (A) Profiles of porosity of a vertical cross section at different times (B and C use the same legend as shown in this figure). Each layer shown represents 8000 layers, and each layer is therefore 8000 times narrower than shown. Porosity profiles in other figures are presented in the same way. (B) The corresponding profiles of homogenized permeability normal to the layers at different times clearly show the formation of a seal. (C) Homogenized pressure profiles at different times, showing the development of a large pressure gradient across the seal as the seal develops through the mechano-chemical instability. (D) Decreasing of fluid flux through the system normal to the layers as the seal develops. Notice the dramatic deceleration of the flux near the end of the simulation caused by the acceleration of the increased rate of the band intensification.

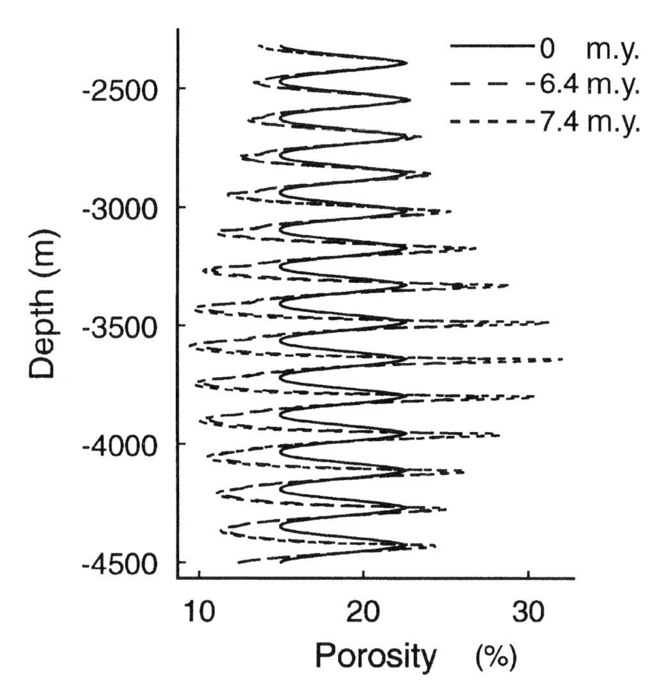

Figure 4. Porosity profiles at different times of a simulation with the same parameters as Figure 3A except that grain size is two times smaller than that in the case of Figure 3. Here it only takes 7.4 m.y. to have a result similar to that of Figure 3A because the grain size is smaller.

Initially the fluid pressure is hydrostatic everywhere. However, fluid pressure at the bottom of the basin was set to increase gradually with time (Figure 5C). After 32 m.y. of diagenesis two seals were formed within the finer grain beds (Figure 5B) across the basin. The texture fluctuation in the finer grain beds had the greatest amplification (Figure 5F). Permeability normal to the layers decreased more rapidly in those regions (Figure 5D). Large pressure gradients were developed across the seal zones while the pressure gradient approaches hydrostatic elsewhere (Figure 5C). This agrees with the observations from the Anadarko basin (Powley, 1990). The fact that diagenetically banded seals appear preferentially in finer grain formations (Powley, 1990; Shepherd et al., this volume) may be explained by this simulation.

Effect of Fluid Pressure on the Rate and Depth of the Formation of a Seal

Increasing fluid pressure decreases the rate of amplification of textural patterns because it tends to cancel the variations in the grain stress associated with variations in texture and the lithostatic stress. In Figure 6A, the boundary conditions at both the top and the bottom of the domain are hydrostatic for the fluid pressure. The porosity deviation amplification at the bottom of the domain is larger than that at the top because the temperature and depth effect overcomes the pressure effect. In Figure 6B, the porosity deviation amplification at the bottom of the domain is smaller than that at the top because the overpressure boundary condition applied to the bottom of the domain represses amplification at greater depth.

From the result of Figure 3C, D and Figure 5C we see that once a seal is formed, it traps the fluid in the compartment and builds up overpressure within it. This suggests that the relatively high pressure (overpressure) within the compartment will decrease the structural change within it, and the relatively low pressure at the top of the overlying seal will speed up the textural differentiation and the formation of a seal. This mechanism enhances the compartmentation process and the macroscopic heterogeneity of the sedimentary basin.

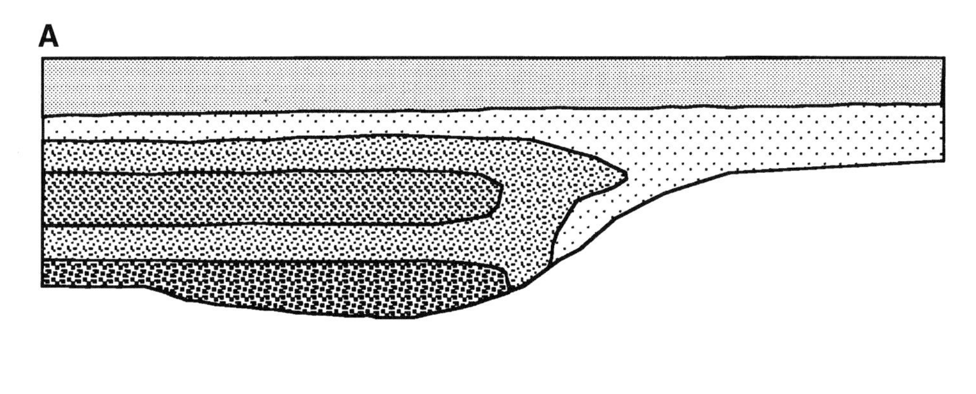

Figure 5. Development of seals within fine grain beds. (A) Initial grain-size distribution.

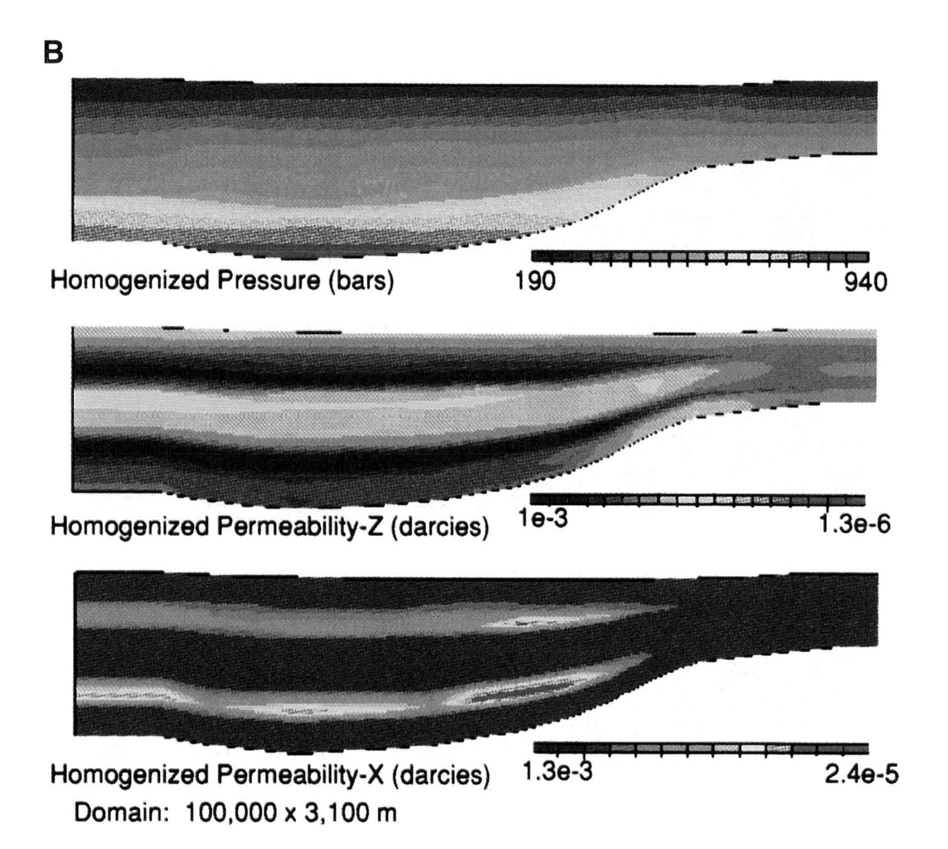

Figure 5. Development of seals within fine grain beds (continued). (B) Contour maps of homogenized pressure, horizontal permeability, and vertical permeability after 32 m.y. of diagenesis.

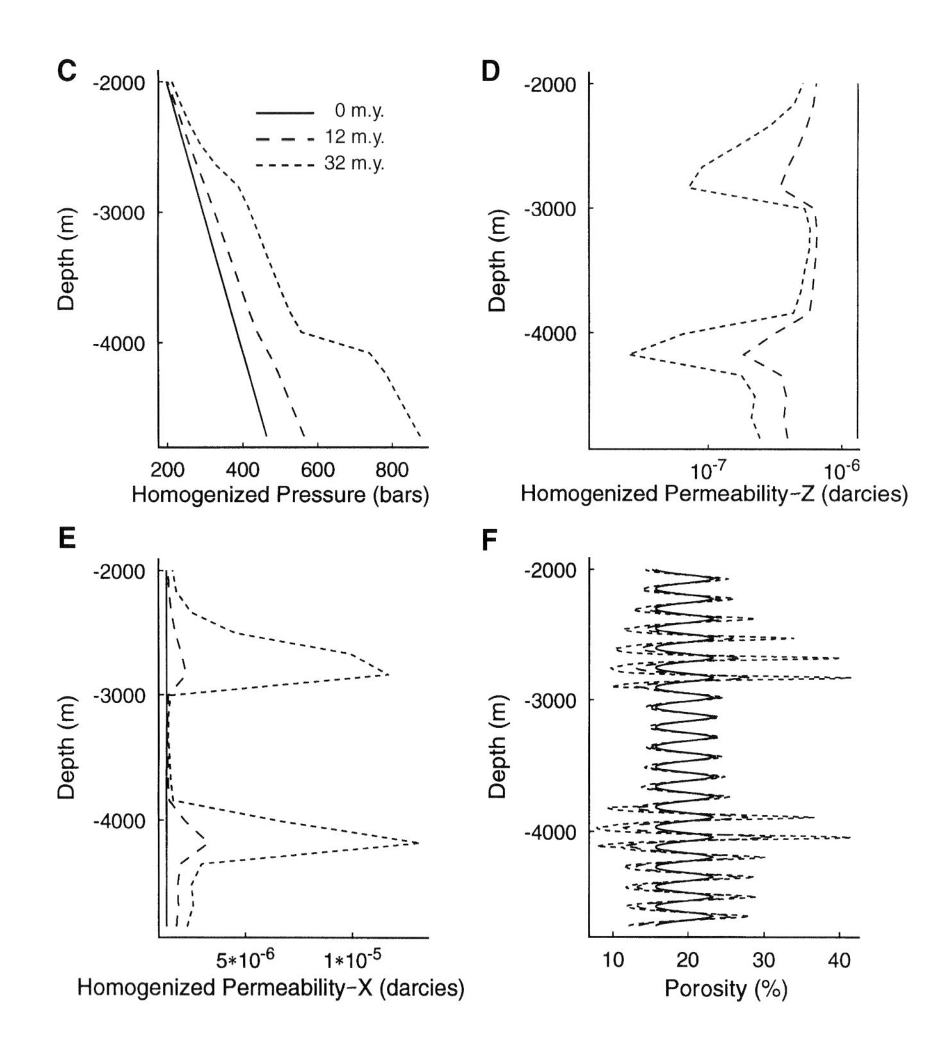

Figure 5 (continued). (C–E) Profiles of homogenized pressure, and vertical and horizontal permeabilities along a transect through (B), respectively at different times (D, E, and F use the same legend as shown in C), and (F) the corresponding porosity profiles at different times. In this simulation the domain is 10 km in the horizontal direction and 3100 m in the vertical direction. The depth of the bottom of the domain is 5000 m. The geothermal gradient is 30°C/km.

A

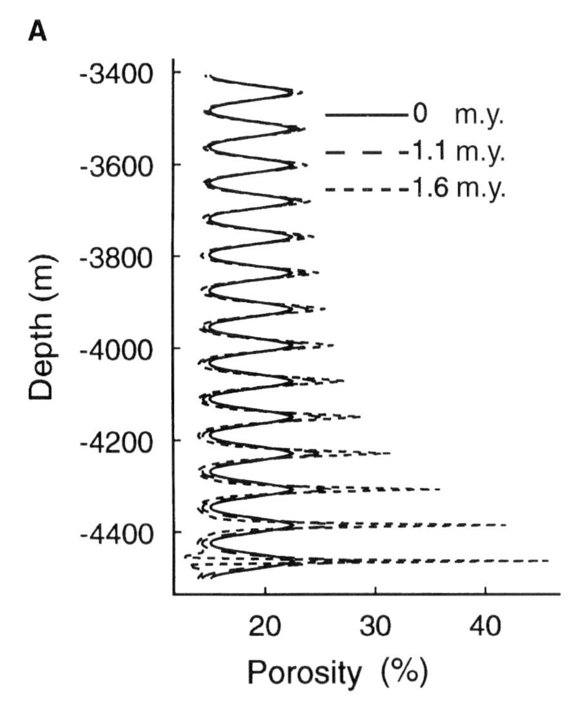

B

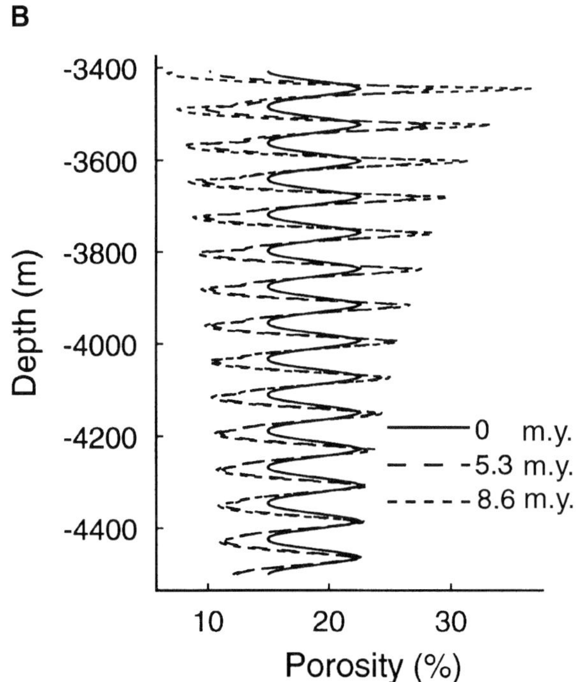

Figure 6. Simulations showing the effect of pressure on seal depth. The initial data and parameters are the same as those used in the case of Figure 3 except for the domain size and boundary conditions on pressure. The domains here are only 1092 m and each layer actually represents 4000 layers and is shown 4000 times wider than it actually is. (A) Hydrostatic pressure conditions are set for both top and bottom of the domain. Depth-related stresses and temperature effects dominate the system. (B) Hydrostatic pressure at the top of the domain is set while overpressure is set at the bottom of the domain. The high pressure depresses the texture differentiation in the bottom half of the domain.

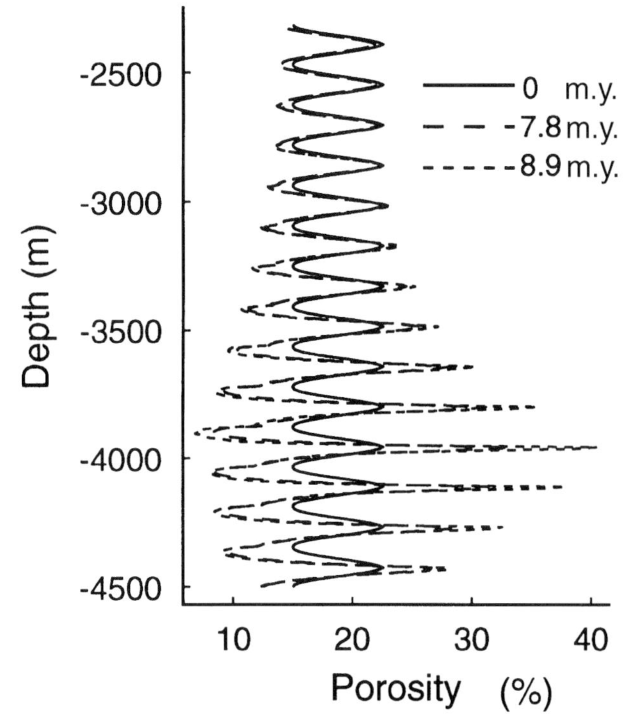

Figure 7. Porosity profiles at different times of a simulation with the same parameters and initial data as those used in Figure 3A except the intensity of overpressure at the bottom of the domain for this case is smaller. The depth of the center of the seal in this case is 4020 m in contrast with the depth of the seal in Figure 3A of 3660 m.

The intensity of overpressure also influences the depth of seal formation. Decreasing the intensity of overpressure increases the depth of the seal as shown in Figures 3 and 7. The intensity of overpressure in the simulation shown in Figure 7 is less than that in Figure 3, so the seal in the case of Figure 7 formed at a depth of 4020 m which is deeper than the depth of the seal, 3660 m, in the case shown in Figure 3. All the other conditions are the same for the cases of Figures 3 and 7. This implies that seals can form in shallower parts of a basin in which there is some mechanism, for example, high rate of sedimentation, to build higher overpressure.

Effect of Temperature on the Rate and Depth of the Formation of a Seal

High temperature considerably speeds up the process of diagenesis because of the dependence of reaction rates, diffusion coefficients, and solubilities on temperature. The simulations shown in Figures 3 and 8 use the same initial texture data and other parameters but have different geothermal gradients. With a temperature gradient of 35°C/km for the case of Figure 3, it takes 12.6 m.y. to increase the porosity deviation from 3.6% to 26.5%. In contrast, it takes about 26 m.y. to have a similar result (porosity deviation 22.2%) for the case with a temperature gradient of 30°C/km shown in Figure 8. This suggests that variation in geothermal gradient may play a very

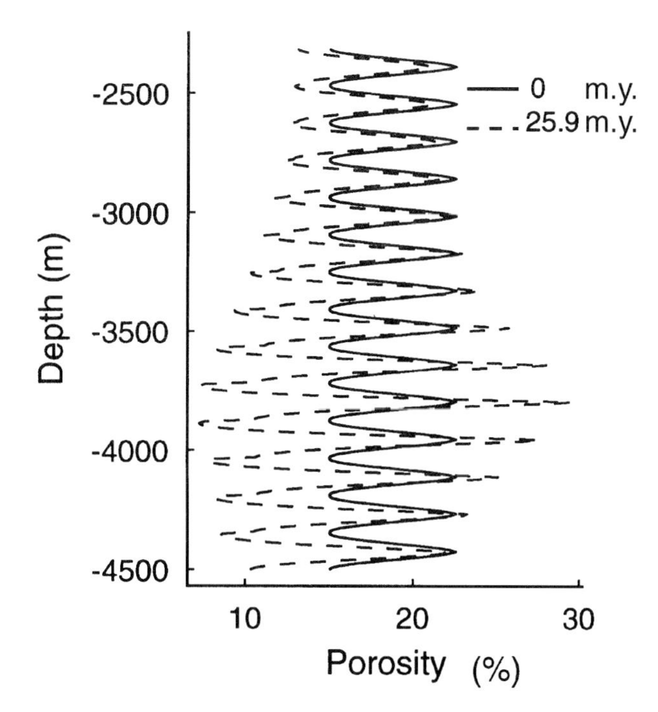

Figure 8. Porosity profiles at different times of a simulation with the same parameters and initial data as those used in the case of Figure 3 except the geothermal gradient for this case is 30°C/km, slightly less than 35°C/km used in Figure 3. The depth of the center of the seal is about 3810 m, which is deeper than that in the case of Figure 3, and it takes about 26 m.y. to increase the porosity deviation to 22% while it only needs 12.6 m.y. in the case of Figure 3.

important role in the diagenesis and formation of seals.

In addition, a lower geothermal gradient increases the depth of the seal as shown in Figures 3 and 8. The depth of the seal in the case of Figure 8 is about 200 m deeper than that in the case of Figure 3.

Fluid Flow and Stress Effects on the Formation of a Seal

Intensive texture differentiation resides within a relatively small interval of depth (Figures 3A, 4, and 7), there being large contrasts between the intense regions and their surroundings. However, intensive texture differentiation tends to spread in a larger interval of depth if the mass transfer by convection is not important in the concentration equation 4. This implies that fluid flow influences the formation of seals and the generation of large-scale heterogeneity in sedimentary basins.

It was also noted during the development of our code that the ratio of vertical stress to horizontal stress strongly affects the formation of a seal. Since this ratio varies greatly from one tectonic environment to another, this may help to explain why compartmentation is well developed in some basins but not in others and gives us some clues for the relationship between diagenetic processes and the tectonics of a basin.

These facts suggest that the fully coupled reaction-transport and mechanical problem needs to be solved and appropriate constraints should be made based on geological observation in order to establish a predictive model.

Development of Homogenized Anisotropy

During the development of the seals in the case shown in Figure 5 the homogenized permeability along the layers (in the horizontal direction) actually increases (Figure 5E), while the homogenized permeability normal to the layers decreases (Figure 5D). This is also true for all other simulations shown in this paper. This indicates that relatively uniform sandstone can become more anisotropic in texture and hydrological properties during diagenesis.

CONCLUSIONS

Origins of banded structures in a sedimentary basin vary from sedimentary to purely diagenetic. It is suggested that while the presence of an early sedimentary prepattern may be documentable, the effect of diagenetic feedback and pattern-forming instability may be profound in determining the ultimate feature of the banded pattern. There are a number of pathways (processes) by which rocks become banded via diagenesis. More detailed petrological, geological, and geochemical work needs to be done to make the mechanisms more convincing, though we do have some simulation results demonstrating some of the mechanisms.

Layered pressure seals involve many layers. The genesis time and location of these seals are affected by local stresses and fluid pressure (functions of local rock structures) but, in turn, they influence the basin-scale distribution of these factors. A computational homogenization technique is proposed to meet this challenging coupled two-scale basin modeling problem. The analysis yields two sets of coupled equations. One set describes the basin-scale dynamics. The other treats local variations of the variables. Although the homogenization calculation is extremely CPU-time intensive, the computations can be easily parallelized due to the mathematical structure of the method.

The fully coupled reaction-transport and mechanical problem needs to be solved and appropriate constraints should be made based on geological observation in order to establish a predictive model for the genesis time and location of seals.

The following were demonstrated by simulations carried out for pure quartz sandstones undergoing stress-induced reactions, diffusion, and fluid flow. Seals are very likely to form within finer grain beds or at some specific depth within macroscopically uniform grain sediments. The depth and the time needed to form a seal depend upon the intensity of overpressure, fluid flow, geothermal gradient, subsidence velocity, grain size, the sediment sequence, and tectonic environment. High overpressure or higher geothermal gradient tend to decrease the depth of the seal. Lower fluid pressure is favorable for the differentiation of the rock texture (and the formation of the seal). Once a seal starts to develop, the process is self-accelerating and the sediments become more heterogeneous and anisotropic. Each of these deserves further research; the investigation being greatly facilitated by the computational homogenization approach proposed here.

ACKNOWLEDGMENTS

The authors benefited considerably from the reviews by L. Cathles and two anonymous reviewers. The parallel code for the homogenization calculation is a combined effort of J. Gotwals and the authors based on a sequential program for this problem developed by the authors. A preconditioned conjugate gradient finite element PDE solver written by J. Mu is used in the parallel code. We also appreciate the discussions with E. Sonnenthal, J. Mu, and M. Maxwell. Research is supported in part by contracts from the Gas Research Institute (No. 5092-260-2443), AMOCO Production Company, and Mobil Exploration and Producing Services, and grants from the Basic Energy Sciences Program of the U.S. Department of Energy (No. DEFG291ER14175), Shell Development Company, Phillips Petroleum Company, and Intel Corporation.

REFERENCES CITED

Alvarez, W., T. Engelder, and W. Lowrie, 1976, Formation of spaced cleavage and folds in brittle limestone by dissolution: Geology, v. 4, p. 698–701.

Amaziane, B., A. Bourgeat, and J. Koebbe, 1992, Numerical simulation of diphasic flow in heterogeneous porous media by homogenization, *in* W.E. Fitzgibbon and M.F. Wheeler, eds., Modeling and Analysis of Diffusive and Advective Processes in Geosciences: Philadelphia, SIAM.

Araki, H., J. Ehlers, K. Hepp, R. Kippenhahn, A.H. Weidenmuller, J. Weiss, and J. Zittartz, 1987, Homogenization Techniques for Composite Media, Proceedings, Udine, Italy, 1985: Berlin, Springer-Verlag, 397 p.

Bathurst, R.G.C., 1986, Diagenetically enhanced bedding in argillaceous platform limestones: stratified cementation and selective compaction: Sedimentology, v. 34, p. 740–778.

Bensoussan, A., J.L. Lyons, and G. Papanicolnon, 1978, Asymptotic Analysis for Periodic Structures: Amsterdam, North-Holland.

Bjorkum, P.A., and O. Walderhaug, 1990, Geometrical arrangement of calcite cementation within shallow marine sandstones, *in* P. Ortoleva, B. Hallet, A. McBirney, I. Meshri, R. Reeder, and P. Williams, eds., Proceedings of the Workshop on Self-Organization in Geological Systems, Santa Barbara, 1988: Earth-Science Reviews, v. 29, p. 145–162.

Boles, J.R., 1987, Six million year diagenetic history, North Coles Levee, San Joaquin Basin, California, *in* J.D. Marshall, ed., Diagenesis of Sedimentary Sequences, Geological Society Special Publication Number 36, p. 191–200.

Bowers, T.S., K.J. Jackson, and H.C. Helgeson, 1984, Equilibrium Activity Diagrams: Springer-Verlag.

Chilingarian, G.W., 1963, Relationship between Porosity, Permeability, and Grain Size Distribution of Sand and Sandstone: Proc. Int. Sedimentl. Congr.

Dewers, T., and P. Ortoleva, 1988, The role of geochemical self-organization in the migration and trapping of hydrocarbons: Applied Geochemistry, v. 3, p. 287–316.

Dewers, T., and P. Ortoleva, 1989a, Mechano–chemical coupling in stressed rock: Geochimica et Cosmochimica Acta, v. 53, p. 1243–1258.

Dewers, T., and P. Ortoleva, 1989b, The self-organization of mineralization patterns in metamorphic rocks through mechano-chemical coupling: Journal of Physical Chemistry, v. 93, p. 2842–2848.

Dewers, T., and P. Ortoleva, 1990a, A coupled reaction/transport/mechanical model for intergranular pressure solution, stylolites, and differential compaction and cementation in clean sandstone: Geochimica et Cosmochimica Acta, v. 54, p. 1609–1625.

Dewers, T., and P. Ortoleva, 1990b, Geochemical self-organization III: a mechano-chemical model of metamorphic differentiation: American Journal of Science, v. 290, p. 473–521.

Dewers, T., and P. Ortoleva, 1990c, The interaction of reaction, mass transport, and rock deformation during diagenesis: mathematical modeling of intergranular pressure solution, stylolites, and differential compaction/cementation, *in* I. Meshri and P. Ortoleva, eds., Prediction of Reservoir Quality Through Chemical Modeling: AAPG Memoir, v. 49, p. 147–160.

Dewers, T., and P. Ortoleva, 1990d, Differentiated structures arising from mechano-chemical feedback in stressed rocks, *in* P. Ortoleva, B. Hallet, A. McBirney, I. Meshri, R. Reeder, and P. Williams, eds., Self-Organization in Geological Systems: Proceedings of a Workshop held 26-30 June 1988, University of California Santa Barbara: Earth-Science Reviews, v. 29, p. 147–160.

Dewers, T., and P. Ortoleva, 1992, Non-linear dynamics in chemically compacting porous media, *in* W.E. Fitzgibbon and M.F. Wheeler, eds., Modeling and Analysis of Diffusive and Advective Processes in Geosciences: SIAM, p. 100–121.

Dewers, T., and P. Ortoleva, 1994, Formation of stylolites, marl/limestone alternations, and clay seams through unstable chemical compaction of argillaceous carbonates, *in* K.H. Wolf and G.V. Chilingarian, eds., Diagenesis Vol. 4: Amsterdam, Elsevier.

Douglas, J., and T. Arbogast, 1990, Dual porosity models for flow in naturally fractured reservoirs, *in* J. Cushman, ed., Dynamics of Fluid in Hierarchical Porous Media: Academic Press.

Ekdale, A.A., and R.G. Bromley, 1988, Diagenetic microlamination in chalk: Journal of Sedimentary Petrology, v. 58, p. 857–861.

Ene, H., 1990, Application of the homogenization method to transport in porous media, *in* J. Cushman, ed., Dynamics of Fluid in Hierarchical Porous Media: Academic Press.

Engelder, T., and S. Marshak, 1985, Disjunctive cleavage formed at shallow depths in sedimentary rocks: Journal of Structural Geology, v. 7, p. 327–343.

Eshelby, J.D., 1957, The determination of the elastic field of an ellipsoidal inclusion and related problems: Proceedings of the Royal Society of London, v. A241, p. 376–396.

Feeney, R., S.L. Schmidt, P. Strickholm, J. Chadam, and P. Ortoleva, 1983, Periodic precipitation and coarsening waves: applications of the competitive particle growth model: Journal of Chemical Physics, v. 78, p. 1293.

Feinn, D., W. Scalf, S. Schmidt, M. Wolff, and P. Ortoleva, 1978, Spontaneous pattern formation in precipitating systems: Journal of Chemical Physics, v. 69, p. 27.

Heald, M.T., and R.C. Anderegg, 1960, Differential cementation in the Tuscarora sandstone: Journal of Sedimentary Petrology, v. 30, p. 568–577.

Hornung, U., 1991, Homogenization of miscible displacement in unsaturated aggregated solids, *in* G.D. Maso and G.F. Dell'Antonio, eds., Composite Media and Homogenization Theory, An International Center for Theoretical Physics Workshop, Trieste, Italy, January 1990: Birkhauser, Boston.

Kamb, W.B., 1961, The theory of preferred crystal orientation developed by crystallization under stress: Journal of Geophysical Research, v. 55, p. 257–271.

Keller, J., 1977, Effective behavior of heterogeneous media, *in* U. Landman, ed., Statistical Mechanics and Statistical Method in Theory and Application, p. 631–644.

Levandowski, D., M.E. Kaley, S.R. Silverman, and R.G. Smalley, 1973, Cementation in Lyons Sandstone and its role in oil accumulation, Denver Basin, Colorado: Bulletin of the American Association of Petroleum Geologists, v. 57, p. 2217.

Liesegang, R.E., 1913, Geologische Diffusionen: Dresden, Steinkopff.

Lovett, R., J. Ross, and P. Ortoleva, 1978, Kinetic instabilities in first order phase transitions: Journal of Chemical Physics, v. 69, p. 947.

Nayfeh, A.H., 1973, Perturbation Methods: New York, John Wiley and Sons.

Ortoleva, P., 1994, Geochemical Self-Organization: New York, Oxford University Press.

Ortoleva, P., Z. Al-Shaieb, and J. Puckette, in press, Genesis and dynamics of basin compartments and seals: American Journal of Science.

Ortoleva, P., E. Merino, J. Chadam, and C.H. Moore, 1987, Geochemical self-organization I: Reaction–transport feedbacks and modeling approach: American Journal of Science, v. 287, p. 979–1007.

Ortoleva, P., T. Dewers, and B. Sauer, 1993, Modeling diagenetic bedding, stylolites, concretions and other mesoscopic mechano-chemical structures, *in* Proceedings of Carbonate Microfabrics Symposium and Workshop: in press.

Ostwald, W., 1925, Organ für das Gesamtgebiet der reinen und angewandten Kolloidchemie und für die Veröffentlichungen der Kolloid-Gesellschaft: Kolloid-Zeitschrift, v. 36, 330 p.

Powley, D., 1990, Pressure and hydrogeology in petroleum basins, *in* P. Ortoleva, B. Hallet, A. McBirney, I. Meshri, R. Reeder, and P. Williams, eds., Self-Organization in Geological Systems: Proceedings of a Workshop held 26–30 June 1988, University of California Santa Barbara: Earth-Science Reviews, v. 29, p. 215–226.

Prager, S., 1956, Periodic precipitation: Journal of Chemical Physics, v. 25, p. 279.

Qin, C., 1992, A fractal interpretation of scale dependence of permeability: GSA Abstracts with Programs, v. 24, p. A303.

Ricken, W., 1986, Diagenetic Bedding: Berlin, Springer-Verlag, 210 p.

Sauer, B., T. Dewers, and P. Ortoleva, 1991, Genesis of pressure seals through differentiated compaction/cementation (DCC): Low porosity authigenic carbonate bands in sandstones: OKINTEX Group, Annual Report, February, 1991 for the Gas Research Institute.

Tada, R., and R. Siever, 1989, Pressure solution during diagenesis: a review: Annual Review of Earth and Planetary Sciences, v. 17, p. 89–118.

Tigert, V., and Z. Al-Shaieb, 1990, Pressure seals: Their diagenetic bedding pattern, *in* P. Ortoleva, B. Hallet, A. McBirney, I. Meshri, R. Reeder, and P. Williams, eds., Self-Organization in Geological Systems: Proceedings of a Workshop held 26–30 June 1988, University of California Santa Barbara: Earth-Science Reviews, v. 29, p. 227–240.

VII. Reaction-Transport-Mechanical Modeling of Seals and Compartments

Chapter 26

Numerical Simulations of Overpressured Compartments in Sedimentary Basins

E. Sonnenthal
Peter J. Ortoleva
Indiana University
Bloomington, Indiana, U.S.A.

ABSTRACT

In many sedimentary basins there are regions where the fluid pressure exceeds the hydrostatic pressure and may approach lithostatic values. These overpressured compartments occur in actively forming basins as well as Paleozoic intracratonic basins. The development and maintenance of fluid overpressures in compartments, over tens to hundreds of millions of years, must require pressure-generating mechanisms and rock-sealing processes to retard the loss of fluid. Several pressurizing processes have been invoked for different basins and include thermal expansion, organic reactions, disequilibrium mechanical compaction, poroelastic deformation, and pressure-solution. Whether the seal is discordant or concordant with bedding, it must be diagenetically altered or mechanically compacted for it to have sufficiently low permeability to retard appreciable fluid loss. Mechanisms of seal formation are therefore dependent on pressure and temperature and thus may be tightly coupled to pressure-generating mechanisms. The thermal, tectonic, and depositional history of the basin directly affects these processes.

This paper describes a model and gives computer simulations in two dimensions of coupled fluid flow, compaction, hydrofracturing, deposition, and subsidence in sedimentary basins. The mechanism of compaction used is a water–film diffusion model of pressure solution for quartz in a periodic array of truncated spheres. Although this process is not the only possible compaction mechanism, it is important in quartz-rich sandstones and illustrates the coupling of overpressuring and diagenesis. Results of simulations show that regions of overpressure can encompass several lithologies with the upper transition to normal pressure cutting across dipping beds, which were deformed by differential tectonic subsidence. Some compartments are enclosed within individual beds, sealed by greater compaction on the margins of the beds. Interiors of these overpressured compartments become fractured and much more permeable as the fluid pressure exceeds the sum of the horizontal stress and rock strength. Compaction and thermal expansion of the fluid combine to cause overpressuring that slows down the rate of compaction by reducing stresses on grain contacts. Rocks surrounding the overpressured compartment continue to compact and further tend to seal it from its surroundings. Application of such coupled models to specific basins would require detailed information on the dominant overpressuring mechanisms and geologic history, yet could yield much insight into the formation of compartments and deep-basin fluid flow patterns.

INTRODUCTION

Many sedimentary basins exhibit abnormal fluid pressures (above or below the hydrostatic gradient; Powley, 1990; Bradley and Powley, this volume). Observations of fluid pressure profiles in basins around the world led to the idea of large pressure compartments (a few to over a hundred kilometers across) that are box-like volumes of abnormal fluid pressure (Powley, 1990; Bradley, 1975; Bradley and Powley, this volume).

The upper transition to abnormal fluid pressure is commonly around 3 km in depth, although it may be less than 2 km or greater than 5 km (Powley, 1990; Negus-de Wys and Dorfman, 1991). Shallow overpressured compartments (<3 km) are often in areas of rapid sedimentation or where horizontal stresses are high (e.g., convergent margins). Fluid overpressures are most common in actively subsiding basins whereas underpressures are found mainly in uplifted onshore basins (Powley, 1990). Overpressures that approach or exceed the lithostatic gradient are characteristic of areas of rapid sedimentation and subsidence, as in some Gulf Coast basins, those under substantial lateral compression such as forearc basins, accretionary wedges (Shi et al., 1989; Berry, 1973), continental thrust systems, and along strike-slip fault boundaries such as the San Andreas (Negus-de Wys and Dorfman, 1991). Interiors of compartments commonly exhibit a network of centimeter-scale fractures, which have been attributed to hydrofracturing, changes in differential stress with depth, or some combination of the fluid pressure and differential stress (Powley, 1990; Althaus, 1975). Zones of highly cemented diagenetically altered rock that cross lithologic contacts have been found in the transition region from normal (hydrostatic) to abnormal pressure and are proposed to be laterally continuous seals (Powley, 1990; Tigert and Al-Shaieb, 1990; Weedman et al., 1991; Al-Shaieb et al., this volume). Seals may be highly impermeable shales, altered paleosols, faults, or diagenetically banded and/or stylolitized rock (Drzewiecki et al., 1991; Ely et al., 1991; Logan and Lin, 1991; Martinsen and Jiao, 1991; Simo et al., 1991; see Qin and Ortoleva, this volume). Nearly vertical faults and shallow-dipping unconformities may also have acted as or promoted the formation of effective seals, especially if they or the adjacent region has undergone diagenetic alteration (Logan, 1991; Al-Shaieb et al., 1991).

The development and maintenance of abnormal fluid pressures require pressure-generating (or reducing) mechanisms and low-permeability features (seals) which retard fluid movement between the compartment and its surroundings. Proposed pressure-generating mechanisms include mechanical compaction (Chapman, 1972), mechano-chemical compaction (pressure solution or lithification; Bradley, 1975), methane generation (Hedberg, 1974; Tissot and Welte, 1984), poroelastic deformation, and thermal expansion (Barker, 1972; Bradley, 1975). In shales and other low-permeability rocks mechanical compaction leads to overpressuring, termed disequilibrium compaction (Hunt, 1979). The latter processes are dependent on rates of sedimentation and subsidence, the thermal gradient and mode(s) of heat transport, the chemical composition of the fluid and sediment, and the sediment texture. Hydrofracturing may lead to a reduction in pressure and localized fluid flow; therefore, the overall stress regime, the rock rheology, and the tectonic history of the basin are important components in the evolution of abnormal pressures and compartments.

The essence of compartmentation is the spatial separation of the low-permeability seal zones from an abnormally pressured interior having high fluid connectivity (Powley, 1990). A bed undergoing compaction (e.g., disequilibrium mechanical compaction) is not a compartment in itself, unless it develops a higher-permeability interior (possibly fractured) with low-permeability margins acting as seals. Compartments are, however, often made of many lithologic units with seals crossing stratigraphic contacts (Al-Shaieb et al., 1991; Powley, 1990). They have beds and compartments inside having differing fluid pressures, hence the concept of a megacompartment complex (Al-Shaieb et al., 1991 and this volume). An important goal of our work is to model the development of compartments as well as the dynamics of overpressuring.

Measurement of pressures in the subsurface, in addition to detailed geological and geochemical studies, has outlined the form and structure of compartments and perhaps the age and origin of seals (Al-Shaieb et al., 1991 and this volume). The time scale of fluid overpressuring and the dynamics of compartmentation are, however, a result of the coupling of several time-dependent processes (sedimentation, fluid flow, compaction, organic reactions, etc.). Thus, a variety of (or the same) phenomena can result from differences in the rates of these competing and reinforcing mechanisms.

Fundamental relationships describing the coupling of compaction and fluid flow were described in detail by McKenzie (1984, 1987). This work provided analytical solutions for fluid flow and compaction using bulk viscosities to calculate rates of strain. In another study, Person and Garven (1992) provided numerical solutions of flow in the Rhine Graben by coupling heat and mass transfer and compaction. The rates of compaction were set by a porosity-depth curve and are not coupled to the fluid pressure or temperature. Coupling of fluid pressure, hydrofracturing, and pressure solution was solved numerically in one-dimensional domains by Gavrilenko and Gueguen (1993). The latter model treated pressure solution only on crack asperities, disregarding compaction in the porous matrix. While this may be more applicable to low-porosity deep crustal rocks, matrix compaction is clearly important in sedimentary basins. In our work we calculate compaction rates based on differences between the stresses on load-supporting grain contacts and the grain faces in equilibrium with the pore fluid pressure. Therefore, rates of compaction are fully coupled to the fluid pressure, temperature, and the grain size (through the effect of differing distances over which diffusion must take place in different-sized grain contacts). Earlier one-dimensional numerical models of sediment compaction by pressure solution reproduced well observations of porosity reduction in sandstones as a function of depth (Angevine and Turcotte, 1983; Dewers and Ortoleva, 1990a); thus, we believe this model gives reasonable compaction rates.

In this paper we give mathematical models for a fully coupled suite of reaction, transport, and mechanical (RTM) processes. Simulations are presented illustrating the development of overpressured compartments through the coupling of mechano-chemical processes, thermal expansion, fluid flow, and the controls of sediment texture, sedimentation rate, and tectonic subsidence.

THE RTM MODEL

The state of the deep-basin fluid-rock system is described in terms of the spatial distribution of a set of macroscopic variables. The rock matrix is characterized by the grain size and shape; the number density of grains; and the number, size, and distribution of fracture nuclei. The pore fluid is described by its composition. Physical parameters include the fluid pressure (P), temperature (T), local average stress, and grain-scale stresses; the latter describe the stresses on grain–grain contacts and free faces (those in contact with pore fluid). The purpose of the model is to predict the evolution in time of the spatial distribution of these variables.

Two distinct sets of processes may take place during compaction. Dominantly mechanical processes include grain sliding, plastic grain deformation, and grain breakage. Mechano–chemical (i.e., pressure solution–mediated) compaction is driven by free energy deviations owing to stress and/or strain. We consider only mechano-chemical compaction in the RTM model described here, because of its greater importance in deep-basin compaction.

The following laws and processes are the basis of simulations discussed in this paper:

1. Local conservation of mass for pore fluid species accounting for aqueous phase and mineral reactions as well as local diffusive transport
2. Conservation of water mass, Darcy flow, fluid thermal expansion, and compressibility
3. Mineral growth/dissolution reactions (compaction by pressure solution and cementation by free face precipitation)
4. Macroscopic-scale force balance
5. Grain-scale force balance
6. Hydrofracture growth and closure
7. Vertical rock flow owing to basement subsidence and compaction

Macro-Stresses

The macroscopic stress is simplified by considering only the principal diagonal components:

$$\begin{bmatrix} \sigma_{xx}^{m} & 0 & 0 \\ 0 & \sigma_{xx}^{m} & 0 \\ 0 & 0 & \sigma_{xx}^{m} \end{bmatrix} \tag{1}$$

Force balance in the vertical (z) direction yields

$$\frac{\partial \sigma_{zz}^{m}}{\partial z} = -g\left[\rho_f \phi + (1 = \phi)\rho_s\right] \tag{2}$$

for mass densities of fluid (ρ_f) and solid matrix (ρ_s), gravitational acceleration (g), and porosity (ϕ). Integration of the previous equation requires $\underline{\underline{\sigma}}^{m}$ at the upper boundary, which is the hydrostatic fluid pressure of the overlying water column.

The horizontal stress σ_{xx}^{m} is set to be a function of depth and the vertical stress. Fracture gradient studies in the Gulf Coast (Althaus, 1975) give a linear variation from 0.7 at the near-surface to 1.0 at approximately 6 km. This relationship was used for all simulations described in this paper.

Truncated Sphere Model

Description

We assume that the rock matrix consists of a single stress-supporting mineral—quartz. Other minerals may reside in the pore space or on quartz grain surfaces. Consider a porous monomineralic rock where every representative volume is filled with a periodic array of truncated spheres (Figure 1; Weyl, 1959). The areas of these contacts (A_x, A_y, A_z, A_f) and the porosity are obtained from the grain dimensions (L_x, L_y, L_z, and L_f; see Dewers and Ortoleva, 1990a). Although hexagonal or other close-packed grain lattices would be closer to a true rock texture, mathematical difficulties arise because of the complicated stress field and the

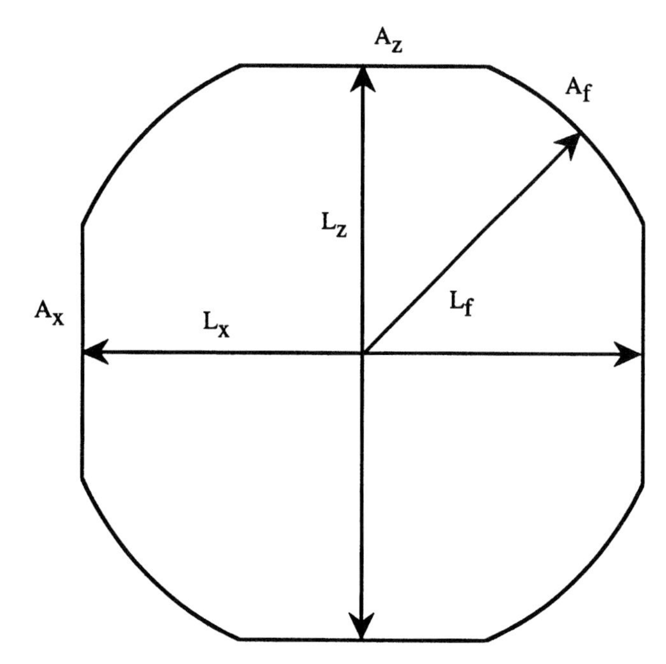

Figure 1. Truncated sphere model (after Weyl, 1959) showing grain dimensions and surface site areas. The grain height, width, and radius are denoted L_z, L_x, and L_f, respectively. Grain–grain contact areas are shown as A_z and A_x with the free-face area exposed to the pore fluid given as A_f. Variables in the third dimension are denoted L_y and A_y (not shown).

effects of grain sliding. However, because rate coefficients for pressure solution are only approximate (plus or minus a few orders of magnitude) employing more sophisticated grain packing and stresses is probably not warranted.

Grain-Scale Stresses

Growth and dissolution of a heterogeneously stressed grain will differ between sites on the grain surface in response to free-energy variations. By definition, σ_{zz}^m is the macroscopic normal stress applied to the z–face of a macrovolume element. Because A_z is the grain–grain contact area and $L_xL_y - A_z$ is the area of the z–side of the unit cell that is exposed to pore fluid, the periodicity of the lattice implies that

$$-L_xL_y\sigma_{zz}^m = A_zP_z + \left(L_xL_y - A_z\right)P. \tag{3}$$

Here P_z is defined to be the normal stress to the z–contact averaged over that contact area (see, however, Rutter, 1976, 1983; Dewers and Ortoleva, 1990b).

Free Energies at Reaction Sites

For porous media in general, the deviation $(\tilde{\mu})$ of the molar Gibbs free energy from its value in the stress-free state (Kamb, 1959) is given by

$$\tilde{\mu} = \left(P_n, \underline{\underline{\sigma}}\right) = \frac{P_n}{\rho_s^m} + \frac{F}{\rho_{su}^m} \tag{4}$$

where ρ_s^m and ρ_{su}^m are the molar densities of quartz in the stressed and stress-free states, F is the Helmholtz strain energy, and P_n is the normal pressure at a site on the grain. In the water–film diffusion model, the strain energy is much less important than the stress on the contacts and is therefore neglected here. The value of P_n differs among the four sites of reaction. For the free face, P_n is the fluid pressure P, while P_n equals P_z for the fluid film on the z–contact.

Pressure Solution Kinetics

Growth/Dissolution Rate Laws

The dynamics of the textural variables are dictated by the kinetics at each of the four grain-site types. For quartz, we consider the reaction:

$$\text{quartz} \Leftrightarrow SiO_2\,(aq) \tag{5}$$

The change in grain height (L_z) over time is determined by the rate of the pressure solution reaction (G_z):

$$\frac{DL_z}{Dt} = G_z \tag{6}$$

where D/Dt is the material derivative (rock-fixed frame of reference) and is defined as (Tritton, 1988):

$$D/Dt = \frac{\partial}{\partial t} + \vec{u} \bullet \vec{\nabla} \tag{7}$$

The first-order kinetic law for the stress-supporting z–face is given by:

$$G_z = k_z\left(c - c_z^{eq}\right) \tag{8}$$

where k_z is the rate constant, c is the pore-fluid concentration of SiO_2 (aqueous), and c_z^{eq} is the equilibrium concentration of SiO_2 in the water film between stress supporting z–contacts (Weyl, 1959; Rutter, 1976, 1983; Dewers and Ortoleva, 1990a, b). Within the water film the equilibrium concentration of $SiO_2\left(c_z^{eq}\right)$ is given by Dewers and Ortoleva (1990a) as

$$c_z^{eq} = \hat{c}\exp\left[\frac{3(P_z - P)}{RT}\left\{\frac{1}{\rho_{su}^m} - \overline{V}\right\}\right]. \tag{9}$$

The equilibrium pore-fluid concentration of SiO_2 $(\hat{c})$ as a function of P and T was obtained by a polynomial fit to the experimental data of Bowers et al. (1984) (from C. Qin, personal communication); $\overline{V}$ is the partial molar volume of aqueous silica (obtained from Millero, 1972).

The rate coefficient k_z depends on the pressure solution mechanism. For water–film diffusion it can be written in terms of the diffusion coefficient D_c for migration along the water film and thickness Δ_c of the contact:

$$k_z = \frac{2\pi D_c\Delta_c}{A_z\rho_s^m}. \tag{10}$$

The thickness of the contact decreases with stress across it and hence with depth (Tada et al., 1987). The diffusion coefficient D_c should increase with temperature (and depth); however, because of the decrease in Δ_c, the depth dependence of the factor $\Delta_c D_c$ is difficult to estimate. In addition, D_c should decrease with increased ordering in the water film, and it may increase as Δ_c decreases. In these simulations a constant value of 10^{-7} cm was chosen for Δ_c (after Tada et al., 1987).

The diffusion coefficient in the water film is obtained by an Arrhenius relation:

$$D_c = A\exp\left\{\frac{-E_{ac}}{RT}\right\} \tag{11}$$

where A equals 2.54×10^{-4} cm^2/sec and E_{ac} is 40 kj/mol (Dewers and Ortoleva, 1990a, in press).

Local Dissolution/Precipitation Kinetics

When fluid flow rates are small, most of the solute created by pressure solution at the contacts precipitates locally at the free face. Under this situation a steady state can be set up where the overall rate of pressure solution–mediated compaction is modified by the free-face kinetics.

When local closure is obtained, conservation of mass implies that the rate of mass added to the free face is equal to the rate of loss of mass at the contacts. This yields

$$A_z G_z + A_f G_f = 0. \tag{12}$$

We assume the same process occurs at both active sites. The form of the G_f equation is

$$G_f = k_f(c - \hat{c}). \tag{13}$$

Combining the above equations we obtain c in terms of $\hat{c}$ and c_z^{eq}. As a result

$$G_z = \tilde{k}_z\left(\hat{c} - c_z^{eq}\right) \tag{14}$$

where

$$\tilde{k}_z = \frac{k_z}{1+\lambda} \quad \text{and} \quad \lambda = \frac{k_z A_z}{k_f A_f}. \tag{15}$$

Free-face growth may be modified by coatings (e.g., clay or petroleum). Assuming a steady-state diffusion-limited process we obtain for k_f

$$k_f = k_{fo} / \left[1 + \rho_s^m k_{fo} \Delta_f / D_f\right] \tag{16}$$

where k_{fo} is the rate coefficient for the clean quartz surface, Δ_f is the thickness of the free-face coating, and D_f is the diffusion coefficient in the coating. D_f was set to 10^{-4} D_c based on the much smaller diffusion rate perpendicular to clay layers than along them (Farver, 1989). The free-face coating thickness was calculated from the clay volume fraction and the total free-face surface area, assuming a uniform thickness over all grain faces.

These relationships allow a variety of interesting diagenetic phenomena. When the free face is kinetically inhibited (i.e., k_f is small) then λ is large and the effective rate coefficient $\tilde{k}_z$ is reduced (also shown by Mullis, 1991). This is a reflection of the common observation of porosity preservation in sandstones with chlorite- or clay-coated grains (Tillman and Almon, 1979). When free-face kinetics is very fast, λ is small and G_z takes on the form one would expect if the pore fluid is in equilibrium with the uncoated free face; then G_z is as in equation 8 and c replaced by $\hat{c}$. Thus, the pore fluid is buffered at $\hat{c}$ by the fast free-face kinetics.

Fracture Kinetics

Development of a fracture network involves the coupling of basin stresses, fluid pressure, rock rheology, and the nucleation, growth, and healing of fractures. Our basic premise in this model of hydrofracturing is that the fractures are small in scale compared to the scale of fluid flow and that the fracture network can be described as varying continuously in space. We also assume that a finite number of fracture nuclei are present as grain-scale flaws and thus fracture nucleation laws are not required.

In the following sections we describe a model for subcritical crack growth similar to that implemented by Dewers and Ortoleva (in press). Modifications were made to the radius (length) growth law to allow for reversible fracture healing and a different formulation is used for fracture aperture (width) which is dependent on the fracture radius (Sleep, 1988). In addition, each representative volume of rock is given a statistical distribution of fracture nuclei, and the tensile strength required for each one to initiate is set as a function of the rock tensile strength. This is a simple way to approximate the effects of rock heterogeneity and fracture tip interaction on the growth of fractures.

Fracture Growth and Healing Kinetics

Fractures are assumed to be vertical penny-shaped discontinuities. Fractures propagate if the crack extension force (Segall, 1984; also termed fracture propagation energy: Pollard and Aydin, 1988) exceeds a critical value. The crack extension force is proportional to the square of the stress intensity factor for dilational (i.e, mode I) fractures, while the propagation energy is related to the surface free energy of an ideal brittle solid (Segall, 1984; Atkinson, 1984). The rate G of the change of radius r of these fractures is given by (Segall, 1984; Atkinson, 1984):

$$G = v_{fr}\left\{1 - \exp\left[\frac{-U}{RT}\right]\right\}. \tag{17}$$

The limiting propagation velocity (v_{fr}) depends on the mechanism of crack generation. For subcritical crack growth and healing, the rate-limiting kinetic mechanism may be surface, grain-boundary or thin-film diffusion, or reaction–kinetic controlled (Atkinson, 1984; Swanson, 1984; Smith and Evans, 1984). The factor U in equation 17 is (Segall, 1984; Atkinson, 1984)

$$U = \frac{1}{\rho_s^m a}\left\{\frac{\pi\left(1-v^2\right)}{E} r\sigma_T^2 - Q\right\} \tag{18}$$

where a is the fracture aperture; v is Poisson's ratio (0.1); E is the Young's Modulus (1.5×10^{11}gcm/s^2); σ_T is the far-field tensile stress, oriented parallel to the direction of the least-compressive stress; and Q is the critical value of the crack extension force (values chosen are typical for sandstones). The first term in equation 18 is the fracture propagation energy. In our model σ_T is equal to $P + \sigma_{xx}^m$. To allow fractures to heal as pressure drops, the term σ_T^2 is replaced by $|\sigma_T|\sigma_T$, thus giving reversible fracture opening and closing. These equations imply that if the basin is in a compressive stress regime, a fluid pressure exceeding a critical value may induce tensile stresses at grain-scale cracks sufficient to initiate crack propagation (Pollard and Aydin, 1988). The critical fluid pressure, p^*, is found from equation 19:

$$p^* = -\sigma_{xx}^m + \tau \tag{19}$$

where

$$\tau = \left[\frac{QE}{\pi\left(1-v^2\right)r}\right]^{1/2} \tag{20}$$

The latter term is the effective tensile strength of the rock in the vicinity of the fracture, and decreases with increasing crack length.

In equation 17, U is smaller than the product RT; therefore, G may be approximated by (Segall, 1984):

$$G = \frac{v_{fr}U}{RT}. \tag{21}$$

The limiting propagation velocity (v_{fr}) can be related to the temperature and fracture aperture (Wang et al., 1983):

$$v_{fr} = \frac{v'}{a^m T}\exp\left\{-E_f / RT\right\} \tag{22}$$

where v', m, and E_f (activation energy) are dependent on the rate-determining mechanism for fracturing. Assuming surface diffusion as the rate-controlling mechanism suggests vfr = 10–12cm6K/erg sec (Dunnington, 1967), $E_f = 8.0 \times 1011$ ergs/mol, and m = 2; Q is taken to be 104 erg/cm^2 (Segall, 1984; Atkinson, 1984).

The fracture aperture (a) for an elastic medium, assuming equalization of the fluid pressure (P) in the matrix and the fracture, is given by:

$$a = r(P - P_1)/E \tag{23}$$

where P_1 is the least principal stress (the minimum of $-\sigma_{xx}^m$ and $-\sigma_{xx}^m$ in our model) and E is the elastic Young's Modulus (Sleep, 1988).

Equation 20 shows that as the fracture length (r) increases, the tensile stress necessary to propagate fractures decreases with the inverse square root of fracture length, suggesting an unstable situation in which fractures grow without bound for constant stress conditions. However, a number of natural limitations on propagation of fractures arise, particularly in the case of hydrofractures. Mechanical interaction between propagating crack tips can limit fracture propagation, so that as a fracture network develops, the average fracture spacing is related to average fracture length (Pollard and Aydin, 1988). For the present analysis we limit the radius of fractures to 10 cm and the aperture to one-tenth of the radius, based on observations of fracture sizes in overpressured compartments (Powley, 1990).

Statistical Model of Fracturing

Heterogeneity of tensile strength and other properties will bring a statistical aspect to the fracture kinetics. We address this point in our macroscopic modeling by considering a distribution of material properties and associated fractures.

Consider a set of fracture classes labeled $i = 1,2,\cdots,N_{fr}$. Then we let r_i and η_i be the radius (length) and number per rock volume of class i fractures. The change in the radius of an i-type fracture over time is given by

$$\frac{DR_j}{Dt} = G_i \tag{24}$$

where G_i is the rate of growth of i-type fractures. We assume that the number of fractures is conserved:

$$\frac{\partial n_i}{\partial t} = -\vec{\nabla} \bullet (\eta_i \vec{u}) \tag{25}$$

Therefore, the identity of each fracture is kept, even if it is closed and has a very small radius (taken to be approximately the grain diameter).

The statistics can be implemented by making some assumptions on the distributions of τ_i and η_i. First define the following set of quantities

$$\eta = \sum_{i=1}^{N_{fr}} \eta_i$$

$$\hat{\eta}_i = \eta_i / \eta, \sum_{i=1}^{N_{fr}} \hat{\eta} = 1. \tag{26}$$

With these definitions and equation 25 we have

$$D\eta_i / Dt = 0 \tag{27}$$

$$\frac{\partial \eta}{\partial t} = -\vec{\nabla} \bullet (\eta \vec{u}). \tag{28}$$

For simplicity, we assume that the $\hat{\eta}_j$ are the same for all sediments initially; hence equation 27 implies that

$$\hat{\eta}_j = \text{constant}. \tag{29}$$

A distribution of tensile strengths for the τ_i is accomplished by giving a linear distribution of the crack extension force (Q_i) from 0.4*Q to Q.

The fracture porosity is given by the sum of the volumes of the fractures multiplied by the proportion of each fracture type.

$$\phi_{fr} = \eta \sum_{i=1}^{N_{fr}} \hat{\eta}_i \pi r_i^2 a_i \quad (30)$$

where a_i is the aperture for class i fractures.

Fluid Flow and Overpressure

The fluid pressure is solved by combining the equation for the conservation of mass of the fluid and Darcy's Law. Conservation of water (mass) is given by:

$$\frac{\partial \rho_f \phi}{\partial t} = -\vec{\nabla} \bullet \left[\rho_f \phi(\vec{v} + \vec{u})\right] \quad (31)$$

where ρ_f is the mass density of water, $\vec{v}$ is the fluid velocity relative to the solid matrix, $\vec{u}$ is the rock velocity, and ϕ is the total porosity (McKenzie, 1984; Palciauskas and Domenico, 1989). The total porosity is the sum of the matrix porosity (ϕ_{mtx}) and the fracture porosity (ϕ_{fr}). Darcy's Law yields the flux relative to the (moving) rock matrix:

$$\phi \vec{v} = -\frac{\kappa}{\mu}\left(\vec{\nabla} P - \rho_f \vec{g}\right) \quad (32)$$

where κ is the permeability, μ is the fluid viscosity, and $\vec{g}$ is the downward-directed gravity vector (assumed constant and of magnitude g). This expression for Darcy's Law is consistent with the form of the Navier–Stokes equation (Tritton, 1988) and its reduction to Darcy's Law by homogenization theory (Keller, 1980). A linearized equation of state for pure water provides the mass density as a function of pressure and temperature:

$$\rho_f = \rho_o\left[1 + \beta(P - P_o) - \alpha(T - T_o)\right] \quad (33)$$

where P_o and T_o are the reference pressure and temperature (1 bar, 293.15K) and ρ_o is the reference density at P_o and T_o. The coefficient of isothermal compressibility for the fluid (β), of thermal expansion (α), and of the fluid viscosity (μ) are assumed constant for all P–T conditions; ($\beta = 3.3 \times 10^{-5}$ bar^{-1}, $\alpha = 0.001$ deg^{-1}, $\mu = 0.01$ poise).

The Lagrangian form (rock-fixed frame of reference, utilizing the material derivative defined in equation 7) can then be written as

$$\frac{D_\rho \phi}{Dt} = \vec{\nabla} \bullet \left[\frac{x_{\rho f}}{\mu}\left(\vec{\nabla} P - \rho_f \vec{g}\right)\right] - \rho_f \phi \vec{\nabla} \bullet \vec{u} \quad (34)$$

The term containing $\vec{\nabla} \bullet \vec{u}$ is the source term for compaction-generated overpressure, and for solely vertical compaction can be written as:

$$\vec{\nabla} \bullet \vec{u} = \frac{G_z}{L_z}. \quad (35)$$

The overpressure (Ψ) is given by the difference of the total fluid pressure and the hydrostatic component P_h:

$$\psi = P - P_h. \quad (36)$$

We define P_h to be the pressure distribution that is required to make the vertical (z) component of the fluid velocity relative to the matrix equal to zero. Hence,

$$\partial P_h / \partial z + g\rho_f(P_h, t) = 0, \quad (37)$$

and thus

$$P_h = P_o + g\int_z^0 \rho_f(P_h, T)\, dz. \quad (38)$$

Permeability

The magnitude of overpressure and the fluid flow rate are strongly dependent on the permeability of the rock. The permeability is a function of the rock texture and the fracture size and connectivity. The overall rock permeability is given by the sum of the matrix and fracture permeabilities:

$$\kappa = \kappa_{mtx} + \kappa_{fr}. \quad (39)$$

The matrix permeability is taken to be a function of the matrix porosity (ϕ_{mtx}), grain size ($L_z/2$), and tortuosity. Permeability variations of several orders of magnitude have been documented from the same lithologic unit over a range of porosities (Blatt et al., 1980). The modified version of the Carmen–Kozeny relationship used here is:

$$\kappa_{mtx} = \frac{\varepsilon(L_z/2)^\gamma \phi_{mtx}^\delta}{\left(1 - \phi_{mtx}\right)^2} \quad (40)$$

where ε incorporates the tortuosity and grain-shape factor ($\varepsilon = 10^{-4} - 10^{-6}$) δ is 4 to 6, and γ equals 2. We have varied these parameters to treat the effects of differing clay fractions and types of clay that have different crystal habits and affect permeability differently (Blatt et al., 1980). For the simulations of the next section this formulation gives matrix permeabilities on the order of 10^{-2} to 10^{-12} darcys, with very low permeabilities reflecting highly compacted and cemented low-porosity rocks. This range encompasses a range of observed and suggested permeabilities of reservoir and seal rocks (Blatt et al., 1980; Smith, 1971).

Fracture permeability is calculated using an implementation of percolation theory (Gueguen et al., 1990; Dienes, 1982), a statistical approach based on the size and proximity of fractures. For a fairly narrow distribution of crack sizes and an isotropic orientation of these fractures, the permeability is given by:

$$\kappa_{fr} = \frac{4\pi}{15} f \frac{\bar{a}^3 \bar{r}^2}{\bar{l}^3} \quad (41)$$

where $\bar{a}$ equals the mean fracture aperture, $\bar{r}$ is the mean fracture radius, and f is the fraction of interconnected cracks. The mean values $\bar{a}$ and $\bar{r}$ are calculated from the a_i and r_i assuming a normal distribution. For a network of cracks where each has four neighbors then

$$f \cong 54(q - q_c)^2, \tag{42}$$

where q is the probability of crack intersection and q_c is the critical value of q, such that if $q \le q_c$ the fracture permeability is zero. Values for q and q_c are:

$$q = \frac{\pi^2 \bar{a}^2}{4\bar{\ell}^3}, q_c = \frac{1}{3}. \tag{43}$$

The average crack spacing is given by

$$\bar{\ell} = \frac{1}{\eta^{1/3}} \tag{44}$$

where η is the crack density. The net result of this formulation is that as fractures grow they increase rock permeability through their greater size and by their higher interconnectivity.

Rates of Sedimentation and Subsidence

Rates of sedimentation can differ by a few orders of magnitude depending on the type of basin and the depositional facies, as well as over time. In the northern Gulf Coast sedimentation rates have been 12–18 cm/100 yrs through the Pleistocene to Holocene and 6 cm/100 yrs in the Miocene (Nunn, 1985). Very high sedimentation rates (~20 cm/100 yrs) are also characteristic of basins formed in active collisional settings. In contrast, deeper-water sedimentation rates may be less than 1 mm/100 years to a few centimeters per hundred years (Blatt et al., 1980). Sedimentation rates of less than 1 cm/100 years to about 5 cm/100 years were chosen to reflect an average of slow and fast depositional settings.

NUMERICAL METHODS

Our simulations have employed two different finite element techniques and coordinate systems to solve the fluid pressure and texture equations. Our earlier simulations were performed on an equally spaced basin-adapted grid that was transformed to a fixed rectangular coordinate system. The first-order texture equations were solved by finite differences and the pressure equation by an interior collocation finite element method from the ELLPACK computational package (Rice and Boisvert, 1985). One advantage of this method was the greater availability of solution techniques for uniform rectangular grids. Disadvantages were some loss of resolution as the basin grew through sedimentation and subsidence, loss of numerical accuracy caused by the grid transformation, and large computer memory requirements.

In later simulations we used a rock-fixed frame of reference with an unequally spaced grid adapted to grain-size variations and the sedimentation rate. The fluid pressure equation was solved using a finite element method with a conjugate gradient squared preconditioner written by Dr. Jun Mu (Laboratory for Computational Geochemistry, Indiana University). This technique had the advantage of solving the pressure equation on a nonrectangular domain, thus eliminating the necessity of the grid transformation.

SIMULATIONS

The following simulations illustrate the effects of differing grain size, sedimentation rate, stratigraphic relations, and basin geometry on the form of overpressured compartments. The coupling of fluid pressure, compaction, and hydrofracturing are shown to have an important role in the development of compartments.

1. *Rift Basin Evolution.* The following pair of simulations show the development of overpressure, hydrofracturing, and the progressive enclosure of a coarse sandstone lens by the preferential compaction of surrounding fine-grained sandstone.
   a. The first simulation is of a 136 km wide basin that reached a maximum depth of 3.76 km after 14 m.y. (Figure 2). The left and right boundaries are impermeable and the upper and lower boundaries were set to the hydrostatic pressure. A simple sedimentation history was given with fine-grained sand and coarse-grained sand lenses surrounded by medium-grained sand (Figure 2A). The sedimentation and subsidence rates were set equal and decrease from left to right. The temperature gradient was set to a constant 30°C/km, as are all of the following simulations.

      Figure 2B shows the development of low-permeability seals in the fine-grained sand lens and along the lower hydrostatic boundary near the left corner. The lowest permeability rock (~10^{-12} darcys) formed in the zone immediately adjacent to the hydrostatic lower boundary because of its greatest depth and the lack of overpressure that would retard compaction. Overpressuring began in the low-permeability rock in the lower left corner and also under, and within, the fine-grained lens, eventually merging into a long zone hugging the basement (Figure 2C). Because of the seal in the fine-grained lens the fluid flow path was directed horizontally around the seal and into the higher-permeability coarse sand lens (shown schematically in Figure 2C).

      At 15 m.y. the basin has subsided to about 3.91 km at its maximum depth (below sea level). Trends of overpressuring and permeability through a vertical section taken about 45 km from the left boundary and through the second overpressure maximum are shown in Figure 3A and B. The top of overpressure followed roughly the downward sinking of the upper seal and is about 100 meters lower at 15 m.y. than at 14 m.y. Permeability minima in

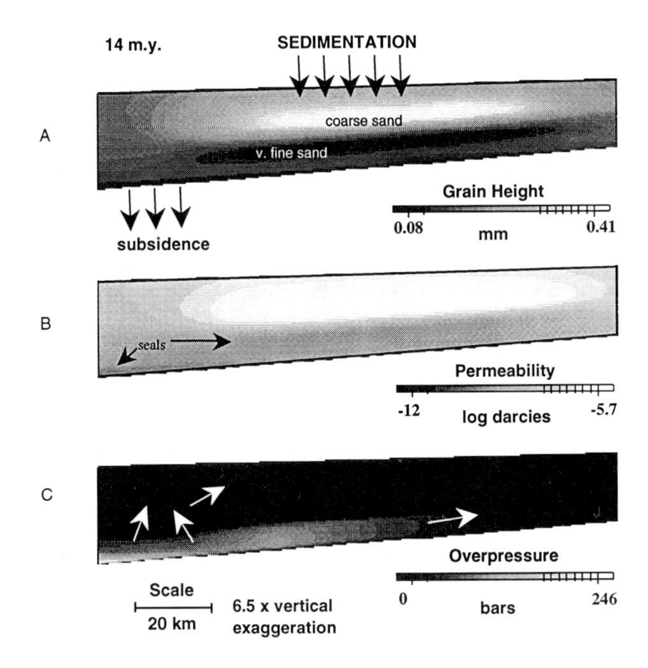

Figure 2. Rift basin simulation after 14 m.y. The basin is 136 km wide and its maximum depth is 3.76 km. (A) Grain height showing sand lenses surrounded by initially uniform sediment, which compacted preferentially with subsidence (see contours near left boundary). (B) Permeability map with seals indicated. (C) Overpressure contours hugging the lower boundary. The highest overpressure is at the lower left corner with a secondary high within the lower part of the fine-sand lens and in the underlying sediment. Arrows denote the predominant fluid flow paths.

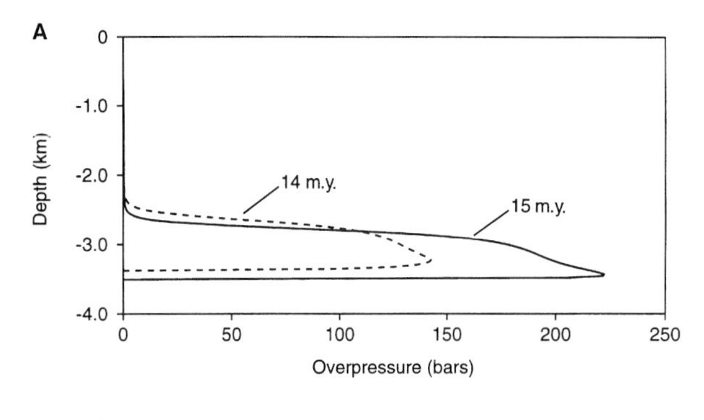

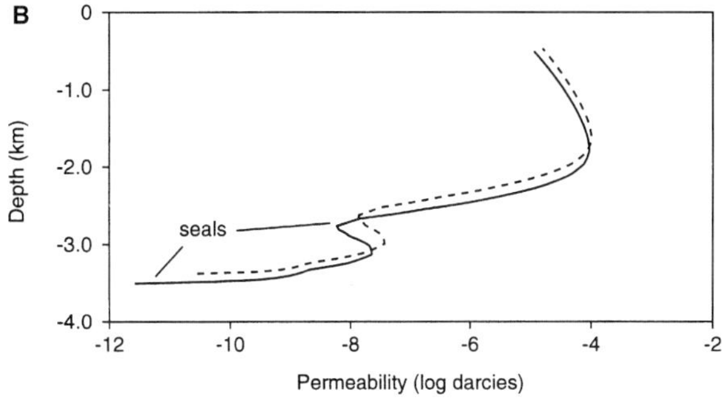

Figure 3. Vertical profiles through rift basin at 14 m.y. and 15 m.y. from simulations shown in Figure 2. (A) Overpressure as a function of depth below sea level. The ocean depth is about 500 m and the top of overpressure is at about 2 km below the sediment-water interface. The bottom boundary is set to hydrostatic pressure, hence the sharp drop in overpressure at the base. (B) Permeability minima in the fine-grained sand lens and along the lower hydrostatic boundary form the upper and lower seals to the zone of overpressure.

Figure 3B correspond to the upper and lower seals. As mentioned previously, the lowest-permeability rock developed at the bottom of the basin adjacent to the hydrostatic boundary, where there is no overpressure to retard compaction, and the stress and temperature are highest.

b. This simulation has a sedimentation history similar to that of the previous basin, but the width is 200 km and the basin reached a maximum depth of 7.3 km over 40 m.y. The porosity and overpressure are shown in Figure 4A–C. The lower boundary is set to the hydrostatic pressure and could represent a highly permeable bed or fractured basement rock (e.g., oceanic crust). The side boundaries are maintained at hydrostatic pressure to simulate permeable faults. Note that the overpressured compartment encompasses the lower third of the sandstone lens and extends into the fine-grained sediment. Most of the compaction and fluid pressuring originated in the fine-grained sediment.

As the sandstone lens subsided, the fine-grained sediment compacted preferentially and closed off the upper tip of the lens, trapping fluid or hydrocarbons that could have been carried into the lens. This illustrates the updip transport of fluid from a fine-grained source region into the coarse sandstone reservoir. The overpressure has increased to nearly 90% of the lithostatic pressure (Figure 4B) and a network of hydrofractures has developed (Figure 4C). The fractures have grown where the fluid pressure exceeds the sum of the horizontal stress and rock tensile strength. At depths greater than about 6.5 km the stress is isotropic (i.e., $\sigma_{xx}^{m} = \sigma_{zz}^{m}$) and we see the termination of fracture growth and the subsequent closing of fractures, because the fluid pressure does not exceed the sum of the horizontal stress and rock tensile strength. The fractures within the overpressured compartment allow for increased connectivity internally even though the fine-grained sediment has lost much of its original porosity there. The migration of high fluid overpressures into the sand lens helps to preserve matrix porosity during further burial. The overpressured compart-

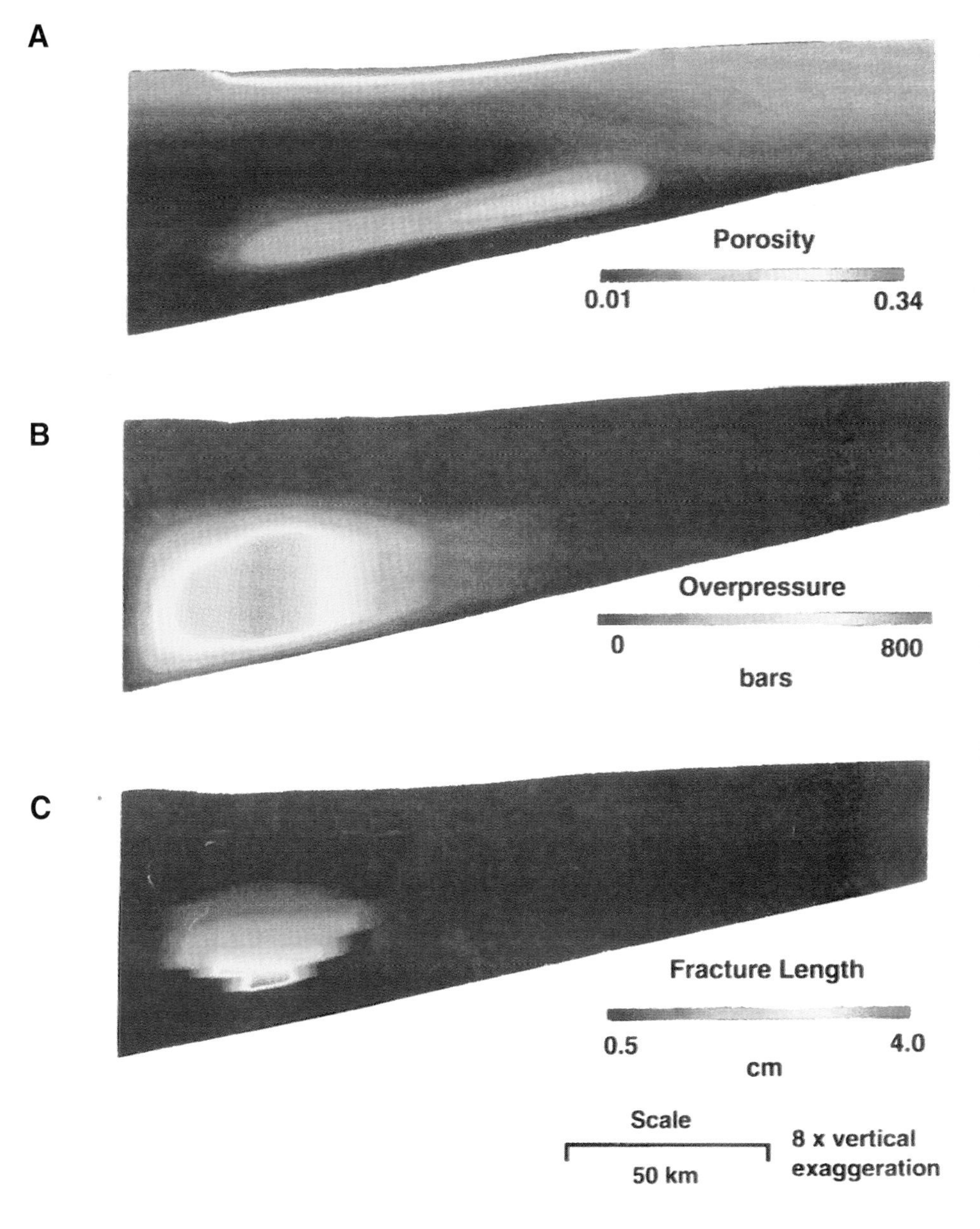

Figure 4. 200 km wide rift basin after 40 m.y. The maximum depth below sea level is 7.3 km. (A) Porosity profile shows the near-complete isolation of the sand lens as finer-grained material surrounding it has compacted preferentially. The porosity in the lens has two local maxima; the updip higher-porosity zone has preserved porosity, whereas the deeper-lying higher-porosity region is due to a combination of fracture porosity and preserved matrix porosity from overpressuring. (B) Note enlargement of overpressured zone compared to simulation in Figure 2 and pressures that approach lithostatic. (C) Contour map of fracture length showing growth of fractures in overpressured zone.

ment has the form of a distorted rectangular box, encompassing both fine-and coarse-grained beds, and is an example of a compartment that is not confined to a single unit.

Another important result was the development of a seal adjacent to the left hydrostatic boundary (fault). Because this rock could not maintain overpressure it compacted more than the nearby overpressured zone, thus tending to seal off the hydrostatic boundary. This is one way in which a fault can generate a seal even though it is not the seal itself.

2. *Intracratonic Basin.* The size and form of overpressured compartments may be influenced by the basin shape and the boundary conditions, as shown in the previous simulations. This simulation is of a 200 km wide basin of form similar to that of an intracratonic basin such as the Anadarko basin and many forearc basins (Powley, 1990; Ingersoll, 1988). A simple sedimentation history of parallel basinwide beds illustrates the effect of basin shape and boundary conditions on the form of overpressured compartments.

The sediment initially deposited into this basin was a very coarse sand, grading upward into a very fine sand, and then overlain by a thick fine sand and finally topped by a very fine sand. The basement contact was set as an impermeable barrier and the side walls at the extreme left set to hydrostatic pressure and impermeable on the right. The maximum sedimentation rate was about 500 m/m.y. and the subsidence rate was set equal to the local sedimentation rate. Therefore, variations in the depth of the sediment-water interface are due solely to differential compaction.

Fluid overpressures exceeding 500 bars are localized within the center of the very fine grained bed at its point of deepest burial (Figure 5A). A

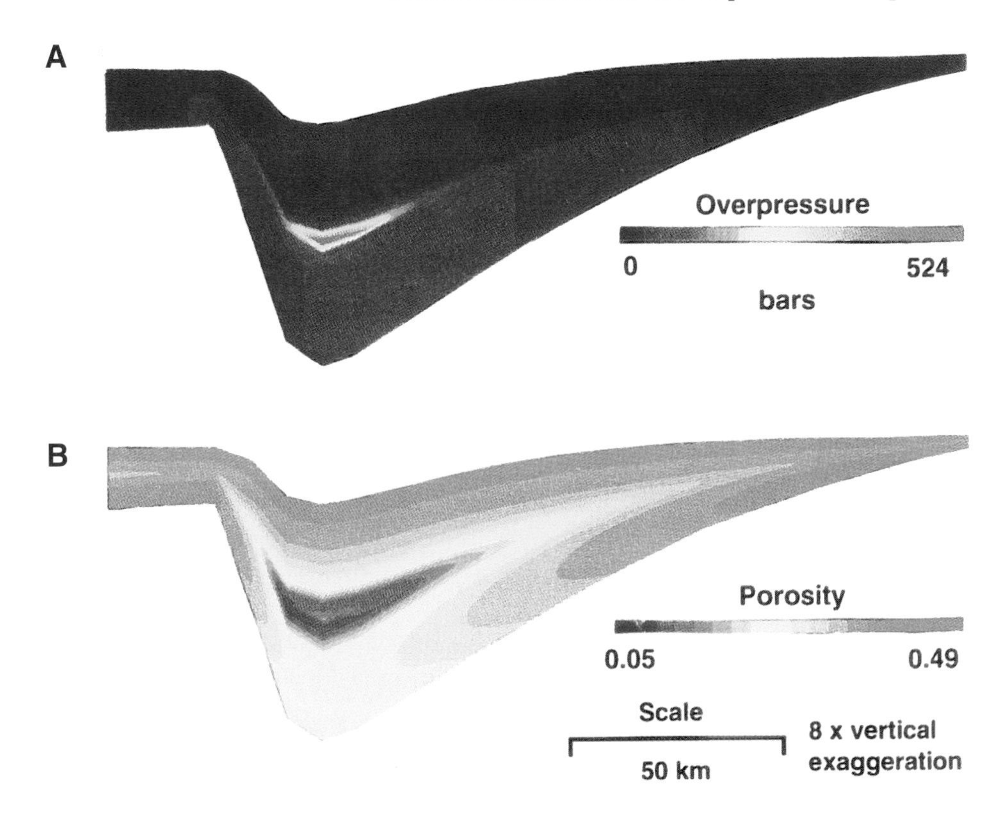

Figure 5. Intracratonic basin after 14 m.y. The maximum depth below sea level is 6.8 km. Beds were deposited parallel to each other and with uniform grain size across the basin. The lower and left boundaries are set to hydrostatic pressure and the right boundary is impermeable. The upper boundary is set to hydrostatic pressure. (A) Shaded contour map of fluid overpressure. Highest pressures are within the fine-grained bed in its deepest point of burial. (B) Shaded contour map of porosity. The fine-grained bed has a very low porosity margin with a region of preserved porosity and high fracture porosity in its interior.

shaded contour map of porosity after 14 m.y. (Figure 5B) shows the enhanced compaction of the lowermost bed in the deepest part of the basin and of the overlying fine-grained bed. In a vertical section of the porosity taken slightly to the right of the deepest part of the basin the porosity reaches a minimum in the very fine grained unit at about 2.5 km below the sediment-water interface (Figure 6A). At the point where the very fine grained bed is deepest the porosity is much higher in the center of the bed than at its margins, a result of fracturing and some porosity preservation due to high overpressures. High fluid overpressure is confined to the center of the fractured interior and is sealed off by the enhanced compaction in the surrounding normally pressured rock. The distribution of grain height shows the localization of the compartment in the fine-grained bed (Figure 6B).

CONCLUSIONS

The simulations described in this paper represent only a small sample of the possible basin histories and the resulting compartmentation/overpressure phenomena. Yet, given the simple mineralogy, stratigraphic relationships, and basin tectonics, several important aspects of overpressured compartments have been demonstrated. Overpressured compartments may form within a single bed or encompass several units and cross stratigraphic boundaries. Fine-grained sand compacts more rapidly than coarser materials and can become overpressured at shallower depths. Therefore, the porosity distribution is not solely a function of depth, but of grain size and other factors that affect the rate of pressure solution, such as temperature, overpressure, and mineral coatings.

Initially acting as fluid sinks, highly permeable faults or beds can allow compaction to progress faster in the adjacent rock than in the overpressured zone. This creates a permeability barrier that further isolates the region of overpressuring (autoisolation; see Ortoleva, this volume) and in combination with an overlying zone of compaction results in a compartment bounded by diagenetic seals in three dimensions. In the fault-adjacent seal all of the cementation needed to make the seal can be derived locally, avoiding the necessity of long-range transport of aqueous species.

Once sealed off from its surroundings, a compartment can develop high fluid pressures as it subsides and thermal expansion of the fluid becomes increasingly important. Hydrofracturing of the compartment interior increases the internal connectivity, creating new fluid pathways that are less controlled by the initial bedding. Apart from their importance in basin diagenesis, the development of compartments must also influence, and in some cases control, the transport and trapping of hydrocarbons.

ACKNOWLEDGMENTS

Research supported in part by contracts from the Gas Research Institute (No. 5092-260-2443), Amoco Production Company, and Mobil Exploration and Producing Services, and grants from the Basic Energy Sciences Program of the U.S. Department of Energy

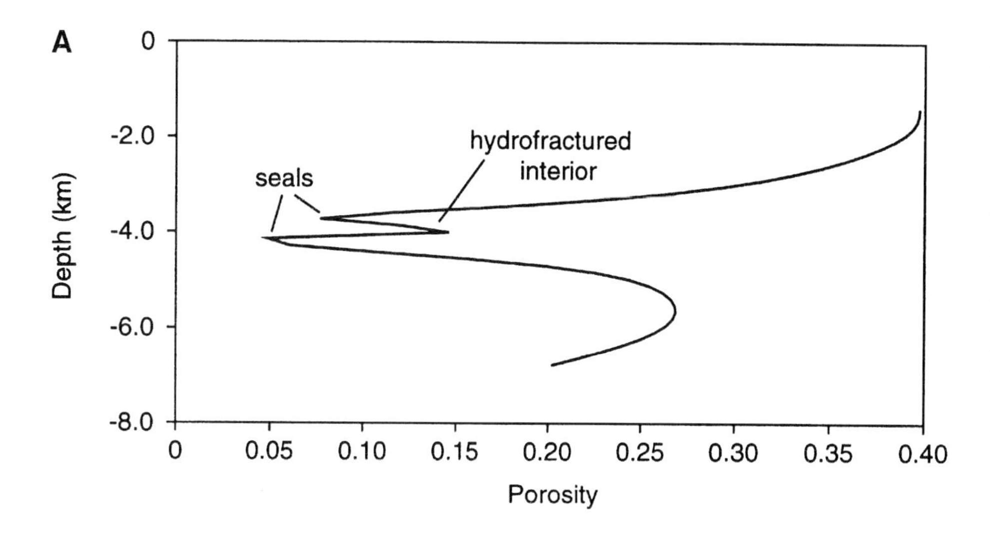

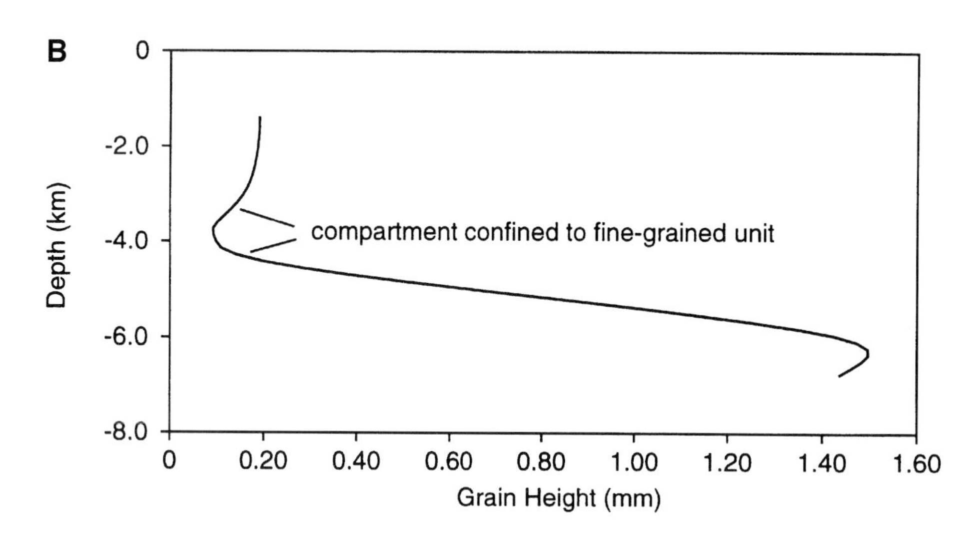

Figure 6. Vertical profiles through center of overpressured compartment in intracratonic basin shown in Figure 5. (A) Porosity minima at top and bottom of bed form seals to hydrofractured higher-porosity interior. (B) Grain height profile showing that the compartment is confined to the fine-grained bed (the top is about 2.5 km below the sediment-water interface).

(No. DEFG291ER14175), Shell Development Company, IBM, and Sun Microsystems. Thanks go to Dr. Jun Mu for his help in the implementation of numerical methods and to Dr. Tom Dewers for help in the initial stages of this project. We also appreciate the perceptive comments of the reviewers.

REFERENCES CITED

Al-Shaieb, Z., J. Puckette, P. Ely, and A. Abdalla, 1991, Megacompartment complex in the Anadarko Basin: a completely sealed overpressured phenomenon: American Association of Petroleum Geologists Annual Meeting Abstracts, p. 69.

Althaus, V.E., 1975, A new model for fracture gradient: SPWLA Sixteenth Annual Logging Symposium, p. 1–19.

Angevine, C. L., and D. L. Turcotte, 1983, Porosity reduction by pressure solution: a theoretical model for quartzarenites: Geological Society of America Bulletin, v. 94, p. 1129–1134.

Atkinson, B.K., 1984, Subcritical crack growth in geological materials: Journal of Geophysical Research, v. 89, p. 4077–4114.

Barker, L., 1972, Aquathermal pressuring—role of temperature in development of abnormal-pressure zones: American Association of Petroleum Geologists Bulletin, v. 56, p. 2068–2071.

Berry, F.A.F., 1973, High fluid potential in California Coast Ranges and their tectonic significance: American Association of Petroleum Geologists Bulletin, v. 57, p. 1219–1249.

Blatt, H., G. Middleton, and R. Murray, 1980, Origin of Sedimentary Rocks: Prentice Hall, New Jersey, 782 p.

Bowers, T.S., K.J. Jackson, and H.C. Helgeson, 1984, Equilibrium Activity Diagrams: Springer-Verlag.

Bradley, J.S., 1975, Abnormal formation pressure: American Association of Petroleum Geologists Bulletin, v. 59, p. 957–973.

Chapman, R.E., 1972, Primary migration of petroleum from clay source rocks: American Association of Petroleum Geologists Bulletin, v. 56, p. 2185–2191.

Dewers, T., and P. Ortoleva, 1990a, A coupled reaction/transport/mechanical model for intergranular pressure solution, stylolites, and differential compaction and cementation in clean sandstones: Geochimica et Cosmochimica Acta, v. 54, p. 1609–1625.

Dewers, T., and P. Ortoleva, 1990b, Geochemical self-organization III: a mean field, pressure solution model of spaced cleavage and metamorphic segregational layering: American Journal of Science, v. 290, p. 473–521.

Dewers, T., and P. Ortoleva, in press, Nonlinear dynamical aspects of deep basin hydrology: fluid compartment formation and episodic fluid release: American Journal of Science.

Dienes, J.K., 1982, Permeability, percolation and statistical crack mechanics, *in* R.E. Goodman and F.E. Heuze, eds., Issues in Rock Mechanics: New York, American Institute of Mining, Metallurgical, and Petroleum Engineers, p. 86–94.

Drzewiecki, P., D.R. de Miranda, G.R. Moline, L.D. Shepherd, and A. Simo, 1991, Stylolitization and diagenetic banding in the St. Peter Sandstone of the Michigan Basin: Symposium on Deep Basin Compartments and Seals, May 15–18, 1991, Stillwater, Oklahoma, Session Abstracts.

Dunnington, H.V., 1967, Aspects of diagenesis and shape change in stylolitic limestone reservoirs, *in* Proceedings of the 7th World Petroleum Congress, Mexico, 1967, v. 2, p. 337–352.

Ely, P., Z. Al-Shaieb, J. Puckette, and A. Abdalla, 1991, Top seals in the Anadarko Basin: their banded character: Symposium on Deep Basin Compartments and Seals, May 15–18, 1991, Stillwater, Oklahoma, Session Abstracts.

Farver, J.R., 1989, Oxygen self-diffusion in diopside with application to cooling rate determinations: Earth and Planetary Science Letters, v. 92, p. 386–396.

Gavrilenko, P., and Y. Gueguen, 1993, Fluid overpressures and pressure solution in the crust: Tectonophysics, v. 217, p. 91–110.

Gueguen, Y., T. Reuschle, and M. Darot, 1990, Single-crack behavior and crack statistics, *in* D.J. Barber and P.J. Meredith, eds., Deformation Processes in Minerals, Ceramics and Rocks: Unwin–Hyman, p. 48–71.

Hedberg, H.D., 1974, Relation of methane generation to undercompacted shales, shale diapirs and mud volcanoes: American Association of Petroleum Geologists Bulletin, v. 58, p. 661–673.

Hunt, J.M., 1979, Petroleum Geochemistry and Geology: San Francisco, W.H. Freeman, 617 p.

Ingersoll, R.V., 1988, Tectonics of sedimentary basins: Geological Society of America Bulletin, v. 100, p. 1704–1719.

Kamb, W.B., 1959, The theory of preferred crystal orientation developed by crystallization under stress: Journal of Geology, v. 67, p. 153–160.

Keller, J.B., 1980, Darcy's Law for flow in porous media and the two-space method, *in* Sternberg et al., eds., Nonlinear Partial Differential Equations in Engineering and Applied Science: Marcel Dekker, Inc., p. 429–443.

Logan, J.M., 1991, The influence of fault zones on crustal-scale fluid transport: American Association of Petroleum Geologists Annual Meeting Abstracts, p. 158.

Logan, J., and C.W. Lin, 1991, Fault zones as seals and conduits to fluid flow: Symposium on Deep Basin Compartments and Seals, May 15–18, 1991, Stillwater, Oklahoma, Session Abstracts.

Martinsen, R., and J. Jiao, 1991, Characteristics of unconformities and their possible role in the development of pressure compartments in clastic reservoir/source rock systems: Symposium on Deep Basin Compartments and Seals, May 15–18, 1991, Stillwater, Oklahoma, Session Abstracts.

McKenzie, D., 1984, The generation and compaction of partially molten rock: Journal of Petrology, v. 25, p. 713–765.

McKenzie, D.P., 1987, The compaction of igneous and sedimentary rocks: Journal of the Geological Society, London, v. 144, p. 299–307.

Millero, F.J., 1972, The partial molar volumes of electrolytes in aqueous solutions, *in* Horne, ed., Water and Aqueous Solutions; Structure, Thermodynamics and Transport Properties: John Wiley and Sons, Inc.

Mullis, A. M., 1991, The role of silica precipitation kinetics in determining the rate of quartz pressure solution.

Negus-de Wys, J., and M. Dorfman, 1991, The geopressured–geothermal resource transition to commercialization: Industrial Consortium for the Utilization of the Geopressured–Geothermal Resource, v. 1 Proceedings, p. 9–17.

Nunn, J.A., 1985, State of stress in the northern Gulf Coast: Geology, v. 13, p. 429–432.

Ortoleva, P., 1994, Geochemical Self-Organization: Oxford University Press, New York.

Palciauskas, V.V., and P.A. Domenico, 1989, Fluid pressures in deforming porous rocks: Water Resources Research, v. 25, p. 203–213.

Person, M., and G. Garvin, 1992, Hydrologic constraints on petroleum generation within continental rift basins: theory and application to the Rhine Graben: American Association of Petroleum Geologists Bulletin, v. 76, p. 468–488.

Pollard, D.D., and A. Aydin, 1988, Progress in understanding jointing over the past century: Geological Society of America Bulletin, v. 100, p. 1181–1204.

Powley, D.E., 1990, Pressures and hydrogeology in petroleum basins: Earth-Science Reviews, v. 29, p. 215–226.

Rice, J.R., and R.F. Boisvert, 1985, Solving Elliptic Problems Using ELLPACK: Springer-Verlag, 497 p.

Rutter, E.H., 1976, The kinetics of rock deformation by pressure solution: Philosophical Transactions of the Royal Society of London, v. A283, p. 203–219.

Rutter, E.H., 1983, Pressure solution in nature, theory and experiment: Journal of the Geological Society, London, v. 140, p. 725–740.

Segall, P., 1984, Rate-dependent extensional deformation resulting from crack growth in rock: Journal of Geophysical Research, v. 89, p. 4185–4196.

Shi, Y., C-Y. Wang, W-T. Hwang, and R. von Huene, 1989, Hydrogeological modeling of porous flow in the Oregon accretionary prism: Geology, v. 17, p. 320–323.

Simo, A., P. Brown, P. Drzewiecki, C. Johnson, J.W. Valley, and M.R. Vandrey, 1991, Petrographic and geochemical characterization of diagenetic seals in the St. Peter Sandstone: Symposium on Deep Basin Compartments and Seals, May 15–18, 1991, Stillwater, Oklahoma, Session Abstracts.

Sleep, N.H., 1988, Tapping of melt by veins and dikes. Journal of Geophysical Research, v. 93, p. 10255–10272.

Smith, D.L., and B. Evans, 1984, Diffusional crack healing in quartz: Journal of Geophysical Research, v. 89, p. 4125–4136.

Smith, J.E., 1971, The dynamics of shale compaction and evolution of pore fluid pressures: Mathematical Geology, v. 3, p. 239–263.

Swanson, P.L., 1984, Subcritical crack growth and other time- and environment-dependent behavior in crustal rocks: Journal of Geophysical Research, v. 89, p. 4137–4152.

Tada, R., R. Maliva, and R. Siever, 1987, A new mechanism for pressure solution in porous quartzose sandstone: Geochimica et Cosmochimica Acta, v. 51, p. 2295–2301.

Tigert, V., and Z. Al-Shaieb, 1990, Pressure seals: Their diagenetic banding patterns, *in* P. Ortoleva, B. Hallet, A. McBirney, I. Meshri, R. Reeder, and P. Williams, eds., Self-Organization in Geological Systems, Proceedings of a Workshop held 26–30 June 1988, University of California at Santa Barbara: Earth-Science Reviews, v. 29, p. 227–240.

Tillman, R.W., and W.R. Almon, 1979, Diagenesis of Frontier Formation offshore bar sandstones, Spearhead Ranch Field, Wyoming: SEPM Special Publication v. 26, p. 337–378.

Tissot, B.P., and D.H. Welte, 1984, Petroleum Formation and Occurrence: Springer-Verlag, 699 p.

Tritton, D.J., 1988, Physical Fluid Dynamics. 2nd Edition: Clarendon Press, Oxford, 519 p.

Wang, J.S.Y., C. Tsang, and R.A. Sternbentz, 1983, The state of the art of numerical modeling of thermohydrologic flow in fractured rock masses: Environmental Geology, v. 4, p. 133–199.

Weedman, S., S.L. Brantley, and W. Albrecht, 1991, Secondary compaction in sandstones in the vicinity of a pressure seal: Geological Society of America, abstracts, p. 410.

Weyl, P.K., 1959, Pressure solution and the force of crystallization: Journal of Geophysical Research, v. 64, p. 2001–2025.

Chapter 27

Role of Pressure-Sensitive Reactions in Seal Formation and Healing: Application of the CIRF.A Reaction-Transport Code

Y. Chen
W. Chen
A. Park
Peter J. Ortoleva
Indiana University
Bloomington, Indiana, U.S.A.

ABSTRACT

Pressure seals exist at depth that have impressively low permeability and are robust, despite evidence of tectonic or other fracturing. We suggest that there likely is some mechanism whereby the quality of a seal improves with increasing pressure gradient to which the seal is subjected. Thus, through diagenesis pressure seals self-enhance and heal, if breached.

A generic mechanism for the development of quality seals is proposed that involves a positive feedback as follows. The pressure dependence of diagenetic reactions can lead to precipitation from a fluid as it moves down a pressure gradient. Precipitation in a given zone will decrease permeability and tend to focus the overall pressure gradient in the zone. However, the augmented local pressure gradient promotes an even greater rate of precipitation in the zone of the original precipitation. Thus there is the tendency toward increased local pressure gradient development.

In the context of compartments, there is an even greater tendency toward seal self-enhancement. As the seal develops about an overpressuring compartment interior, the overall pressure drop across the developing seal increases—i.e., fluid escape is increasingly retarded as the seal develops. Thus there is a greater pressure head localized to the region where the pressure gradient–induced precipitation is taking place.

It is shown that one way this may occur is via the pressure dependence of equilibria of aqueous and mineral reactions. As an illustrative example, we consider the carbonate-quartz system, as layers of such cements constitute an important contribution to the banded seals in sandstone observed in a number of basins. We find that this system allows for the rather efficient development or healing of seals. This phenomenon is evaluated by using the quantitative, reaction-transport code CIRF.A. Numerical simulations show the time scales for the development and healing processes and their dependence on the salinity, temperature, depth, applied pressure head, and fluid chemistry to which the seal is subjected. The existence of such self-enhancing sealing mechanisms gives us new insights into the nature of seals and compartments. Including the processes underlying this phenomenon should aid in the prediction of the location and properties of compartments.

SEALING ROCKS AS SMART MATERIALS

We conjecture that regions of rock in a sedimentary basin can become hydrologically isolated solely through the development of diagenetic seals (Dewers and Ortoleva, 1988; Powley, 1990). Seals are observed to sustain very large pressure gradients (see Bradley and Powley, this volume) and we investigate the possibility that they can heal themselves once ruptured by fracturing. The purpose of the present study is to show that the development and healing of these seals may involve a process of self-enhancement whereby the efficiency of the barrier to flow increases with the pressure gradient involved due to pressure-dependent reaction equilibria.

The notion that a sedimentary basin can be compartmented into kilometer-scale domains of hydrologically isolated rock has been introduced by Powley and Bradley (PB) (Powley, 1975, 1990; Bradley, 1975). The compartment interior is separated from its environment by a shell of low-permeability rock (or seal).

To explain compartments we must address the following observations:

1. Seals must be able to sustain large pressure gradients over tens and hundreds of millions of years.
2. Tectonic activity and overpressuring will breach seals so that there likely is a mechanism of healing.
3. Horizontal seals may cut across strata—i.e., they need not be stratum-bound (Bradley and Powley, this volume; Al-Shaieb et al., this volume).
4. The three-dimensional isolation of compartment interiors is often accomplished through vertical seals.

These factors suggest to us that there must be a diagenetic component to the sealing process—i.e., stratigraphic or structural trapping does not explain (1) rupture healing, (2) the fact that seals can cut across lithology, and (3) the fact that sedimentary features or folding do not always form a complete three-dimensional shell around observed abnormally pressured zones.

We suggest that the above factors can be explained by the existence of a self-enhancing seal-forming and -healing process. In particular, we consider a class of mechanisms wherein a decrease in pressure causes precipitation from the fluid. Thus pressures exerted across a seal will cause precipitation leading to decrease in permeability within the seal. Such a process makes a sealing rock a "smart material" able to evolve in response to the conditions to which it is subjected in a way that enhances its sealing properties.

While there could be a variety of mechanisms that improve seal hydraulic resistivity or heal punctures, some mechanism involving fluid pressure seems most attractive. Temperature or fluid composition differences across seals could cause precipitation as fluids traverse a leaky seal and thereby decrease its permeability. However, temperature differences across vertical seals are not typically large. Because compartment interiors have a variety of fluid compositions—saline, acidic, CO_2-rich, etc.—fluid composition does not seem to be an important factor.

The fluid pressure difference that can develop across a seal can be hundreds of bars (Bradley and Powley, this volume; Al-Shaieb et al., this volume). The magnitude of these pressures and their potential effect on pore fluid mineral and organic solid reactions suggest the following theme. A decrease in fluid pressure can cause the precipitation of a given mineral. Then as fluid in equilibrium with that mineral on the high-pressure side of the seal passes through the seal, it will cause precipitation within the high-pressure gradient (intraseal) zone, improving on seal quality.

Recent studies on pressure seals in the Anadarko basin (Tigert and Al-Shaieb, 1990; Ortoleva and Al-Shaieb, in press; Al-Shaieb et al., this volume) and the Michigan basin (Shepherd et al., this volume) have shown their layered structure. The layering (or banding) has arisen through diagenesis—more specifically via the precipitation of silicate and carbonate cementation in an initially relatively uniform sandstone. Observations and theory suggest a number of types of bands (see Qin and Ortoleva, this volume). The actual pressure seal consists of a complex mixture of these banding types. Much of the banding apparently has its origins in the spatially unstable mechano-chemical dynamics of the compacting sandstone (see Qin and Ortoleva, this volume).

One type of band in the above-mentioned seals consists of intense calcite or dolomite cementation in a host sandstone. These bands are rather perplexing. That the carbonate cementation is localized (and not pervasive throughout the sandstone) might be related to local conditions (particular clays, for example; Boles, 1991, private communication) favorable to carbonate mineral nucleation. As we shall show in more detail below, localization could also occur in layers with some early local decreased permeability (and hence augmented pressure drop) when the sandstone is subjected to an overall pressure gradient.

The partial molar volume of ionic species in water tends to be negative (the charge attracts a shell of associated water overcoming the bare ion volume for smaller ions). Thus dissolution reactions resulting in ionic aqueous solutes often have a negative change in system volume upon reaction. For example, the reaction

$$\text{calcite} \Leftrightarrow Ca^{2+} + CO_3^{2-} \tag{1}$$

is favored to the right as pressure (p) increases. Hence the equilibrium constant (solubility) increases with p. If a calcite-saturated fluid interacting with a rock reacting according to equation 1 only flows down a gradient in p, calcite would be precipitated if the fluid entered the system saturated with respect to calcite at the (elevated) inlet pressure. This scenario would be favored at relatively high pH and ionic strength so that complexes of CO_3^{2-} with H^+ are not favored (although complexes of Ca^{2+} with OH^- would complicate the picture at very high pH).

However, breaking up ion complexes creates more higher-charged species and hence results in decreases of volume. Thus a drop in pressure favors complexing and—for calcite, for example—would make it possible to have more total dissolved solutes in a low-pressure solution than in a high-pressure one. Therefore the pressure gradient–induced sealing mechanism proposed above would not be favored in a system of low pH and salinity.

The equilibrium constants for the various aqueous and mineral dissolution reactions differ. Elevation of temperature tends to decrease the dielectric constant for water and hence favors ion complexing. Thus with other factors held constant there should be a maximum temperature (and hence a related depth) beyond which the pressure gradient–induced sealing mechanism can operate.

In summary, the pressure gradient sealing mechanism can likely operate only in a well-defined window in fluid composition, temperature, and mineral content space. We show in detail for the case of calcite that this window is within typical diagenetic conditions and demonstrate that the spatial and temporal relations for the resulting precipitation allow for the creation and robustness of quality seals. In both this and the carbonate layering noted in the previous paragraph, the relatively high solubility and fast kinetics of carbonate minerals are found to be key to the feasibility of the mechanism to operate on a relevant time scale.

Our approach is to use a quantitative reaction-transport model. In particular, we augmented the code CIRF.A (Chen et al., in preparation) to include pressure corrections to the equilibrium constants of the mineral and aqueous reactions.

Engineers sometimes refer to materials that respond favorably to stimuli as being "smart." Pressure gradient accumulation systems may in this context be termed "smart." Pressure gradient accumulation systems may therefore be termed "smart seals." They develop as better permeability barriers the greater the applied pressure drop across them. Considering the potential chemical diversity of water-rock-petroleum systems, it seems to us that there is likely a variety of pressure gradient accumulation-type smart seals.

One of the implications of the present work is to underscore again the importance of "banded" (i.e., layered) diagenetic pressure seals. The present mechanism operates most efficiently for precipitation in very localized regions and not in a broad domain. In a rock layered by some precursor such as depositional features or diagenetic bedding (Ricken, 1986; Dewers and Ortoleva, 1990a, b, c, 1992; Qin and Ortoleva, this volume) the present mechanism can localize cements or organic precipitates in the relatively few tight regions over the entire thickness of a putative composite (layered) seal. Thereby these relatively few tight layers are made even more impermeable, making the overall package a very effective seal.

THE RTM MODEL AND CIRF.A CODE

Simple Pressure Corrections to the Equilibrium Conditions and Rate Laws

The reaction-transport model adopted accounts for fluid flow, solute transport, and fast (i.e., equilibrated) aqueous phase reactions and grain growth/dissolution kinetics as set forth in Chen et al. (1991). Here we show how the model was supplemented to include pressure effects.

The assumption was made that pressure entered the reaction dynamics only through the equilibrium constant (i.e., not through the rate coefficient). Consider a reaction of the form

$$v_0(\text{mineral}) + \sum_{\alpha=1}^{N} v_\alpha \text{ (aqueous species } \alpha) = 0 \tag{2}$$

for stoichiometric coefficients $v_0, v_1, \ldots, v_v$ involving a mineral (when $v_0 \neq 0$) and N aqueous species α $+1,2,\ldots,N$. Letting $g(p,T)$ be the molar free energy of the mineral and μ_α be the chemical potential of α we have (Prigogine and Defay, 1967)

$$v_0 g + \sum_{\alpha=1}^{N} v_\alpha \mu_\alpha = 0. \tag{3}$$

Introducing the partial molar volume V_α and reference pressure $\overline{p}$ we assume

$$\mu_\alpha = \mu_\alpha(\overline{p}, T, \underline{c}) + (p - \overline{p})V \tag{4}$$

for the system at temperature T and $\underline{c}(=\{c_1, c_2, \ldots, c_N)$ denotes the concentrations of species in the pore fluid. Next we write

$$\mu_\alpha(\overline{p}, T, \underline{c}) = \mu_\alpha^*(\overline{p}, T) + RT \ln a_\alpha \tag{5}$$

for reference potential μ_α* and activity a_α. The Debye Huckle equation (Helgeson et al., 1981) is used to evaluate the ionic strength of the activity. With this equation 3 becomes

$$\prod_{\alpha=1}^{N} a_\alpha^{v_\alpha} = K(\bar{p},T)\exp\{-(p-\bar{p})\Delta V/RT\} \equiv K^* \quad (6)$$

$$RT \ln K(\bar{p},T) = -v_0 g - \sum_{a=1}^{N} v_a \mu_a^*(\bar{p},T) \quad (7)$$

$$\Delta V = v_0 V_0 + \sum_{\alpha=1}^{N} v_\alpha V_\alpha \quad (8)$$

Here ΔV is the change of volume due to reaction and V_0 is the solid molar volume. As a simplification we have neglected the p-dependence of V_0 and the V_α and will also neglect the $\underline{c}$-dependence of the V_α.

For the fast (equilibrated) aqueous reactions, equation 6 was used for the pressure corrections. The rate of mineral reaction G(cm/sec) was written in the form

$$G = k\left(\frac{1}{K^*}\prod_{\substack{\alpha \\ v_\alpha<0}} a_\alpha^{v_\alpha} - \prod_{\substack{\alpha \\ v_\alpha<0}} a_\alpha^{-v\alpha}\right) \quad (9)$$

No pressure correction to the rate coefficient k was made. In the above we assumed that $v_\alpha > 0$ for a dissolution product.

Partial Molar Volume Data

Millero (1971, 1972) gives the conventional partial molar volumes of some species in water at 25°C. From these values we have obtained an effective radius for each species by using a semi-empirical equation suggested by Hepler (Millero, 1971)

$$V_\alpha = Ar_\alpha^3 - BZ_\alpha^2 / r_\alpha \quad (10)$$

where A and B are factors and Z_α is the valence of species α. We assume that the radius r_α of a species does not change with temperature and pressure. However, A and B change with temperature (Millero, 1972). By using least mean square fitting, we have obtained polynomial expressions for A and B as a function of temperature so that the partial molar volumes for a species at a desired temperature may be calculated according to equation 10.

For cations and anions, A and B have different values (Millero, 1972). The average of these two values of A is used for neutral species. The polynomial expressions we get are listed in Table 1.

If the volume change of a reaction and the partial molar volumes of all but one species involved in the reaction are known, the partial molar volume and hence the radius of that species can be obtained according to

$$\bar{V}_u^o = \Delta V - \sum_{\alpha \neq u} v_\alpha \bar{V}_\alpha^o. \quad (11)$$

We calculated the radius of some species by this method based on the volume reaction data in SOLMINEQ.88 (Kharaka et al., 1988). This radius data has been put in the CIRF.A database.

NUMERICAL SIMULATIONS OF SEAL GENESIS AND HEALING

Numerical simulations of the pressure gradient accumulation effect serve to illustrate its potential as a sealing mechanism and to set it in the context of competing and reinforcing mechanisms. In this section we consider several examples of smart seals that can arise in carbonate-cemented sandstones.

Carbonate Cement Band Self-Enhancement

Consider the development of a tight band of calcite cementation in a sandstone. The reactions considered are listed in Table 2. They involve aqueous species and complexes for the H-Ca-C-O system and the calcite grain growth/dissolution reaction.

First we simulated a one-dimensional system consisting of a uniform sandstone and a single horizon with a relatively small amount of calcite cement as seen in Figure 1. As the system evolves, the higher pressure gradient in the "proto-seal region" promotes calcite precipitation there. The latter cementation decreases permeability in that horizon and hence the applied pressure gradient is increasingly localized to the cementing horizon as seen in Figure 2. This

Table 1. Polynomial expressions for A and B.

	A	B
Cation	$22.015 - 0.31829T + 1.9897 \times 10^{-3}T^2$ $-5.6455 \times 10^{-6}T^3 + 7.3997 \times 10^{-9}T^4$ $-36103 \times 10^{-12}T^5$	$53.350 - 0.27633T + 4.228 \times 10^{-4}T^2$
Anion	$0.94377 + 1.5053 \times 10^{-2}T - 1.0147$ $\times 10^{-5}T^2$	$-109.45 + 1.1088T - 3.5502 \times 10^{-3}T^2$ $+ 3.8896 \times 10^{-6}T^3$
Neutral species	$-7.6469 + 9.0906 \times 10^{-2}T - 2.2533$ $\times 10^{-4}T^2 + 1.8909 \times 10^{-7}T^3$	

Table 2. Reactions involved in our simulations.

quartz = $SiO_2(aq)$
calcite = $Ca^{2+} + CO_3^{2-}$
CO_2 (aq) + $H_2O = H^+ + HCO_3^-$
$HCO_3^- = H^+ + CO_3^{2-}$
$CaCO_3 + H^+ = Ca^{2+} HCO_3^-$
$H_2O = H^+ + OH^-$
$NaCl = Na^+ + Cl^-$

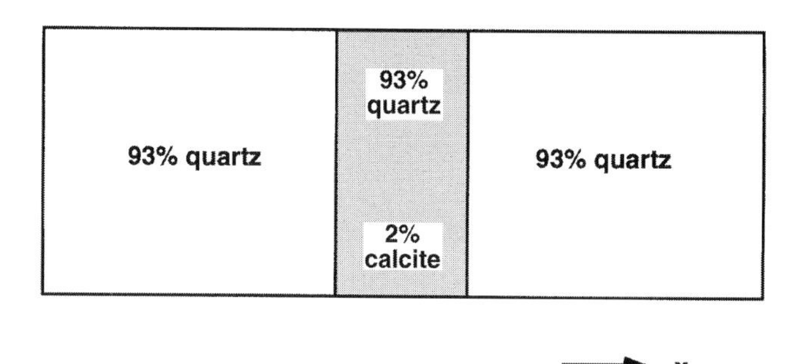

Figure 1. Initial configuration of the system. Domain size: 150 m × 100 m.

process continuously accelerates until porosity and permeability within the calcite-cemented horizon become very low.

It is important to note the following:

- As the system evolves the flux through the seal becomes arrested (Figure 3).
- The dynamics of the pressure gradient accumulation are self-accelerating; both the porosity minima ϕ_{min} and the flux (Figure 3) decrease at a rapidly accelerating rate.
- The dependence of ϕ_{min} achieved by a given time is a rapidly increasing function of the imposed pressure gradient; this strong dependence arises because of the applied gradient-driven flow as well as the greater rate of precipitation fostered by the increasing pressure gradient within the calcite-cemented horizon (Figure 4).

This study illustrates the basic pressure gradient effect in a calcite-cemented sandstone.

Sandstone Heterogeneity Promotes Local Calcite Precipitation

In the above case study, calcite was allowed to precipitate only in the horizon with the original cementation—i.e., it was assumed that a nucleation barrier prevented calcite cementation elsewhere. Here we drop this restriction; calcite is allowed to grow everywhere. In this case, a single horizon has lower initial porosity, but there is uniform, small calcite precipitation everywhere in this system initially (Figure 5). The pressure difference of 150 bars applied across the

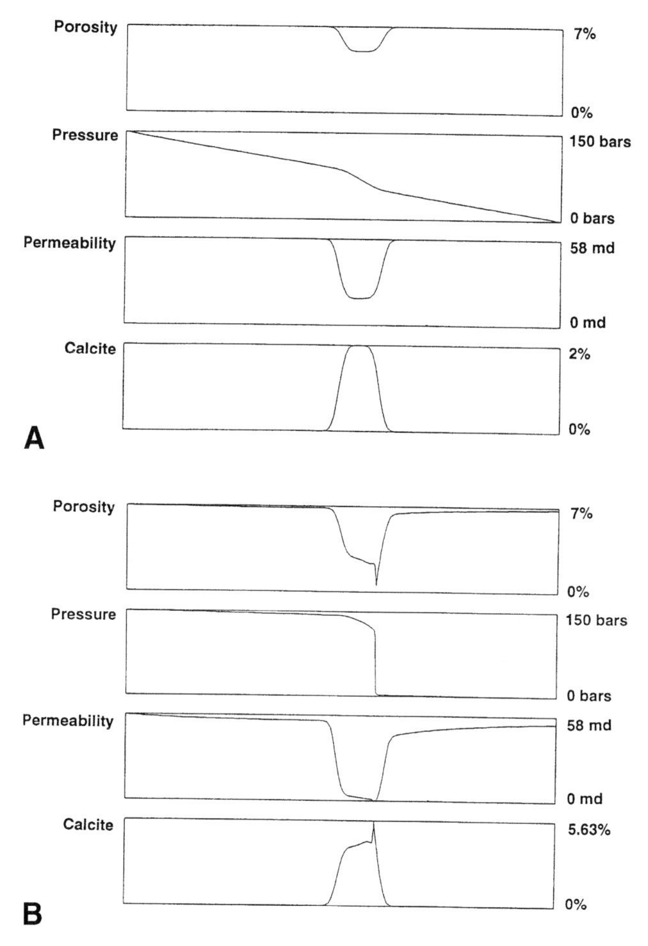

Figure 2. 1-D simulation results. (A) At time 0. (B) After 2280 years, calcite precipitates in the "proto-seal region." Permeability decreases there and pressure drop is localized.

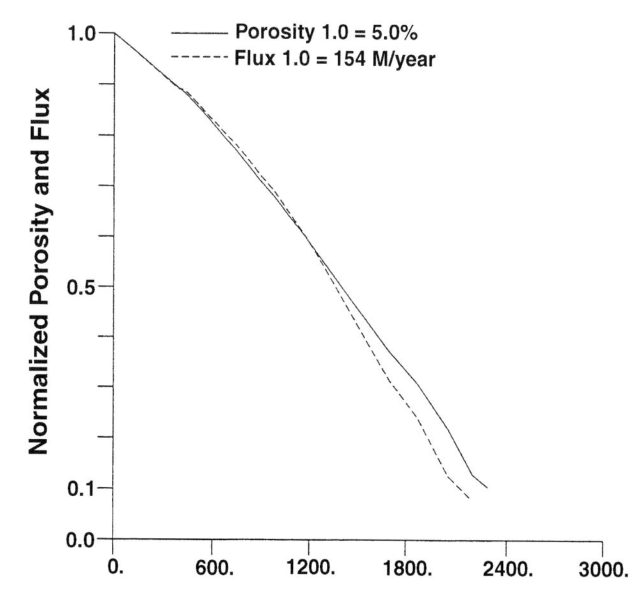

Figure 3. Flux and minimum porosity for the simulation of Figure 2 decrease with time.

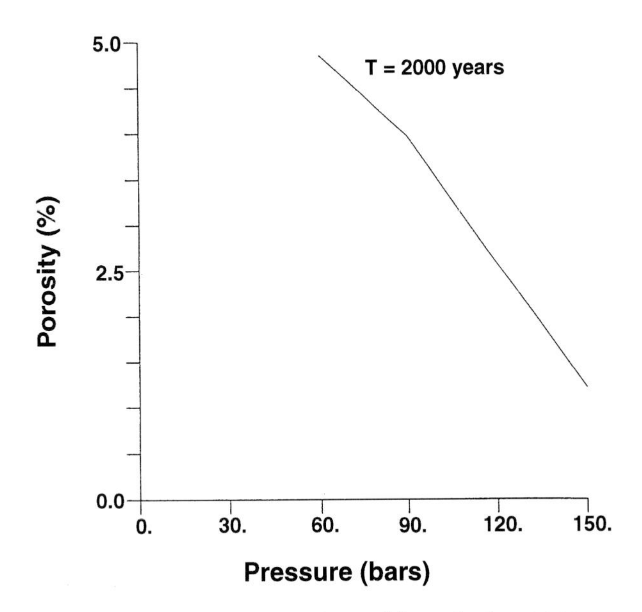

Figure 4. Minimum porosity achieved after 2000 years depends strongly on the imposed pressure difference.

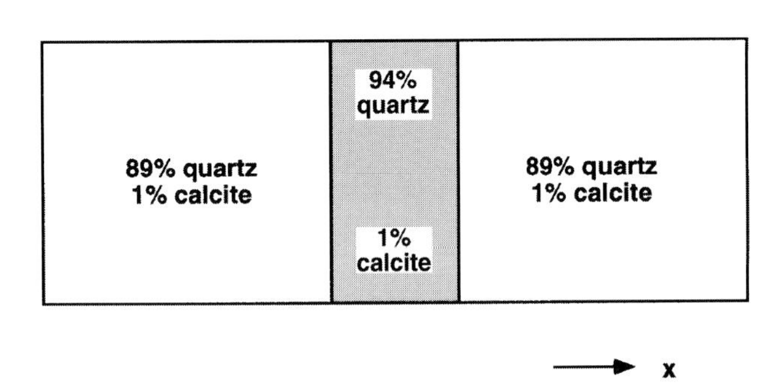

Figure 5. Initial configuration of a system where calcite is allowed to precipitate everywhere. Domain size: 150 m ×100 m.

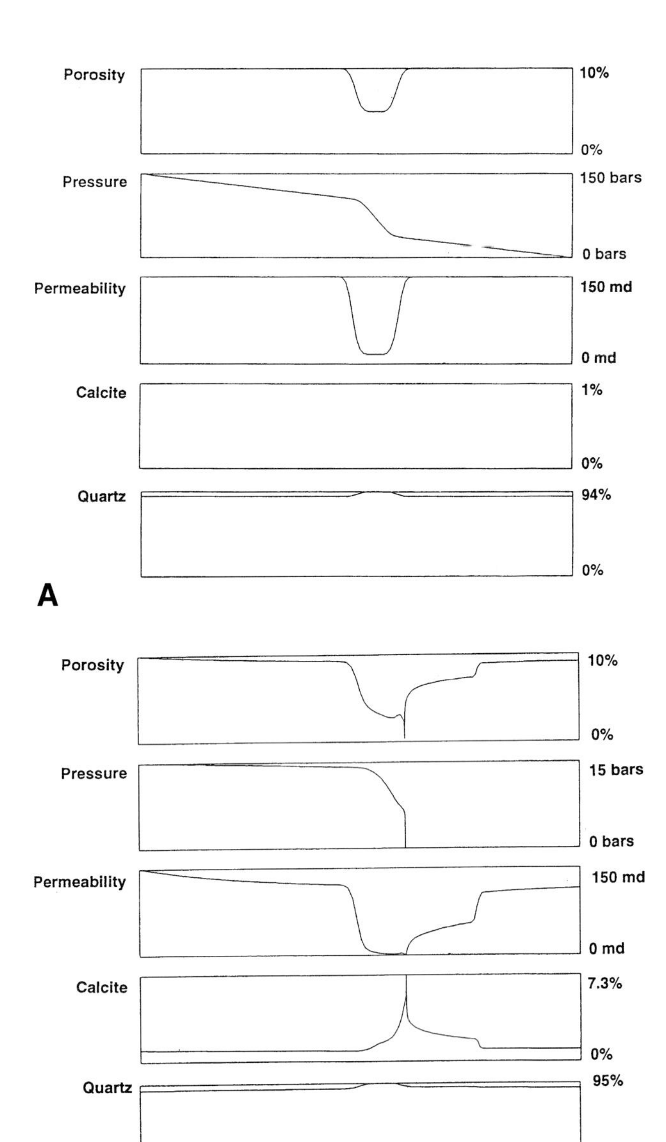

Figure 6. 1-D simulation results. (A) At time 0. (B) After 3780 years. Calcite precipitates everywhere but is by far greatest in the initially lower permeability region.

150 m domain is typical of what would exist across a pressure seal in a sandstone (see Tigert and Al-Shaieb, 1990). However, the result (Figure 6) shows that textural (and hence permeability) heterogeneity in the sandstone can promote local calcite precipitation.

The underlying dynamics are quite transparent. Heterogeneity in the sandstone leads to variation of the local fluid pressure gradient when an overall pressure gradient is imposed. The regions of initially lower permeability, if layered or lenticular, have larger local pressure gradients. Hence there is the accompanying enhanced local calcite precipitation and the pressure gradient accumulation effect takes off.

We have carried out a number of such simulations for the same time but under different applied pressure heads. There are two distinct effects:

- The minimum porosity obtained at a given time decreases with increased imposed pressure head.
- The precipitation profile has an increasingly broad tail and skewing to the region downstream from the original lower porosity/permeability sandstone. We suggest that this bears a strong resemblance to observed situations wherein carbonate precipitation overlies the overpressured zone.

In conclusion, the pressure gradient accumulation process can provide a mechanism for the development and enhancement of seals.

The latter may be the case for carbonate cementation banding in the diagenetically banded sandstone seals observed in the Anadarko basin (Tigert and Al-Shaieb, 1990; Al-Shaieb et al., this volume) and the Michigan basin (Shepherd et al., this volume)—see also Qin and Ortoleva (this volume). These observed banded sandstone seals involve a number of distinct types of bands, one of which is calcite cemented. Mechanisms for the formation of these various types of bands are discussed in Qin and Ortoleva (this volume). The mechanism discussed in the present work could enhance the sealing capacity of any of these other types of bands in that another type of band could serve as an initial nonuniformity as in the present simulation. Alternatively, the chemical environment of a given type of band may be more favorable than others for the nucleation of calcite, which then would lead to the strong localization of calcite in analogy with the simulations of Figures 1 and 2.

Healing of Punctured Seals and Development of Lateral Seal Continuity

A key property of a smart seal should be its ability to heal itself once breached. Here we show how local breaches (punctures) can be healed. Furthermore, the mechanism can give seals lateral continuity.

In Figure 7, we see a schematic view of a horizontal seal with a puncture or with broken lateral continuity. The "seal" is a horizon of diminished porosity in the sandstone. In the puncture the sealing capacity is compromised by either a depositional feature or a breaching process such as hydrofracturing. If an overall pressure gradient is imposed there will be a region of augmented pressure gradient within the puncture accompanying the bottleneck from flow through the puncture. As a result, we see in Figure 8 that the compromised region is cemented and the puncture is healed. Alternatively, this scenario can be understood to be the development of lateral continuity in the sealing capacity associated with depositional or diagenetic nonuniformity in the precursor sealing horizon.

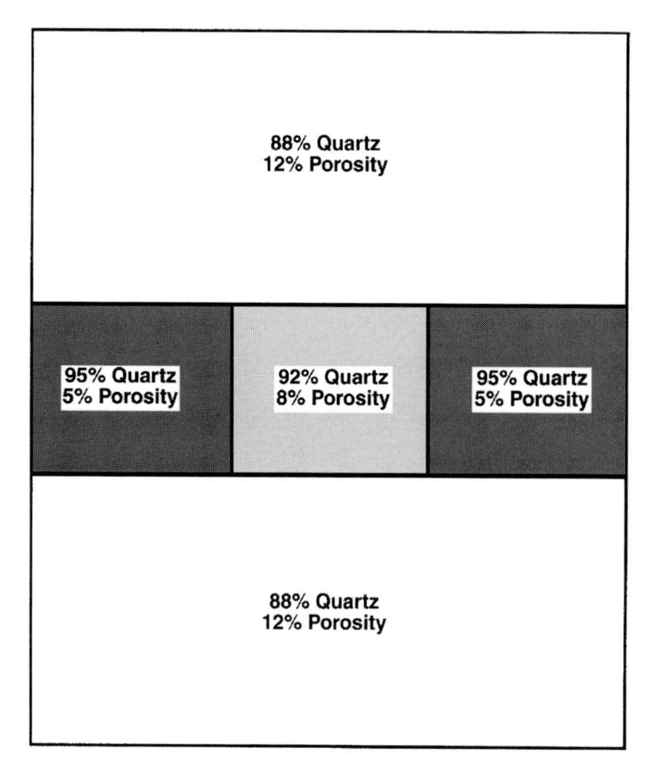

Figure 7. 2-D initial system configuration: a horizontal seal with a puncture. Domain size: 150 m × 100 m. 150 bars of pressure difference is imposed.

Initially Uniform System–Self-Developed Seal

In the above experiments, the systems initially are heterogeneous. Now let us consider an initially homogeneous system with a length of 1 km and porosity of 10%.

An overall pressure difference of 200 bars was imposed on the system. The result (Figure 9) shows that a lower-permeability region appears gradually and the permeability within that region goes to zero finally. In other words, a local pressure seal developed from the uniform system. This phenomenon shows that pressure gradients can play an important role not only in enhancement and healing of seals, but also in seal genesis from initially uniform sediment. Similar experiments but with different overall pressure gradients show that this factor affects the positions of the self-developed seals. The higher the pressure difference the larger the distance between the inlet and seal.

Influence of Fluid Chemistry

Inlet Fluid pH

Experiments similar to those described in Figures 1–6 indicate that when the inlet pH equals 6.5 the seal enhancement and healing process goes faster than when pH is higher or lower than this value. We believe this is because of the dual effects of H^+ in the process. On the one hand, low pH will let the inlet fluid carry more free calcium ions into the domain by reducing the amount of calcium complexed to CO_3^{-2}, speeding up the precipitation of calcite. On the other hand, a high concentration of H^+ makes more CO_2 in the form of H_2CO_3 and HCO_3^- and reduces the concentration of CO_3^{-2}, thus the precipitation rate of calcite becomes lower.

Salinity

The above experiments use seawater (i.e., the concentration of NaCl is about 0.6 m) as inlet fluid. In order to determine the effects of the salinity of the inlet fluid, another two experiments, one with fresh water and the other with 1.0 M NaCl as inlet fluid, were carried out. Experiments with fresh water take about 200 years longer to obtain a porosity and permeability somewhere in the system that is several orders of magnitude lower than their original values. Experiments with seawater take about 4200 years to finish. The seawater case takes a shorter time because in seawater activity coefficients of ions become less than 1.0, which in turn makes fluid able to carry more calcium and carbonate when the fluid is in equilibrium with calcite. The experiment with higher salinity does not show much difference from the case of seawater. These results must be viewed with some caution, however,

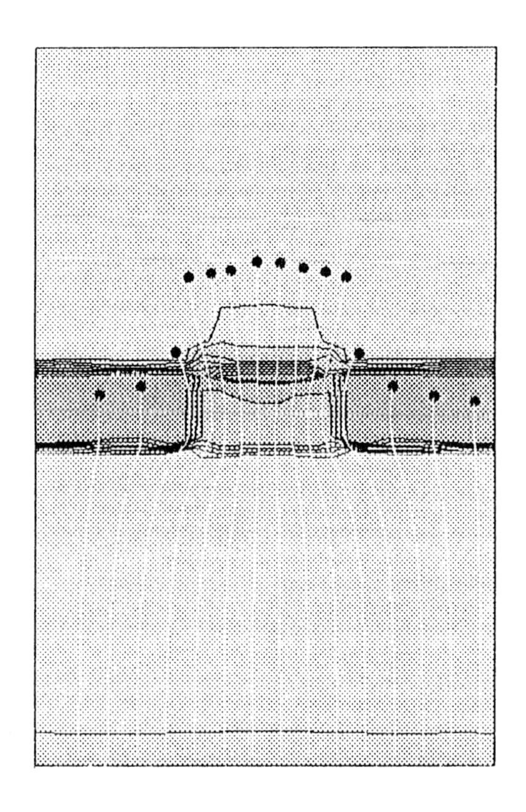

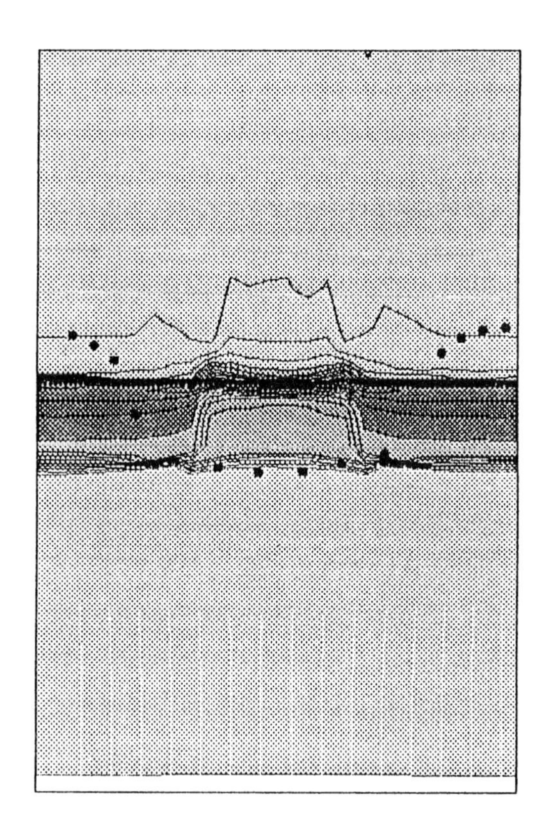

Porosity **1.1E+4 years**

Porosity **2.1E+4 years**

A **B**

Figure 8. Self-healing of a punctured horizontal seal. Porosity gray-level map after (A) 11,000 years, (B) 21,000 years. Calcite precipitates mainly in the punctured region, and the puncture has been healed after 21,000 years. Black lines are contour lines. Black dots and their trailing white lines show the trajectory of fluid particles. After 21,000 years the overall resistance to fluid flow in the puncture is a little higher than in the original unpunctured region. Irregularities in the contours are graphical artifacts and do not reflect inaccuracy in the calculation.

because the ionic strength corrections to the activities used are not very accurate at these salinities.

Complexing Reactions

We have added more inorganic calcium complexes in the fluid chemistry to see whether the process of calcite precipitation will speed up. This does not appear to be the case. This is due to the fact that 99% of the calcium is transported in the form of free ions. This result verifies that the fluid chemistry we chose is complete.

Though inorganic complexes of calcium do not appear to be important under diagenetic conditions in the present context, the importance of organic complexes has not been established here.

Permeability-Texture Relation

A key link in the pressure gradient autocatalytic feedback is the dependence of the permeability on texture. The feedback is strengthened when the permeability decreases more rapidly with the amount of precipitation. Therefore any predictions of pressure gradient feedback phenomena will be sensitive to the permeability-texture relation.

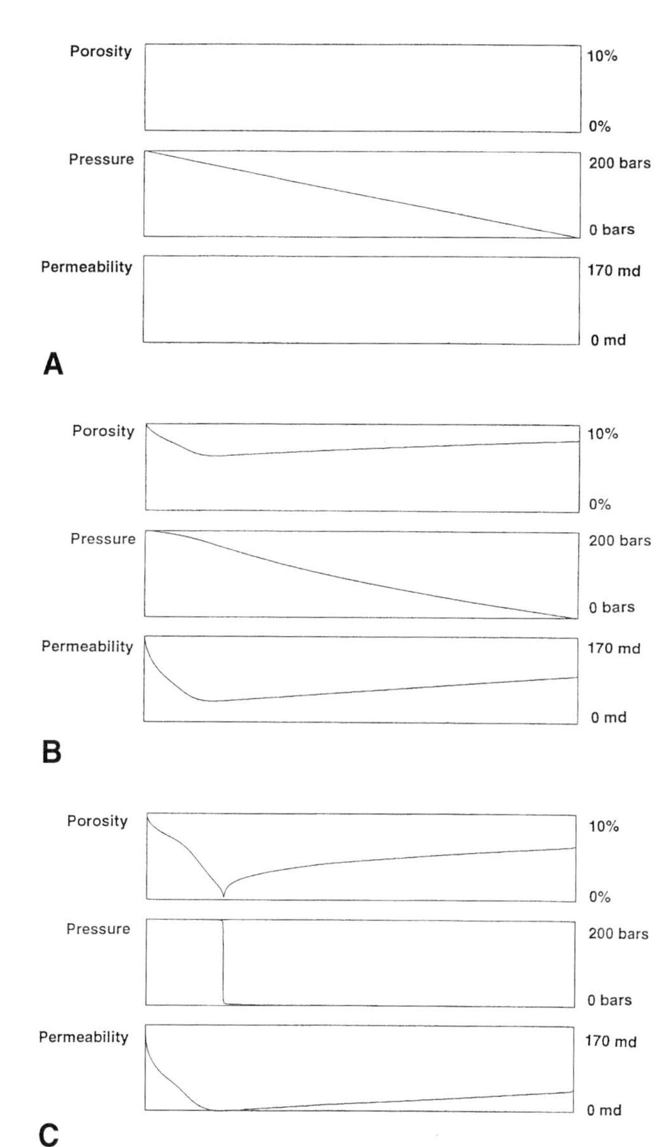

Figure 9. 200 bars of pressure difference was imposed on an initially uniform system with a length of 1 km. A modest seal developed from the (A) original uniform state by (B) 38,500 years and then to a well-developed seal at (C) 210,000 years.

A number of permeability relations have been discussed in the literature. The simplest case is a porosity power law dependence (McKenzie, 1984; Sleep, 1988):

$$\kappa = \bar{\kappa}\phi^{\alpha} \tag{12}$$

for constants $\bar{\kappa}$ and α (α>0) where κ is permeability and ϕ is porosity. For such a system the strength of the feedback increases with α.

A more dramatic effect is expected when the permeability vanishes at some nonzero porosity, because pore throats become cemented closed. One formula capturing the effect is the following:

$$\kappa = \bar{\kappa}\phi^{\alpha}\left\{\exp\left[-\left(\phi^{*}/\phi\right)^{\beta}\right]\right\} \tag{13}$$

for constants $\bar{\kappa}$, α, β, and ϕ.

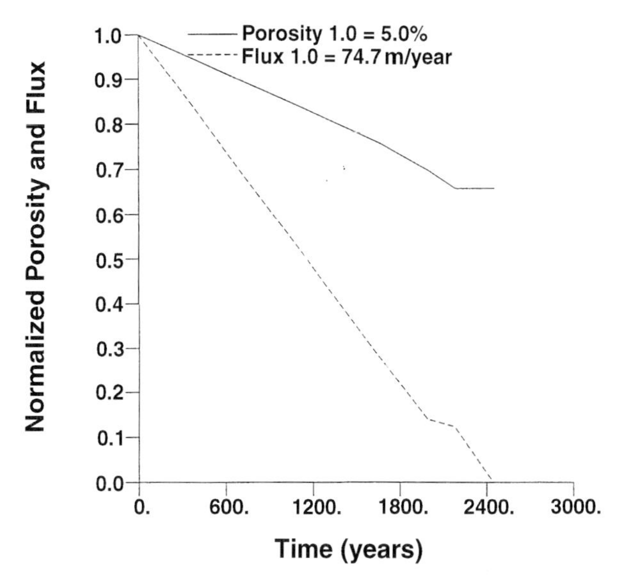

Figure 10. Evolution of flux and minimum system porosity using permeability law (13). Flux decreases more rapidly and is arrested earlier because the decrease in permeability with calcite precipitation is stronger here than in the simulation of Figure 6 where permeability law (12) was used.

An experiment was carried out with equation 13 as the permeability relation. All other conditions remain the same as in the case described in Figure 6. ϕ^* and β were set to be 6% and 3, respectively; α was 3 as in Figure 6. The result is shown in Figure 10. The flux through the system decreases more rapidly than for the case of Figure 6.

From this result we draw the conclusion that, as expected, the permeability-texture relation has a strong effect on our experiments. Accurate permeability relations are crucial for the prediction of sealing phenomena.

Positive Feedback Between Overpressuring and Sealing

Now consider the additional feedback arising when the pressure gradient accumulation effect is operating in a seal bounding an actively overpressuring compartment. For simplicity, we study this effect by considering the imposition of a constant flux through the developing sealing region (driven by compaction or other overpressuring mechanism in the underlying compartment interior). In other words, we impose a constant flux boundary condition rather than impose a constant pressure difference on the system.

Figures 11 and 12 show the results of an experiment that has the same conditions as those in Figures 5 and 6 except that the constant flux boundary condition was used. Figure 11 indicates that under such conditions porosity decreases in an accelerated way. This is because a decrease in porosity will enhance the pressure difference across the domain (Figure 12). The

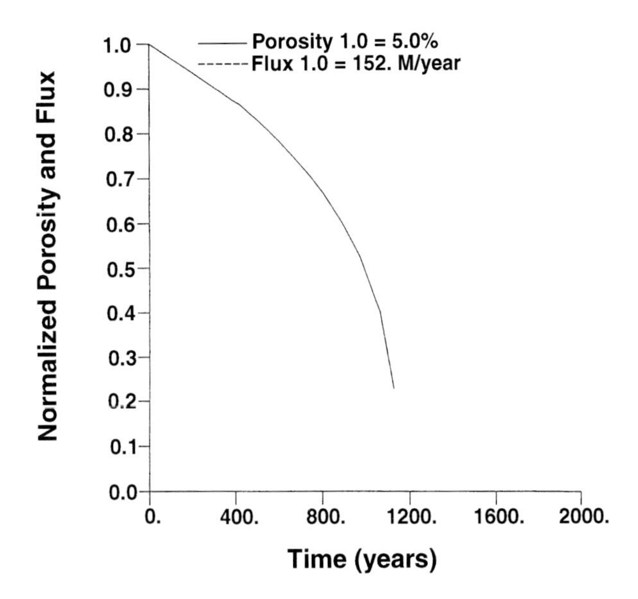

Figure 11. Minimum porosity change with time under constant flux of 152 m/yr. The rate of porosity decrease accelerates. Other conditions are the same as in Figure 6.

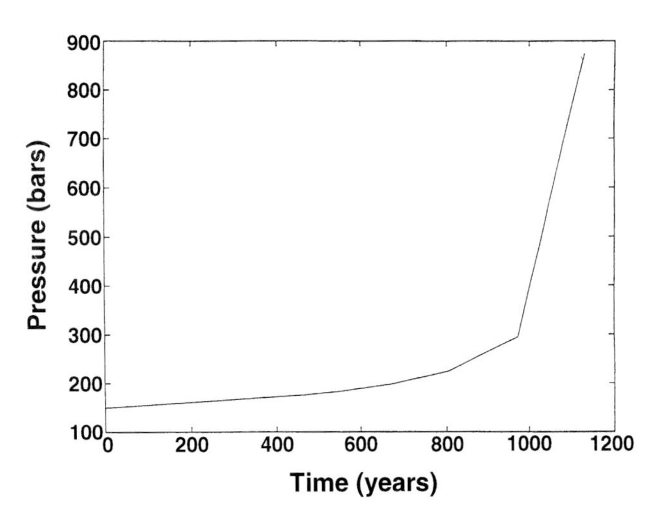

Figure 12. Pressure difference across the system increases with time due to the development of a seal when the flux through the system is held constant.

higher pressure difference will speed up the precipitation of calcite. Thus a positive feedback loop is active. Note that at the end of the experiment, the pressure difference across the system is about five times greater than that initially imposed.

FURTHER CONSIDERATIONS

The pressure gradient feedback mechanism can, in principle, operate for a broad range of chemistries. In future work we shall investigate more thoroughly the effects of fluid composition and, in particular, the possible influence of organic or other complexes.

The possibility exists for the present phenomenon to operate through the precipitation of asphaltines or paraffins from petroleum pore fluids. Thereby overpressured oil-filled compartments could automatically limit leakage. Such precipitation is in fact commonly observed during petroleum production, where it can be a cause of formation damage or pipeline clogging; such precipitation is induced by either pressure or temperature changes.

For the present mechanism to be effective in forming, enhancing, or healing seals, it must operate efficiently. The volumes of flow-through must not be so great that they would deplete the compartment of its fluids. Consider a compartment of volume V_c and porosity ϕ_c while the region of the top seal being formed or healed has area A_s and porosity ϕ_s. Let t^* be the time to achieve the closure and, for simplicity, take the flow-through velocity v and seal-porosity to decrease linearly:

$$v = v(0)\left(1 - t/t^*\right) \tag{14}$$

$$\phi_s = \phi_s(0)\left(1 - t/t^*\right) \tag{15}$$

Then the volume of fluid required to achieve closure, v_f, is

$$V_f = A_s v(0)\phi_s(0)\int_0^{t^*} dt(1 - t/t^*) = \frac{1}{3}A_s v(0)\phi_s(0)t^* \tag{16}$$

Let the completion efficiency E_{comp} be defined by $E_{comp} = 1 - V_f/V_c$. Then we get

$$E_{comp} = 1 - \frac{1}{3}A_s v(0)\phi_s(0)t^*/v_c. \tag{17}$$

For a simulation as in Figure 3 we have $v(0) = 150$ m/yr and $\phi(0) = .05$. Considering a compartment of 10 km × 10 km × 1 km = 100 km^3 in volume and an incomplete seal area A of 100 m × 100 m, we have, roughly,

$$E_{comp} = .9995. \tag{18}$$

Thus the seal is completed or healed with negligible loss of fluid ($E_{comp} = 1$ for no loss).

If the permeability decreases very rapidly with calcite precipitation via the clogging of pore throats then the above efficiencies for completion could be improved. The above results suggest that the present mechanism could even operate to heal a compartment-wide breached horizontal seal in such a case. The mechanism could also cement hydrofractures in a breached seal when the fracture porosity was small. Thus we believe that the pressure gradient mechanism can be an important factor in compartment genesis and preservation.

ACKNOWLEDGMENTS

Research supported in part by contracts from the Gas Research Institute (No. 5092-260-2443), Mobil Exploration and Producing Services, and Amoco Production Company, and grants from the Basic

Energy Sciences Program of the U.S. Department of Energy (No. DEFG291ER14175), Shell Development Company, IBM, and Sun Microsystems.

REFERENCES CITED

Bradley, J.S., 1975, Abnormal formation pressure: AAPG Bulletin, v. 12, p. 328–329.

Dewers, T., and P. Ortoleva, 1988, The role of geochemical self-organization in the migration and trapping of hydrocarbons: Applied Geochemistry, v. 3, p. 287–316.

Dewers, T. and P. Ortoleva, 1990a, A coupled reaction/transport/mechanical model for intergranular pressure solution, stylolites, and differential compaction and cementation in clean sandstones: Geochimica et Cosmochimica Acta, v. 54, p. 1609–1625.

Dewers, T., and P. Ortoleva, 1990b, Differentiated structures arising from mechano-chemical feedback in stressed rocks, *in* P. Ortoleva, B. Hallet, A. McBirney, I. Meshri, R. Reeder, and P. Williams, eds., Self-Organization in Geological Systems: Proceedings of a Workshop held 26-30 June 1988, University of California Santa Barbara: Earth-Science Reviews, v. 29, p. 283–298.

Dewers, T., and P. Ortoleva, 1990c, The interaction of reaction, mass transport, and rock deformation during diagenesis: Mathematical modeling of intergranular pressure solution, stylolites, and differential compaction/cementation, *in* I. Meshri and P. Ortoleva, eds., Prediction of Reservoir Quality Through Chemical Modeling: AAPG Memoir 49, p. 147–160.

Dewers, T., and P. Ortoleva, 1992, Non-linear dynamics in chemically compacting porous media, *in* W.E. Fitzgibbon and M.F. Wheeler, eds., Modeling and Analysis of Diffusive and Advective Processes in Geosciences: SIAM, p. 100–121.

Helgeson, H., D. David, and G. Flowers, 1981, Theoretical prediction of the thermodynamic behavior of aqueous electrolytes at high pressures and temperatures: IV. Calculation of activity coefficients, osmotic coefficients, and apparent molal and standard and relative partial molal properties to 600°C and 5KB: American Journal of Science, v. 281, p. 1249–1516.

Kharaka, Y.K., W.D. Gunter, K. Aggarwalp, E.H. Perkins, and J.D. DeBraal, 1988, SOLMINEQ.88—A computer program for geochemical modeling of water-rock interactions: USGS Water Resources Investigations Report 88-4227, 420 p.

McKenzie, D.P., 1984, The generation and compaction of partially molten rock: Journal of Petrology, v. 25, p. 713–765.

Millero, F., 1971, The molal volumes of electrolytes: Chemical Reviews, v. 71, p. 147–176.

Millero, F., 1972, The partial molal volumes of electrolytes in aqueous solutions, *in* R. Horne, ed., Water and Aqueous Solutions: Structure, Thermodynamics and Transport Properties: John Wiley and Sons, Inc.

Ortoleva, P., and Z. Al-Shaieb, in press, Genesis and dynamics of basin compartments and seals: American Journal of Science.

Powley, D., 1975, Normal and abnormal pressures: Lecture presented to AAPG Advanced Exploration Schools, 1980–1987.

Powley, D., 1990, Pressures, hydrogeology and large scale seals in petroleum basins, *in* P. Ortoleva, A. McBirney, I. Meshri, R. Reeder, and P. Williams, eds., Self-Organization in Geological Systems, Proceedings of a Workshop held 26–30 June 1988, University of California Santa Barbara: Earth-Science Reviews, v. 29, p. 215–226.

Prigogine, I., and R. Defay, 1967, Chemical Thermodynamics: London, Longmans, Green and Co. Ltd., 543 p.

Qin, C., M. Maxwell, and P. Ortoleva, 1991, Layered diagenetic pressure seals: a mechano-chemical computational homogenization modeling study: OKINTEX Group, Quarterly Report June–August 1991, for the Gas Research Institute, p. 24–56.

Ricken, W., 1986, Diagenetic Bedding: Berlin, Springer-Verlag.

Sleep, N.H., 1988, Tapping of melt by veins and dikes. Journal of Geophysical Research, v. 93, p. 10255–10272.

Tigert, V., and Z. Al-Shaieb, 1990, Pressure seals: Their diagenetic banding patterns, *in* P. Ortoleva, A. McBirney, I. Meshri, R. Reeder, and P. Williams, eds., Self-Organization in Geological Systems, Proceedings of a Workshop held 26–30 June 1988, University of California Santa Barbara: Earth-Science Reviews, v. 29, p. 227–240.

Chapter 28

Simulating the Development of a Three-Dimensional Basinwide Overpressured Compartment

M. Maxwell
Peter J. Ortoleva
Indiana University
Bloomington, Indiana, U.S.A.

ABSTRACT

Basinwide overpressured compartments have been observed in the Anadarko and other basins. Results are presented on a mechano-chemical model of this phenomenon that illustrates how the imposed pattern of sedimentary features and faulting can produce such a three-dimensionally isolated compartment.

The model is based on the solution of equations describing pressure solution, stresses, conservation of pore fluid solutes, and the evolution of pore fluid pressure. The model accounts only for quartz chemistry; fluid overpressuring is solely through thermal expansion of the fluid and compaction.

INTRODUCTION

A basin-scale, overpressured compartment has been shown to exist in the Anadarko basin. The goal of the present study is to show how such an overpressured "megacompartment" (Al-Shaieb et al., this volume) arises in three spatial dimensions naturally from the sedimentary and tectonic history of such an aulacogen under the influence of the network of inorganic diagenetic reaction, transport, and mechanical (RTM) processes.

Here we start such a study via a minimal, three-spatial-dimensional RTM model. In work in progress we are developing three-dimensional models of increasingly greater chemical, transport, textural, and mechanical complexity.

The megacompartment of the Anadarko basin is defined roughly as follows (Al-Shaieb et al., this volume). A basal seal follows the Woodford Shale. A side seal is adjacent to the Wichita fault in poorly sorted granite wash. The top seal is fairly horizontal, lying within the Red Fork, Atoka, Morrow, Springer, Chester, and Mississippi formations. The three-dimensional closure of the compartment appears to occur by the convergence of the top and basal seals. Over- and underlying the compartment are normally pressured rock. The ultimate goal of the present work is to illustrate how this can take place through specific diagenetic processes and, in so doing, enable one to make quantitative predictions for exploration and field development in the Anadarko and other basins.

The temporal development of the three-dimensionally isolated megacompartment is primarily the history of the genesis of the surrounding seal. Timing is important in that the formation of the surrounding seal must precede the destruction of the interior porosity—otherwise the putative compartment would simply be a region of low porosity. The collection of seals (top, basal, and side) must fit together exactly to constitute complete three-dimensional closure. Furthermore, seals themselves may develop as or from smaller-scale compartments—i.e., a closed array of compartments can constitute a larger-scale compartment. The timing of these various events relative to that of petroleum generation and migration is of interest for exploration and production considerations.

Here the temporal development of a megacompartment in an RTM model is illustrated. The RTM model used is similar to that adopted in our earlier two-dimensional study (Sonnenthal and Ortoleva, this volume). In the next section the model is briefly reviewed while in the final section some preliminary results are presented.

THE MODEL

RTM Model

The RTM model consists of a set of equations describing the diagenetic processes and the assumed sedimentological and tectonic history. The model is simulated in three dimensions via our RTM code developed in this project.

The underlying processes driving the system are assumed to be the following:

1. Darcy flow with a texture-dependent permeability
2. Conservation of fluid mass
3. Pressure solution-mediated compaction via a water–film diffusion/truncated sphere model (Dewers and Ortoleva, 1990)
4. Linear equation of state
5. Macroscopic vertical stress calculated from overburden and horizontal stress imposed by the tectonic regime
6. Grain-overgrowth kinetics using a fluid pressure-mediated equilibrium constant and the diffusion limitation induced by clay coatings
7. Thermal effects accounted for by an imposed geothermal gradient and temperature-dependent rate and equilibrium constants for the quartz reaction in the assumed monomineralic system

These effects are incorporated into equations for conservation of water and pore fluid solute mass, grain geometry and packing variables, and macro- and grain-scale force balance. The equations are solved on a three-dimensional spatial grid that moves with the rock velocity as the basin deforms, compaction occurs, and sediment is added. New grid points are added to follow the input sediment. Grid addition is done in an adaptive way so that sharp transitions of texture between different beds are captured.

The sedimentation history is fed into the program in terms of the velocity of sediment mass input and the size and geometry of the input grains and the variation of these quantities over the top surface of the sediment pile. The sediment is deposited in a marine environment so that a sea level is assumed. The time course of the bottom of the sediment pile's geometry is imposed by the assumed tectonic history.

The output of the computer code is the simulated time course of the distribution in three dimensions of the texture, porosity, permeability, fluid pressure and flow, and the geometry of the various formations and the entire basin.

Specific Sedimentation and Tectonic History

We chose a simple sedimentary and tectonic history as suggested in Figures 1 and 2. The basin is bounded on one side by a fault. The fault is assumed to be highly permeable. The basin is underlain by an asymmetric bowl-shaped impermeable basement. The time course of the basement-sediment interface is a key control variable.

The sediment history leads to bedding as suggested in the cross section of Figure 1. The bed interfacing with the basement is assumed to be fractured and thereby highly permeable. It is modeled as a coarse-grained sandstone.

Above the coarse-grained underlayer is a "shaly" formation that we model as a very fine grained sandstone. This formation will be referred to as the "shale."

Overlying the shale is a graded formation that is poorly sorted adjacent to the fault, a medium sandstone further into the basin (away from the fault), and fine in the periphery of the basin away from the fault. For simplicity we model the poorly sorted (granite wash) via its most pressure–solution reactive component—the fine-grained material.

The basin is topped with a relatively fine grained formation, reflecting a period of slower subsidence. The subsidence history of the sediment-basement interface is taken to be a period of relatively constant subsidence velocity (of about 12 Ma) followed by a longer quiescent phase.

Banded Seals

Observations from the Anadarko (Tigert and Al-Shaieb, 1990; Al-Shaieb et al., this volume) and Michigan (Shepherd et al., this volume) basins show that diagenetic processes have led to banded patterns of cementation that yield rock of high sealing quality. In a series of modeling studies we have shown that such banding can arise through mechano-chemical processes and are of a range of distinct types (Dewers and Ortoleva, 1989, 1990; Qin and Ortoleva, this volume). Here we focus on one type of banding observed in seals—compaction/cementation alternations.

A full mechano-chemical model capturing the submeter-scale banding in a basin-scale calculation is only feasible via a homogenization approach (Qin and Ortoleva, this volume). Here we consider a simpler model. A sequence of bands is assumed involving an alternation of A- and B-type layers. The thickness of the layers is assumed equal but may vary slowly on the supra-band (say 10 m or greater) scale. Assuming a steady state diffusional exchange between bands and letting the concentration of silica be denoted a and b in the A and B bands, respectively, we have

$$\gamma(b-a) - A_f^A G_f^A - A_z^A G_z^A = 0 \qquad (1)$$

$$-\gamma(b-a) - A_f^B G_f^B - A_z^B G_z^B = 0 \tag{2}$$

where γ is an inverse time associated with the diffusive exchange. As a compromise we write

$$\gamma = \frac{2}{\rho n \Delta^2}\left\{\frac{1}{\phi^A D^A} + \frac{1}{\phi^B D^B}\right\}^{-1} \tag{3}$$

where Δ is the thickness of an A, B couple; ρ is the molar density of quartz; n is the number of quartz grains per unit volume; ϕ is porosity; and D is the macrodiffusion coefficient. Here G_f^A is the rate of growth of the free face radius of a truncated sphere grain in the A-band and G_z^A is the rate of growth of the vertical (z) dimension while A_f^A and A_z^A are the grain facet areas (see Dewers and Ortoleva, 1990; Sonnenthal and Ortoleva, this volume). The above two equations provide a pair of linear equations for a and b in terms of the texture and stress in the A- and B-bands. Separate evolution equations for the grain geometry parameters L_f^A and L_f^B (and similarly for L_z^A and L_z^B) are written for each band.

The key effect of the banding is to minimize the permeability in the direction normal to the bands (here the vertical direction). While a treatment of banded seal genesis requires a fuller homogenization approach (Qin and Ortoleva, this volume), we simulate this effect by considering the vertical and horizontal permeabilities (κ_v, κ_h, respectively) in the form

$$\kappa_v = 2\left(\frac{1}{\kappa_A} + \frac{1}{\kappa_B}\right)^{-1} \tag{4}$$

$$\kappa_h = \frac{1}{2}(\kappa_A + \kappa_B) \tag{5}$$

and by treating the permeability as a diagonal tensor.

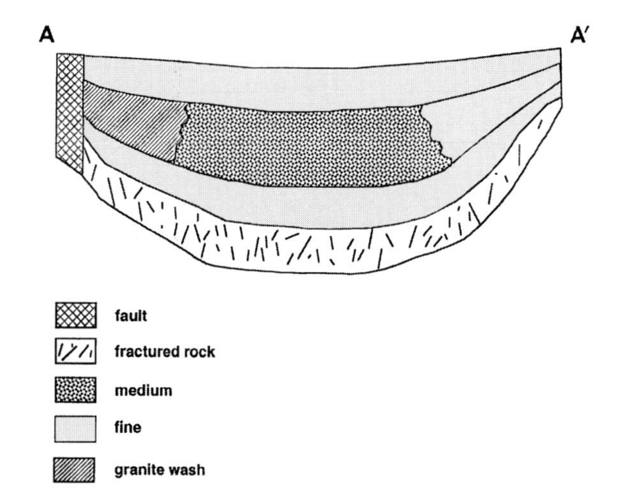

Figure 1. Schematic basin cross section used for modeling study. A–A′ transect as in Figure 2.

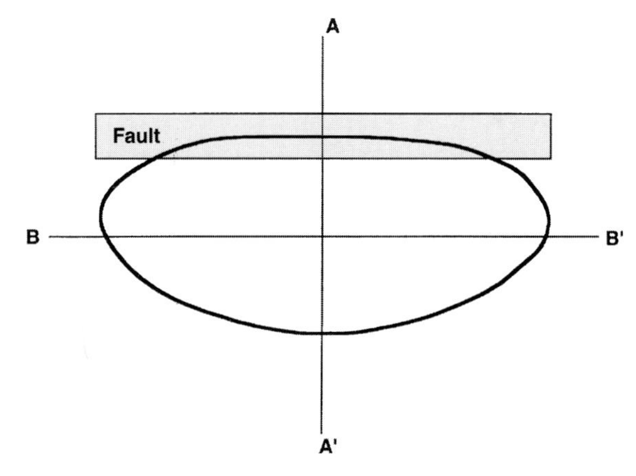

Figure 2. Map view of an Anadarko basin-like model showing transects A–A′ and B–B′. Figure 1 shows cross section corresponding to A–A′.

SAMPLE HISTORIES

To illustrate the model, we investigate the following sedimentological history. The basin is assumed for simplicity to have only quartz sediment. The qualitative aspects of the sedimentology are very crudely associated with that of the (admittedly mineralogically complex) Anadarko basin.

The number and range of the control variables (sedimentology and tectonics) that affect basin evolution, and in particular diagenesis, give each basin its individual character. Evolution of the megacompartment of the Anadarko basin seems to have proceeded in several stages that for the preliminary study we caricaturize as follows (see Figure 1). As sediment is added, the Devonian shale (Woodford Formation) achieves sufficient depth that it compacts. Compaction elevates pressure within the shale and that developed overpressure eventually represses further compaction in the interior of the shale. However, at the periphery of the shale fluid is lost to the surrounding more-permeable beds. This causes a rind of more highly compacted rock bounding the shale on top and bottom. Thereby, in this early stage the shale becomes a bed-localized compartment.

With further burial the medium-grain–size material and its more active near-fault and far-from-fault finer sediment are added. Further burial adds the finer overburden.

With time the medium material becomes enclosed in faster compacting sediment. This compacting envelope eventually develops significant overpressure and serves as a shell with complete three-dimensional closure.

In the third stage fluid squeezed from the surrounding seals and compaction of the interior lead to overpressuring of the interior. Finally, a steady state is obtained wherein fluid leakage from the interior is just balanced by compaction. This steady state is long-lived (i.e., on the 100 million year time frame) if the surrounding seals are of sub-nanodarcy permeability (Ortoleva et al., in press).

To test this scenario, we carried out a preliminary simulation. The domain and sedimentary history are suggested in Figures 1 and 2. A view of the simula-

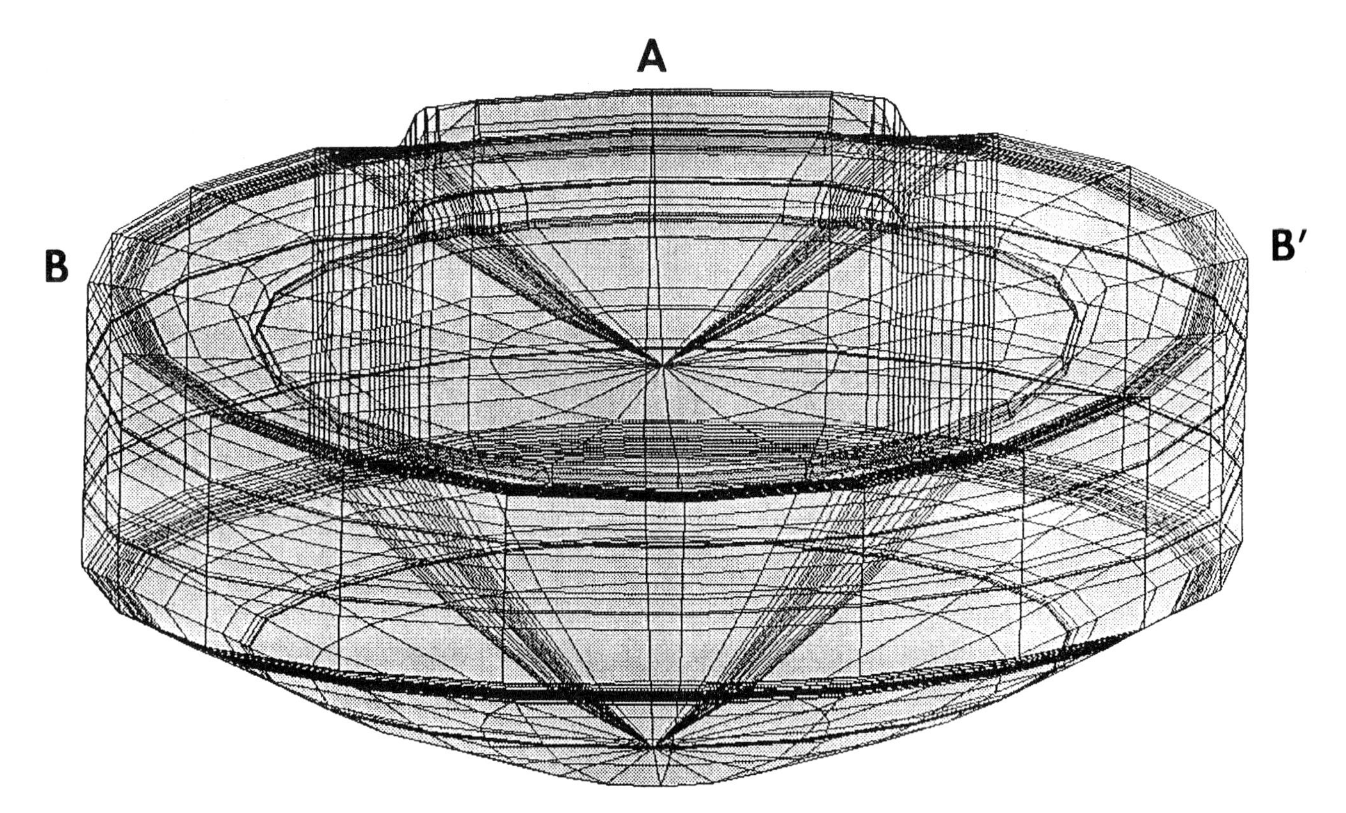

Figure 3. Three-dimensional computational grid at 35 Ma showing approximately 1/15 of grid nodes. For clarity only grid nodes lying on the surface of the domain are shown. Note adaptive grid spacing used to capture regions where properties change rapidly is dynamically readjusted with sedimentation and diagenesis.

tion grid at 35 Ma is given in Figure 3. The state of the system after 35 million years is seen in Figure 4A and 4B. At the time shown, the basal, top, and side seals have completed. The minimum permeability in the basal seal is 0.1 nanodarcy, while in the top and lateral seals, it is 0.2 nanodarcy; this is in stark contrast to the permeability in the interior which is 2×10^{-3} darcy and that of the uncompacted fine material at the top of the basin, where it is 10^{-4} darcy. Three-dimensional closure to form the megacompartment has been obtained with an overpressure of at least 125 bars. In future work we shall attempt to extend these (computationally intensive) simulations to longer times to observe the elevation of pressure in the megacompartment interior and its longevity on the 100 Ma time scale.

The effects of hydrofracturing, mineralogy, more accurate basin stresses, nonconstant geothermal gradient, and diagenetic banding as discussed above, as well as specific tectonic history, will lead to many variations on the above theme. These effects are presently being incorporated into our model and code. The present study does illustrate the feasibility of three-dimensional diagenetic RTM modeling. We believe that this type of modeling will be of great value in identifying new types of reservoirs and predicting their quality, characteristics, and locations in a basin.

ACKNOWLEDGMENTS

Research was supported in part by contracts from the Gas Research Institute (5092-260-2443), Mobil Exploration and Producing Services, and Amoco Production Company, and grants from the Basic Energy Sciences Program of the U.S. Department of Energy (DEFG291ER14175), Shell Development Company, Intel Supercomputers, and Sun Microsystems.

REFERENCES

Dewers, T., and P. Ortoleva, 1989, The role of geochemical self-organization in the migration and trapping of hydrocarbons: Applied Geochemistry, v. 3, p. 287–316.

Dewers, T., and P. Ortoleva, 1990, A coupled reaction/transport/mechanical model for intergranular pressure solution, stylolites and differential compaction and cementation in clean sandstones:

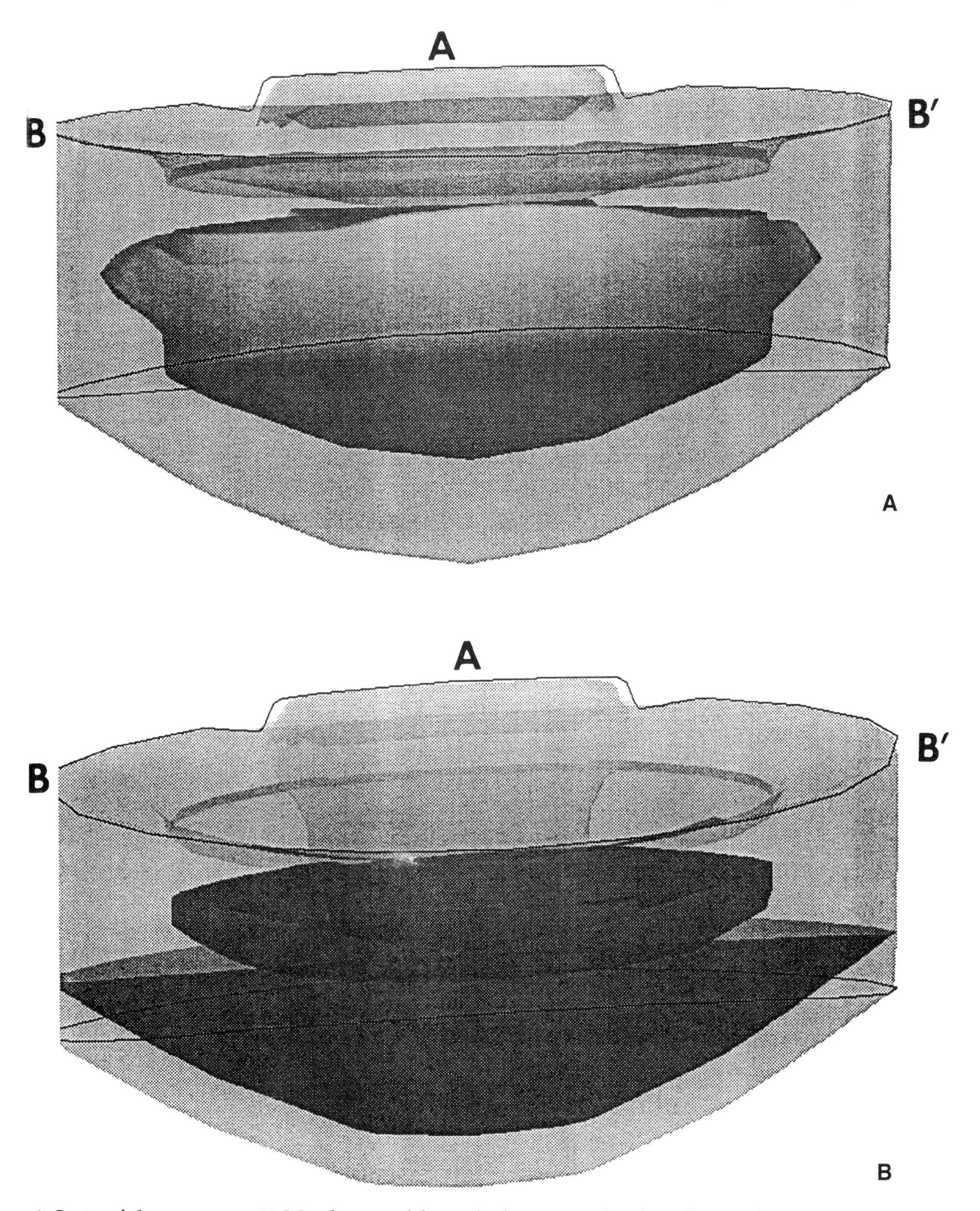

Figure 4. State of the system at 35 Ma shown with vertical exaggeration by a factor of 30. (A) Isosurface of overpressure at 125 bars showing megacompartment interior. (B) Porosity isosurface of 25%; the upper body indicates the porosity-preserved interior whereas below the bottom surface there is a zone where porosity exceeds 25% modeling underlying fractured zone (see Figure 1).

Geochimica et Cosmochimica Acta, v. 54, p. 1609–1625.

Ortoleva, P., Z. Al-Shaieb, and J. Puckette, in press, Genesis and dynamics of basin compartments and seals: American Journal of Science.

Tigert, V., and Z. Al-Shaieb, 1990, Pressure seals: their diagenetic banding patterns: Earth-Science Reviews, v. 29, p. 227–240.

VIII. Other Considerations

Chapter 29

Comparative Hydrodynamic and Thermal Characteristics of Sedimentary Basins and Geothermal Systems in Sediment-Filled Rift Valleys

Wolfgang Polster
H. L. Barnes
The Pennsylvania State University
University Park, Pennsylvania, U.S.A.

ABSTRACT

Detailed geophysical, hydraulic, and geochemical data were compiled from the literature for sedimentary basins and for liquid-dominated geothermal fields of sediment-filled rift valleys. The objective was to use the geothermal data as a guide to the effects to be expected in sedimentary basins from the upflow along growth faults and other high-permeability zones.

These *geothermal reservoirs* usually lie 0.6–0.8 km below the surface. Representative conditions in such reservoirs are temperatures of 120–370°C, thermal gradients of 15–80°C/km, salinity of 3–27 g/l, pH of about 4.5–5.5, pressure from hydrostatic to lithostatic, average porosity of 10–20%, and permeability of 0.1–600 md, typically. Comparable zones in *sedimentary basins* with similar thermal, geochemical, hydrodynamic, and lithological conditions are, for example: (1) upflow zones along growth faults (e.g., Wilcox trend, northern Gulf of Mexico Basin, up to 60°C/km) or piercement structures (e.g., Danish Central Graben, North Sea Basin, up to 50°C/km), (2) upflow zones along deep, but permeable strata of sedimentary basins (e.g., Alberta basin, ~40°C/km), and (3) sediments near ancient rift zones (e.g., Gabon basin).

Average bulk permeability and porosity are typically reduced by hydrothermal upflow in both environments. For example, the bulk porosity at a depth of 0.6–2.5 km in unaltered sediments is up to 10% higher than in an adjacent, moderate-temperature (120–200°C) geothermal reservoir. Apparently, the upflow of hydrothermal fluids into sedimentary strata can cause significant reduction of porosity and permeability in a short time (<16,000 yr) even at moderate temperatures and geothermal gradients (15–60°C/km). Representative bulk porosities of hydrothermally altered sediments at temperatures exceeding 250°C are about 3–10%, comparable to the porosity of basins commonly found at depth below about 4–5 km. Such hydrothermal processes may easily form basinal seals that are effective traps for hydrocarbon fluids.

Sedimentary basins have temperatures and gradients similar to those in geothermal systems but at greater depth. For example, a temperature range from 120–190°C and a gradient of 45°C/km are typical at 2.2–3.8 km depth in sedimentary basins and 0.6–2.2 km depth in geothermal reservoirs. This means that, given similarities in geochemical conditions, similar water/rock interactions can occur. For instance, diagenetic alterations as a result of the influx of hot brines into clastic sediments are often similar in both cases.

Geothermal systems differ markedly from basins in having higher surface-heat flow, much higher near-surface gradients, and much lower gradients in the presence of cold water recharge. Additionally, organic solutes are absent in this type of geothermal reservoir while organic and biological reactions play an important role in sedimentary basins.

INTRODUCTION

Interest in hydrocarbon exploration and the evolution of sedimentary basins has stimulated many studies on the influence of water/rock interactions on rock permeability and porosity. A detailed knowledge of the hydraulic and geochemical conditions within deep sedimentary basins is necessary to allow exploration and production of gas from deep reservoirs. Studies of the local and regional geothermal regimes in the Gulf Coast Basin (e.g., Tyler et al., 1985; Bodner and Sharp, 1988; Pfeifer and Sharp, 1989), the sedimentary basins of the Rocky Mountains (Meyer and McGee, 1985), the North Sea Basin (e.g., Jensenius, 1987; Liewig et al., 1987; Jensenius and Munkgaard, 1989), the Alberta basin (Tilley and Longstaffe, 1989; Tilley et al., 1990), and the Angola rift basins (e.g., Walgenwitz et al., 1985; Girard et al., 1989) provide examples of the migration and upflow of hydrothermal fluids and their effects on the maturity of basin sediments. However, the plethora of pertinent variables (e.g., water chemistry, fluid flow, fluid pressure, and thermal conditions) makes it difficult to identify the mechanisms causing time-dependent changes in the reservoir properties of permeability and porosity.

Liquid-dominated geothermal systems in sedimentary environments provide unique opportunities to examine the influences of hot brines on the evolution of reservoir properties of clastic sediments over a large temperature interval (100–350°C). By comparing the characteristics of sedimentary basins with those of active geothermal systems, many of these effects can be individually evaluated. Their environments are similar during several stages of evolution. In the centers of geothermal systems a temperature of 120–250°C is typically encountered at a depth of about 500–1000 m. In sedimentary basins a temperature above 120°C is usually reached below 2.5 km. However, the conditions found in sedimentary basins and the reservoir zones of geothermal systems are often similar, when zones with comparable temperature are compared, except that the pertinent geochemical reactions are predominantly biological-inorganic and organic-inorganic in sedimentary basins up to 160°C.

All geothermal systems discussed here are typically found within continental rifts or where oceanic spreading ridges extend onto the continent. Representative examples are the East Mesa, Westmoreland, Heber, Dunes, Salton Sea (all USA), and Cerro Prieto (Mexico) geothermal fields, which are located in deltaic sedimentary sequences of upper Cenozoic age (Muffler and White, 1969). All of these geothermal fields are found in the Salton Trough (Southern California and Mexico), a complex rift valley between the North American and Pacific plates (Elders and Cohen, 1983). A generalized cross section through a geothermal field in a sediment-filled rift valley is given in Figure 1. Descriptions of the tectonics and the lithologic and stratigraphic relationships of the Salton Trough are given elsewhere (e.g., Randall, 1974; Elders, 1979; Elders and Cohen, 1983; Newmark et al., 1986).

We have compiled detailed geophysical, hydraulic, and geochemical data on this common type of geothermal field to help understand basinal evolution. The objectives are a comparative evaluation of (1) similarities and differences in thermal structure, hydrodynamic properties, and basic differences in water chemistry; (2) cementation, dissolution, and alteration histories; (3) mixing effects between aquifers with different compositions; and (4) changes in reservoir properties due to the upward flow of hot brines.

Water/rock interactions in both environments are time-dependent functions of thermal (e.g., temperature gradient, fluid temperature), hydrodynamic (e.g.,

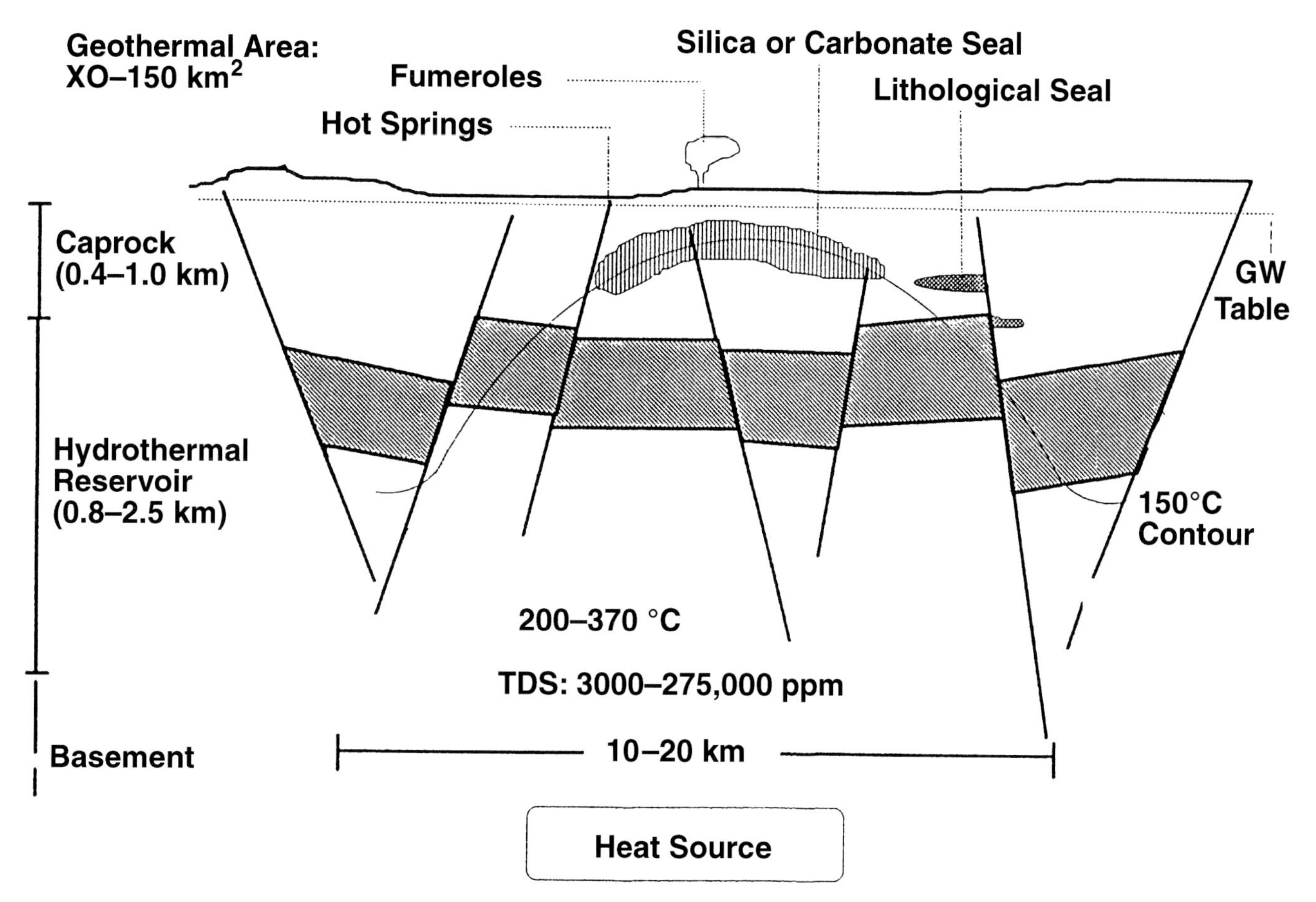

Figure 1. Generalized cross section through a geothermal system in a sediment-filled rift valley.

fluid pressure, rock permeability and porosity, fluid flow), lithological (e.g., sedimentary history, rock and facies type), and geochemical (e.g., water chemistry, mineralogy) conditions. For effective comparison of the development of geothermal systems and sedimentary basins, it is necessary to define comparable zones in both environments in which the key parameters mentioned above are analogous during the evolutionary stages.

We will summarize the characteristics of rift valley–associated geothermal systems to demonstrate that the constraints in geothermal fields and in certain regions of sedimentary basins are similar during several stages of development of their thermal, hydrodynamic, and geochemical characteristics.

THERMAL CHARACTERISTICS OF SEDIMENTARY BASINS AND GEOTHERMAL SYSTEMS

Although the mean surface heat flow of sedimentary basins, continental rifts, and rift valley–associated geothermal fields is different, their temperature gradients are quite similar over large temperature and depth intervals (see Figures 2, 3, 4).

There is an inverse correlation between the mean thickness of the continental crust and the mean surface heat flow of both sedimentary basins and rifts (Figure 3) even when tectonic events or age differences are neglected. This implies that the mean surface heat flow in both environments is primarily controlled by the depth to the mantle.

In sedimentary basins, the mean surface heat flow is usually smaller, about 85 mW/m^2, and varies less than within continental rifts (e.g., Pannonian basin: 69–130 mW/m^2 ; Horvath et al., 1979). The mean surface heat flow in younger continental rifts is usually above 95 mW/m^2 (Figure 3) and can reach extreme values in regions with anomalous high heat flow (>2000 mW/m^2). Rift valley–associated geothermal fields are commonly located in regions with anomalous high surface heat flow. All known geothermal fields of the Salton Trough have a surface heat flow above 200 mW/m^2 (Lachenbruch et al., 1985).

Sedimentary Basins

The temperature distribution in both sedimentary basins and geothermal systems is critical for the kinetics of alteration, precipitation, and dissolution that control seal formation. Therefore, we shall examine

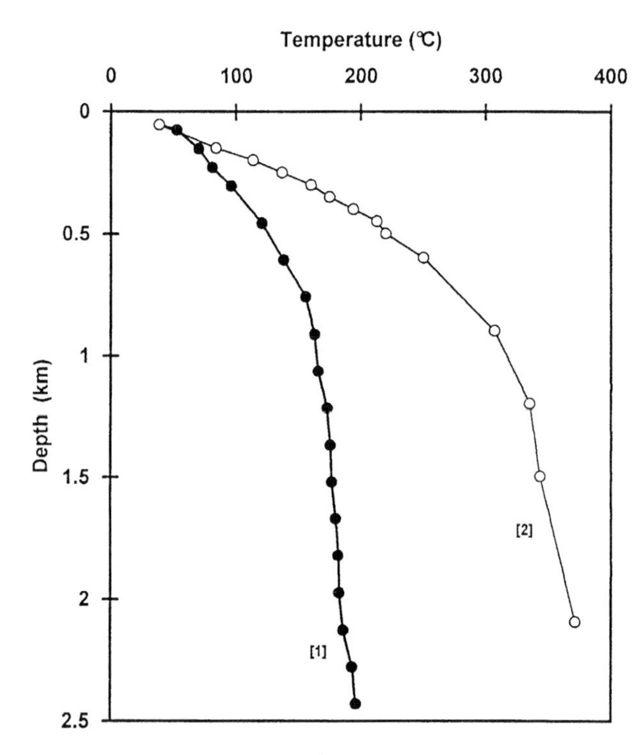

Figure 2. The dependence of temperature on depth in representative geothermal systems: [1] moderate-temperature geothermal fields (e.g., East Mesa, well 6-1; Pearson, 1976); [2] maximum in high-temperature fields (e.g., Salton Sea, well Elmor # 1; Elders and Cohen, 1983).

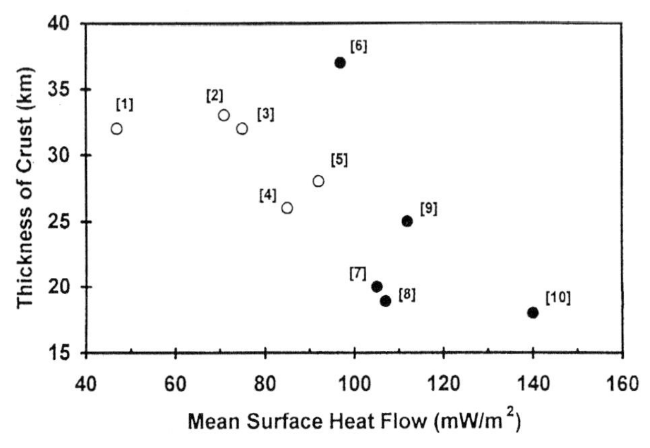

Figure 3. Variation in mean surface heat flow and of the crustal thickness in sedimentary basins (●) and continental rifts (O): [1] Transylvanian basin (Czechoslovakia) (Veliciu and Visario, 1980); [2] Songliao basin (China) (Ma et al., 1989); [3] Norwegian-Danish basin (Balling, 1979); [4] Pannonian basin (Hungary) (Bodri and Bodri, 1980); [5] Basin and Range Province (USA) (Hermance, 1982); [6] Baikal Rift (USSR) (Barberi et al., 1982; Morgan, 1982); [7] East African Rift (Hermance, 1982; Barberi et al., 1982); [8] Rio Grande Rift (USA) (Hermance, 1982; Barberi et al., 1982); [9] Rhine Graben (Germany) (Barberi et al., 1982; Morgan, 1982); [10] Salton Trough (USA) (Fuis et al., 1982; Lachenbruch et al., 1985).

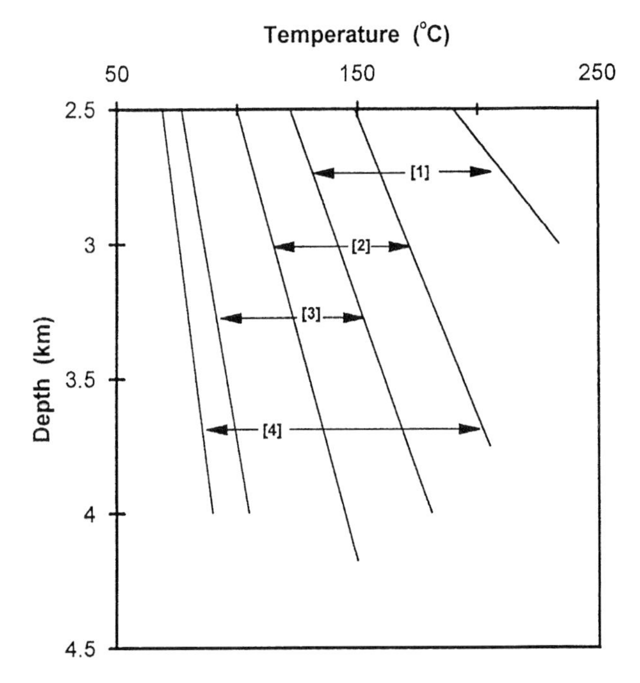

Figure 4. Representative geothermal gradients in sedimentary basins: [1] Pannonian basin (Hungary) (Horvath et al., 1979); [2] Cooper basin (Australia) (Kantsler et al., 1978); [3] Gulf Coast Basin (USA) (McBride and Sharp, 1989; Jones, 1975); [4] Niger Delta Basin (Nigeria) (Robert, 1988).

the temperatures and their lateral and vertical variation in basins and then in geothermal systems.

The temperatures in a sedimentary basin vary considerably depending on the position in the basin (Figure 4). Along the Gulf Coast Basin, the temperatures encountered at a depth of 2.5 km may be as low as 80°C or as high as 150°C. In the center of geothermal systems a temperature of 120–250°C is typical of a depth between 0.5 and 1.0 km (Figure 2). The large differences in the temperature distribution between sedimentary basins, as well as in the thermal regimes of a single basin, are indicated by a wide range of mean geothermal gradients of 10 to 70°C/km (averaged over several km), as shown in Figure 4.

Within most sedimentary basins, there are broad thermal anomalies that are commonly caused by the following four conditions.

1. Upflow of hot brines from deeper sediments into overlying sediments commonly affects the temperature distribution. For example, along the Wilcox growth-fault trend (Gulf Coast Basin, southern Texas), a major perturbation of temperatures below 2 to 3 km is apparently caused by the upflow of deeper hot brines along growth faults (Bodner and Sharp, 1988; Pfeifer and Sharp, 1989). Here, the temperature distribution indicates that anomalously high temperatures and geothermal gradients do not coincide with the top of the geopressured zone in the Wilcox trend. This is evidence that anomalously warm fluids are escaping upward from the overpressured zone into

overlying, hydrostatically pressured sediments (Pfeifer and Sharp, 1989). In these two zones, the geothermal gradient is about 20–30°C/km above the upflow zone, but below a depth of about 2–3 km, within the thermal anomaly, the geothermal gradient increases significantly to 45–60°C/km (Bodner and Sharp, 1988). The regional geothermal gradient, outside the approximately 20–35 km wide fault zone, at a depth below 2.0 km, is only about 33–38°C/km (Bebout et al., 1982; Bodner et al., 1985). Additional indications of the upwelling of hydrothermal fluids along the Wilcox fault-trend are an increase in secondary porosity (McBride and Sharp, 1989) and feldspar alteration (Land et al., 1987) in the overlying sediments, the local and regional salinity distribution (Morton et al., 1983), and the presence of near-surface uranium deposits (Galloway, 1982). A similar steepening of geothermal gradients occurs at upflow zones in the Pannonian basin (Hungary), which are revealed by hot springs (Horvath et al., 1979).

There are other basins showing similar thermal effects by hydrothermal flows. In another example in the deep Alberta basin, hot, moderately saline (2–3 wt %), and methane-rich brines migrated upward along permeable strata (conglomerates, sandstones) and fractures. During maximum burial of the basin, an upflow of hot brines into the Cretaceous Fahler and Cadotte formations produced an extensive quartz cementation (Tilley and Longstaffe, 1989; Tilley et al., 1990). Petrographic and fluid-inclusion data from these formations document a significant geothermal anomaly (170–200°C) with an increased geothermal gradient of about 40°C/km (Tilley and Longstaffe, 1989; Tilley et al., 1990). Other evidence of ancient thermal anomalies and of increased thermal maturity of sediments is found in the South Gabon rift basin due to the upflow or lateral circulation of hydrothermal fluids (Giroir et al., 1988). Similarly, cross-formational flow of hot brines is reported from near salt domes of the North Sea Basin (e.g., Liewig et al., 1987; Jensenius, 1987).

2. Contrasts in thermal conductance between overlying, low-conductivity sediments and deeper, high-conductivity sediments, such as salt diapirs, can decrease the geothermal gradient with depth from 30–50°C/km to 10°C/km in salts (e.g., Gabon basin; Robert, 1988, p. 48). In most cases, the lateral gradient near the salt diapir is large but the normal gradient returns within a couple of kilometers laterally away from the diapir (Jensen, 1990). Similarly, conductivity contrasts are responsible for thermal anomalies in the southern Aquitaine region, France. The normal geothermal gradient is about 10–20°C/km in the southern Aquitaine region but increases sharply to 60°C/km above very-conductive reef sediments of the upper Lacq oil reservoirs (Coustau et al., 1969).

3. Transitions between hydrostatically pressured formations and deeper, overpressured strata are often accompanied by significant increases in the geothermal gradient, generally from a regional gradient to about 35–70°C/km within geopressured zones (Jones, 1975; Peng, 1980; Kharaka et al., 1980).

4. Troughs within sedimentary basins are often characterized by anomalously high geothermal gradients. Examples are the Patchwarra and Nappamerri Trough in the Cooper basin, Australia (Kantsler et al., 1978) and the Danish Central Trough in the North Sea Basin (Jensenius, 1987; Jensenius and Munkgaard, 1989). In both basins, outside the troughs, the geothermal gradient is usually about 30°C/km. In contrast, within these troughs geothermal gradients are commonly higher, up to 50°C/km.

Geothermal Systems

The shape of geothermal systems is largely controlled by horizontal and vertical variations in lithology, mineralogy, and hydraulic properties of the host rock. The area of geothermal fields in sediment-filled rift valleys can be up to about 150 km^2 (Meidav et al., 1975; Mañón et al., 1977; DiPippo, 1987). The volume of hot fluid (>150°C) in moderate-temperature fields may be as small as 2.6 km^3 (e.g., East Mesa; Davies and Sanyal, 1979) but can be up to 123 km^3 (e.g., Westmoreland; DiPippo, 1987). For easier comparison of geothermal systems and sedimentary basins, it is useful here to divide geothermal systems into three depth zones: caprock, reservoir, and basement (Figure 1).

The lateral and vertical variation of geothermal gradients in geothermal systems depends primarily on the thickness of the caprock where there is conductive heat flow, and on the depth to the reservoir where there is convective heat and fluid flow. Both influxes of cool recharge water and the permeability distribution have additional influence on the temperature distribution in a geothermal field. As examples, temperature gradients of geothermal fields of the Salton Trough are idealized and summarized in Figure 2.

Above geothermal systems are usually unconsolidated, clastic sediments within which the near surface gradient can vary widely. The maximum temperature gradient of geothermal fields at the surface (depth of <10 m) is commonly between 180°C/km (e.g., East Mesa; Combs, 1971) and 830°C/km (e.g., Salton Sea; Newmark et al., 1988).

The caprock defining the top of geothermal systems consists typically of low-permeability sediments (lithological seals) and cemented horizons (chemical seals) that restrict convective fluid and heat flow (e.g., Younker et al., 1982; Adams, 1985; Elders, 1982; Moore and Adams, 1988). The caprocks of geothermal fields in the Salton Trough are mainly in sandstones with interbedded thick shale horizons. The depth of the caprock lies between 0.35 km (Dunes field; Elders and Bird, 1974) and approximately 2.5 km (Cerro Prieto field; Elders, 1979). In the Cerro Prieto field, the cap is not continuous and the reservoir consists of leaky aquifers where deeper shale units act as local caprocks (Halfman et al., 1982). However, an average caprock thickness of about 0.4–1.0 km, based on the Dunes, East Mesa, Heber, and Salton Sea geothermal fields, is representative (e.g., Randall, 1974; Hoagland, 1976; Bird and Elders, 1976; Elders and Cohen, 1983).

In the caprock, the temperature can reach 120–150°C in moderate-temperature geothermal fields and approximately 200–240°C in high-temperature fields (Pearson, 1976; Younker et al., 1982; Lippmann and Bodvarsson, 1985; Moore and Adams, 1988). Within the thermal cap above geothermal reservoirs, temperature gradients are usually highly variable, for example, from averages of about 135°C/km for the Heber field (Salveson and Cooper, 1981) to 380°C/km for the Salton Sea field (Elders and Cohen, 1983).

The caprock acts either as a barrier (e.g., Salton Sea field) or as a mixing zone (e.g., Dunes field) between shallow groundwater and deeper warm reservoir fluids. At the periphery of geothermal fields, there is often a reversed or reduced temperature gradient due to the lateral influx of cooler water from adjacent local aquifers. The temperature in this zone is typically 100–200°C. In the caprock, heat transfer is conductive but changes predominantly to convective in the underlying reservoir as shown by a rapid decrease in the geothermal gradient with depth (Figure 2). Fluid flow is mostly restricted to fractures and fault systems because of sealing by precipitation to form low permeability horizons, thereby preventing convective fluid overturn (Elders, 1982).

Below the caprock, the reservoir has different characteristics. The thickness of the reservoir ranges typically from 1.0 to 2.5 km (Figure 1). In moderate-temperature fields such as Heber (Salveson and Cooper, 1981), the average reservoir brine temperature is below 190°C but is up to 297°C in high-temperature geothermal fields (e.g., Cerro Prieto; Williams and Elders, 1984). Convective fluid flow within sandstone- and siltstone-dominated reservoirs is commonly assumed, based upon the small, linear geothermal gradient of most geothermal fields, as in Figure 2 (Younker et al., 1982; Lachenbruch et al., 1985; Newmark et al., 1988).

Hot brines for power generation are normally produced from reservoir depths between about 0.7 and 1.8 km. The bulk permeability is comparatively large at the upper and middle depths of reservoirs in the main production zones. Here, in the Salton Sea field, the sandstone permeability varies between 100 and 500 md (Morse and Thorsen, 1978), values representative of this type of geothermal field. Some geothermal reservoirs of the Salton Trough (e.g., Salton Sea, Cerro Prieto) are divided by shale barriers (Schroeder, 1976; Halfman et al., 1982) that may separate geochemically distinct reservoirs, as for example at the Cerro Prieto field (Grant et al., 1984; Truesdell and Lippmann, 1986).

Within the center of the reservoir, the temperature gradient is usually constant. Relatively high values consistent over large depth intervals occur in or near upflow zones of hydrothermal fluids if no major amount of cooler water is flowing laterally into the reservoir. Examples of representative temperature gradients typical for this zone are given in Figure 2 and Table 1. At the center of moderate-temperature reservoirs (170–200°C), the gradient is usually between 45°C/km (e.g., Heber field; Lippmann and Bodvarsson, 1985) and about 60°C/km (e.g., East Mesa field; Elders and Cohen, 1983). The temperature gradient is up to about 80–100°C/km in the central zone of the high-temperature geothermal fields at Cerro Prieto (Younker et al., 1982) and at the Salton Sea (Williams and Elders, 1984). However, lower gradients of about 40°C/km (Fournier, 1988) are representative and more common, for example, as observed at the Salton Sea field (e.g., Well State 2-14; Sass et al., 1987).

Comparison of Basins and Geothermal Systems

Comparison of thermal regimes of sedimentary basins and geothermal systems (Figures 2, 4, Table 1) provides the following conclusions.

1. The mean surface heat flow and subsurface temperatures at comparable depth are greater in geothermal systems than in sedimentary basins, but the change of temperature with depth is very similar in both environments if comparing gradients below the caprock of geothermal systems with gradients of sedimentary basins. Therefore, the temperature and geothermal gradient in geothermal reservoirs and sedimentary basins is often similar when the thermal structures of geothermal reservoir rock sequences at shallow depths are compared.

2. The mean temperature gradients of sedimentary basins, 10–70°C/km, are comparable to the temperature gradients found in the center of shallow, moderate-temperature geothermal reservoirs, <80°C/km. For example, a temperature interval of about 120–190°C corresponds to a depth of 2.2–3.8 km in sedimentary basins and to a depth of 0.6–2.2 km in the upflow region of moderate-temperature geothermal reservoirs. Here, the temperature gradient is 45°C/km within the 120–190°C interval of a warm sedimentary basin and also in the center of a moderate-temperature geothermal system. Only the depth to this temperature interval is different in the two environments. The difference in temperatures is, therefore, caused only by a significantly higher temperature gradient in the caprock of the geothermal system.

3. The temperature gradients of geothermal reservoirs, where large amounts of cool water are recharged, are often very low or negative. Comparable thermal structures are apparently rare or absent in sedimentary basins.

4. The temperatures and gradients found in geothermal reservoirs are especially similar to thermal conditions found in the vicinity of salt domes (e.g., Danish Trough), to upflow zones along faults in basins (e.g., Wilcox trend), and to deep basins where dewatering takes place along permeable strata (e.g., Alberta basin).

5. The temperature gradients of active geothermal fields are similar to gradients that are estimated for hydrothermally altered sediments of ancient sedimentary rift basins. Hydrothermally altered sediments of the Gabon rift basin indicate a gradient of 100–200°C/km. This range is comparable to present gradi-

Table 1. Representative modern mean temperature gradients of geothermal systems and sedimentary basins.

Geothermal Systems	Depth (km)	Temperature gradient (°C/km) Min.	Max.	Source
Near surface	<10	180	830	[1]
Caprock	<0.4–1.0	135	380	[2]
Reservoir	>0.4–1.0	15–60	<80–100	[3]
Reservoir temp. (avg.)		170–210°C	>250°C	

Basins	Temperature Gradients (°C/km) Basinwide	Maximum at Anomaly	Possible Cause	Source
Pannonian basin	39–69	69	upflow	[4]
Gulf Coast Basin	20–40	70	upflow/overpr.	[5]
North Sea Basin	18–40	50	diapirs/overpr.	[6]
Niger Delta	16–35	—	—	[7]
Cooper basin	30–50	55	troughs	[8]
Mugland rift basin	13–33	—	—	[9]
Aquitaine basin	10–20	50	very conductive reef	[10]

Data sources: [1] Combs (1971), Newmark et al. (1988); [2] Salveson and Cooper (1981), Elders and Cohen (1983); [3] Lippmann and Bodvarsson (1985), Elders and Cohen (1983), Younker et al. (1982), Williams and Elders (1984); [4] Horvath et al. (1979); [5] Jones (1975), Kharaka et al. (1980), Bodner and Sharp (1988); [6] Robert (1988, p. 35), Liewig et al. (1987), Jensenius (1987); [7] Robert (1988, p.58); [8] Kantsler et al. (1978); [9] Schull (1988); [10] Robert (1988, p. 34).

ents of 250°C/km (Salton Sea field) and about 80°C/km (East Mesa field) in geothermal upflow zones averaged over the drilled depth range.

Fluid Composition

Geothermal fluids of the Salton Trough are predominantly meteoric water flowing into the systems along deep aquifers. Any fluids from magmatic sources contribute insignificantly to the composition of the geothermal fluid, as indicated by isotopic data (Elders and Cohen, 1983). In contrast, the fluids of sedimentary basins may be a mixture of recent or older meteoric water and diagenetic and connate waters (Kharaka and Thordsen, 1992) with residence times, in general, several orders of magnitude longer than those of geothermal fluids.

The compositions of the solutions in the two types of systems are related. The inorganic solute content of the Salton reservoir is strongly influenced by evaporation processes and dissolution of evaporates. Both processes are often also important for the evolution of basinal brines. For example, highly saline fluids that are found in the Salton Sea geothermal field and in various sedimentary basins are predominantly the result of halite dissolution (McKibben et al., 1988a; Kharaka and Thordsen, 1992). However, there may also be large differences between geothermal and basinal fluid chemistries. For example, reactive organic species up to 10,000 mg/l contribute significantly to the total dissolved species in basinal fluids (Kharaka et al., 1986; Surdam et al., 1989). Therefore, biological and organic reactions play an important role in sedimentary basins. In contrast, except for methane, hydrocarbons are absent in this type of geothermal system. The pH of geothermal fluids and basinal brines is often similar (4.5–5.5), but the buffer systems that fix the pH may be different in the two environments. The pH in geothermal systems is often controlled by mineral buffer systems, like the feldspar-quartz-muscovite buffer. In contrast, organic buffer systems may control the pH of the fluid in sedimentary basins (Surdam et al., 1989; Lundegard and Kharaka, 1990).

The discussion of the salinity distribution in geothermal systems and sedimentary basins illustrates the great variability of many geochemical parameters in both environments. Both maximum reservoir temperatures and correlated average salinities are plotted in Figure 5 for geothermal systems. Included are salinities from six different geothermal systems in sediments and from ten geothermal fields located in volcanics-dominated rift valleys of the Taupo Fault Zone, New Zealand. A representative geothermal

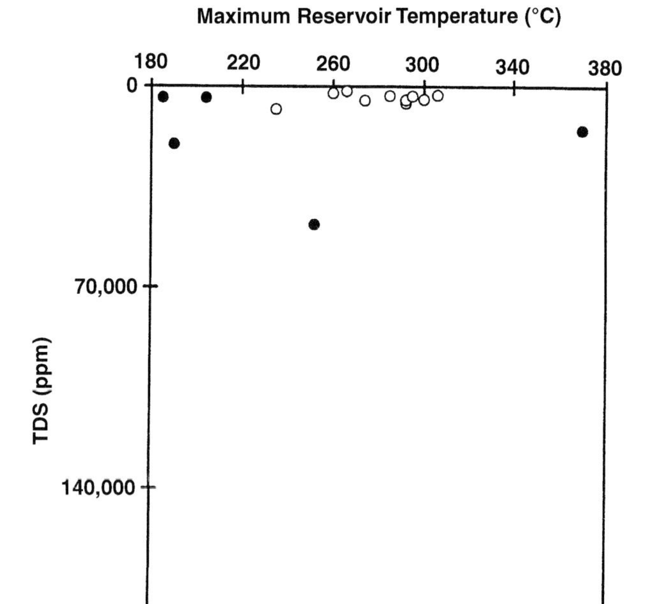

Figure 5. Average salinity and maximum reservoir temperature in sediment-associated (●) (Coplen, 1976; Olsen and Matlick, 1978; Ershaghi and Abdassah, 1983; Elders and Cohen, 1983; Elders et al., 1984) and volcanics-dominated (O) (Ellis, 1979) geothermal fields.

range for the average of total dissolved solids (TDS) in the sediment-dominated type is 4000 ppm (e.g., East Mesa; Coplen, 1976) to 200,000 ppm (e.g., Salton Sea; Ermak, 1977). It is likely that the deep, hypersaline brines which are typical of most of the geothermal fields of the Salton Trough originated from partially evaporated and downward-percolated water. The water source was either the formerly existing lake or the local groundwater (Crosby et al., 1972; Rex, 1985). Additionally, dissolution of evaporites from the deltaic sediments contributed to the high salinity of the deep brines (McKibben et al., 1986, 1988a).

The large variability in salinity of high- and low-temperature geothermal reservoirs is illustrated by the Salton Sea (TDS: 50,000–270,000 ppm) and East Mesa systems (TDS: 1,500–24,000 ppm) (Coplen, 1976; McKibben et al., 1988a). The content of dissolved solids in basinal fluids is also highly variable. In general, the fluid salinity increases with depth until a maximum value is reached. The salinity usually stays constant below this depth. The salinity of fluids in deep sedimentary basins can be as low as X,000 ppm TDS. In contrast, in the Michigan basin deep saline brines contain up to 400,000 ppm TDS. Interfaces where large salinity changes occur over a small depth interval are common in both geothermal systems and sedimentary basins. Typical examples are at the top of overpressured compartments in the Gulf Coast Basin (Jones, 1975; Morton et al., 1981; Morton and Land, 1987) and in the Salton Sea field where at a depth of 1–1.5 km, rapid changes in salinity from X,000 to 200,000 ppm TDS have been observed (McKibben et al., 1988a, 1988b; Williams and McKibben, 1989). This large variation is typical of both sedimentary basins and geothermal fields that are located in sedimentary environments. It is considerably larger than within the fields of the Taupo Fault Zone (Ellis, 1979), where total dissolved solids are in general below 10,000 ppm.

FLUID FLOW IN GEOTHERMAL SYSTEMS

Fluid flow in geothermal fields is predominantly controlled by the distribution of the porosity, permeability, fractures, faults, temperature, and fluid density within the caprock, reservoir, and basement. Characteristics of fluid flow, including fluid and reservoir volume, specific discharge, and distance to the recharge area are summarized in Figures 6 and 7.

The main recharge area of the buried deltaic sediments of the Salton Trough is probably near where the Colorado River enters the rift valley (Elders et al., 1984). This water replenishes the deep aquifers that are the main water supply of the geothermal fields (Figure 6). The distance from the recharge area to the different hydrothermal systems of the trough is about 50–80 km. Near thermal anomalies, the deep aquifer water is heated and forms a water plume rising due to the expansion of the fluid. At least in some of the geothermal reservoirs of the Salton Trough (e.g., Cerro Prieto, East Mesa), the upflow of hot brines into the reservoir takes place preferentially where the vertical permeability of the basement and lower reservoir is enhanced by faults and fractures (e.g., Bailey, 1977; Halfman et al., 1986).

Significant convective heat transport seems to be required in reservoirs to account for the high heat flux through caprocks. However, the shape and scale of convection in the Salton Sea field, for example, is uncertain and the proposed convection models differ considerably (e.g., Younker et al., 1982; Kasameyer et al., 1984; Rex, 1985; Fournier, 1988).

The size of the flow through the reservoirs (Figures 6 and 7) is indicated by a water/rock volume ratio of about 3:1 to 2:1 for Cerro Prieto (Williams and Elders, 1984) and a ratio of about 1:1 for the Salton Sea (Clayton et al., 1968) and East Mesa (Coplen, 1976) fields. These water/rock ratios imply that the flow through the geothermal reservoirs was fast enough to allow the reservoir rock to react with up to three times larger volume of hot geothermal fluid (e. g., Cerro Prieto) within the relatively short lifetime of the system. The integrated fluid flow gives a specific discharge through the reservoir of about 0.6 m/yr for the Cerro Prieto (Elders, 1982; Elders et al., 1984) and East Mesa fields (Kassoy, 1975). This estimate of the specific discharge of the Cerro Prieto field is based on a reservoir area of 6 km^2 and a flow of 3.6×10^6 m^3/yr

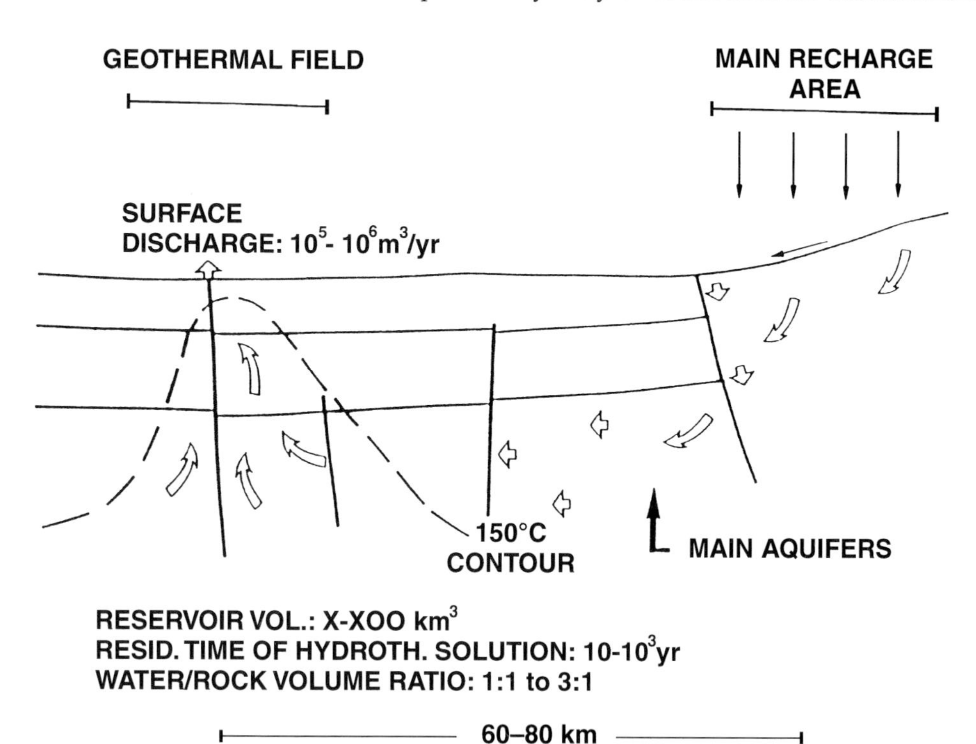

Figure 6. Representative flow data for a recharge area and a geothermal field.

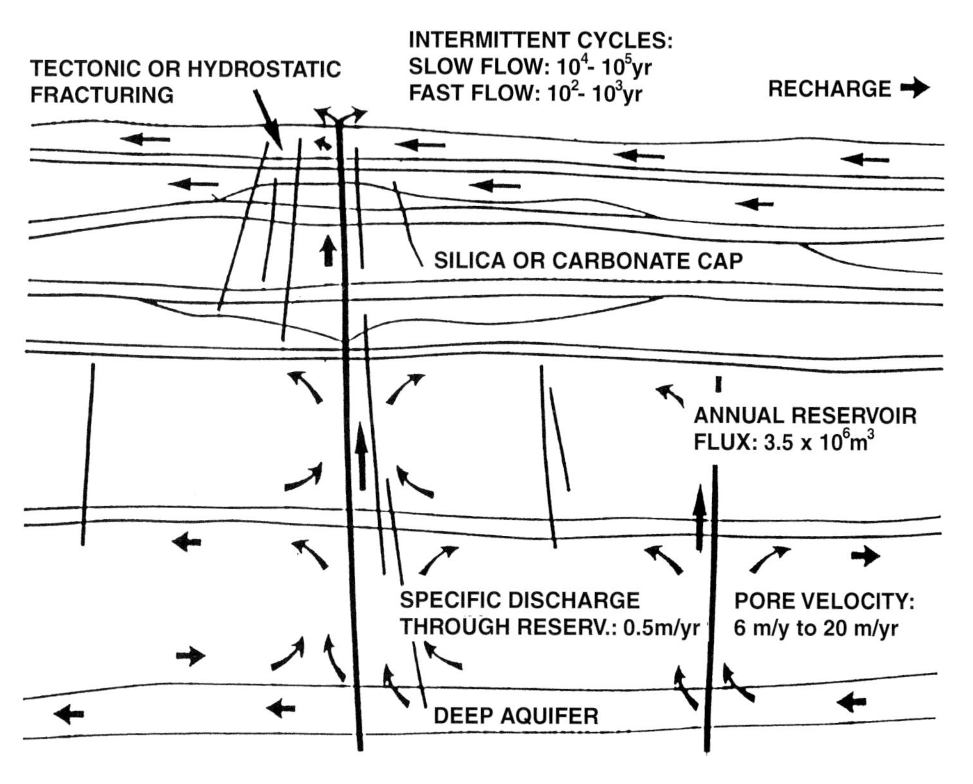

Figure 7. Schematic flow relationships in geothermal fields.

through the reservoir, lasting 10,000 yr (Elders et al., 1984). McKibben and Williams (1990) estimate that the upflow volume of metalliferous brine is about 2×10^8 m^3/yr in the Salton Sea geothermal field. The pore velocity of the geothermal fluid is about 6 m/yr at Cerro Prieto and up to about 20 m/yr at the East Mesa field, using the estimated average reservoir porosities listed in Table 2. The observed range of 6–20 m/yr for the reservoir pore velocity may also be representative of other fields in the Salton Trough.

The upflow of reservoir fluid into the caprock zones is highly restricted by low-permeability shales

Table 2. Porosity (%) in geothermal systems.

	Salton Sea	Cerro Prieto	Heber	East Mesa
Caprock	—	—	15–30 [1]	33 [2]
Base of caprock	15–20 [3]	—	—	—
Upper reservoir	20 [4]	—	—	—
Lower reservoir	11 [5]	—	10 [6]	—
Reservoir (avg.)	14 [7]	10 [8]	23–30 [9]	23 [10]

Data sources: [1] Ershagi and Abdassah (1983); [2] calculated here from wire data given in Pearson (1976); [3] McDowell and Elders (1980); [4] Schroeder (1976); [5], [7] calculated here from data given in Younker et al. (1982), McDowell and Elders (1980), Tewhey (1977); [6] Ershagi and Abdassah (1983); [8] Elders et al. (1981); [9] Bodvarrsson and Witherspoon (1989); [10] calculated here from wire data given in Pearson (1976).

and cemented horizons. Therefore, heated water from shallow regional aquifers contributes most to the surface discharge of fluid, while only a small fraction results from upflow of reservoir fluid. A comparison of recent fluid discharges at the surfaces of the geothermal fields of the Salton Trough reveals large differences in flow volumes. The largest discharge, about 5×10^6 m^3/yr, is reported from the Cerro Prieto geothermal field (Truesdell et al., 1984). Moderate-temperature systems, like the Westmoreland or Dunes fields, usually have no surface discharge. However, the surface discharge can change quickly. For example, the present discharge of about 3.9×10^4 m^3/yr at the Salton Sea geothermal field was considerably larger in the last century (Muffler and White, 1969).

The Duration of Geothermal Systems

Investigations of the history of the continental crust beneath the Salton Trough show that the thermal anomalies of the Salton Trough might be at least 20,000 yr old (Kasameyer et al., 1984; Kasameyer and Hearst, 1988). However, the present systems seem to be younger as indicated by several types of evidence. Fission-track-annealing studies of apatite show that a reservoir temperature of 160–180°C was reached less than 10,000 yr ago at Cerro Prieto (Stanford, 1981; Elders et al., 1984). The estimated age of the Salton Sea geothermal system is about 16,000 yr (Muffler and White, 1969; White, 1981) and of the East Mesa field about 10,000 yr. There is strong evidence (e.g., heating rates, cementation events, brine residence time, change of surface discharge) that the hydraulic, thermal, and geochemical conditions changed considerably during the life of these geothermal systems (e.g., Skinner et al., 1967; Andes and McKibben, 1987). A change of fluid flow and fluid chemistry is reflected by several cementation events and fracture-filling periods in most fields of the Salton Trough. In the vicinity of the reservoir, temperature changes can be especially fast due to variation in volume and temperature of the discharged or recharged water. On the west side of the Cerro Prieto field, the reservoir temperature increased by 50–100°C over 1–10 yr, for example (Stanford, 1981).

Detailed studies of fractures and their mineralogy in the Dunes (Bird and Elders, 1976), Cerro Prieto (Elders, 1982), and Salton Sea fields (McKibben et al., 1988b) reveal that they have been episodically reopened, allowing spurts of chemically and thermally unequilibrated fluid into the upper reservoir (e.g., Salton Sea field) or caprock (e.g., Dunes field). The mixing of upper reservoir water with upflowing deeper water causes precipitation, especially of carbonates and quartz, and favors resealing of fractures. Subsequently, upward fluid flow shrinks due to a gradual decrease of fracture and matrix permeability. The cause of the episodic fracturing that periodically increases the permeability of geothermal systems (Facca and Tonani, 1967; Elders, 1982) is uncertain. Periods of increased seismic activity and fracture propagation may reflect intrusive events, compaction, or release of overpressure in the reservoir (Elders, 1982).

Both hydrostatic (e.g., Salton Sea, East Mesa, and Heber field) and overpressured (Cerro Prieto) reservoirs occur in the Salton Trough (Helgeson, 1968; Elders, 1982; Halfman et al., 1986). At present, fluid flow is relatively fast due to open fractures in the Salton Sea geothermal field. Consistent with episodic self-sealing and subsequent fracturing, this fast fluid flow governs a short brine residence time of about 10^2–10^3 yr (Zukin et al., 1987) and is producing fracture-filling precipitates that are significantly different from older precipitated mineral assemblages (McKibben et al., 1988a, 1988b; Williams and McKibben, 1989). In contrast, the Cerro Prieto field has lithostatic pressure at a shallow depth of 3.6 km (Bermejo et al., 1979; Lippmann and Bodvarsson, 1983; Ayuso, 1984; Halfman et al., 1986), and fluid flow is limited by partially closed fractures.

Comparison of Fluid Flow in Sedimentary Basins and Geothermal Systems

Modeling studies of the subsurface flow in sedimentary basins show that the flow rates typical of sedimentary basins are normal for geothermal systems. Flow rates in sedimentary basins, where compaction-driven flow is dominant, are usually lower than 0.01 m/yr (Bethke, 1986). In contrast, flow rates in sedimentary basins vary over a wide range where fluid flow is predominantly gravity driven (Garven and Freeze, 1984a, b; Bethke, 1986). Within parts of a sedimentary basin, gravity-driven flow may reach a velocity of up to several m/yr. Therefore, the residence time of the fluid in sedimentary basins is in general several orders of magnitude longer than in geothermal fields due to the large recharge and discharge distances and the overall slower flow rates in sedimentary basins.

Overpressured compartments are reported from both geothermal fields and sedimentary basins. In deep, overpressured compartments of the Gulf Coast, pressures up to 1000 bars have been measured (Kharaka et al., 1985). The pressure is lithostatic at a depth of 3.6 km in the Cerro Prieto geothermal field (Bermejo et al., 1979; Lippmann and Bodvarsson, 1983).

The fluid flow characteristics of geothermal systems are similar to those in the upflow regions of sedimentary basins. Similar pore velocities of up to several m/yr can be found in regions with (1) thermohaline-induced convection and upflow along salt diapirs (e.g., North Sea and Gulf Coast basins) (Rabinowicz et al., 1985; Hanor, 1987; Bethke, 1989), (2) upflow along normal and growth faults (e.g., Gulf Coast; Tyler et al., 1985; Liewig et al., 1987), and (3) discharge regions of continental basins (Cathles and Smith, 1983; Garven and Freeze, 1984a).

Flow rates associated with convection cells near salt diapirs are of magnitudes similar to the flow rates observed in geothermal fields. For example, the flow rate is about 10^6 m^3/yr in an upflow zone associated with a salt diapir (Rabinowicz et al., 1985) and is about 3.6×10^6 m^3/yr in the Cerro Prieto geothermal field (Elders et al., 1984). The fluid flow along faults during episodic dewatering of overpressured compartments, as proposed by Cathles and Smith (1983), may have an order of magnitude similar to the fluid flow during the episodic fracturing of the caprock of geothermal systems. The upflow of deep, unequilibrated brines in sedimentary basins can result in the cementation of faults by quartz or carbonates similarly to the episodic self-sealing processes observed in geothermal systems.

PERMEABILITY AND POROSITY IN GEOTHERMAL FIELDS

Bulk porosity and permeability of these fields reflect mainly lithologic variations and the effects of fracturing, alteration, dissolution, and cementation mostly in pores of rocks. Therefore, a comparison of porosities and permeabilities of unaltered sediments around geothermal systems with those of hydrothermally altered sediments reveals the effects of hot brines in cooler sedimentary environments. Especially informative are investigations of the hydrothermally altered sediments of the geothermal systems in the Salton Trough. Averaged values of the porosity and permeability of the caprock, reservoir, and basement of several geothermal fields are given in Tables 2 and 3.

All geothermal fields of the Salton Trough are associated with small, positive gravity anomalies (Rex, 1966). The anomalies are a few kilometers deep and reflect densification of the sediments by water/rock interactions (Meidav and Rex, 1970). These anomalies occur in the Cerro Prieto (Lyons and van de Kamp, 1980), Salton Sea (Seamount and Elders, 1981; Sass et al., 1988) and East Mesa fields (Biehler, 1971; Goldstein and Carle, 1986), for example. At East

Table 3. Permeability and transmissivity in geothermal reservoirs.

	Permeability (md)			
	horiz.	vert.	Transmissivity (Dm)	Source
Salton Sea				
Upper reservoir sandstones	~500	—	—	[1]
Reservoir shales	—	0.1–1.0	—	[2]
Reservoir	100–500	—	—	[2]
East Mesa				
Reservoir (avg.)	170	45	—	[3]
Reservoir	—	—	3–12	[4]
Cerro Prieto				
Reservoir sandstones	~100	~1–10	—	[5]
Reservoir	—	—	4–40	[4]
Heber				
Reservoir	5–125	—	12–60	[4]

Data sources: [1] Schroeder (1976), Riney et al. (1979); [2] Morse and Thorsen (1978); [3] calculated here from wire log data given in Pearson (1976); [4] Bodvarsson and Witherspoon (1989); [5] Halfman et al. (1986).

Mesa, the excess mass is approximately 10^{12}–10^{13} kg (Meidav et al., 1975). In detail, the local gravity anomalies in the geothermal fields of the Salton Trough are due to the precipitation of minerals and progressive metamorphism of the clastic sediments. The hydrothermal alteration of clastic sediments was carefully studied for the Salton Sea (e.g., Muffler and White, 1969, McDowell and Elders, 1980, 1983; Bird et al., 1984; McKibben and Elders, 1985; Yau et al., 1986, 1988), Cerro Prieto (e.g., Elders, 1979, 1984; Schiffmann et al., 1984), and East Mesa geothermal fields (Hoagland, 1976; Fournier, 1976). Petrologic studies show that the extent of rock alteration is predominantly a function of initial permeability, temperature, and bulk rock composition. Therefore, distinct mineral zones that are defined by dominant minerals (e.g., illite, chlorite) often follow the isothermal surfaces which characterize the temperature distribution in each geothermal field. Rock permeability has a significant influence on the progress of mineral alteration. In permeable metasandstones and metasiltstones, reactions between phyllosilicates and changes in their textures and composition all take place at lower temperatures than the alteration of shale sequences (Hoagland, 1976; McDowell and Elders, 1979, 1980).

Caprocks

In the Salton Trough, the caprocks of the geothermal fields commonly consist of sandstone and siltstone sequences that have a relatively large average porosity (>30%) and thick shale horizons. In most geothermal systems of the Salton Trough (e.g., Salton Sea and Cerro Prieto) the vertical bulk permeability of the caprocks is initially low, because of thick, clay-rich horizons. Further decrease in permeability is caused by hydrothermal carbonate or quartz cementation that can readily seal regional aquifers contributing to the low vertical bulk permeability of the caprocks.

For example, in the upper 0.3 km of the Dunes geothermal field, seven thoroughly quartz-cemented horizons form low-porosity, impermeable barriers in the elsewhere loosely consolidated caprock. This advanced sealing of the initially permeable aquifer horizons causes an increase of the rock density from 2.2 to 2.6 g/cm^3 and a decrease in porosity to as low as 3% (Elders, 1982). The extensive cementation is caused by episodic precipitation of quartz from hot brines that flowed horizontally along aquifers into cooler sections of the caprock (Elders and Bird, 1974; Coplen, 1976). The sealing of the Dunes caprock is so effective that no hot springs reveal the higher-temperature fluids that are present in the reservoir.

Not all caprocks are highly impermeable, however. At the East Mesa field, the caprock consists mainly of relatively uncemented and unconsolidated clay-rich sand and siltstones (Hoagland, 1976; Coplen, 1976) down as deep as 0.85 km (Pearson, 1976). Detailed porosity and permeability data, derived from well logs and core logs of eight wells of the East Mesa field (in Pearson, 1976) give an average bulk porosity of about 33% (Figure 8 and Table 2) and an average vertical bulk permeability of about 354 md (Figure 10 and Table 2). The petrographic study of well 6-1 in the East Mesa field by Hoagland (1976) shows minor cementation of the caprock sediments consistent with the small decrease in porosity, but significant decrease in both horizontal and vertical permeabilities as shown by Figures 9 and 10.

The zone between the caprock and reservoir (0.4–1.0 km, ~200–240°C) in high-temperature fields is usually extensively cemented due to the mixing of hydrothermal reservoir fluid and cooler aquifer fluids that laterally recharge the reservoir. In the Salton Sea field at a depth of about 0.4–0.9 km (Randall, 1974), a narrow zone between the cap and reservoir rock is highly cemented with a porosity now less than 15–20%. Below this cemented zone in the reservoir, the deeper porosity remains at approximately 30%, a value typical of the uppermost part of the reservoir (McDowell and Elders, 1979, 1980). Here in this narrow zone intense carbonate cementation (calcite, dolomite, and ankerite) prevents fluid circulation and preserves biotite, muscovite, and chlorite, minerals that otherwise are destroyed by reactions above and below this zone (McDowell and Elders, 1983). The upflow of warm hydrothermal fluid from the reservoir into upper parts of the caprock is highly dependent on the presence of open faults and fractures because of the zones of low vertical caprock permeability (Elders, 1982). Therefore, surface discharge through hot springs is present above only some geothermal fields (e.g., Salton Sea, Cerro Prieto).

A comparison of the clay alteration within sedimentary basins and geothermal fields can contribute significantly to our understanding of diagenesis. For example, there is textural evidence that the mechanism for the smectite-to-illite reaction is different in the caprock and reservoir shales due to the different degree of water/rock interactions (Yau et al., 1986). Illite in shales of low-permeability caprock seems to be a product of direct smectite replacement, which is similar to that found in the low-permeability shales of the Gulf Coast. In contrast, within the more-permeable reservoir shales, smectite is being dissolved and illite seems to result from a direct precipitation from solution which may reflect the significant higher fluid flow and water/rock ratio in the reservoir (Yau et al., 1986, 1988).

Reservoirs

The porosity and permeability of geothermal reservoirs decrease markedly with increasing depth in all geothermal fields of the Salton Trough (Ershaghi and Abdassah, 1983). Data on the bulk porosity of several geothermal reservoirs show that a range from 30% to 3% is typical. The average reservoir porosity is usually between 10 and 23% (Table 2). High-temperature reservoirs like that at the Salton Sea ($T_{avg.}$ = 285°C) or Cerro Prieto ($T_{avg.}$ = 297°C) have a slightly lower average reservoir porosity of about 10–15% (Tewhey, 1977; McDowell and Elders, 1980; Elders, 1982; Elders

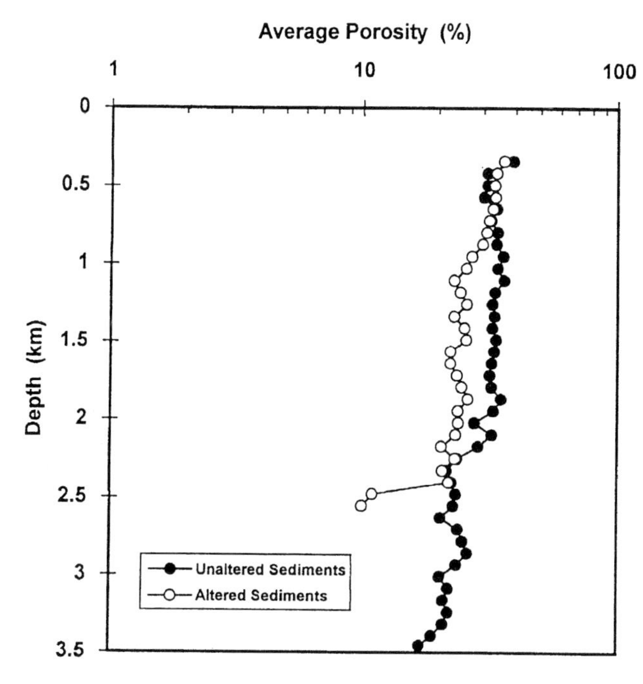

Figure 8. Porosity of the East Mesa field (based on 8 wells) and adjacent sediments (based on 4 wells). After Polster and Barnes (1990). Data source: Pearson (1976).

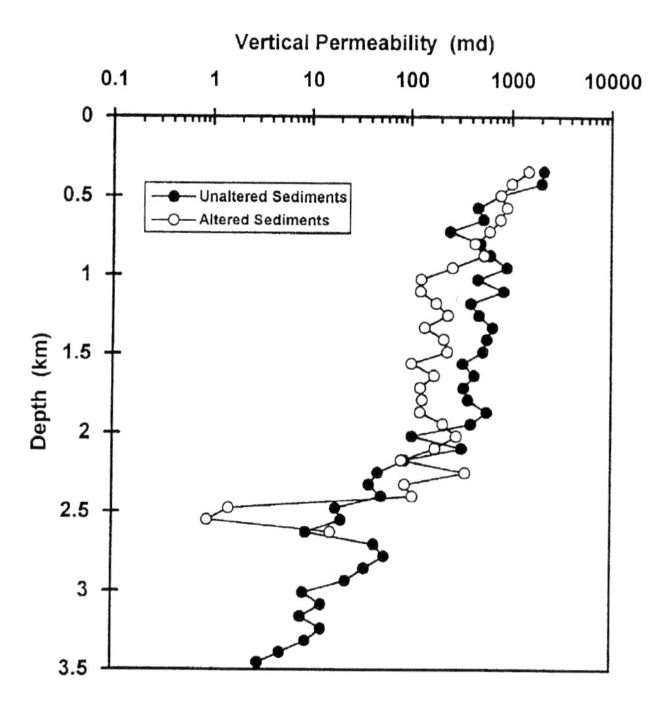

Figure 10. Vertical permeability of the East Mesa field (based on 8 wells) and adjacent sediments (based on 4 wells). After Polster and Barnes (1990). Data source: Pearson (1976).

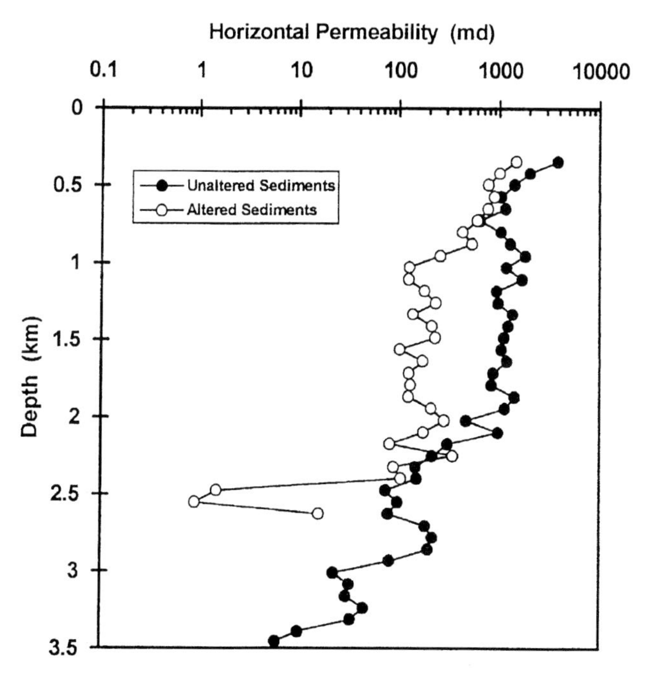

Figure 9. Horizontal permeability of the East Mesa field (based on 8 wells) and adjacent sediments (based on 4 wells). After Polster and Barnes (1990). Data source: Pearson (1976).

and Cohen, 1983), which is lower than the average of about 23% for the moderate-temperature systems ($T_{max.}$= 204°C; Smith, 1979) at East Mesa. There is apparently a greater reduction of pore space in reservoir rocks of high-temperature fields resulting from the combined effects of alteration, precipitation, and dissolution reactions.

Representative bulk rock permeabilities of geothermal reservoirs vary within 5–500 md (Table 4). Highly permeable reservoir rocks, like the sandstone units at the Salton Sea or East Mesa, have a horizontal permeability up to about 500 md (Schroeder, 1976; Pearson, 1976; Riney et al., 1979). Pressure-transient tests in the East Mesa field show that the lateral hydraulic continuity of producing aquifers can be poor even over short distances of X00 m (Howard et al., 1978). Such compartmentation indicates that horizontal flow is restricted in parts of the reservoir by faults or lithological variations.

The vertical permeability of reservoir rocks is especially low in shales as shown by Table 3. Morse and Thorsen (1978) estimate the vertical permeability of the shales of the Salton Sea reservoir to be 0.1–1.0 md, and Halfman et al. (1986) find a corresponding vertical permeability of 0.005–0.1 md for the shales of the Cerro Prieto reservoir. Large-scale convection of hydrothermal fluids in geothermal reservoirs is, therefore, unlikely in clastic sediments with high shale contents or thick, continuous shale layers. Vertical fluid flow may be even more restricted in reservoir rocks with a relatively low bulk permeability such as Cerro Prieto (10–30 md) or Heber (5–125 md) (Lippmann and Bodvarsson, 1985; Bodvarsson and Witherspoon, 1989), especially in zones where fractures contribute little to the upflow of fluids. Locally, precipitation due to boiling may further

Table 4. Horizontal and vertical permeability (k, in md) and porosity (Φ, in %) of the East Mesa field (based on 8 wells) and adjacent wells (based on 4 wells).

Permeability (md):	Average			Average Min. /Max.		
	(k_h)	(k_v)	(Φ)	(k_h)	(k_v)	(Φ)
Depth: 0.3–0.84 km						
East Mesa: (caprock)	808	354	33	431–1469	166–785	31–36
Sediments: (adjacent)	1270	714	33	643–3850	246–2106	30–39
Depth: 0.84–2.59 km						
East Mesa: (reservoir)	167	44	23	0.9–532	0.4–231	10–31
Sediments: (adjacent)	957	404	30	6–1827	3–898	17–36

Data source: Pearson (1976).

reduce rock permeability in the deep aquifers of the Cerro Prieto geothermal field (Truesdell et al., 1984).

A comparison of the extent of diagenetic reactions in geothermal fields and surrounding nonthermal sediments can give valuable information about the kinetics of mineral reactions. For example, the conversion of smectite to illite in interstratified clays is almost complete at about 180–200°C (90–95% illite) in the Salton Sea geothermal field and also in the adjacent nonthermal sediments (McDowell and Elders, 1980, 1983; Yau et al., 1986). This clearly shows that the time necessary for the smectite-to-illite conversion is shorter than 16,000 yr under the hydraulic and geochemical conditions that exist in the Salton Sea geothermal field.

Faults and fractures, either tectonic or hydraulic, can enhance the rates of upflow of geothermal fluid considerably (Helgeson, 1968; Morse and Thorsen, 1978; Schroeder, 1976; Younker et al., 1982). For example, the size of the upflow zone and the shape of the hot water plume of East Mesa (Combs and Hadley, 1977; Riney et al., 1980; Goyal and Kassoy, 1981; Goldstein and Carle, 1986) and of the Cerro Prieto field (Lippmann, 1982; Truesdell and Lippmann, 1986) are highly influenced by the locations of regional faults.

HYDROTHERMAL EFFECTS IN MODERATE-TEMPERATURE GEOTHERMAL FIELDS: IMPLICATIONS FOR SEDIMENTARY BASINS

Detailed porosity and permeability data are available for useful comparisons. For example, Table 4 and Figures 8, 9, and 10 are porosity and permeability data calculated from wire logs from 12 wells in the East Mesa area. These porosity and permeability data are given in Pearson (1976) for eight different hydrothermal and four nonthermal wells recalculated here into averages over 76 m intervals. The type of wire logs used and methods of sandstone/shale discrimination are given in Pearson (1976) and are discussed in Howard et al. (1978). The caprock of the East Mesa field, 0.61–0.76 km thick ($T < 120$°C), consists mainly of shales and unconsolidated sandstones and extends to a depth of 0.85 km (Pearson, 1976). The average porosity is about 33% in the caprock and at the adjacent depth in the nearby sediments. The influence of the hydrothermal fluid on the caprock is small, causing a slightly smaller maximum porosity (36%) compared to the surrounding, nonthermal sediments (39%), consistent with intensive cementation of sandstones only at greater depths below 0.64 km (Hoagland, 1976). However, this small degree of cementation of the caprock results in a significant decrease in the vertical permeability, as we have documented above.

The difference in the average reservoir porosity of the East Mesa field (~23%) compared to the adjacent nonthermal wells (~30%) indicates a significant hydrothermal pore-space reduction in this geothermal reservoir (Table 4). The porosity versus depth trend in the surrounding sediments does not show a significant decrease of porosity due to compaction within the upper 1.8 km (Figure 8). Especially remarkable is the reduction of the average vertical permeability in the reservoir to about 44 md when compared to that of the surrounding sediments (~404 md). The average vertical permeability within the reservoir (Figure 10) drops from about 15 md at 1.0 km depth to values below 0.01 md in the deeper reservoir zones (below 2.4 km), limiting major fluid transport to horizontal flow or flow along fractures. In the East Mesa reservoir, the increased permeability due to fractures is confirmed by hydraulic well tests. The 60–120 m thick aquifer below a depth of 1.25 km has a permeability of 110–220 md when estimated from hydraulic well tests, which is considerably larger than the permeability of about 30 md expected from logs (Pearson, 1976; Howard et al., 1978). The observed difference in permeabilities can be interpreted as fracture permeability because the logging tools used take only the matrix permeability into account. Porosity and permeability trends in both geothermal and nonthermal wells show that geothermal fluids at moderate temperatures (~100–200°C) dramatically reduce reservoir rock permeability and porosity. The reservoir of the East Mesa geothermal field has an initial silicate–sulfide alteration followed by a later carbonate precipitation (Hoagland, 1976). The initial hydrothermal event reduced pore space by precipitation of quartz in sandstones and the precipitation of illite and chlorite in shales from a reducing fluid also depositing minor pyrite. Pre-existing carbonates were partly dissolved during this stage. The alteration of sandstones is usually more intense than the alteration in shales due to differences in their initial permeabilities. The degree of alteration is, therefore, highly dependent on access of the hydrothermal fluid to the sedimentary section. An extensive precipitation of authigenic calcite postdates the silicate–sulfide alteration and preferentially reduces the pore space where the initial quartz cementation had only a minor effect on the initial porosity.

SUMMARY AND CONCLUSIONS

Geothermal gradients of 10–70°C/km in various sedimentary basins are similar to those of geothermal reservoirs, 15–80°C/km, although the mean surface heat flow is considerably larger in geothermal systems than in sedimentary basins. The same temperatures are reached at shallower depths in geothermal systems because of the higher temperature gradients in their caprocks. The gradient is very similar in both environments when comparing zones below the caprock near the center of geothermal systems with equivalent zones in sedimentary basins. For example, a temperature gradient of 45°C/km and a temperature interval of about 120–190°C are characteristic of both a depth interval in warm sedimentary basins of 2.2–3.8 km and a corresponding depth interval of 0.6–2.2 km in the upflow regions of moderate-temperature geothermal reservoirs. The thermal regimes in these comparable zones are similar, but the hydrostatic fluid pressure is considerably higher in the equivalent section of sedimentary basins. Both geothermal systems and compartments in sedimentary basins can be considerably overpressured.

Similar to upflow zones in geothermal systems, upflow zones of hot brines are found in sedimentary basins (1) along tectonic and growth faults, (2) in regions with thermohaline-convection and upflow along piercing salt diapirs, and (3) along deep and permeable strata. Hydrodynamic conditions are comparable in upflow zones of sedimentary basins and geothermal systems as indicated by similar flow rates and pore velocities.

Upflow zones in sedimentary basins are usually associated with regional thermal anomalies. There, the temperatures and temperature gradients (up to 70°C/km) are similar to those found in the center of geothermal reservoirs. Geothermal fields with moderate reservoir temperatures (~120–200°C) are analogs to upflow regions in deeper sedimentary basins in terms of fluid flow and thermal state. Therefore, diagenetic alteration in clastic sediments that are a result of the influx of hot brines is often very similar in both environments.

An influx of hot brines in both geothermal systems and sedimentary basins at moderate temperatures, 120–200°C, is commonly revealed by (1) increased sediment maturity, such as advanced illitization, (2) precipitation or dissolution of quartz and/or carbonates, (3) fluid inclusion data proving high geothermal gradients (up to 60°C/km) and a change in water chemistry, and (4) a change of rock permeability.

There is convincing evidence from reservoirs in moderate-temperature geothermal systems and from upflow zones in sedimentary basins that an influx of hot brines into clastic sediments can result in a significant reduction of rock porosity and permeability. In the East Mesa geothermal reservoir, the average bulk porosity is reduced by precipitation of quartz and carbonates from the ascending fluids, a decrease quantified by comparison with nearby hydrothermally unaltered sediments. The decrease of reservoir porosity in the geothermal system is accompanied by a dramatic decrease in vertical bulk permeability. This clearly shows that reservoir rock properties of clastic sediments can change dramatically within a few thousand years as a result of an influx of moderate-temperature brines. Similarly, a reduction of pore space by quartz precipitation has been observed in the upflow zones of the Alberta basin (e.g., Tilley and Longstaffe, 1989; Tilley et al., 1990) and North Sea Basin (e.g., Liewig et al., 1987).

A detailed comparison of basin diagenesis with the hydrothermally enhanced diagenesis of geothermal systems provides insight on (1) the influence of permeability variations on diagenesis, (2) the kinetics of mineral reactions, and (3) the reaction path. As demonstrated for the smectite-to-illite transformation, alteration sequences reach consistent states of advancement along isothermal surfaces in geothermal systems as well as in unaltered surrounding sediments. The conversion of smectite to illite in interstratified clays is almost complete at about 180–200°C (90–95% illite) in the Salton Sea geothermal field and also in the adjacent unaltered nonthermal sediments (McDowell and Elders, 1980, 1983; Yau et al., 1986) although the temperature intervals occur at very different depths. Therefore, the smectite-to-illite conversion is not a simple function of burial depth, at least not in the Salton Trough. The age of the Salton Sea geothermal system, ~16,000 yr, is an upper limit for the time necessary to complete the smectite-to-illite conversion under the hydraulic and geochemical conditions of the system.

The influence of rock permeability on the progress of mineral alteration in geothermal fields can be dominant. In geothermal systems, clays in sandstone sequences are more intensely altered than clays within shales. Reduction of initial permeability by intense cementation, and the subsequent decrease in water/rock ratio, is often reflected by the preservation of otherwise unstable mineral phases. The reaction process of the smectite-to-illite conversion seems dependent also on the rock permeability. In the permeable Salton Sea reservoir shales, illite seems to be precipitated directly from solution. In contrast, illite replaces smectite directly in low-permeability caprock shales, an important process in the low-permeability shales of the Gulf Coast Basin (Yau et al., 1986, 1988).

This comparison of sediment-hosted geothermal fields and upflow zones of sedimentary basins shows clearly that analogous conditions exist in terms of temperature, temperature gradient, salinity distribution, and fluid flow in both environments. Detailed petrographic and geochemical studies of clastic sediments in moderate-temperature geothermal systems and basinal upflow zones may contribute significantly to our understanding of shale diagenesis and porosity and permeability distribution in both environments.

ACKNOWLEDGMENTS

This manuscript was greatly improved through reviews provided by Drs. T. Engelder and L. R. Kump of the Pennsylvania State University and Y. K. Kharaka of the U.S. Geological Survey. This research was supported by the Gas Research Institute under Contract Number 5088-260-1746, and by the National Science Foundation under Grant Number EAR-8903750.

REFERENCES CITED

Adams, M. C., 1985, Tracer stability and chemical changes in an injected geothermal fluid during injection-backflow at the East Mesa geothermal field: Proceedings of the 10th Symposium on Geothermal Reservoir Engineering, Stanford University, Stanford, California, p. 247–252.

Andes, J. P., and M. A. McKibben, 1987, Thermal and chemical history of mineralized fractures in cores from the Salton Sea Scientific Drilling Project (abst.): EOS, Transactions, American Geophysical Union, v. 68, p. 439.

Ayuso, M. A., 1984, Gerencia de proyectos geothermoeléctricos: Com. Fed. de Electr., internal report, no. 1384-027 and 028, Departamento de Evaluación y Yacimientos, 15 p.

Bailey, 1977, A hydrogeological and subsurface study of Imperial Valley geothermal anomalies, Imperial Valley, California: M.S. Thesis, University of California, Riverside, California, 100 p.

Balling, N., 1979, Subsurface heat flow in Europa, *in* V. Cermak and R. Haenel, eds., Geothermics and geothermal energy: Stuttgart, Schweizerbartsche Verlagsbuchhandlung (Nagele u. Obermiller), p. 161–171.

Barberi, F., R. Santacroce, and J. Varet, 1982, Chemical aspects of rift magma, *in* G. Palmason, ed., Continental and oceanic rifts: Washington, D. C., American Geophysical Union, Geodynamic series, v. 8, p. 223–225.

Bebout, D. G., B. R. Weise, A. R. Gregory, and M. B. Edwards, 1982, Wilcox sandstone reservoirs in the deep subsurface along the Texas Gulf Coast, their potential for production of geopressured energy: The University of Texas, Bureau of Economic Geology, Report no. 117, 125 p.

Bermejo, M. F., F. X. Navarro, O. F. Castillo, B. C. Esquer, and C. A. Cortez, 1979, Pressure variation at the Cerro Prieto reservoir during production: Proceedings of the 2nd Symposium on the Cerro Prieto geothermal field, p. 473–493.

Bethke, C. M., 1989, Modeling subsurface flow in sedimentary basins: Geologische Rundschau, v. 78, p. 129–154.

Bethke, C. M., 1986, Inverse hydrologic analysis of the distribution and origin of Gulf Coast-type geopressured zones: Journal of Geophysical Research, v. 91, p. 6535–6545.

Biehler, S., 1971, Gravity studies in the Imperial Valley, *in* R. W. Rex, ed., Cooperative geological-geophysical-geochemical investigations of geothermal resources in the Imperial Valley of California: U.S. Bureau of Reclamation, Contract no. 14-06-300-2149, University of California, Riverside, California, p. 29–41.

Bird, D. K., and W. A. Elders, 1976, Hydrothermal alteration and mass transfer in the discharge portion of the Dunes geothermal system, Imperial Valley, California, U.S.A.: Proceedings of the 2nd United Nations Symposium on Geothermal Resources, San Francisco, v. 1, p. 285–295.

Bird, D. K., P. S. Schiffman, W. A. Elders, A. E. Williams, and S. D. McDowell, 1984, Calc–silicate mineralization in active geothermal systems: Economic Geology, v. 79, p. 671–695.

Bodner, D. P., and J. M. Sharp, Jr., 1988, Temperature variations in South Texas subsurface: AAPG Bulletin, v. 72, p. 21–32.

Bodner, D. P., P. E. Blandchard, and J. M. Sharp, Jr., 1985, Variations in Gulf Coast heat flow created by groundwater flow: Gulf Coast Association of Geological Societies Transactions, v. 35, p. 19–28.

Bodri, L., and B. Bodri, 1980, Geothermal model of the heat anomaly of the Pannonian basin, *in* V. Cermak and R. Haenel, eds., Geothermics and geothermal energy: Stuttgart, Schweizerbartsche Verlagsbuchhandlung (Nagele u. Obermiller), p. 37–43.

Bodvarsson, G. S., and P. A. Witherspoon, 1989, Geothermal reservoir engineering part I: Geothermal Science and Technology, v. 2, p. 1–68.

Cathles, L. M., and A. T. Smith, 1983, Thermal constraints on the formation of Mississippi Valley-type lead-zinc deposits and their implications for episodic basin dewatering and deposit genesis: Economic Geology, v. 78, p. 983–1002.

Clayton, R. N., J. P. Muffler, and D. E. White, 1968, Oxygen isotope study of calcite and silicates of the River Ranch #1 well, Salton Sea geothermal field, California: American Journal of Science, v. 266, p. 968–979.

Combs, J., 1971, Heat flow and geothermal resource estimates for the Imperial Valley area of California: U.S. Bureau of Reclamation, Final report no. 14-06-

300-2194, p. 5–28.
Combs, J., and O. Hadley, 1977, Microearthquake investigations of the East Mesa geothermal anomaly, Imperial Valley, California: Geophysics, v. 42, p. 29–41.
Coplen, B., 1976, Cooperative geochemical resource assessment of the Mesa geothermal system: U.S. Bureau of Reclamation, Final report no. 14-06-300-2479, 97 p.
Coustau, H., J. Gauthier, G. Kulbicki, and E. Winnock, 1969, Hydrocarbon distribution in the Aquitaine Basin of southwest France, *in* K. Brighton, ed., The exploration of petroleum in Europe and North Africa: Institute of Petroleum and American Association of Petroleum Geology, p. 73–85.
Crosby, J. W., R. M. Chatters, J. V. Anderson, and R. L. Fenton, 1972, Hydrothermal evaluation of the Cerro Prieto geothermal system utilizing isotopic techniques: Washington State University, College of Engineering, Research Division, no. 27/11-5.
Davies, D. G., and S. K. Sanyal, 1979, Case history report on East Mesa and Cerro Prieto geothermal fields: Scientific Software Co. and Los Alamos Scientific Laboratory Report, LA-7889-MS, 182 p.
DiPippo, R., 1987, Geothermal power generation from liquid dominated reservoirs: Geothermal Science and Technology, v. 1, p. 63–124.
Elders, W. A., 1979, The geological background of the geothermal fields of the Salton Trough, geology and geothermics of the Salton Trough, *in* W.A. Elders, ed., Field trip guide for the 92th annual GSA meeting, San Diego, University of California, Riverside: University of California, Riverside, California, Campus museum contributions, no. 5, 1–19 p.
Elders, W. A., 1982, Determination of fracture history in geothermal reservoirs through study of minerals: Geothermal Resources Council, Special Report no. 12, p. 62–66.
Elders, W. A., and D. K. Bird, 1974, Investigations of the Dunes anomaly, Imperial Valley, California: Active formation of silicified caprocks in arenaceous sands a low-temperature geothermal environment in the Salton Trough of California, U.S.A.: Proceedings of the Symposium on water-rock reaction of the International Union of Geochemistry and Cosmochemistry, Prague, Czechoslovakia, p. 150–157.
Elders, W. A., and L. H. Cohen, 1983, The Salton Sea geothermal field, California, as a nearfield natural analog of a radioactive waste repository in salt: Institute of Geophysics and Planetary Physics, University of California, Riverside, California, Report UCR/IGPP-83/10, 139 p.
Elders, W. A., A. E. Williams, and J. R. Hoagland, 1981, An integrated model for the natural flow regime in the Cerro Prieto hydrothermal system based upon petrological and isotope geochemical criteria: Proceedings of the 3rd Symposium on the Cerro Prieto geothermal field, San Francisco, California, Lawrence Berkeley Laboratory Report LBL-1197 p. 102–109.
Elders, W. A., D. K. Bird, A. E. Williams, and P. S. Schiffmann, 1984, Hydrothermal flow regime and magmatic heat source of the Cerro Prieto geothermal system, Baja California, Mexico: Geothermics, v. 13, p. 27–47.
Ellis, A. J., 1979, Explored geothermal systems, *in* H. L. Barnes, ed., Geochemistry of Hydrothermal Ore Deposits, 2nd. ed.: New York, J. Wiley & Sons, p. 632–683.
Ermak D. L., 1977, Potential growth of electrical power production from Imperial Valley geothermal resources: Lawrence Livermore Laboratory, University of California, Livermore, California, Report URL-52252, 27 p.
Ershaghi, I., and D. Abdassah, 1983, Interpretation of some wireline logs in geothermal fields of the Imperial Valley, California: California Regional Meeting of the Society of Petroleum Engineers, Ventura, California.
Facca, G., and F. Tonani, 1967, The self-sealing geothermal field: Bulletin Volcanoligique, p. 271–273.
Fournier, R. O., 1976, A study of the mineralogy and lithology of cuttings from the U.S. Bureau of Reclamation Mesa 6-2 drillhole, Imperial County, California, including comparisons with East Mesa 6-1 drillhole: USGS Open File Report 76–88, 57 p.
Fournier, R. O., 1988, Double-diffusive convection as a mechanism for transferring heat and mass within the Salton Sea geothermal brine: Proceedings of the 13th Symposium on Geothermal Reservoir Engineering, Stanford University, Stanford, California, p. 101–106.
Fuis, G. S., W. D. Mooney, J. H. Healey, G. A. McMechan, and W. J. Lutter, 1982, "Crustal structure of the Imperial Valley," The Imperial Valley, California, Earthquake of October 15, 1979: USGS Professional Paper no. 1254, p. 25–49.
Galloway, W. E., 1982, Epigenetic zonation and fluid flow history of uranium-bearing fluvial aquifer systems, South Texas uranium province: Bureau of Economic Geology, Report of Investigation no. 119, 31 p.
Garven, G., and A. Freeze, 1984a, Theoretical analysis of the role of groundwater flow in the genesis of stratabound ore deposits: 1. Mathematical and numerical model: American Journal of Science, v. 284, p. 1085–1124.
Garven, G., and A. Freeze, 1984b, Theoretical analysis of the role of groundwater flow in the genesis of stratabound ore deposits: 2. Quantitative results: American Journal of Science, v. 284, p. 1125–1174.
Girard, J. P., S. M. Savin, and J. L. Aronson, 1989, Diagenesis of the Lower Cretaceous arcoses of the Angola Margin: Petrologic, K/Ar dating, and $^{18}O/^{16}O$ evidence: Journal of Sediment Petrology, v. 59, p. 519–538.
Giroir, G., E. Merino, and D. Nahon, 1988, Diagenesis of Cretaceous sandstone reservoirs of the South Gabon Rift, West Africa: mineralogy, mass transfer, and thermal evolution: Journal of Sedimentary Geology, v. 59, p. 482–493.
Goldstein, N. E., and S. Carle, 1986, Faults and gravity

anomalies over the East Mesa hydrothermal–geothermal field: Geothermal Resources Council, Transactions, v. 10, p. 223–228.

Goyal, K. P., and D. R. Kassoy, 1981, A plausible two-dimensional vertical model of the East Mesa geothermal field, California: Journal of Geophysical Research, v. 86, no. B11, p. 10,719–10,733.

Grant M. A., A. H. Truesdell, and A. Mañón, 1984, Production-induced boiling and cold water entry in the Cerro Prieto geothermal reservoir indicated by chemical and physical measurements: Geothermics, v. 13, p. 117–140.

Halfman, S. E., M. J. Lippmann, and R. Zelwer, 1982, The movement of geothermal fluid in the Cerro Prieto field as determined from well log and reservoir engineering data: Proceedings of the 8th Symposium on Geothermal Reservoir Engineering, Stanford University, Stanford, California, p. 171–176.

Halfman, S. E., M. J. Lippmann, and G. S. Bodvarsson, 1986, Quantitative model of the Cerro Prieto field: Proceedings of the 11th Symposium on Geothermal Reservoir Engineering, Stanford University, Stanford, California, p. 127–134.

Hanor, J. S., 1987, Kilometer-scale thermohaline overturn of pore waters in the Louisiana Gulf Coast: Nature, v. 327, p. 501–503.

Helgeson, H. C., 1968, Geology and thermodynamic characteristics of the Salton Sea geothermal system: American Journal of Science, v. 266, p. 129–166.

Hermance, J. F., 1982, Magnetotelluric and geomagnetic deep-sounding studies in rifts and adjacent basins: Constraints on physical processes in the crust and upper mantle, *in* G. Palmason, ed., Continental and oceanic rifts: Washington, D. C., American Geophysical Union, Geodynamic series, v. 8, p. 169–192.

Hoagland, J. R., 1976, Petrology and geochemistry of hydrothermal alteration in borehole Mesa 6-2, East Mesa geothermal area, Imperial Valley, California: Department of Earth Sciences and Institute of Geophysics and Planetary Physics, University of California, Riverside, California, IGPP-UCR-76-12, 90 p.

Horvath, F., L. Bodri, and P. Ottlic, 1979, Geothermics of Hungary and the tectonophysics of the Pannonian Basin "Red Spot," *in* V. Vermak and L. Rybach, eds., Terrestial heat flow in Europa: New York, Springer-Verlag, p. 206–217.

Howard, J., J. A. Apps, S. Benson, N. E. Goldstein, A. N. Graf, J. Haney, D. Jackson, S. Juprasert, E. Majer, D. McEdwards, T. V. McEvilly, T. N. Narasimhan, B. Schechter, R. Schroeder, R. Taylor, P. van de Kamp, and T. Wolery, 1978, Geothermal resource and reservoir investigations of U.S. Bureau of Reclamation leaseholds at East Mesa, Imperial Valley, California: Lawrence Berkeley Laboratory, University of California, Berkeley, California, LBL-7094, p. 305.

Jensen, P. K., 1990, Analysis of the temperature field around salt diapirs: Geothermics, v. 19, p. 273–283.

Jensenius, J., 1987, High-temperature diagenesis in shallow chalk reservoir, Skjold oil field, Danish North Sea: Evidence from fluid inclusions and oxygen isotopes: AAPG Bulletin, v. 71, p. 1378–1386.

Jensenius, J., and Munkgaard, N. C., 1989, Large-scale hot water migration systems around salt diapirs in the Danish Central Trough and their impact on diagenesis of chalk reservoirs: Geochimica et Cosmochimica Acta, v. 53, p. 79–88.

Jones, P. H., 1975, Geothermal and hydrocarbon regimes, northern Gulf of Mexico: Proceedings of the 1st Geopressured–Geothermal Energy Conference, Center of Energy Studies, University of Texas, Austin, Texas, p. 15–89.

Kantsler, A. J., A. C. Cook, and G. C. Smith, 1978, Lateral and vertical rank variation: implications for hydrocarbon exploration: Journal of Australian Petrology Exploration Association, p. 143–156.

Kasameyer, P. W., and J. R. Hearst, 1988, Borehole gravity measurement in the Salton Sea Scientific Drilling Project State Well 2-14: Journal of Geophysical Research, v. 93, no. B11, p. 13,037–13,045.

Kasameyer, P. W., L. W. Younker, and J. M. Hansen, 1984, Development and application of a hydrothermal model for the Salton Sea geothermal field, California: Geological Society of America Bulletin, v. 95, p. 1,242–1,252.

Kassoy, D. R., 1975, Heat and mass transfer in models of undeveloped geothermal fields: Proceedings of the United Nations Geothermal Symposium, San Francisco, v. 1, p. 1707–1713.

Kharaka, Y. K., and J. T. Thordsen, 1992, Stable isotope geochemistry and the origin of waters in sedimentary basins, *in* H. Claver and S. Chauduri, eds., Isotopic Signatures and Sedimentary Basins: New York, Springer-Verlag, p. 441–466.

Kharaka, Y. K., M. S. Lico, and W. W. Carothers, 1980, Predicted corrosion and scale-formation properties of geopressured–geothermal waters from the northern Gulf of Mexico Basin: Journal of Petroleum Technology, v. 32, p. 319–324.

Kharaka, Y. K., R. W. Hull, and W. W. Carothers, 1985, Water-rock interactions in sedimentary basins, *in* D. L. Gautier, J. K. Kharaka, and R. C. Surdam, eds., Relationship of organic matter and mineral diagenesis: Society of Economic Paleontologists and Mineralogists, Short Course, v. 17, p. 79–176.

Kharaka, Y. K., L. M. Law, W. W. Carothers, and D. F. Goerlitz, 1986, Role of organic species dissolved in formation waters in mineral diagenesis, *in* D. L. Gautier, ed., Relationship of organic matter and mineral diagenesis: Society of Economic Paleontologists and Mineralogists, v. 38, p. 111–122.

Lachenbruch, A. H., J. H. Sass, and S. P. Galanis, Jr., 1985, Heat flow in southernmost California and the origin of the Salton Trough: Geophysical Research, v. 90, p. 6,709–6,736.

Land, L. S., K. L. Milliken, and E. F. McBride, 1987, Diagenetic evolution of Cenozoic sandstones, Gulf of Mexico sedimentary basin: Sedimentary

Geology, v. 50, p. 195–225.

Liewig, N., N. Clauer, and F. Sommer, 1987, Rb-Sr and K-Ar dating of clay diagenesis in Jurassic sandstone oil reservoir, North Sea: AAPG Bulletin, v. 71, p. 1467–1474.

Lippmann, M. J., 1982, Overview of Cerro Prieto Studies: Proceedings of the 8th Symposium on Geothermal Reservoir Engineering, Stanford University, Stanford, California, p. 49–66.

Lippmann, M. J., and G. S. Bodvarsson, 1983, Numerical studies of the heat and mass transport in the Cerro Prieto geothermal field, Mexico: Water Resources Research, v. 19, p. 753–767.

Lippmann, M. J., and G. S. Bodvarsson, 1985, The Heber geothermal field, California: Natural state and exploitation modeling studies: Geophysical Research, v. 90, p. 745–758.

Lundegard, P. D., and J. K. Kharaka, 1990, Geochemistry of organic acids in subsurface waters, *in* D. C. Melchior and R. L. Basset, eds., Chemical modeling of aqueous systems II: American Chemical Society Symposium Series, v. 416, p. 169–189.

Lyons, D. J., and P. C. van de Kamp, 1980, Subsurface geological and geophysical study of the Cerro Prieto geothermal field, Baja California, Mexico: University of California Lawrence Berkeley Laboratory, Berkeley, California, Report LBL-10540.

Ma, L., J. Yang, and Z. Ding, 1989, Songliao basin—an intracratonic continental sedimentary basin of combination type, *in* X. Zhu, ed., Chinese sedimentary basins: New York, Elsevier, p. 77–87.

Mañón, A., E. Mazor, M. Jiménez, A. Sánchez, J. Fausto, and C. Zenizo, 1977, Extensive geochemical studies in the geothermal field of Cerro Prieto, Mexico: Lawrence Livermore National Laboratory, Livermore, California, Report for the Department of Energy, Contract no. W-7405-Eng-48, LBL-7019, 113 p.

McBride, E. F., and J. M. Sharpe, Jr., 1989, Sediment petrology, a guide to paleohydrologic analysis, example of sandstones from the northwestern Gulf of Mexico: Journal of Hydrology, v. 108, p. 367–386.

McDowell, S. D., and W. A. Elders, 1979, "Geothermal metamorphoses of sandstone in the Salton Sea geothermal system," geology and geothermics of the Salton Sea, *in* W. A. Elders, ed., Field trip guide for the 92th annual GSA meeting, San Diego, University of California, Riverside: University of California, Riverside, California, Campus museum contributions, no. 5, p. 70–76.

McDowell, S. D., and W. A. Elders, 1980, Authigenic layer silicate minerals in Borehole Elmore # 1, Salton Sea geothermal field, California, U.S.A.: Contributions to Mineralogy and Petrology, v. 74, p. 293–310.

McDowell, S. D., and W. A. Elders, 1983, Allogenic layer silicate minerals in borehole Elmore # 1, Salton Sea field, California: American Mineralogist, v. 68, p. 1146–1159.

McKibben, M. A., and W. A. Elders, 1985, Fe-Zn-Pb mineralization in the Salton Sea geothermal system, Imperial Valley, California: Economic Geology, v. 80, p. 539–559.

McKibben, M. A., and W. E. Williams, 1990, Intrusion-driven brine diapirism and ore diagenesis in the Salton Trough (abst.): V. M. Goldschmidt Conference, 1990, p. 65.

McKibben, M. A., A. E. Williams, W. A. Elders, and L. S. Eldridge, 1986, Metamorphosed Plio-Pleistocene evaporites and the origins of the sulphur and salinity in the Salton Sea geothermal brines (abst.): EOS, Transactions, American Geophysical Union, v. 67, p. 1258.

McKibben, M. A., W. E. Williams, and S. Okubo, 1988a, Metamorphosed Plio-Pleistocene evaporites and the origin of hypersaline brines in the Salton Sea geothermal system, California: fluid inclusion evidence: Geochimica et Cosmochimica Acta, v. 52, p. 1057–1067.

McKibben, M. A., J. P. Andes, Jr., and A. E. Williams, 1988b, Active ore formation at a brine interface in metamorphosed deltaic lacostrine sediment: the Salton Sea geothermal system, California: Economic Geology, v. 83, p. 511–523.

Meidav, T, R. James, and R. W. Rex, 1970, Investigation of geothermal resources in the Imperial Valley and their potential value of desalination of water and electricity production: Institute of Geophysics and Planetary Physics, University of California, Riverside, California, 54 p.

Meidav, T., R. James, and S. Sanyal, 1975, Utilization of gravimetric data for estimation of hydrothermal reservoir characteristics in the East Mesa geothermal field, Imperial Valley, California: Proceedings of the 1st Symposium on Geothermal Reservoir Engineering, University of Stanford, Stanford, California, p. 1–8.

Meyer, H. J., and H. W. McGee, 1985, Oil and gas fields accompanied by geothermal anomalies in the Rocky Mountain region: AAPG Bulletin, v. 69, p. 933–945.

Moore, J. N., and M. C. Adams, 1988, Evolution of the thermal cap in two wells from the Salton Sea geothermal system, California: Proceedings of the 13th Symposium on Geothermal Reservoir Engineering, Stanford University, Stanford, California, p. 107–112.

Morgan, P., 1982, Heatflow in rift zones, *in* G. Palmason, ed., Continental and oceanic rifts: Geodynamic Series, v. 8, p. 107–122.

Morse J. G., and L. D. Thorsen, 1978, Reservoir engineering study of a portion of the Salton Sea geothermal field: Geothermal Resources Council, Transactions, v. 2, p. 471–474.

Morton, R. A., and L. S. Land, 1987, Regional variations in formation water chemistry, Frio Formation (Oligocene), Texas Gulf Coast: AAPG Bulletin, v. 71, p. 191–206.

Morton, R. A., C. M. Garret, J. S. Posey, Jr., J. H. Han, and L. A. Jirik, 1981, Salinity variations and chemical compositions of waters in the Frio Formation, Texas Gulf Coast: The University of Texas, Austin,

Texas, Bureau of Economic Geology, Report prepared for U.S. Dep. of Energy, Contract no. DE-AC08-79ET2711, 96 p.

Morton, R. A., J. H. Han, and J. S. Posey, Jr., 1983, Variations in chemical compositions of Tertiary Formation waters, Texas Gulf Coast, *in* Consolidation of Geological Studies of Geopressured Geothermal Resources in Texas: The University of Texas, Austin, Texas, Bureau of Economic Geology, Report prepared for U.S. Dep. of Energy, Contract no. DE-AC08-79ET2711, p. 63–135.

Muffler, L. J. P., and D. E. White, 1969, Active metamorphism of upper Cenozoic sediments in the Salton Sea geothermal field and the Salton Trough, southeastern California: Geological Society of America Bulletin, v. 80, p. 157–182.

Newmark, R. L., P. W. Kasameyer, L. W. Younker, and P. C. Lysne, 1986, Research drilling at the Salton Sea geothermal field, California: the shallow thermal gradient project: EOS, Transactions, American Geophysical Union, v. 67, p. 698–707.

Newmark, R. L., P. W. Kasameyer, and L. W. Younker, 1988, Shallow drilling in the Salton Sea region: the thermal anomaly: Journal of Geophysical Research, v. 93, p. 13,005–13,023.

Olsen, E. R., and J. S. Matlick, 1978, A flow-through model for the Westmoreland geothermal system, Imperial Valley, California: Institute of Geophysics and Planetary Physics, University of California, Riverside, California, Report UCR/IGPP-78/7, 37 p.

Pearson, R. O., 1976, Planning and design of additional East Mesa geothermal test facilities (Phase 1b): Energy Research and Development Administration, Division of Geothermal Energy, Report SAN/1140-1/1, v. 1, p. 1–1 to 5–7.

Peng, D.-J., 1980, Geopressured geothermal resources: Proceedings of the New Zealand geothermal workshop, University of Auckland, Auckland, New Zealand, p. 181–187.

Pfeifer, D. S., and J. M. Sharp, Jr., 1989, Subsurface temperature distributions in South Texas: Gulf Coast Association of Geological Societies Transactions, v. 39, p. 231–245.

Polster, W., and H. L. Barnes, 1990, Geothermal systems in sediment-filled rift valleys: hydrodynamic and geochemical characteristics: Proceedings of the 15th Symposium on Geothermal Reservoir Engineering, University of Stanford, Stanford, California, p. 89–95.

Rabinowicz, M., J.- L. Danduarand, M. J. Jakubowski, J. Schott, and J.- P. Cassan, 1985, Convection in a North Sea oil reservoir: interferences on the diagenesis and hydrocarbon migration: Earth and Planetary Science Letters, v. 74, p. 387–404.

Randall, W., 1974, An analysis of the subsurface structure and stratigraphy of the Salton Sea geothermal anomaly, Imperial Valley, California: Ph.D. Dissertation, University of California, Riverside, California, 62 p.

Rex, R. W., 1966, Heat flow in the Imperial Valley of California (abst.): American Geophysical Union, Transactions, v. 47, p. 181.

Rex, R. W., 1985, Temperature-chlorinity balance in the hypersaline brines of the Imperial Valley, California: International Symposium on Geothermal Energy, Geothermal Resources Council, Hawaii, International volume, p. 351–356.

Riney, T. D., J. W. Pritchett, and L. F. Price, 1980, Modeling of the East Mesa hydrothermal system: Geothermal Resources Council, Transactions, v. 4, p. 476–480.

Riney, T. D., J. W. Pritchett, L. F. Price, and S. K. Gary, 1979, A preliminary model of the East Mesa hydrothermal system: Proceedings of the 5th Symposium on Geothermal Reservoir Engineering, University of Stanford, Stanford, California, p. 211–214.

Robert, P., 1988, Organic metamorphism and geothermal history: Microscopic study of organic matter and thermal evolution of sedimentary basins: Elf-Aquitane and D. Reidel Publishing Co., p. 307.

Salveson, J. O., and A. M. Cooper, 1981, Exploration and development of the Heber geothermal field, Imperial Valley, California: Proceedings of the New Zealand Geothermal Workshop, University of Auckland, Auckland, New Zealand, p. 3–6.

Sass, J. H., J. D. Hendricks, S. S. Priest, and L. C. Robinson, 1987, Temperature and heat flow in the State 2-14 well, Salton Sea Scientific Drilling Program (abs.): EOS, Transactions, American Geophysical Union, v. 68, p. 454.

Sass, J. H., S. S. Priest, L. E. Duda, C. C. Carson, J. D. Hendricks, and L. C. Robinson, 1988, Thermal regime of the State 2-14 well, Salton Sea Scientific Drilling Project: Journal of Geophysical Research, v. 68, no. B11, p. 12,995–13,004.

Schiffman, P. S., W. A. Elders, A. E. Williams, S. D. McDowell, and D. K. Bird, 1984, Active metasomatism in the Cerro Prieto geothermal system, Baja California, Mexico: a telescoped low-pressure, low-temperature metamorphic facies series: Geology, v. 12, p. 12–15.

Schroeder, R. C., 1976, Reservoir engineering report for the Magma-SDG&E geothermal experimental site near the Salton Sea, California: Lawrence Livermore National Laboratory, Livermore, California, Report CRL-52094, 62 p.

Schull, T. J., 1988, Rift basins of Interior Sudan: Petroleum exploration and discovery: AAPG Bulletin, v. 72, p. 1128–1142.

Seamount, D. T. , and W. A. Elders, 1981, Use of wireline logs at the Cerro Prieto in the identification of the distribution of the hydrothermally altered zones and dike location: 3rd Symposium on the Cerro Prieto field, p. 123–133.

Skinner, B. J., D. E. White, H. J. Rose, and R. E. Mays, 1967, Sulfides associated with the Salton Sea geothermal brine: Economic Geology, v. 62, p. 316–330.

Smith, J. L., 1979, Geology and geothermics of the Salton Trough, *in* W. A. Elders, ed., Field trip guide for the 92th annual GSA meeting, San Diego, University of California, Riverside: University of California, Riverside, California, Campus museum contributions, no. 5, 94 p.

Stanford, S. J., 1981, Dating thermal events by fission track annealing, Cerro Prieto geothermal field, Baja California, Mexico: M.S. Thesis, Institute of Geophysics and Planetary Sciences, University of California, Riverside, California, 113 p.

Surdam. R. C., L. J. Crossey, E. S. Hagen, and H. P. Hessler, 1989, Organic–inorganic interactions and sandstone diagenesis: AAPG Bulletin, v. 73, p. 1–23.

Tewhey, J. D., 1977, Geologic characteristics of a portion of the Salton Sea geothermal field: Lawrence Livermore Laboratory, Livermore, California, Report UCRL-52267, 51 p.

Tilley, B. J., and F. J. Longstaffe, 1989, Diagenesis and isotopic evolution of porewaters in the Alberta Deep Basin: The Falher Member and the Cadomin Formation: Geochimica et Cosmochimica Acta, v. 53, p. 2529–2546.

Tilley, B. J., B. N. Nesbitt, and F. J. Longstaffe, 1990, Thermal history of the Alberta Deep Basin: comparative study of fluid inclusions and vitrinite reflectance data: AAPG Bulletin, v. 73, p. 1206–1222.

Truesdell, A. H., and M. J. Lippmann, 1986, The lack of immediate effects from the 1979–80 Imperial and Victoria earthquakes on the exploited Cerro Prieto geothermal reservoir: Geothermal Research Council, Transactions, v. 10, p. 405–411.

Truesdell, A. H., N. L. Nehring, J. M. Thompson, and C. J. Janic, 1984, A review of progress in understanding the fluid geochemistry of the Cerro Prieto geothermal system: Geothermics., v. 13, p. 65–74.

Tyler, N., M. P. R. Lighth, and T. E. Ewing, 1985, Saline fluid flow and hydrocarbon migration and maturation as related to geopressure, Frio Formation, Brazoria County, Texas: Proceedings 6th Conference on Geopressured Geothermal Energy, The University of Texas, Austin, Texas, p. 83–92.

Veliciu, S., and M. Visario, 1980, On the low heat flow in the Transylvanian basin, *in* V. Cermak and R. Haenel, eds., Geothermics and geothermal energy: Stuttgart, Schweizerbartsche Verlagsbuchhandlung (Nagele u. Obermiller), p. 91–100.

Walgenwitz, F., M. Pagel, and E. George, 1985, Paléothermomètrie, rapport de phase no. 2; calcul des paléotemperatures par microthermomètrie sur inclusions aqueuses et hydrocarbures synchrones: Sociéte National Elf Aquitaine, rapport interne, Dir. Expl. DRAG, Pan, 34 p.

White, D. E., 1981, Active geothermal systems and hydrothermal ore deposits, *in* B. J. Skinner, ed., 75th anniversary volume of Economic Geology: El Paso, Mexico, Economic Publishing Co., p. 392–423.

Williams, A. E., and W. A. Elders, 1984, Stable isotope systematics of oxygen and carbon isotopes in rocks and minerals from the Cerro Prieto geothermal anomaly, Baja California, Mexico: Geothermics, v. 13, p. 49–63.

Williams, A. E., and M. A. McKibben, 1989, A brine interface in the Salton Sea geothermal system, California: Fluid inclusion and isotopic characteristics: Geochimica et Cosmochimica Acta, v. 53, p. 1905–1920.

Yau, Y.-C., D. R. Peacor, and S. D. McDowell, 1986, Smectite-to-illite reactions in Salton Sea shales: a transmission and analytical electron microscopy study: Journal of Sedimentary Petrology, v. 57, p. 335–342.

Yau, Y.-C., D. R. Peacor, R. E. Beane, and E. J. Essene, 1988, Microstructures, formation mechanism, and depth-zoning of phyllosilicates in geothermally altered shales, Salton Sea, California: Clays and Clay Minerals, v. 36, p. 1–10.

Younker, L. W., P. W. Kasameyer, and J. D. Tewhey, 1982, Geological, geophysical and thermal characteristics of the Salton Sea geothermal field, California: Journal of Volcanology and Geothermal Research, v. 12, p. 221–258.

Zukin, J. G., D. E. Hammond, T.-L. Ku, and W. A. Elders, 1987, Uranium–thorium series radionucleides in brines and reservoir rocks from two deep geothermal boreholes in the Salton Sea geothermal field, California: Geochimica et Cosmochimica Acta, v. 51, p 2719–2732.

Chapter 30

The Mechanical Properties of Rock through an Ancient Transition Zone in the Appalachian Basin

Irene Meglis*
Terry Engelder
The Pennsylvania State University
University Park, Pennsylvania, U.S.A.

ABSTRACT

Hydrostatic compression tests were used to measure the variation in mechanical properties of Devonian core cut through an ancient pressure transition zone in the Appalachian Plateau of western New York. The properties of this core are strongly influenced by low aspect ratio microcracks developed during stress relief. In the siltstones, linear compressibility decreases with increasing confining pressure to a minimum value between 14 and 16 $\times 10^{-6}$ MPa^{-1} in all directions. In the shales, linear compressibility parallel to bedding is approximately 10 $\times 10^{-6}$ MPa^{-1} for all confining pressures, but normal to bedding it decreases with increasing confining pressure throughout the range of test pressures (140 MPa), reaching minimum values between 25 and 40 $\times 10^{-6}$ MPa^{-1} at high pressure. The high linear compressibility normal to bedding reflects compression and closure of a dense population of bedding-parallel microcracks. Microcrack porosity is lowest in the core 50 m above the ancient transition zone and is higher further above and below the zone. In conclusion, rocks within the ancient transition zone have lower intrinsic compressibilities than overlying rocks (8–12 $\times 10^{-6}$ MPa^{-1} versus 9–18 $\times 10^{-6}$ MPa^{-1}). Such a marked change in elastic properties near the transition zone may serve as a seismic reflector.

INTRODUCTION

In searching for pressure compartments one seismic exploration strategy involves the assumption that seal zones above pressure compartments have mechanical properties that differ from those of the overlying rock. The boundary between rocks of different mechanical properties will appear as a seismic reflector in seismic sections. To further examine the possibility that seal zones have unique mechanical properties, we chose to study cores** of Devonian sedimentary rocks from a well that intersects a transition

* Present address: Queen's University, Kingston, Ontario, Canada.

** The core was taken as part of a major drilling program by the Eastern Gas Shales Project (E.G.S.P.) of the Department of Energy.

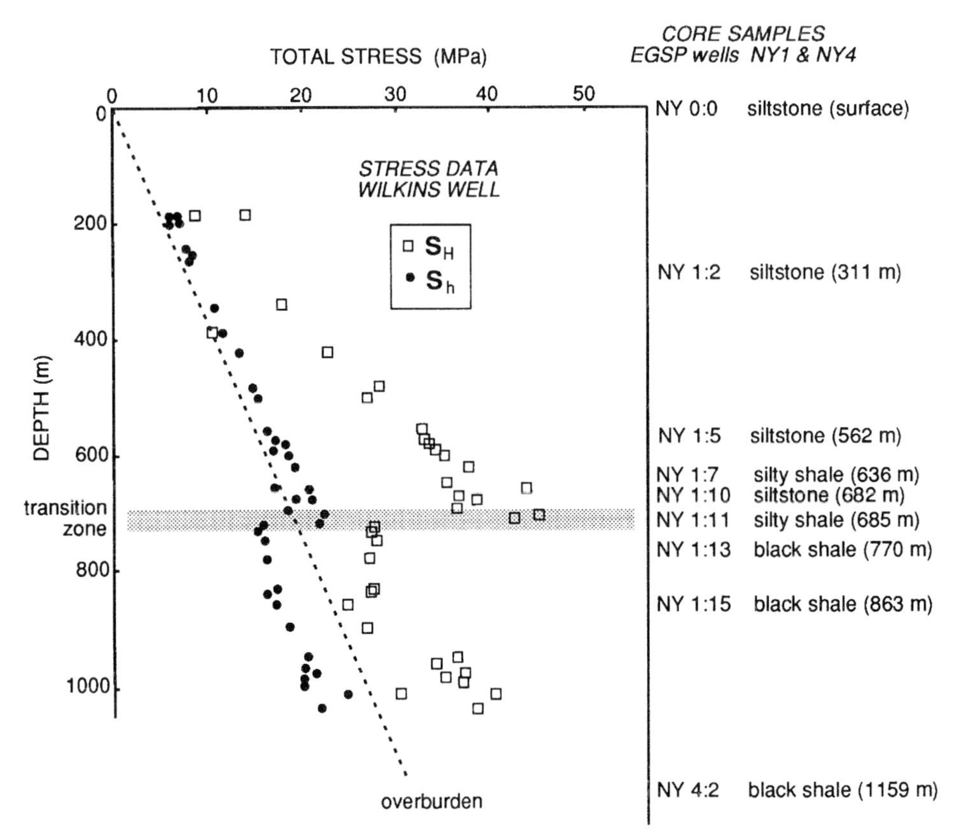

Figure 1. In situ stress profile measured by Evans et al. (1989a) in the Wilkins, Appleton, and O'Dell wells. S_H and S_h are the maximum and minimum horizontal stress components, respectively. The depth and generalized lithology of the core samples from E.G.S.P. wells NY-1 and NY-4 are also indicated.

zone, possibly an ancient seal zone once separating normally and abnormally pressured rocks. The core came from a portion of the Appalachian basin where a compaction-depth trend and the regional distribution of joints suggest that the deeper shales were once overpressured (Engelder and Oertel, 1985; Lacazette and Engelder, 1992).

Further evidence that part of the Appalachian basin was once overpressured comes from an unusual pattern of in situ stress as a function of depth (Figure 1). In three wells near South Canisteo, New York, the horizontal components of stress, measured by hydraulic fracturing, increase systematically with depth to approximately 700 m (Evans et al., 1989a). Below this depth the horizontal components of stress are less by as much as 20 MPa but continue to increase with depth. Evans et al. (1989b) interpret this stress drop as an indication that the section of shale below 700 m was previously overpressured and that the lower stress below this depth is a consequence of poroelastic contraction accompanying the long-term dissipation of high pore pressure. The transition zone refers to the rock in the vicinity of the top boundary of the previously abnormally pressured rock. This zone has a unique geochemical signature consistent with a pressure seal (Albrecht, 1992). Evans et al. (1989a) found that the stress drop, independent of the magnitude of overburden stress, occurred at the base of the Rhinestreet Formation within the Upper Devonian section of the Appalachian Plateau.

The immediate objective of this study is to determine if the mechanical properties of samples from the transition zone are different from overlying rocks, since this condition may be useful in the seismic exploration for transition zones in other basins. To do so, we reloaded core under hydrostatic compression to achieve conditions found in situ. Then we used both static and dynamic techniques to measure the elastic properties of the core. However, we had to test the possibility that in situ stress controlled the mechanical properties of the rock through the development of low aspect ratio microcrack porosity upon stress relief. If in situ stress plays a role in the development of mechanical properties, then the hydrostatic pressure at microcrack closure will correlate with the magnitude of stress normal to the microcracks in question.

Stress Relief Microcrack Porosity

Microcrack porosity in cores of crystalline rock is proportional to mean stress at depth (e.g., Carlson and Wang, 1986; Meglis et al., 1991). In sedimentary rocks, the directions and magnitudes of maximum, intermediate, and minimum strains resulting from crack closure under hydrostatic compression in the laboratory correlate with inferred directions and magnitudes of the horizontal and vertical components of in situ stresses (Strickland and Ren, 1980). Based on these observations a large proportion of the microcrack porosity in core is interpreted as opening or propagating during stress relief. Furthermore, the elastic properties of core differ from in situ rock properties largely because of the presence of stress relief microcracks in the core. Thus, any core-based study aimed at assessing in situ properties of rock within a pressure transition zone must account for the effect of stress relief microcracking.

There are three potential problems in using core to infer the in situ elastic properties in sedimentary rocks, and these are the same problems arising from the use of crack porosity or crack closure pressure to infer the in situ stress. First, a pre-existing rock anisotropy will contribute to a preferred orientation of stress relief microcracks, making it difficult to use microcrack closure to infer the ratio of vertical and horizontal components of in situ stress (Meglis et al., 1991). Second, a large population of horizontal microcracks develops as a consequence of the stress concentrations responsible for incipient core disking upon stress relief by coring (Jaeger and Cook, 1963; Meglis, 1992). Third, a bedding-parallel weakness (i.e., bedding planes) can contribute to the microcrack fabric. If the maximum stress is vertical, a large horizontal crack population is expected from stress relief in samples (Strickland and Ren, 1980; Ren and Roegiers, 1983). However, the influence of both bedding-parallel weakness and incipient disking on the formation of microcracks is difficult to resolve. In brief, tectonic fabric, horizontal bedding, and incipient core disking in sedimentary rocks may all contribute to a preferred orientation of microcracks that in turn influences the elastic properties of the rock.

Samples

The suite of core used in this study was recovered from Eastern Gas Shales Project (E.G.S.P.) wells NY-1 and NY-4, located approximately 30 km from the South Canisteo study area (Figure 2). The lithologies intersected in the E.G.S.P. wells NY-1 and NY-4 correlate with those in the Canisteo wells as described by Evans et al. (1989b). The E.G.S.P. wells intersect a coarsening-upward sequence ranging from black, strongly laminated shale near the bottom (NY1:13, NY1:15, and NY4:2) to siltstones (NY1:2, NY1:5, and NY1:10) and interbedded fine siltstone to mudstone that we call a silty shale (NY1:7, NY1:11) (Figure 1). An outcrop sample of Ithaca siltstone from Watkins Glen, New York, was studied as well (NY:0) (Scott et al., 1992). Based on fission track analysis in Devonian sediments of the western Appalachian basin, Miller and Duddy (1989) concluded that rocks currently at the surface have been uplifted and eroded by 2–3 km since the Early Cretaceous (120–140 Ma). Bedding is horizontal in all samples. Cores NY1:10 and NY1:11 come from within the transition zone. Sample NY1:10 corresponds to the K sand of the Rhinestreet Formation (Plumb et al., 1991).

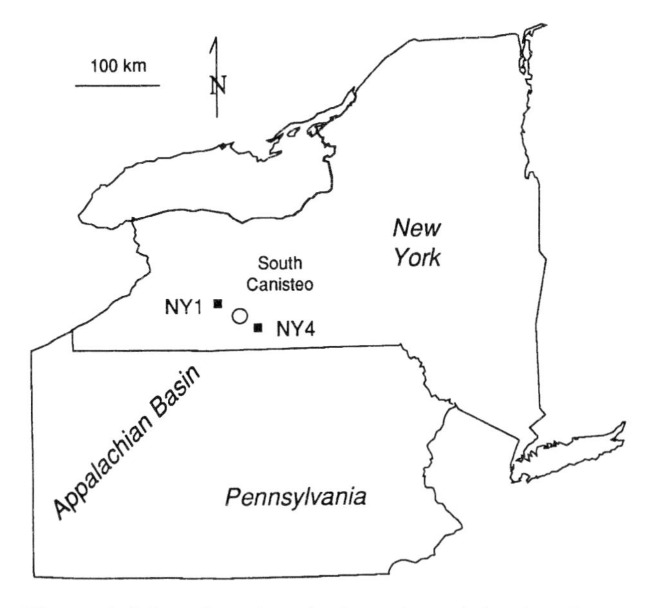

Figure 2. Map showing the location of the South Canisteo wells, where stress measurements were made, and wells NY-1 and NY-4, where core samples were taken, in the Appalachian basin of western New York.

PROCEDURE

Samples were cut into 50 cm blocks oriented horizontally with vertical faces parallel and perpendicular to N65°E. This is the approximate orientation given in a number of references for the contemporary tectonic stress field in upstate New York (Zoback and Zoback, 1989) and is the approximate orientation of fractures induced by hydraulic fracturing in the Wilkins and Appleton wells (Evans et al., 1989a). Samples were dried under a vacuum at room temperature and then fitted with three strain gauge rosettes and four transducer pairs, as shown in Figure 3. Dow-Corning RTV 3140 silicone sealant was applied to exclude the pressurizing medium (hydraulic oil). Measurements were made at approximately 20 confining pressure intervals during loading, and the same during unloading.

Three compressional and one shear wave time-of-flight measurements were made using transducers with 1 MHz resonant frequency. Source transducers were excited with a 20V pulse, and the output of the receivers was captured on a LeCroy 9400A digitizing oscilloscope.

Microcrack strain was calculated following the procedure of Siegfried and Simmons (1978). A least-squares fit to the linear strain curves from 100 to 140 MPa determines the intrinsic compressibility, and the microcrack strain is calculated from the difference between the intrinsic and the measured strain at ambient pressure. When redundant measurements of strain were available for a given direction, the average was used.

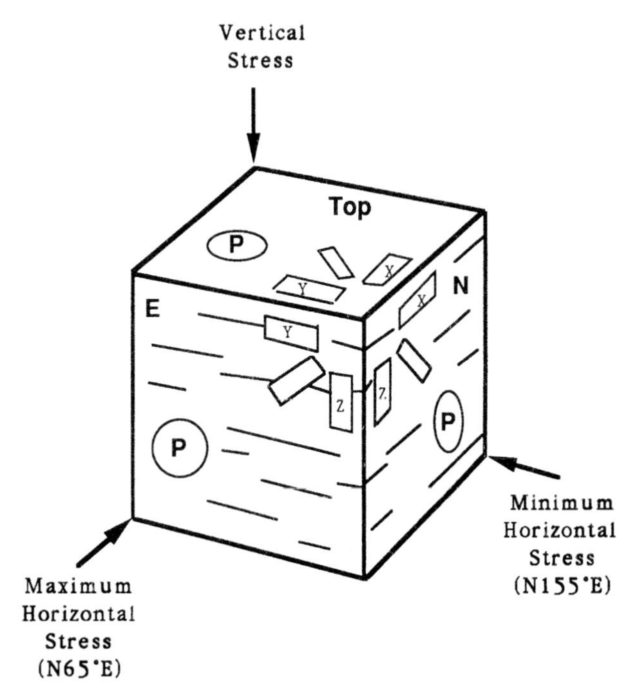

Figure 3. Schematic showing sample orientation relative to principal horizontal in situ stress directions and horizontal bedding. Velocities were measured in three directions: N65°E (X), N155°E (Y), and vertically along the core axis (Z). Six strain measurements were made in these three directions, with redundant measurements on adjacent faces, and three measurements at 45° to these directions.

The strain measurements can be resolved to approximately 2 $\mu\varepsilon$. Redundant gauges generally gave values that differed by less than 100–200 $\mu\varepsilon$ for total linear microcrack strain. However, redundant gauges on the highly fissile shales differed by 600–1600 $\mu\varepsilon$ out of nearly 4000 $\mu\varepsilon$. The small differences between redundant measurements in the less cracked samples arise from a number of sources, including sample inhomogeneity, strain gauge bond thickness, and electrical contact resistance. The very large differences in the shale samples are likely to reflect primarily sample inhomogeneity.

Pulse travel times can be resolved to 0.005–0.010 μsec, and sample dimensions were measured to 0.01 mm. Velocities were calculated from the first break of the arrival and are generally repeatable to within ±0.02 km/sec. In the highly cracked shale specimens, the vertical wave arrivals are barely detectable at ambient pressure and the precision is less than 0.50 km/sec.

OBSERVATIONS

Linear Strain and P-Wave Velocity

The behavior of our nine samples clusters into three groups depending on lithology: the siltstones, the black shales, and the silty shales. We discuss one example from each lithology in the following paragraphs. In general, the slope of the strain curve between 100 and 140 MPa is taken as β, the "intrinsic" linear compressibility of the rock. Extrapolating the linear portion of the strain-pressure curve to ambient pressure yields the linear strain due to the closure of microcracks oriented approximately normal to the strain gauge (i.e., the crack strain). Summing the crack strain in three perpendicular directions yields the volumetric microcrack strain, or the microcrack porosity (Simmons et al., 1974).

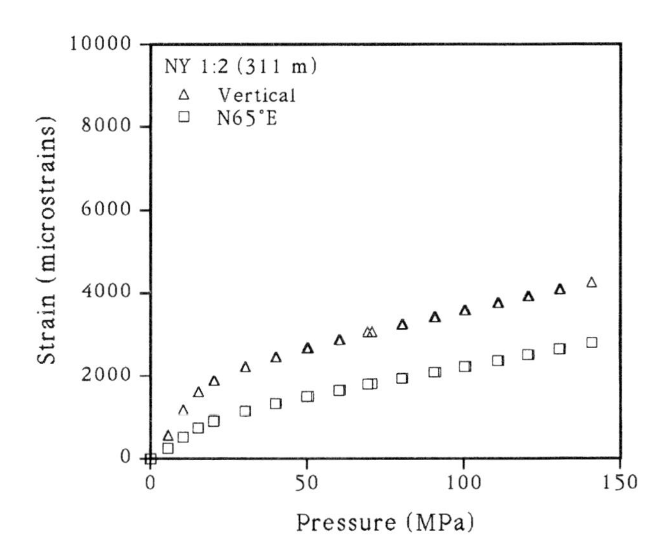

Figure 4. Example of linear strain as a function of confining pressure in sample NY1:2, one of the shallow siltstones (311 m). Strains were measured on a vertical face with strike N65°E, measured in the vertical and horizontal (N65°E) directions.

Siltstone

Curves for linear strain and P-wave velocities as a function of confining pressure for a siltstone sample, NY1:2 (311 m depth), reflect the compression and closure of microcracks (Figures 4 and 5). As confining pressure is applied to the sample, velocities increase and linear strain is large. Above 30 to 50 MPa the strain versus pressure curve becomes linear, and the P-wave velocity reaches its maximum value.

Black Shale

In a black shale sample (Marcellus shale from 1159 m depth), the horizontal strain is extremely low up to the maximum pressure, whereas the vertical strain is much larger (Figure 6). Although the slope of the vertical strain-pressure curve decreases significantly with pressure, it does not become linear even at the maximum confining pressure. Velocity in the horizontal direction is almost constant with confining pressure, but in the vertical direction both compressional and shear wave velocities increase from low values at ambient pressure to approximately 4.0 and 2.5 km/sec, respectively, above 10–30 MPa (Figure 7).

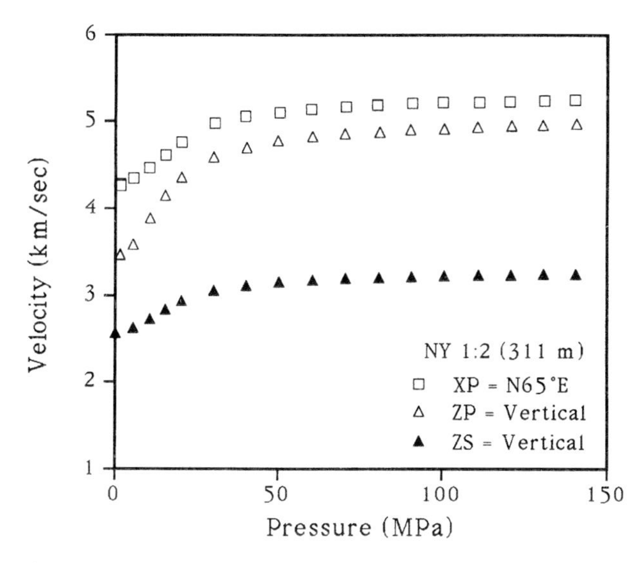

Figure 5. Example of compressional and shear wave velocities as functions of confining pressure in sample NY1:2 (311 m).

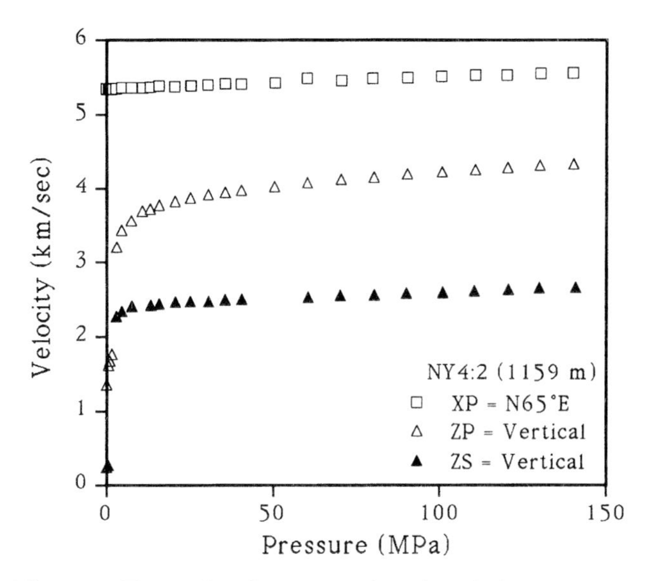

Figure 7. Example of compressional and shear wave velocities as functions of confining pressure in sample NY4:2 (1159 m).

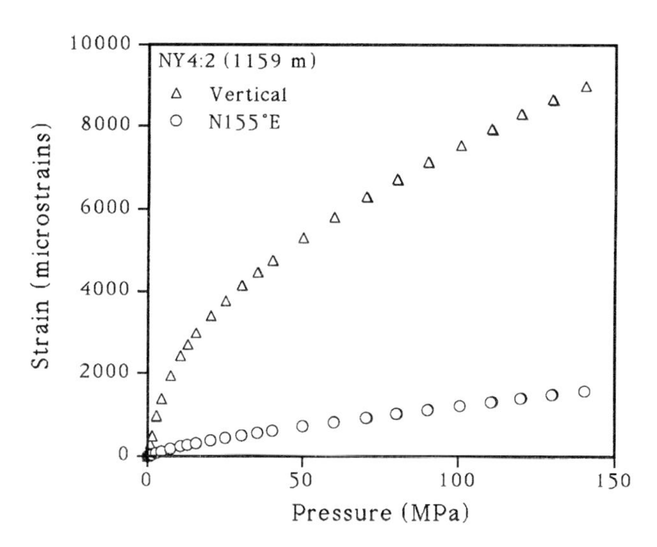

Figure 6. Example of linear strain as a function of confining pressure in sample NY4:2, one of the deep black shales (1159 m). Strains were measured on a vertical face with strike N155°E, measured in the vertical and horizontal (N155°E) directions.

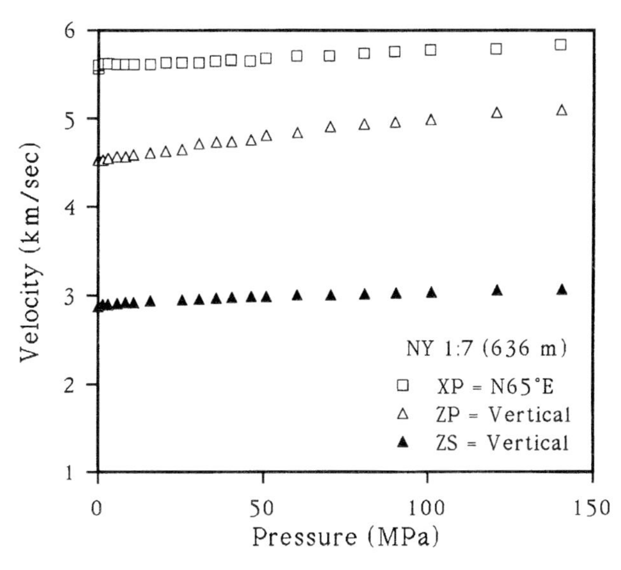

Figure 8. Example of linear strain as a function of confining pressure in sample NY1:7, one of the silty shales (636 m). Strains were measured on a vertical face with strike N155°E, measured in the vertical and horizontal (N155°E) directions.

The strong increase in P-wave velocity below 30 MPa confining pressure likely reflects closure of the prominent bedding-parallel partings visible in the core specimens. Consistent with the nonlinear nature of the vertical strain-pressure curve, the vertical velocities do not reach a constant value but continue to increase slightly up to the maximum pressure.

Silty Shale

The rocks described as "silty shales" have characteristics that are intermediate between those of the siltstones (NY1:2) and the black shales (NY4:2) (Figure 8). As in the black shales, microcrack strain measured parallel to bedding in silty shale sample NY1:7 (from 636 m depth) is essentially zero, and microcrack strain in the vertical direction is higher. However, the microcrack strains in the vertical direction are significantly lower than those measured in the black shales. As in the siltstones, velocities measured at ambient pressure in the silty shales are high in both the horizontal and vertical directions, without the very strong anisotropy of the shales (Figure 9).

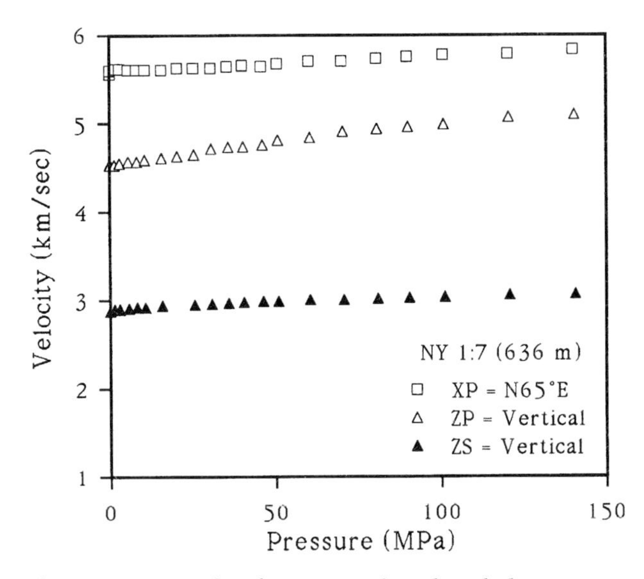

Figure 9. Example of compressional and shear wave velocities as functions of confining pressure in sample NY1:7 (636 m).

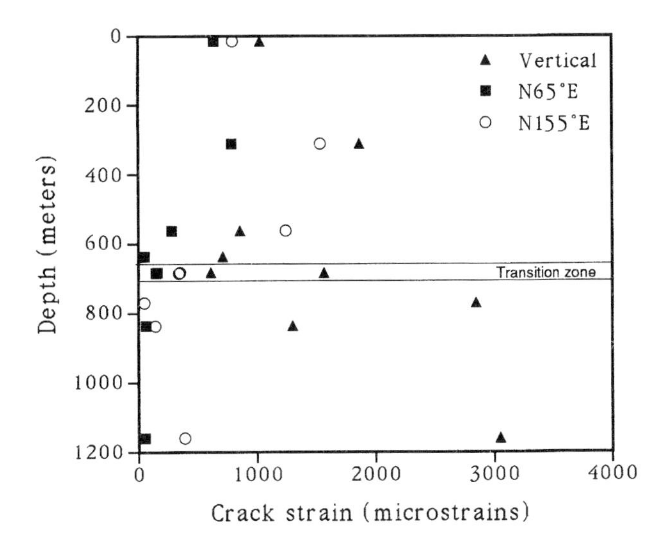

Figure 10. Linear crack strain measured in cores from E.G.S.P. wells NY-1 and NY-4 in the directions of the vertical, minimum horizontal (N155°E), and maximum horizontal (N65°E) stress directions. The transition zone at the base of the Rhinestreet Formation is marked (685 m).

Unlike in the siltstones, however, velocities in the silty shales increase very gradually with confining pressure and show no clearly defined break in slope reflecting the closure of a set of microcracks.

Microcrack Porosity

Microcrack porosity is calculated from the sum of the individual linear microcrack strains in three mutually perpendicular directions (Figure 10). Vertical strain is the largest in all but one sample. The crack strain is less anisotropic in the coarser siltstones and silty shales than in the deeper black shales. Low aspect ratio microcrack porosity reaches a minimum value in a silty shale recovered just above the transition zone (Figure 11).

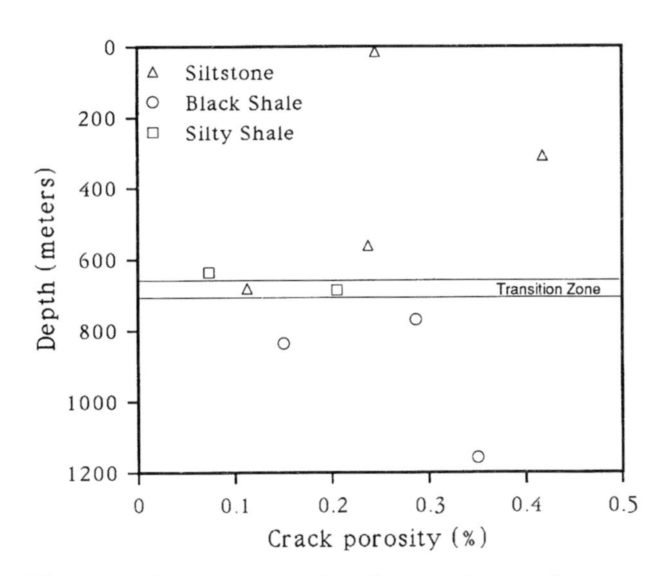

Figure 11. Low aspect ratio microcrack porosity as a function of initial sample depth in the cores from E.G.S.P. wells NY-1 and NY-4.

Compressibility

The linear compressibilities measured between 100 and 140 MPa confining pressure are plotted as functions of sample depth for three mutually perpendicular directions (Figure 12A) and again at a larger scale for the two horizontal directions which are maximum stress (S_H) at N65°E and minimum stress (S_h) at N155°E (Figure 12B). Although we use the term "intrinsic" compressibilities, the rocks may yet contain open cracks with aspect ratios greater than 0.003. This aspect ratio is approximately equal to the ratio of the closure pressure to the intrinsic Young's Modulus, assuming a Young's Modulus of 30 GPa (Walsh, 1965). The intrinsic compressibilities measured vertically are highest in all samples; compressibility in the direction of minimum horizontal stress is larger than compressibility in the direction of maximum horizontal stress for all but two of the samples. Only sample NY1:15 (836 m) has a higher compressibility in the N65°E direction for both static and dynamic measurements.

The variation in compressibility from one sample to the next correlates roughly with lithology (Figure 13). The siltstones are somewhat less anisotropic than the shales and the silty shales and generally have a decreasing intrinsic compressibility with depth. The silty shales are slightly more anisotropic, with lower compressibility parallel to bedding. The black shales have even lower compressibilities parallel to bedding;

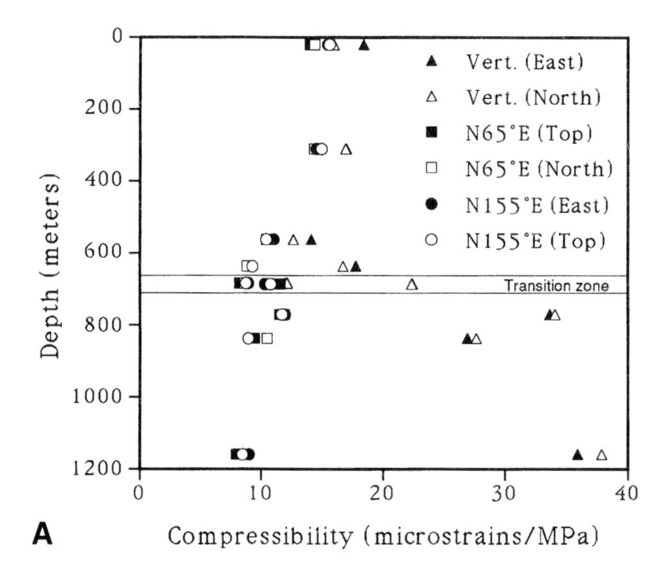

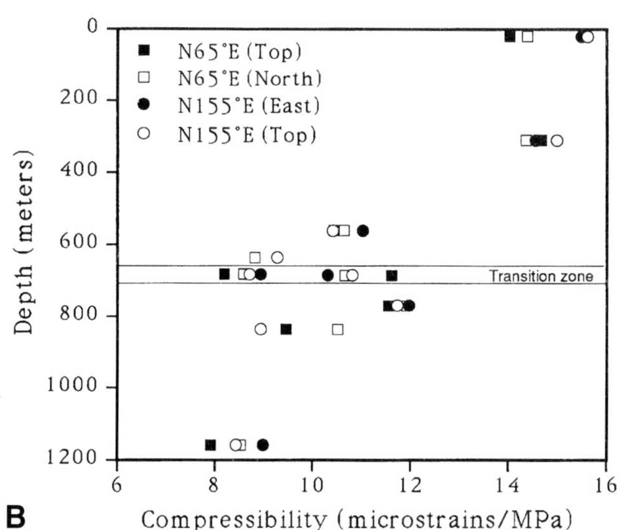

Figure 12. Linear compressibility measured in cores from E.G.S.P. wells NY-1 and NY-4 at maximum pressure in directions corresponding to known stress directions. (A) Measurements in a given direction from redundant gauges are distinguished as solid and outline symbols. The direction normal to the face on which the strain was measured is given in parentheses, where north is the direction N155°E and east is the direction N65°E. (B) Linear compressibility in the horizontal principal stress directions shown at a larger scale.

however, in the direction normal to bedding, compressibility is relatively high.

Velocity

Compressional wave velocities measured at ambient pressure vary as functions of initial sample depth and propagation direction (Figure 14A). The cores taken from above the transition zone are generally less anisotropic than those from below, reflecting primarily the contrast in lithology between the siltstones

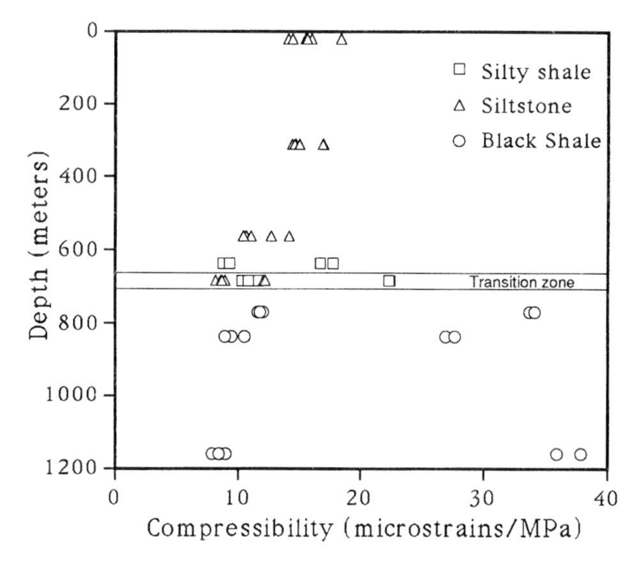

Figure 13. Linear compressibility measured in three mutually perpendicular directions at maximum pressure in cores from E.G.S.P. wells NY-1 and NY-4, distinguished by general lithology.

and silty shales above and the black shales below the transition zone. P-wave velocities measured normal to bedding (vertical propagation direction) are lowest in every sample. At ambient pressure, P-waves propagating horizontally in the direction of S_h (N155°E) have lower velocities than do P-waves propagating in the direction of S_H (N65°E) (Figure 14B).

At the maximum confining pressure (140 MPa), velocities increase in all directions relative to the ambient velocities and show a smaller anisotropy (Figure 15A). Velocities of P- and S-waves propagating vertically in the siltstones and shales increase significantly more than do those measured in the silty shales. However, the waves propagating vertically are still the slowest at maximum pressure in all samples. In general, P-waves propagating in the direction N155°E are slower than those propagating N65°E, a result consistent with the compressibilities derived from the strain measurements (Figure 15B). Note the higher velocities in the vicinity of the transition zone.

Vertical P- and S-wave velocities measured at approximately in situ lithostatic stress levels (using a density of 2710 kg/m^3) are plotted as a function of the sample's initial depth in Figure 16. We find very good agreement between the core sample data and the sonic log data from the equivalent formations in the Wilkins well (Plumb et al., 1991). Again, the highest velocities are found in the vicinity of the transition zone.

Dynamic Versus Static Moduli

We compute the dynamic compressibility, β, for the E.G.S.P. samples assuming a density ρ = 2710 kg/m^3,

$$\beta = K^{-1} = \left[\rho\left(V_p^2 - \frac{4}{3}V_s^2\right)\right]^{-1}$$

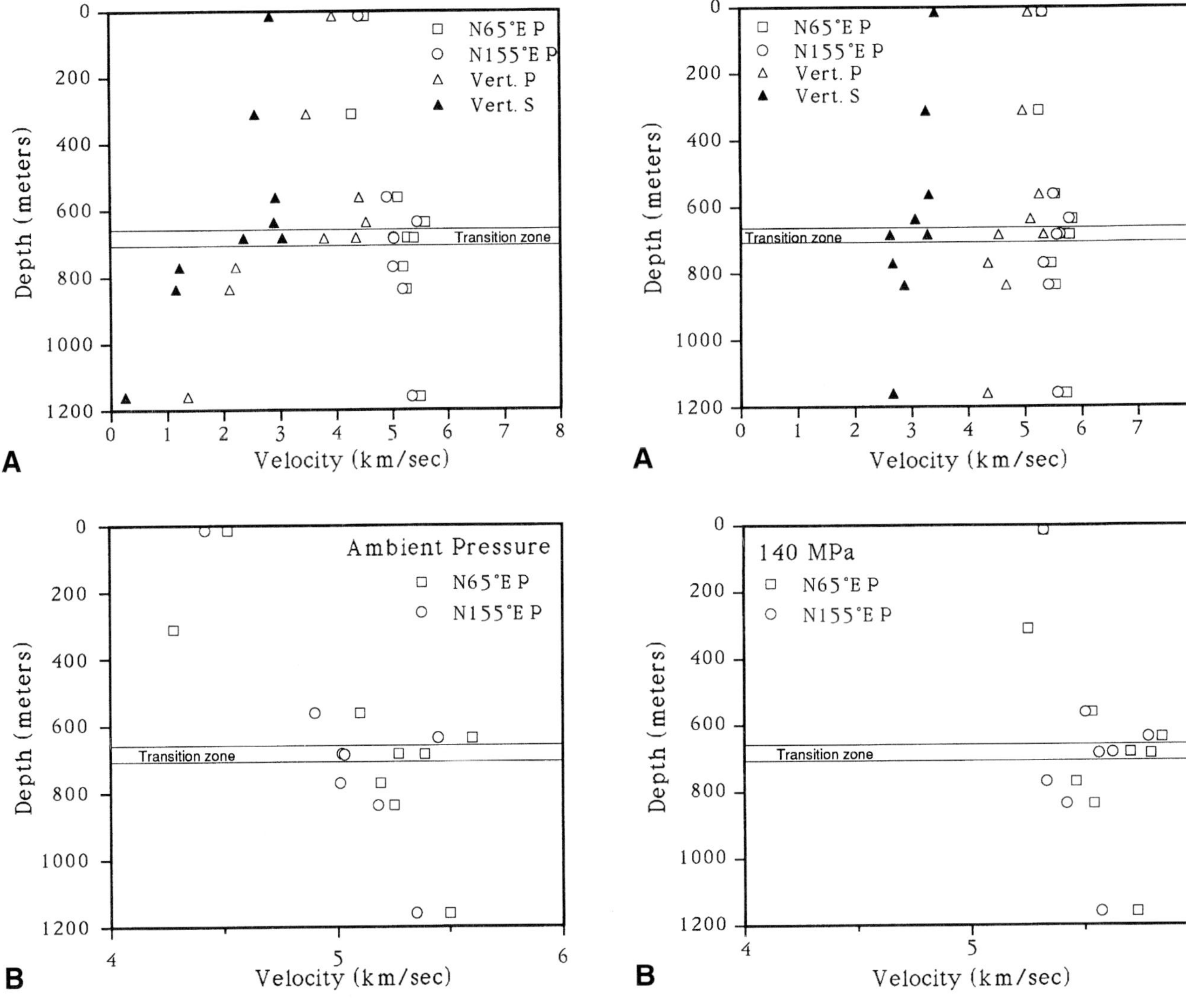

Figure 14. Compressional and shear wave velocities measured in cores from E.G.S.P. wells NY-1 and NY-4 at ambient pressure. (A) Compressional and shear wave velocity measurements made in directions corresponding to the vertical, horizontal minimum (N155°E), and horizontal maximum (N65°E) stress directions. (B) Compressional wave velocities measured at ambient pressure in the horizontal maximum (N65°E) and horizontal minimum (N155°E) stress directions.

Figure 15. Compressional and shear wave velocities measured in cores from E.G.S.P. wells NY-1 and NY-4 at maximum pressure (140 MPa). (A) Measurements in the vertical, horizontal minimum (N155°E), and horizontal maximum (N65°E) stress directions. (B) Compressional wave velocities measured at maximum pressure (140 MPa) in directions corresponding to the horizontal minimum (N155°E) and maximum (N65°E) stress.

where V_p and V_s are the velocities of vertically propagating P- and S-waves. To compute the static volumetric compressibility, we sum the linear compressibilities in three directions. Static volumetric compressibilities are lower in the siltstones and silty shales (less than or equal to 47 με/MPa) than in the deeper shales (greater than or equal to 47 με/MPa) (Figure 17). In general the dynamic compressibilities are lower than the static. The lowest compressibilities, both static and dynamic, are observed in the low-porosity siltstones and silty shales in the vicinity of the transition zone at the base of the Rhinestreet Formation. These compressibilities are consistent with in situ moduli derived from log data taken in wells from the South Canisteo study area (Plumb et al., 1991).

DISCUSSION

The Correlation Between Elastic Properties and the Transition Zone

A comparison of trends in sample microcrack porosity and stress in the Wilkins well shows that the lower stress found below the transition zone (Figure

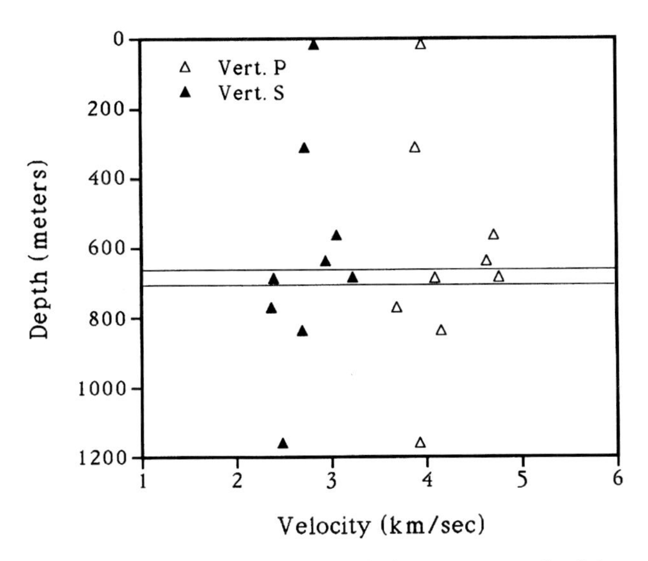

Figure 16. Compressional and shear wave velocities measured in the vertical direction in cores from E.G.S.P. wells NY-1 and NY-4 at confining pressures approximately equal to in situ lithostatic stress.

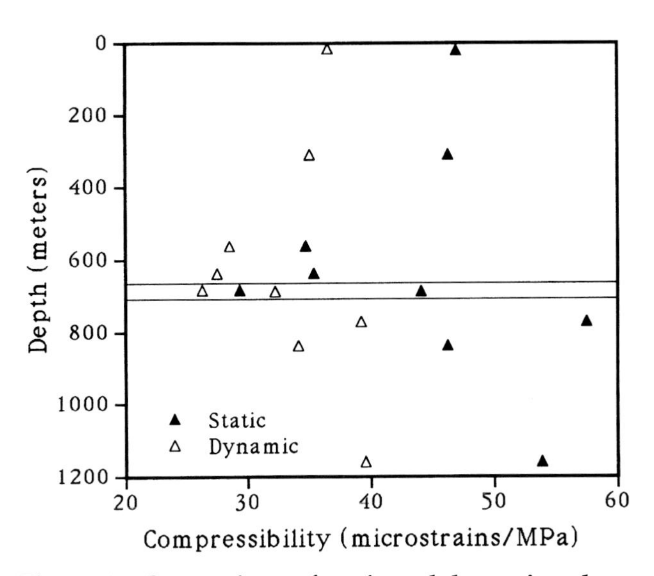

Figure 17. Comparison of static and dynamic volumetric compressibilities in the cores from E.G.S.P. wells NY-1 and NY-4.

1) does not correlate with lower microcrack porosity in the corresponding core samples (Figure 11). The lowest crack porosity is measured in a sample from approximately 50 m *above* the transition zone, where horizontal stress is relatively high.

Seismic methods for detecting overpressured zones rely on a contrast in properties between the transition zone and the surrounding rock. Therefore, differences in the mechanical properties of the transition zone may be useful in differentiating this zone from the normally pressured rocks above and the overpressured rocks below. Although the E.G.S.P. wells are no longer abnormally pressured, the lower microcrack porosity in the siltstones and silty shales in the vicinity of the transition zone is consistent with a more thoroughly indurated rock, as might be expected for a sealed zone. This low porosity may be due to the addition of carbonate cement documented by Albrecht (1992). The low porosity in these samples is reflected in high velocities and relatively low compressibilities which, if characteristic of the transition zone, may be detectable by seismic exploration methods. In fact, we note that data from Plumb et al. (1991) show slightly higher sonic velocities within the transition zone than within the overlying rock.

Microcrack Fabric

Several investigators have used stress relief microcracks to infer the magnitude and orientation of in situ stress components (e.g., Strickland and Ren, 1980; Ren and Roegiers, 1983), but they must assume that all the microcracks form on relief of in situ stress by coring. Several factors may complicate the correlation between low aspect ratio microcrack porosity and in situ stress. First, the substantial difference between the properties of the highly laminated black shales and the less anisotropic siltstones, seen in the linear microcrack strains, indicates that lithology has a very strong influence on the orientation and amount of cracking. In the deeper black shales, the increasing crack porosity reflects primarily bedding-parallel cracks, which may result from a number of factors other than increasing overburden stress. Batzle and Simmons (1976) reported very large compressibilities and large crack strains measured normal to bedding in claystones from the Raft River, Idaho, geothermal area. They attribute these large values in part to crushing and rotation of clay grains during decompression. Core disking may also affect samples of the undercompacted E.G.S.P. black shales. Finally, several samples of the E.G.S.P. black shales split parallel to bedding after being cut into blocks. This is believed to reflect the influence of dehydration and indicates that in the black shales, fractures do not form solely as a result of stress relief during coring.

In the Devonian sequence of New York, the maximum stress is horizontal in the shallow part of the well, providing a better opportunity to separate the effect of core disking from other stress relief microcracks. Yet, like our study, other studies of microcrack strain in sedimentary cores indicate crack strains measured normal to bedding are often larger than those measured parallel to bedding. Batzle and Simmons (1976) reported larger axial (vertical) than radial (horizontal) microcrack strains in four of six core samples from geothermal areas. Strickland and Ren (1980) also found vertical crack strains to be the highest in most of the cores of sandstone and shale they studied from Texas, Louisiana, and Pennsylvania. Neither of these studies reported in situ stress magnitudes. Ren and Hudson (1983) presented two case histories in which microcrack strain ratios were compared with ratios of in situ stress magnitudes using cores of sandstone, and in one case found the maximum stress predicted by microcrack strain was vertical, whereas hydraulic fracture data indicated a horizontal maximum stress.

Enough factors contribute to the development of horizontal microcracks upon stress relief that they are unreliable as a measure of in situ stress. However, in both case histories the ratios of *horizontal* crack strains were in good agreement with the ratios of horizontal stress magnitudes (Ren and Hudson, 1985).

Stress Relief Fabric Versus Tectonic Fabric

Because the siltstone samples are relatively isotropic at high pressure in both velocity and microcrack strain, it might appear that they are good candidates for the estimation of in situ stress using techniques that measure microcrack anisotropy. However, in most samples of the E.G.S.P. core, the maximum horizontal stress direction is parallel to the direction of lowest microcrack strain, and vertical crack strains (measured normal to bedding) are the highest in all but one sample.

In the E.G.S.P. cores, the static compressibility measurements indicate that fabric anisotropy persists even up to 140 MPa, making it unlikely that the patterns in microcrack strain result from the release of in situ stress. The velocity measurements made both at ambient and maximum confining pressure are consistent with the strain data; the more compressible directions, N155°E and vertical, have the lowest velocities, even at maximum pressure. In contrast, for a velocity anisotropy developed as a consequence of stress relief microcracking, the direction N65°E (parallel to the maximum stress direction) should correlate with the direction of lowest velocity and highest compressibility.

Sedimentary rock subjected to foreland fold-thrust tectonics, as is the case for the Devonian rocks of the Appalachian Plateau, carries an anisotropic tectonic fabric (Engelder and Marshak, 1985). Evans et al. (1989b) measured the preferred orientation of chlorite in cores from NY-1 and NY-4 and found a secondary preferred orientation of chlorite normal to N33°W. They interpreted this preferred orientation as resulting from northwest-directed Alleghanian compression. The relatively high closure pressures of the cracks in the siltstone samples are consistent with a population of microcracks in situ aligned parallel to the chlorite fabric (Cliffs Minerals, 1982, p. 27). If some stress relief microcracking has occurred as well, the majority of the microcracks have apparently aligned preferentially parallel to the chlorite grains. Hence, horizontal microcrack strains do not correlate directly with the relative magnitudes and orientations of in situ stress within the Appalachian Plateau.

CONCLUSIONS

A study of cores of Devonian shales and siltstones from western New York shows that the low microcrack porosity in the siltstones and silty shales within an ancient transition zone are consistent with a more thoroughly indurated rock, as might be expected for a sealed zone. The low porosity in these samples is reflected in high velocities and relatively low compressibilities which, if characteristic of the transition zone, may be detectable by seismic exploration methods. However, the magnitudes of low aspect ratio microcrack strains reflect primarily the degree of bedding-parallel microcracking. A weak correlation was also observed between the distribution of crack strains and a secondary mineral fabric developed during Alleghanian compression. Lithology strongly influences the distribution of cracks. Measurements of velocity and linear strain under hydrostatic confining pressure up to 140 MPa indicate that low aspect ratio microcracks in the highly fissile black shales are almost exclusively subhorizontal partings, whereas in the coarser siltstones and silty shales microcracks are distributed less anisotropically. The direction of maximum horizontal in situ stress correlates with the direction of the relative minimum microcrack strain. These results indicate that the relative magnitudes of in situ stress of relatively minor importance in controlling the porosity and preferred orientation of low aspect ratio cracks in these samples. Therefore, methods for inferring stress *a priori* from measurements of stress relief microcracking in cores must be applied cautiously in bedded sedimentary rocks and in basins affected by a tectonic fabric.

ACKNOWLEDGMENTS

We thank E.K. Graham, R.J. Greenfield, S. Mackwell, and Z.T. Bieniawski for reviewing an early version of this paper. This work was supported by GRI contract #5088-260-1746 and NSF grant #EAR87-20592.

REFERENCES CITED

Albrecht, W., 1992, Geochemistry of diagenetic pressure seal formation: Ph.D. Thesis, The Pennsylvania State University, University Park, Pennsylvania, 265 p.

Batzle, M. L., and G. Simmons, 1976, Microfractures in rocks from two geothermal areas: Earth Planet. Sci. Lett., v. 30, p. 71–93.

Carlson, S. R., and H. F. Wang, 1986, Microcrack porosity and in situ stress in Illinois Borehole UPH-3: Journal of Geophysical Research, v. 91, p. 10421–10428.

Cliffs Minerals Inc., 1982, Analysis of the Devonian shales in the Appalachian Basin, Vol. 1: Unpublished Report, Cliffs Minerals Inc., Granville, West Virginia.

Engelder, T., and S. Marshak, 1985, Disjunctive cleavage formed at shallow depths in sedimentary rocks: Journal of Structural Geology, v. 7, p. 327–343.

Engelder, T., and G. Oertel, 1985, Correlation between abnormal pore pressure and tectonic jointing in the Devonian Catskill Delta: Geology, v. 13, p. 863–866.

Evans, K., T. Engelder, and R. Plumb, 1989a, Appalachian Stress Study, 1, a detailed description of in situ stress variations in Devonian Shales of the Appalachian Plateau: Journal of Geophysical Research, v. 94, p. 7129–7154.

Evans, K., G. Oertel, and T. Engelder, 1989b, Appalachian stress study, 2, analyses of Devonian shale core: some implications for the nature of contemporary stress variations and Alleghanian deformation in Devonian rocks: Journal of Geophysical Research, v. 94, p. 7155–7170.

Jaeger, J. C., and N. G. W. Cook, 1963, Pinching-off and disking of rocks: Journal of Geophysical Research, v. 68, p. 1759–1765.

Lacazette, A., and T. Engelder, 1992, Fluid-driven cyclic propagation of a joint in the Ithaca siltstone, Appalachian Basin, New York, *in* B. Evans and T. F. Wong, eds., Fault mechanics and transport properties of rocks: Academic Press Ltd., London, p. 297–324.

Meglis., I.L., 1992, The influence of stress relief on the distribution of microcracks in anisotropic rocks: a laboratory study of compressibility, ultrasonic velocity, and attenuation in two suites of core: Ph.D. Thesis, The Pennsylvania State University, University Park, Pennsylvania, 185 p.

Meglis, I., T. Engelder, and E. K. Graham, 1991, The effect of stress relief on ambient microcrack porosity in core samples from the Kent Cliffs (New York) and Moodus (Connecticut) scientific research boreholes: Tectonophysics, v. 186, p. 163–173.

Miller, D.S., and I. R. Duddy, 1989, Early Cretaceous uplift and erosion of the northern Appalachian Basin, New York, based on apatite fission track analysis: Earth and Planetary Science Letters, v. 93, p. 35–49.

Plumb, R., K. Evans, and T. Engelder, 1991, Geophysical log responses and their correlation with bed-to-bed stress contrasts in Paleozoic rocks, Appalachian Plateau, New York: Journal of Geophysical Research, v. 96, p. 14509–14528.

Ren, N.-K., and P. J. Hudson, 1985, Predicting the in-situ state of stress using Differential Wave Velocity Analysis: Proc. U.S. Symp. on Rock Mech., 26th, p. 1235–1244.

Ren, N.-K., and J.-C. Roegiers, 1983, Differential Strain Curve Analysis—A new method for determining the pre-existing in-situ stress state from rock core measurements: 5th Int. Congress on Rock Mech., v. 2, p. F117–F127.

Scott, P. A., T. Engelder, and J. J. Mecholsky, 1992, The correlation between fracture toughness anisotropy and surface morphology of the siltstones in the Ithaca Formation, Appalachian Basin, *in* B. Evans and T. F. Wong, eds., Fault mechanics and transport properties of rocks: Academic Press Ltd., London, p. 341–370.

Siegfried, R., and G. Simmons, 1978, Characterization of oriented cracks with Differential Strain Analysis: Journal of Geophysical Research, v. 83, p. 1269–1278.

Simmons, G., R. Siegfried, and M. Feves, 1974, Differential Strain Analysis: a new method for examining cracks in rocks: Journal of Geophysical Research, v. 79, p. 4383–4385.

Strickland, F. G., and N.-K. Ren, 1980, Use of Differential Strain Curve Analysis in predicting in-situ stress state for deep wells: Proc. U.S. Symp. Rock Mech., 21st, p. 523–532.

Walsh, J. B., 1965, The effect of cracks on the compressibility of rock: Journal of Geophysical Research, v. 70, p. 381–389.

Zoback, M. L., and M. D. Zoback, 1989, Tectonic stress field of the Continental United States, *in* L. Pakiser and W. Mooney, eds., Geophysical Framework of the Continental United States: Boulder, Colorado, Geological Society of America Memoir 172, p. 523–539.

Index